# Classic Cordilleran Concepts:
# A View from California

Edited by

Eldridge M. Moores
Department of Geology
University of California
Davis, California 95616-8605

Doris Sloan
Department of Geology and Geophysics
University of California
Berkeley, California 94720-4767

Dorothy L. Stout
Cypress College
Cypress, California 90630

1999

Copyright © 1999, The Geological Society of America, Inc. (GSA). All rights reserved. GSA grants permission to individual scientists to make unlimited photocopies of one or more items from this volume for noncommercial purposes advancing science or education, including classroom use. Permission is granted to individuals to make photocopies of any item in this volume for other noncommercial, nonprofit purposes provided that the appropriate fee ($0.25 per page) is paid directly to the Copyright Clearance Center, 222 Rosewood Drive, Danvers, MA 01923, USA, phone (978) 750-8400, http://www.copyright.com (include title and ISBN when paying). Written permission is required from GSA for all other forms of capture or reproduction of any item in the volume including, but not limited to, all types of electronic or digital scanning or other digital or manual transformation of articles or any portion thereof, such as abstracts, into computer-readable and/or transmittable form for personal or corporate use, either noncommercial or commercial, for-profit or otherwise. Send permission requests to GSA Copyrights.

Copyright is not claimed on any material prepared wholly by government employees within the scope of their employment.

Published by The Geological Society of America, Inc.
3300 Penrose Place, P.O. Box 9140, Boulder, Colorado 80301

Printed in U.S.A.

GSA Books Science Editor Abhijit Basu

**Library of Congress Cataloging-in-Publication Data**

Classic Cordilleran concepts : a view from California / edited by
   Eldridge M. Moores, Doris Sloan, Dorothy L. Stout.
      p.    cm. -- (Special paper ; 338)
   Includes bibliographical references.
   ISBN 0-8137-2338-8
   1. Geology--California.   I. Moores, Eldridge M., 1938-   .
II. Sloan, Doris, 1930-   .  III. Stout, Dorothy L., 1941-   .  Series:
Special papers (Geological Society of America) : 338.
QE89.C54   1999
557.94--dc21                                                99-23409
                                                              CIP

**Cover:** Specimen of high-grade gold in quartz from the Oriental mine, Alleghany, California. Harold and Erica Van Pelt, photographers.

10 9 8 7 6 5 4 3 2 1

# Contents

*Preface* .................................................................... *vii*

**Section I. One hundred years of geology**
   *One hundred years: Introduction* ........................................ *3*
   *Chapter 1. A century of Cordilleran research* ............................ *5*
      W. R. Dickinson
   *Chapter 2. Early history of the Cordilleran section* .................... *15*
      D. L. Stout
   *Chapter 3. Some salient European contributions to Cordilleran tectonics* ... *31*
      A. M. C. Şengör

**Section II. Where it started**
   *Where it started: Introduction* ......................................... *39*
   *Chapter 4. California gold* ............................................. *41*
      **Classic Paper:** W. Lindgren, 1895, *Characteristic features
      of California gold-quartz veins* ...................................... *43*
      **Commentary:** J. K. Böhlke, *Mother Lode gold* ........................ *55*
   *Chapter 5. The great San Francisco earthquake of 1906* .................. *69*
      **Classic Paper:** A. C. Lawson, 1908, *The California earthquake of April 18, 1906–
      Report of the State Earthquake Investigation Commission (excerpts)* ... *71*
      **Commentary:** C. S. Prentice, *San Andreas fault: The 1906 earthquake
      and subsequent evolution of ideas* .................................... *79*
   *Chapter 6. The San Andreas and related fault systems* ................... *87*
      **Classic Paper:** M. L. Hill and T. W. Dibblee, Jr., 1963, *San Andreas, Garlock,
      and Big Pine faults, California* ...................................... *89*
      **Commentary:** T. W. Dibblee, Jr., *A lifetime of field work along the
      San Andreas fault system, California* ................................ *101*

**Section III. Plate Tectonics—Coast to Mountains**
   *Plate tectonics: Introduction* .......................................... *107*
   *Chapter 7. The Franciscan: California's classic subduction complex* ..... *111*
      **Classic Paper:** W. G. Ernst, 1970, *Tectonic contact between the Franciscan mélange
      and the Great Valley Sequence–crustal expression of a Late Mesozoic Benioff zone* ... *113*
      **Commentary:** J. Wakabayashi, *Subduction and the rock record:
      Concepts developed in the Franciscan Complex, California* ............ *123*
   *Chapter 8. The Great Valley Group: The arc-trench gap* .................. *135*
      **Classic Paper:** R. W. Ojakangas, 1968, *Cretaceous sedimentation,
      Sacramento Valley, California (excerpts)* ............................. *137*
      **Commentary:** R. V. Ingersoll, *Post-1968 Research on the Great Valley Group* ... *155*
   *Chapter 9. The Sierra Nevada: Central California's arc* ................. *161*
      **Classic Paper:** P. C. Bateman and C. Wahrhaftig, 1966, *Geology
      of the Sierra Nevada (excerpts)* ...................................... *165*
      **Commentary:** J. B. Saleeby, *On some aspects of the geology of the Sierra Nevada* ... *173*
   *Chapter 10. Geochronology of California's Arc* .......................... *185*
      **Classic Paper:** J. F. Evernden and R. W. Kistler, 1970, *Chronology of emplacement
      of Mesozoic batholithic complexes in California and western Nevada (excerpts)* ... *189*
      **Commentary:** W. D. Sharp and P. R. Renne, *Commentary on "Chronology of
      emplacement of Mesozoic batholithic complexes in California and western Nevada"
      by J. F. Evernden and R. W. Kistler* .................................. *193*

*Chapter 11. The southern California batholith* .................................................. *201*
    **Classic Paper:** E. S. Larsen, Jr., 1943, *Batholith and associated rocks
of Corona, Elsinore, and San Luis Rey quadrangles, southern California (excerpts)* ..... 204
    **Commentary:** G. R. Gastil, *Esper S. Larsen, Jr. on the batholith
of Southern California revisited* .................................................. 217
*Chapter 12. The assembly of California* .................................................. *219*
    **Classic Paper:** E. M. Moores, 1970, *Ultramafics and orogeny, with models
of the U. S. Cordillera and the Tethys* .................................................. 221
    **Commentary:** E. M. Moores, Y. Dilek, and J. Wakabayashi, *California terranes* ........ 227

## Section IV. Water and Oil–the Vital Fluids of California

*Water and oil: Introduction* .................................................. *237*
*Chapter 13. Gilbert's hydraulic experiments* .................................................. *241*
    **Classic Paper:** G. K. Gilbert, 1914, *The transportation of débris
by running water (excerpts)* .................................................. 243
    **Commentary:** E. A. Keller, *Transportation of débris by running water
(Professional Paper 86) by Grove Karl Gilbert* .................................................. 253
*Chapter 14. Applied stratigraphy in the Coast Ranges* .................................................. *257*
    **Classic Paper:** R. M. Kleinpell, 1938, *Miocene stratigraphy of California (excerpts)* .... 259
    **Commentary:** W. B. N. Berry, *Stratigraphic paleontology:
From oil patch to academia* .................................................. 267
*Chapter 15. The Monterey Formation: The source of oil* .................................................. *273*
    **Classic Paper:** M. N. Bramlette, 1946, *The Monterey Formation of California
and the origin of its siliceous rocks (excerpts)* .................................................. 275
    **Commentary:** R. J. Behl, *Since Bramlette (1946): The Miocene
Monterey Formation of California revisited* .................................................. 301
*Chapter 16. Water management: Slaking California's thirst* .................................................. *315*
    **Classic Paper:** *The California Water Plan, 1957 (excerpts)* .................................................. *317*
    **Commentary:** R. T. Bean and E. M. Weber, *The California Water Plan* .................. 323
    **Commentary:** C. J. Hauge, *Water in California–1998: a brief update* ................. 333

## Section V. Tectonics in the Desert

*Tectonics in the desert: Introduction* .................................................. *339*
*Chapter 17. Basin and Range extension* .................................................. *341*
    **Classic Paper:** R. L. Armstrong, 1972, *Low-angle (denudation) faults,
hinterland of the Sevier Orogenic Belt, eastern Nevada and western Utah* .............. 343
    **Commentary:** B. Wernicke and J. Spencer, *Retrospective on
"Low-angle (denudation) faults, hinterland of the Sevier Orogenic Belt,
eastern Nevada and western Utah" by Richard Lee Armstrong* ...................... 357
*Chapter 18. Mojave—Basin and Range boundary* .................................................. *363*
    **Classic Paper:** G. A. Davis and B. C. Burchfiel, 1973, *Garlock fault:
An intracontinental transform structure, southern California* ....................... 365
    **Commentary:** J. D. Walker and A. F. Glazner, *Tectonic development
of the southern California deserts* .................................................. 375
*Chapter 19. Death Valley: The ultimate desert valley* .................................................. *381*
    **Classic Paper:** L. F. Noble, 1941, *Structural features of the Virgin Spring area,
Death Valley, California (excerpts)* .................................................. 385
    **Commentary:** L. A. Wright and B. W. Troxel, *Levi Noble's Death Valley,
a 58-year perspective* .................................................. 399

**Section VI. The Modern Landscape—Mountains to Coast**
    *The modern landscape: Introduction* ................................................. *415*
    *Chapter 20. Modern and ancient volcanoes* ........................................ *419*
        **Classic Paper:** Howel Williams, 1929, *The volcanic domes
        of Lassen Peak and vicinity, California* ............................................. *421*
        **Commentary:** R. L. Christensen and M. A. Clynne, *Volcanism in California* .......... *431*
    *Chapter 21. Mountains shaped by ice* ................................................ *439*
        **Classic Paper:** E. Blackwelder, 1931, *Pleistocene glaciation
        in the Sierra Nevada and Basin Ranges* ............................................. *441*
        **Commentary:** A. R. Gillespie, M. M. Clark, and R. M. Burke, *Eliot Blackwelder
        and the alpine glaciations of the Sierra Nevada* ................................... *443*
    *Chapter 22. California's shifting slopes* ............................................ *453*
        **Classic Paper:** K. Terzaghi, 1950, *Mechanism of landslides (excerpts)* .............. *455*
        **Commentary:** R. J. Shlemon, *Commentary on Karl Terzaghi's 1950
        "Mechanism of landslides"* ......................................................... *471*

*Epilogue: Past, present, and future context of Cordilleran geology* ..................... *475*
*Index* ................................................................................ *483*

# Preface

The publication of this volume coincides with the Centennial Meeting of the GSA Cordilleran Section, at the University of California, Berkeley, June 2–4, 1999, the symbolic birthplace of the Section (though the first meeting was at the California Academy of Sciences in San Francisco). The volume had its conception at a salubrious Cordilleran Management Luncheon in an idyllic setting at the 1997 Cordilleran Section meeting in Kona, Hawaii. At the Kona meeting, the phrase "The Past as Prologue to the Future: The Century of the Pacific Rim" was officially announced as the theme of the Centennial meeting.

During planning for the Berkeley meeting, the conversation turned to the many classic geologic concepts developed during the past century's geologic studies in the Cordillera of western North America. The idea arose of a special publication outlining some of these concepts, how they originated, and the present status of knowledge about them. This volume is the result.

As is evident from the Table of Contents, this volume consists of a series of entries on important topics of Cordilleran geology. Each entry typically includes a classic article, a short explanation of why the article is classic, a short biographic sketch and contemporary picture of the author or authors, and an update or commentary about the present status of the subject by an active worker or workers. Classic articles are variously reprinted in their entirety, as excerpts, or incorporated into the commentaries.

The list of entries is by no means complete. It concentrates on California, rather than the entire Cordilleran Section region. Even California is too vast a subject to be covered adequately in a volume of manageable size. We hope that the volume will give, however, some flavor of the development of classic geologic concepts from the Cordillera, as viewed from California, and that it will serve as encouragement for future Cordilleran Section meeting organizers to develop comparable publications for other regions as appropriate or for other GSA sections to do it for their own anniversaries (although for centennials, we are going to have to wait 48 years for the next one!—e.g., Rocky Mountain Section, 2047; Southeastern, 2052; Northeastern, 2065, North and Central, 2066, respectively).

We have clustered these entries into a series of thematic sections, each with its own short introduction. We recognize that no thematic grouping is perfect—there is considerable overlap, some duplication, and some cycling from one group to the other. Still, we hope that the clustering will enhance the coherence of the widely diverse offerings. The organization of papers varies from one section to another, as outlined in the section introductions. We hope that the reader will find the total product interesting and useful. Thematic groups include:

- Introduction
- Where it started
- Plate tectonics—from coast to mountains
- Water and oil—the vital fluids of California
- The desert
- The modern landscape—mountains back to coast.

Throughout this volume, we have included selected photographs of Cordilleran geology or geoscientists from the past, as well as pen-and-ink sketches by Clyde Wahrhaftig of the U.S. Geological Survey and University of California–Berkeley. Clyde was never in the field or traveling without a sketch pad. Over the years he filled several small volumes, and he used many of his sketches to illustrate his informal publications. We also have included several reminiscences of Cordilleran geologists to try to give the reader an added flavor of the Section and its geology. These boxed quotes are the responsibility of the editors, not the authors of the chapters in which they appear.

*Note regarding the CD*

A compact disk (CD) is included in the pocket on the inside back cover of this volume. On it we have reproduced features that enhance the quality of the printed chapters. There are four items:
1. A collection of color photographs and images, edited by Carol Prentice, of the San Andreas fault, to accompany her chapter on the San Andreas fault.
2. An animation of the plate tectonic evolution of the western North American margin, by Tanya Atwater. This animation supplements the chapters on the San Andreas fault and the section on Plate Tectonics.
3. An animation of the southeast Asian region by Robert Hall, which may serve as a model of the possible evolution of the western North American region in Mesozoic–early Cenozoic time, and which is intended to accompany the chapter on California Terranes by E. M. Moores, Y. Dilek, and J. Wakabayashi.
4. The Cordilleran Section and Pacific Rim from space; NASA images edited by William R. Muehlberger.

## ACKNOWLEDGMENTS

We acknowledge with thanks the many authors of the contributions who labored under severe editorial pressure to submit their contributions in a timely manner for publication by the Meeting in June 1999. We also thank GSA Books Editor Abhijit Basu, Executive Director Donald M. Davidson, Managing Editor Faith Rogers, former Maps and Books Managing Editor Sharon Schwoch, and current Maps and Books Managing Editor Naomi Horii for their hard work and overall support to the success of this project.

For help with the classic papers, we wish to thank Jan Haskell, Don Bain, director of the Computing Facility of the Geography Department at UC Berkeley, and Gan Golan and A'lan Abruzzo, who spent many hours at the scanner and OCR to reproduce the original papers.

## NOTE TO THE READER:

The classic articles included in this volume were scanned from the original publications. In some cases the original type or page layout was of uneven quality, which was not possible to change in our reproductions. Some of the articles included typographical errors which we were unable to correct. In addition, a number of the original articles were too long and had to be excerpted, with consequent loss of continuity. Length was obviously a problem with books. In several cases we could not avoid omitting many figures and tables, although they convey so much of the important meaning of the original. We have tried to include enough to convey the "feeling" of the classic, but we encourage the reader to seek out the original for the full treatment.

Symbols for excerpting:
Sections missing: * * * * *
Minor text missing: . . .

Geological Society of America
Special Paper 338
1999

# Section I
# ONE HUNDRED YEARS OF GEOLOGY

# *One hundred years: Introduction*

The history of the Cordilleran Section of the Geological Society of America (GSA) and the development of geologic concepts during its life have unfolded during a century of momentous scientific and societal changes. Five wars, a major revolution in geology, enormous societal upheavals, mind-boggling and ever-accelerating technological changes in transportation, communication, and information have marked the past century. Developments in the Cordilleran Section have occurred within this wider framework.

We begin "at the beginning" with the three essays comprising this section. William Dickinson's essay, entitled "A Century of Cordilleran Research" chronicles the geologic effort and the growth of knowledge over the past century. It conveys a sense of the excitement that has swept Cordilleran geology, particularly with the advent of the plate tectonic revolution, and it outlines the inherent conflicts and challenges in the geoscientific community as we enter the twenty-first century. Dorothy L. Stout's essay "Early History of the Cordilleran Section" chronicles the establishment of the section at a small meeting in Berkeley in 1899 and its development through its early years. It also gives, through a series of quotations, a flavor of what the section and the geology of the Cordillera mean to a representative cross section of Cordilleran geoscientists. A. M. C. Şengör's essay highlights the European contribution to Cordilleran geology. Şengör's comments may come as a surprise to many Cordilleran geologists and should be the cause of some reflection and humility. He documents the fact that several German geologists in the early part of this century had considerable insight into the North American geology, but their work was mostly published in German, and went largely unread and unquoted on this side of the Atlantic. One can only imagine what greater progress might have been made if things had been different!

The story is unfinished. There is much work to do, many unanswered questions, many exciting avenues of inquiry to pursue. These three essays set the stage and provide a historical basis in which to view the essays that follow.

MANUSCRIPT ACCEPTED BY THE SOCIETY ON NOVEMBER 23, 1998.

Since its inception the Cordilleran Section of the Geological Society of America (GSA) has always recognized that a field trip is an essential feature of any geological meeting. The discussions, explanations, and arm waving in the field will often solidify into new concepts and insights into geological events. Friendships and connections are made that extend far beyond the classroom and workplace. Meeting organizers have taken care to record routes, prepare written descriptions of outcrops, and encourage the participation of graduate students to share and highlight recent thesis research with fellow geologists.

Field guidebooks are a unique genre of the geologic literature because often, except for the thesis, there may not be a description of a specific location written again. They provide a legacy and a future reference to enable other geologists to retrace an excursion. UC Berkeley, Stanford, and Caltech have been frequent meeting sites for the Cordilleran Section, but other meetings at Cal State Los Angeles and Cal State Long Beach have all produced guidebooks that have added to the geologic knowledge of our state. Guidebooks published for the annual meeting of the Cordilleran Section in California have covered topics as diverse as engineering geology, gold-bearing Tertiary gravels, hydrogeology, landslides, and neotectonics, and geographically have ranged from the offshore islands to the Mojave Desert, the Sierras, Death Valley, and the Klamath Mountains. By ensuring that libraries are aware of guidebooks produced by each meeting, students and members of the public become aware of the contribution of earth scientists to the understanding and economic benefits of the world in which they live.

This volume commemorates an important anniversary in the history of the geosciences. It provides a concise account of the development of the Cordilleran Section in one place. The legacy of the founders is also being passed on by geologists who have expanded on their early ideas and classic papers presenting the latest ideas and theories on the original concepts. By creating a Centennial Volume the editors and authors have ensured that libraries will acquire this work, archive it, and provide access for future generations to classic geological literature written by geoscientists who have used their powers of observation for the benefit and education of the people, especially those who live in California and the western United States of America.

Barbara E. Haner, Librarian
Science & Engineering Library
Geology/Geophysics Collection, UCLA

Geological Society of America
Special Paper 338
1999

# Chapter 1
# A CENTURY OF CORDILLERAN RESEARCH

**William R. Dickinson**
*Department of Geosciences, University of Arizona*
*P.O. Box 210077*
*Tucson, Arizona 85718*
*E-mail: wrdickin@ccit.arizona.edu*

## WILLIAM R. DICKINSON

It is most fitting that William R. Dickinson be the author of the Introduction to this volume, because he is one of the pioneer Cordilleran practitioners of Circumpacific tectonics and of plate tectonics. Dickinson was born in Tennessee, but he received all his degrees from Stanford University (B.S., 1952; M.S., 1956; Ph.D., 1958). His professional life has been spent as faculty member at Stanford (1958–1979), and at the University of Arizona, where he is currently Emeritus Professor of Geosciences. A GSA member and/or fellow since 1958, Dickinson has served as 1994 GSA President, Chair of the Cordilleran Section (1974–1975), General Chair for the 1987 Annual Meeting in Phoenix, and the 1990 Cordilleran Section Meeting in Tucson. Dickinson's honors include a Guggenheim Fellowship (1965), the GSA Penrose Medal (1991), and election to the National Academy of Sciences (1992). He was the first Chair of the Geological Sciences Board (now the Board on Earth Sciences) of the National Research Council (1981–1984).

Dickinson's manifold scientific interests and contributions have focused mainly on the relationship between tectonics and sedimentation, including the nature of depositional systems associated with active tectonism, the plate setting and evolution of sedimentary basins, the relationship between sand composition and provenance, and the history of Circumpacific orogenic belts. Dickinson's work in the late 1960s with the New Zealand geologist Trevor Hatherton (e.g., Hatherton and Dickinson, 1969), about the relationship between composition of active volcanism and depth to Wadati-Benioff zones around the Pacific, was one of the cornerstones of the plate tectonic revolution. In 1967, Dickinson was co-chair, with Arthur Grantz, of a "Conference on Geologic Problems of the San Andreas Fault System" (Dickinson and Grantz, 1968). During this meeting he enthusiastically reassured a young upstart assistant professor (myself) who presented, to a decidedly mixed reception, an attempt at global analysis of San Andreas–like faults. More important, Dickinson was convenor of the second Penrose Conference, held at Asilomar, California in 1969, entitled "The Meaning of the New Global Tectonics for Geology" (Dickinson, 1970), which was truly a watershed event in the history of geology. At that meeting, the full import of the plate tectonic revolution burst upon the participants like a dam failure. Dickinson's final-day summation of the relationship between active tectonic environments and sedimentation (subsequently published as Dickinson, 1971, 1972) administered what seemed at the time to be the final *coup de grace* to the old geosynclinal concept. I remember that as one of the most exciting scientific moments of my life!

Dickinson continues to be active in research. His current interests, with his faculty colleagues, include the isotopic signature of the Ouachita system, the interpretation of paleomagnetic data from the Cordilleran margin, ages of detrital zircons in key Cordilleran sedimentary assemblages, the structural evolution of the Great Valley forearc basin, and the geologic context of archaeological sites in the South Pacific. He has had many students over the years who have gone on to productive careers in geology (including authorship of contributions to this volume). Bill Dickinson is truly a living Cordilleran legend.

Eldridge M. Moores

## REFERENCES CITED

Dickinson, W. R., 1970, The new global tectonics - 2nd Penrose Conference: Geotimes, v. 15, p. 18–20.
Dickinson, W. R., 1971, Plate tectonic models of geosynclines: Earth and Planetary Science Letters, v. 10, p. 165–174.
Dickinson, W. R., 1972, Evidence for plate-tectonic regimes in the rock record: American Journal of Science, v. 272, p. 551–576.
Dickinson, W. R., and Grantz, A., 1968, Proceedings of Conference on Geologic Problems of San Andreas Fault System: Stanford University Publications in the Geological Sciences, v. XI, 374 p.
Hatherton, T., and Dickinson, W. R., 1969, The relationship between andesitic volcanics and seismicity in Indonesia, the Lesser Antilles, and other island arcs: Journal of Geophysical Research, v. 74, p. 5301–5310.

# *A century of Cordilleran research*

**William R. Dickinson**
*Department of Geosciences, University of Arizona, P.O. Box 210077, Tucson, Arizona 85718; e-mail: wrdickin@ccit.arizona.edu*

Looking back over a century of geoscience research in the Cordillera is both an inspiring and a humbling exercise. So much has been learned that could not have been imagined in 1899! Other contributions to this volume discuss some of the classic papers in various fields that have led the way to fundamental new insights. In this opening chapter, I prefer to highlight the largely unsung but crucial contributions to the overall research effort that have been made by thousands of faces in the crowd laboring steadily outside the limelight to advance our understanding of Cordilleran geoscience. It is easy to forget that the pacesetting perceptions of the giants among us would be impossible without the information gained, day to day, by uncelebrated colleagues who work assiduously, by their own best lights, to dispel the scientific darkness bit by bit. In our admiration for intellectual leaders, we should always remember that science is the most supremely collegial and collaborative of all human endeavors. They also make their marks who never win an award, but must be content with the plaudits of their own consciences.

## LEADERS AND FOLLOWERS

The symbiosis between leaders and followers in geoscience is an intricate and at times topsy-turvy relationship. Intellectual leadership may be achieved only at the price of attention to a narrow bandwidth of the full spectrum of geoscience, or may be attained only during a brief span of a full career. The same person honored for insight into one nook or cranny of our field may be as dependent as anyone else on the thoughts of others for understanding other facets of geoscience. As no one is a leader in every aspect of our joint investigations, or throughout professional life, everyone is a follower in some matters at all times, and in all matters at some times. We are much more a company of scientific brethren, universally plagued with weaknesses as well as blessed with strengths, than the popular science press, which loves a hero, is wont to posit.

Perhaps the inevitable mingling of leader and follower stems partly from the almost-arrogant breadth of what we are pleased to claim as part and parcel of geoscience. Ours is fundamentally a problem-oriented science, and our chosen problem of understanding the Earth is so broad as to admit virtually every scientific discipline ever devised as grist for our intellectual mills. Rare is the paleontologist, no matter how innovative and effective at the study of fossil life, who can comprehend more than the rudiments of seismology, and the reverse is just as surely true.

For the ultimate synthesis of our ideas, we thus tend to stand not only on the shoulders of our predecessors, as all scientists do, but also lean more than most on the shoulders of contemporaries who solve pieces of the general puzzle for the benefit of all. For many of our central concerns, such as paleogeography or geochronology, we cannot make overall progress without combining the labors of people with quite disparate talents and interests. In the Cordillera, we learned in the middle years of the century that microfossils and isotopes were as important for the development of a regional geochronology as the classic ammonites and clamshells of the great biostratigraphers of the past. In recent years, we have learned that envisioning the evolving paleogeography of the Cordillera through time requires blending insights gleaned in almost equal measure from paleomagnetism, basin analysis, biogeography, and plate tectonics. None of those subdisciplines even existed in modern guise at the middle of the twentieth century.

## TIME AND MOTION

A challenge that many had to overcome in the early days of Cordilleran research was the sheer logistical difficulty of pursuing field studies before the advent of aircraft, paved highways, and four-wheel drive vehicles. Chester Longwell once recounted for me his first visit to his beloved Las Vegas country near the turn of the century. After riding a train for several

---

Dickinson, W. R., 1999, A century of Cordilleran research, *in* Moores, E. M., Sloan, D., and Stout, D. L., eds., Classic Cordilleran Concepts: A View from California: Boulder, Colorado, Geological Society of America Special Paper 338.

days from the East Coast to Salt Lake City, he made his way, presumably on horseback, to the San Pete Valley of central Utah, where the old U.S. Geological Survey (USGS) evidently maintained a field depot. There he picked up a wagon and team that had been allotted to him, and drove his camp outfit down the west flank of the Wasatch to Las Vegas, returning by the same means at the end of his field season.

A host of less-famous geologists doubtless had to face the same general kind of physical obstacles to accomplish their work through the years. One can only admire the pluck and the grit they needed to dredge solid scientific information out of the brush-covered or forested coastal mountains and the isolated desert ranges of the Great Basin. Well into mid-century, mapping projects in the more remote ranges of the Canadian Cordillera often entailed disappearing into snowy mountains by pack train for several months at a time. Using the technology of helicopter and float plane brought to fruition for other purposes during the war years, postwar Canadian researchers were able to mount a coordinated geologic assault on their segment of the Cordillera that became the envy of their counterparts across the border, where exploratory research was less systematically supported.

Early workers also lacked key technological aids we all use now. Few of us can imagine, for example, undertaking any regional mapping project without aerial photos or satellite images, which give us a bird's-eye view of the terrain in a form we can ponder at our leisure. In many areas, the subtle structural and stratigraphic trends visible on airborne or space imagery make the difference between an accurate geologic map and an educated guess. One can only marvel at people who somehow constructed geologic maps of poorly exposed areas, like much of the coastal California ranges, without any roadcuts at all to examine! In the laboratory, of course, every technique the old-timers used has been vastly improved, and we now have ten techniques, or maybe even a hundred, for every one they could call upon.

## CHANGING RESEARCH STYLES

Early in the twentieth century, field research was essentially an exploratory enterprise. No one knew what lay just over almost any hill, and entire mountain ranges were total mysteries, the nature and ages of the rocks within them virtually unknown. By mid-century, much scorn was heaped on the previous practice of awarding advanced degrees for the simple mapping of quadrangles. Opinion had by then developed the perspective that a research project should be designed to answer or test a specific question or hypothesis. In the early going, however, the question was simply what was out there, and many new relationships were perceived by thoughtful but largely unsung geologists following their noses up unmapped canyons and along previously unstudied ridges. Even as late as mid-century, for example, the number of quadrangles that had been adequately mapped geologically in the whole of Oregon east of the Cascades could be counted on the fingers of one hand. Faced with such wholesale ignorance, the research community of the interwar years quite properly regarded basic geologic mapping as the highest priority: without at least rudimentary knowledge, relevant questions cannot be asked and useful hypotheses cannot be advanced.

It is interesting in retrospect to observe that each major new investigative tool has fostered a fresh round of exploratory fishing expeditions to test the waters. When isotopic dating became

---

**Musings, what the Cordillera has meant to me**

The Earth affects all creatures, everywhere, and thus geologists no matter their origin are rightly proud of the work they do. But for this exercise we will myopically tread a parochial pathway, and try to imagine how the Cordillera has exerted a special control in chiseling out privileged careers. For me it all started when I arrived for graduate school at the University of California, Santa Barbara, in 1967, well before plate tectonics had gone prime time. I came from New York where we calculated dips in field class by measuring the drop in elevation per mile of a marker bed from one quarry to another. Even the high-grade metamorphic rocks of the Adirondack Mountains seemed to have flat lying foliations. How my eyes popped during an orientation field trip when John Crowell showed us overturned Quaternary strata along the San Andreas fault!

In the years since, our comprehension of the complexity and richness of the Cordillera has continued to expand. More than once I have heard colleagues boast that "it's all here." A bit exaggerated? Sure, but why not. What is particularly nice about the "all" is that so much of it is tied together, from craton to oceanic crust, in myriad configurations. And where the expected relations do not fit, we have come to appreciate terranes, those strange bedfellows evincing a disparate and wandering history.

The connectednesses and unconnectednesses of the Cordillera foster multidisciplinary studies, team work in the most wonderful sense. Of course, competition amongst different teams has been spirited, with the human side of research playing its own role. But the ebb and flow of Cordilleran ideas remains a fertile playground for learning, for interacting, and for solving riddles about how the Earth works.

David G. Howell, USGS, 1998

---

feasible on a systematic scale, for example, rock masses of any suitable character were dated in a frenzy of activity, different laboratories competing to produce the most voluminous as well as the most reliable results. The wisest heads focused their attention on rock bodies having field relations that were thought to be already well understood, but isotopic ages time and again showed extant understanding to be faulty. In California, for example, knowledge that the Sierran granites are largely Cretaceous rather than Jurassic, and that Salinian granites are largely Mesozoic rather than Precambrian, forced wrenching reevaluation of fundamental aspects of California geology.

A pattern of successive exploratory probes of analogous nature has provided a sort of rhythmic beat for Cordilleran research through the years. One can view the initial study of microfaunas, first of foraminifers and later of coccoliths and radiolarians, in much the same light as applications of isotopic dating. The list of explorations can be expanded beyond geochronology to include, noting just a few salient examples, the study of radiogenic isotopes as markers of crustal origins, observations of paleomagnetism for both stratigraphic and tectonic purposes, stratal measurements of paleocurrents to delineate patterns of sediment dispersal, subsurface reflection profiling at multiple scales, and diverse compilations of mineral facies and paleothermic and paleobarometric indicators as guides to the conditions of magmatism and metamorphism. The field appraisal of mineral physics and geochemistry nicely illustrates the crossover interaction between experimental and observational earth science, for only limited field observations could be made until the behavior of mineral systems had been calibrated in the laboratory.

Each exploratory effort, including field mapping and a host of other more recent avenues of progress not mentioned specifically here, began with a ground-breaking phase, during which rather diffuse research programs were pursued in serendipitous or shotgun fashion. Later in each case came a mature phase during which work was ever more tightly constrained by previously established guideposts, and by more sophisticated appraisal of the types of issues that each methodology or technique could address effectively. Every new departure tends to start as a diffuse program that draws often-justifiable criticism as being poorly designed and unlikely to yield definitive results. Crude beginnings are a function, however, of having to walk before running. New research avenues that prove fruitful soon develop the logical controls that amplify their impact, and broaden to include the yeoman efforts of everyday geoscientists whose participation is vital for full success.

Throughout the century of research, an atmosphere of tension has always seemed to exist between those who place their trust exclusively in field mapping by tried and true methods, and those who strive to bring new applications to the fore as a means of augmenting the analysis of field relations. There is potential gain from softening this aura of tension, and there are signs that we are overcoming our proclivity for that particular brand of pointless controversy. Those who take their inspiration from sophisticated instrumentation have learned that the power of any technique is improved by close attention to field relations, and those whose love is establishing the relationships of rocks in the field have learned that multiple laboratory investigations of well-selected samples can vastly improve their insights. The blending of the two approaches to geoscience has also occurred within the minds of individual geoscientists, as specialists in this or that pet methodology have come to do their own field work with the care and attention to detail that allows them to gain maximum ground with their demanding and time-consuming procedures.

Having said all that, it remains a continuing truism that Cordilleran geoscientists tend to maintain a prime allegiance to lessons from the field, in a quasireligious manner. Perhaps the sublime grandeur of so many Cordilleran landscapes and the great complexity of most Cordilleran rock assemblages are partly responsible for this durable geophilosophy. Whatever the reasons that nearly all of us hold so steadfastly to the time-honored concept that field relations are the ultimate test of any hypothesis, we should bless them, for we do study the Earth, and have no business indulging ourselves in any kind of virtual reality.

Many fear, however, that instruction in field methods and practice in field observations have lost some of their steam in competition with more sophisticated technology for the attention of young geoscientists. Those who plot educational strategy for the future need to ponder how best to preserve the kernel of our hard-earned ways of addressing basic field work as we move inevitably into an increasingly complex research environment. Just as budding young medical doctors continue to dissect cadavers, simply to learn how the human body is put together, all geoscientists need direct experience with rocks in their natural state. Even a working career spent interpreting seismic reflection profiles inside an office building needs the insights provided by personal knowledge of instructive field relations.

## CHANGING RESEARCH GOALS

For much of the century in review, the chief driver of both private and public funding for geoscience research was the need for improved knowledge in the service of exploration for and development of mineral resources, whether the target was metals, nonmetals, or petroleum and natural gas. Until quite recently, also, the employment opportunities for geoscientists lay almost entirely within the mineral industries, apart from the comparatively few slots available in academia. There have always been a few fortunate individuals able to cater almost exclusively to their own intellectual curiosity in designing research programs. The loosing of Tom Dibblee on the California Coast Ranges by Mason Hill comes to mind immediately as a case in point. Dibblee's mapping led demonstrably to the earliest documentation of dramatic strike slip along the San Andreas fault. It is well to remember, however, that his funding during the halcyon years came directly from the perception, well borne out by events, that an improved picture of regional geology would lead to fresh ideas for exploration. It is not for naught that the main producing horizon of the Cuyama oilfield was dubbed the Dibblee sand.

For most practitioners through many decades of the century, a much closer attention to mineral occurrences was a professional imperative. Even such luminaries as Jim Gilluly were funded during their formative years by topical investigations of mining districts, or for regional reconnaissance intended to improve the climate for new mineral discoveries. That so much fundamental geoscience was pioneered along the way is a tribute to one of two factors. Some might hold that attention to practical problems is a superb way to focus the mind on basic issues. Others might argue that an alert mind will generate valuable geoscience even when assigned the most banal of ostensible tasks. In either case, the annals of Cordilleran geoscience give the lie to the petulant claim that the funding of strategic research frustrates the imagination and inevitably stymies conceptual advances. One of my most cherished memories concerns an evening spent with Tom Thayer, watching him burn the midnight oil in an effort to ensure that the samples he was to collect by helicopter the next day, as mandated by legal edict to satisfy the dictates of the law regarding geochemical evaluation of potential wilderness areas, would also advance our petrologic knowledge of his treasured John Day ophiolite. We work as we must for our livelihoods, but no one can order us to stop thinking.

Over the past quarter of the century, environmental concerns, linked in part to ground-water resources and geohazards, have become an additional prime driver for geoscience research, and a major factor in the employment arena as well. Environmental geoscience has brought a wider perspective to our field as a whole, and we can only lament the fact that so many biologists and ecologists still do not fully appreciate the crucial role that geoscience must play in paleolandscape analysis and historical ecology. The inevitable tension between environmental preservation and resource exploitation unfortunately also raises the specter of a brand of intellectual and programmatic conflict that could well damage geoscience research in the service of either rubric if not resisted firmly by geoscientists themselves. Unless we hammer home the point that fundamental geoscience rests on the same core concepts, regardless of application, we risk getting caught in political whirlwinds.

## EXPLORATION AND ENVIRONMENTALISM

Citizen groups, industry lobbyists, legislative staffers, and agency administrators, few of whom have any roots in geoscience, tend to view funding for geoscience research from one of two disparate and mutually contradictory viewpoints. On the one hand are those whose interests and inclinations lead them to view research in generic support of exploration or other economic activities quite favorably, but at the same time to view research into environmental geoscience as a largely mischievous activity that can only erect unwanted roadblocks in their path. On the other hand are those whose dedication to environmental concerns leads them to view environmental geoscience in a favorable light, but anything that improves our picture of mineral distribution as a potential threat to ground they prefer to keep sacrosanct.

The inherent conflict between the two viewpoints is exacerbated by the fact that some in the first group have direct pecuniary interests to protect, and that some in the second group take an almost mystical approach to the Earth as the sacred product of creation, whether that act be seen as willful or no. Devotees of the latter team thus tend to regard kingpins of the "extractive industries," a deliberately pejorative term, as despoilers of our common heritage, whereas champions of the former team tend to view professional "environmentalists," a term equally as pejorative in some circles, as starry-eyed utopianists who will lead society astray if given the chance. With such antagonisms in the air, any debate that begins as a discussion about geoscience can easily degenerate into a dispute about morals and political philosophy.

The problem for geoscience, of course, is that Earth processes have never divided themselves into extractive and environmental compartments. The same geochemistry that informs us about the transport of metallic elements in ore-forming solutions also informs us about the behavior of contaminant plumes. Much of our knowledge concerning the sedimentary dynamics of fluvial systems and strandline associations and coral reefs comes from research sponsored by industry to learn more about fossil analogues that serve as subsurface reservoirs for fluid hydrocarbons. The underlying unity of all geoscience is self-evident to most geoscientists, but we must never tire of trying to make that point, over and over again, to nongeoscientists. If we fail to do so successfully, we may find that the inevitable conflict between people with an emphasis on exploration and those with a focus on the environment will conspire to suck resources from both geoscience areas into less-controversial allied fields. Legislators and program managers are notoriously sensitive to criticism, and it is all too easy for them to avoid getting a black eye from either side of the fence by placing a plague on both houses, forgetting that the Earth is really all the same house and that geoscience research cannot readily predict the specific applicability of improved information and insight.

## LEGIONS OF GEOSCIENTISTS

Such concerns aside, the most overt characteristic of the Cordillera-wide research community in geoscience has been its explosive growth in sheer numbers through the past century. There are probably more competent geoscientists working in the Cordillera today than there were in the entire nation, or perhaps

Snow Peak from Lundy Pond, August 23, 1963. Sketch by Clyde Wahrhaftig.

even in the world, at the turn of the century. This gives us now the capacity to attack a host of outstanding problems in detail across a broad front of disciplines and subdisciplines, many of them hardly a gleam in any eye only a scant few decades ago. As our numbers have grown, the ratio of followers to leaders has also grown exponentially, and the significance of the patient grunt work done by the followers looms ever larger in the overall research equation.

One index to the astonishing multiplication of effort that we have witnessed over the past century is the number of reputable departments of geoscience that now exist in the Cordilleran region. In 1899, there were only a handful of institutions up and down the coast where one could seek a solid education with any depth at all in the geosciences. At present, within our region alone, there are at least 40 four-year academic institutions with able staffs large enough to support strong programs with real breadth in geoscience, and there are many more two-year institutions that offer introductory and entry-level geoscience courses that are as carefully designed and as effectively taught as those at four-year schools. Essentially all academic institutions today assume the responsibility of doing their part to educate general student populations about geoscience, as well as to recruit young talent into the field. This existing educational base, when compared to the past, is an amazing resource for the future, and can only be enhanced by further development of the world-wide computer web and other communication mechanisms that will soon make the libraries of the world as accessible to students pursuing their studies far from traditional academic centers as anyone on the most prestigious campus. Determination and drive can now generate academic excellence anywhere, despite vagaries of fortune and opportunity.

Our growing numbers and the increasing variety of our specialties pose, however, some unfamiliar intellectual challenges that we must learn to overcome. When we were fewer, personal friendships and frequent contacts tended to weave the Cordilleran research community together into a tight network through which fresh ideas could propagate at surprising speed. We have lost that easy collegiality, and with it some of the flexibility to respond quickly to new information or concepts. Personal interchange at regional meetings and imaginative use of the Internet fortunately have the potential to overcome the geographic isolation that threatens constantly to thwart our highest aims.

Even more difficult to overcome is the tendency for geoscientists to plow the furrows of narrow disciplinary interest without adequate attention to the work of those in other fields which may have broad import. With so many people active in each subfield of geoscience, it is all too easy to imagine that we are getting the needed outside stimulus from our professional contacts, while actually working firmly within the limiting confines set by our disciplinary blinders. The history of geoscience shows repeatedly, however, that valuable synthesis must draw on a range of disciplinary roots. We simplify the scope of our problems to make any progress at all, but sacrificing breadth for depth, if taken to extremes, is a sure way to guarantee failure in our aspiration to construct an integrative geology that meets the Earth on its own terms. The revolutionary impact of plate tectonics had the effect for several years of dismantling the barriers between specialties, but that impulse to integrative thinking has largely spent itself. We need constantly to remind newcomers to geoscience, as well as ourselves, that the Earth is far too complex for any single discipline to comprehend, and that our disciplinary boundaries are divisions of convenience, not principle.

> Though I am one of the elders, I often cross the hall to a concurrent session of another group, our *avantgarde*, where there is an almost evangelical zeal to quantify, and if this means abandoning the classical geological methods of inquiry, so much the better; where there are some who think of W. M. Davis as an old duffer with a butterfly-catcher's sort of interest in scenery; where there is likely to be, once in a while, an expression of anger for the oldsters, who through their control of jobs, research funds, honors, and access to the journals, seem to be bent on sabotaging all efforts to raise geology to the stature of a science; where, in the urgency of change, it seems that nothing old is good.
>
> J. Hoover Mackin, chair
> Cordilleran Section, 1951
> *in* Rational and Empirical Methods of Investigation in Geology

## GEOLOGY AND TECHNOLOGY

At once the means of multiplying subdisciplines, and also of reunifying them, is the development of new techniques and methodologies. It requires a conscious effort of remembrance to recall the immense range of technological innovation that has spawned so much of the progress in Cordilleran geoscience over the years. Consider, for example, the few exploratory wells that had been drilled for petroleum and natural gas within the Cordillera at the time of World War I. Our knowledge of the subsurface was then almost nil except in local mining districts, and there our information extended to shallow levels only.

Even as late as World War II, the use of microfossils for biostratigraphy was in its infancy, isotope geology of any kind was ponderous and rudimentary owing to the scarcity and imprecision of existing instrumentation, and all geophysical investigations were shockingly primitive by modern standards. Conceptual impediments were equally severe, because we cannot perceive what we cannot imagine any more than we can measure what we cannot perceive. Everyone of my generation is painfully aware, of course, that some of our most respected ideas were laughed at with the advent of plate tectonics and the parallel acceptance of continental drift and sea-floor spreading, but our conceptual shortcomings went much deeper than that.

At mid-century, we were still debating whether granite was an igneous or metamorphic rock, had just begun to appreciate the reality of turbidite sedimentation, still tended to describe welded tuffs as lava flows, and had no clear idea that ophiolite, if indeed we used the word at all, had anything to do with the oceanic realm.

The historic record shows plainly that openness to innovative technology and fresh thinking, even when one or the other overturns cherished theory, is essential for continued progress in geoscience. The record shows as well that followers are as necessary as leaders when a new technology or methodology is brought into play. They do not serve us well who reject, in the name of tradition or past personal successes, the timely leadership of others with a better mouse trap in hand or a promising wild idea in mind.

## FUNDAMENTAL AND APPLIED GEOSCIENCE

One of the lasting strengths of Cordilleran research throughout the century has been the stimulating flow of people and ideas between fundamental and applied fields of geoscience, academia and industry and government, and private and public interests. Even while subdisciplines as varied as engineering geology and paleoseismology have been invented and brought to flower, and as time-honored professions such as petroleum and mining geology have been guided into unfamiliar directions by unexpected opportunities, there has never been any way to foresee whether the next exciting avenue of research would emerge from the seemingly driest scholarly concerns or from the most practical of investigations.

Nor is it clear even yet whether our greatest contributions to the larger community of our fellow citizens stem from the mundane considerations that surround selection of dam sites, alignment of highways, designation of hazard zones, production of mineral resources, and ground-water management, or whether the intellectual vision embodied in our hard-won understanding of the tectonic evolution of the Cordillera, and how it has conditioned the environment in which we live, is the most crucial product of our work through the years. The geologic context of public life is more lasting than any specific adaptation to geologic knowledge, for the needs and aspirations of society change.

In either case, the free flow of concepts and techniques from fundamental to applied research is an indispensable key to progress in geoscience. Think of all the once-esoteric methods of downhole logging that are now standard practice for so many applications of practical concern. Or think of the augering and trenching and other inherently simple approaches, used first in applied fields to answer equally simple and specific questions, that have been adapted for quite open-ended inquiries into basic phenomena of geoscience. Once again, the roles of leader and follower are inherently muddy in the constant interplay between fundamental and applied geoscience.

## THE WORLD SCENE

Cordilleran geoscience is proud of its global impact on geologic thinking, and rightly so. We are fortunate that so many outstanding practitioners of our science have lived out their research careers working in our region. The things we do and the things we think as we address problems of Cordilleran geoscience have wide impact, and of that fact we can be justly proud.

At the same time, we need to guard always against Cordilleran provinciality. We are so many, here within the Cordillera, that it is all too easy to think we need never look outward for inspiration. Sitting on the Pacific periphery of a huge continent, examining a segment of the immense Circumpacific orogenic belt in ever-increasing detail, there is a great temptation for us to suppose that all the key lessons of Earth history can be learned in our own backyard. We can be surprisingly resistant to ideas that emerge from studies of the Alpine-Himalayan orogenic belt, or even from the other side of our own ocean. To be fair to our own central enterprises within the Cordillera, we need to make sure that we meld the thoughts we develop ourselves with advances in thinking that emerge elsewhere.

## GUIDES TO SUCCESS

Although my focus has been on the pedestrian underpinnings to innovative breakthroughs in Cordilleran geoscience, implicit in that stance is the belief that many of us can become intellectual leaders, at least for a while or within a restricted arena of investigation. What converts a follower into a leader? No one should suppose that native talent or force of personality alone can do the trick. It seems to me, however, that several traits are essential and have been common to all the research leaders highlighted in this volume.

First and foremost, all have been diligent. Some of the men and women you will read about were jolly folk, cracking jokes on every occasion and laughing their way through life. Others were regular sobersides, never accused by anyone as having a sense of humor. All have in common, however, that they have been deadly serious about their science, yet uniformly motivated to pursue it with wild abandon. For leadership, there is no substitute for unwavering determination to persevere, nor for absolute allegiance to the integrity of the investigative process.

Second, nearly all have had open minds wherever and however science impinged on their thinking, able to suspect error in self as well as in others, and willing to retract or adjust their own most tenaciously held prior opinions as additional data pile up and new concepts arise. Those lacking the habit of mind that embraces needed revision of thought have wound up their careers as lovable old dinosaurs, likely as not with their central contributions overturned and discredited by the flow of events. An alert mind covers many failings, however, and a lively and adaptable approach to geoscience can be maintained by people whose private lives present the starkest contrasts, from those who can be described almost as libertines to the most steadfast and moralistic souls that anyone could ever expect to encounter.

Third, the aspiration to leadership requires the intellectual courage, and entails the personal willingness, to be unpopular, at least temporarily or in some circles. No one needs a guide to reinforce conventional wisdom, and few gain a lasting reputation for leadership by shoring it up, although plaudits of a tamer sort come readily enough from riding the bandwagons that pass our way. Fresh ideas are almost always unsettling, and the leaders we most admire had the guts at some point in the development of their ideas to upset and annoy their fellows. Those with grace, however, never embarked on personal vendettas for the sake of scientific triumph, and those with lasting reputations conquered the seductive temptation to push data beyond the bounds of relevance for the sake of personal prominence as a goal in itself.

Lasting prominence as a leader in the geoscience community requires instead a firm and uncompromising allegiance to valid data and sound logic, which are now as ever the sole pillars of authority in science. The buoyant and contentious spirit of youth is ever our strongest bulwark against slavish deference to self-styled leaders without the data and logic to hold their ground, and the future belongs to currently unknown geoscientists whose still unplumbed careers will offer unexpected insights that we cannot yet imagine as we embark on the next century of Cordilleran research.

MANUSCRIPT ACCEPTED BY THE SOCIETY NOVEMBER 23, 1998

Geological Society of America
Special Paper 338
1999

# Chapter 2
# EARLY HISTORY
# OF THE CORDILLERAN SECTION

**Dorothy L. Stout**
*Cypress College*
*Cypress, California 90630*
*E-mail: gaea@deltanet.com*

## DOROTHY L. STOUT

Dorothy L. Stout is a professor of geoscience at Cypress College in California. An early interest in history led to her eventually majoring in the ultimate history course—geology. After raising three daughters and teaching for many years she went back to school in her forties to Claremont Graduate School to develop a dissertation on "The development of geologic knowledge and education and its applications in California before 1934" at the suggestion of one of California's legends, Mason L. Hill. He had participated in a GSA–National Association of Geoscience Teachers field trip that she had led to examine classic geologic sites in Britain and suggested the theme of her dissertation.

Stout is a councilor with the Geological Society of America, a past president of the National Association of Geoscience Teachers (NAGT), serves on the American Geological Institute's Higher Education Committee, has served on the American Geophysical Union's Education and Human Resources Committee, and has been involved in developing and running National Science Foundation-funded workshops with geological organizations for teachers and professors oriented toward the use of technology in the classroom. Stout received her B.A. (1963) and M.A. (1965) degrees in geology from Bowling Green State University, and her doctoral degree (1987) from Claremont Graduate School in geoscience education.

Eldridge M. Moores

---

A meeting of the Cordilleran Section of the Geological Society of America was the first professional meeting I ever attended. My mind has lost the year, but the meeting was in Portland, Oregon. I was working in the Baker area of northeastern Oregon. The meeting featured Howard Brooks and Norm Wagner expounding on what they had learned about the part of Oregon I was just beginning to understand. The list of Cordilleran Section meeting revelations has lengthened over the years and my mind has been stretched over much of the Western United States and Canada.

This section was one of the first to embrace the importance of improving our teaching of geosciences. I am grateful for the seedbed it has provided future educators.

Greg Wheeler, President
National Association of Geology Teachers

Geological Society of America
Special Paper 338
1999

# *Early history of the Cordilleran section*

**Dorothy L. Stout**
*Cypress College, Cypress, California 90630; e-mail gaea@deltanet.com*

> On all sides they clamor: Of what interest to us are
> the sayings or deeds of old?
> We are self-taught; our youth has learned for itself.
> Our band does not accept the dogmas of the ancients.
> We do not burden ourselves in following their utterances.
> Rome may cherish the authors of Greece.
> I dwell on the Petit-Pont and am a new authority.
> It may have been discovered before; I boast it is mine.
> —*John of Salisbury*
> *The Entheticus, thirteenth century*

> I have happily lived during the plate tectonic revolution that has so successfully revealed how Earth's crust wrinkles and breaks and pieces move about, when we have learned that most of crustal history is recorded in blocks and slices caught up in continental terranes, and only parsimoniously preserved. I have witnessed a second revolution coincident with the growth of plate tectonics: a huge increase in understanding of what we know about Earth history during more than a half century of mapping, dating, geochemistry, geophysics, and other approaches, as geologists have scurried over our planet. I have also lived at a place near mountains and deserts where rocks revealing crustal history are magnificently exposed. How fortunate I have been!
> —*John Crowell (1996), 1995 Penrose Medalist,*
> *California student, teacher, researcher, and resident*

## INTRODUCTION

The Centennial of the Cordilleran Section of the Geological Society of America (GSA) was celebrated June 2–4, 1999, during the ninety-fifth annual meeting. The first meeting of the Cordilleran Section was held December 29–30, 1899, at the California Academy of Sciences in San Francisco and the University of California (in Berkeley) in response to a call issued by Professor Andrew C. Lawson of the University of California.

In its early years, GSA, which was established in 1888, had good representation from geologists working in the western United States at annual meetings on the East Coast. G. K. Gilbert of the U.S. Geological Survey (USGS) was the fourth president in 1892 (and served again in 1909), and Joseph LeConte, Professor of Geology at the University of California, became the eighth president in 1896. However, by the late 1890s travel back to the East Coast for the annual GSA meetings between Christmas and New Years had become increasingly difficult for western geologists. Largely because of this travel burden, a regional geological club was founded in early 1899, which later that year became the Cordilleran Section, the first section of GSA, 13 years after its establishment.

---

Stout, D. L., 1999, Early history of the Cordilleran section, *in* Moores, E. M., Sloan, D., and Stout, D. L., eds., Classic Cordilleran Concepts: A View from California: Boulder, Colorado, Geological Society of America Special Paper 338.

Table 1 lists the officers of the Cordilleran Section and the location of the annual meetings. Since its inception, the section has provided a forum along the North American Pacific Coast for discourse on geologic issues, problems, and discoveries (see Dickinson, this volume). The minutes of meetings (preserved at GSA headquarters) are an often unadorned and unembellished record of leadership, changes, and events. In this paper I attempt to capture the setting of the section's establishment and early development.

The Cordilleran Section originated in heady times. By 1900 Clarence King's early (1880s) vision of a USGS aiding the industrial progress of the country was in full swing. Completion of the great surveys of the western territories had provided massive amounts of new scientific data. These earlier exploratory expeditions led to a need for more geological detail. The USGS met this need with its geological folios as good topographic maps became available.

## Initial Members

At its inception the Cordilleran Section was dominated by geologists at a few geological departments which had evolved in the late nineteenth century, principally at the University of California (now the Berkeley Campus) and Stanford University. These departments were initially staffed by imported geologists, mostly trained in the eastern United States or in Europe. However, by the 1890s rigorous programs in geological training were established in California.

The premier Cordilleran Section member was its first chair, Joseph LeConte. He had arrived in California in the 1860s, the same decade as King (and the transcontinental railroad). LeConte, the eighth president of GSA (in 1896) and the first chair of the Cordilleran Section, spent the first half of his life in the southern United States. Initially trained as a medical doctor,

John Muir. Bancroft Library, University of California, Berkeley

he went on to pursue an interest in geology with Louis Agassiz at Harvard. In 1869 at age 46, he left the war-torn South to take up a position teaching natural history at the newly established University of California. After his first year of teaching he accompanied his students to Yosemite, where he met John Muir, with whom he later collaborated on a number of subjects, including glaciation. Muir, LeConte, and the first president of Stanford University, David Starr Jordan, helped found the Sierra Club in 1892.

LeConte was a self-taught scientific generalist, who wrote papers on such diverse subjects as medicine, geology, mining, philosophy, and evolution. His book *Elements of Geology* (1878) quickly became the leading American textbook in geology and, largely through LeConte's efforts, by 1900 the University of California had become known as the leading center for the study of science in western North America (Steller–Stout, 1987).

Andrew C. Lawson came to the University of California as an assistant professor of mineralogy and geology in the early 1890s. Lawson, who had recently received his doctorate from The Johns Hopkins University, published his interpretation of the complex chronology of the Precambrian shield in Canada in the first volume of the GSA *Bulletin* (see also Lawson, 1893, 1895). At the University of California Lawson taught the laboratory and field courses and handled the modern scientific aspects of geology, while LeConte concentrated on the philosophical bases of science. Lawson's field work in the San Francisco area

Clarence King with Cotter (Muleskinner) and Gardner. Bancroft Library, University of California, Berkeley

> We crack the rocks and make them ring,
> And many a heavy pack we sling,
> We run our lines and tie them in,
> We measure strata thick and thin,
> And Sunday work is never sin,
> By thought and dint of hammering.
> Andrew C. Lawson

David S. Jordan, Bancroft Library, University of California, Berkeley

led to the first publications on the San Andreas fault (Hill, 1981; see also Hill and Dibblee, 1953).

Other founding members included John C. Merriam (GSA president, 1919), who was hired by Lawson in 1895 to teach paleontology, and Ernest W. Hilgard, who had been professor of agriculture and botany since 1875 and was a geologist and expert on soils. George D. Louderback received his doctorate from Berkeley in 1899 and in 1906 returned as a faculty member, eventually taking over the subject of petrology from Lawson.

Founding members from Stanford included John C. Branner and James P. Smith. New competition in geological education had begun with the California State Senate's rebuff of Governor Leland Stanford's proposal to build another university. In 1891 David Starr Jordan assumed the presidency of Leland Stanford Junior University with the aim of making it the "Harvard of the West." Jordan hired Branner, an established geologist with experience as State Geologist of Arkansas, and at Cornell University and Indiana University, to develop Stanford's Department of Geology and Mining. Branner and Smith, a paleontologist, taught all of the geology courses for a number of years, inspiring such students as Herbert C. Hoover (who was hired in the department because he could type). In 1898 Waldemar Lindgren (GSA president, 1924) came to Stanford from the USGS to serve as associate professor of mining and metallurgy.

Branner and Lawson were firm believers in field experience for students learning geology. Although Lawson introduced field courses at Berkeley, it was through Branner's efforts that Stanford became the first American university to require a summer field geology course as part of the baccalaureate degree program (Norris, 1981). At Stanford all geology majors had to spend two summers in the field, usually compiling a geologic map of a quadrangle. Soon most American universities were requiring field geology courses. The USGS's employment of geologists from Berkeley and Stanford in the gold fields and in other mapping projects was significant. The leadership and education emanating from professors at both institutions who were actively engaged in teaching and research provided important steps in the development of geological training in California. Berkeley and Stanford had a particular advantage over students being trained in the eastern United States. Most of the completed geology in the western United States was generalized, and detailed studies were needed. The

Stanford's first graduating class. Seated to the left is J. P. Smith; seated to his left is J. C. Branner. Standing on the far right is J. J. Hollister. Herbert Hoover is standing second from left. Photograph from the Geohistory Archives, School of Earth Sciences, Stanford University.

**TABLE 1 (on this and facing page). LIST OF CORDILLERAN SECTION OFFICERS AND MEETINGS, 1899 TO PRESENT.**

| YEAR | CHAIR | SECRETARY | TREASURER | MEETING SITE |
|---|---|---|---|---|
| 1899 (1) | J. LeConte | J.E. Talmage | A. Lawson | Cal Acad. Sci. & UC Berkeley |
| 1900 (2) | J. LeConte TC<br>W. Blake | J.E. Talmage | A. Lawson | Cal Acad. Sci. & UC Berkeley |
| 1901 (3) | W.C. Knight TC<br>H.W. Turner | A.S. Eakle | A. Lawson | Cal Acad. Sci. |
| 1902 (4) | H. W. Turner TC<br>H.W. Fairbanks | J.C. Merriam | A. Lawson | Cal Acad. Sci. |
| 1903 (5) | H.W. Fairbanks TC<br>E.W. Hilgard | W.C. Knight | A. Lawson | UC Berkeley |
| 1904 (6) | E.W. Hilgard | G.D. Louderback | A. Lawson | UC Berkeley |
| 1905 (7) | W.G. Tight | G.D. Louderback | A. Lawson | UC Berkeley |
| 1906 (8) | J.C. Branner | W.C. Mendenhall | G.D. Louderback | Stanford |
| 1907 (9) | A. Lawson | G.K. Gilbert | G.D. Louderback | Albuquerque, N.M. with GSA |
| 1908 (10) | A. Lawson TC<br>J.C. Branner | G.K. Gilbert | G.D. Louderback | Stanford |
| 1910 (11) | A. Lawson | G.K. Gilbert | G.D. Louderback | UC Berkeley |
| 1911 (12) | A. Lawson | G.K. Gilbert | G.D. Louderback | UC Berkeley |
| 1912 (13) | A. Lawson | W.S. Tangier Smith | G.D. Louderback | Stanford |
| 1913 (14) | J.C. Branner TC<br>A. Lawson | W.S. Tangier Smith | G.D. Louderback | UC Berkeley |
| 1914 (15) | J.C. Branner TC<br>A. Lawson | W.S. Tangier Smith | G.D. Louderback | U Washington |
| 1915 (16) | H. Foster Bain TC<br>C. F. Tolman | C. E. Weaver | G. D. Louderback TC<br>J. A. Taff | Stanford & UC Berkeley with GSA |
| 1916 (17) | C.F. Tolman TC<br>G.D. Louderback | C.E. Weaver | J.A. Taff | San Diego |
| 1917 (18) | C.F. Tolman | | J. A. Taff | Stanford |
| 1918 | No Meeting | | | |
| 1919 (19) | B.L. Clark TC | | G.D. Louderback AS | Throop (Caltech) |
| 1920 | No Meeting | | | |
| 1921 (20) | G.D. Louderback | C.F. Tolman | A.F. Rogers | UC Berkeley |
| 1922 (21) | G.D. Louderback | C.F. Tolman | A.F. Rogers | Stanford |
| 1923 (22) | W.S.W. Kew | E. Blackwelder | J.P. Buwalda | UC Berkeley |
| 1924 (23) | F.M. Anderson<br>E. Blackwelder TC | | J.P. Buwalda | Stanford |
| 1925 (24) | W.D. Smith | E. Blackwelder | J.P. Buwalda | UC Berkeley |
| 1926 (25) | E. Blackwelder | W.J. Miller | J.P. Buwalda | Stanford |
| 1927 (26) | W.J. Miller | W.S.W. Kew | J.P. Buwalda | UCLA (Southern Branch) |
| 1928 (27) | R. Anderson | A.F. Rogers | R.W. Chaney | UC Berkeley |
| 1929 (28) | B.L. Clark | F.L. Ransome | R.W. Chaney | Stanford |
| 1930 (29) | T.W. Vaughan | E. Blackwelder | R.W. Chaney AS<br>H. Schenck | UC Berkeley |
| 1931 (30) | J.P. Buwalda | T.W. Vaughan | R.W. Chaney | Caltech |
| 1932 (31) | C.E. Weaver | E. Blackwelder | R.W. Chaney | Stanford |
| 1933 (32) | R.T. Hill | A.C. Lawson | A.O. Woodford | UCLA |
| 1934 (33) | R.W. Chaney | | A.O. Woodford | UC Berkeley |
| 1935 (34) | B. Willis | W.S.T. Smith | A.O. Woodford | Stanford |
| 1936 (35) | A.O. Woodford | C. Stock, AS<br>A. Water | C.A. Anderson | Caltech |
| 1937 (36) | G.E. Goodspeed | A. C. Waters | C.A. Anderson | UC Berkeley |
| 1938 (37) | A.F. Rogers | H. S. Gale | C.A. Anderson | Stanford |
| 1939 (38) | R.D. Reed | H. Williams | C.A. Anderson | UC Berkeley with GSA |
| 1940 (39) | J. Gilluly | V.P. Gianella | C.A. Anderson | UCLA |
| 1941 (40) | C. Stock | H.G. Schenck | C.A. Anderson | Stanford |
| 1942 (41) | N.L. Taliaferro | H.S. Gale | C.A. Anderson | Caltech |
| 1943 | No Meeting | | | |

| YEAR | CHAIR | SECRETARY | TREASURER | MEETING SITE |
|---|---|---|---|---|
| 1944 | No Meeting | | | |
| 1945 | No Meeting | | | |
| 1946 (42) | H. G. Schenck | U.S. Grant | C.A. Anderson | UC Berkeley |
| 1947 (43) | V.P. Gianella | S.W. Muller | C.M. Gilbert | Stanford |
| 1948 (44) | H. Williams | I. Campbell | C.M. Gilbert | Caltech |
| 1949 (45) | S.W. Muller | E.D. McKee | C.M. Gilbert | UC Berkeley |
| 1950 (46) | B. Gutenberg | J.H. Mackin | C.M. Gilbert | U Washington |
| 1951 (47) | J.H. Mackin | D.T. Griggs | V.L. VanderHoof | U Southern California |
| 1952 (48) | I. Campbell | C. Gilbert | V.L. VanderHoof | U Arizona |
| 1953 (49) | W. C. Putnam | H. Coombs | V.L. VanderHoof | Stanford |
| 1954 (50) | C. M. Gilbert | L. Staples | V.L. VanderHoof | U Washington |
| 1955 (51) | M. L. Hill | T. Clements | V.L. VanderHoof | UC Berkeley |
| 1956 (52) | H.A. Coombs | W.F. Barbat | V.L. VanderHoof | U Nevada, Reno |
| 1957 (53) | A.O. Woodford | C.W. Merriam | V.L. VanderHoof | UCLA |
| 1958 (54) | F.J. Turner | K.O. Emery | V.L. VanderHoof | U Oregon |
| 1959 (55) | B. Page | G. Oakeshott | V.L. VanderHoof | U Arizona |
| 1960 (56) | G. Oakeshott | P. Misch | V.I. VanderHoof | U British Columbia |
| 1961 (57) | C. Campbell | E.M. Baldwin | V.I. VanderHoof | San Diego State College |
| 1962 (58) | J. Noble | E. Bailey | R.V. Fisher | U Southern California |
| 1963 (59) | K.B. Krauskopf | W.C. Smith | R.V. Fisher | UC Berkeley |
| 1964 (60) | C.W. Merriam | V.J. Okulitch | R.V. Fisher | U Washington |
| 1965 (61) | J. Verhoogen | G. Tunnel | R.V. Fisher | Fresno State College |
| 1966 (62) | W. Easton | E.R. Larson | B. McKee | U Nevada, Reno |
| 1967 (63) | P. Misch | G. Gastil | B. McKee | UC Santa Barbara |
| 1968 (64) | R. Wallace | D. McIntyre | B. McKee | U Arizona |
| 1969 (65) | A. Hietanen-Makela | J. Crowell | B. McKee | U Oregon |
| 1970 (66) | J. Crowell | W. Gussow | B. McKee | Cal State, Hayward |
| 1971 (67) | D.B. Slemmons | T.W. Dibblee, Jr. | B. McKee | UC Riverside |
| 1972 (68) | J. Hazzard | G.A. Macdonald | B. McKee | U Hawaii |
| 1973 (69) | A. Waters | P.D. Snavely | B. McKee | Portland State U |
| 1974 (70) | P.D. Snavely | W.R. Dickinson | M.L. Stout | U Nevada, Las Vegas |
| 1975 (71) | W.R. Dickinson | C.A. Hopson | M.L. Stout | Cal State, Los Angeles |
| 1976 (72) | C.A. Hopson | R.G. Coleman | M.L. Stout | Washington State U Pullman |
| 1977 (73) | R.G. Coleman | W.G. Ernst | M.L. Stout | Sacramento State U |
| 1978 (74) | W.G. Ernst | A. McBirney | M.L. Stout | Arizona State U |
| 1979 (75) | A. McBirney | H.P. Taylor | M.L. Stout | San Jose State U |
| 1980 (76) | T.L. Péwé | E.J. Dasch | M.L. Stout | Oregon State U |
| 1981 (77) | E.J. Dasch | W.H. Matthews | M.L. Stout | Hermosillo, Mexico |
| 1982 (78) | W.H. Matthews | A. Navarro-Galindo | M.L. Stout | Cal State, Fullerton |
| 1983 (79) | G.A. Davis | M.C. Blake, Jr. | M.L. Stout | U Utah |
| 1984 (80) | M.C. Blake, Jr. | J.W. Schopf | M.L. Stout | U Alaska |
| 1985 (81) | J.W. Schopf | R. Armstrong | M.L. Stout | U British Columbia |
| 1986 (82) | R. Armstrong | L. Anderson | B.A. Blackerby | Cal State, Los Angeles |
| 1987 (83) | L. Anderson | W. Hamilton | B.A. Blackerby | U Hawaii |
| 1988 (84) | W. Hamilton | R. Yeats | B.A. Blackerby | U Nevada, Las Vegas |
| 1989 (85) | R. Yeats | M. Zoback | B.A. Blackerby | Eastern Washington U U Idaho |
| 1990 (86) | M. Zoback | M. Stout | B.A. Blackerby | U Arizona |
| 1991 (87) | M. Stout | P. Abbott | B.A. Blackerby | San Francisco State U |
| 1992 (88) | P. Abbott | R. Schweickert | B.A. Blackerby | U Oregon |
| 1993 (89) | R. Schweickert | D. S. Cowan | B.A. Blackerby | U Nevada - Boise State U |
| 1994 (90) | D.S. Cowan | E.M. Moores | B.A. Blackerby | Cal State, San Bernardino |
| 1995 (91) | E.M. Moores | C.J. Hickson | B.A. Blackerby | U Alaska |
| 1996 (92) | C.J. Hickson | J.C. Moore | B.A. Blackerby | Portland State U |
| 1997 (93) | J.C. Moore | A.B. Till | B.A. Blackerby | U Hawaii |
| 1998 (94) | A.B. Till | M. Rusmore | B.A. Blackerby | Cal State, Long Beach |
| 1999 (95) | G.A. Davis | M. Rusmore | B.A. Blackerby | UC Berkeley |

Notes: TC = temporary chair; AS = Assistant Secretary

Andrew C. Lawson's first field class. Near Carmel Mission, Thanksgiving, 1891. Department of Geology and Geophysics, University of California, Berkeley

climate was conducive to year-round study of the complex and varied geology, which posed intriguing problems. These two universities emphasized field geology and turned out young men well trained for leadership and geologic work.

## ORGANIZATION OF THE CORDILLERAN SECTION

When the travel burden imposed on West Coast geologists prompted Lawson in 1899 to meet with Joseph LeConte, J. C. Merriam, J. P. Smith, E. W. Hilgard, G. D. Louderback, and others to discuss starting a West Coast geological organization, the result was the formation of the Cordilleran Geological Club. The first meeting was held in Berkeley on February 4, 1899, and a committee was appointed to look into affiliation with either the California Academy of Sciences or the 13-year-old Geological Society of America. At the second meeting, held April 29, 1899, at the California Academy of Sciences in San Francisco, the committee decided to affiliate with the academy under the name Geological Section of the California Academy of Sciences. This connection continued until the seventh meeting (April 12, 1902), when it was voted to sever connections with the academy and change the name to The Le Conte Club. The club had always been open to anyone with a degree in geology, but was largely supported by geologists from Stanford, UC Berkeley, and the USGS; hence it met alternately each year at the campuses of Stanford and Berkeley. (The club met until 1971, when, owing to several factors—the ease of attending the annual Cordilleran Section meetings, competition from the newly formed Peninsula Geological Society, and from talks given at the USGS in Menlo Park—it was decided to terminate the club.)

The same geologists who organized the Cordilleran Geological Club had continued their efforts to affiliate with GSA, which at that time had a total membership of 239 members (Eckel, 1982). The annual meeting that year in Washington, D.C., had 70 registrants. In response to Lawson's petition and request to the Council of the Society to establish a West Coast section, GSA Secretary H. W. Fairchild stated that such an organization would be within the constitution and consistent with the purpose of the society.

Formal authorization of the Cordilleran Section as part of the Geological Society of America occurred at 10:30 a.m. on December 29, 1899, in the council room of the California Academy of Sciences in San Francisco. LeConte and Lawson were elected chairman and secretary of the section for the first year. These two officers, together with Professor J. E. Talmage, acted as an executive council for the year. The GSA Council's approval of the section averted what might well have come to secession of a large and influential segment of the fellowship (Eckel, 1982, p. 41).

Secretary Lawson's handwritten notes from this first annual meeting identify the following members as present: Professors Joseph LeConte, J. C. Branner, E. W. Hilgard, A. C. Lawson, J. E. Talmage (University of Utah), J. C. Merriam, and H. W. Fairbanks, and visitors George D. Louderback and J. W. Sinclair (paleontology student at Berkeley).

Minutes of the first meeting in Lawson's handwriting.

John C. Merriam
Geological Society of America

Ernest W. Hilgard
Bancroft Library, University
of California, Berkeley

George D. Louderback
Department of Geology and
Geophysics, University of
California, Berkeley

Joseph LeConte
Geological Society of America

John C. Branner
Geological Society of America

Andrew C. Lawson
Geological Society of America

J. E. Talmage
Geological Society of America

J. W. Sinclair
Geological Society of America

Photos of attendees at the first Cordilleran Section meeting (except H. W. Fairchild, for whom no photo was available).

The Cordilleran Section was formally approved as a part of the Geological Society of America, with the understanding that for scientific and social purposes the section should be open to any fellow of the society, but for business purposes, the membership should comprise only fellows resident in the Cordilleran belt of North America.

Below is the text of the telegram received by Cordilleran Secretary Andrew Lawson from GSA Secretary Fairchild formally recognizing the new section. A postal card followed.

A two-day meeting ensued—the first day in the council room of the California Academy of Sciences in San Francisco and the second day at the Geology Department of the University of California at Berkeley. The papers presented at the meeting reflect interest in a wide variety of subjects.

• F. W. Cragin, "The discovery of a goat antelope in the cave fauna of Pike's Peak region."

• J. C. Merriam, "On the occurrence of ground-sloths in the Quaternary of Middle California" and "Classification of the John Day beds."

• J. E. Talmage, "Notes concerning erosion forms and exposures in the deserts of South Central Utah," "On certain peculiar markings on sandstones from the vicinity of Elen Cañon, Arizona," and "Conglomerate puddings from the Paria River, Utah."

• H. W. Fairbanks, "The peneplain question upon the Pacific Coast."

• W. S. Tangier Smith, "A topographic study of the islands of Southern California."

• Joseph LeConte, "An early geological excursion (informal narrative of a camping trip in 1844 to Lake Superior)."

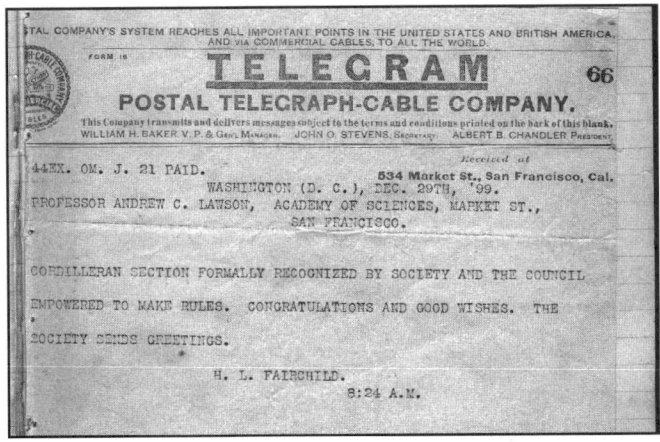

> Have wired you that the Society has officially authorized your Cordilleran Section and referred to the Council matters of conditions or regulations. Will write fully later. Please send me at least two clean samples of all your printing for our records, — several copies is possible.
>
> Dec. 29 1899.
>
> Herman LeRoy Fairchild, Secretary.

- Bailey Willis, "Some coast migrations, Southern California."
- J. C. Branner, "The sandstone reefs of Brazil."
- E. W. Hilgard, "The geological significance of soil study."
- E. W. Claypole, "The American Devonian placoderms."
- Andrew C. Lawson, "The Berkeley Hills—A detail of Coast Range geology."

## FIRST YEARS OF THE CORDILLERAN SECTION

The second annual meeting of the Cordilleran Section was again at the California Academy of Sciences, San Francisco, December 28–29, 1900. Wilbur C. Knight (Wyoming) was elected chairman; Lawson was again elected secretary, a position he held until 1905. Arthur S. Eakle, Berkeley mineralogy professor, was elected councilor. A committee established to formulate regulations for the governance of the section consisted of Andrew C. Lawson, John C. Branner, and E. W. Claypole. Papers read include those by William P. Blake (a geologist on the Pacific Railroad Survey), Andrew C. Lawson, Wilbur C. Knight, and John C. Merriam. On the following day papers given included those by H. W. Turner of the USGS (chair of the Cordilleran Section, 1902), H. W. Fairbanks (chair, 1903), E. W. Hilgard (chair, 1904), and Andrew C. Lawson.

The third meeting, the first held after rules of operation had been adopted, was again held in the Academy of Sciences, San Francisco on December 30–31, 1901. As secretary of the organization, Lawson wrote a brief report on the rules adopted for the section by the council of the society at Denver, on August 26, 1901. These rules related to officers, geographic limits, membership, dates of meetings, expenses, and publications. The section was finally official!

### Early meetings, rules, and awards

From time to time, rules and regulations of the section were reexamined. For example, in 1915 the GSA Council

1915 GSA Banquet. Photograph from the archives of the Department of Geology and Geophysics, University of California, Berkeley.

noted some laxity in presentations and adopted the following rules.

1. All titles of papers are to be accompanied by abstracts, suitable for printing in the proceedings, in case the full paper is not published.

2. If the length of time for presentation of a paper is not given in the program, the time shall be limited to 15 minutes.

3. The presiding officer will enforce the time limit on each paper, unless the meeting extends the time by special vote.

4. Papers whose authors are not present when the title is called will go to the end of the program, unless held in place by special vote.

The section minutes in 1928 indicate concern about the quality and timing of papers being presented: A motion from Chair Blackwelder and seconded by Lawson reflects "the Council was instructed to regulate the time assigned to papers for presentation on the program and to reject papers not suitable for the program."

Also in 1928, Blackwelder suggested the desirability of the society offering a prize for the best paper presented, and A. C. Lawson offered to contribute $50 for such a prize, "to be awarded annually to a young man or woman who is under 30 years of age, who shall have presented to the section the most satisfactory and most important paper setting forth the results of his own research in geology." The award was presented to Howel Williams, a student of Lawson, for his paper entitled "The Geology of the Marysville Buttes." Unfortunately, the following year, even though a number of members each contributed $5 for the award, they found no one deserving of the award.

### Expansion south

In 1916 the seventeenth section meeting was held in San Diego, the first in southern California. A true integration of the sciences took place, because many other science organizations were represented at the meeting: the Pacific Division of the American Association for the Advancement of Science (an organization started as a result of breaking off from GSA); the Astronomical Society of the Pacific; the San Francisco Section of the American Mathematical Society; the California Section of the American Chemical Society; the Puget Sound Section of the American Chemical Society; the Technical Society of the Pacific Coast; the Seismological Society of America; the Western Society of Naturalists; the Cooper Ornithological Club, Pacific Slope Branch; the American Association of Economic Entomologists; the San Francisco Society; the Archaeological Institute of America; and the California Academy of Sciences.

Expansion of geology programs into southern California included geology departments founded at the University of Southern California in 1909, the University of California, Southern Branch (now UCLA), in 1919, and Throop College of Technology, which became the California Institute of

A. O. Woodford, Geological Society of America

> While I was a student at Pomona College, my professor, A. O. Woodford, was secretary of the section. The meetings alternated between northern and southern California and also in the north between Berkeley and Stanford and in the south between UCLA and Caltech. Woodford attended all of them and never failed to return with a summary for us students, of what was presented. Because of finances, my attendance was limited to those at UCLA and Caltech. The great thing about attending was that ALL the important western geologists were there and a college student could hear and meet them if he wanted to. I remember Bailey Willis, A.C. Lawson, Eliot Blackwelder, George Louderback, Nicholas Taliaferro, John Buwalda, Chester Stock, John Maxson, Ian Campbell, W. J. Miller, Ralph Arnold, Harry O. Woods, Norman E.A. Hinds, W. M. Davis, F. L. Ransome, and toward the end of the period, the friendly Beno Gutenberg and the others of the new seismology group, C. F. Richter and Hugo Benioff. As an undergraduate and graduate student and later as a teacher, I expected that I must learn all there was to know about all aspects of geology. Outside of a good geology library, the Cordilleran Section meetings were wonderful chances to be exposed to all phases of geology and the experts in those fields.
>
> H. Stanton Hill, 1998

Technology in 1920. Prominent Cordilleran geologists from southern California included Throop professors John Buwalda (Cordilleran Section secretary, 1923–1927 and chair, 1931) and Chester Stock (Section chair, 1941). The 1919 Cordilleran Section meeting was held at Throop in Pasadena. A. O. Woodford (secretary, 1933–1935; chair, 1936 and 1957) started the department at Pomona College, and two of Woodford's students were Mason Hill (section chair, 1954) and C. A. Anderson (Section Secretary, 1936–1946).

*Branner Club*

In 1922, the year preceding John C. Branner's passing, the southern California counterpart of the LeConte Club was organized to promote good fellowship among geologists and to foster the study of geology and cooperation with kindred scientific associations in southern California. The club was named for the mentor who had trained many of the geologists working in southern California. The first Branner Club meeting in conjunction with the Cordilleran Section meeting occurred in 1927. This meeting, hosted at UCLA, had William Morris Davis (GSA president, 1911) as the speaker. Secretary John Buwalda described the speech as "a most interesting anecdotal narrative regarding two eminent geologists, who had contributed greatly to geologic knowledge of the West and the founding of the U.S. Geological Survey, G. K. Gilbert and J. W. Powell." The Branner Club continues today with quarterly meetings held at the Athenaeum at Caltech.

*Field trips*

From foot to horseback to airplane, field trips have highlighted section meetings since the beginning. For example, early field excursions included Hunter's Point, Point San Pedro, Santa Cruz ocean beaches, Mt. Diablo, and a five-day trip to Yosemite Valley. Yosemite field-trip leaders F. E. Matthes and F. C. Calkins of the USGS described in the 1915 meeting announcement features in part as follows: "Unusual glacial effects are exposed for inspection in the Tuolumne Meadows locality and from this camp excursions can be made by saddle or on foot to study glacial and other geology in the canyons and high Sierras. The details of the geology are familiar to Messrs. Matthes and Calkins, who kindly agree to direct us to those most interesting." Round trip fare, San Francisco to Yosemite, is $23..."

Accommodations ranged from spartan to quite luxurious. For example, the field trip offered by Ralph Reed of Texas Oil from Ventura to Taft in conjunction with the Caltech meeting of 1936 advertised in the meeting announcement that participants could stay at the Hotel Ventura for "$1.00 a day each person with bath, $1.50 each person with shower, or $2.00 each person with tub bath."

The forty-seventh meeting in 1951, hosted by the University of Southern California, offered for $10.00 an aerial trip over the Los Angeles basin along the San Andreas fault from Cajon Pass to the Devil's Punchbowl, returning by way of the San Gabriel Range and the northern portion the Los Angeles basin.

The effects of World War II took their toll on the section. Field trips scheduled for the 1942 meeting at Caltech were canceled because of the rubber shortage. The 1943 Berkeley meeting was canceled because of war-related activities and transportation problems.

*Semicentennial Cordilleran Section meeting*

The forty-fifth annual meeting was held in Berkeley on April 15–16, 1949, to celebrate the semicentennial of the Cordilleran Section. The Paleontological Society, Pacific Coast Branch, and the Seismological Society of America met at the same time. The technical sessions, which were held at the Claremont Hotel in Oakland, included sessions on petrology, structure, and physiography, mineralogy and engineering geology, and stratigraphy and sedimentation.

## WESTERN GEOLOGIC ISSUES AT THE TURN OF THE CENTURY

The quest for gold and petroleum; the imbalance of precipitation to population; the threat from hazards such as volcanic eruptions, landslides, and earthquakes; insight from observations of climate change; paleontology's growing importance as a tool in petroleum exploration and in the debate on the age of the Earth, all combined to enhance comprehension of how the Earth works. These subjects were critical for turn of the century geologists and residents on the West Coast.

During this time, the USGS, established in 1879, and the California Mining Bureau (now California Division of Mines and Geology), established in 1880, had made strides in systematically examining the geology of California. Mining was to provide the basis for subsequent industrialization, and the USGS project to map the entire United States meant expanding efforts in California.

1895 Stanford geological expedition to the Klamath Mountains near Redding, California. Photograph from the Geohistory Archives, School of Earth Sciences, Stanford University.

Cover of the 1949 semicentennial program

Mapping of the northern and southern portions of the Sierra Nevada in the 1880s and 1890s by USGS geologists H. W. Turner (1902 Cordilleran chair), assisted by F. L. Ransome, and of the east-central Sierra Nevada by Waldemar Lindgren, assisted by Stanford student Herbert Hoover, resulted in publication in the late 1890s of the *Gold Belt Folios*. These publications received great attention because they appeared at a time when President McKinley's emphasis on the gold standard had sharply increased interest in gold. Development of dredging technology and improvement in mining and milling methods made lower grade deposits profitable. In 1898 Lindgren joined the faculty at Stanford and in 1924 became GSA president.

The Sierra Nevada and the Peninsular Ranges contained some of California's greatest mineral wealth. Their daunting geologic complexity led to detailed field and laboratory studies to try to unravel the geological relationships controlling economic deposits.

Also crucial to the development of California was water. A lingering drought in the latter half of the 1890s caused the USGS to initiate stream measurements in California. In central California the USGS surveyed potential reservoir sites, principally Hetch Hetchy, for capacity and cost and initiated systematic measurements on the San Joaquin, Kings, and Salinas Rivers. In southern California, with an ever-increasing population, water was running dry at the site of El Pueblo de Nuestra Senora la Reina de Los Angeles de Porciuncula. Lt. Gaspar de Portola had chosen this site in 1769 for a mission and a large settlement because of the presence there of a perennial stream. By 1899, Los Angeles required new sources of water, and in 1900 on Alta California's southern border, George Chaffee developed a canal to bring Colorado River water to the Salton trough, which he named the Imperial Valley.

Petroleum production also was increasing (Hoots and Bear, 1954). In 1900 some 4 million barrels of oil were pumped in California by ~2400 incorporated oil companies. By 1905, California's annual petroleum output was 34 million barrels pumped from 2450 producing wells; Kern River, the largest field, produced 15 million barrels.

Geologic knowledge acquired in universities was being applied to oil exploration, because surface mapping of anticlines had led to successful development of fields. Random prospecting augmented the search for oil, but knowledge of surface geology began to dominate discovery. Stanford and Berkeley increasingly concentrated on training petroleum geologists. Ralph Arnold, a 1902 graduate of Stanford, was an unusually effective early petroleum geologist. The simultaneous evolution of the petroleum industry and the burgeoning geology departments at Stanford and Berkeley led to a virtual monopoly by those departments in training geologists.

In the 1880s and 1890s USGS geologist J. S. Diller mapped the junction of the Klamath Mountains, the Sierra Nevada, and the Cascade Range. His work resulted in USGS folios covering parts of northern California and Lassen Peak in 1889 and Mount Shasta in 1895. I. C. Russell of the USGS worked out the chronological sequencing of glaciation, Pleistocene lake levels, and volcanism in the area of Mono Lake during his work in the Sierra Nevada and the western Great Basin. The eruption of Lassen Peak in 1914 led to widespread recognition of the need to study volcanism in California, a recognition reflected in subsequent Cordilleran section meetings.

At the turn of the century, engineering geology did not exist as a separate field. It required a larger population, some severe problems that needed an understanding of geology, and some

Oil fields in Los Angeles, 1908. Bancroft Library, University of California, Berkeley

dramatic events to establish this field. These events soon occurred, with rapid population growth, drought, increasing need for importing water, earthquakes (e.g., 1906 in San Francisco and 1933 in Long Beach), and the eruption of Lassen Peak.

*San Andreas, tectonics, and seismicity*

The San Andreas fault has been the object of many Cordilleran papers, field trips, and Penrose Conferences, all starting with the impact of the San Francisco earthquake of 1906. The combined talents of John C. Branner, Andrew Lawson, George Louderback, Harry F. Reid (elastic rebound theory), G. K. Gilbert, and others converged. (G. K. Gilbert, who was at Berkeley at the request of President Roosevelt to study the effects of hydraulic mining on agriculture in the Sacramento Valley, had previously stated that "It is the natural and legitimate ambition of a properly constituted geologist to see a glacier, witness an eruption and feel an earthquake," but until 1906 Gilbert had not "felt" a major earthquake in a major way!) These events, of course, tested all their theories of faulting, earthquakes, and mountain building.

> From California's subsurface come those recurring violent actions known as earthquakes, several of which have dealt rather harshly with man and his works during the period of historic record. These shocks have originated along faults, or breaks in the earth's crust that represent repeated slippage over very long spans of time. Thousands of faults are known within the state, and many of them can be classed as large in terms of their total displacements. Many of them also are geologically active in the sense of having moved within the past 10,000 years, and more than a few have been active in historic times.
>
> Richard Jahns, Geologic Jeopardy
> GSA President, 1971

At the meeting in 1906 Chair Branner initiated a tradition still continued today of presentations of "breaking events," for the program announced events surrounding the April 18, 1906, San Francisco earthquake. Branner exhibited a collection of photographs, showing the geologic effects of the earthquake and planned an excursion to the locus of the earthquake rift.

Other early papers were presented on faulting and earthquakes, including "Recent faulting in Owens Valley," by W. D. Johnson; "California earthquakes: A synthetic study of the recorded shocks" by H. O. Wood in 1910; "Mountain producing forces" by H. F. Reid in 1911. Also in 1910, with Lawson's influence, the section expressed its strong interest in "establishment of a National Bureau of Seismology organized under the Smithsonian Institution with power to collect seismology data, establish an observing station, study and investigate special earthquake regions within the national domain, and cooperate with other scientific bodies and organizations and individual scientists in forwarding the development and dissemination of seismological knowledge."

The aftermath of the San Francisco earthquake led to the establishment of the Seismological Society of America. Lawson led the efforts for it and served as the first vice chair. Subsequently, many meetings of the Cordilleran Section were held jointly with the Seismological Society. While Stanford and Berkeley continued as the most frequently selected sites for section meetings in the first half of the century, in 1914 the meeting was held at the University of Washington.

Secretary Buwalda of Caltech noted in the minutes that at the 1927 meeting at Stanford, "An event probably without precedent in any geological meeting held in North America occurred on Saturday afternoon at about 3:21. While Mr. J. P. Fox was presenting a paper dealing in part with the active fault line known as the Hayward Rift, a fairly strong earthquake, consisting of three or four impulses apparently in a N-W or NW-SE direction, shook the meeting room and slightly disturbed the assembled geologists."

At the 1929 meeting, M. L. Hill presented a paper entitled "A contribution on the structure of the San Gabriel Mountains,

Professor Bailey Willis with an early earthquake engineering model of the Alexander Building in San Francisco. The model was mounted on a shaking table driven by a cam cut from a strong motion seismogram. Photograph from the Geohistory Archives, School of Earth Sciences, Stanford University.

California," in which he recognized active reverse faults along the mountain front. In the 1930 meeting H. O. Wood and J. F. Buwalda presented a paper entitled "Horizontal displacement along the San Andreas fault in the Carrizo Plain, California." Partly from study of aerial photographs, they presented evidence for large-scale (miles to tens of miles) of horizontal displacement based on offset streams and other geologic features. Thus was born the concept of large-scale horizontal displacement on the San Andreas fault.

*Awareness of climate change and engineering issues*

Many meetings in the section have concentrated on climatic conditions and the relationship between glaciation and pluvial lake levels. For example, Eliot Blackwelder gave two talks in 1927 entitled "Evidence of a Third Glacial Epoch in the Sierra Nevada" and "Pleistocene Lakes of the Basin Range Province," and a third in 1929 on the still-remote Death Valley, including evidence for pluvial lakes there.

As the century wore on, the ever-growing population of southern California led to the development of more aqueducts. F. L. Ransome of Caltech (vice chair, 1929) and G. D. Louderback of Berkeley had written the authoritative geologic report on the 1928 St. Francis dam disaster (the second-highest number of fatalities in a natural event in California history). Ransome's and Blackwelder's 1930 talks, entitled "Geologic considerations affecting the choice of a route for the Colorado River aqueduct," and "History of the Colorado River," respectively, emphasized recognition of the role of geology in making informed decisions on engineering projects.

## CONCLUSIONS

By and large the leadership of the Cordilleran Section has been from academia, with notable exceptions from industry, for example, Ralph Reed (1933; Texas Oil) and Mason Hill (Atlantic-Richfield), or from government, for example, H. W. Turner (USGS), W. S. W. Kew (USGS), and Gordon Oakeshott (CDMG). Several members were chair of the section more than once (George Louderback, C. F. Tolman, and Greg Davis; Lawson holds the record, five times). Several secretaries have been very important contributors to the integrity and continuity of the section. They include Andrew Lawson, the prime mover in organizing the section (seven one-year terms), George Louderback (ten terms), Charles Anderson (nine terms), V. L. VanderHoof (eleven terms), Bates McKee (eight terms), Martin Stout (twelve terms), and Bruce Blackerby (fourteen terms). Four of these secretaries went on to become chairs (Lawson, Louderback, Chaney, and Stout).

The early influence of UC Berkeley and Stanford is reflected in the large number of meetings held at these institutions (Berkeley, 23; and Stanford, 15), primarily before 1950. Since then there has been much broader participation of institutions from the rest of the Cordilleran Section with meetings

---

Since its founding a century ago, it is doubtful whether any of GSA's geographically defined sections have had a greater influence on the scientific development of tectonics. I can only recall the last third or so of this century and the Cordilleran Section's role in it, but section meetings have often been the initial forum for the presentation of groundbreaking research in structure and tectonics–in some cases with worldwide implications. One example comes immediately to mind. I still remember a paper by Stephen Bezore at a meeting in Eugene, Oregon, in 1969. Bezore, a Master's student of Eldridge Moores at UC Davis, presented the results of his thesis research in the Mount St. Helena area near Napa Valley. He concluded that the enigmatic Coast Range ultramafic-mafic complex in this area was quite possibly the oceanic mantle and crust unconformably overlain by sedimentary rocks of the Knoxville Formation. To me, having puzzled over the origin of an ultramafic-mafic complex in the Klamath Mountains, it was the most exciting paper I had ever heard at any meeting! In the 1970s, Cordilleran Section meetings became the leading venue for the presentations of other discoveries of ophiolitic sequences from Baja California to Alaska. Recognition of ophiolites as remnants of oceanic lithosphere aided Cordilleran workers in proposing some of the first plate tectonic interpretations of any orogenic belt; many of the initial presentations of these ideas were at section meetings.

I personally regard the decade from the mid-1970s to the mid-1980s as the most exciting time in Cordilleran tectonic studies, with the section at center stage in the presentation of such studies. This was the decade when Cordilleran geology and research by Cordilleran geologists profoundly influenced tectonic thinking worldwide in two major topics—accretionary tectonics along convergent plate boundaries, and intraplate extensional tectonics involving low-angle normal faults and their footwall metamorphic core complexes. These Cordilleran-originated concepts proved contagious as geologists around the world began to reexamine their own orogens—and discovered examples of "suspect terranes" and extensional detachment faults in many of them.

Ophiolite emplacement, continental plate tectonics, accretionary tectonics, and extensional metamorphic core complexes: The record is clear that Cordilleran Section meetings played a profound and global role in the development and dissemination of seminal ideas on these major topics of the plate tectonics era.

Greg Davis
University of Southern California, 1998

in Alaska, British Columbia, Hawaii, Nevada, Arizona, and elsewhere in California. Five section chairs from Stanford also became presidents of the society (John C. Branner, Bailey Willis, Eliot Blackwelder, Konrad Krauskopf, and William Dickinson [although he was from the University of Arizona when he was 1994 GSA president]), three from UC Berkeley (Joseph LeConte, Andrew Lawson, John C. Merriam), two from Caltech (Ian Campbell and Chester Stock), two from UCLA (James Gilluly and W. Gary Ernst), one from UC San Diego (T. Wayland Vaughan), and one from UC Davis (Eldridge Moores).

Section members who served as GSA president but not section chair include Waldemar Lindgren (1924), Charles Palache (one of Lawson's first students, 1939), Adolf Knopf (1944), William Rubey (1950), Ernst Cloos (1954), Thomas B. Nolan (1961), Richard Jahns (1971), Luna Leopold (1972), Clarence Allen (1974), Leon Silver (1979), and George Thompson (1997).

As the Cordilleran Section prepares to enter a new century and communication increases and the "world grows smaller," the thoughts by Greg Davis on p. 29 on the influence of the section on the development of just one area, tectonics, emphasize the vast impact the section has had on the geosciences.

## REFERENCES CITED

Crowell, J. C., 1996, Medals and awards for 1995: GSA Today, March, p. 13.

Eckel, E. B., 1982, The Geological Society of America: A life history of a learned society: Geological Society of America Memoir 155, 167 p.

Gilbert, G. K., 1906, The investigation of the San Francisco earthquake: Popular Science Monthly, p. 97.

Hill, M. L., 1981, San Andreas fault: history of concepts: Geological Society of America Bulletin, v. 92, p. 1122-1131.

Hill, M. L., and Dibblee, T. W., Jr., 1953, San Andreas, Garlock, and Big Pine faults, California: Geological Society of America Bulletin, v. 64, p. 443–458.

Hoots, H. W., and Bear, T. L., 1954, History of oil exploration and discovery in California, in Jahns, R., ed., Geology of southern California: California Division of Mines and Geology, Bulletin 170, p. 5–9.

Lawson, A. C., 1893, Post-Pliocene diastrophism of southern California: California University Publications in Geological Sciences, v. 1, p. 115–160.

Lawson, A. C., 1895, Sketch of the geology of the San Francisco Peninsula, California: U.S. Geological Survey, 15th Annual Report, p. 399–476.

Norris, R. M., 1981, Early geologic education in California: Berkeley and Stanford show the way: Journal of Geological Education, v. 29, p. 169–175.

Steller-Stout, D., 1987, The development of geologic knowledge and education and its applications in California before 1934 [Ph.D. thesis]: Claremont, California, Claremont Graduate School, p. 413.

MANUSCRIPT ACCEPTED BY THE SOCIETY NOVEMBER 23, 1998

Printed in U.S.A.

Chapter 3
# SOME SALIENT EUROPEAN CONTRIBUTIONS TO CORDILLERAN TECTONICS

*A. M. C. Şengör*
*ITU Maden Fakultesi*
*Jeoloji Bolumu*
*Ayazaga, Istanbul 80626*
*Turkey*

## A. M. CELÂL ŞENGÖR

Celâl Şengör is a structural geologist whose research interest is in large-scale tectonics. He was born in Istanbul, Turkey, on 24th March, 1955, into a family of merchants and industrialists. His interest in geology began in primary school, under the influence of Jules Verne, and later was much encouraged by the two leading figures in the Turkish earth sciences, Professors Ihsan Ketin and Sirri Erinç, with whom he had come into contact. After a year in Germany, Şengör moved to the United States to continue his education in geology, which he began in the University of Houston in 1974. In his third year, he moved to the State University of New York at Albany where he completed his bachelor's, master's, and doctoral degrees, the latter two under the direction of John F. Dewey. His doctorate work concerned the geology of the Albula Pass in eastern Switzerland. After completing his doctorate, Şengör returned to Istanbul and joined Ketin's department at the Istanbul Technical University where he currently is a professor. He has held visiting posts in a number of European universities, including Collége de France and University of Oxford.

Most of Şengör's work has been in Eurasia, mainly along the Alpine-Himalayan ranges, though he also did field work in Central Asia. His contribution to North American geology has been the discovery of the Grenville-age Llano suture in 1976, while he was still in Houston. He has published some 140 research papers and several popular science articles in over nine languages. Şengör has been editor, associate editor, or editorial board member of numerous international earth science journals. He is a member of Academia Europaea (1990) and a founding member of the Turkish Academy of Sciences. His honors include the Prix Lutaud of the Académie des Sciences (Paris), Rammal Medal of the Physical Society of France and the Foundation of the Ecole Normale Supériure, and the Bigsby Medal of the Geological Society of London. Şengör is an honorary fellow of the Geological Society of America, the Geological Society of France and the Austrian Geological Society.

Eldridge M. Moores

# Some salient European contributions to Cordilleran tectonics

**A. M. C. Şengör**
*ITU Maden Fakültesi, Jeoloji Bolumu, Ayazaga, Istanbul 80626, Turkey*

The large-scale structure of the Cordilleran orogen was synthesized in some detail in terms of global tectonic models earlier by European geologists than by Americans. This work led to some further conceptual models and terminology that dominated tectonics until the rise of plate tectonics and in some cases beyond, shaping some plate tectonic interpretations. Eduard Suess (1831–1914) was the first to study the structure of the entire North American Cordillera between Alaska and Mexico. The main result of his study was that the Cordillera had, like "the Caledonides and the Saharides," (Suess, 1909a, 1909b, p. 443; Note that all page numbers to Suess refer to his English translation) a symmetric structure consisting of two oppositely verging flanks and a central "Zwischengebirge" (translated in the English edition of *The Face of the Earth* as the "Intermediate Range"). The Zwischengebirge included "the batholiths of Idaho and the Sierra Nevada, in addition to the batholiths of Columbia, among which we may reckon the whole series of the Cascade range." To the west of the intermediate range, structural vergence was westward: "In the north, so far as any direction of folding is perceptible, they are driven towards the west or south-west, and in lat. 49°N this movement may apparently be recognized in the most westerly parts of the Rocky Mountains. In the Sierra Nevada, overfolding towards the west-south-west still occurs" (Suess, 1909b, p. 443). Through the work of Hershey (1906), Suess knew that the Klamaths could be divided into "four flakes dipping to the east and to some extent imbricate" (p. 421), and therefore belonged to the western, west-vergent flank. The east-vergent flank was very clear in Mexico and could be followed northward along the western border of the Colorado Plateau, the Wasatch, and the "gneiss masses on the west side of the Rocky Mountains" (p. 443). In addition, Suess (1909b, p. 443) recognized that "Strike-faults cut through the structure, which is thus often divided into long strips, or often also let down to form deep troughs (Owens Valley, Death Valley, Bolson de Mapimi). Eruptive rocks crop out along these faults."

This picture of the Cordillera made a deep impression on Leopold Kober (1883–1970) another Viennese tectonician, whose writings made Suess's term "Zwischengebirge" so well known that its introduction began to be credited to him. It led him to develop a theory of orogenic structure that portrayed all orogens to be symmetric, just like the North American Cordillera. When Kober published his theory using the structure of the Alpine System as illustration, he hastened to emphasize its universality, and the first extra-Alpine example he chose was precisely Suess's double-sided Cordillera: "A grandiose example [of the double-sided orogenic structure] is provided by North America: the Rocky Mountains, the Zwischengebirge with the granodioritic scar, and the Elias Mountains on the Pacific coast. The Rocky Mountains are the easterly-moved mountains that moved onto the Cretaceous of the Laurentian foreland. The Elias Mountains were overfolded towards the west. In between is the Zwischengebirge (E. Suess), a long depression, filled with marine Mesozoic, occupied by volcanoes, and along the zone extends the immense granodioritic scar" (Kober, 1914a, p. 256; also see Kober, 1914b, p. 203). Later, in both editions of his famous textbook, Kober not only reemphasized the double-sided nature of the North American Cordillera, but he fortified the interpretation by using the new data that had appeared since the publication of Suess' classic (Kober, 1921, 1928).

Kober's interpretation of the double-sided nature of the Cordillera was later much improved by the great German geologist Hans Stille (1876–1966). In addition to his many smaller publications, in which Stille discussed the structure and tectonic evolution of the North American Cordillera, his *Einführung in den Bau Amerikas* (Stille, 1940) constitutes a real milestone in understanding not only the tectonics of the North American Cordillera, but the entire double continent of America. (Despite its fame, this great book reached very few readers, because the store of its publisher was bombed shortly after publication;

---

Şengör, A.M.C., 1999, Some salient European contributions to Cordilleran tectonics, *in* Moores, E. M., Sloan, D., and Stout, D. L., eds., Classic Cordilleran Concepts: A View from California: Boulder, Colorado, Geological Society of America Special Paper 338.

according to Marshall Kay there were only four copies in the United States as of 1974). Based on an exhaustive survey of the American literature and some field checks in the Canadian Cordillera (1913), Texas (1933), and California (1931), Stille elaborated the tectonic evolution in terms of the east-west asymmetry of the paleogeographic realms summarized in the eugeosyncline-miogeosyncline couple, outlined the temporal evolution of magmatism and its coordination with deformation as mafic (commonly ophiolitic) initial volcanism, felsic synorogenic plutonism, felsic subsequent volcanism and mafic final volcanism, and emphasized the change of style of deformation from early alpinotype (penetrative deformation with nappes) to later germanotype (nonpenetrative, fault-block dominated). Stille's paleogeographic concepts and geosynclinal nomenclature found a loud echo in the United States through the work of Marshall Kay.

### HANS WILHELM STILLE

Hans Wilhelm Stille, a member of an old farming family in Lower Saxony, was born on October 10, 1876, in Hannover, the second of the five children of Eduard and Meta (Hanckes) Stille. After reading chemistry for one year at the Technical University of Hannover, he moved to the University of Göttingen, where he studied under the noted German stratigrapher Adolf von Koenen. Von Koenen gave him a region in the Teutoburg Forest, between Detmold and Altenbecken, to map for his doctoral degree. Here Stille documented the existence of Mesozoic faults. After obtaining his degree in 1899 he was hired by the Geological Survey of Prussia, where he continued mapping around his dissertation area.

In 1908 Stille moved to the Technical University of Hanover as professor of geology and mineralogy and founded the Geological Society of Lower Saxony. His first papers revealing his tectonic views were published in the journal of this society. These papers showed him to be a "conservative" tectonician, having roots in the central European mining tradition. After his move to Göttingen, his alma mater, via a six-month sojourn in Leipzig as professor, Stille's global tectonic research activity unfolded. Göttingen was the classical area of the central European block mosaic, where Stille believed the dependence of the intensity of folding on geosynclinal subsidence was demonstrated. He also believed that the area demonstrated that folding took place in short and worldwide "phases" separated in time by long intervals of slow vertical movements free of faulting. These slow movements he called "epeirogenic," after G. K. Gilbert, and the rapid movements accompanied by intense folding and faulting he called "orogenic." In 1924 he published his influential book *Grundfragen der Vergleichenden Tektonik* (*Basic Questions of Comparative Tectonics*), which largely determined the course of twentieth century tectonics before the rise of plate tectonics.

Stille was called to Berlin as the head of the prestigious Geological Institute and the Museum of Natural History of the University of Berlin in 1932. Here his research activity was extended outside Europe, and it was in Berlin that Stille became interested in the tectonics of the Americas. Already in 1907 he had published the results of his field studies in the Rio Magdalena area in South America. In 1913 he saw the Canadian Cordillera during an excursion of the International Geological Congress. In 1931 and 1933, Ralph D. Reed led him through southern California and Philip B. King led him through the Marathon region in Texas. He held numerous seminars on regional tectonics devoted to the Americas in Berlin in the 1930s. The outcome of all these studies was the great book *Einführung in den Bau Amerikas* (*Introduction to the Structure of America*). Published in 1940, most of the printed copies were destroyed shortly afterward during an air raid on Berlin, which accounts for its rarity despite its tremendous fame. Although only four copies exist in the United States, this book nevertheless exercised much influence in the United States through Stille's numerous pupils and friends, the foremost among the latter being Marshall Kay. After the war, Stille stayed in East Berlin (because the university was there) until his retirement, and there founded an Institute of Geotectonics attached to the German Academy of Sciences (formerly Prussian Academy of Sciences). He retired in 1949 and returned to his hometown of Hannover, then in West Germany. He continued publishing until 1960 and died on December 26, 1966. Stille did see the initial lights of plate tectonics, discussed it with his former students, reportedly did not much like it, but had the greatness to acknowledge that his time had passed. He once told his former student Andreas Pilger that he was still writing papers after 1960 "just for fun; afterwards I throw them into the dustbin."

A. M. C. Şengör

Another significant interpretation by a European geologist of the tectonics of the North American Cordillera was the interpretation by Becker (1934, p. 119) of the Basin and Range extensional system in terms of secondary extension in a broad right-lateral shear zone, of which the San Andreas fault was considered a part.

> Through the northwestward drift of the Pacific block, the Great Basin is placed under a rotating stress field, which creates extensional tears with north-south strikes. These are the faults that cut out the horsts and the tilted blocks of the Basin Ranges.
> On the basis of what has been said, I wish to define the concept of "Basin Range Structure," which plays a significant role in the American literature: "Under Basin Range structure, we understand the phenomenon that an older folded mountain is disrupted into a system of horsts, grabens, and tilted blocks through parallel extensional tears and which have no relation to the older mountain structure." I shall leave undecided, whether it is also necessary to add the following as well: "These extensional tears form as a consequence of two crustal blocks sliding past one another."

Recent studies have shown again the young age of the Basin and Range structure and its direct genetic relation to the relative motion of North American and the Pacific plates (Atwater and Stock, 1998), thereby rehabilitating the once-popular hypothesis of Becker (often credited to others).

A popular text book of internal geodynamics in the later 1930s in Europe was the *Einführung in die Geologie* (*Introduction to Geology*); its author, the even more popular Hans Cloos (1885-1951), the man to whom the invading allied armies in 1944 were to offer the position of mayor of Bonn (which he declined), and who received the Penrose Medal of the Geological Society of America in 1948. There was much talk of the geology of the Cordillera in that famous book. Cloos described with some care its double-sided structure in a cross-section from the California shelf to the Wasatch front (with many references to his European and American colleagues; Cloos, 1936, p. 418-419), partly on the basis of his own field observations in the summer of 1927 together with those of his student Robert Balk (cf. Cloos, 1928, 1947, esp. p. 396). The San Andreas fault is portrayed, as a matter of fact, as a srike slip fault with some 40 km of offset and possibly more, but with due regard to Bucher's objections (Cloos, 1936, p. 251). For the "enormous extension" (p. 604) in the Basin and Range, Cloos thought that "uplift for many 1000s of metres" (p. 405) was mainly responsible, but he could not do without adding that "in the western part, there are structures indicating sideways crustal movement, and most recently earthquakes have confirmed this assumption" (p. 405).

Cloos' text book shows how commonplace was the knowledge of the double-sided structure of the North American Cordillera, the strike slip nature of the San Andreas and its possible influence on the Basin and Range in Europe by the 1930s. Apart from Stille's paleogeographic work, North American geologists had taken slight notice of these interpretations and not uncommonly they reinvented the wheel. This reinvention was sometimes necessary. In 1966 B. C. Burchfiel and G. A. Davis submitted to Science a paper entitled "Two-sided nature of the Cordilleran orogen and its tectonic implications" with a view to combating the prevalent opinion that the entire Cordillera was an asymmetric, east-vergent orogen, as James Dwight Dana had portrayed it a century earlier. It was rejected twice with such typical comments as: "the idea proposed does not merit publication *except as an unfounded but interesting speculation*" (from an unpublished referee's comment, courtesy of Professor G. A. Davis; italics are mine). As a European, I am amazed that in 1966 anybody could call the symmetric structure of the North American Cordillera an unfounded speculation! The referees might have told Burchfiel and Davis that what they were talking about was old hat, but none of the four referees who recommended rejection knew that it was widely known—at least to the Europeans—that the North American Cordillera was structurally symmetric. When I discussed this with Professor Davis, he confessed that they had not known it either.

My superficial knowledge of the history of the study of Cordilleran geology tells me that a more extensive and perhaps more careful reading of the European literature might have avoided much unnecessary controversy in western North America (and do not let us forget that this paper contains only a small, but important, fraction of what was described and discussed of the American geology by European geologists!). There were some attempts to bring this about (e.g., Teichert, 1931), but evidently they proved insufficient. Witness what Curt Teichert wrote half a century later in an unpublished letter to A. M. C. Şengör dated September 7, 1982.

> This [the 1931 paper] has quite an interesting history. When I was a student in the mid- and late 1920's, there were great stirrings in geological thought in Central Europe. Book after book was being published by Stille, Staub, Kober, Argand, Wegener, Haarmann, and others. We students read them avidly and they were discussed in endless seminars and colloquia. I had the great good luck to spend the year 1930 on a post-doctoral in the United States, mostly in Washington, where I observed that this great stream of ideas seemed to have completely bypassed the American geological community. In the spring of 1930 I was asked to give a talk at a meeting of the Geological Society of Washington and instead of talking about cephalopods on which I was working under my fellowship, I chose to enlighten my colleagues about some of these things that had excited us so much in Europe. The secretary then suggested that my talk should be published and the result is the attached paper, which to the best of my knowledge has never been cited or referred to by anybody anywhere.

A final note on this sort of provincialism refers to the little-known debate between James Gilluly (1886–1980) and Hans Stille in 1950, following Gilluly's (unjustly) famous 1948 Presidential address to the Geological Society of America (Gilluly,

1949). In my experience, few American geologists seem to have taken notice of the 1950 debate published in Geologische Rundschau (Gilluly, 1950a, 1950b). When I was a student in America in the 1970s, I was recommended to read Gilluly to see how Stille's theory of episodic orogeny was discarded. But when I then also read the ensuing debate, I noted to my amazement that Stille had driven Gilluly literally to the ground on United States examples (California and Utah), simply because Gilluly either had not taken the time to read carefully Stille's theory and especially his orogeny-epeirogeny distinction, or he had not understood them. As a consequence, the European geologist who read the debate then would have found little reason to take Gilluly seriously. Those few who did take him seriously, such as Rudolf Trümpy, did so only because they thought the American fashion chic, and changed their minds when they considered the matter seriously (Trümpy, 1973, p. 247). Only plate tectonics really killed Stillean episodicity and exposed that Gilluly had poorly defended a strong case out of ignorance of the prevailing theoretical corpus.

## REFERENCES CITED

(All page references to Suess' book are to the English edition).

Atwater, T., and Stock, J., 1998, Pacific–North America plate tectonics of the Neogene southwestern United States; an update: International Geology Review, v. 40, p. 375–402.

Becker, H., 1934, Die Beziehungen zwischen Felsengebirge und Großem Becken im westlichen Nordamerika—Ergebnisse einer Studienreise durch die Vereinigten Staaten III: Zeitschrift der Deutschen Geologischen Gesellshaft, v. 86, p. 115–120.

Cloos, H., 1928, Bau und Bewegung der Gebirge in Nordamerika, Skandinavien und Mitteleuropa, *in* Soergel, W., ed., Fortschritte der Geologie und Paläontologie: v. VII, Berlin, Gebrüder Borntraeger, 87 p.

Cloos, H., 1936, Einführung in die Geologie–Ein Lehrbuch der Inneren Dynamic: Berlin, Gebrüder Borntraeger, XII+503 p. (reprinted by the same publisher in 1963).

Cloos, H., 1947, Gespräch mit der Erde–Geologische Welt and Lebensfahrt, R. Piper & Co., München, 410 p.

Cloos, H., 1954, Conversation with the Earth, Routledge & Kegan Paul Ltd., London, 440 p.

Gilluly, J., 1949, Distribution of mountain building in geologic time: Geological Society of America Bulletin, v. 60, p. 561–590.

Gilluly, J., 1950a, Distribution of mountain building in geologic time: Geologische Rundschau, v. 38, p. 89–91.

Gilluly, J., 1950b, Reply to Discussion by Hans Stille: Geologische Rundschau, v. 38, p. 103–107.

Hershey, O. H., 1906, Some West Klamath stratigraphy: American Journal of Science, ser. 4, v. 21, p. 58–66.

Kober, R. L., 1914a, Die Bewegungsrichtung der alpinen Deckengebirge des Mittelmeers: Dr. A. Petermanns Mitteilungen aus Justus Perthes' Geographischer Anstalt, v. 60, part I, p. 250–256.

Kober, L., 1914b, Alpen und Dinariden: Geologische Rundschau, v. 5, p. 175–204.

Kober, L., 1921, Der Bau der Erde: Berlin, Gebrüder Borntraeger, 324 p.

Kober, L., 1928, Der Bau der Erde, zweite neubearbeitete und vermehrte Auflage: Berlin, Gebrüder Borntraeger, 500 p.

Stille, H., 1940, Einführung in den Bau Amerikas: Berlin, Gebrüder Borntraeger, 717 p.

Stille, H., 1950a, Bemerkungen zu James Gillulys "Distribution of Mountain Building in Geologic Time": Geologische Rundschau, v. 38, p. 91–102.

Stille, H., 1950b, Nochmals die Frage der Episodizität und Gleichzeitigkeit der orogenen Vorgänge: Geologische Rundschau, v. 38, p. 108–111.

Suess, E., 1909a, Das Antlitz der Erde, Volume III/2: Wien, F. Tempsky; and Leipzig, G. Freytag, 789 p.

Suess, E., 1909b, The face of the Earth (Das Antlitz der Erde), translated by Hertha B. C. Sollas under the direction of W. C. Sollas, Volume 4: Oxford, Clarendon Press, 673 p.

Teichert, C., 1931, Recent German theories about structural geology: Washington Academy of Sciences Journal, v. 21, p. 1–12.

Trümpy, R., 1973, The timing of orogenic events in the Central Alps, *in* DeJong, K. A., and Scholten, R., eds., Gravity and tectonics: New York, John Wiley & Sons, p. 229–251.

MANUSCRIPT ACCEPTED BY THE SOCIETY NOVEMBER 23, 1998.

Geological Society of America
Special Paper 338
1999

# Section II
# WHERE IT STARTED

# Where it started: Introduction

The study of California geology is founded on two principal features—gold, and the San Andreas fault. Although a few prior observations of California geology existed (principally those of the Frémont expedition of 1843–1844), the discovery of gold at Sutter's Mill in 1848 and the ensuing gold rush gave powerful impetus to the study of California geology. The account of the California gold rush has been chronicled many times, most recently by McPhee (1993, 1998). John Marshall's discovery led to one of the great human migrations in history. California gold kept the Union afloat during the Civil War, and San Francisco became the first truly international city on the eastern Pacific Rim. In 1860, the California legislature formed a geological survey under Josiah D. Whitney. The Whitney Survey began a geological study of California, but when its first reports appeared with "academic," rather than "practical geology," it was slowly starved for funds and its work soon ceased (McPhee, 1998, p. 474).

The initial documentation of the geology of gold in California rests with the work of the U.S. Geological Survey in the 1880s and 1890s. Several workers, principally W. Lindgren, H. W. Turner, and J. S. Diller, mapped much of the Sierra Nevada. It has been said that these geologists worked mostly from horseback, completing a 30 minute quadrangle in two field seasons—a formidable achievement. The state of nearly complete denudation that existed in the Sierra Nevada at the end of the nineteenth century made the geology much better exposed that at the present time. The folio maps are remarkably accurate, and in many places still stand as the best available geologic information.

Using these quadrangle folios as a basis, W. Lindgren wrote his classic article "Characteristic features of California gold-quartz veins," reprinted here, which forms the basis of J. K. Böhlke's chapter on Mother Lode gold. It is interesting to note Böhlke's observations that the "ultimate origin of the hydrothermal fluids continues to be problematic," and that "the origin of the gold is essentially unknown, as it was a century ago."

It is also important to note that Mother Lode quartz veins were not the only sources of California gold. In addition to placer deposits in modern streams, which were the source of the first discoveries, the Auriferous Gravels, stream deposits of Eocene age, formed an important source of gold mined hydraulically from 1853 to 1884, when hydraulic mining in California was banned by U.S. Court order and subsequent congressional action (Averill, 1946, p. 144). These early Tertiary gravels were derived from a highland east of the present crest of the Sierra Nevada (Yeend, 1974), and may have been derived from a Tibet-like plateau east of the present Sierra Nevada (Dilek and Moores, 1987; 1999).

The San Andreas fault constitutes the second major impetus to the study of California geology. By 1900, California was well established as a prosperous, vigorous place. San Francisco was the largest city on the eastern Pacific Rim, and the most international, and it also had a somewhat "racy" reputation. Its initial growth came from the California gold rush, but it was the silver of the Comstock Lode of Virginia City, Nevada, just east of the Sierra Nevada, that made San Francisco rich (e.g., Glasscock, 1931). The 1906 San Francisco earthquake literally shook San Francisco, and its geologic community, to its very core.

Throughout most of historical time, before the invention of seismographs, earthquakes had been thought to occur only in populated regions and were considered manifestations of the ultimate wrath of an angry god or gods. Moralizing was especially virulent in England, leading up to and following the Great Lisbon earthquake of 1755 (Kendrick, 1957). Some of this moral attitude persisted through to the twentieth century. San Francisco's "sinful" reputation was enhanced as a result of the 1906 earthquake. Some observers have noted an undercurrent of this moral stance still running through some non-Californian press accounts of California earthquakes.

We include two chapters on the San Andreas fault. The first chapter reprints part of the initial report on the 1906 earthquake, written by A. C. Lawson and H. F. Reid, and an update by Carol S. Prentice. The second includes a reprint of the article by Hill and Dibblee (1953) which first presented evidence for large-scale horizontal movement on the San Andreas fault, and a personal reminiscence by Thomas W. Dibblee, Jr. Together they give a good impression of past and present work on this important geologic feature.

The San Andreas fault is by no means the only fault in California. The Pacific–North American plate boundary is a large family of faults and associated structures extending from offshore to well inland.

**Note also that an animation of San Andreas evolution by T. Atwater is included on a CD in the pocket in the inside back cover of this volume.**

## REFERENCES CITED

Averill, C.V., 1946, Placer mining for gold in California: California Divison of Mines and Geology Bulletin 135, 357 p.

Dilek, Y., and Moores, E. M., 1987, Tibetan model for the late Mesozoic–early Tertiary tectonics of the western U.S.: Eos (Transactions, American Geophysical Union) v. 68, p. 1474.

Dilek, Y., and Moores, E. M., 1999, A Tibetan model for the early Tertiary Western U.S.: Geological Society of London Journal, in press.

Glasscock, C. B., 1931, The big bonanza; the story of the Comstock Lode: Indianapolis, Bobbs-Merrill Company, 368 p.

Kendrick, T. D., 1957, The Lisbon earthquake: Philadelphia, Lippincott, 255 p.

McPhee, J., 1993, Assembling California: New York, Farrar, Straus, and Giroux, 304 p.

McPhee, J., 1998, Annals of the former world: New York, Farrar, Straus, and Giroux, 696 p.

Yeend, W. E., 1974, Gold-bearing gravel of the Ancestral Yuba River, Sierra Nevada, California: U.S. Geological Survey Professional Paper 722, p. 44.

MANUSCRIPT ACCEPTED BY THE SOCIETY NOVEMBER 23, 1998.

Participants in a conference on the future of seismological studies at Caltech in 1929. Front row, left to right: unknown assistant, Leason Adams, Hugo Benioff, Beno Gutenberg, Harold Jeffreys, Charles Richter, Arthur Day, Harry Wood, Ralph Arnold, John P. Buwalda. Back row: unknown assistant, Perry Byerly, Harry Fielding Reid, John Anderson, Father J. B. Macelwane. Courtesy of the Berkeley Seismological Laboratory.

Printed in U.S.A.

Geological Society of America
Special Paper 338
1999

# Chapter 4
# CALIFORNIA GOLD

**J. K. Böhlke**
*U. S. Geological Survey*
*431 National Center*
*Reston, Virginia 20192*
*E-mail: jkbohlke@usgs.gov*

## WALDEMAR LINDGREN

Waldemar Lindgren (1860-1939) was one of the preeminent geologists in the field of ore deposits around the end of the nineteenth and beginning of the twentieth centuries. Lindgren was born in Sweden in 1860 and studied in Freiberg, Germany, starting in 1878. He moved to the United States in 1883 and worked for the U.S. Geological Survey from 1884 to 1915, where he held the position of Chief Geologist from 1911 to 1912. Lindgren moved to the Massachusetts Institute of Technology in 1912 to become Professor of Geology and Head of the Department of Geology, then held an emeritus position from 1933 until his death in 1939. Two of Lindgren's major early assignments at the U.S. Geological Survey were studies of the Pacific Coast Range mercury deposits and the Sierra Nevada gold-quartz veins, with some assistance from later U.S. President Herbert Hoover. Those studies had a significant influence on Lindgren's global interest in hydrothermal fluids, water-rock interactions, and classification of hydrothermal ore deposits based on physical and chemical conditions of formation. Lindgren is perhaps best known for his textbook on *Mineral Deposits*, first published in 1913, which featured the "Lindgren classification" of ore deposits based on genetic criteria. Lindgren received numerous awards, including the Penrose Gold Medal from the Society of Economic Geologists (1928), and the Penrose Medal from the Geological Society of America (1933). He was president of GSA in 1924.

## CLASSIC PAPER (LINDGREN, 1895)

This chapter features a reproduction and discussion of Lindgren's classic paper on the geology, geochemistry, and origin of gold-quartz veins in California, published in the Geological Society of America *Bulletin* in 1895. That paper reviewed the major findings of Lindgren and coworkers during the preceding decades of study in the region, summarizing years of observations made when large numbers of mines were active. Although more than a century old, this paper was selected for inclusion in this volume because it contains a brief yet comprehensive account of the state of knowledge about the gold-quartz veins of the western Sierra Nevada foothills metamorphic belt by one of the foremost ore deposit geologists of all time, and because it anticipates some of the difficult issues that have continued to challenge subsequent investigators to the present time (e.g., Robert et al., 1991; this study). Lindgren's paper is a rewarding read, as much for its casual tone as for its critical overview of the characteristics and origin of the veins.

Source: Graton (1939); Butler (1950)

J. K. Böhlke

---

> It is chiefly through the intimate connection with the art of mining and the development of mineral resources of the country that geology has acquired the importance which it is now has and especially in its relation to the state.
>
> J. D. Whitney<br>California Geological Survey

# CHARACTERISTIC FEATURES OF CALIFORNIA GOLD-QUARTZ VEINS*

BY WALDEMAR LINDGREN

(Read before the Society December 29, 1894)

## CONTENTS

| | Page |
|---|---|
| Introduction | 221 |
| Geographic distribution | 222 |
| Geologic relations | 222 |
| Age | 225 |
| Differing types of gold deposits | 226 |
| Structural relations | 226 |
| The filling of the veins | 229 |
| Association of minerals in gangue and ores | 230 |
| Distribution of the gold in the veins | 231 |
| The alteration of the country rock | 232 |
| Genetic conclusions | 236 |
| Comparison with quicksilver deposits | 238 |
| Origin of the gold | 239 |
| Summary | 240 |

*Published by permission of the Director of the U. S. Geological Survey

THE CENTRAL GOLDBELT OF CALIFORNIA.

GOLD QUARTZ VEINS OF CALIFORNIA.

W. Lindgren

## Introduction.

The gold-quartz veins of California, in spite of many local variations, form a remarkably well defined type of mineral deposits, the salient characteristics of which it is intended to portray in this paper. The results, indicated in brief outlines, have been obtained during general and detailed mapping for the United States Geological Survey in the gold-bearing region of California.

Referring to the map of the distribution of veins, it should be stated that Inyo county, as well as the central and eastern part of San Bernardino, contains many gold deposits which have not been indicated. In many cases they carry both silver and gold like the Comstock mines, or differ in other respects from the normal gold-quartz veins, though the latter are not without representatives. For many notes and valuable suggestions I am under obligation to Messrs G. F. Becker, H. W. Turner and J. S. Diller. The reports of the state mineralogist of California have also been frequently consulted in the preparation of the maps.

## Geographic Distribution.

The general map of California accompanying this paper indicates the extent and distribution of the gold-quartz veins. Beginning in the peninsular range south of the Mexican boundary, the deposits continue in scattered form and with many intermissions up to Fresno county, a few of them also occurring at isolated points along the coast ranges south of San Francisco. In Fresno county they become more abundant, and in Mariposa county the auriferous belt rapidly widens. From here northward to the point where they are covered by the great lava fields of northeastern California the maximum development is obtained. In latitude 40° the gold deposits extend from the great valley on the west to the summits of the Sierra Nevada on the east. In a northwesterly direction the continuation of the gold-bearing area is found in Shasta, Trinity, Siskiyou and Del Norte counties in California, and its northerly end occupies the counties of Jackson, Josephine and Curry in southwestern Oregon. Volcanic flows and more recent superjacent formations cover the gold-bearing area toward the east and north.

A smaller auriferous belt of less importance runs along the eastern slope of the Sierra Nevada, beginning in Alpine county and continuing southward through Mono, Inyo and San Bernardino counties. Most of the deposits along this line differ more or less from the normal type of the western slope.

## Geologic Relations.

In the northern part of the Mexican peninsula and in San Diego county granitic rocks prevail, but in them are imbedded numerous more or less contact-metamorphosed areas of slates and schists of uncertain age. The gold-quartz veins usually occur in, or at least close to, these areas. The principal mining districts in San Diego county are Julian and Banner, in the central part, and Pinacate, near the northern boundary.[1]

Granitic rocks, with smaller schist areas, continue through San Bernardino and Los Angeles counties. Placer deposits and smaller veins are found around San Bernardino mountain, as well as at several places in clay-slate near the summit of the range,[2] in the central and northern part of Los Angeles county. Very scattered and isolated deposits occur in Ventura, Santa Barbara, San Luis Obispo, Monterey and Santa Cruz counties. In Monterey paying veins have been found near the coast at Los Burros, sandstone being mentioned as the country-rock. A short distance north of Santa Cruz a few gold-quartz veins are said to occur in unaltered sedimentary formations. In Kern county there is a line of paying veins with a northeasterly strike, extending from Kernville to Tehachipi pass. Granitic rocks predominate, but contain a number of smaller schist areas, with which the gold deposits appear to be associated. The locality is of interest on account of the number of hot springs occurring near the veins. Tulare county contains but few quartz-veins, but placer diggings are found along several of the rivers.

In Fresno county, again, several streaks and smaller areas of schists and slates occur in the main granitic mass; again, the quartz-veins, which here attain greater importance, are closely associated with the former, though not exclusively occurring in them. Continuing northward for about fifteen miles to Mariposa, these belts of schists and slates suddenly widen, and at the same time begin to contain numerous and rich quartz veins. Between this region and the lava fields of the north lie the most productive gold-mining regions of California.

The western slope of the Sierra Nevada is from here northward occupied by a gradually widening belt of rocks, to which the name "metamorphic series" is usually given. It attains its maximum width in Butte and Plumas counties and continues across northwestern California and southwestern Oregon to the Pacific ocean. The eastern part and the summit of the Sierra Nevada are still occupied by the continuation of the southern granitic area, bordering upon the "metamorphic series," with an irregular contact-line, along which evidence of

---

[1] Mr W. H. Storms has described interesting lenticular veins from the former locality, which, according to his explanation, doubtless correct, are only modifications of normal fissure veins. Eleventh Ann. Rep. State Mineralogist of California.
[2] Acton mining district

the later origin and intrusive character of the granite may be frequently observed. This contact-line is indicated on the map. The "metamorphic series," sometimes also referred to collectively as the "auriferous slates," is a very complex mass of rocks. It consists largely of more or less altered and highly compressed sediments, of an age ranging from early Paleozoic to late Jurassic, and bearing evidence of having been subjected to several mountain-building disturbances. Associated with these sediments are igneous masses—augite-porphyrite, diabase, serpentine, etcetera—also ranging in age from Paleozoic to late Mesozoic, though the greater mass of them appear to date from late Jurassic or early Cretaceous time. To a considerable extent these igneous rocks have been acted on by the dynamo-metamorphic processes which also affected the sedimentary rocks, and are largely converted into crystalline schists. It may be said in general that the sedimentary rocks prevail in the eastern part of the metamorphic belt, while along the great valley basic, igneous rocks are found in the greatest abundance. The granitic rocks of the high Sierra Nevada are to a large extent *granodiorites*, the name adopted on the survey maps for a quartz-mica-diorite containing more or less orthoclase. In the metamorphic series there are many smaller masses of the same rock—the latest intrusions—which are usually but little affected by dynamo-metamorphic processes.

The intimate connection of the gold deposits with the metamorphic series or the auriferous slates has been recognized for a long time, and Professor Whitney emphasizes it repeatedly in his works. The auriferous region, indeed, corresponds closely with the extent of the metamorphic series. Even in the south, where the granitic rocks predominate, it has been shown that the gold deposits are usually connected with the scattered schist areas. Few gold-quartz veins are found in the granitic area, and then usually near the contact. Within the typical gold-bearing region the veins are distributed with remarkable impartiality, and occur in almost any of the great variety of rocks which make up the metamorphic series. They are found in granite, diorite, granodiorite, gabbro and serpentine; in quartz-porphyrite, augite, or hornblende-porphyrite and diabase; in amphibolite and other dynamo-metamorphosed rocks; in sedimentary, more or less altered slates, sandstones and limestones. In Tertiary volcanic rocks gold deposits are only found on the eastern slope of the range. It is apparently impossible to formulate any law as to their lithologic occurrence or to say that they prevail in any one kind of rock in the metamorphic series.

Regarding the quartz-veins of California F. von Richthofen has made a frequently quoted statement which in a certain sense may be correct, but which unless qualified is apt to lead to grave errors. It is as follows:[3]

The auriferous quartz veins "have in their occurrence clearly discernible connection with the extension of the granite. They are crowded closely at its contact with the metamorphic rocks, and occur here partly in the former, partly in the latter. The greater the distance from the granite, the rarer they become in the metamorphic rocks, and only occur as an exception where the influence of the outcropping granite would not be expected on account of its distance. In the same way they become less frequent in the granitic regions as the distance from the contact increases, and are, as a rule, entirely lacking in the interior of the large granite masses."

This statement cannot be accepted for the main granitic contact, which, on the contrary, except near Sonora, is remarkably barren of important deposits. In the larger part of the gold region a wide belt of Paleozoic slates comparatively poor in gold deposits separates this contact from the principal gold-producing districts. In very many places, however, the contact clearly marks the abrupt beginning of auriferous deposits, though perhaps poor and of small extent. The sudden change of recent and Tertiary river-beds from barren to auriferous when cutting across the contact is often very noticeable.

Though not applicable to the main granitic contact, the statement quoted is to a certain degree true of the smaller masses of granodiorite scattered through the metamorphic series, for it is very common to find the gold-quartz veins clustered near their contacts in the manner indicated. It is not so general, however, as to be called a rule or a law, for there are many included granitic masses the contacts of which are in no way remarkable for abundant deposits.

Dr W. Moericke, who has recently published several very interesting papers on the gold deposits of Chile, has come to the conclusion that they are closely associated with acid, igneous rocks, and drawn a comparison between the occurrences of that country and California.[4] In view of this, it may be well to emphasize the fact that the gold-quartz veins of California do not in their surface relation show any remarkable dependence on acid, igneous rocks. The great mother-lode, for instance, is in location and occurrence of its ores in no way related to such rocks, they being, on the contrary, as a rule, distant from it. Normal gold-quartz veins in diabase and augite-porphyrite sometimes occur far away from other rocks, although the larger areas of the former are, on the whole, rather barren.

## AGE.

Before beginning the discussion of the characteristics of the deposits, their age may be briefly touched upon. It has long been apparent and insisted upon by Whitney, von Richthofen and others that the quartz-veins of California are

---

[3] Zeitschrift der deutschen Geol. Gesell., B. xxi, 1869, p. 727.

[4] Zeitschrift fur prakt. Geol. Jahrgang, 1894, p. 28.

first of these types of impregnation is not of great economic importance, But, the second sometimes affords large masses of low grade ores.[7]

## STRUCTURAL RELATIONS.

Regarding the structural relations of the normal gold-quartz veins it should first be stated that they are fissure veins, and emphatically not so-called segregated[8] veins or "lenticular masses" in the auriferous slates.

It is everywhere plain and evident that the fissures have been broken open subsequently to the metamorphism of the rocks. These post-Jurassic and post-granitic quartz-veins form the latest chapter in the Mesozoic revolution in the Sierra Nevada.

Neither Whitney nor von Richthofen commit themselves to an expression of the "segregated" nature of the veins. A. Phillips, in his book on mineral deposits, mentions their affinity to fissure veins, although classing them as "segregated veins." All these writers, however, state that the veins nearly always conform in strike and dip to the inclosing slates. This has evidently led authors of recent text books to class the California veins as "segregated." Thus Professor J. F. Kemp, in his "Ore deposits of the United States," classes them as such with some doubt, while Professor R. S. Tarr, in his "Economic geology of the United States," thinks that "in spite of the recent observations (by H. W. Fairbanks) it still seems as though these quartz-veins must be of segregation origin."

Quartz-veins like those Professor Tarr has in mind, formed by a sort of dynamo-metamorphic process, I am quite sure do not exist in the gold-belt. The somewhat auriferous "fahlbands" in certain amphibolites approach nearest his conception. I am by no means prepared to deny, however, that there may be some minor ore-bodies deposited in openings in the slate from silicious solutions derived from the immediately surrounding rocks, but if they occur, they are surely exceptions to the general rule. In altered quartzose slates nodules and lenses of quartz seemingly of such origin frequently occur on a small scale.

This rule of "parallelism with inclosing slates" must unquestionably be rejected in a general description of the veins. It should first be pointed out that a very large number of veins, especially in the northern part of the gold-belt,

---

[7] See later, under "The alteration of the country-rock," page 235, line 4 from bottom.

[8] The term "segregated vein" is not quite clear and has been variously interpreted. A. Phillips evidently considered the only criterion of a segregated vein to be in its parallelism with inclosing slaty or schistose rocks, admitting motion along the walls and filling by foreign material, while R. S. Tarr, in a recently published volume, regards a segregated vein as the result of dynamo-metamorphism and a concentration of material from surrounding rocks, preexisting cavities not being necessary. I have used it as meaning more or less lenticular openings in the mass of slates and schists, parallel to strike and dip, produced by longitudinal compression and filled by a sort of lateral secretion or exudation from the surrounding rock.

---

of late Jurassic or early Cretaceous age, and the same authors have suggested that they probably owe their origin to thermal action following the granitic intrusion. For the larger number of the quartz-veins this is undoubtedly true. It is certain that the majority were formed subsequent to the latest dynamo-metamorphism of the sedimentary and old eruptive rocks of the Sierra Nevada, subsequent also to the granitic intrusion. It is, however, also certain that some deposits antedate this period, for in the latest sedimentary member of the bed-rock series there are conglomerates containing quartz pebbles and free gold,[5] which appears to have been concentrated as placer gold at the time the conglomerates were formed. It does not appear easy to separate the earlier deposits from the later, but it is probable that they were neither very numerous nor very rich.

Again, the eruptive activity of late Tertiary time which was centered along the summit and on the eastern slope of the Sierra Nevada was followed by another period of thermal activity, and another line of gold deposits was formed. This intermittently recurring action confirms von Richthofen's generalization that a region once metalliferous is always metalliferous. Successive eruptions in such vicinity produce successive mineral deposits, while other eruptive centers are wholly barren of them.

## DIFFERING TYPES OF GOLD DEPOSITS

It is desirable to eliminate a few deposits of a different type from the prevailing one. Most important among them are the *impregnations*,[6] of which several examples occur in the Sierra Nevada and which may be of two types: First, zones containing grains of iron pyrites disseminated in fresh dynamo-metamorphic amphibolitic schists. These zones are seldom strongly auriferous, but may enrich quartz-veins passing through them, and are apparently similar to the so-called "fahlbands" in crystalline schists. These deposits are distinctly older than the principal quartz-veins and contemporaneous with the dynamo-metamorphism which produced the schists from the diabases and other rocks. Second, impregnations of later date forming irregular zones, in which the massive rocks or schists have been decomposed and filled with secondary auriferous sulphides. These deposits are probably contemporaneous with the principal period of vein-filling and only a phase of it, in which the solutions, instead of following distinct fissures, permeated whole masses of rocks. The

---

[5] W. Lindgren: Am. Jour. Sci., October, 1894.
[6] This word is here used in its general sense, and not confined to the filling of interstitial spaces in porous rocks.

from Placer to Butte county, do not occur in slates or schists, but in massive rocks, such as diabase, granodiorite or gabbro, and among these a predominating direction of dip and strike does not exist. In slates and schists the veins often strike about parallel to the slaty cleavage—that is, northerly or northwesterly—but other directions are nearly as common. Only very exceptionally is there a strict parallelism in both strike and dip. The great mother-lode, for instance, is parallel to the strike of inclosing rocks, but differs not inconsiderably from them in dip. Its character of fissure vein is clear and unquestionable, and has been justly insisted upon by H. W. Fairbanks.[9] All directions and all dips are in fact represented among the California quartz veins, only dips below 20° and above 70° are comparatively rare. A general rule for strike and dip cannot be given; different laws guide them in different mining districts. The quartz-veins are the expression of the greater and minor strains to which the Sierra Nevada has been subjected, and a study of the former will, to a considerable degree, illustrate the latter, which have certainly varied in intensity and direction from point to point. Thus, to pick out a few illustrating examples, the veins of Ophir, Placer county, consist of two principal systems, one set of veins running west-northwest and dipping south, while the other has a west-southwest strike and southerly dip, both cutting the surrounding schists obliquely to their strike and dip. At Grass Valley and Nevada City there is one system with a general northerly direction and dipping either east or west; another system courses east and west and dips north or south at varying angles. The surrounding rocks are here mostly massive. The veins in the vicinity of Sierra Buttes, Sierra county, show the greatest divergencies in strike and dip. Equally variable are the veins about Sonora, Tuolumne county.

The force producing these fissures appears in most cases to have been a compressive stress acting at an angle more or less oblique to the horizontal. In some cases this force produced one large and prominent fracture, but far more commonly one or several series of fractures, or a sheeting[10] of the country-rock along which the auriferous solutions could circulate. Along the larger fissures considerable movement has taken place, but when the country-rock has been sheeted the motion along the individual joints has probably not been very great. In many cases, when the direction of the movement could be proved, it has been found that a relative upward movement of the hanging wall has taken place. The force did not produce a single, sudden and catastrophic movement; on the contrary, it continued for long time, resulting in repeated dislocations, as proved by the reopening and refilling of some veins and by a sheeting of some veins, producing what is usually described as "ribbon rock." Recemented quartz-breccias are also of common occurrence.

I should here like to mention one misleading circumstance relating to parallelism of vein and country-rock. When larger fissures are opened in massive rocks it is not at all uncommon to find the immediately adjoining wall-rock converted entirely locally into schists parallel to the fissure, under the influence of the enormous shearing stress to which it has been subjected. Such veins would have the appearance of cropping in preëxisting schist-masses, and of parallelism in strike and dip with these. The conclusion to be derived from the relation of the veins to the larger, regionally metamorphosed schist-masses is that the schistose structure antedates the formation of the vein fissures; and that the forces to which these fissures are due, while bearing a general similarity to those manifested in the cleavage, often differed from them in direction to a sensible extent.

Different rocks influence the character of the fissures to some extent. In massive rocks[11] they are apt to be straight, clear cut and well defined; in slates and serpentines there is often a tendency to splinter into a network of cracks and fissures, extremely small, but often very rich. In such cases the whole mass, country-rock and vein, may be extracted and milled. Linked veins are common and chambered veins sometimes occur.

Very long and continuous veins are not common, and in this respect the mother-lode is rather an exception. Only rarely can a quartz-vein be traced more than a few miles, and many important veins crop out only for a short distance.[12]

### THE FILLING OF THE VEINS.

The typical gold-quartz veins cannot be considered as anything but fissures and fractures filled with quartz, accompanied by small amounts of native gold and metallic sulphides. Replacement proper of the minerals of the country-

---

[9] Tenth Ann. Rep. State Mineralogist of California, 1890, p. 86.
[10] The relation of the forces and the sheeting has been discussed by Mr G. F. Becker: Bull. Geol. Soc. Am., vol. 4, p. 13. See G.F. Becker: "Geology of the Comstock Lode." Mon. III. U. S. Geol. Survey, p. 182, and S. F. Emmons, "Structural Relations of Ore Deposits," Trans. Am. Inst. Min. Eng., vol. xvi, p. 814.
[11] G. F. Becker: Quicksilver Deposits of the Pacific Slope, p. 409.
[12] It is not true that every fissure vein holds out to indefinite depth, though it is probable that most of the larger veins of the gold-belt will continue to a depth exceeding the limit of practicable exploitation. As a rule, the probable permanence of a fissure vein will be in direct proportion to its traceable length and to its width. In regions where strong sheeting of the rocks has taken place it is quite probable that many of the smaller fissures and joint-planes will pinch out and disappear in depth. Fissuring, after all, is most intense near the surface, and probably comparatively few of the fissures reach down to deep seated regions; when the rocks become plastic by pressure and heat, or by a suitable relation between the applied stresses, such as must prevail below a certain level, all fissures must cease to exist. The smaller fissures probably received their quartz-filling by communication with the larger ones, which must be regarded as the principal conduits for the solutions.

rock along the fissure by quartz I have never been able to observe, and cases supposed to be of such nature have always proved to be due to the shattering of the country-rock and the filling of it by silica along narrow cracks. The clean quartz usually forming the vein I cannot account for in any other way than by filling of cavities, as it does not seem possible that a replacement of the ferro-magnesian silicates and other minerals could occur without leaving chloritic stains or other signs in the resulting mass. In all quartz-veins of this type it seems unavoidable to admit the existence of open spaces along the vein, supported at frequent intervals by the contact of the two walls or by rock fragments. Even the heaviest veins show in the underground workings frequent places where the walls "shut down." Such fillings of clean quartz may vary in width from a few inches to several feet. "Horses," of course, frequently appear in the larger veins, separating them in two or more parts. The heaviest veins appear to be found along the mother lode in Tuolumne and Mariposa counties, where the clean quartz often reaches a width of 10 to 15 feet, and in isolated cases even more. This extreme thickness seldom continues for any great distance. It may probably be safely assumed that gold-quartz veins of this type cannot be formed at any extreme depths below the surface, probably not below 10,000 feet, for at such depth open spaces could hardly exist. These very large fissure veins are, however, not very abundant; a moderate width of one to three feet is far more common. In many cases, indeed, there are no large open spaces at all, but a network of smaller cracks and fractures, in which the solutions deposited their contents.

## Association of Minerals in Gangue and Ores.

In the predominating milky white quartz of the veins but few other gangue minerals are found. Calcite, more rarely magnesium carbonate, or a mixture of both, occur occasionally, but always in subordinate quantities and usually concentrated near the walls. The quartz is ordinarily massive, but excellent examples of comb-structure may be found. Barite[13] and fluorite are conspicuously absent. A white mica with pearly luster is sometimes found in the quartz at some of the mines along the mother-lode, and a green potassium-mica, colored by chromium, and which Professor Silliman has called *mariposite*,[14] occurs abundantly in places, though usually not in the quartz itself. Roscoelite, a vanadium-potassium-mica has been found in one place, and albite occurs in isolated cases.[15] Rhodonite has been found in Plumas county. Titanium minerals, such as titanite, ilmenite and anatase, occur occasionally.

The native gold is distributed through the quartz-gangue in an irregular manner. The particles may be of microscopic size, or coarser, and visible to the naked eye as scales, threads and smaller masses. Occasionally large pieces of all weights up to fifty pounds or more, will be found. In the ores from the larger mines it is, however, rare to find the gold visible to the naked eye. The gold always contains a little silver, in rarer cases as much as 30 per cent.

A variable but always comparatively small quantity of metallic minerals accompanies the gold. It varies from a fraction of 1 per cent to 5 or 6 per cent, but ordinarily makes up from 2 to 3 per cent of the mass of the quartz. Sulfides are most common, but compounds of arsenic, antimony and tellurium also occur. A list of the associated minerals in the quartz-veins would include the following species:[16]

Iron-pyrites (universally present).
Pyrrhotite (not common).
Copper-pyrites (common).
Zinkblende (common).
Galena (common).
Molybdenite.
Arsenical pyrites (common).
Tetrahedrite
Antimonal lead sulphides (rare)
Cinnabar (rare)
Tellurium minerals - Hessite, Altaite, Calaverite, Sylvanite, Petzite, Melonite (frequent, in small quantities)
Nickel and cobalt minerals (very rare).

Marcasite is noticeably absent from gold deposits as noted by Mr Louis. I have once seen it, however, from a mine at Grass Valley.

These metallic minerals, usually referred to as "sulphurets," contain more or less gold and silver and are frequently very rich, the concentrates ranging from thirty to several hundred dollars per ton.

Bismuth and cadmium have been found in small quantities in the concentrates from the Nevada City mines, the former also in Shasta county.[17] Compounds of tin, wolframium, uranium, boron,[18] phosphorus and fluorine appear to be entirely absent. Cuprite, bornite and chalcocite are also lacking. Cobalt and nickel minerals are occasionally present. Titanium occurs sparingly.

A slight influence of the wall-rock upon the character of the mineral association cannot be denied. It appears to be a fact that veins in granodiorite contain more sulphurets than those in other rocks. Pyrrhotite appears to be entirely confined to veins in this rock. It is known only from the vicinity of

---

[13] For rare occurences of barite see W. Lindgren, Am. Jour. Sci., vol. xliv, 1892, p. 92, and H. W. Turner, Am. Jour. Sci., vol. xlvii, 1895.

[14] For analyses see H. W. Turner: "Further notes of the gold ores of California," Am. Jour. Sci., vol. xlvii, 1895.

[15] See the interesting paper by H. W. Turner: Am. Jour. Sci., May, 1894, vol. xlvii, p. 467.

[16] Compare a paper by Henry Louis in Min. Magazine, vol. x, 1893, p. 241, on the minerals associated with gold deposits in general.

[17] R. Pearce, in Trans. Am. Inst. Min. Eng. vol. xvii, p. 447.

[18] Tourmaline has been found in the abnormal veins of Meadow Lake, Nevada county. Am. Jour. Sci., vol. xlvi, 1893, article 30.

Washington, in Nevada county, Sonora, in Tuolumne county, and Westpoint, in Calaveras county. Veins in black sedimentary slate or on the contact between greenstone and slate seldom contain much besides iron-pyrites, and perhaps arsenical pyrites; neither are veins in augite-porphyrite or diabase usually rich in sulphurets. Veins in gabbro often contain copper. These are no strict rules, however, and the influence of the wall-rock may, on the whole, be considered as remarkably small.

## Distribution of the Gold in the Veins.

Gold is universally distributed in the quartz-veins of California. The definition of what is ore, or quartz paying for exploitation and metallurgic treatment, will necessarily vary at different times and in different places. Under exceptional circumstances rock containing as little as one or two dollars of gold to the ton will pay. In the deep mines the tenor of the extracted ore is usually from five to twenty dollars.

In wider veins a small streak near one of the walls will sometimes contain the pay, while the rest is comparatively barren. Equal distribution of value in cross-section is, however, common enough. Considered in projection on the plane of the vein, there is rarely an equal distribution of the gold over large surfaces. The richer ore is concentrated in bodies and masses, which sometimes may be wholly irregular, but which usually show more or less regular outlines. These richer masses are called chutes or chimneys, and appear on the plane of the vein in long-drawn linear or elliptic form, with a dip which usually is above 45 degrees. Flat ore-chutes occur, however, as, for instance, in the Idaho mine, Grass Valley. Their width ranges from a few feet up to several hundred, and their length may exceed 2,000 feet. It is not uncommon to find one of these ore-chutes give out in depth, but another chute will then probably be found in some place below it, if a thorough exploration is carried out. It is a practical rule in many districts, and one which holds good in a remarkable number of cases, that the chutes dip to the left when one is standing on the apex and looking down along the dip. The explanation of the ore-chutes is difficult. They may, as F. Posepny and others have suggested, simply indicate the direction of least resistance for the gold-bearing solutions. This explanation is not entirely satisfactory, for in the intervals between the chutes it is by no means the rule to find the walls shut down tight. On the contrary, it is common to find the barren vein between them as wide if not wider than the rich vein in them. An increase in the quantity of the sulphurets always accompanies the increase of gold in the ore-chutes.

No gradual decrease in the tenor of the ore takes place with increasing depth; on the whole, the character remains constant. Individual ore-chutes may be exhausted, but others, as a rule, are found below them.

Certain veins show no large bodies of milling ore at all, but coarse gold concentrated at certain points; such deposits are called "pocket veins." Small seams may sometimes carry a surprisingly large amount of gold. Intersection of seams or veins often, but by no means always, produce pockets or ore-chutes.

## The Alteration of the Country-rock.

The study of the changes and alterations which the rocks adjoining the fissures have undergone is a subject of the highest importance, for in this way a closer insight into the genetic processes of the vein may often be obtained.

It would at first glance seem more likely that the rock in the vicinity of the quartz-filled veins would have undergone a silicification. Such is not the case. Instead of a silicification there is, as a rule, a most marked carbonatization, or a conversion of the country-rock to carbonates. Most intense next to the vein, the alteration gradually decreases at a distance from it, the width of the altered zone varying according to the width of the vein. The carbonate zone, surrounding the quartz-vein on both sides, may often be studied to great advantage in small veinlets cutting through hand specimens.

This action upon the adjoining country-rock is in itself, to my mind, the strongest possible evidence against the application of lateral secretion in its narrower sense to these veins. It appears to completely refute the theory of the veins being formed by percolating surface waters, and prove the existence of an agency active in the fissures and gradually extending outward.

The solutions circulating in the fissures acted with different intensity on different rocks. Nearly all igneous rocks, acid or basic, are profoundly altered, the latter more than the former, and serpentine more than any other. Only extremely silicious rocks, and especially certain carbonaceous slates, appear to successfully withstand the action of these solutions. The process of carbonatization has not in all cases been carried out to its full extent; in some veins it is more marked than in others; occasionally fresh rock may lie close up to the vein.[19] Crushing of the rock next to the vein facilitates the process and increases the width of the altered zone, which may vary from a few inches up to twenty feet and even more in exceptional cases. With all variations, there is no doubt that the process is a general one, and characteristic for the type.[20]

The result of the process, when it has been thoroughly carried out, is the conversion of the country-rock by replacement to a mixture of carbonates, white potassium-micas (sericite), a small amount of chloritic minerals and

---

[19] Such cases are perhaps due to layers of impervious clay-like detritus on the wall.
[20] A good instance has been described by the author in the Fourteenth Ann. Rep. U. S. Geological Survey, in a paper entitled "The Gold-silver Veins of Ophir, California," now in press.

residuary quartz; besides, there is always a large amount of iron pyrites,[21] usually more than in the vein; arsenical pyrites[21] is also frequently present, but never, as far as I know, any other sulphides in noticeable amounts. Calcium carbonate usually prevails, but the carbonates of magnesium, iron and manganese are also present. According to numerous analyses, calcium is always added, while nearly all of the sodium is carried away. The potassium of the orthoclase remains transferred to the sericite. As abundant potassic micas are often found in wall-rocks originally very poor in this metal, it is probable that some potassium was also added. In silicious rocks the quartz is often attacked, but never completely carried away. The iron ores and partly also the bisilicates of the original rock appear to have been converted into pyrites,[22] while the titanium in it was transformed to leucoxene.[23]

In the case of clean-cut fissures, with well defined quartz-veins, it is usual to find by far the largest amount of gold in the quartz and in the sulphides associated with it. The altered country-rock is not entirely barren, but it does not often contain native gold, and its sulphides are much poorer than those in the quartz.[24] This is not entirely without exceptions, for in several places, usually adjoining rich chutes, the altered country-rock will pay for milling, and may, in isolated cases, go as high as $12 per ton. And again, there are cases in which the altered country-rock is traversed by a great number of minute quartz seams, in which the gold is concentrated. Such a case is the Rawhide mine, Tuolumne county, in which this altered and fissured country-rock is far richer than the main quartz-vein. At the same place the gold sometimes also penetrates and coats the cleavage faces of the adjoining talcose or serpentinoid schistose rock. One frequently hears of native gold in talc, slate or other rocks. I have always found such occurrences to be more or less altered rocks from the immediate vicinity of some vein. The gold occurs on minute, sometimes hardly visible, seams traversing them. Indeed, many fissures are absolutely microscopic.

It has been stated above that serpentine[25] is peculiarly liable to alteration by the auriferous solutions. The zones of altered rock are in this case often very large and always very characteristic. They may be twenty or thirty feet wide, or in the case of branching veins a whole area, several hundred feet across, may be more or less completely altered. The serpentine is converted into a mixture of magnesic and calcic carbonates, a green micaceous mineral containing potassium and colored by chromium, to which the name of mariposite has been given by Professor Silliman, together with more or less iron pyrites. The altered mass is frequently shattered and traversed by seams of mixed quartz and carbonates. It has a rather coarse, crystalline structure, and a bright green color from the disseminated mariposite. The carbonates referred to as ankerite by Professor Silliman are in reality, as indicated by H. W. Fairbanks,[26] a mixture of varying composition, ranging fromn calcite to magnesite, and often containing considerable iron. Magnesic carbonate, on the whole, predominates. The mineral mariposite is, as Silliman observes, only associated with magnesian and chloritic rocks. Fairbanks[27] states that it is particularly eminently characteristic of the mother-lode. This is not correct. It is, however, characteristic of all quartz-veins in or at the contact of serpentine, though occasionally occurring in very small quantities in diabase and other basic rocks. The writer has noticed the same characteristic mixture of carbonates and mariposite from a great many places in the gold-belt besides the mother-lode; thus, for instance, at the Phoenix and Red Chief mines in Sierra county, and also near Washington, Nevada county. It appears at the mother-lode wherever that great quartz-vein breaks through serpentine. Quartz mountain, Tuolumne county, is an excellent place to study it.

Along the mother-lode the altered serpentine has been variously interpreted. Whitney inclined to the belief that the vein represented a stratum of silicified dolomite, a theory that has not been supported by more detailed investigation. Fairbanks, who some years ago carefully examined the mother-lode,[28] considered it at first as vein-matter deposited in open fissures, but regarded it subsequently (as the needed, once open space would manifestly have been too large, in places several hundred feet) as an altered, coarsely crystalline basic rock. The latter theory, while nearer the truth, is unnecessary. A careful investigation will not fail to disclose the fact that the mixture of carbonates and mariposite is nothing but an altered serpentine, and abundant transitions may be found to prove this. A locality showing this plainly is the App mine at Quartz mountain, Tuolumne county. This conversion is not astonishing when the facility is considered with which the serpentine is decomposed by carbonated waters into magnesite and chalcedonic quartz. Experiments by C. Doelter[29] show that while at ordinary temperature and pressure water

---

[21] Both occur as small but extremely sharp crystals, while the sulphurets in the quartz are usually massive.

[22] A similar alteration has been shown to have taken place in the country-rock of the Comstock lode by G. F. Becker, Monograph III, U. S. Geol. Survey, p. 210.

[23] The alteration and replacement of the wall-rocks has been emphasized by S. F. Emmons in regard to the fissure-veins of Colorado and Montana, and he points out that, especially where extensive sheeting has taken place, the fillings of open spaces are often small compared with the alteration and impregnation of the adjoining country-rock. "Structural relations of ore-deposits," Trans. Am. Inst. Min. Eng., xvi, p. 808.

[24] This fact, as well as many others, of course, speaks strongly against the derivation of the gold in the vein from the decomposed zone adjoining it.

[25] As well as talc-schist and other slaty magnesian rocks derived from serpentine.

[26] Tenth Ann. Rep. State Mineralogist, p. 85.

[27] Loc. cit.

[28] Loc. cit.

[29] Allgemine chemische geologie, Leipzig, 1890, p. 190.

containing carbon-dioxide will, with simultaneous decomposition and formation of carbonates, dissolve 0.3 per cent orthoclase and 0.5 per cent oligoclase, serpentine will be dissolved at the rate of 1.24 per cent.

The large bodies of decomposed rock referred to on page 226 as containing impregnations of auriferous pyrites and rarely free gold are in many respects interesting. In the ferruginous outcrops the iron-pyrites is usually converted into ferric hydroxide and the gold set free; the whole mass can then sometimes be profitably mined and milled, though it is of very low grade. Veins and seams of quartz are often entirely absent in these impregnated zones. In the cases which have come under my observation the action on the rock is much the same as in the decomposed wall-rocks of the veins—that is, there is an abundance of carbonates and iron-pyrites in sharp edged, little crystals. While there is abundant evidence of replacement by carbonates, I have not yet seen anything proving a replacement by quartz, though the possibility of such a process cannot be denied. However, in these deposits the action of the solution on the rock-forming minerals must have produced much free silica in solution and probably also much sodic silicate; in fact, there are in these deposits occasional masses of granular, grayish quartz very different from the ordinary vein-quartz and probably partly chalcedonic. This quartz often contains iron-pyrites in small scattered crystals, and appears to represent in part residual masses from leaching, in part deposition from the supersaturated silicious waters. H. W. Fairbanks has recently described two deposits in El Dorado county, the Big Canyon and the Shaw mines,[30] as showing in marked degree a replacement of the rocks by silica. Though the latter mine was not worked during my examination of the Placerville sheet, I have, through the kindness of Mr H. W. Turner, had occasion to examine an excellent suite of specimens, lately collected. The vein is partly in black slate, partly in a feldspathic dike. Both rocks contain an abundance of stringers and seams of quartz rock by the former mineral. On the contrary, the porphyritic dike is to a very marked degree converted into carbonates in the vicinity of the veins. Regarding the Big Canyon mine, I have seen only two specimens of greenstone impregnated by pyrite from this mine, and collected by Mr H. W. Turner. These specimens show carbonatization to a considerable extent, but no evidence of replacement by silica. It is not intended to deny that such a process may take place, but only to point out that it is something requiring more and more detailed investigation. Calcite is found pseudomorphic after an enormously large number of minerals, while pseudomorphs of quartz after other minerals are much less common.

## GENETIC CONCLUSIONS.

The country-rock altered to carbonates, standing in strong contrast to the vein filled nearly exclusively by quartz, affords a much-needed key to the genetic processes of the deposits. It shows, first, that besides silica, the water circulating in the fissures contained large amounts of carbon-dioxide, as well as dissolved calcic carbonate. It certainly contained sodium as carbonate taken up from the feldspars of the adjoining rocks, probably also as silicate and chloride. It further contained sulphur, in what form is not certain, but most probably as sulphuretted hydrogen or as sulpho-salts. The presence of large quantities of sulphates does not appear probable.

Waters of this composition, containing abundant carbon-dioxide, are only known in nature as ascending, usually hot springs. The process of deposition took place as follows:

At first the carbonated waters began to act with great energy on the soluble minerals in the wall-rocks of the fissures, converting them more or less completely into a mixture of carbonates, potassium-micas and pyrites, adding calcium-carbonate and sulphur, probably also potassium, to them, and abstracting sodium. Finally, this process being completed, and the walls usually coated with crystals of carbonates, the formation of the latter ceased, and in this surrounding of carbonates the silica now began to be deposited, and with it the gold and the rest of the metallic sulphides.

A most interesting question in connection with this subject is, why the walls should, to such a large extent, act as a separating barrier for the gold and most of the sulphides. Mr G. F. Becker, in discussing the quicksilver deposits of the Pacific coast,[31] has suggested that this may be due to an osmotic action, transmitting through the septum only the chemically active solutions.

Admitting that the gold-quartz veins were deposited by such mineral waters, the next question is, in what form the gold and other metals were in solution. While not intending to enter into a detailed discussion of the difficult problems associated with the question,[32] I would like to call attention to a few general facts connected with them. Gold is soluble at 200° centigrade in a 10 per cent solution of carbonate of sodium to the extent of 1.23 per cent (Doelter), while silver is hardly attacked. Silicates of alcalies dissolve gold at 250° centigrade to the smaller extent of 0.101 (Doelter and Liversidge). Besides, gold is more or less soluble in a great many other salts (T. Egleston). G. F. Becker has shown the solubility of gold in alkaline sulphides, and the

---

[30] Twelfth Ann. Rep. State Mineralogist, 1894, pp. 103 and 114.

[31] Mineral Resources of the United States, 1892, p. 21.

[32] Mr A. Liversidge has recently given an interesting historic résumé of the experiments regarding the solubility of gold, as well as many original experiments, in the Proc. Roy. Soc., New South Wales, vol. xxvii, 1893, p. 303.

Nevada the association of minerals is native gold with predominating quartzose gangue; carbonates in the wall-rocks; next in importance, iron-pyrites with smaller quantities of the minerals of copper, lead, zinc, arsenic, and antimony; quicksilver-ores are occasionally present. In the Coast ranges we have quicksilver in predominating quartzose, and to some extent carbonate gangue; next in importance, iron-pyrites with smaller quantities or copper, antimony, arsenic, and nickel; gold is very commonly present.

Regarding the rocks adjoining the deposits, Mr. Becker says[34] they—

"have in many cases been greatly modified. Metamorphic rocks often appear to have been converted into or replaced by more or less dolomitic carbonates by the action of solutions. . . . Both serpentine and the metamorphic rocks seem to be subject to this conversion."

Containing a similar association of metals, similar gangue and similar altered country-rock, it seems justifiable to express the conviction that similar mineral solutions have circulated in both classes of deposits; and in fact, the still abundant thermal waters found in intimate connection with the ore deposits in the quicksilver region closely correspond in character to the inferred composition of the once existing hot springs of the gold-belt. They all show free carbonic acid, as well as abundant carbonates (sodic, calcic and magnesic); silica is always present; usually also sulphuretted hydrogen or alkaline sulphides.

Carbonatization of the wall-rocks of fissure veins has neither been described by A. v. Groddeck nor by F. v. Sandberger in their researches though the former has found abundant sericite in many. The wall-rocks of the Comstock lode, according to Mr. Becker, are rich in iron-pyrites, but do not contain much carbonates. J. H. L. Vogt has shown that along certain veins of Norway the granite and gneiss are altered to products resembling the "greisen" of the tin deposits.

## Origin of the Gold.

Regarding the origin of the hot, auriferous solutions which have produced the gold-quartz veins it is best, at this stage of our knowledge, to speak with great reserve. Even the results of assays or analyses of country-rock must be received with the greatest caution, to make sure that the percentage discovered is primary constituent and not later impregnation. It is not to be denied that many reasons speak strongly against a derivation from the surrounding rocks. Thus, for instance, the diorites of Nevada City and Grass Valley contain an

---

[34] Monograph XIII, U.S. Geol. Survey, p. 392.

---

solubility of the sulphides of Hg, Fe, Cu and Zn in either sodic sulphide, sodic sulph-hydrate or sodic carbonate, partly saturated with sulphuretted hydrogen. Silicate of gold, the existence of which was first suggested by G. Bischof, has been frequently mentioned as probably contained in mineral waters; but it should be borne in mind that the existence of this salt has never been proved. It appears that the mentioned facts are sufficient to show that the mineral waters, once circulating in the quartz veins of California may easily have held gold in solution.[33] It seems of questionable use to speculate on the particular combination in which the gold is contained in the water, for, according to the views of modern chemistry, watery solutions, when sufficiently diluted, contain the solids in a state of dissociation, so that it is uncertain whether salts of gold could exist as such in the always much diluted natural solutions.

The precipitation of the solids contained in the solution could have been brought about by many means, such as diminution of pressure, dilution, etcetera. The reducing influence of carbonaceous slates, so often maintained as the probable cause of the precipitation of the gold, appears of questionable importance. Veins entirely in massive rocks and far away from any sedimentary areas show too much similarity with those in such areas to attribute a paramount weight to this argument.

## Comparison with Quicksilver Deposits.

There are certain interesting analogies between the gold-quartz veins of the Sierra Nevada and the quicksilver deposits of the Coast ranges. In the Sierra

---

[33] These reactions are of course by no means the only ones which are likely to take place. It is thus very likely that the reactions established by C. Newbery (Trans. Roy. Soc. Victoria, vol. ix, p. 754) have taken place. According to him the iron is contained as ferrous carbonate with sulphates; chlorides of gold can be held in such very diluted solutions in presence of alkaline carbonates and excess of $CO_2$. "This is true of chloride of gold, and if the sulphide is required in solution, it is only necessary to charge the solution with an excess of $H_2S$. In this manner both sulphides may be retained in the same solution, depositing gradually with the escape of the carbonic acid." It does not seem probable, however, that sulphates have played a very important part in the chemistry of the gold veins. The explanation of Phillips for the contemporaneous deposition of gold and pyrites (Proc. Roy. Soc. London, vol. xvi, 1868, p. 294) was that as gold is soluble to some extent in ferric sulphate, solution of this salt containing gold was transformed by a reducing agent into pyrites, the gold at the same time being reduced to the metallic state. The presence of a ferric salt in deep-seated waters would be a very unusual occurrence. The presence of ferrous sulphate on the other hand, in solution carrying gold does not appear possible, for the latter would be immediately precipitated. The fact that in the gold-quartz veins silver occupies such a subordinate position would seem to lend strength to the view that the solutions once circulating in them were not adapted for the dissolving of silver compounds. While thus G. F. Becker found that PbS and $Ag_2S$ were insoluble in sodic sulphide, sodic sulphydrate or in solution of sodic carbonate partly saturated with hydrogen sulphide, these salts or metallic silver may be soluble to a very slight degree when in combination with other compounds. An alloy of much gold with slight amount of silver may thus be soluble. I do not know of any experiments on this subject. Doelter, referring to the dissolving action of sodic silicate and carbonate on silver, remarks that it is "hardly" attacked, thus implying some action.

appreciable amount of barium, and still there is no trace of barite in the veins of those localities. In another instance the diabasic rocks of the same region contain copper, and yet the gold veins passing through these rocks are remarkably poor in copper minerals.

In discussing this difficult question there are several broad facts which must be borne in mind:

First, that the gold-quartz veins throughout the state of California are closely connected in extent with the above-described metamorphic series, and that the large granite areas are almost wholly void of veins, though fissures and fractures are not absent from them.

Second, that in the metamorphic series the gold-quartz veins occur in almost any kind of rock, and that if the country-rock exerts an influence on the contents of the veins, it is, at best, very slight.

Third, that the principal contact of the metamorphic series and the granitic rocks is in no particular way distinguished by rich or frequent deposits.

It is further apparent that gold deposits have been formed at different periods, though by far most abundantly in later Mesozoic times. Some of these later veins may have been locally enriched by passing through earlier impregnations in schist or old concentrations in the sandstones and conglomerates of the metamorphic series, the gold contents of which have, however, only been proved in isolated cases.

These considerations, though involving many most difficult questions, strengthen the belief that the origin of the gold must be sought below the rocks which now make up the surface of the Sierra Nevada, possibly in granitic masses underlying the metamorphic series.[35]

### SUMMARY.

The auriferous deposits extend through the state of California from north to south in an irregular and broken line.

The gold-quartz veins occur predominantly in the metamorphic series, while the large granitic areas are nearly barren. The contact of the two formations is not distinguished by rich or frequent deposits.

The gold-quartz veins are fissure-veins, largely filled by silica along open spaces, and may dip or strike in any direction.

The gangue is quartz, with a smaller amount of calcite; the ores are native gold and small amounts of metallic sulphides. Adjoining the veins the wall-rock is usually altered to carbonates and potassium-micas by metasomatic processes.

The veins are independent of the character of the country-rock, and have been filled by ascending thermal waters, charged with silica, carbonates and carbon-dioxide.

Most of the veins have been formed subsequent to the regional metamorphism which affected the auriferous slates and the older igneous rocks associated with them, and also subsequent to the granitic intrusions which closed the Mesozoic igneous activity in the Sierra Nevada.

---

[35] Mr Becker, reasoning from analogy, has some time ago suggested such a derivation: "Quicksilver Deposits," p. 449. Militating against this view is the general absence of compounds of boron and fluorine so often occurring in ore deposits in granitic rocks.

---

...the nozzles are sixteen feet long, and are called dictators, monitors, or giants. They require ever more ditches and flumes.... Where two men working a rocker can wash a cubic yard a day, two men working a mountainside with a dictator can bring down and drive through a sluice box fifteen hundred tons in twelve hours...

Although the nozzle has the appearance of a naval cannon, it is mounted on a ball socket and is so delicately counterbalanced with a "jockey box" full of small boulders that, for all its power, it can be controlled with one hand. Every vestige of what has lain before it—forest, soil, gravel—is driven asunder, washed over, piled high, and flushed away. At a hundred and twenty-five pounds of pressure per square inch, the column of shooting water seems to subdivide into braided pulses hypnotic to the eye, and where it crashes at the end of its parabola it sounds like a storm sea hammering a beach.

John McPhee, Assembling California, p. 64-65

# Mother Lode gold

### J. K. Böhlke
*U.S. Geological Survey, 431 National Center, Reston, Virginia 20192; e-mail: jkbohlke@usgs.gov*

## INTRODUCTION

Mother Lode gold is of interest not only because of its important role in the history and economics of California, but also because it is a significant representative of a distinctive and globally important ore-deposit type. Similar clusters of low-sulfide gold-quartz vein deposits have formed near major fault zones in tectonically active metamorphic complexes on all continents from the Archean to the Cenozoic. Those deposits commonly are associated with younger placer deposits formed by subsequent exposure and weathering of the veins. The amount of gold extracted in California through 1995 was ~115 million ounces (equivalent to about 3600 metric tons or 190 cubic meters), of which roughly 75% came from veins and placer deposits in the metamorphic complexes of the western Sierra Nevada foothills (Clark, 1969, 1985; U.S. Bureau of Mines, 1983–1995).

The California gold rush was initiated by the discovery of placer gold in 1848, and the placer deposits have yielded most of the gold extracted from the Sierra Nevada foothills since then. Some of the classic geologic studies in the region were done on the distribution of the placers, including maps of the buried and partially exhumed Tertiary river systems (e.g., Whitney, 1879; Lindgren, 1911). Early rapid development of placer mines was followed almost immediately by discovery and development of hard-rock vein deposits (as early as 1849), some of which were first exposed in the floors of placer drift mines. The veins have outlived the placers as important producers, aided by closure of the hydraulic placer operations in 1884. Partly because of their economic persistence, but also because of their more enigmatic origins, the vein deposits also have been the subject of more study and speculation in the twentieth century. Major studies of the gold-quartz veins of the Sierra Nevada foothills region were done while mining activity flourished from the late 1800s until the 1940s, when World War II caused the mines to close. Relatively little mining and research activity occurred in the first few decades after the war, but interest increased abruptly after the price of gold was deregulated in 1968, and several new studies were completed in the 1970s and 1980s. Attention shifted in the 1990s toward the environmental consequences of the mines and whether they pose a threat to residential development in the region. Particular concerns include erosion and sedimentation of old placer workings and tailings piles, arsenic and mercury weathering from sulfides in the ores and tailings, and mercury left from ore-treatment practices (Kattlemann, 1995; Ashley, 1997).

Since Lindgren's classic summary of the state of knowledge in 1895, there have been a number of other major contributions to the scientific literature on the Sierran gold-quartz veins. Among those are a set of U.S. Geological Survey Professional Papers published in the early part of the twentieth century that gave more detailed descriptions of veins in specific subregions of the metamorphic belt. Knopf (1929) described the gold deposits of the "Mother Lode" district, which was defined in that study as a narrow linear north—south–trending mineralized zone (~200 km long) along the Melones fault zone in the southern and central part of the western Sierra Nevada metamorphic belt. This included an area of mineralization somewhat larger than what was included in the Mother Lode as described by Ransome (1900). Ferguson and Gannett (1932) described the Alleghany district of California, which is in the northern foothills near a major fault zone that they correlated with the Melones fault zone farther south. Johnston (1940) described the Grass Valley district, which included the most productive veins in the northwest part of the metamorphic belt. These three important publications, along with the earlier studies by Lindgren, Turner, Becker, Ransome, and others, contain irreplaceable documentation of the subsurface geology and structure of the major vein systems. The

---

Böhlke, J. K., 1999, Mother Lode gold, *in* Moores, E. M., Sloan, D., and Stout, D. L., eds., Classic Cordilleran Concepts: A View from California: Boulder, Colorado, Geological Society of America Special Paper 338.

more recent of those studies also contain detailed descriptions and photomicrographs of mineral textures. Because of similarities in the geologic setting, chemistry, age, and mineralogy among the different districts of the Sierran foothills, studies in all of those areas are included, along with mention of similar deposits in the Klamath Mountains, in the following discussion of Mother Lode gold. Other major studies and compilations of mine workings, mineralogy, and production data, and field trip guides for these deposits were published by the California Division of Mines and Geology and the U.S. Geological Survey, and include those of Logan (1934), Jenkins et al. (1948), Clark (1969), Hotz (1971), and many others. A brief outline of the history of California gold mining through the 1960s was given by Clark (1969).

## UPDATING LINDGREN'S 1895 PAPER

Some of the important general conclusions summarized by Lindgren (1895; see also 1896, 1933) include the following. (1) Regionally, the veins are widely distributed in a variety of metamorphic rocks and in some minor igneous bodies within the metamorphic belt, but they are not common in the major batholiths. (2) Locally, the veins occupy discordant structures and were formed by intermittent open-space filling in tectonically active fault systems in Late Jurassic or Early Cretaceous time, subsequent to the last regional penetrative metamorphic deformation. (3) Aqueous fluids responsible for the mineralization were far from chemical equilibrium with their host rocks, having ascended from significant depths bearing large quantities of $CO_2$, along with gold and other constituents. Massive bodies of bright green mariposite (Cr-bearing mica), carbonate, and quartz formed by hydrothermal alteration (metasomatism) of ultramafic rocks are among the most conspicuous products of the interaction between the $CO_2$-rich vein fluids and wall rocks, but more subtle analogous effects are ubiquitous in other rock types. (4) Although not stated firmly, it was implied (and restated in Lindgren, 1933) that the veins were in some way genetically related to deep-seated magmatic activity.

Many of Lindgren's (1895) observations and conclusions have been confirmed by subsequent work, but some difficult issues remain unresolved despite the application of new technologies. This essay continues with a brief updated discussion of some of the major topics addressed by Lindgren concerning the origin of the deposits, focusing on results of recent studies that include methods and data that were not available when the early work was done. Some types of information that have had significant influence on the subject in recent decades include plate tectonic models of regional geology; geochronologic and isotopic data bearing on the ages of the deposits and origins of the mineralizing fluids; microthermometric and chemical analyses of fluid inclusions; and thermodynamic and mass-transfer analyses of wall-rock alteration and mineralization reactions. Studies completed in the twentieth century also contain more detailed mineralogic data owing to the use of the petrographic microscope and other microanalytical tools. The results of this review exercise demonstrate the importance of basic geologic observations and testify to the perception of the early geologists; however, they also support Lindgren's reluctance to state firm conclusions about the genesis of the ores.

## AGE AND TECTONIC ENVIRONMENT OF GOLD MINERALIZATION

> Within the typical gold-bearing region, the veins are distributed with remarkable impartiality, and occur in almost any of the great variety of rocks which make up the metamorphic series.
>
> Lindgren, 1895, p. 224

> Regarding the structural relations of the normal gold-quartz veins, it should first be stated that they are fissure veins, and emphatically not so-called segregated veins or "lenticular masses"
>
> Lindgren, 1895, p. 226

> It is everywhere plain and evident that the fissures have been broken open subsequently to the metamorphism of the rocks. These post-Jurassic and post-granitic quartz-veins form the latest chapter in the Mesozoic revolution in the Sierra Nevada.
>
> Lindgren, 1895, p. 227

The major gold-quartz veins of the western Sierra Nevada foothills occur within pervasively metamorphosed, sheared, folded, and faulted rocks of widely varying lithologies and ages. In the Alleghany district, host-rock lithologies of mineralized veins range from serpentinite to granite, metamorphic grades range from blueschist to amphibolite, and apparent ages of protolith deposition, metamorphism, and igneous intrusion range from about 140 Ma to older than 390 Ma (Ferguson and Gannett, 1932; Coveney, 1981; Böhlke and McKee, 1984; Böhlke and Kistler, 1986). Similar variations may be observed on a larger scale throughout the gold-bearing region.

Lindgren's (1895) Plate 1, showing the regional extent of gold-quartz mineralization within the "metamorphic series" west of the Sierran batholith, features some linear arrays of mine locations corresponding to major structural trends. An updated version of this figure compiled by the U.S. Geological Survey in the 1980s confirms the concentration of deposits along several major northwest-southeast–trending structures, some of which separate contrasting metamorphic complexes (Fig. 1). Those major structures are known in part as the foothills fault system, which includes the Melones fault zone, near which many of the most productive gold districts are located (Clark, 1960). Displacements along some of these structures were large and complex. Some of the faults in the region may have originated as suture zones along which geologically diverse terranes were emplaced tectonically, but the history of Paleozoic through

Jurassic subduction, accretion, and deformation in the Sierran region is controversial (e.g., reviews by Burchfiel and Davis, 1981; Saleeby, 1981; Schweickert, 1981; Schweickert et al., 1988; Sharp, 1988; Day et al., 1988). Most of the large fault displacements in the Foothills Metamorphic Belt apparently ended in the Late Jurassic around the time of the Nevadan compressional event, which has been dated as 155 ± 3 Ma (Schweickert et al., 1984). At about that time, subduction ceased in the foothills region and was initiated farther west in the vicinity of the Coast Ranges. Minor deformations have been reported in Cretaceous intrusive rocks just south of the metamorphic belt indicating movements between about 113 and 147 Ma (Bateman et al., 1983; Paterson et al., 1987), but their significance to the foothills fault system is uncertain.

Almost all studies have concluded that the Sierran gold-quartz veins formed after the regional pervasive deformation of the Nevadan compressional event, but the critical tectonic elements responsible for the mineralization have been difficult to resolve. Underground mapping has demonstrated that many of the vein systems were characterized by minor reverse and reverse-oblique faulting that apparently was active intermittently during the mineralization (Lindgren, 1895; Knopf, 1929; Ferguson and Gannett, 1932; Cooke, 1947; Böhlke, 1989). Common vein textures such as brecciation, open-space vug fillings, "ribbon structures," and pressure-solution features referred to as "crinkly banding" indicate an active tectonic environment and complex episodic development of the mineralization. These observations have been confirmed consistently by researchers studying the veins, and have led in part to their inclusion in a global class of "orogenic (metamorphic-hosted) gold-quartz veins" (Böhlke, 1982; Groves et al., 1998). Veins of this type have also been termed "mesothermal" (Lindgren, 1933) for the moderately high temperatures and pressures of formation. Ferguson and Gannett (1932) derived a depth of around 6 km for ore deposition in the Alleghany district from estimates of subsequent erosion and placer accumulation. Isotope data and fluid-inclusion analyses indicate that temperatures and pressures probably were within the range of about 250–400 °C and 0.5–3 kbar, respectively (Coveney, 1981; Böhlke and Kistler, 1986; Weir and Kerrick, 1987). Those pressures would correspond to depths of about 2–10 km under lithostatic conditions.

Sibson et al. (1988) proposed that veins with these general characteristics might represent episodic escape routes for deep fluids trapped near the brittle-ductile transition, consistent with ore deposition occurring at depths of about 10–15 km or less, depending on the geothermal gradient. Several authors have called upon the voluminous Cretaceous magmatism in the Sierra Nevada as a source for various combinations of heat, fluids, and/or dissolved constituents (Lindgren, 1895; Knopf, 1929; Johnston, 1940; Albers, 1981; Böhlke and Kistler, 1986; Weir and Kerrick, 1987). Cloos (1935) compared the gold-quartz vein faults to marginal thrusts formed around magmatic intrusions. Others have deemphasized the significance of magmatic activity (Landefeld, 1988; Elder and Cashman, 1992), and some have suggested larger components of lateral movement on the major regional structures during the mineralization (Landefeld, 1988).

The difficult issues relating to the tectonic environment of the mineralization and origin of the mineralizing fluids largely involve the ages of the veins. Direct evidence for the ages of the Sierran veins has been provided by concordant Rb-Sr, K-Ar, and laser microprobe $^{40}Ar/^{39}Ar$ dating of hydrothermal mica, mostly mariposite (Evans and Bowen, 1977; Kistler et al., 1983; Böhlke and Kistler, 1986; Böhlke et al., 1989). The selection of mariposite for this purpose is important because it is one of the most clearly identifiable products of the carbonate metasomatism associated with the gold-bearing veins, and because it formed in ultramafic rocks that had essentially no relict mica and no K or Rb other than that added by the metasomatic fluids. Data summarized by Böhlke and Kistler (1986) indicate that mariposite formed in the vicinity of the Melones fault zone throughout much of its north-south extent in Early Cretaceous time, between about 108 and 127 Ma. An anomalous age of ca. 140–145 Ma was obtained for a specimen from the Brunswick mine in the northeast part of the Grass Valley district, though the intrusive host for some other veins in the southwest part of the district yielded an apparent age of 127 Ma (Böhlke and Kistler, 1986). Landefeld (1988) suggested that the bulk of the geochronologic data indicating Cretaceous mineralization ages in the Sierran deposits are misleading and that the veins formed earlier than 127 Ma. Although K-Ar and Rb-Sr radiometric ages may be younger by varying amounts than initial crystallization ages for minerals subjected to high temperatures, there is no evidence yet for such an effect in the vein micas, which formed at relatively low temperatures (probably around 350 °C or less) and which yield concordant results from both dating methods. Elder and Cashman (1992) reported a K-Ar age of 147 Ma for hydrothermal mica from the Quartz Hill mine in the Klamath Mountains, and they summarized evidence for younger gold mineralization limited by intrusive ages to less than 136 Ma in other Klamath districts. Lindgren (1895) also cited evidence for multiple ages of vein mineralization in the Sierran region.

Accepting the radiometric ages of the metasomatic assemblages, several authors have suggested correlations of mineralization with some of the major tectonic-plutonic events that occurred around the same times. Albers (1981), Böhlke and McKee (1984), Böhlke and Kistler (1986), and Weir and Kerrick (1987) emphasized the overlap in apparent ages of the majority of the Mother Lode veins (108–127 Ma) with the ages of pervasive low-grade blueschist metamorphism in the Coast Ranges and of a major pulse of igneous intrusive activity in the Sierra Nevada. Those observations suggested a genetic link between the veins and the post-Nevadan east-dipping subduction and arc magmatism. At that time, the major vein systems would have been deep within the proximal Sierran forearc above the subducting Farallon plate, perhaps about 200 km east

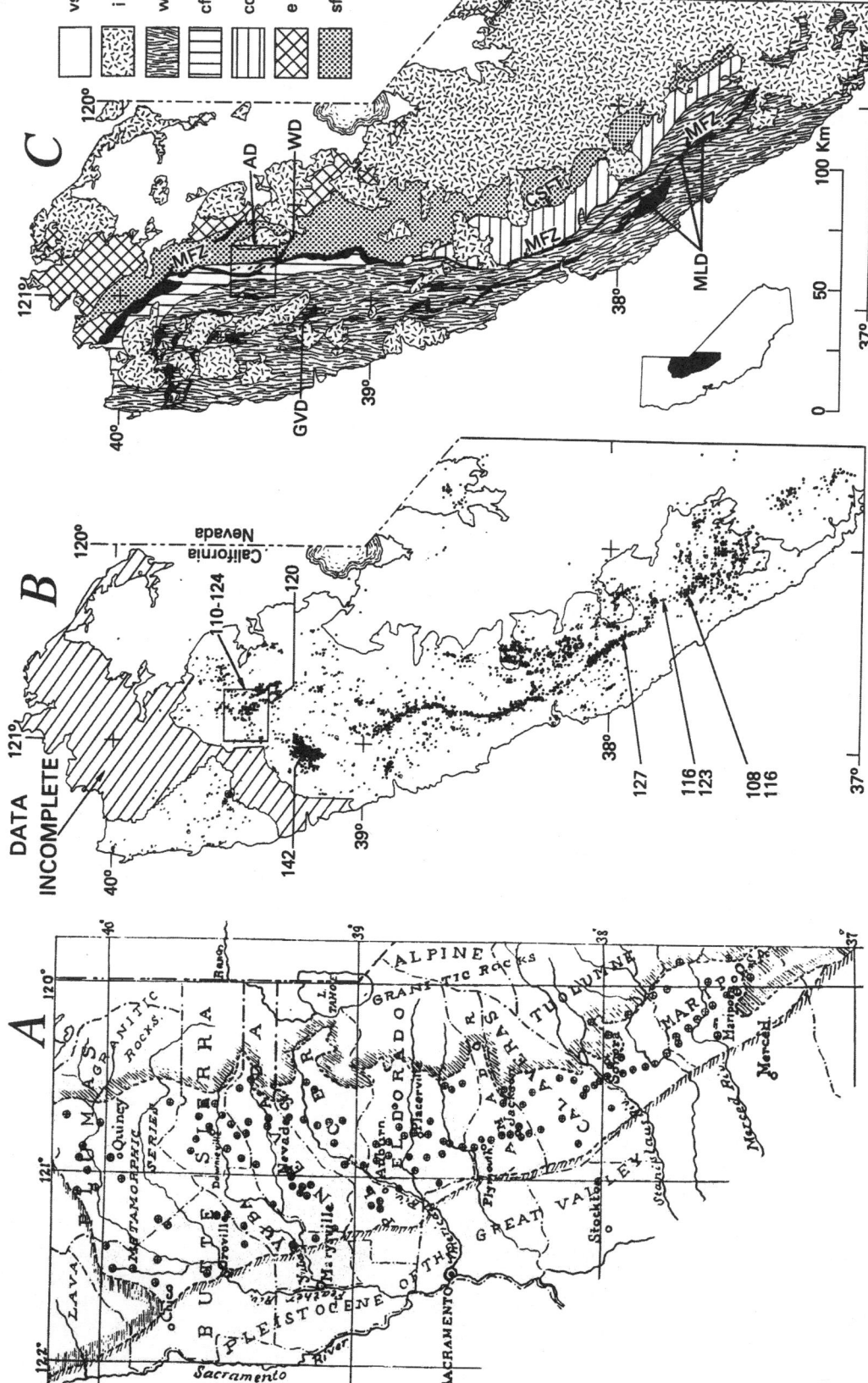

Figure 1. Distribution of gold-quartz veins in the Sierra Nevada foothills metamorphic belt. A: Gold quartz veins of the "central gold belt" of California (from Lindgren, 1895; Plate 1). B: Lode gold deposits of the Sierra Nevada foothills region from the U.S. Geological Survey Computer Resources Information Bank (through 1984) with radiometric ages of hydrothermal minerals (in Ma; from Böhlke and Kistler, 1986). C: Selected geologic features (from Böhlke and Kistler, 1986): black areas—ultramafic rocks; vs—unmetamorphosed volcanic and sedimentary rocks (post-Cretaceous); i—intrusive rocks (mostly Mesozoic); w—metavolcanic and metasedimentary rocks of the western foothills; cf—Calaveras Formation; cc—Calaveras Complex; e—eastern Paleozoic and Mesozoic melange and arc rocks; sf—Shoo Fly Complex; MFZ—Melones fault zone; CSFT—Calaveras–Shoo Fly thrust; MLD—Mother Lode district; AD—Alleghany district; GVD—Grass Valley district; WD—Washington district. Despite almost a century between compilations and two orders of magnitude difference in the number of sites located, both maps indicate the same overall distribution of mineralized veins within the metamorphic belt, but not in the major batholiths, and a tendency for veins to be concentrated along major northwest-southeast–trending structural features.

of the plate boundary, 50–100 km west of the magmatic arc axis, 50–100 km east of the shoreline of the Great Valley forearc depression, at about lat 45–55°N (Ingersoll, 1982; Debiche et al., 1987; Hamilton, 1988; Fig. 2). Most workers, including Lindgren (1895, p. 224–225), commented on the lack of evidence for a direct connection between exposed plutons and gold-quartz vein fluid generation. Nevertheless, it is possible that a rapid increase in the rate of regional intrusive activity beginning about 125–130 Ma might have been associated with minor reactivation of preexisting fault zones and accelerated deep-crustal devolatilization and/or fluid circulation within the proximal forearc, resulting in a pulse of vein mineralization during the earlier stages of the magmatic pulse. The magmas need not have supplied fluids or even heat directly to the vein-forming systems, but the magmas and the hydrothermal fluids both could represent responses to a significant thermal-tectonic event that postdated the Nevadan compressional event by about 30–40 m.y. (or less for possible earlier episodes of mineralization).

Other authors have emphasized correlations between the timing of mineralization and of regional shifts in plate tectonic interactions derived from paleomagnetic studies. Data summarized by Engebretson et al. (1985) indicate several significant changes in the rates and directions of post-Nevadan relative motions between the Farallon and North American plates. Page and Engebretson (1984) suggested that the Sierran gold-quartz vein faults may have formed during a period of nearly normal convergence, bracketed by times of more oblique convergence. Elder and Cashman (1992) suggested that two episodes of vein mineralization in the Klamath Mountains may have occurred in response to shifts in the convergence rates and directions at about 150 and 133 Ma, on the basis of a revised set of convergence vectors. Similarly, Goldfarb et al. (1991) proposed that an episode of Tertiary gold mineralization in the Alaska-Juneau gold belt at 55–56 Ma occurred in response to changes in relative motions between the Kula and North American plates.

Goldfarb et al. (1991) and Elder and Cashman (1992) suggested that fluids generated by metamorphic devolatilization reactions in subducted materials were released through the crust in relatively short bursts at times when transcrustal fault openings were generated by major changes in the regional stress fields. If it could be proven that the fluids came from subducted materials, then this model would differ somewhat from those that call upon crustal fluid sources in the earlier-accreted materials, but those fluid sources are not currently distinguishable. Goldfarb et al. (1998) concluded that generation of thermal anomalies within accreted marginal terranes was a common requirement for generation of orogenic gold-quartz veins in several regions around the Pacific basin at various times during the Phanerozoic. Regional thermal anomalies or events within the accreted terranes could be the result of magmatic activity, thermal relaxation above deceased subduction zones, or other processes. Similarly, recent models of Archean orogenic gold-quartz veins involve crustal accretion, rising geoisotherms, and crustal-scale fluid flow, in part as a result of recent studies of Phanerozoic deposits like the Sierran veins (Kerrich and Wyman, 1990; Groves et al., 1998). However, in the Sierra Nevada foothills and elsewhere, it remains to be proven whether vein-related faulting was caused more commonly by regional uplift, magma displacements, or external shifts in plate tectonic interactions, and whether the timing of mineralization was more dependent on specific episodes of fluid generation (e.g., thermal events), fluid release (faulting events), or possibly fluid convection. Understanding more specifically the genetic environment of the Sierran veins could still benefit from further geochronologic studies, which have been limited in scope.

## ORIGIN OF FLUIDS AND GOLD

> The country-rock altered to carbonate, standing in strong contrast to the vein filled nearly exclusively by quartz, affords a much-needed key to the genetic processes of these deposits.
> Lindgren, 1895, p. 236

> These considerations, though involving many most difficult questions, strengthen the belief that the origin of the gold must be sought below the rocks which now make up the surface of the Sierra Nevada, possibly in granitic masses underlying the metamorphic series.
> Lindgren, 1895, p. 240

> Regarding the origin of the hot, auriferous solutions which have produced the gold-quartz veins it is best, at this stage of our knowledge, to speak with great reserve.
> Lindgren, 1895, p. 239

The ultimate origin of the hydrothermal fluids responsible for the gold mineralization in the Sierra Nevada foothills metamorphic belt, as in similar deposits around the world, continues

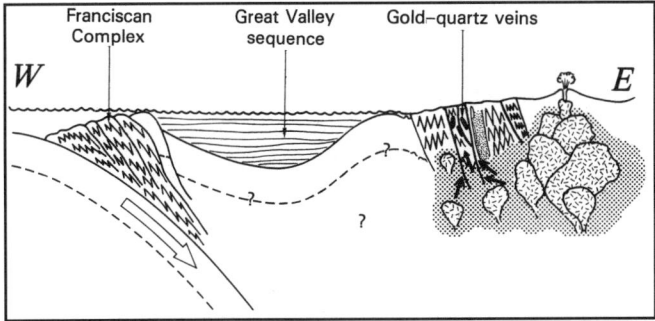

Figure 2. Generalized schematic cross section of a continental-margin subduction system representing the western margin of North America in Early Cretaceous time (from Böhlke and Kistler, 1986) (not to scale). The supposed location of the Sierra Nevada foothills gold-bearing region is shown in the proximal forearc complex of previously accreted metamorphic rocks above the subducting Farallon plate. Ore-bearing fluids are of unspecified origin, but are shown ascending along major fault zones from deep metamorphic reaction zones within the crust (arrows).

to be problematic. Perhaps each constituent of the fluids should be considered separately; possibly even the oxygen and the hydrogen in the water represent predominantly different sources. Some components, like K, Na, Sr, Ba, B, and possibly silica and sulfur, may have had relatively short residence times in the fluids as they were transferred from one type of wall rock to another by divergent metasomatic reactions and fluid flow along discordant structures, whereas some, like $CO_2$ and possibly gold, may have traveled relatively far (Lindgren, 1895, 1896; Knopf, 1929; Ferguson and Gannett, 1932; Coveney, 1981; Böhlke, 1988, 1989). It is important to recognize that the requisite geochemical conditions for generating a low-sulfide gold-quartz vein deposit may be fairly simple. Requirements for a focused fluid-flow path may be more demanding the greater the depth, requiring major seismic zones deep in the crust and only minor joints and fractures nearer the surface; but the fluid source and delivery mechanism may be relatively free to vary.

One of the most important major ingredients defining the fluids responsible for gold deposits of this general type is carbonic acid. Most observers of the Mother Lode, especially Lindgren, emphasized the importance of carbonate metasomatism of the wall rocks adjacent to the veins (Fig. 3). Recent studies aided by electron microprobe and X-ray diffraction techniques have provided details of the mineralogy and compositions of the carbonates. In addition to calcite, the metasomatic carbonates include dolomite-ankerite and magnesite-siderite solid solutions, the abundances and compositions of which vary according to host-rock lithology and proximity to the veins (Böhlke, 1988, 1989). Carbonate metasomatism is most abundant and conspicuous in the mafic and ultramafic rocks because of the higher initial concentrations of divalent cations (Ca, Mg, Fe). Böhlke (1989) argued on the basis of thermodynamic evidence that carbonatization reactions involving dolomite-ankerite and magnesite-siderite solid solutions in a variety of wall rocks precluded a local source of fluid $CO_2$, in agreement with Lindgren and others who concluded that the $CO_2$ must have come from greater depths.

Courtis (1890) apparently was the first to report evidence of liquid $CO_2$ in fluid inclusions from the Mother Lode. The most detailed descriptions of $CO_2$-rich fluid inclusions were given by Coveney (1981) and Coveney and Kelly (1971) for samples from the Alleghany district, in which dawsonite (Na-Al carbonate) was reported as a common daughter mineral. Additional observations have been reported for samples from Alleghany (Weir and Kerrick, 1987; Böhlke et al., 1989; Böhlke and Irwin, 1992) and from the Klamath Mountains (Elder and Cashman, 1992). Studies in Grass Valley (Johnston, 1940) and in the southern Mother Lode (Weir and Kerrick, 1987) did not identify $CO_2$-rich fluid inclusions, despite similar carbonate-bearing metasomatic assemblages. Maximum concentrations of $CO_2$ reported so far are about 11 mole% in Alleghany and more than 20 mole% in the Klamaths. $H_2O$-$CO_2$ fluids with those compositions could be consistent with metamorphic devolatilization reactions occurring in mixed carbonate-silicate rocks at depth.

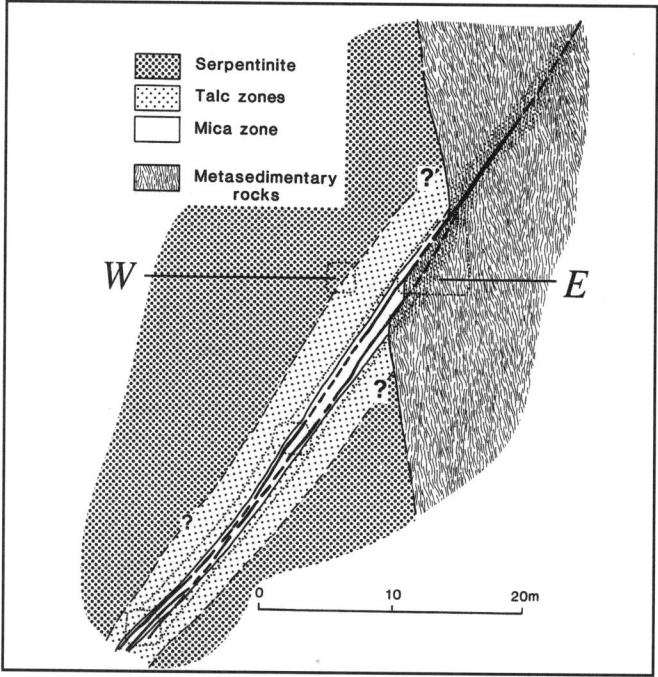

Figure 3. Vertical section through a portion of the Gold Crown mine, Alleghany district, illustrating relations between vein structure, host-rock lithology, and wall-rock alteration (from Böhlke, 1989). The east-dipping contact between serpentinite and metasedimentary rocks is cut by a west-dipping gold-bearing quartz vein with minor reverse offset. Metasomatic wall-rock alteration haloes are parallel to the vein. Proximal alteration assemblages in serpentinite include dolomite-ankerite, magnesite-siderite, quartz, and mariposite, with crystalline native gold in quartz veinlets. Proximal alteration assemblages in metasedimentary rocks include dolomite-ankerite, magnesite-siderite, quartz, albite, sericite, and pyrite, with disseminated gold in pyrite. Details of the alteration zoning were given in Böhlke (1989).

$CO_2$-rich springs and gaseous emissions are associated with active tectonics throughout the world, including subduction zones and magmatic arcs (Barnes et al., 1978). Ascending $CO_2$-rich fluids may be common expressions of underlying regional metamorphism and/or partial melting in those environments (Barnes et al., 1978; Sano and Marty, 1995; Kerrick and Caldiera, 1998).

There are other important features of the vein-forming fluids that could provide clues about the sources of specific constituents. Several studies have reported stable isotopic compositions of hydrothermal minerals and fluid inclusions in the Sierran veins (Marshall and Taylor, 1981; Böhlke and Kistler, 1986; Taylor, 1987; Weir and Kerrick, 1987; Böhlke et al., 1988). Fluid $^2H/^1H$ ratios estimated from analyses of mica are relatively high ($\delta^2H$ = −30 per mil ± 20 per mil) and fluid $^{18}O/^{16}O$ ratios estimated from analyses of quartz, carbonate, and mica are generally high but regionally variable ($\delta^{18}O$ = +5 per mil to +14 per mil). Those data are consistent with equilibration between the fluids and a variety of common metamorphic rocks at metamorphic temperatures. The $^{13}C/^{12}C$ ratios of hydrothermal carbonate minerals are somewhat variable but generally slightly less than those of marine carbonate ($\delta^{13}C = -4$ per mil ± 3 per mil). Similar $\delta^{13}C$ values have been accepted by some workers as evidence for a magmatic source of $CO_2$ in some Archean gold-quartz veins (e.g., Burrows et al., 1986), as they are not much different from values characteristic of the Earth's mantle. But they also are similar to average values of mixed sedimentary sources containing both reduced and oxidized carbon, from which fluids of various sources could have obtained the carbon. Phase relations in fluid inclusions indicate that $CH_4$ generally was not a significant species, although it may have been present locally (Coveney, 1981; Weir and Kerrick, 1987; Böhlke and Irwin, 1992). The $^{34}S/^{32}S$ ratios of hydrothermal sulfide minerals are somewhat variable; with average $\delta^{34}S$ values are around +2 per mil ± 3 per mil. Those data are consistent with S having been derived from various combinations of metavolcanic and metasedimentary rocks, or from magmas, and transported largely in the form of aqueous sulfide. Vein carbonates have regionally variable initial $^{87}Sr/^{86}Sr$ ratios (0.704–0.718), indicating that Sr was acquired by the hydrothermal fluids from various sources ranging from mafic igneous rocks to continent-derived clastic sedimentary rocks (Böhlke and Kistler, 1986). Fluid inclusions in the veins generally have chloride concentrations less than that of seawater, in some cases as low as 0.1 mol/L (Coveney and Kelly, 1971; Coveney, 1981; Weir and Kerrick, 1987; Böhlke et al., 1989; Böhlke and Irwin, 1992). The low chlorinity values eliminate seawater or connate brines as potential major fluid sources. The I/Cl ratios in fluid inclusions are higher than those of seawater by almost three orders of magnitude, possibly as a result of interaction between the fluids and marine sedimentary rocks (Böhlke and Irwin, 1992). Near neutral or slightly alkaline fluid pH values are indicated by carbonate daughter minerals, including dawsonite, in fluid inclusions (Coveney and Kelly, 1971), bicarbonate/chloride ratios greater than 1 in fluid inclusions (Böhlke et al., 1989; Böhlke and Shock, 1990), and equilibrium thermodynamic calculations for observed mineral assemblages (Weir and Kerrick, 1987; Böhlke, 1989). All of these features could be consistent with a metamorphic fluid formed by devolatilization reactions in a heterogeneous lithologic terrain, although some could also be interpreted in different ways. However, with the possible exception of the chlorinity and H isotope values, these features cannot reliably be assumed to represent characteristics acquired by the fluids at their source, because they may have been susceptible to continuous changes along the fluid-flow paths to the sites of ore deposition. The H isotope data are difficult to interpret because the composition of meteoric water in the region during Cretaceous time is not precisely known and may not have been much different from that of metamorphic fluids (Böhlke et al., 1988).

The possibility that deeply circulating meteoric waters participated in the Mother Lode mineralization was suggested by Ransome (1900), Lindgren (1895), Knopf (1929), and others, generally without much evidence other than that the vein-forming fluids were believed to be aqueous, and by analogy with certain types of modern geothermal waters. Nesbitt et al. (1986) noted that $\delta^2H$ values of fluids extracted from gold-quartz veins of the western North American Cordillera, including the Mother Lode, are low and tend to decrease with latitude, qualitatively in much the same way that the isotope ratios vary in modern precipitation. This observation indicates that the sampled fluids were meteoric in origin and could provide support for deep circulation of meteoric water in the vein faults (Nesbitt and Muehlenbachs, 1989). However, it has also been observed that fluids extracted from bulk samples of vein quartz for H isotope analysis may be largely secondary (Böhlke and Kistler, 1986; Pickthorn et al., 1987; Taylor et al., 1991). At Alleghany, there is evidence for a positive correlation between $\delta_2H$ and $CO_2$ contents of fluid inclusions in bulk samples (Marshall and Taylor, 1981; Böhlke et al., 1988), which could imply that the fluids with low meteoric isotope values were not the same as the $CO_2$-rich metasomatic fluids. This is also indicated by the observation that fluid $\delta^2H$ values inferred from analyses of hydrous hydrothermal minerals such as mariposite in the veins commonly are higher than the bulk fluid inclusion values (Böhlke and Kistler, 1986; Taylor et al., 1991). The $\delta^2H$ values of serpentine minerals may be less useful for this purpose than those of mariposite because serpentinites apparently interacted with a variety of fluids, not all associated with the gold mineralization. Some support for a high-temperature component of meteoric water was derived from laser-microprobe analyses of noble gases in primary hydrothermal fluid inclusions in quartz from Alleghany (Böhlke et al., 1989; Böhlke and Irwin, 1991). Those inclusions were estimated to have concentrations of $^{36}Ar$ and $^{84}Kr$ between about 25% and 100% of those expected in meteoric waters. They also have moderately elevated and variable $^{40}Ar/^{36}Ar$ ratios (~400 to 900) that are within the range of some modern geothermal waters of meteoric origin. While it seems certain that meteoric water entered the veins at some time, it is not well established whether it constituted a significant fraction of an ascending $CO_2$-rich ore-bearing fluid, mixed with an ascending nonmeteoric $CO_2$-rich fluid during the mineralization, or entered only much later after the deposits were cooled and uplifted.

The origin of the gold is essentially unknown, as it was a century ago. Magmatic sources were suggested by Lindgren (1895, 1933), Knopf (1929), Johnston (1940), and others. Lindgren's (1895) paper contains an enlightened discussion about gold solubility and concludes that hot aqueous fluids could have transported the gold in sufficient quantities to form the ores. The probable aqueous species involved in the transport were uncertain because the experimental data available at the time were limited to fairly low temperatures and unrealistic fluid compositions. Subsequent hydrothermal experiments combined with thermodynamic analyses and fluid-inclusion studies indicate that aqueous sulfide species may have been largely responsible for the solubility of the gold, and that the low chlorinity and low acidity of the fluids may have precluded transport of much larger quantities of base metals such as Fe, Cu, Zn, and Pb (Helgeson, 1969; Seward, 1991; Coveney, 1981; Böhlke, 1989). Fluids from any of the proposed sources might have acquired gold from a variety of rocks within or beneath the metamorphic belt before or during their ascent along the vein faults, provided the fluids were capable of reacting with gold host minerals in the rocks. Speculations have been made regarding favorable source rocks for leaching of the gold, including ultramafic and mafic rocks of accreted oceanic and island-arc terranes (Albers, 1981), and preconcentrated sources such as gold-bearing volcanogenic massive sulfide deposits (Dodge and Bateman, 1977; Smith, 1981), but the evidence has not been conclusive. Böhlke (1988) noted that gold could be released by reaction of Mother Lode–type $CO_2$-rich fluids with gold-bearing sulfide minerals in rocks with lower than average Fe/Mg ratios.

## ORE DEPOSITION

> The precipitation of the solids contained in the solution could have been brought about by many means...
> Lindgren, 1895, p. 238

> The causes to which the occurrence, form, and direction of the ore shoots are due are very obscure, and to extend the discussion to them would be to go farther in the realms of hypothesis than is here desirable.
> Lindgren, 1896, p. 184

Considering the cause and distribution of ore deposition, it is important to note variations in scale and to recognize that there are different types of ore occurrences within the vein systems, ranging from high-grade vein pockets to low-grade disseminated wall-rock ores, that may require somewhat different approaches. Regionally, the spatial association between the gold-quartz veins and some of the major northwest-southeast–trending fault zones in the metamorphic belt was clear from the earliest studies, and this has been confirmed and refined by more recent data (Fig. 1). As the major fault zones commonly contain slivers and lenses of ultramafic rocks including sheared serpentinite, there is also a significant, though far from ubiquitous, regional association of gold-quartz veins with serpentinite.

Although gold may be "universally distributed in the quartz-veins of California" (Lindgren, 1895, p. 231), it clearly is not uniformly distributed at the mine scale. Within the major vein systems, there is no documented evidence for systematic spatial variations in fluid temperatures or compositions over vertical distances of at least 2 km; however, there is evidence for temporal variations in both the temperatures and compositions of fluids locally within individual ore shoots or samples. Ore shoots commonly are said to be located preferentially in areas of intense and late fracturing, near vein intersections, and along certain types of lithologic boundaries. Petrographic studies of mineral textures commonly indicate that gold and its most closely associated sulfides (commonly arsenopyrite or galena) were deposited relatively late in the sequence of vein filling phases (Ferguson and Gannett, 1932; Johnston, 1940; Cooke, 1947; Coveney, 1981). However, the distinguishing fluid properties characteristic of those times and places are not precisely known. If the gold was transported in the form of aqueous sulfide complexes, it is theoretically possible for it to have precipitated from solution in response to various combinations of cooling, mixing, degassing, sulfide precipitation, and increasing or decreasing fluid pH or oxidation state (Seward, 1991). More than one of these mechanisms may have been important in the formation of ore bodies.

The most difficult ore bodies to locate are small high-grade pockets in the veins that produce spectacular specimens (Fig. 4) and short-term profits, but may be too few and far between to support long-term confidence. Ferguson and Gannett's (1932) description of the Alleghany district contains several interesting statements on the subject of high-grade ores, ranging from the obvious: "The most reliable mineralogic guide is the actual presence of free gold..." (p. 58) to the more perceptive: "It is estimated that 80% of the production of the district has been derived from those portions of the veins where serpentinite either formed one wall or was less than 100 feet from the ore" (p. 57). The latter statement has proven to be quite useful in subsequent development at Alleghany. Cooke (1947) determined "association indices" relating various vein characteristics with high-grade gold occurrences within the Sixteen-to-One Mine in Alleghany. The features that appeared to be most commonly associated with vein gold included proximity to serpentinite, crushing and brecciation in the quartz, and some minerals such as mariposite, galena, and arsenopyrite. Cooke (1947, p. 246) suggested that serpentinite was "the dominant control of bends in the vein, and therefore of high-grade" perhaps by focusing flow of large amounts of gold-bearing vein fluids, possibly also by promoting mixing of fluids with slightly different compositions. To explain high-grade ores in the Oriental mine at Alleghany, Coveney (1981) proposed mixing of the gold-bearing vein fluids with more reduced fluids emanating from serpentinite. Fluids carrying dissolved $H_2$ gas produced by serpentinization reactions forming magnetite in the ultramafic rocks could have reacted with aqueous gold species in the veins to precipitate the gold. Hydrogen isotope data have been interpreted as indicating that serpentinization and carbonate metasomatism were accomplished independently by fluids from different sources (Böhlke and Kistler, 1986; Böhlke et al., 1988). If serpentinization and carbonate metasomatism could be shown to have occurred both simultaneously and independently, then mixing of the respective fluids could be an explanation for Ferguson and Gannett's (1932) observation. However, reaction of $CO_2$-rich ore-bearing fluids with previously formed serpentinite would be more likely to cause fluid oxidation than fluid reduction, owing to the conversion of magnetite to Fe-bearing carbonates (Böhlke, 1989). Wittkopp (1980) suggested that the gold was transported by fluids that moved along serpentinite contacts and was deposited where those fluids came into contact

Figure 4. Specimen of high-grade gold in quartz from the Oriental mine, Alleghany, California (Harold and Erica Van Pelt, photographers). "One hears rumors that there is no gold... We have decided to investigate. Everybody lies here." (This is from a letter written by Arnold Brandt to his brother, sent from the Klondike in July, 1898; (transcribed by Catherine M. Böhlke from the collection of Eugenia Brandt Böhlke; though this letter was not written from California, it probably represents many that were.) The evidence belies the rumors.

with preexisting quartz veins. Böhlke and Irwin (1991, 1992) reported evidence for fluid-inclusion populations with differing $^{40}Ar/^{36}Ar$ and $CH_4/CO_2$ ratios in the Oriental mine, but it is not clear whether they mixed to cause gold deposition.

Evidence for fluid unmixing ($H_2O$-$CO_2$ phase separation) as a mechanism for precipitating gold has been reported in some similar deposits elsewhere in the world (e.g., Robert and Kelly, 1987). Coexisting $CO_2$-rich and $H_2O$-rich fluid inclusions were reported from the Quartz Hill deposits in the Klamath Mountains (Elder and Cashman, 1992). However, fluid inclusions bearing such evidence from the Mother Lode region have not been found despite considerable effort by several investigators (e.g., Coveney, 1981; Weir and Kerrick, 1987; my own data). Relatively constant $^{36}Ar/Cl$ ratios in fluid inclusions with varying $CO_2$ concentrations also indicate that unmixing was not the major process causing changes in the $CO_2$ contents of the fluids (Böhlke and Irwin, 1991).

Much of the accumulated evidence cited here could be consistent with concentrated deposition of high-grade gold ores in veins where ascending ore-bearing fluids mixed with other (meteoric?) fluids in areas of prolonged or intense fracturing, but other processes also are possible. Evidence for episodic fault movement and fluid flow indicates that chemistry, temperature, and pressure of the ore-bearing fluids may have changed cyclically within the vein segments. Metasomatic reactions along anastomosing fluid flow paths in heterogeneous host-rock lithologies could have produced subtle variations in the compositions of different ore-bearing fluid parcels that subsequently remixed, causing ore deposition. Some metasomatic reactions, such as sulfidization of iron minerals or oxidation of reduced carbon, might also have caused local precipitation of gold in veins more directly without subsequent mixing.

In addition to the gold ore shoots within the quartz veins, some of the gold in the Sierran vein systems occurs as disseminated deposits associated with pyrite, apparently formed in the wall rocks by metasomatism by the same types of fluids that formed the veins. Lindgren (1895) mentioned some such occurrences (p. 226 and 235), but then wondered "why the [vein] walls should, to such a large extent, act as a separating barrier for the gold and most of the sulfides" (p. 237), implying that disseminated ores were atypical. Nevertheless, significant ores of the disseminated pyritic type include the "gray ores" and mineralized schists in metavolcanic and metasedimentary rocks in the southern Mother Lode (Knopf, 1929) and the "episyenite" ores in granite at Alleghany (Coveney, 1981; Böhlke, 1989). The close association of the disseminated gold with pyrite (commonly as microscopic inclusions within pyrite grains) indicates that rock sulfidization and consequent fluid desulfidization may have caused gold to precipitate locally from saturated solutions containing gold-bisulfide complexes. Assuming this to be the case, Böhlke (1988) proposed that favorable host rocks for disseminated gold-pyrite mineralization could be predicted on the basis of their Fe/Mg ratios. It was suggested that a gold-bearing fluid reacting with the dominant wall rocks along a flow path might approach a state in which the fluid S/C ratio was buffered by coexisting pyrite and carbonates with intermediate Fe/Mg ratios. If that fluid encountered rocks with low Fe/Mg ratios, Fe would tend to be sequestered as a minor component in Mg-rich carbonate phases and pyrite would be less likely to form. If the same fluid encountered rocks with high Fe/Mg ratios, pyrite would be more likely to form at the expense of Fe-rich carbonate and those rocks would tend to be sulfidized and mineralized. This model is consistent with occurrences of disseminated gold in granite, ferruginous quartzite, and some types of metavolcanic and metasedimentary rocks. It is possible that other low-grade ores of this type have been overlooked in the region.

## SIMILARITIES TO COAST RANGE MERCURY DEPOSITS

> There are certain interesting analogies between the gold-quartz veins of the Sierra Nevada and the quicksilver deposits of the Coast Ranges.
> 
> Lindgren, 1895, p. 238

Comparing the Sierran gold-quartz veins with the Coast Range mercury deposits, Lindgren noted similarities in the local enrichments of certain elements (e.g., Au, As, Hg), in the occurrence of some of the gangue minerals associated with the ores (e.g., quartz, carbonate, pyrite), and in some features of wall-rock alteration (carbonatization of serpentinite and other metamorphic rocks). These would be useful observations indeed if "...in fact, the still abundant thermal waters found in intimate connection with the ore deposits in the quicksilver region closely correspond in character to the inferred composition of the once existing hot springs of the gold-belt" (Lindgren, 1895, p. 239). The most obvious difference between the two types of deposits is the much greater depth at which most of the Sierran veins are believed to have been deposited. It would be interesting to know if Coast Range–type deposits forming within a few hundred meters of the surface could represent the upper parts of Mother Lode–type hydrothermal systems extending many kilometers into the crust. Subsequent studies of Coast Range thermal waters and of Sierran gold-quartz fluid inclusions have confirmed and expanded the list of similarities in fluid composition, but genetic analogies between the two types of fluids are less clear.

Thermal waters in the vicinity of mercury ± gold deposits and silica-carbonate alteration in the Coast Ranges commonly have large amounts of both neutral $CO_2$ and bicarbonate, relatively high (nonmeteoric) $\delta^2H$ and $\delta^{18}O$ values, slightly negative $\delta^{13}C$ values, and unusually high I/Cl ratios (White, 1957; Barnes, 1970; White et al., 1973), as did the Sierran vein-forming fluids. These characteristics could be interpreted as normal signatures of metamorphic fluids derived in part from thermal decomposition of mineral and organic constituents of sedimentary rocks (White, 1957; Barnes, 1970). However, some of the Coast Range thermal waters with high $\delta^2H$ values may also include varying amounts of modified older marine connate waters buried with the Great Valley sequence (White et al., 1973; Peters, 1993). Volcanism driving the recent hydrothermal circulation and mineralization in the Coast Ranges may have been related in part to the northward passage of the Mendocino triple junction as Pacific plate subduction was replaced by strike-slip movement along the San Andreas fault zone, possibly with minor rifting (Dickinson and Snyder, 1979; Page and Engebretson, 1984; Rytuba, 1993). It is possible that analogous conditions of strike-slip faulting and/or rifting were met when the veins formed in the Sierran foothills region during the period of Cretaceous subduction in the Coast Range region; however, the veins appear to have been relatively close to the active magmatic arc at the time, and the common reverse sense of movement and pressure-solution features in the vein faults do not suggest rifting. Nevertheless, even if the tectonic environments were not the same, it is reasonable to suppose that fluids with similar compositions ascending in and near fault zones in orogenic regions could generate ore deposits with some overlapping characteristics at different depths (Lindgren, 1933). This point was reemphasized by several authors, including Nesbitt et al. (1989), who compared gold, antimony, and mercury deposits in the Canadian Cordillera, and Groves (1993), who compared deposits in Archean metamorphic complexes of western Australia. In the context of Groves' (1993) review, the major deposits of the Mother Lode may represent only the middle part of a continuum of hydrothermal mineralization that extends over much of the range of crustal temperatures and pressures in active orogenic belts.

## UNFINISHED WORK

> Any visitor to the [Alleghany] district whose imagination is at all active must be intrigued by the number of interesting problems without apparent direct economic bearing which present themselves in even a cursory study of the mines.
>
> Ferguson and Gannett, 1932, p. 60

It is unfortunate that more work has not been done on the Sierran gold deposits in recent years, because important questions about their origins remain unanswered. As indicated by Ferguson and Gannett (1932), more detailed understanding of the place of the veins in the geologic history of the region could transcend the particular concerns of local mining interests and provide useful information about Sierran tectonic-igneous-metamorphic processes as well as insights into the origins of similar types of gold deposits worldwide. Documentation of the significance of plate tectonic convergent margins in generating Phanerozoic gold-quartz veins (including the Sierran veins) has had an influence on similar models for some of the more voluminous Archean deposits (e.g., Kerrich and Wyman, 1990; Groves et al., 1998).

It is clear that more work on the local and regional distributions of mineralization ages is warranted. Additional dating methods (e.g., stepwise $^{40}Ar$-$^{39}Ar$, Re-Os) could be applied to some of the deposits to determine whether the existing Rb-Sr and K-Ar data should be interpreted differently. Regional age patterns or cooling histories might emerge that would permit more precise correlations to be made between the mineralization and either post-Nevadan regional migration of deep-seated plutonism, terrane boundary adjustments, postsubduction rebound, or other geologic activities. Measurements of helium isotopes in fluid inclusions might be useful in assessing contributions of juvenile gases, which might be relevant for the origins of some other fluid constituents. More detailed evaluation of stable isotope systematics could limit further debate about fluid sources and flow paths. With respect to ore prospecting and mine development, it remains to establish clearly the processes responsible for localization of the ores within the veins. Despite various indications that different fluids occupied the veins at different times, it has not been demonstrated conclusively whether documented changes in temperature, $CO_2$ content, $^{40}Ar/^{36}Ar$, $\delta^2H$, or other properties were related to gold deposition. Numerous attempts have proven the difficulty of relating specific generations of fluid-inclusions in quartz to events concerning gold deposition, but perhaps more effort with fluid-inclusion microanalytical techniques on both the quartz and the gold could yield useful new results. Fluid inclusions from Alleghany are among the largest and best preserved in deposits of this type.

## CONCLUSIONS

As a summary of the state of knowledge at the time, Lindgren's (1895) paper was wonderfully comprehensive and yet selective in the sense that certain topics were presented as being well understood and others clearly not so. Lindgren was most emphatic when describing the formation of the gold-quartz veins by precipitation from aqueous solutions in open spaces resulting from late faulting after the last regional pervasive deformation and metamorphism of the host rocks. These conclusions have been supported by subsequent work. Advances in tectonic theory, field mapping, and geochronology have added new information about the geologic history of the foothills region and its relation to global processes, but have not resolved the essential cause of the formation of the veins. Lindgren also was clear in his interpretation of water-rock disequilibrium and the significance of fluid evolution and wall-rock alteration. Recent detailed studies of wall-rock alteration assemblages have provided more detailed information on properties of the fluids and some potential clues about processes that might have controlled the solubility and transport of gold.

Lindgren apparently was more reluctant about stating the origins of the fluids and the gold, and he left open the question of what controlled the local distribution of the ores. New data derived from stable isotope analyses and fluid-inclusion studies, experiments on metamorphic and hydrothermal processes, and thermodynamic analyses have created new debates but have not answered these questions conclusively. Whether the aqueous ore-bearing fluids owe their existence largely to metamorphic devolatilization reactions or began as descending meteoric waters, it is clear that they obtained many of their chemical and isotopic characteristics by interactions with metamorphic rocks beneath the sites of ore deposition, and it is possible that some constituents were added from magmatic sources. It seems to be established that meteoric waters entered the veins at some time, but it is not clear when this first occurred or if it contributed to focused deposition of the gold in the veins. Lindgren apparently was not impressed by occurrences of disseminated ores in wall rocks, but subsequent mining and study have shown that they may be locally important and possibly predictable. Despite a significant amount of recent work, it seems that the available tools and approaches have not been exhausted yet; further studies of the veins could make contributions to understanding not only the distribution of Mother Lode gold, but also the geologic history of the Sierra Nevada region.

## ACKNOWLEDGMENTS

Many important contributions have been made to the study of Mother Lode gold by Donald R. Dickey, whose family has owned and operated the Oriental Mine in Alleghany since the 1930s. Mr. Dickey has trained some of the best miners in the region, has supported research efforts by scientists in academia, industry, and the government, has made his property available for field excursions and sampling, has donated valuable specimens to museums, and has shared unsurpassed knowledge gained by experience in the search for gold. My work on California gold was supported by D. R. Dickey and others in and around the Alleghany district, by A. S. Radtke, N. J. Page, R. P. Ashley, E. H. McKee, and W. C. Shanks, III, of the U.S. Geological Survey, by G. H Brimhall, Jr., H. C. Helgeson, and I. S. E. Carmichael of the University of California at Berkeley, and by N. C. Sturchio of Argonne National Laboratory. Helpful discussions, guidance, and review comments on this paper from P. B. Barton, B. M. Böhlke, R. M. Coveney, Jr., J. H. Dilles, E. M. Moores, and D. Stout are much appreciated. R. P. Ashley provided a compilation of gold production data.

## REFERENCES CITED

Albers, J. P., 1981, A lithologic-tectonic framework for the metallogenic provinces of California: Economic Geology, v. 76, p. 765–790.

Ashley, R. P., 1997, Environmental geochemistry of gold deposits in the Mother Lode belt, California: U.S. Geological Survey Open-File Report 97–496, p. 7.

Barnes, I., 1970, Metamorphic waters from the Pacific tectonic belt of the west coast of the United States: Science, v. 168, p. 973–975.

Barnes, I., Irwin, W. P., and White, D. E., 1978, Global distribution of carbon dioxide discharges, and major zones of seismicity: U.S. Geological Survey Water-Resources Investigations Open-File Report 78–39, 12 p.

Bateman, P. C., Busacca, A. J., and Wawka, W. N., 1983, Cretaceous deformation in the western foothills of the Sierra Nevada, California: Geological Society of America Bulletin, v. 94, p. 30–42.

Böhlke, J. K., 1982, Orogenic (metamorphic-hosted) gold quartz veins, in Erickson, R. L., compiler, Characteristics of mineral deposit occurrences: U.S. Geological Survey Open-File Report 82–795, p. 70–76.

Böhlke, J. K., 1988, Carbonate-sulfide equilibria and "stratabound" disseminated epigenetic gold mineralization: A proposal based on examples from Alleghany, California: Applied Geochemistry, v. 3, p. 499–516.

Böhlke, J. K., 1989, Comparison of metasomatic reactions between a common $CO_2$-rich vein fluid and diverse wall rocks: Intensive variables, mass transfers, and gold mineralization at Alleghany, California: Economic Geology, v. 84, p. 291–327.

Böhlke, J. K., and Irwin, J. J., 1991, Noble gases and halides in gold quartz veins as indicators of fluid sources, salt sources, mixing, and unmixing, in Robert, F., Sheahan, P. A., and Green, S. B., eds., Nuna Research Conference on Greenstone Gold and Crustal Evolution: Val d'Or, Quebec, Geological Association of Canada, Society of Economic Geologists, p. 15.

Böhlke, J. K., and Irwin, J. J., 1992, Laser microprobe analyses of Cl, Br, I, and K in fluid inclusions: Implications for sources of salinity in some ancient hydrothermal fluids: Geochimica et Cosmochimica Acta, v. 56, p. 203–225.

Böhlke, J. K., and Kistler, R. W., 1986, Rb-Sr, K-Ar, and stable isotope evidence for the ages and sources of fluid components in gold-quartz veins of the northern Sierra Nevada foothills metamorphic belt, California: Economic Geology, v. 81, p. 296–322.

Böhlke, J. K., and McKee, E. H., 1984, K-Ar ages relating to metamorphism, plutonism, and gold quartz vein mineralization near Alleghany, Sierra County, California: Isochron/West, no. 39, p. 3–7.

Böhlke, J. K., and Shock, E. L., 1990, Carbonic acid dissociation in aqueous metamorphic fluids: Geological Society of America Abstracts with Programs, v. 22, no. 7, p. A348.

Böhlke, J. K., Coveney, R. M., Jr., Rye, R. O., and Barnes, I., 1988, Stable isotope investigation of gold quartz veins at the Oriental mine, Alleghany district,

California: U.S. Geological Survey Open-File Report 88–279, 24 p.

Böhlke, J. K., Kirschbaum, C., and Irwin, J. J., 1989, Simultaneous analyses of noble gas isotopes and halogens in fluid inclusions in neutron-irradiated quartz veins using a laser microprobe noble gas mass spectrometer: U.S. Geological Survey Bulletin 1890, p. 61–88.

Burchfiel, B. C., and Davis, G. A., 1981, Triassic and Jurassic tectonic evolution of the Klamath Mountains–Sierra Nevada geologic terrane, in Ernst, W. G., ed., The geotectonic development of California (Rubey Volume I): Englewood Cliffs, New Jersey, Prentice-Hall, p. 50–70.

Burrows, D. R., Wood, P. C., and Spooner, E. T. C., 1986, Carbon isotope evidence for a magmatic origin of Archean lode gold deposits: Nature, v. 321, p. 851–854.

Butler, B. S., 1950, Memorial to Waldemar Lindgren: Geological Society of America Proceedings, Annual Report for 1949, p. 177–196.

Clark, L. D., 1960, Foothills fault system, western Sierra Nevada, California: Geological Society of America Bulletin, v. 71, p. 483–496.

Clark, W. B., 1969, Gold districts of California: California Division of Mines and Geology Bulletin 193, 186 p.

Clark, W. B., 1985, Gold districts of California: An update: California Geology, v. 38, p. 3–4.

Cloos, E., 1935, Mother Lode and Sierra Nevada batholith: Journal of Geology, v. 43, p. 225-249.

Cooke, H. R., 1947, The original Sixteen-to-One gold quartz vein, Alleghany, California: Economic Geology, v. 42, p. 211–250.

Courtis, W. M., 1890, Gold-quartz: American Institute of Mining Engineers Transactions, v. 18, p. 639–644.

Coveney, R. M., Jr., 1981, Gold quartz veins and auriferous granite at the Oriental mine, Alleghany, California: Economic Geology, v. 76, p. 2176–2199.

Coveney, R. M., Jr., and Kelly, W. C., 1971, Dawsonite as a daughter mineral in hydrothermal fluid inclusions: Contributions to Mineralogy and Petrology, v. 32, p. 334–342.

Day, H. W., Schiffman, P., and Moores, E. M., 1988, Metamorphism and tectonics of the northern Sierra Nevada, in Ernst, W. G., ed., Metamorphism and crustal evolution of the western United States (Rubey Volume VII): Englewood Cliffs, New Jersey, Prentice-Hall, Inc., p. 737–763.

Debiche, M. G., Cox, A., and Engebretson, D., 1987, The motion of allochthonous terranes across the North Pacific basin: Geological Society of America Special Paper 207, 49p.

Dickinson, W. R., and Snyder, W. S., 1979, Geometry of triple junctions related to San Andreas transform: Journal of Geophysical Research, v. 84, p. 561–572.

Dodge, F. C. W., and Bateman, P. C., 1977, The Sierra Nevada batholith, California, U.S.A., and spatially related mineral deposits, in The relations between granitoids and associated ore deposits of the Circum-Pacific region: Geological Society of Malaysia Bulletin 9, p. 17–29.

Elder, D., and Cashman, S. M., 1992, Tectonic control and fluid evolution in the Quartz Hill, California, lode gold deposits: Economic Geology, v. 87, p. 1795–1812.

Engebretsen, D. C., Cox, A., and Gordon, R. G., 1985, Relative motions between oceanic and continental plates in the Pacific basin: Geological Society of America Special Paper 206, 59 p.

Evans, J. R., and Bowen, O. E., 1977, Geology of the southern Mother-Lode, Tuolumne and Mariposa Counties, California: California Division of Mines and Geology Map Sheet 36, scale 1:24000.

Ferguson, H. G., and Gannett, R. W., 1932, Gold quartz veins of the Alleghany district, California: U.S. Geological Survey Professional Paper 172, 139 p.

Goldfarb, R. J., Snee, L. W., Miller, L. D., and Newberry, R. J., 1991, Rapid dewatering of the crust deduced from ages of mesothermal gold deposits: Nature, v. 354, p. 296–298.

Goldfarb, R. J., Phillips, G. N., and Nokleberg, W. J., 1998, Tectonic setting of synorogenic gold deposits of the Pacific Rim: Ore Geology Reviews, v. 13, p. 185–218.

Graton, L. C., 1939, Waldemar Lindgren: Economic Geology, v. 34, p. 850a–850f.

Groves, D. I., 1993, The crustal continuum model for late-Archaean lode-gold deposits of the Yilgarn Block, Western Australia: Mineralium Deposita, v. 28, p. 366–374.

Groves, D. I., Goldfarb, R. J., Gebre-Meriam, M., Hagemann, S. G., and Robert, F., 1998, Orogenic gold deposits: A proposed classification in the context of their crustal distribution and relationship to other gold deposits: Ore Geology Reviews, v. 13, p. 7–28.

Hamilton, W., 1988, Tectonic setting and variations with depth of some Cretaceous and Cenozoic structural and magmatic systems in the western United States, in Ernst, W. G., ed., Metamorphism and crustal evolution of the western United States (Rubey Volume II): Englewood Cliffs, New Jersey, Prentice-Hall, p. 1–40.

Helgeson, H. C., 1969, Thermodynamics of hydrothermal systems at elevated temperatures and pressures: American Journal of Science, v. 267, p. 729–804.

Hotz, P. E., 1971, Geology of lode gold districts in the Klamath Mountains, California and Oregon. U.S. Geological Survey Bulletin 1290, 97 p.

Ingersoll, R. V., 1982, Initiation and evolution of the Great Valley forearc basin of northern and central California, USA., in Leggett, J. K., ed., Trench-forearc geology: Sedimentation and tectonics of modern and ancient active plate margins: Geological Society of London Special Publication 10, p. 459–466.

Jenkins, O. P., 1948, Geologic guidebook along Highway 49–Sierran gold belt–the Mother Lode country: California Division of Mines Bulletin 141, 164 p.

Johnston, W. D., Jr., 1940, The gold-quartz veins of Grass Valley, California: U.S. Geological Survey Professional Paper 194, 101 p.

Kattelmann, R., 1995, Impacts of gold mining on water resources of the Sierra Nevada, in Hotchkiss, W. R., Downey, J. S., Gutentag, E. D., and Moore, J. E., eds., Water resources at risk: American Institute of Hydrology, p. LL-66 to LL-74.

Kerrich, R., and Wyman, D. A., 1990, The geodynamic setting of mesothermal gold deposits: An association with accretionary tectonic regimes: Geology, v. 18, p. 882–885.

Kerrick, D. M., and Caldeira, K., 1998, Metamorphic $CO_2$ degassing from orogenic belts: Chemical Geology, v. 145, p. 213–232.

Kistler, R. W., Dodge, F. C. W., and Silberman, M. L., 1983, Isotopic studies of mariposite-bearing rocks from the south-central Mother Lode, California: California Geology, September, p. 201–203.

Knopf, A., 1929, The Mother Lode system of California: U.S. Geological Survey Professional Paper 157, 88 p.

Landefeld, L. A., 1988, The geology of the Mother Lode gold belt, Sierra Nevada foothills metamorphic belt, California: Bicentennial Gold 88, Melbourne, May, 1988, p. 167–172.

Lindgren, W., 1895, Characteristic features of California gold-quartz veins: Geological Society of America Bulletin, v. 6, p. 221–240.

Lindgren, W., 1896, The gold-quartz veins of Nevada City and Grass Valley districts, California: U.S. Geological Survey Annual Report 17, p. 13–262.

Lindgren, W., 1911, The Tertiary gravels of the Sierra Nevada of California: U.S. Geological Survey Professional Paper 73, 226 p.

Lindgren, W., 1933, Mineral deposits (fourth edition): New York, McGraw-Hill Book Co., 930 p.

Logan, C. A., 1934, Mother Lode gold belt of California: California State Division of Mines Bulletin 108, 240 p.

Marshall, B., and Taylor, B. E., 1981, Origin of hydrothermal fluids responsible for gold deposition, Alleghany district, Sierra Nevada, California: U.S. Geological Survey Open-File Report 81-355, p. 280–293.

Nesbitt, B. E., and Muehlenbachs, K., 1989, Origins and movement of fluids during deformation and metamorphism in the Canadian cordillera: Science, v. 245, p. 733–736.

Nesbitt, B. E., Murowchick, J. B., and Muehlenbachs, K., 1986, Dual origins of lode gold deposits in the Canadian Cordillera: Geology, v. 14, p. 506–509.

Nesbitt, B. E., Muehlenbachs, K., and Murowchick, J. B., 1989, Genetic implications of stable isotope characteristics of mesothermal Au deposits and related Sb and Hg deposits in the Canadian Cordillera: Economic Geology, v. 84, p. 1489–1506.

Page, B. M., and Engebretson, D. C., 1984, Correlation between the geologic record and computed plate motions for central California: Tectonics, v. 3, p. 133–155.

Paterson, S. R., Tobisch, O. T., and Radloff, J. K., 1987, Post-Nevadan deformation along the Bear Mountains fault zone: Implications for the foothills terrane, central Sierra Nevada, California: Geology, v. 15, p. 513–516.

Peters, E. K., 1993, $\delta^{18}O$ enriched waters of the Coast Range Mountains, northern California: Connate and ore-forming fluids: Geochimica et Cosmochimica Acta, v. 57, p. 1093–1104.

Pickthorn, W. J., Goldfarb, R. J., and Leach, D. L., 1987, Dual origins of lode gold deposits in the Canadian Cordillera: Comment: Geology, v. 15, p. 471–472.

Ransome, F. L., 1900, Mother Lode district folio: U.S. Geological Survey Atlas of the United States, Folio 63, 11 p.

Robert, F., and Kelly, W. C., 1987, Ore-forming fluids in Archean gold-bearing quartz veins at the Sigma mine, Abitibi greenstone belt, Quebec, Canada: Economic Geology, v. 82, p. 1464–1482.

Robert, F., Sheahan, P. A., and Green, S. B., eds., 1991, Greenstone gold and crustal evolution: Geological Association of Canada, NUNA Conference Volume, 252 p.

Rytuba, J. J., ed., 1993, Active geothermal systems and gold-mercury deposits in the Sonoma–Clear Lake volcanic fields, California: Society for Economic Geology (SEPM) Guidebook Series, v. 16, 361 p.

Saleeby, J., 1981, Ocean floor accretion and volcanoplutonic arc evolution of the Mesozoic Sierra Nevada, in Ernst, W. G., ed., The geotectonic development of California (Rubey Volume I): Englewood Cliffs, New Jersey, Prentice-Hall, Inc., p. 132–181.

Sano, Y., and Marty, B., 1995, Origin of carbon in fumarolic gas from island arcs: Chemical Geology, v. 199, p. 265–274.

Schweickert, R. A., 1981, Tectonic evolution of the Sierra Nevada Range, in Ernst, W. G., ed., The geotectonic development of California (Rubey Volume I): Englewood Cliffs, New Jersey, Prentice-Hall, Inc., p. 87–131.

Schweickert, R. A., Bogen, N. I., Girty, G. H., Hanson, R. E., and Merguerian, C., 1984, Timing and structural expression of the Nevadan orogeny, Sierra Nevada, California: Geological Society of America Bulletin, v. 95, p. 967–979.

Schweickert, R. A., Bogen, N. L., and Merguerian, C., 1988, Deformational and metamorphic history of Paleozoic and Mesozoic basement terranes in the western Sierra Nevada metamorphic belt, in Ernst, W. G., ed., Metamorphism and crustal evolution of the western United States (Rubey Volume VII): Englewood Cliffs, New Jersey, Prentice-Hall, Inc., p. 789–822.

Seward, T. M., 1991, The hydrothermal geochemistry of gold, in Foster, R. P., ed., Gold metallogeny and exploration: Glasgow, Blackie, p. 37–62.

Sharp, W. D., 1988, Pre-Cretaceous crustal evolution in the Sierra Nevada region, California, in Ernst, W. G., ed., Metamorphism and crustal evolution of the western United States (Rubey Volume VII): Englewood Cliffs, New Jersey, Prentice-Hall, Inc., p. 823–864.

Sibson, R. H., Robert, F., and Poulsen, K. H., 1988, High-angle reverse faults, fluid-pressure cycling, and mesothermal gold-quartz deposits: Geology, v. 16, p. 551–555.

Smith, R. M., 1981, Source of Mother Lode gold: California Geology, v. 34, no. 5, p. 99–103.

Taylor, B. E., 1987, Stable isotope geochemistry of ore-forming fluids, in Kyser, T. K., ed., Short course in stable isotope geochemistry of low temperature fluids: Mineralogical Association of Canada Shortcourse Handbook, v. 13, p. 337–445.

Taylor, B. E., Robert, F., Ball, M., and Leitch, C. H. B., 1991, Mesozoic "Mother Lode Type" gold deposits in North America: Primary vs secondary (meteoric) fluids: Geological Society of America Abstracts with Programs, v. 23, p. A174.

United States Bureau of Mines, 1983-1995, U.S. Bureau of Mines, Minerals Yearbook: Washington, D. C., U.S. Government Printing Office.

Weir, R. H., and Kerrick, D. M., 1987, Mineralogic, fluid inclusion, and stable isotope studies of several gold mines in the Mother Lode, Tuolumne and Mariposa Counties, California: Economic Geology, v. 82, p. 328–344.

White, D. E., 1957, Magmatic, connate, and metamorphic waters: Geological Society of America Bulletin, v. 68, p. 1659–1682.

White, D. E., Barnes, I., and O'Neil, J. R., 1973, Thermal and mineral waters of nonmeteoric origin, California Coast Ranges: Geological Society of America Bulletin, v. 84, p. 547–560.

Whitney, J. D., 1879, The auriferous gravels of the Sierra Nevada: Harvard College Museum of Comparative Zoology, Memoir 6, 569 p.

Wittkopp, R. W., 1980, A hypothesis for the localization of gold in quartz veins, Alleghany district, California, in Lloyd, R. C., and Rapp, J. S., eds., Mineral resource potential of California: American Institute of Mining Engineers, Sierra Nevada Section, p. 9–22.

MANUSCRIPT ACCEPTED BY THE SOCIETY NOVEMBER 23, 1998.

Geological Society of America
Special Paper 338
1999

Chapter 5

# THE GREAT SAN FRANCISCO EARTHQUAKE OF 1906 AND SUBSEQUENT EVOLUTION OF IDEAS

**Carol S. Prentice**
*U.S. Geological Survey*
*345 Middlefield Road, MS 977*
*Menlo Park, California 94025*
*E-mail: cprentice@usgs.gov*

### ANDREW COWPER LAWSON

Born in Scotland in 1861, Andrew Lawson emigrated with his family to Ontario, Canada, during his childhood. Lawson was educated at the University of Toronto, joined the Geological Survey of Canada, and later pursued graduate studies at The Johns Hopkins University, where he received his doctorate. After a few years as a consulting geologist, he was persuaded to join the faculty of the Geology Department of the University of California at Berkeley in 1890, and remained a member of this department for the rest of his life.

Lawson's geological research included some of the first geologic mapping of northern California, and in 1893 he named one of the faults he mapped along the San Francisco peninsula as the San Andreas fault. Although formally named after San Andreas Lake, a reservoir within the fault zone, Lawson's temperament was such that many have speculated that he actually named the fault after himself, since "Andreas" is Spanish for "Andrew."

It was Lawson who recognized the need for an organized scientific study of the 1906 San Francisco earthquake, which ruptured his San Andreas fault, and he convinced the Governor of California to establish and appoint the State Earthquake Investigation Commission. Lawson was elected chairman of the commission, and coordinated and edited the subsequent report.

Lawson did not restrict his work to California: his geological research took him from the Archean terranes of Canada to ore deposits in Brazil, and his published papers range in discipline from petrography, to ore genesis, to isostasy. He created and edited for 30 years the first scientific publication series at the University of California at Berkeley, the *Bulletin of the Department of Geology, University of California.* The many stories about him told by his former students and colleagues indicate that although he was often fiercely argumentative and often difficult to interact with, his sharp intellect and provocative ideas place him among the finest of California geologists.

Nearly 100 years after its publication, *The Report of the State Earthquake Investigation Commission* (Lawson, 1908, volume I; Reid, 1910, volume II) remains a model of excellence in postearthquake investigations, and an example of multidisciplinary study and cooperation among scientists. Much of our modern knowledge of strike-slip faulting, transform plate margins, strain accumulation and release, and earthquake occurrence is the result of studies initiated along the San Andreas fault, all of which are built upon the foundation provided by the report. The passages reproduced here reflect only a small fraction of this exceptional 643-page document. They were chosen to represent some of the most important observations and conclusions, including descriptions of geomorphic features associated with the San Andreas fault, descriptions of the 1906 ruptures, and a summary of the theory of elastic rebound.

(Source: Byerly, P., and Louderback, G. D., 1964.)

Carol S. Prentice

# THE CALIFORNIA EARTHQUAKE OF APRIL 18, 1906

## REPORT

OF THE

## STATE EARTHQUAKE INVESTIGATION COMMISSION

IN TWO VOLUMES AND ATLAS

VOLUME I

BY

ANDREW C. LAWSON, Chairman

IN COLLABORATION WITH G. K. GILBERT, H. F. REID, J. C. BRANNER, H. W. FAIRBANKS, H. O. WOOD, J. F. HAYFORD AND A. L. BALDWIN, F. OMORI, A. O. LEUSCHNER, GEORGE DAVIDSON, F. E. MATTHES, R. ANDERSON, G. D. LOUDERBACK, R. S. HOLWAY, A. S. EAKLE, R. CRANDALL, G. F. HOFFMAN, G. A. WARRING, E. HUGHES, F. J. ROGERS, A. BAIRD, AND MANY OTHERS

VOLUME I, PART I

WASHINGTON, D. C.
PUBLISHED BY THE CARNEGIE INSTITUTION OF WASHINGTON
1908

# THE CALIFORNIA EARTHQUAKE OF APRIL 18, 1906.

## INTRODUCTION.

On the morning of April 18, 1906, the coastal region of Middle California was shaken by an earthquake of unusual severity. The time of the shock and its duration varied slightly in different localities, depending upon their position with reference to the seat of the disturbance in the earth's crust; but in general the time of the occurrence may be stated to be $5^h$ $12^m$ A. M. Pacific standard time, or the time of the meridian of longitude 120° west of Greenwich; and the sensible duration of the shock was about one minute.

The shock was violent in the region about the Bay of San Francisco, and with few exceptions inspired all who felt it with alarm and consternation. In the cities many people were injured or killed, and in some cases persons became mentally deranged, as a result of the disasters which immediately ensued from the commotion of the earth. The manifestations of the earthquake were numerous and varied. It resulted in the general awakening of all people asleep, and many were thrown from their beds. In the zone of maximum disturbance persons who were awake and attending to their affairs were in many cases thrown to the ground. Many persons heard rumbling sounds immediately before feeling the shock. Some who were in the fields report having seen the violent swaying of trees so that their top branches seemed to touch the ground, and others saw the passage of undulations of the soil. Several cases are reported in which persons suffered from nausea as a result of the swaying of the ground. Many cattle were thrown to the ground, and in some instances horses with riders in the saddle were similarly thrown. Animals in general seem to have been affected with terror.

In the inanimate world the most common and characteristic effects were the rattling of windows, the swaying of doors, and the rocking and shaking of houses. Pendant fixtures were caused to swing to and fro or in more or less elliptical orbits. Pendulum clocks were stopt. Furniture and other loose objects in rooms were suddenly displaced. Brick chimneys fell very generally. Buildings were in many instances partially or completely wrecked; others were shifted on their foundations without being otherwise seriously damaged. Water or milk in vessels was very commonly caused to slop over or to be wholly thrown from the vessel. Many water-tanks were thrown to the ground. Springs were affected either temporarily or permanently, some being diminished, others increased in flow. Landslides were caused on steep slopes, and on the bottom lands of the streams the soft alluvium was in many places caused to crack and to lurch, producing often very considerable deformations of the surface. This deformation of the soil was an important cause of damage and wreckage of buildings situated in such tracts. Railway tracks were buckled and broken. In timbered areas in the zone of maximum disturbance many large trees were thrown to the ground and in some cases they were snapt off above the ground.

The most disastrous of the effects of the earthquake were the breaking out of fires and, at the same time, the destruction of the pipe systems which supplied the water necessary to combat them. Such fires caused the destruction of a large portion of San Francisco, as all the world knows; and they also intensified the calamity due to the earthquake at Santa Rosa and Fort Bragg. The degree of intensity with which the earthquake made itself felt by these various manifestations diminished with the distance from the seat of disturbance, and at the more remote points near the limits of its sensibility it was perceived only by a feeble vibration of buildings during a brief period.

The area over which the shock was perceptible to the senses extends from Coos Bay, Oregon, on the north, to Los Angeles on the south, a distance of about 730 miles; and easterly as far as Winnemucca, Nevada, a distance of about 300 miles from the coast. The territory thus affected has an extent, inland from the coast, of probably 175,000 square miles. If we assume that the sea-bottom to the west of the coast was similarly affected, which is very probably true, the total area which was caused to vibrate to such an extent as to be perceptible to the senses was 372,700 square miles. Beyond the limits at which the vibrations were sufficiently sharp to appeal to the senses, earth waves were propagated entirely around the globe and were recorded instrumentally at all the more important seismological stations in civilized countries.

The various manifestations of the earthquake above cited, including the cracking and deformation of the soil and incoherent surface formations, were the results of the earth jar, or commotion in the earth's crust. The cause of the earthquake, as will be more fully set forth in the body of this report, was the sudden rupture of the earth's crust along a line or lines extending from the vicinity of Point Delgada to a point in San Benito County near San Juan; a distance, in a nearly straight course, of about 270 miles. For a distance of 190 miles from Point Arena to San Juan, the fissure formed by this rupture is known to be practically continuous. Beyond Point Arena it passes out to sea, so that its continuity with the similar crack near Point Delgada is open to doubt; and the latter may possibly be an independent, tho associated, rupture parallel to the main one south of Point Arena. It is most probable, however, that there is but one continuous rupture. The course of this fissure for the 190 miles thru which it has been followed is nearly straight, with a bearing of from N. 30° to 40° W., but with a slight general curvature, the concavity being toward the northeast, and minor local curvatures. The fissure for the extent indicated follows an old line of seismic disturbance which extends thru California from Humboldt County to San Benito County, and thence southerly obliquely across the Coast Ranges thru the Tejon Pass and the Cajon Pass into the Colorado Desert. This line is marked by features due to former earth movements and will be referred to in a general way as a *rift*, the term being adopted from the usage for analogous features in Palestine and Africa.[1] To distinguish it from other rifts of similar origin, it will be referred to more specifically as the San Andreas Rift, the name being taken from the San Andreas Valley on the peninsula of San Francisco, where it exhibits a strongly pronounced character and where its diastrophic origin was first recognized in literature.

The plane or zone on which the rupture took place is, so far as can be determined from a study of the surface phenomena, nearly vertical; and upon this vertical plane there occurred a horizontal displacement of the earth's crust or at least of its upper part. The displacement was such as to cause the country to the southwest of the rift line to be moved northwesterly relatively to the country on the northeast side of that line. The differential displacement in a horizontal direction was probably not less than 10 feet for the greater part of the Rift; in many places it measured over 15 feet, and in one place as much as 21 feet.

[1] Roy. Geograph. Soc. vol. iv, 4, 1894. The Great Rift Valley, by J. W. Gregory, London, 1896.

## INTRODUCTION.

This differential displacement of the earth's crust along the plane of rupture constitutes a *fault*, and will be so referred to in the text of the report. It is named the San Andreas fault. The intersection of the fault plane or narrow zone with the surface of the ground is manifested by cracks, heaved sod, scarps, etc., and these manifestations are designated the *fault-trace*. As a result of this fault, all the fences, roads, railways, bridges, tunnels, dams, pipes, and other structures which crost its path were dislocated. All property lines and other survey lines which were intersected by it were offset. Inasmuch as the movement of the earth which caused the fault was not confined to its immediate vicinity, but was distributed over a considerable belt of country on either side of the trace of the rupture, the latitudes and longitudes of a large portion of the Coast Ranges of California were changed, and the triangles established by the Coast and Geodetic Survey in its triangulation of the region were distorted.

In addition to the horizontal displacement there was, particularly toward the northern end of the fault, a vertical displacement probably nowhere exceeding 2 to 3 feet, whereby the country to the southwest was raised relatively to that to the northeast. In many places, however, particularly toward the southern end of the fault, no vertical displacement can be detected; and there is some indication that, if there was vertical displacement in this region, it was the reverse of that observed in the northern portion of the fault. This rupture of the earth's crust gave rise to certain manifestations at the surface which resemble those described above as a result of the vibratory commotion of the earth, due to the sudden displacement. The cracking and rending of the surface along the line of the fault is a direct expression of the rupture and displacement which originated the earthquake, whereas the cracks, fissures, and lurching of the soft bottom lands and the landslide cracks on the hillsides, whether near the fault line or remote from it, are referable to the oscillation of the crust. The two classes of phenomena must, therefore, be discriminated, particularly as there has been a tendency on the part of some observers to class the secondary phenomena with the primary and interpret the former as indicative of fault lines in the earth's crust, when in reality they are merely superficial phenomena.

While the shock was perceptible to the senses to the extent above indicated in California, Nevada, and Oregon, the distribution of the higher grades of intensity was remarkably linear and was definitely related to the fault line, and to the general trend of the coast of California. This may be brought out in a preliminary way by stating that a zone of destructive effects extends parallel to the Rift from Humboldt Bay, in Humboldt County, to the vicinity of King City in Monterey County, a distance of 350 miles. If we take the throw of brick chimneys and allied phenomena as indicating the limits of what may be called destructive effects, the width of this zone may be fairly approximated at about 70 miles, or about 35 miles on either side of the fault, or its prolongation where no actual fault is observable at the surface. The length of this zone of destruction is thus five times greater than its width, and the total area within which the shock was sufficiently severe to throw brick chimneys may be placed at something over 25,000 square miles; it being assumed that the severity to the southwest of the fault, beneath the waters of the Pacific, was equal to that on the land. If the fault near Point Delgada be regarded as distinct from that extending from Point Arena southeasterly, then the total area of these high intensities would be considerably larger in the direction of the Pacific.

Within this outer limit of destructive effects the intensity increased toward the fault. But proximity to the fault was not the only factor determining the degree of intensity. The soft, more or less incoherent, and water-saturated alluvial formations of the valley-bottoms were much more severely shaken than the rocky slopes of the intervening ridges, and the structures upon them were consequently more commonly and more completely wrecked. It is not understood by this excessive damage on the valley-

## 4  REPORT OF THE CALIFORNIA EARTHQUAKE COMMISSION.

bottoms that the vibratory movement due to the passage of the earth-wave was characterized by greater energy than where it traversed elastic rocks; but that this energy was manifested in a form of movement more destructive to structures upon the surface. The intensity of the shock upon the valley-bottoms, as inferred from damage, seemed abnormally high. In terms of energy it was probably not abnormal. It thus became necessary to discriminate between *apparent intensity* and *real intensity*. Inasmuch as we have to deal primarily with observable effects and record these as a basis for inference, it has been found convenient to use the term "apparent intensity" in a technical sense thruout the report; and all the grades of intensity specified, even when the qualification "apparent" is omitted because of the wearisomeness of its reiteration, are grades of "apparent intensity" arrived at by applying literally the criteria of the Rossi-Forel scale.

City Hall, San Francisco (Lawson, 1908, Plate 82).

# THE SAN ANDREAS RIFT AS A GEOMORPHIC FEATURE.

## GENERAL.

Extending thru the greater part of the Coast System of mountains from Humboldt County to the Colorado Desert, a distance of over 600 miles, is a line or narrow zone characterized by peculiar geomorphic features, referable either directly to the modern deformation of the surface of the ground or to erosion controlled by the lines upon which such deformation has taken place. This peculiar feature has been known, both to Californian geologists and to residents of the sections where its characters are most prominent, but its extent and importance were not fully appreciated until after the earthquake of April 18, 1906. It is commonly reported among the residents of the southern interior Coast Ranges, particularly in San Benito, Monterey, and San Luis Obispo Counties, that displacement of the ground occurred on this line in the earthquake of 1857 and in certain later earthquakes. The first reference in scientific literature to this feature appears to have been in the year 1893, in a paper entitled "The Post-Pliocene Diastrophism of the Coast of Southern California," by Andrew C. Lawson, which is quoted in the sequel. The next reference to this peculiar line is in the eighteenth annual report of the U. S. Geological Survey for 1896–1897, Part IV, in a paper by Schuyler on "Reservoirs for Irrigation," where, pp. 711–713, the significance of the line is fully recognized in the following words quoted in full:

This reservoir has especial interest, not only as the first one of any magnitude completed on the Mojave Desert or Antelope Valley side of the Sierra Madre in southern California, but because it lies directly in the line of what is known as "the great earthquake crack" of this region, which is marked by a series of similar basins behind a distinct ridge that appears to have been the result of the great seismic disturbance.

This remarkable line of fracture can be traced for nearly 200 miles thru San Bernardino, Los Angeles, Kern, and San Luis Obispo Counties, and deviates but slightly here and there from a direct course of about N. 60° to 65° W. There appears to have been a distinct "fault" along the line, the portion lying south of the line having sunken and that to the north of it being raised in a well-defined ridge. In many places along the great crack, ponds and springs make their appearance, and water can be had in wells at little depth anywhere on the south side of the break, in this portion of its course at least; and where the line crosses Little Rock Creek, the blue clay has formed a submerged dam, which has forced the underflow near the surface and created a "cienega" immediately above it. After crossing the line, the water of the creek drops quickly away into the deep gravel and sand of the wash. The same effect is noticeable at other streams, and it has been suggested as the probable cause of the very distinct rim marking the lower margin of the San Bernardino Valley artesian basin and confining its waters within well-defined limits, as this rim is nearly on a prolongation of the line that is traceable on the north side of the mountains — the break having crost the mountains thru the Cajon Pass on the line of Swartout Canyon.

In 1899 the essential features of the same line in the region north of the Golden Gate were recognized and discust by F. M. Anderson.[1] In later years Dr. H. W. Fairbanks has traced out the line in various field trips and has given several public lectures descriptive of its features and its significance, but has published no systematic account of his studies.

The fact that the earthquake of April 18, 1906, was caused by a rupture and displacement of the earth's crust along this line for a distance of about 190 miles, immediately focussed the attention of local geologists upon it. Among those engaged upon

[1] The Geology of the Point Reyes Peninsula, Bull. Dept. Geol. Univ. Cal, vol. 2, No. 5, p. 143 *et seq.* Anderson, however, supposed, as is indicated by the last paragraph of his paper, that the faulting antedates entirely the Pleistocene terrace formations.

its investigation, it became known as the "rift line." Since the earthquake it has been traced as a geomorphic or physiographic feature from Humboldt County to the Colorado Desert, with a possible gap between Shelter Cove and Point Arena, where, if continuous, it lies beneath the Pacific. Its continuity has, however, been satisfactorily established from Point Arena to Whitewater Canyon, at the northern end of the Colorado Desert, a distance of 530 miles. Thruout this entire distance it lies along depressions or at the base of steep slopes which are either the direct result of crustal displacement or of stream erosion, operating with exceptional facility along lines of displacement. There can be no doubt that the displacements have been recurrent thru a considerable part, if not the whole of Pleistocene time, and that in parts of its extent, at least, the movements have taken place on fault-lines which originated in pre-Miocene time. The later movements on this line have given rise to minor features which subaerial and stream erosion have not yet obliterated, and it is these minor features chiefly which have attracted attention to the Rift by reason of their striking contrast with more common geomorphic forms due to erosion. These minor features are chiefly low scarps and troughs bounded on one or both sides by low, abrupt ridges in which frequently lie ponds or swamps of quite small extent.

A summary account will now be given of this rift line as a geomorphic feature.

Fault-trace a mile northwest of Olema. Looking northwest. Illustrates ridge phase. G.K.G. (After Lawson 1908, Plate 40A).

## REVIEW OF SALIENT FEATURES.

It will be of advantage briefly to review the salient features of the San Andreas Rift, in the light of the facts presented in the foregoing detailed description of its extent and character, and of other facts to which attention will be directed.

The San Andreas Rift has been traced with three interruptions from a point in Humboldt County, between Point Delgada and Punta Gorda, to the north end of the Colorado Desert, a distance of over 600 miles. These three interruptions are: The stretch between Shelter Cove and the mouth of Alder Creek, where for a distance of about 72 statute miles it traverses the bottom of the Pacific Ocean; the stretch from the vicinity of Fort Ross to Bodega Head, where for 13 miles it is similarly on the ocean bottom; and the stretch from Bolinas Lagoon to Mussel Rock, where it lies beneath the Gulf of the Farallones for about 19 miles. Of these interruptions only the first involves any doubt as to the continuity of the feature, and this doubt is in large measure removed by the evidence cited hereafter as to the position of the trace of the fault of April 18, 1906.

Thruout its extent the Rift presents a variable relation to the major geomorphic features of the region traversed by it. In Humboldt County it lies within the mountainous tract inland from the coast but to the seaward side of the higher land. From Shelter Cove to Alder Creek it lies to the west of a steep, terraced, coastal slope. From Alder Creek to Fort Ross, it finds its expression in a series of rectilinear, sharply incised valleys, the alinement of which converges upon the coast line to the south at a very acute angle. But near Fort Ross the Rift, without deviation of its general trend, crosses the divide to the coastal side of the ridge which separates these valleys from the ocean, and traverses the terraced coastal slope. Beyond Fort Ross it again lies to the west of a steep coastal slope. From Bodega Head to Bolinas Lagoon the Rift is a remarkably pronounced depression, lying between the main coastal slope and the rather high and precipitous easterly side of the Point Reyes Peninsula. About 0.6 of this depression is below sealevel, forming Tomales and Bodega Bays. This defile is one of the most remarkable and interesting phases of the Rift. It has been a line of repeated faulting in past geological time, and evidently separates a well-marked and probably relatively mobile crustal block from the main continental land mass.

South of Mussel Rock the Rift traverses for a few miles a rolling upland, marked by ponds and old scarps, but with no very marked contrast in relief, and then passes into the very marked and rectilinear San Andreas Valley, along the base of the northeast flank of the Santa Cruz Range. From here to the gap at Wright Station it lies along the base of the range at a distance nowhere greater than 2 miles from the crest. Passing thru the gap at Wright, it crosses from the northeast flank of the range to the southwest flank. Similarly passing thru the gap between the Santa Cruz and Gavilan Ranges at Chittenden, it is again found on the northeast flank of the latter. In effecting this last-mentioned change of position relatively to the mountain crests, a distinct deviation in the trend of the Rift is observable (see map No. 5) as if the path of the Rift accommodated itself to the mass of the mountain blocks. Farther south, near Bitterwater, the Rift leaves the northeast flank of the Gavilan, and lies along the southwest base of a straight ridge of the Mount Hamilton Range. Still farther south in Cholame Valley it follows the northeast base of the ridge which separates Cholame Valley from San Juan Valley. In the Carissa Plain it hugs the southwest flank of the Temblor Range. But the most noticeable reversal of the relative position of the Rift to the adjacent mountain slopes is beyond Tejon Pass. From Tejon Pass to near Cajon Pass, the Rift lies along the steep northerly flank of the San Rafael and San Gabriel Ranges, on the southern edge of Mojave Desert; but at Cajon Pass it passes thru between the San Gabriel and San Bernardino Ranges, and thence easterly lies on the south side of the latter range. Thus from the San Francisco Peninsula to its southern end, so far as the extent of the Rift is at present known, there is a fairly regular and rather remarkable alternation of the relative positions of the Rift and the mountains adjacent to it.

## THE EARTH MOVEMENT ON THE FAULT OF APRIL 18, 1906.

### THE FAULT-TRACE.

The successive movements which in the past have given rise to the peculiar geomorphic features of the Rift, either directly or by control of erosion, have with little question been attended in every case by an earthquake of greater or less violence. The earthquake of April 18, 1906, was due to a recurrence of movement along this line. The movement on that day was of the nature of a horizontal displacement on an approximately vertical fault plane or zone, whereby the country on the southwest side was moved to the northwest and the country on the northeast side to the southeast. This displacement was manifested at the surface by the dislocation and offsetting of fences, roads, dams, bridges, railways, tunnels, pipes, and other structures which crost the line of the fault. The surface of the ground was torn and heaved in furrow-like ridges. Where the surface consisted of grass sward, this was usually found to be traversed by a network of rupture lines diagonal in their orientation to the general trend of the fault. Small streams flowing transverse to the line of the fault had their trenches dislocated so that their waters became impounded. These and similar phenomena of disruption constitute the *fault-trace*.

The width of the zone of surface rupturing varied usually from a few feet up to 50 feet or more. Not uncommonly there were auxiliary cracks either branching from the main fault-trace obliquely for a few hundred feet or yards, or lying subparallel to it and not, so far as disturbance of the soil indicated, directly connected with it. Where these auxiliary cracks were features of the fault-trace, the zone of surface disturbance appears to have included them frequently had a width of several hundred feet. The displacement appears thus not always to have been confined to a single line of rupture, but to have been distributed over a zone of varying width. Generally, however, the greater part of the dislocation within this zone was confined to the main line of rupture, usually marked by a narrow ridge of heaved and torn sod.

The amount of the horizontal displacement, as measured on dislocated fences, roads, etc., at numerous points along the fault-trace, was commonly from 8 to 15 feet. In some places it exceeded this and at one place it was as much as 21 feet. Toward the south end of the fault the amount of displacement was notably less and finally became inappreciable. Nearly all attempts at the measurement of the displacement were concerned with horizontal offsets on fences, roads, and other surface structures at the point of their intersection by the principal rupture plane, and ignore for the most part any displacement that may be distributed on either side of this in the zone of movement. The figures thus obtained may, therefore, in general be considered as representing a minimum for the amount of differential movement. In one or two cases, however, when the displacement has been measured on soft ground subject to slumping, and the measured offset is higher than usual, the results may be in excess of the true crustal displacement.

Besides this horizontal displacement of the crust, there was also, particularly in the region north of the Golden Gate, a distinct uplift of the country to the southwest of the Rift, relatively to that on the northeast. This differential vertical movement was made

# THE CALIFORNIA EARTHQUAKE OF APRIL 18, 1906

## REPORT

### OF THE

## STATE EARTHQUAKE INVESTIGATION COMMISSION

IN TWO VOLUMES AND ATLAS

---

VOLUME II

---

THE MECHANICS OF THE EARTHQUAKE

BY

HARRY FIELDING REID

WASHINGTON, D. C.

Published by the Carnegie Institution of Washington

1910

---

manifest by the appearance of low, abrupt fault-scarps, ranging from less than a foot up to 3 feet. Many of these occurred along the slope of somewhat degraded fault-scarps due to former movements, and served to revivify them. In other cases the new scarps have been developed on slopes where no trace of a previous scarp can be detected. The low scarp which was formed on April 18 is by no means a continuous feature, but appears at a great many places not widely spaced along the fault-trace, extending often for hundreds of yards at a stretch, with intervals where no abrupt scarp can be detected. In the latter places it is probable that the differential vertical movement has been distributed over a zone of some width, underlain by formations in which the deeper seated fracture would be taken up by plastic deformation. The scarp almost invariably faces the northeast, but a few cases have been noted in which a fresh scarp on the fault-trace faced the southwest for a short distance. These will be mentioned more particularly in the detailed descriptions which follow. Associated with the fault-trace, it is quite common to find secondary or induced movements of the soil, particularly on steep slopes. These partake of the nature of landslides, and very commonly exhibit the characteristic landslide scarp. This is usually, however, easy to distinguish from the scarp on the fault proper, or on the auxiliary cracks, since it lacks evidence of horizontal displacement, and the broken sod is not traversed by diagonal, torsional cracks.

The differential displacement of the earth's crust above indicated occurred only on the northern portion of the Rift. South of San Juan, in Benito County, there is no indication along the Rift in the shape of rupture of the soil, or the dislocation of transverse structures, which points to the displacement of the underlying formations. It is not, however, to be certainly inferred from this that there was no deep-seated rupture south of that point. Many earthquakes are known which are referable to sudden slips in the earth's crust for which there is no corresponding rupture at the surface. It is probable that the slip, which is so manifest as a surface rupture to the north of San Juan, was continued as a subsurface movement for many miles south of that point.

North of San Juan the displacement on the fault-trace has been followed practically continuously to a point on the northern coast of California a little beyond Point Arena, a distance of 190 miles. At this point the fault-trace as a continuous feature passes out to sea, and the evidence of displacement is lost. At Shelter Cove, in southern Humboldt County, however, where as previously stated the Rift features appear again, evidence of displacement due to movement on April 18 is also found. The doubt as to whether the Rift in Humboldt County is continuous with that which leaves the coast near Point Arena, of course also applies to the question of the continuity of the rupture on the day of the earthquake. If we assume that the line of rupture is continuous thruout, then its full extent from San Juan to Telegraph Hill is about 270 miles.

Beginning with southern Humboldt County, a somewhat detailed account will now be given of the phenomena of the displacement which occurred on April 18, 1906.

\* \* \* \* \* \*

## THE NATURE OF THE FORCES ACTING.

We know that the displacements which took place near the fault-line occurred suddenly, and it is a matter of much interest to determine what was the origin of the forces which could act in this way. Gravity can not be invoked as the direct cause, for the movements were practically horizontal; the only other forces strong enough to bring about such sudden displacements are elastic forces. These forces could not have been brought into play suddenly and have set up an elastic distortion; but external forces must have produced an elastic strain in the region about the fault-line, and the stresses thus induced were the forces which caused the sudden displacements, or elastic rebounds, when the rupture occurred.[1] The only way in which the indicated strains could have been set up is by a relative displacement of the land on opposite sides of the fault and at some distance from it. This is shown by the northerly displacement of the Farallon Islands of 1.8 meters between the surveys of 1874–1892 and 1906–1907, but the surveys do not decide whether this displacement occurred suddenly at the time of the earthquake, or grew gradually in the interval between them; there are valid reasons, however, for accepting the latter alternative, as the following considerations show: The Farallon Islands are far beyond the limits of the elastic distortion revealed by the surveys, so that we can not ascribe their displacement to elastic rebound; and we have seen that this is the only kind of force which could have produced a sudden movement; and what

* * *

is still more convincing, we shall shortly see that not only was the displacement of 1.8 meters of the Farallons between the survey of 1874–1892 and 1906–1907 insufficient to account for the slip on the fault, but the additional displacement of 1.4 meters which they experienced between the surveys of 1851–1865 and 1874–1892 leaves this quantity still too small.

We must therefore conclude that the strains were set up by a slow relative displacement of the land on opposite sides of the fault and practically parallel with it; and that these displacements extended to a considerable distance from the fault. Let us consider this process; suppose we start with an unstrained region, fig. 6, in which the line $AOC$ is straight; suppose forces parallel to $B''D'$ to act on the regions on opposite sides of the line $B'D'$ so as to displace $A$ and $C$ to $A''$ and $C''$; the straight line $AOC$ will be distorted into the line $A''OC''$; if the distortion is beyond the strength of the rock, a rupture will occur along $B''D'$; the line $A''OC''$ will be broken and the two parts will become straight again and will take the positions $A''O'$ and $C''Q'$; and $O'Q'$ will represent the relative slip at the line of rupture, which will be equal to $A''A''$, the sum of the opposite displacements which $A$ and $C$ gradually experienced when they were brought to $A''$ and $C''$. All points on the western face of the fault will move a distance $OO'$ to the north, and all points on the eastern face a distance $OQ'$ to the south. The straight line which occupied the positions $A''O'$ and $C''Q'$ just before the rupture will be distorted to $A''B''$ and $C''D'$, these lines being exactly like $A''O$ and $C''O$, but turned in opposite directions. The sum of $O'B''$ and $Q'D'$ will exactly equal $O'Q'$, the total slip.

When we examine the actual displacements about the fault-line, we find that the slip $B'D'$, fig. 5, about 6 meters, is fully 4 meters greater than the relative displacement of $A'$ and $C'$ since the survey of 1874–1892; this means that the region was not unstrained at that time, but that $A'$ and $C'$ had already suffered a relative displacement of about 4 meters from their unstrained positions; that is, two-thirds of the stress which caused the rupture had already accumulated 25 years ago. Going still further back to the surveys of 1851–1865, we find that the total relative displacement of distant points on

* * *

opposite sides of the fault since that date amounts to about 3.2 meters, a little more than half enough to account for the slip on the fault-plane; therefore 50 years ago the elastic strain, which caused the rupture in 1906, had already accumulated to nearly half its final amount. It seems not improbable, therefore, that the strain was accumulating for 100 years, altho there is no satisfactory reason to suppose that it accumulated at a uniform rate.

We can picture to ourselves the displacements and the strains which the region has experienced as follows: let $AOC$ (fig. 6) be a straight line at some early date when the region was unstrained. By 1874–1892, $A$ had been moved to $A'$ and $C$ to $C'$, and $AOC$ had been distorted into $A'OC'$; by the beginning of 1906, $A$ had been further displaced to $A''$ and $C$ to $C''$, the sum of the distances $AA''$ and $CC''$ being about 6 meters; and $AOC$ had been distorted into $A''OC''$. When the rupture came, the opposite sides of the fault slipt about 6 meters past each other; $A''O$ and $C''O$ straightened out to $A''O'$ and $C''Q'$; and the straight lines which occupied the positions $A''O'$ and $C''Q'$ just before the rupture, were distorted afterward into the lines $A''B''$ and $C''D'$, these lines being exactly like the lines $A''O$ and $C''O$ but turned in opposite directions. The straight lines, which occupied the positions $A'O'$ and $C'Q'$ in 1874–1892, were distorted into $A''O'$ and $C''Q'$ in the beginning of 1906; at the time of the rupture their extremities on the fault-line had the same movements as other points on that line; $O'$ moved to $B''$ and $Q'$ to $D'$. If we should move the left half of our figure so as to make $A'O'$ continuous with $C'Q'$, fig. 6 would then be practically similar to fig. 5 and similar letters in the two figures would refer to the same points; in fig. 5, however, we have supposed $C'$ to remain stationary and have attributed all the relative movement to $A'$, whereas in fig. 6 we have divided the movement equally between $A'$ and $C'$; as we do not know the actual, but only the relative, movement this difference has no significance.

What was actually determined by the two surveys were the distances of points on the line $C'D'$ and $A''B''$ in fig. 5 measured from the line $C'A'$; and this is equivalent in fig. 6 to the distances of the line $C''D'$ from $C''Q'$, and $A''B''$ from $Q'A''$ less the distance $OO'$. The divergence of the lines $A''B''$ and $C''D'$ from straight lines does not represent the strains which existed in the region just before the rupture, but only the strains accumulated before 1874–1892; we have seen that the total strains set up by 1906 are represented by the divergence from straight lines of the lines $A''O$ and $C''O$, or their counterparts, $A''B''$ and $C''D'$.

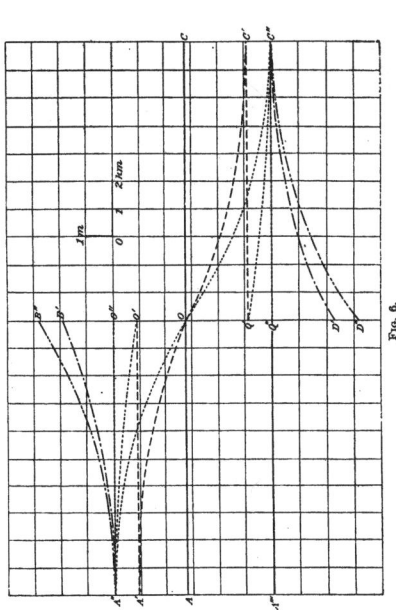

Fig. 6.

# San Andreas fault: The 1906 earthquake and subsequent evolution of ideas

**Carol S. Prentice**
*U.S. Geological Survey, 345 Middlefield Road, MS 977, Menlo Park, California, 94025; e-mail: cprentice@usgs.gov.*

## THE 1906 EARTHQUAKE AND THE REPORT OF THE STATE EARTHQUAKE INVESTIGATION COMMISSION

The San Andreas fault first gained the attention of California's citizenry, as well as that of the international scientific community, during and immediately after the great San Francisco earthquake of 18 April 1906. This earthquake, which occurred when the northern 435–470 km of the San Andreas fault ruptured (Fig. 1), devastated the city of San Francisco and caused widespread damage throughout much of coastal northern California. Because of this earthquake and the subsequent geologic investigation, the San Andreas fault is by far the most famous fault in North America, and perhaps the most famous fault in the world. An icon of popular culture, portrayed in books, movies, and popular songs, it is as much a symbol of California as "golden" hills or surfers riding Pacific Ocean waves. It is also one of the most thoroughly studied faults on Earth.

The scientific inquiry conducted in the aftermath of the earthquake was among the world's earliest comprehensive postearthquake investigations and was the first conducted in the western United States. The results of this investigation were published as *The Report of the State Earthquake Investigation Commission* (referred to herein as the 1906 report), which consists of two volumes, the first edited by Andrew Lawson, and the second by Harry Fielding Reid. After nearly a century, the 1906 report still provides invaluable data and insights for modern research on earthquakes and faults.

A. C. Lawson, professor of geology at the University of California, Berkeley, was largely responsible for the 1906 report. He was chairman of the State Earthquake Investigation Commission (SEIC), editor of volume I, and a contributor of field observations (Byerly and Louderback, 1964). Other contributors to the 1906 report include some of the most prominent geoscientists of the time, as well as many who achieved prominence later in their careers. Among the contributors were G. K. Gilbert, whose remarkably perceptive observations along the San Andreas surface rupture near Olema laid the foundation for the field of tec-

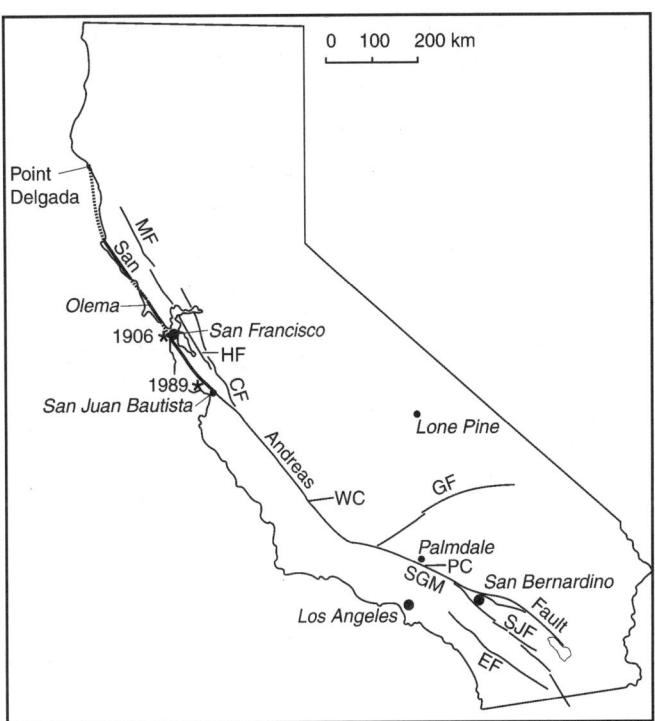

Figure 1. Location map showing San Andreas and related strike-slip faults. Boldest part of San Andreas represents the 435-km-long 1906 rupture (rupture may or may not have continued north or northwest of Point Delgada for as much as 35 additional km). Stars represent approximate locations of 1906 and 1989 earthquake epicenters. CF, Calaveras fault; EF, Elsinore fault; GF, Garlock fault; HF, Hayward fault; MF, Maacama fault; PC, Pallett Creek; SGM, San Gabriel Mountains; SJF, San Jacinto fault; WC, Wallace Creek (modified from Wallace, 1990).

Prentice, C. S., 1999, San Andreas fault: The 1906 earthquake and subsequent evolution of ideas, *in* Moores, E. M., Sloan, D., and Stout, D. L., eds., Classic Cordilleran Concepts: A View from California: Boulder, Colorado, Geological Society of America Special Paper 338.

tonic geomorphology, and H. F. Reid, whose development of the elastic rebound theory in volume II of the 1906 report remains the fundamental model for understanding the origin of earthquakes (Lawson and Byerly, 1951). Also among the contributors were: G. Davidson, the first president of the Seismological Society of America (Townley, 1922); J. C. Branner, founding chairman of the Stanford University Geology Department, and later president of Stanford University (Townley, 1922); H. O. Wood, later the founder and first director of the Seismological Laboratory at Caltech (Richter, 1959); F. E. Matthes, preeminent geomorphologist and topographer of Yosemite Valley and other national parks (Fryxell, 1956); and F. Omori, one of Japan's most celebrated seismologists (Debew, 1968). The coordinated effort of the SEIC was essential to achieving a scientific understanding of the 1906 San Andreas fault rupture, and Lawson, as chairman of this group, demonstrated vision and leadership seldom equaled in the history of geoscience investigations.

Geologic investigation of the 1906 rupture presented formidable challenges. The 1906 earthquake produced rupture from near San Juan Bautista to at least Point Delgada (Fig. 1), and ~240 km of this rupture length was onshore. Much of the onshore rupture occurred in remote regions, few maps were available, and widespread earthquake damage hindered travel. At the time, the relationship between faults and earthquakes was understood by few scientists (G. K. Gilbert was an exception; he clearly understood this relationship, as demonstrated in Gilbert, 1884). Horizontal slip on a fault had been reported prior to 1906 (Koto, 1893; McCay, 1890), and Gilbert had recognized and described it in his field notes during an August 1883 visit to the 1872 rupture in Lone Pine (Beanland and Clark, 1994). However, strike slip on a fault was not a widely accepted concept, so most of the 1906 earthquake investigators were studying a phenomenon of which they had no prior knowledge.

The 1906 report is important both as the authoritative source that documents observations associated with the 1906 earthquake, and as the pioneering effort to understand earthquakes using geologic, geodetic, and seismic data. The 1906 report contains the first integrated description of the San Andreas fault, extending from Point Delgada, in the north, to Whitewater Canyon, southeast of San Bernardino. Although several segments of the fault had been recognized independently in reports published prior to 1906, none of these workers recognized the connection between their independently mapped faults (Hill, 1981). The detailed description of the 1906 San Andreas fault rupture from Point Delgada to San Juan Bautista is outstanding for its scientific insight and level of documentation. The 1906 report also contains a description of the San Jacinto and Hayward faults, and a summary of the then-current knowledge of the geology of the California Coast Ranges. Also included are maps and photographs, descriptions of other geologic phenomena such as coseismic landslides, geodetic measurements, descriptions of damage used to construct isoseismal maps, descriptions and analyses of seismograms, and accounts of several large nineteenth-century California earthquakes.

# EVOLUTION OF SAN ANDREAS FAULT CONCEPTS

## *Large-scale horizontal displacement*

In spite of the well-documented occurrence of horizontal slip associated with the 1906 earthquake, geologists continued to explain the dramatically contrasting rock types they found juxtaposed across the San Andreas fault in terms of dip-slip displacement (Hill, 1981). The concept of large-scale strike-slip movement took several decades to develop after the occurrence of the 1906 earthquake and did not become well accepted for several more decades. The recognition of hundreds of kilometers of strike-slip displacement is one of the most important results to come out of studies of the San Andreas fault. Yet, the paradigm of vertical fault slip was so deeply ingrained that even after the 1906 report provided indisputable evidence of its occurrence, most scientists (including many of the contributors to the 1906 report) did not consider horizontal slip to be a geologically important mode of behavior for some 50 years after the earthquake.

One of the first geologists to suggest substantial amounts of strike-slip motion was Levi Noble, who proposed about 38 km of strike-slip displacement on Tertiary rocks across the San Andreas fault in reports on the geology of the northern San Gabriel Mountains (Noble, 1926, 1927). This proposition, radical at the time, was not generally accepted by the geologic community, and Noble had reservations about it. Subsequent workers continued to explain the discordant rocks juxtaposed across the San Andreas in terms of dip-slip displacement, and in spite of Noble's suggestion, many continued to believe that total horizontal slip amounted to less than 1 or 2 km (Hill, 1981).

While the geologic community argued over whether or not the several tens of kilometers of horizontal displacement proposed by Noble were possible, results of further regional geologic mapping began to provide data suggesting a much more shocking and revolutionary idea—that hundreds of kilometers of strike slip had taken place across the San Andreas fault over geologic time. Wood and Buwalda (1930) reported 3 m to more than 1000 m of offset on streams in the Carrizo Plain and suggested the possibility of many kilometers of offset. Wallace (1949) recognized right-lateral offset of Quaternary terraces across the San Andreas fault northwest of Palmdale and extrapolated this back into the Tertiary to suggest that displacement of 120 km was possible. In a very influential paper, Hill and Dibblee (1953) not only suggested several hundred kilometers of strike slip across the San Andreas fault, they also presented data showing that older units are laterally offset by a greater distance than are younger units. Although some of the specific correlations and offsets Hill and Dibblee (1953) proposed have been questioned, the basic thesis of their paper has more than stood the test of time and has fundamentally altered the way geologists think about strike-slip faults.

The ideas and data presented in Hill and Dibblee (1953) stimulated much interest and new work along the San Andreas fault. Scenarios that explained the field relations using dip-slip offset became ever more complicated and cumbersome as additional mapping was done, and eventually, in the face of mounting evidence, the concept of large strike-slip displacements was accepted as a superior explanation of the geologic relations by most members of the geologic community (Hill, 1981). The best constrained offsets across the San Andreas suggest a total post-early Miocene right-lateral offset of 300–330 km (Wallace, 1990; Powell et al., 1993) (Fig. 2).

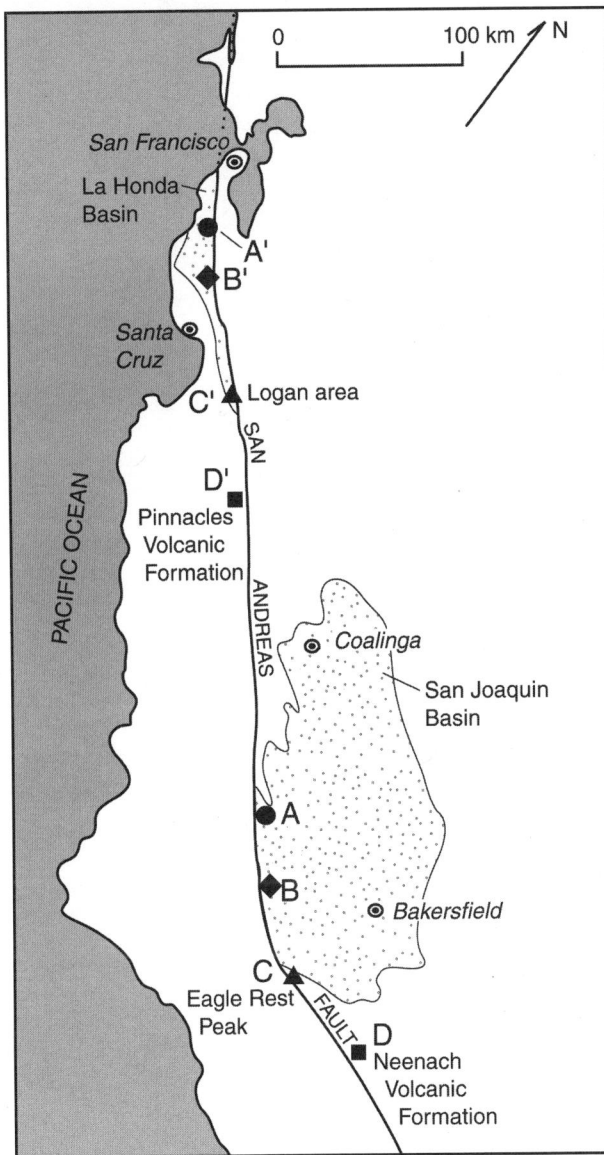

Figure 2. Map of part of central California showing some of the strongest correlations of units offset across the San Andreas fault. Offsets are about 305–330 km. Stippled regions are correlative Tertiary sedimentary basins, offset across the fault (Stanley, 1987). A and A′ represent the Butano Formation and the Point of Rocks Sandstone Member (of the Kreyenhagen Formation), respectively, which are offset parts of an Eocene deep-sea fan. B and B′ are the deepest points within the La Honda and San Joaquin basins, respectively. Triangles represent exposures of unusual quartz-bearing mafic rocks near Logan (C) and Eagle Rest Peak (C′) (Ross et al., 1973). Squares represent uniquely correlated volcanic rocks, The Pinnacles (D) and the Neenach (D′) (Mathews, 1976). Modified from Irwin (1990).

### San Andreas fault and the plate tectonics revolution

That strike-slip displacement was not accepted as a significant geologic phenomenon for so many decades after it had been documented in the 1906 report was probably because no horizontal driving force was known to the scientific community. The advent of plate tectonics in the 1960s provided a new model for the large-scale deformation of Earth's crust and provided a mechanism for producing hundreds of kilometers of horizontal slip. The concept of the San Andreas as a transform fault (Wilson, 1965) provided the key for subsequent plate tectonics analysis of California geology (e.g., Dickinson and Snyder, 1979a, 1979b), which in turn advanced the understanding of global plate tectonics.

A major breakthrough in understanding the role of the San Andreas fault in global plate tectonics was presented in Atwater (1970) and Atwater and Molnar (1973). These studies used marine geophysical data to model the interactions of the Pacific, Farallon, and North American plates since about 30 Ma (Fig. 3). The San Andreas fault represents the Pacific–North American transform plate boundary, which began to evolve when the spreading center between the Pacific and Farallon plates encountered the Farallon–North American subduction zone. This model remains a useful framework for understanding the evolution of the San Andreas fault, although subsequent geologic studies have provided a more complete and complex view (Wallace, 1990; Powell et al., 1993).

### Seismic hazards analysis

Many of the important ideas that underpin current seismic hazards analysis can be traced to the 1906 report. Wood (*in* Lawson, 1908, p. 220–245) observed that the amount of damage in different districts within San Francisco was distinctly related to the nature of the surficial materials, and similar observations were made by workers in other regions along the rupture. Gilbert pointed out that the geomorphic features he observed along the San Andreas fault provided clear evidence of repeated earthquakes, expanding on his previous observations of evidence for recurrent faulting along the Wasatch fault in Utah (Gilbert, 1884). Reid (1910) explained the relevance of geodetic measurements, and, with his description of elastic rebound theory, provided a conceptual model for understanding earthquake recurrence.

Three of the disciplines typically used in assessing seismic hazards, seismology, earthquake geology, and tectonic geodesy, were all significantly advanced by the studies documented in the 1906 report. Subsequently, much of the progress in understanding seismic hazards has resulted from studies initiated on the San Andreas fault. Seismology has contributed much to seismic hazard analyses (e.g., Ben-Menahem, 1995), and many new developments in understanding seismic hazards in the past 10–20 yr have resulted from advances in earthquake geology (often referred to as paleoseismology) and geodesy, both of which have significant roots in the 1906 report.

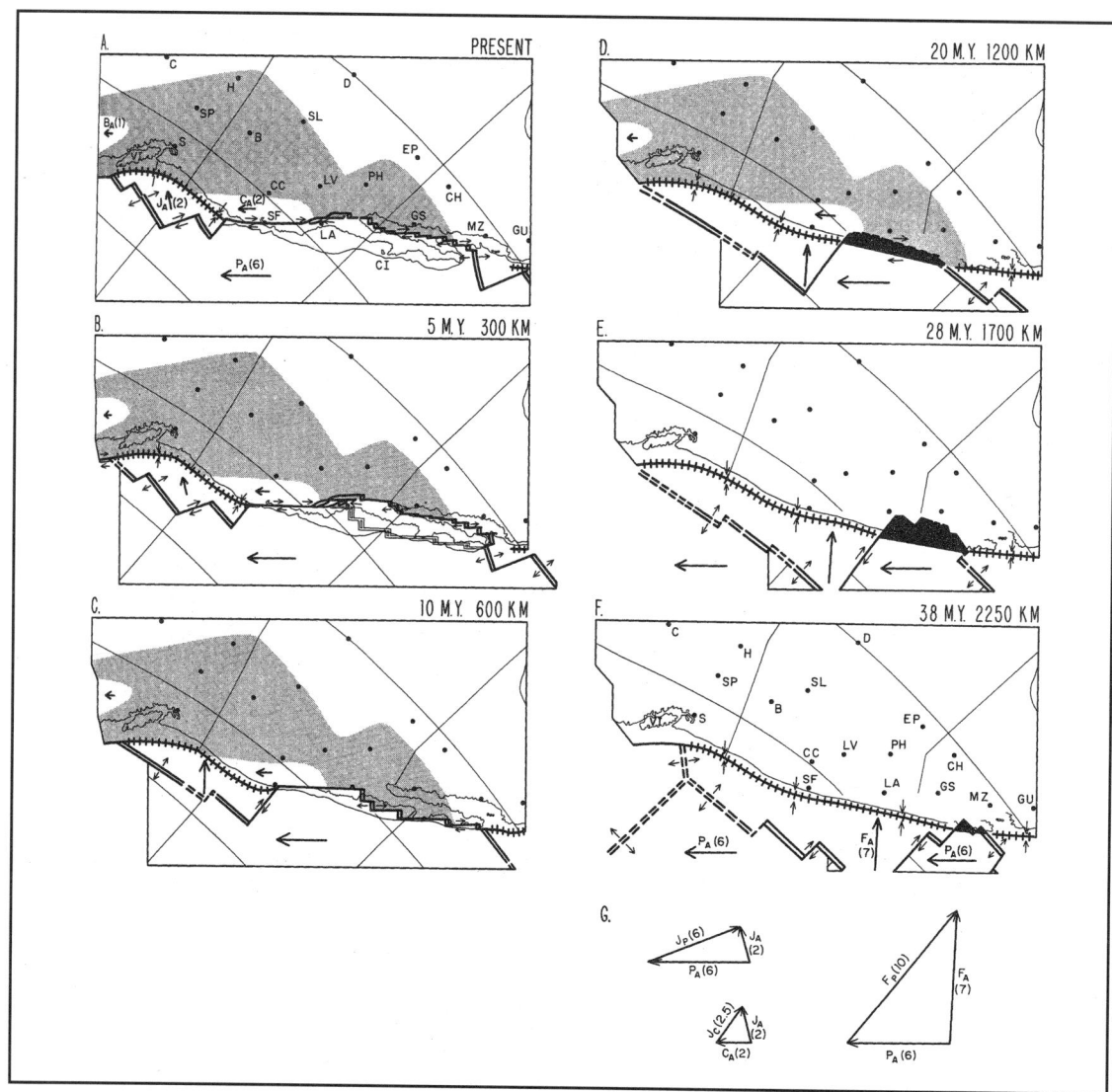

Figure 3. Evolution of the San Andreas transform (from Atwater, 1970, Fig. 16). B—Boise; C—Calgary; CC—Carson City; CH—Chihuahua; CI—Cedros Island; D—Denver; EP—El Paso; GS—Guaymas; GU—Guadalajara; H—Helena; LA—Los Angeles; LV—Las Vegas; MZ—Mazatlan; PH—Phoenix; S—Seattle; SF—San Francisco; SL—Salt Lake City; SP—Spokane; VI—Vancouver Island.

*Earthquake geology.* The geologic study of active faults for seismic hazards analysis can also be traced to the 1906 report. This is especially evident in the sections written by Gilbert and Matthes, who elegantly described the relations between earthquake recurrence, strike-slip faulting, and topography that they observed along the San Andreas fault. Gilbert's earlier observations of repeated fault slip recorded in landforms (Gilbert, 1884) were developed for normal faults, and he first applied them to strike-slip faults in the 1906 report. He later expanded these ideas into a discussion of earthquake forecasting that provided an integrated overview of the entire seismic hazard problem and anticipated many of the elements of modern seismic hazard analysis (Gilbert, 1909).

Geologic interpretation of aerial photography facilitated the widespread use of geomorphic features to delineate active faults. A series of "strip maps" produced during the 1960s and 1970s showing geomorphic features associated with active faults (e.g., Brown and Wolfe, 1972; Ross, 1969; Vedder and Wallace, 1970; Sarna-Wojcicki et al., 1975) had an important impact on the understanding of seismic hazard in California and led to state legislation prohibiting building on active faults (Wallace, 1996).

The development of radiocarbon age dating for late Quaternary materials allowed geologists to begin to quantify the behavior of active faults. Slip rates and dates of prehistoric earthquakes are now some of the most important types of data used in modern seismic hazard analysis. Significant studies in advancing these new quantitative techniques were conducted along the San Andreas fault in central and southern California by K. E. Sieh (Sieh, 1978; Sieh and Jahns, 1984; Sieh et al., 1989). Sieh and Jahns (1984) used the offset of Wallace Creek to estimate a late Holocene slip rate of about 34 mm/yr across the San Andreas fault in the Carrizo Plain. Studies of excavations across the fault at Pallett Creek in southern California (Sieh, 1978; Sieh et al., 1989) provided radiocarbon age estimates of prehistoric earthquakes and a basis for estimating earthquake recurrence intervals (Fig. 4). These studies, and similar work by other investigators, advanced the understanding of Holocene San Andreas fault behavior and developed one of the most important new techniques of earthquake geology since publication of the 1906 report (e.g., McCalpin, 1996; Yeats et al., 1997).

Many fundamental questions about earthquake recurrence remain to be answered. Are earthquakes essentially random events, or are they highly periodic, as several models suggest? Or, as some data indicate, are they clustered in time and/or space? Paleoseismology, identifying and dating prehistoric earthquakes from evidence recorded in deformed late Quaternary strata, can answer these questions and supply a quantitative understanding of the behavior of the faults responsible for generating large, destructive earthquakes (Yeats and Prentice, 1996).

*Geodesy.* The understanding of active faults has also advanced through the contributions of geodetic studies. The 1906 report described one of the earliest geodetic studies of crustal deformation preceding and accompanying an earthquake, and this prototype has led to more detailed and precise geodetic studies of active faults. Surveys done in the nineteenth century during the periods 1851–1865 and 1874–1892 were compared to a postearthquake survey done in 1906–1907. The elastic rebound theory developed by Reid using these data appears in volume II of the 1906 report and provides the underlying theory for the geodetic modeling of strain accumulation and release across active fault zones. In addition, two surveyed arrays were set up across the fault to measure future movement.

Modern geodesy has been transformed as a result of advances in measurement techniques unimaginable at the time the 1906 report was published. Advances such as laser geodimeters, very long baseline interferometry (VLBI), the global positioning system (GPS), and radar interferometry have enabled scientists to make increasingly precise and rapid determinations of crustal deformation and are now yielding very detailed information on strain accumulation and strain release in tectonically active regions (e.g., Carter and Robertson, 1986; Lambeck, 1988; Prescott et al., 1989; Dixon, 1991; Massonnet et al., 1993).

## CONCLUDING REMARKS

Only 1,000 copies of the 1906 report were produced initially, an unfortunate consequence of insufficient funding. This situation was remedied in 1969, when the Carnegie Institution, at the urging of W. W. Rubey and earth scientists (Haskins, 1969) reprinted the 1906 report in its entirety and made it available at low cost. That there was such strong interest in and demand for what at that time was a 60 yr old scientific publication speaks volumes for its importance and place in the history of the earth sciences.

The 1906 report continues to supply information for modern studies in geology, geodesy, and seismology. For example, data contained in the 1906 report concerning the offset of the Wrights tunnel, located near the epicenter of the 1989 Loma Prieta earthquake, have recently been used to show that ridge-top fractures produced in this region in 1906 were not of tectonic origin and to suggest that similar fractures observed in 1989 were also nontectonic. Data from the Wrights tunnel also permit an estimate of the amount of total coseismic near-surface displacement in 1906, at least 1.7–1.8 m (Prentice and Ponti, 1997). Historical research into archival documents and photographs associated with the 1906 report has provided a better understanding of the San Andreas fault near Loma Prieta, allowing comparison between the 1906 and 1989 earthquakes (Prentice and Schwartz, 1991; Prentice and Ponti, 1999). Similar historical data associated with the 1906 report are providing a better understanding of the fault along the San Francisco peninsula (Prentice and Hall, 1996) (Fig. 5). A long-standing controversy over the San Andreas fault near Point Delgada has recently been addressed using data from the 1906 report and associated archival documents and photographs (Brown, 1995; Prentice et al., 1999). A recent geodetic study of the entire 1906 rupture length relies on data found in the 1906 report (Thatcher et al., 1997). A recent analysis of teleseismic

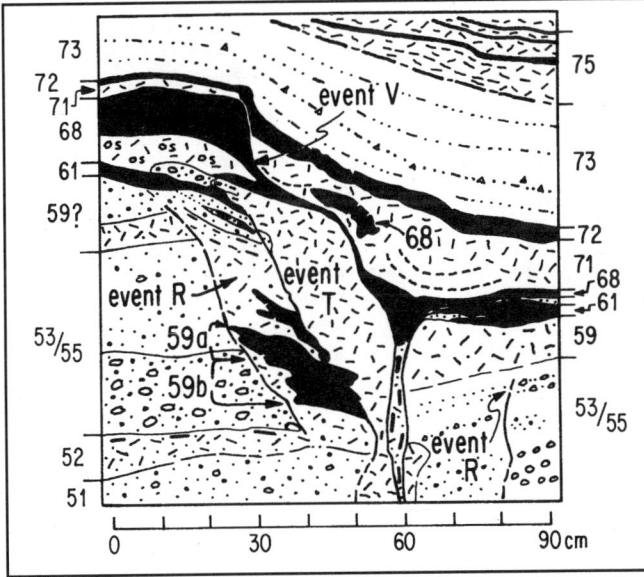

Figure 4. Drawing of part of an exposure in an excavation across the San Andreas fault near Pallett Creek, southern California, showing evidence for paleoearthquakes (from Sieh et al., 1989, Fig. 5). Black beds are peats (radiocarbon dated); stipples represent sand; short-line pattern indicates silt. Feature located above 60 cm mark represents a fissure that opened during earthquake R. Peat deposited in this fissure provides the best minimum age for this event. Fissure reopened during earthquakes T and V (see Sieh et al., 1989, for additional details).

Figure 5. Previously unpublished photograph taken by H. O. Wood in 1906, showing surface rupture and offset fence along the San Francisco peninsula. Photograph was made from a glass-plate negative stored in the Bancroft Library of the University of California, Berkeley (1957.007.110), and is reprinted by permission. Note that there are three fault traces, shown by arrows. The person shown in the photo provides a scale, allowing an estimate of less than 2 m of slip on the main trace. This location is now within a housing development (Prentice and Hall, 1996).

wave forms recorded in 1906 and published in the 1906 report has provided a detailed source model suggesting two regions of localized strong seismic radiation for the 1906 earthquake (Wald et al., 1993).

There is still much to be gained from study of the 1906 report, in spite of the fact that it is nearly a century old and in spite of the great increases in our understanding of the San Andreas fault since the time of its publication. It is exceptionally well written and worth reading for the quality of its prose alone. The 1906 report provides an invaluable lesson about the impact of careful, well-documented scientific investigation. Perhaps more important, it provides a glimpse into the origin of most of the modern concepts about the San Andreas fault and earthquake recurrence, and a sense of the magnitude of the accomplishments during the twentieth century in understanding this truly remarkable feature of the Earth's crust.

## REFERENCES CITED

Atwater, T., 1970, Implications of plate tectonics for the Cenozoic tectonic evolution of western North America: Geological Society of America Bulletin, v. 81, p. 3513–3536.

Atwater, T., and Molnar, P., 1973, Relative motion of the Pacific and North American plates deduced from sea-floor spreading in the Atlantic, Indian, and South Pacific Oceans, in Kovach, R. L., and Nur, A., eds., Proceedings of the conference on tectonic problems of the San Andreas faults system: Stanford University Publications in the Geological Sciences, v. 13, p. 136–148.

Beanland, S., and Clark, M. M., 1994, The Owens Valley fault zone, eastern California, and surface faulting associated with the 1872 earthquake: U.S. Geological Survey Bulletin 1982, 29 p.

Ben-Menahem, A., 1995, A concise history of mainstream seismology; origins, legacy, and perspectives: Seismological Society of America Bulletin, v. 85, p. 1202–1225.

Brown, R. D., 1995, 1906 surface faulting on the San Andreas fault near Point Delgada, California: Seismological Society of America Bulletin, v. 85, p. 100–110.

Brown, R. D., and Wolfe, E. W., 1972, Map showing recently active breaks along the San Andreas fault between Point Delgada and Bolinas Bay, California: U.S. Geological Survey Miscellaneous Investigations Map I-692, scale: 1:24,000.

Byerly, P., and Louderback, G. D., 1964, Andrew Cowper Lawson: National Academy of Sciences Biographical Memoirs, v. 37, p. 185–204.

Carter, W. E., and Robertson, D. S., 1986, Studying the Earth by very-long-baseline interferometry: Scientific American, v. 255, p. 46–54.

Debew, A. G., 1968, Who's Who in science from antiquity to the present, in Marquis Who's who: Hannibal, Missouri, Western Publishing Co., p. 1284–1285.

Dickinson, W. R., and Snyder, W. S., 1979a, Geometry of subducted slabs related to San Andreas transform: Journal of Geology, v. 87, p. 609–627.

Dickinson, W. R., and Snyder, W. S., 1979b, Geometry of triple junctions related to San Andreas transform: Journal of Geophysical Research, v. 84, no. B2, p. 561–572.

Dixon, T. H., 1991, An introduction to the Global Positioning System and some geological applications: Reviews of Geophysics, v. 29, p. 249–276.

Fryxell, F., 1956, Memorial to Francois Emile Matthes (1874–1948): Proceedings Volume of the Geological Society of America, Annual Report for 1955, p. 153–168.

Gilbert, G. K., 1884, A theory of the earthquakes of the Great Basin, with a practical application, Salt Lake City Tribune, Sept. 30, 1883 (reprint): American Journal of Science, ser. 3, v. 27, p. 49–53.

Gilbert, G. K., 1909, Earthquake forecasts: Science, v. 29, p. 121–138.

Haskins, C. P., 1969, Forward to reprinted edition, in Lawson, A. C., ed., The California earthquake of April 18, 1906, Report of the State Earthquake Investigation Commission: Washington, D. C., Carnegie Institute of Washington, p. 451.

Hill, M. L., 1981, San Andreas fault: History of concepts: Geological Society of America Bulletin, v. 92, p. 112–131.

Hill, M. L., and Dibblee, T. W., 1953, San Andreas, Garlock, and Big Pine faults, California: Geological Society of America Bulletin, v. 64, p. 443–458.

Irwin, W. P., 1990, Geology and plate-tectonic development, in Wallace, R. E., ed., The San Andreas Fault System, California: U.S. Geological Survey Professional Paper 1515, p. 61–80.

Koto, B., 1893, On the cause of the great earthquake in central Japan, 1891: Journal of the College of Science, Imperial University of Japan, v. 5, p. 296–353.

Lambeck, K., 1988, Geophysical geodesy: The slow deformation of the Earth: Oxford, Clarendon Press, 718 p.

Lawson, A. C., ed., 1908, The California earthquake of April 18, 1906, Report of the State Earthquake Investigation Commission: Washington, D. C., Carnegie Institute of Washington Publication 87, 451 p.

Lawson, A. C., and Byerly, P., 1951, Harry Fielding Reid: National Academy of Sciences Biographical Memoirs, v. 26, p. 1–12.

Massonnet, D., Rossi, M., Carmona, C., Adragna, F., Peltzer, G., Feigl, K., and Rabaute, T., 1993, The displacement field of the Landers earthquake mapped by radar interferometry: Nature, v. 364, p. 138–142.

Matthews, V., 1976, Correlation of Pinnacles and Neenach Volcanic Formations and their bearing on San Andreas fault problem: American Association of Petroleum Geologists Bulletin, v. 60, p. 2128–2141.

McCalpin, J. P., 1996, Paleoseismology: San Diego, Academic Press, 588 p.

McCay, S., 1890, On the earthquake of September 1888, in the Amuri and Marlborough districts of the South Island, Reports of Geological Exploration 1885–1889: Wellington, New Zealand Geological Survey, p. 1–16.

Noble, L. F., 1926, The San Andreas rift and some other active faults in the desert region of southeastern California: Carnegie Institution of Washington Year Book, v. 25, p. 415–422.

Noble, L. F., 1927, The San Andreas rift and some other active faults in the desert region of southeastern California: Seismological Society of America Bulletin, v. 17, p. 25–39.

Powell, R. E., Weldon, R. J., II., and Matti, J. C., 1993, The San Andreas fault system: Displacement, palinspastic reconstruction, and geologic evolution: Geological Society of America Memoir, 332 p.

Prentice, C. S., and Hall, N. T., 1996, Using historical research to locate the 1906 rupture of the San Andreas fault on the San Francisco Peninsula: Geological Society of America Abstracts with Programs, v. 28, no. 5, p. 102.

Prentice, C. S., and Ponti, D. J., 1997, Coseismic deformation of the Wrights tunnel during the 1906 San Francisco earthquake: A key to understanding 1906 fault slip and 1989 surface ruptures in the southern Santa Cruz Mountains, California: Journal of Geophysical Research, v. 102, p. 635–648.

Prentice, C. S., and Ponti, D. J., in press, 1906 and 1989 coseismic ground fractures in the Santa Cruz Mountains, in Holzer, T. L. and Ponti, D. J., eds., The Loma Prieta, California earthquake of October 17, 1989: U.S. Geological Survey Professional Paper 1551H.

Prentice, C. S., and Schwartz, D. P., 1991, Re-evaluation of 1906 surface faulting, geomorphic expression, and seismic hazard along the San Andreas fault in the southern Santa Cruz Mountains: Seismological Society of America Bulletin, v. 81, p. 1424–1479.

Prentice, C. S., Merritts, D. J., Beutner, E., Bodin, P., Schill, A., and Muller, J., 1999, The northern San Andreas fault near Shelter Cove, California: Geological Society of America Bulletin (in press).

Prescott, W. H., Davis, J. L., and Svarc, J. L., 1989, Global positioning system measurements for crustal deformation: Precision and accuracy: Science, v. 244, p. 1337–1340.

Reid, H. F., 1910, The California earthquake of April 18, 1906, Report of the State Earthquake Investigation Commission, Volume II, The Mechanics of the Earthquake: Carnegie Institute of Washington Publication 87, v. 2, 192 p.

Richter, C. F., 1959, Memorial to Harry Oscar Wood (1879–1958): Proceedings Volume of the Geological Society of America, Annual Report for 1958, p. 219–224.

Ross, D. C., 1969, Map showing recently active breaks along the San Andreas fault between Tejon Pass and Cajon Pass, southern California: U.S. Geological Survey Miscellaneous Investigations Series Map I-553, scale 1:24,000.

Ross, D. C., Wentworth, C. M., and KcKee, E. H., 1973, Cretaceous mafic conglomerate near Gualala offset 350 miles by San Andreas fault from oceanic crustal source near Eagle Rest Peak, California: U.S. Geological Survey Journal of Research, v. 1, p. 45–52.

Sarna-Wojcicki, A. M., Pampeyan, E. H., and Hall, N. T., 1975, Map showing recently active breaks along the San Andreas fault between the central Santa Cruz Mountains and the northern Gabilan Range, California: U.S. Geological Survey Miscellaneous Field Studies Map, MF-650, scale 1:24,000.

Sieh, K. E., 1978, Prehistoric large earthquakes produced by slip on the San Andreas fault at Pallett Creek, California: Journal of Geophysical Research, v. 83, no. B8, p. 3907–3939.

Sieh, K. E., and Jahns, R. H., 1984, Holocene activity of the San Andreas fault at Wallace Creek, California: Geological Society of America Bulletin, v. 95, p. 883–896.

Sieh, K., Stuiver, M., and Brillinger, D., 1989, A more precise chronology of earthquakes produced by the San Andreas fault in southern California: Journal of Geophysical Research, v. 94, no. B1, p. 603–623.

Stanley, R. G., 1987, New estimates of displacement along the San Andreas fault in central California based on paleobathymetry and paleogeography: Geology, v. 15, p. 171–174.

Thatcher, W., Marshall, G., and Lisowski, M., 1997, Resolution of fault slip along the 470-km-long rupture of the great 1906 San Francisco earthquake: Journal of Geophysical Research, v. 102, no. B3, p. 5353–5367.

Townley, S. D., 1922, John Caspar Branner: Seismological Society of America Bulletin, v. 12, p. 1–11.

Vedder, J. C., and Wallace, R. E., 1970, Recent active breaks along the San Andreas fault between Cholame Valley and Tejon Pass, California: U.S. Geological Survey Miscellaneous Investigation Map I-574, scale 1:24,000.

Wald, D. J., Kanamori, H., Helmberger, D. V., and Heaton, T. H., 1993, Source study of the 1906 San Francisco earthquake: Seismological Society of America Bulletin, v. 83, p. 981–1019.

Wallace, R. E., 1949, Structure of a portion of the San Andreas rift in California: Geological Society of America Bulletin, v. 60, p. 781–806.

Wallace, R. E., 1990, The San Andreas Fault System, California: U.S. Geological Survey Professional Paper 1515, 283 p.

Wallace, R. E., 1996, Earthquakes, minerals and me: With the USGS, 1942–1995: U.S. Geological Survey Open-File Report 96-260, 201 p.

Wilson, J. T., 1965, A new class of faults and their bearing on continental drift: Nature, v. 207, p. 343–347.

Wood, H. O., and Buwalda, J. P., 1930, Horizontal displacement along the San Andreas fault in Carrizo Plain, California [abs.]: Pan-American Geology, v. 54, p. 75.

Yeats, R. S., and Prentice, C. S., 1996, Introduction to special section: Paleoseismology: Journal of Geophysical Research, v. 101, no. B3, p. 5847–5853.

Yeats, R. S., Sieh, K. E., and Allen, C. R., 1997, The geology of earthquakes: New York, Oxford University Press, 568 p.

MANUSCRIPT ACCEPTED BY THE SOCIETY ON NOVEMBER 23, 1998.

# Chapter 6
# THE SAN ANDREAS FAULT AND RELATED FAULT SYSTEMS

**Thomas W. Dibblee, Jr.**
*Dibblee Geological Foundation*
*P. O. Box 60560*
*Santa Barbara, California 93160*

Mason L. Hill

Thomas W. Dibblee, Jr.

## MASON L. HILL AND THOMAS W. DIBBLEE, JR.

The convergence and influence of academia, industry, and government in this section is illustrated with the descendant of one of the original Spanish land grantees in California, Thomas Dibblee, Sr. Dibblee took classes with Joseph LeConte at Berkeley at the turn of the century. In the 1920s A. O. Woodford, who had completed his doctoral degree at Berkeley, returned to start a geology department at Pomona and encouraged his student, Mason Hill, to take a course in geology from Lawson at Berkeley. In 1929 Thomas Dibblee Sr., recognizing the anticlinal structure of the Santa Barbara County ranch and oil potential, hired geologist Harold Johnson to map the San Julian Ranch and assess the likelihood of oil. Although no oil was realized, the mapping process stimulated his high-school age son, Thomas Dibblee, Jr., to attend Stanford. Soon after graduation he joined Mason Hill at Richfield Oil, mapping extensively on both sides of the San Andreas fault. In 1952 Dibblee, Jr. joined the U.S. Geological Survey and went on to map more than a quarter of the state of California. In 1953, after many years of collaboration, Hill and Dibblee wrote their paper entitled "San Andreas, Garlock and Big Pine faults," reprinted here. Mason Hill served his entire career at Richfield Oil and at its successor, Arco. Tom Dibblee, Jr. has become a living legend, having personally mapped the geology of more of California than anyone else in history. A foundation named for him, the Dibblee Geological Foundation, was established to provide an outlet for the dozens of geologic quadrangle maps he has produced throughout his career.

Dorothy L. Stout

## SAN ANDREAS, GARLOCK, AND BIG PINE FAULTS, CALIFORNIA

A Study of the Character, History, and Tectonic Significance of their Displacements

By Mason L. Hill and T. W. Dibblee, Jr.

### Abstract

The Big Pine left lateral fault extends northeastward from Big Pine Mountain to the right lateral San Andreas fault, while the left lateral Garlock fault extends northeast from the San Andreas, but from a point 5 miles to the southeast. The Big Pine fault is considered the western segment of the Garlock fault as offset by the San Andreas. Movement on this Garlock-Big Pine fault zone appears to have caused the anomalous east-west trend of the San Andreas fault in this vicinity.

Tens of miles of lateral movement have probably occurred on these faults with the possibility of a cumulative movement on the San Andreas of hundreds of miles since Jurassic time. Such distances are important elements in reconstructing paleogeologic conditions.

The three concurrently active, long, steep, and deep faults are considered major conjugate shears which define a primary strain pattern of relative east-west extension and north-south shortening of an area of approximately 120,000 square miles. A northeast-southwest counterclockwise compressive couple, possibly set up by drag due to the deep-seated movement of rock material from the Pacific region, is tentatively postulated as causing the deformation in this large region.

### CONTENTS

#### TEXT

| | Page |
|---|---|
| Introduction | 443 |
| San Andreas fault | 444 |
| Problems | 444 |
| Description and interpretations | 445 |
| Discussion and conclusions | 449 |
| Garlock fault | 451 |
| Big Pine fault | 451 |
| Relationships of the faults | 452 |
| Mechanics of the faulting | 454 |
| Introduction | 454 |
| Strain pattern | 455 |
| Stress pattern | 456 |
| Genetic concepts and conclusions | 457 |
| References cited | 458 |

#### ILLUSTRATIONS

| Figure | | Page |
|---|---|---|
| 1. | Right lateral offset of drainage lines by the San Andreas fault | 446 |
| 2. | Offsets of Pleistocene and upper Miocene facies by the San Andreas fault | 447 |
| 3. | Possible offsets on the San Andreas fault since Jurassic time | 448 |
| 4. | Sketch section across the Bear Valley fault | 449 |
| 5. | Left lateral offset of drainage lines by the Big Pine fault | 452 |
| 6. | San Andreas and Big Pine-Garlock strain system | 456 |
| 7. | Stress systems for San Andreas and Big Pine-Garlock faults | 457 |

| Plate | | Facing page |
|---|---|---|
| 1. | Lateral fault map of California | 443 |
| 2. | San Andreas, Garlock and Big Pine faults | 450 |
| 3. | Garlock fault | 451 |
| 4. | Map of Big Pine fault and associated structures | 454 |

### Introduction

Recent work shows, contrary to previously published maps, that the Big Pine fault, on the south slope of Big Pine Mountain, extends northeastward to the San Andreas. This work was done by the junior author (Dibblee) in 1947, on photographic and topographic base maps, as part of an oil-exploration program. The characteristics of the Big Pine fault were recognized as possibly of regional tectonic significance which stimulated further studies by Dibblee and Hill (1948) upon which this presentation is founded.

The first part of the discussion presents evidence on the character and amount of the displacements of the faults with particular emphasis on interpretations of the history of the San Andreas fault; the second part deals with interpretations of relationships between these faults and their bearing on some other structural features of the region; and the third part analyzes the mechanics of the deformation and timidly suggests some genetic concepts and conclusions.

Most of the facts concerning the faults discussed are generally known, but some new data and interpretations lead to conclusions, particularly concerning the San Andreas fault, that are not now orthodox. These conclusions are mainly tentative and are offered for the criticism of other workers.

Acknowledgment is made to the Richfield Oil Corporation, for permission to publish, and to A. O. Woodford, Duncan A. McNaughton, John C. Crowell, Rollin Eckis, John Shelton, and John H. Wiese for criticism of the manuscript.

## SAN ANDREAS FAULT

### Problems

This great Right Lateral[1] fault became widely publicized after the San Francisco earthquake of 1906. At that time a maximum of 21 feet of right lateral offset occurred on the San Andreas fault (Lawson, 1908). Much geologic study has since been devoted to this fault, and it has been mapped from Point Arena southeastward almost to Mexico, a distance of about 600 miles; however, the history of its displacements is still controversial. After nearly 50 years of research, workers disagree in their answers to the following questions:

1. What is the San Andreas fault?

Noble (1933, p. 11) says:

"It is marked by a curiously straight and almost continuously traceable chain of scarps, ridges and trough-like depressions.... This line of recent topographic features upon the San Andreas fault

---

[1] The block opposite the observer, as viewed in horizontal section, is offset to the right (Hill, 1947). The terms *right* and *left lateral* are as critical in fault nomenclature as *normal* and *reverse*; they are descriptive of *separation* rather than definitive of *slip*; they are geometric rather than generic; and they are becoming widely used. Some English workers (Anderson, 1942, p. 55) use *dextral* and *sinistral transcurrent* for these types but with a *slip* instead of *separation* connotation.

---

is commonly referred to as the fault trace. Bordering the fault is a belt of roughly parallel branching and interlacing fractures which at some places attains a width of several miles. This belt, the San Andreas fault zone, is a mosaic of elongated sliver-like blocks or wedges whose longer axes trend parallel with the main fault, so that the dominant structure is a sort of slicing...."

On the other hand, Taliaferro (1943, p. 160) says:

"Nearly everywhere along the San Andreas there is abundant physiographic evidence of recent faulting, such as true sag ponds and offset ridges and drainage lines. However, along the earlier mid-Pleistocene faults (which the San Andreas partially follows) there is no similar direct physiographic evidence of faulting.... It (the San Andreas) has produced no important modifications of either the structures or topography formed by these (earlier) diastrophisms.... The supposed branches or 'barbs' are actually earlier faults which were formed by a very different type of movement and which may be traced across the San Andreas."

2. What is the age of the San Andreas fault? All workers agree, on the firm basis of earthquake activity and topographic expression, that this fault is active. However, there is a distinct difference of opinion, in respect to its antiquity. Noble (1933, p. 11) says:

"The fault is a very old line of weakness, upon which movements have recurred through Tertiary and Quaternary time and perhaps through much of pre-Tertiary time."

Taliaferro (1943, p. 161) contends that the San Andreas fault had its inception in late Pleistocene time and that it, with its unique characteristics, should not be confused with older faults which are followed, and occasionally crossed, by the San Andreas.

3. What is known about the orientation (sense) of displacements of the San Andreas fault? All authors have agreed that the latest movements have been actually strike-slip with the southwest block relatively moving to the northwest (right lateral), but here again agreement stops. Those workers who believe that the San Andreas is an old fault differ in opinion on the relative importance of vertical and lateral components of these earlier movements. Most authors, for example Buwalda (1926), Clark (1929), Weaver (1949), and Willis (1929), believe that these earlier movements were predominantly vertical and that reversals of throw have taken place on this and parallel faults.

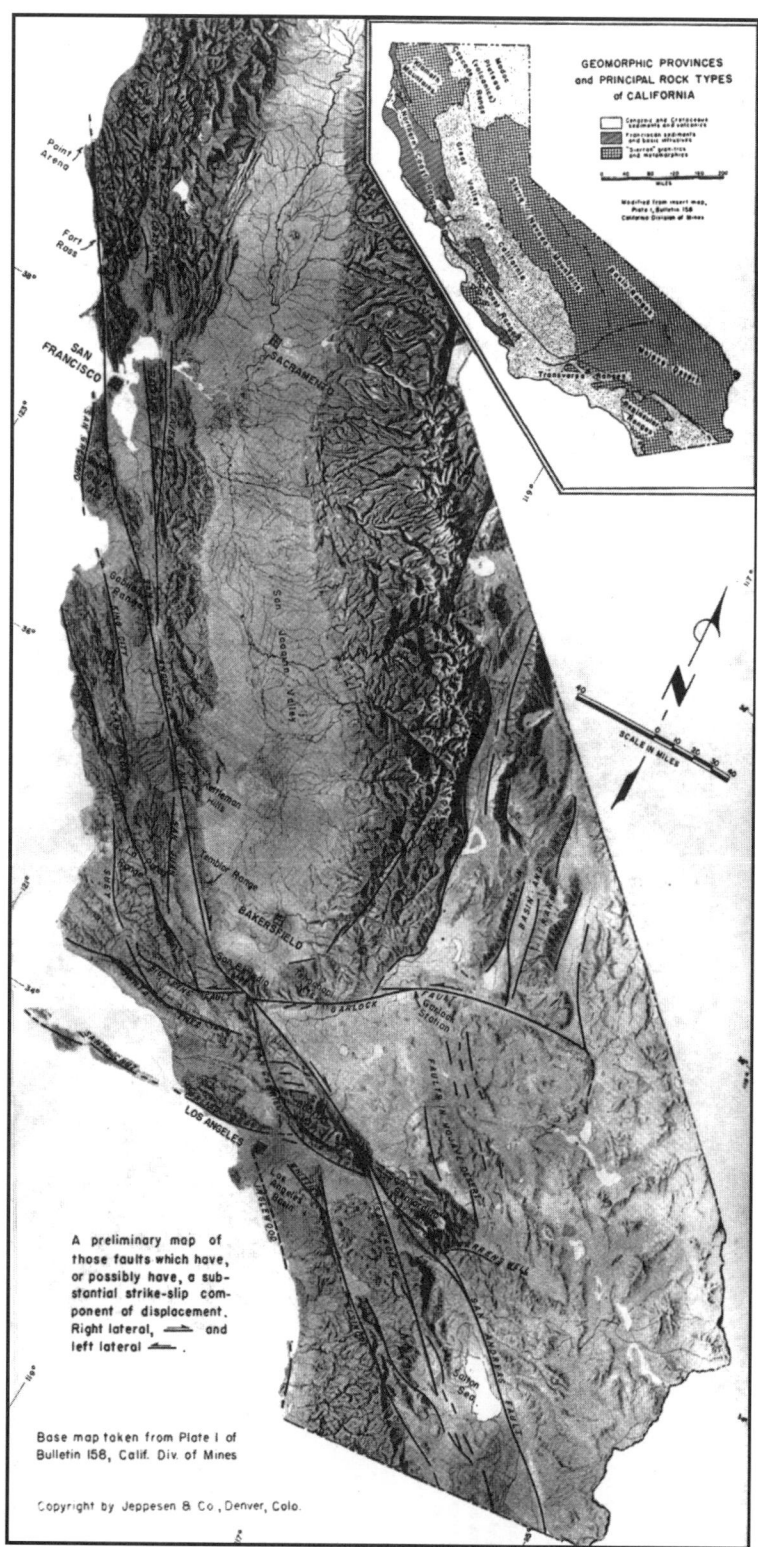

Plate 1: Lateral fault map of California

## SAN ANDREAS FAULT

4. Does the San Andreas fault zone mark a prominent contact between unlike rocks? Noble (1933, p. 11–12) says:

"The profound difference in the rocks on opposite sides of the San Andreas fault shows that the fault movements have been of great magnitude.... Nowhere in the 30 mile sector (north side of San Gabriel Mountains) are the rocks on opposite sides of the fault similar."

Taliaferro (1943, p. 160) says:

"The San Andreas fault, however, is rarely the boundary between these two very diverse types, but usually is either wholly within crystalline (granitic) rocks or Mesozoic (Franciscan) rocks, except where it cuts through either Miocene, Pliocene or Pleistocene sediments."

5. What is the order of magnitude of the cumulative offset on the San Andreas fault? Noble (1925–1926, p. 420) says early Tertiary strata may have been offset by the San Andreas (in a right lateral sense), a total distance of approximately 24 miles.

Taliaferro (1943, p. 161) says:

"....The horizontal shift (on the San Andreas) has been small, and has not been greater than one mile and probably even less."

The present authors believe that: (1) the San Andreas is a steep fault zone of variable width consisting of one or several nearly parallel faults; (2) its inception was quite likely pre-Tertiary, and it is now active; (3) it has probably been characterized by right lateral displacements throughout its history; (4) it marks such an important contact that rarely can it be crossed, except in Recent alluvium, without passing into significantly different rocks; and (5) its cumulative displacement of some rock units is at least tens of miles, and older rocks may have been displaced a few hundred miles.

### Description and Interpretations

The San Andreas fault is so long that no one worker can have a first-hand detailed knowledge of its characteristics throughout its length. The junior author, however, has mapped a 100-mile stretch of the fault and associated geology in the Salton Sea region and most of the 300-mile stretch from the San Emigdio Mountains to San Francisco. This work and available maps and discussions by others are used in the following outline of significant fault characteristics and their bearing on the history of the San Andreas fault.

1. The trace of the San Andreas zone is typically continuous and straight (Pl. 1). Topographic and geologic features clearly show its continuity for about 600 miles. There is evidence of recent activity along its entire course. Excepting a 30-mile segment trending eastward in the San Emigdio Mountains, and another stretch of similar trend 100 miles to the southeast, the zone is remarkably straight from Point Arena southeastward nearly to Mexico. These aspects of continuity and straightness are considered typical of strike-slip faults.

2. The San Andreas is a steep fault which transects major topographic features but develops all along its course one or several parallel trenches, sag ponds, low ridges, saddles, and/or scarps. Its steepness is indicated by the straight trace, the fact that mapped fault planes are nearly vertical, and the failure of near-by drill holes to penetrate the zone. These characteristics are typical of strike-slip faults. The development of fresh topographic features, many of which are in unconsolidated recent sediments, and the common lack of appreciable vertical or consistent vertical components of offset clearly indicate the recency of lateral movements. Seismic evidence for recent right lateral movements on the San Andreas, as summarized by Wallace (1949), comprises the following maximum displacements at the time of earthquakes: 30 feet (San Emigdio Mountains, 1857), 10 feet (San Francisco area, 1868), 21 feet (San Francisco area, 1906), and 10 feet (Salton Sea area, 1940). Geodetic surveys in the San Francisco (Whitten, 1948) and Salton Sea (Meade, 1948) areas also show this type of recent movement.

3. The San Andreas fault zone ranges from a few feet to a few miles in width. Locally a single recent trace may be irregular, with 15-degree variations in strike within a few hundred feet, or it may disappear and be replaced, en echelon, by another. Occasionally two or three parallel traces widen the zone of recent traces to a maximum of about half a mile. Wider segments of the zone consist of several faults (not necessarily active) which are usually steep and nearly parallel to the trend of the zone. These characteristics are considered typical of strike-slip fault zones along which recurring movements have taken place.

4. The apparent throw is commonly reversed along the San Andreas fault as indicated by

FIGURE 1.—RIGHT LATERAL OFFSET OF DRAINAGE LINES BY THE SAN ANDREAS FAULT Elkhorn Hills quadrangle

topographic and geologic relationships. These throws are probably due to the major strike-slip component which places in juxtaposition unlike topographic elevations and geologic sections, and thus the reversals of dip-slip are mainly illusory.

5. Drainage lines are consistently offset in a right lateral sense. These offsets are especially clear on the southwest side of the Temblor Range (Fig. 1) where a maximum of 3000 feet of displacement has occurred through recent movements on the fault. Wallace (1949, p. 805) reports a probable drainage offset of 1½ miles on the north side of the San Gabriel Mountains, and Allen (1946, p. 50) reports 3800-foot offsets of drainage lines near the Gabilan Range, also in a right lateral sense.

6. Recently developed trenches which trend southward into the fault have been observed in aerial reconnaissance on the southwest side of the Temblor Range. These are oriented correctly to be tensional in origin and due to right lateral movement on the San Andreas.

7. Locally developed west-northwest trending folds adjacent to the San Andreas are obviously drag folds resulting from the right lateral movement on the San Andreas. Such drag folds are especially clear in the Salton Sea Region, and, besides indicating the right lateral sense of movement on the fault, many of them show by their discordance with topographic form that the fault was active before the present physiographic features were developed.

8. Wallace (1949) reports a probable 6-mile right lateral offset of terrace deposits on the north side of the San Gabriel Mountains, and L. F. Noble (personal communication) describes similar late offsets in that area of several miles.

9. Between the San Emigdio Mountains and the Temblor Range, there are two facies of Pleistocene gravels. On the southwest side of the San Andreas, the pebbles are granite, gneiss, quartzite, limestone, black shale, and sandstone which undoubtedly came from the San Emigdio Mountains. On the other side of the fault, the pebbles are almost exclusively white siliceous shale which probably came from the Miocene shale of the Temblor Range. These two facies are in direct contact along the San Andreas for several miles. Furthermore, the northwest end of the crystalline clast facies is about 14 miles northwest of the crystalline rocks of the San Emigdio Mountains. These relationships thus indicate a right lateral displacement of approximately 10 miles on the San Andreas fault since Pleistocene deposition in this area (Fig. 2).

10. In the Caliente Range, marine sediments of upper and middle Miocene age grade laterally eastward into continental red beds which strike into the San Andreas fault, whereas strata of the same age are marine shales on the other side of the fault. This juxtaposition of unlike facies again demonstrates substantial lateral movement. In this case the general trend of the western margin of the continental facies in the Caliente Range is northward across the Carrizo Plain toward the San Andreas, whereas possibly the same transition line may be extrapolated southward from along the east side of the San Joaquin Valley to the fault. Thus, by the simple projections the right lateral offset on the

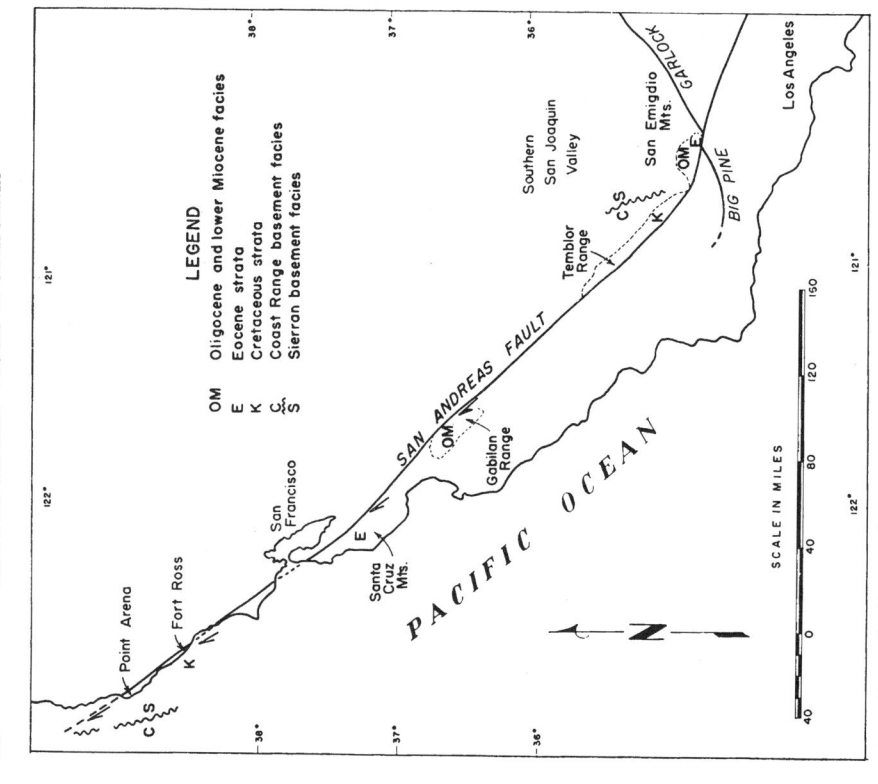

FIGURE 3.—POSSIBLE OFFSETS ON THE SAN ANDREAS FAULT SINCE JURASSIC TIME

These time-rock facies appear to be offset from the southern San Joaquin Valley area to the positions indicated on the west side of the fault. Thus, for example, during Cretaceous time the Fort Ross section may have been adjacent to the Temblor Range section.

fault since upper Miocene time would be about 65 miles, although the probability of irregularities in trend of this facies contact precludes a strictly quantitative solution of that cumulative shift (Fig. 2). Note the comparable offset of the upper Miocene "Pancho Rico," "Santa Margarita" shale, shown in the same figure.

11. Going back only slightly farther in the geologic record, approximately 175 miles of right lateral offset may have accumulated on the San Andreas fault since early Miocene time. This is suggested by the unique similarities of rock types and sequences in the San Emigdio Mountains, as described by Wagner and Schil-

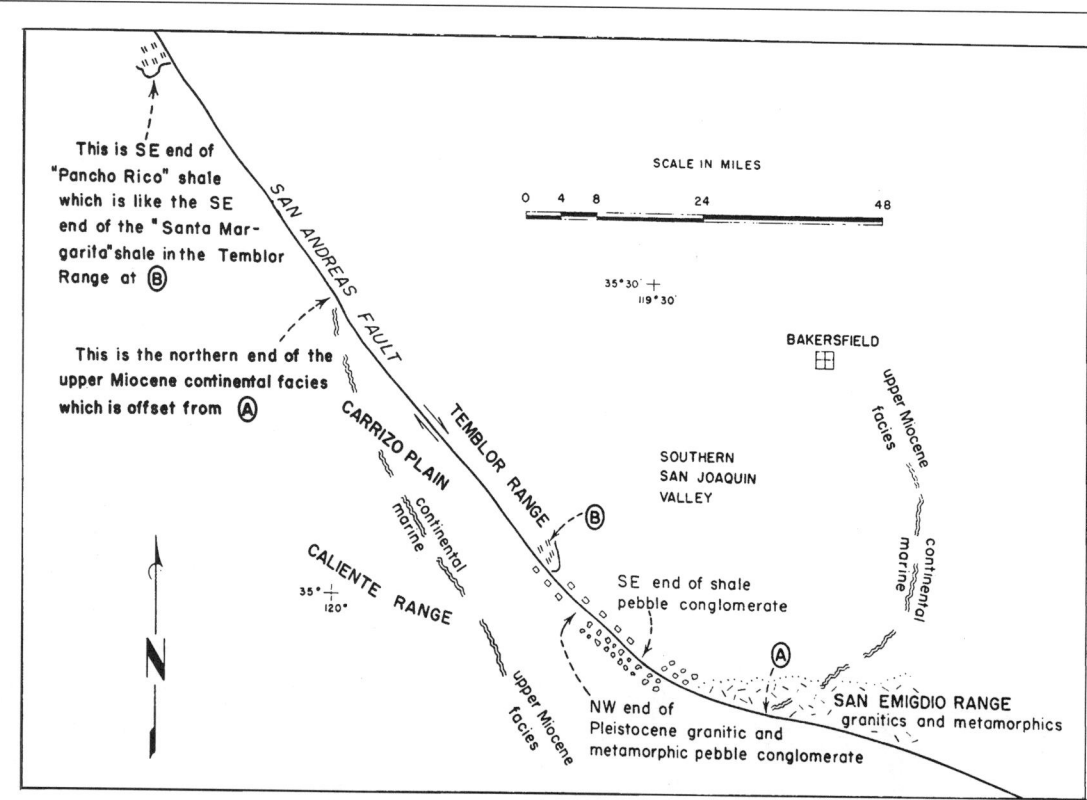

FIGURE 2.—OFFSETS OF PLEISTOCENE AND UPPER MIOCENE FACIES BY THE SAN ANDREAS FAULT

ling (1923), and the Gabilan Range as described by Kerr and Schenck (1925), and Allen (1946). In each of these areas, a section of lower Miocene volcanics, red beds, and marine lower Miocene and Oligocene strata occurs (Fig. 3). A similar relationship is suggested by some lithologic and faunal similarities between the Eocene formations of the Temblor-San Emigdio and the Santa Cruz Mountains which indicate the possibility of an offset of approximately 225 miles since late Eocene time (Fig. 3).

13. Also the southern limit of Cretaceous strata in the Temblor Range may match with the southern limit of Cretaceous beds near Fort Ross which would indicate an offset of approximately 320 miles (Fig. 3). This possibility is strengthened by the improbability that vertical movements were consistent enough to remove the Cretaceous strata from such a long belt on the southwest side of the San Andreas fault zone. Furthermore, the Cretaceous sediments which do occur as close as 12 miles southwest of the San Andreas in the La Panza Range (Pl. 1) are west of other important faults and rest on granitic instead of Franciscan basement.

14. Of great interest would be good evidence for offset of the pre-Cretaceous rocks by the San Andreas. Perhaps the southward-trending contact between the Sierran complex and the Coast Range Franciscan complex which is concealed by the sediments of the Great Valley is truncated by the fault in the southern Temblor Range with its offset equivalent concealed by the sea north of Point Arena. This would amount to at least 350 miles of right lateral displacement since Jurassic time (Fig. 3).

### Discussion and Conclusions

The antiquity and amount of movement on the San Andreas fault is evidenced by the increased offset of older rock facies as matched across the fault zone. The possibility of a pre-Cretaceous age and a cumulative right lateral displacement of several hundred miles is therefore suggested. These conclusions are both speculative and qualitative but tenable enough to justify further critical work to test their correctness and to determine more exactly their values. There is, however, no reasonable doubt that the fault is at least as old as early Pliocene and that the total right lateral offset is at least tens of miles (Figs. 2, 3).

Of perhaps great significance is the increased offset of older rock facies on the west side of the San Andreas from their probable counterparts at the southern end of the San Joaquin

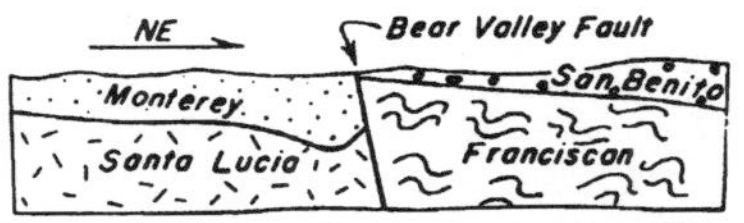

FIGURE 4.—SKETCH SECTION ACROSS THE BEAR VALLEY FAULT
After Wilson (1943, p. 227)

Valley (Fig. 3). This situation appears to indicate that one block was actually moving past the other. If this were not true the counterparts would also be progressively shifted in the opposite direction. A conclusion regarding this movement will be presented under the subject heading of "Mechanics of the faulting."

Those workers who maintain that the San Andreas refers only to a fault which has been active recently and which had its inception in late Pleistocene time and those others who believe in earlier dip-slip and reversals of dip-slip displacement on the San Andreas may be skeptical of these conclusions. However, these differences in opinion appear to result from definitions of the fault and interpretations of displacement from cross-section relationships. The present authors believe that: (1) in many places the San Andreas is a wide zone comprising several important faults which are parallel or nearly parallel to the recent trace, and (2) the apparent dip-slip displacements are mainly the result of substantial lateral movements which put unlike geologic sections in juxtaposition.

The Bear Valley fault, east of the Gabilan Range (Wilson, 1943, p. 251), illustrates these differences in opinion and interpretation. This fault is mapped about half a mile southwest of the recent trace of the San Andreas and dips steeply northeast. It separates middle Miocene (Monterey) sediments resting on the Santa Lucia granite on the southwest side from a section of Plio-Pleistocene (San Benito) gravels resting on the Franciscan complex (Fig. 4). Wilson's interpretation is that this "ancestral" San Andreas fault is not the San Andreas because it shows no evidence of recent movement, it is not vertical, and it is supposed to be characterized by dip-slip rather than strike-slip displacement. According to Wilson the first movement, probably Eocene normal faulting, elevated the west side causing the Franciscan to be removed by erosion. The second period of movement was post-Miocene, and, by reverse faulting, the east side was elevated with accompanying erosion of all the Miocene strata on that block. The third period of movement occurred in late Pleistocene or Recent times as normal faulting which again elevated the west block causing the removal of the San Benito gravels.

The present authors suggest, based on the same data and by pinning a little faith on the law of uniformitarianism, the following tentative interpretations:

The juxtaposition of granitic and Franciscan basement may be the result of a great many miles of cumulative lateral movement on the fault. The absence of Miocene sediments on the northeast side of the fault may be due to many miles of right lateral movement so that the facies of the southwest block in this area were offset far to the southeast. The absence of the Plio-Pleistocene gravels on the southwest side of the fault in this area may also be due to right lateral offset with the possibility that they could be found several miles to the northwest on that side of the San Andreas. Although no recent movement has occurred on the Bear Valley fault, it is probably an inactive and slightly deformed fracture within the San Andreas zone which was characterized by the same type of movement as is evidenced along the recent trace of that fault.

It appears then that the significance of the Bear Valley fault is debatable. It is discussed here only as an example. An alternate interpretation is suggested in order to focus attention on a possibility that seems to have been overlooked. If the Bear Valley fault or other structures similarly related to the San Andreas, such as the Pilarcitos fault in the San Francisco area, can be shown to effect substantial right lateral offset, it will be reasonable to include them in the San Andreas fault zone as a partial manifestation of the important history of that zone. Furthermore, it would eliminate the not too probable interpretations of reversals of dip-slip displacements from cross-section relationships and the postulation of a unique and new strain-stress environment to explain the recent trace of the San Andreas fault.

The type of evidence used in this paper for the cumulative offset on lateral faults is not wholly new, nor are the suggested offset distances unprecedented. Kennedy (1946) has demonstrated 65 miles of strike-slip displacement on the Great Glen fault in Scotland, and the Alpine fault in New Zealand may have 300 miles of lateral displacement according to Wellman (1949 and personal communication). Even in California Vickery (1925) determined that the relatively minor Sunol (Calaveras) fault has had 13 miles of right lateral displacement since early Miocene time by what appears to be a unique match of strata across this fault. Furthermore, the cumulative amount of lateral movements, which are tangential to the earth's surface, need not be as limited as for high-angle dip-slip faults which are necessarily limited by gravitational forces.

Much more work is obviously needed to verify qualitatively some of these tentative conclusions. Many more possible offsets must be analyzed for accuracy and compatibility with respect to geologic time and geographic distribution before sound quantitative conclusions can be reached. In general, the younger rocks should be more easily matched across the fault, but the offsets of older units, particularly the pre-Cretaceous crystalline rocks, will be more significant. For example, is the north end of the Pelona schist on the north side of the San Gabriel Mountains (Wallace, 1949) offset 160 miles from the north end of the Orocopia schist near the Salton Sea (Miller, 1944)? Then, is such an offset commensurate with the pos-

PLATE 2. SAN ANDREAS, GARLOCK, AND BIG PINE FAULTS
FIGURE 1. TERMINATION OF BIG PINE FAULT AT THE SAN ANDREAS
View north

FIGURE 2. TRENCHES ADJACENT TO GARLOCK FAULT

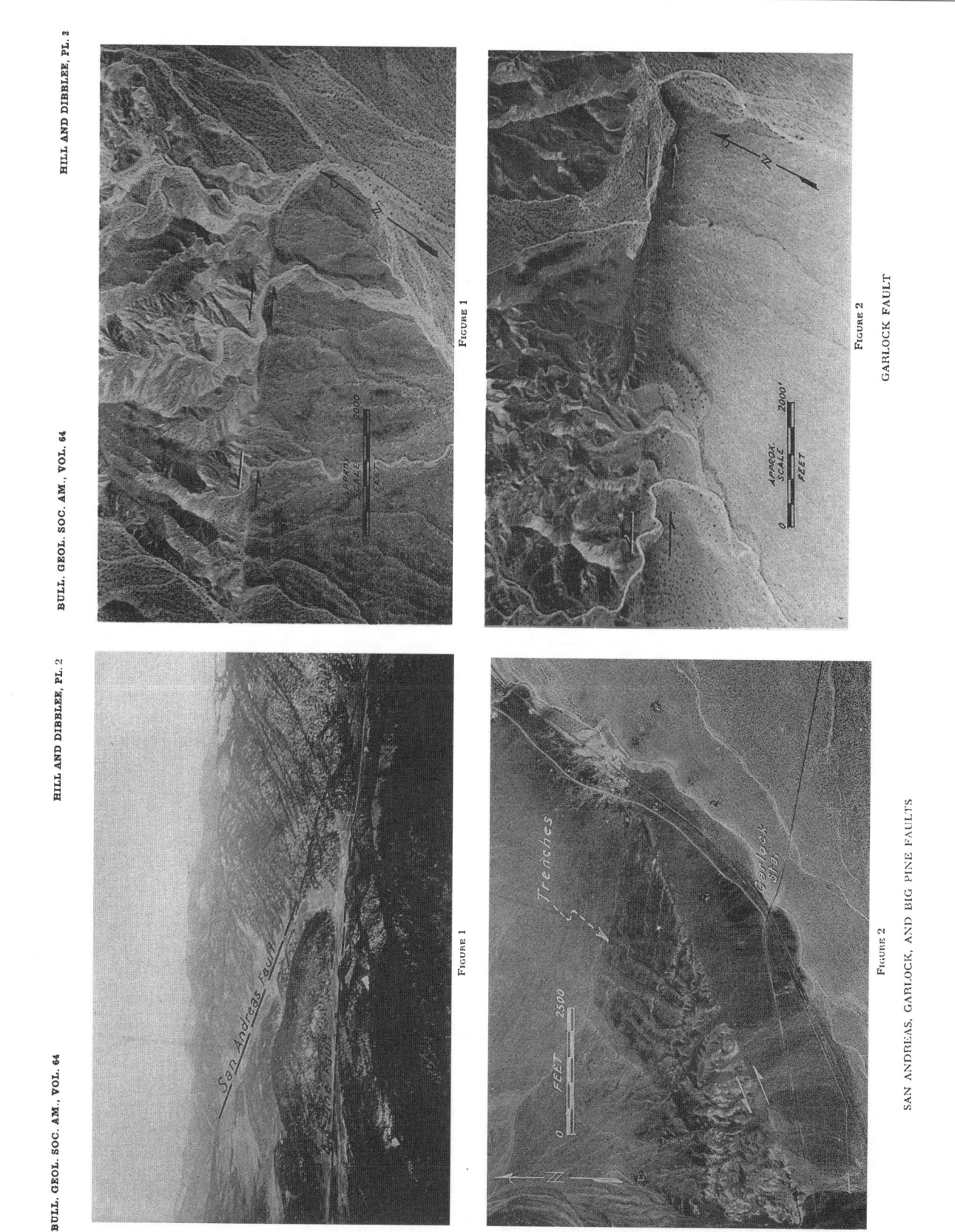

## SAN ANDREAS FAULT

sible offset on the San Andreas north of the Garlock-Big Pine fault zone?

Another worth-while line of research is the study of rates of movement on the San Andreas fault. Geodetic work has demonstrated a right lateral movement in the San Francisco Bay region averaging 2 inches a year since the survey of 1880 (Whitten, 1948). Another method, based on offsets at the time of earthquakes, suggests an average movement along the entire length of the fault of about 0.2 inch a year since 1857 (Wallace, 1949). Now, from the probable and possible offsets described above and the available calculations of geologic time in years, the rates of movement come out in fractions of an inch per year, as follows: 0.25 since Miocene, 0.2 since lower Miocene, 0.3 since Eocene, 0.29 since Cretaceous, and 0.2+ since Jurassic time. These figures are surprisingly consistent and therefore may possibly be taken as evidence that the rate of movement has been rather uniform for some 100 million years. On the other hand, the geodetically determined rate is nearly tenfold greater than these geologically controlled figures. However this great rate might be anomalous because of the short time involved, the short length of the fault involved, the inclusion of the large displacement of 1906, and/or movements outside the fault zone.

## GARLOCK FAULT

The Garlock fault extends from the San Andreas northeast and east-northeast for 150 miles (Pl. 1). It, like the San Andreas fault, transects topographic features, dips steeply, and has apparent variations of throw along its course.

Hess (1910, p. 25) first described and named the fault and indicated its type locality as just north of Garlock Station. Hulin (1925) was the first to suggest lateral displacement. He used the offset of a contact as possible evidence for left lateral displacement of approximately 6 miles.

PLATE 3. GARLOCK FAULT
FIGURE 1. LEFT LATERAL OFFSET OF DRAINAGE LINES Seven miles northeast of Garlock Station.
FIGURE 2. LEFT LATERAL OFFSET OF DRAINAGE LINES Ten miles northeast of Garlock Station.

The present authors followed this fault zone from its junction with the San Andreas northeastward for 150 miles by airplane and saw many well-defined offsets of drainage lines. These are all in a left lateral sense and commonly a distance of 2000 feet (Pl. 3). Also seen on this air reconnaissance were trenches in old alluvium in the vicinity of Garlock Station, immediately north of the fault trace (Pl. 2, fig. 2). These parallel trenches, approximately 40 feet deep, 150 feet wide, and 2000 feet long, trend S. 30°W. into the N. 60°E. trending Garlock fault trace. Their characteristics, including orientation and proximity to the Garlock fault, suggest gash fractures produced by tensional stresses adjacent to a major shear zone.

These striking evidences of recent and actual left lateral movement have, surprisingly, not hitherto been emphasized in the geologic literature. However, Wiese and Fine (1950, p. 1652) report at least 2 miles of left lateral offset on the Pinon Hill fault (a subparallel branch of the Garlock on the south side of the Tehachapi Mountains), and D. F. Hewett (1950) describes new evidence for recent offset on the Garlock fault zone east of Garlock Station. Thus the sense and recency of movement on this great fault are considered firmly established. The time of initiation and the total amount of left lateral offset on the Garlock fault are not yet known.

## BIG PINE FAULT

Nelson (1925, p. 379-380) first mapped and described the Big Pine fault as it occurs on the south side of Big Pine Mountain. He sketchily mapped the fault from here some 15 miles eastward. Later, Reed and Hollister (1936, p. 94-96) extended the fault to the southeast, as it is likewise shown on the Geologic map of California (Jenkins, 1938). The present mapping shows that this southeastward extension is in error, but that the fault does extend in a clearly revealed trace northeastward to the San Andreas (Pl. 2, fig. 1). Now it is evident that Nelson's east-west portion of the fault and this

---

## HILL AND DIBBLEE—FAULTS IN CALIFORNIA

newly mapped and equally long northeast-trending portion gives the Big Pine fault a length of 50 miles and establishes it as a major fault of the region.

The extended portion of the fault trends N. 65°E., with steep dips varying from north-

FIGURE 5.—LEFT LATERAL OFFSET OF DRAINAGE LINES BY THE BIG PINE FAULT
Morro Hill and Reyes Peak quadrangles

west to southeast and with the throw apparently reversing along its course. At its type locality,[2] 9 miles southwest of Mt. Pinos, the fault separates Miocene continental beds from Eocene marine strata to the south (Pl. 4). Here the fault strikes N. 60°E. and dips 70°S. Striations have been found on slickensided planes of the fault zone which are 70° clockwise from the direction of dip.

This left lateral fault has a prominent strike-slip component of displacement as indicated by the following field observations:

1. The oblique-slip striations in the fault zone indicate a predominant strike-slip component of displacement and their orientation combined with the orientation of apparent throw indicates that the southeast side has moved northeastward and up the 70-degree fault plane. On this evidence alone the fault should be classified as a "left lateral reverse," but, since reversals of dip and apparent reversals of throw occur along the fault trace, the appearance of reverse faulting at the type locality is not significant.

2. Left lateral offsets of drainage lines along the Big Pine fault indicate the sense and re-

[2] A type locality is picked as recommended by Buddenhagen et al. (1930).

cency of displacements. Figure 5 shows three such offsets, two of which are displaced approximately 3000 feet.

3. East-west trending drag folds adjacent to the Big Pine fault indicate left lateral displacement (Pl. 4).

4. Left lateral offset of approximately 8 miles since Miocene time is suggested by the possible displacement of a major syncline and a northwest-trending fault contact between Eocene and Miocene strata (Pl. 4).

5. The continental facies of the Miocene, together with its fanglomerate members extends farther west on the north side of the Big Pine fault than on the south side (Pl. 4).

6. A well 14 miles southwest of Mt. Pinos and just north of the fault (Pl. 4) encountered continental Miocene conglomerate on granitic basement which matches a similar sequence on the south side of the fault 14 miles to the east.

The time of initiation of the Big Pine fault is unknown, but its age must be substantial, considering the probable minimum cumulative displacement of approximately 8 miles.

## RELATIONSHIPS OF THE FAULTS

The more important aspects of the San Andreas, Garlock, and Big Pine faults, for the present analysis, are as follows:

1. The faults are essentially contemporaneous, as attested especially by the recency of their displacements.

2. The faults are major structures of the re-

gion, as attested by their great lengths and probable great cumulative displacements.

Of the three fault zones, the San Andreas is dominant, and its trend predominates in the regional faulting. However, the Big Pine and subparallel Santa Ynez faults separate the northwest-trending Coast Ranges from the east-west trending Transverse Ranges, and the Garlock fault separates the Basin and Range region from the Mojave Desert province.

4. The Big Pine and Garlock faults are on the same trend and are alike in the character of their displacements, as attested by the evidence for left lateral movement on each.

5. The San Andreas right lateral fault zone intersects, at a substantial angle, the Garlock-Big Pine left lateral fault zone.

Based on these data, the more important relationships between the faults appear to be as follows:

1. The Big Pine fault is probably the offset extension of the Garlock fault since its essential characteristics are the same as the Garlock, since it is the only major left lateral fault in the area, and since it abuts the San Andreas at a point northwest of the west end of the Garlock fault, as is consistent with the right lateral movement on the San Andreas. Nolan (1943) expressed the opinion that the San Gabriel fault should be found about where the Big Pine fault is now mapped but to his knowledge it did not occur. He concluded that, if the Garlock fault did exist west of the San Andreas, it might be offset to the southeast and if so would indicate an anomalous (left lateral) movement on the San Andreas. Finding what was reasonably expected has fortunately then eliminated the possibility of an unlikely reversal of movement on the San Andreas. Furthermore, this shift of 5 miles is in accord with the 5- or 6-mile offset of terrace deposits by the fault 50 miles to the southeast (Wallace, 1949).

2. The big bend in the trace of the San Andreas, that east-west trending portion (Pl. 1), occurs where abutted by the Garlock and Big Pine faults. This anomalous trend is now readily explained by the left lateral movement on the Garlock-Big Pine fault zone. Even the total amount of roughly 25 miles of offset due to the bend is not, in the opinion of the writers, out of line with the possible amount of left lateral offset on the Garlock-Big Pine zone.

3. Another big bend of the San Andreas occurs in the San Bernardino Mountains, approximately 100 miles southeast of the bend described above (Pl. 1). Here possible left lateral movement on the "Warrens Well" fault may be responsible for this other anomalous trend of the San Andreas fault. Critical work on the "Warrens Well" fault would quite likely establish the merit of this inference.

4. Another speculation is that the San Gabriel fault (Pl. 1) is an ancestral portion of the San Andreas. This fault, which occurs south of the Big Pine, extends southeastward on trend with that portion of the San Andreas which lies north of the Big Pine fault. Thus the San Gabriel fault may have been the southeastern continuation of the San Andreas before the big bend was developed. The fact that, according to Crowell (1950), the San Gabriel fault shows evidence of less recent activity than the present San Andreas is a reasonable consequence of such a possibility. Along this line of thinking, the San Jacinto fault (Pl. 1) could be conceived of as also being an earlier representative of the San Jacinto[3] fault would be younger than the San Gabriel fault and perhaps older (although still active) than the present trace of the San Andreas. Likewise the Calaveras and Hayward faults might perhaps be the earlier initiated manifestations of the San Andreas fault zone in the San Francisco Bay region (Pl. 1). It is further suggested that the cumulative right lateral displacement on this greater San Andreas fault system could be of a magnitude of several hundreds of miles. This concept is strengthened by Crowell's report (1952) of 15 to 25 miles of right lateral movement on the San Gabriel fault alone, and during a short interval of geologic time (between late Miocene and late Pliocene). Such a possibility is obviously shocking, particularly when considering its effect on paleogeographic maps which straddle this zone.

Substantiation of the interpretation that the Big Pine and Garlock faults comprise one fault zone is not necessary to the probability that the bend of the San Andreas is related to their left lateral movements or vital to the following

---

[3] Five miles of right lateral movement is indicated on the San Jacinto fault (Arnett, 1949), and the localization of Jurassic (?) alaskite dikes in the fault zone indicates its early origin (Sprotte, 1949).

---

analysis of the mechanics of the regional deformation. It is, however, essential that the Big Pine and Garlock once comprised one continuous fault if used as evidence of 5 miles of right lateral offset on the San Andreas. However, since the sense and magnitude of the displacement on the San Andreas from other evidence is compatible with this offset of the Garlock-Big Pine trend, the probability is that a continuous line of faulting actually did at one time exist. Also, the interpretations concerning the other big bend of the San Andreas or the possible historical connections between either the San Gabriel or San Jacinto faults and the San Andreas are not essential to the primary theses of this presentation.

## Mechanics of the Faulting

### Introduction

The geomorphology of California ranges from high precipitous mountains to low plains (Pl. 1). The larger so-called valleys are usually structural depressions (basins), and many are separated from the mountains by important faults. The geomorphic provinces are also usually geologic provinces, with the main mountain ranges characterized by highly deformed older sediments, metamorphic and plutonic rocks, and the basins characterized by thick, younger and less deformed sedimentary sections (Pl. 1, insert map). Although there are many exceptions, owing in part to the intense and wide-spread late Pleistocene tectonic activity, some of these geomorphic-geologic provinces have persisted through at least most of the Tertiary period, and therefore some of the bounding faults are probably of considerable age.

The principal rock types of California, classified according to characteristic response to deformational stresses, are as follows:

1. Sierran basement complex (pre-Cretaceous): metasedimentary and metavolcanic rocks, intensely deformed and widely invaded by granitic rocks. Because of physical similarity, the Santa Lucia granites and metamorphics of the southern Coast Ranges and the complexes of the Transverse and Peninsular ranges belong in this group. These are relatively rigid rocks which fail locally by fracturing and, since they or rocks like them are extensively exposed and are presumably of state-wide occurrence at depth, their mechanical behavior is tectonically important.

2. Franciscan basement (pre-Cretaceous): sedimentary and volcanic rocks, regionally unmetamorphosed but highly indurated, commonly intruded by basic igneous rocks which are usually altered to serpentine and have caused local metamorphism. These rocks are exposed in large areas in the Coast Ranges; on the northeast side of the San Andreas fault, and also on the west side of the Nacimiento fault zone. They presumably underlie a much greater area but are probably in turn underlain by granitic rocks. The Franciscan, unlike the granitic basement, is typically incompetent. Although in places intensely fractured, often before being covered by later Jurassic or Cretaceous strata, and usually in fault contact with the other principal rock types, its response to deformational forces has been characterized by folding.

3. Cretaceous and Cenozoic sedimentary and volcanic formations: mainly marine clastic sediments with local volcanics and nonmarine deposits, not strongly lithified and of extremely variable thicknesses and facies. Deposited in large and small basins; locally highly deformed, especially during the late Pliocene-Pleistocene revolution in the Coast and Transverse Ranges, and in uplifts in the Mojave Desert and Salton Sea regions. These rocks form a pliable mantle on the above-described complexes and have therefore responded to tectonic forces primarily by folding, particularly where the sedimentary section is thick or where underlain by Franciscan basement.

The San Andreas, Garlock-Big Pine, and other similar faults are obviously at odds with the geomorphic-geologic provinces and the structural characteristics of the rock types of the region. The unique nature of these faults is shown by their transection of the geomorphic-geologic provinces and by their transection of all the principal rock types, without significant variations.

There are many important northwest-trending steep faults in California which are approximately parallel to the San Andreas, and which are probably also characterized by major right lateral components of displacement (Pl. 1). Several of these are present in the northern

## MECHANICS OF THE FAULTING

Coast Ranges, but they have not been described in detail. In the southern Coast Ranges, they include the Calaveras, Hayward, Pilarcitos, San Gregorio, San Marcos, Reliz Canyon, San Juan, and Suey faults, and the long Nacimiento fault zone; in the Transverse Ranges, the San Gabriel and San Jacinto faults; in the Los Angeles Basin, the Whittier and Inglewood faults; and, in the Peninsular Ranges, the San Jacinto and Elsinore faults. The several northwest-trending faults in the Mojave Desert and in the Basin and Range province, some of which are known to affect right lateral offsets, probably also belong in this group. The faults of this set are much more common than the east-northeast set of lateral faults.

The important east-northeast trending lateral faults which, like the Garlock-Big Pine zone, probably have significant left lateral displacements, include the Santa Ynez, White Wolf, possibly the Malibu Beach, faults in the northwestern portion of the San Gabriel Mountains, the "Warrens Well," and faults on Santa Rosa and Santa Cruz Islands. The faults of this set are most common in the Transverse Ranges. Other faults of both sets are quite likely present on the continental shelf.

Confusion should not exist, although it surely does, between these sets of steep lateral faults and those many other faults, both large and small, which indicate vertical relief or local adjustments due to secondary or superficially acting forces.

Large west-northwest trending thrust and reverse faults are especially common in the Transverse Ranges but they can, perhaps usually, be differentiated from the east-northeastward trending Garlock-Big Pine set by degree of dip and orientation of trend. Further confusion will undoubtedly likewise continue to exist between the thrust and reverse faults of the Coast Ranges and the San Andreas type faults. This is particularly true because superficially and locally acting forces are certain to modify the dips and strikes of the steep northwest-trending right lateral faults.

Strong folding of Cretaceous and Cenozoic strata is very common in both the Coast and Transverse ranges. In the Coast Ranges, the main folds and associated faults typically trend a little westward from the northwest San Andreas trend. In the Transverse Ranges, the main fold axes typically trend a little north of west in contrast to the east-northeast trend of the Garlock-Big Pine set and again the major thrust and reverse faults are approximately parallel to the folds.

It is here hypothesized that the folds and accompanying thrust and reverse faults of the Coast and Transverse ranges are shallow structures compared to the steep, and surely deep, lateral fault system. This concept furthermore suggests that these folds and faults which affect upward relief are secondarily related to the deep-seated forces which presumably developed the lateral fault system of nearly horizontal relief. Other minor faults and folds of diverse trends are even less directly related to the primary regional forces. As an example, the Kettleman Hills anticline of the southern Coast Ranges is a minor structure compared to the San Andreas fault zone, but the faults on it are still more local and superficial and thus more remotely related to the regional tectonic forces.

To ascertain the orientation of deformational forces which develop even minor structures is fraught with difficulties, but in such analyses, based on good mapping and sound principles of mechanics, lies the possibility of taking the Coast Ranges of California out of the defeatist category of a "heterogeneous mobile belt".

### Strain Pattern

Since the Garlock-Big Pine and San Andreas fault zones are the major fracture features of the region, and since they are concurrently active, they are probably genetically related. Then since the displacements of these intersecting fault zones show a systematic relative east-west extension and north-south shortening, their common origin seems to be an inevitable conclusion. Therefore, these fault zones are considered to comprise a "strain pattern" or system which is the result of a particular set of deforming forces.

This strain pattern is considered to be the primary fracture system of the entire region because these are the long faults (measured in hundreds of miles); they are deep faults (at least 10 miles as indicated by the usual calculated depth of foci of earthquakes in this region); and they are probably the faults of

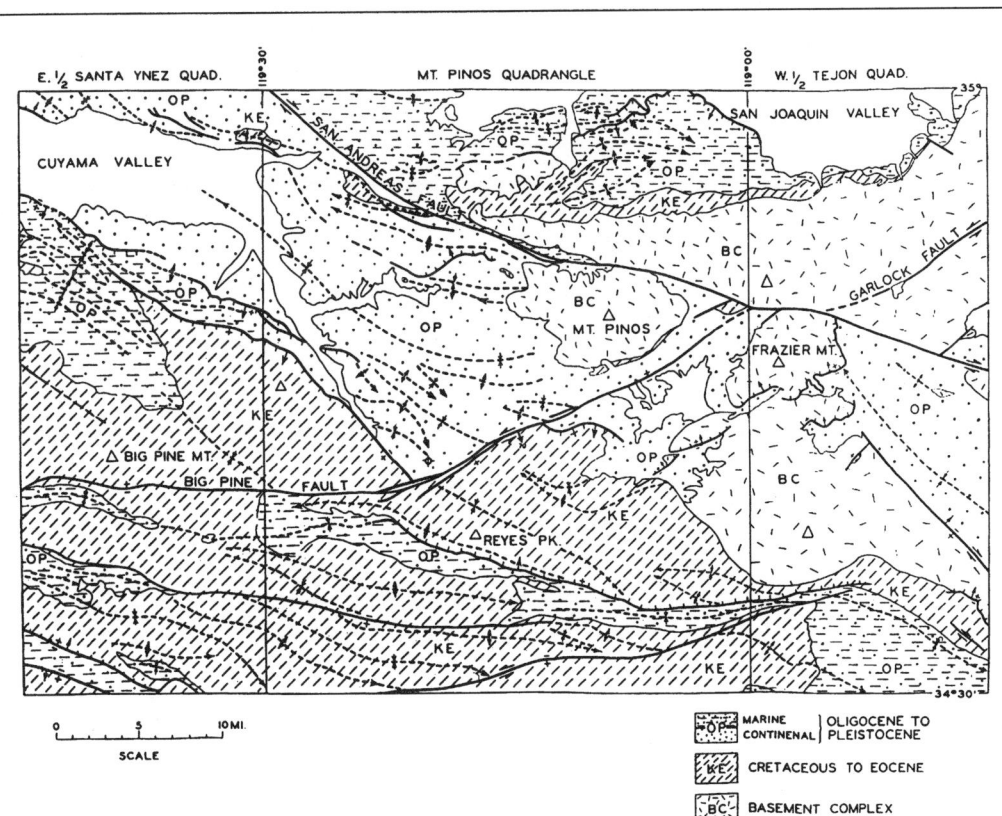

maximum cumulative displacement. Other large faults such as the San Gabriel, Nacimiento, and Santa Ynez (Pl. 1) may be nearly as directly related to the primary regional stresses, but the multitude of smaller faults and folds are undoubtedly in secondary or tertiary order strain systems which are only indirectly related to the primary stress pattern.

The diagram (Fig. 6) of the northwest-trending San Andreas right lateral shear zone and the northeast-trending Garlock-Big Pine left lateral shear zone shows the $CC_1$ axis of shortening and $AA_1$ axis of relative extension. This strain pattern is geotectonically important because its size of approximately 600 x 200 miles makes it a respectable sample of the earth's crust. Even this area of systematic deformation may be, however, only a portion of a much larger unit or even of secondary or tertiary order in relation to the primary strain systems of the crust. The amount of deformation since the inception of this shear system is probably also geotectonically important, but no conclusive evidence is now known by which to date the first displacements or determine the total cumulative movement on either of the fault zones.

Supposing, however, as now seems tenable, that the right lateral displacement on the San Andreas amounts to several hundred miles and the left lateral displacement on the Garlock-Big Pine zone totals several tens of miles, some of the consequences should be as follows: (1) The San Andreas was initiated first and much movement occurred on it before the inception of the Garlock-Big Pine fault; (2) After the Garlock-Big Pine was established, it caused the big bend in the San Andreas and was offset by the San Andreas; and (3) The Garlock-Big Pine zone interrupted the continuity of displacement on the San Andreas and thereafter the north-south wedges actually moved toward each other.

If one side of the San Andreas has actually done most of the moving, indicated by the progressive offsets of Figures 2 and 3, there now appears to be a basis for determining which was the mobile block. Thus after the inception of the Garlock-Big Pine zone and assuming crustal shortening, the north-south wedges have been moving toward each other. Therefore it seems probable that the east side of the San Andreas north of the Garlock fault has actually been moving, as has the west side south of the Big Pine fault (Fig. 6).

### Stress Pattern

The character and orientation of the forces responsible for the primary strain pattern of this large region are not readily definable. Even in this fortunate case where the fault characteristics and relationships clearly show a pattern of deformation, there are many orientations of deforming forces which could cause the observed deformation. An example in which the character and orientation of deforming forces are more obvious is in the case of "feather joints" (Cloos, 1932) where the fractures can be related to secondary forces which arise from known movements. In the case of the subject region, however, we are dealing with the major known deformation and thus cannot know the character and orientation of the causal forces. On the other hand, by knowing the deformational pattern of this region, it may be possible to resolve its relationship to a larger strain pattern which may in turn indicate the stress pattern in this region.

At present, however, only general deductions may be made. The first of these is that the deformation is probably the result of compressive stresses because: (1) experimental work shows that rocks are brittle enough so that tensional stresses would be apt to cause tensional fractures instead of this shear pattern, and (2) regionally important thrust and reverse faults and folds show prominent crustal shortening (and prominent upward relief with the $AA_1$ axis of extension locally oriented vertically). Thus, assuming compressive stresses,

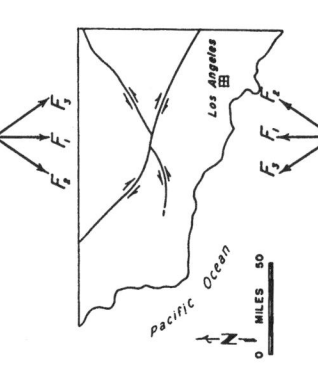

FIGURE 6.—SAN ANDREAS AND BIG PINE-GARLOCK STRAIN SYSTEM
$AA'$ axis of relative extension; $CC'$ axis of maximum shortening

the possible effective orientations (Fig. 7) are north-south oppositional forces, or a northeast-southwest counterclockwise couple, or a northwest-southeast clockwise couple. The northeast-southwest counterclockwise couple is favored because this orientation could perhaps more logically cause both the prominent northwest-trending folds of the Coast Ranges and the east-west folds and reverse faults of the Transverse Ranges (by upward relief in combination with the north-south shortening of the major shear pattern). With directly acting north-south compression, a more nearly equal manifestation of the shear sets would be expected, whereas the northwest right lateral set greatly predominates in this region. Likewise a northwest-southeast clockwise couple is unlikely because, where local upward relief occurred, the folds would be likely to trend northeast-southwest, whereas the general trend of folds in the Coast Ranges is northwestward.

### Genetic Concepts and Conclusions

Because the San Andreas and Garlock-Big Pine fault zones are major structures of great length, depth, displacement, and age; because they constitute a strain pattern of north-south shortening; because the probable orientation of causal forces is northeast-southwest; and because these forces are operative at great depth, a drag mechanism due to flowage of rock material from the Pacific basin to the continent of North America may be the primary reason for deformation in this region.

The mechanical effect of subcrustal convection currents, a presently favored concept (Gilluly *et al*., 1951), may have caused this deformation. Such a drag effect near oceanic margins could, then, be responsible for the permanence and growth of continents, development of geosynclines, deformation of geosynclinal deposits, association of metamorphic and igneous rock facies with orogeny, development of mountain roots, and possibly, the development of major lateral fault systems.

If the major lateral fault system described here is really so directly related to primary forces of deformation, much advancement in the understanding of geotectonics may be expected by the world-wide study of such faults. Perhaps faults with lateral displacements of hundreds of miles could even be responsible for some geologic relationships which seem to require land bridges.

Entirely apart from these speculations, the following conclusions are attained. First, the mapping of the faults and the evidence for the sense of the displacements are considered reliable. Second, the essential contemporaneity of the faults is considered to be proved. Third, this pattern of conjugate shears with opposing centripetal and centrifugal moving wedges is believed to be mechanically sound. Fourth, this strain system shows definite north-south shortening and relative east-west elongation of, during its activity, at least tens of miles and affecting a crustal area of at least 120,000 square miles; as such it is of significant proportions with respect to earth deformation. Fifth, this strain system is, as far as known, compatible with lesser structural features in the region and has in itself set up stress conditions which have developed lower-order strain systems. This major tectonic pattern in California may thus provide a satisfactory framework for the understanding of many elements of deformation within this region. Sixth, if deformation in other large areas proves to be compatible, it may ultimately be possible to develop a system of primary geotectonic forces. Hopefully this approach could, in conjunction with other lines of research, culminate in a usable theory of the causes of earth deformation.

Before the geologic history of California is satisfactorily established, much more critical mapping is needed, especially to determine the sense, duration, and amount of lateral movements on faults. Displacements which are measured in miles will have significant effects on paleogeologic and paleogeographic maps, and even those which are measured only in thousands of feet may, for example, be very critical in studies of the distribution of sedimentary facies. As these data accumulate, we can expect to attain a more nearly complete story of the dynamic geologic history of the region.

### References Cited

Allen, J. E. (1946) *Geology of the San Juan Bautista quadrangle, California*, Calif. Div. Mines, Bull. 133, p. 11–75.

Anderson, E. M. (1942) *The dynamics of faulting*, Oliver and Boyd, London.

Arnett, G. R. (1949) *Geology of the Lytle Creek area, California*, The Compass of Sigma Gamma Epsilon, vol. 126, p. 294–305.

Buddenhagen, H. J., et al. (1930) *Type locality of faults*, Am. Assoc. Petrol. Geol, Bull., vol. 14, p. 797–798.

Buwalda, J. P. (1926) *Certain inferences regarding the nature of movements along the Hayward fault zone, California* (Abstract), Geol. Soc. Am., Bull., vol. 37, p. 212.

Clark, B. L. (1929) *Tectonics of the Valle Grande of California*, Am. Assoc. Petrol. Geol, Bull., vol. 13, p. 199–238.

Cloos, E. (1932) *"Feather joints" as indicators of the direction of movement on faults, thrusts, joints and magmatic contacts*, Am. Jour. Sci., 5th ser., vol. 23, p. 289–304.

Crowell, J. C. (1950) *Geology of the Hungry Valley area, southern California*, Am. Assoc. Petrol. Geol., Bull., vol. 34, p. 1632–1640.

———(1952) *Probable large lateral displacement on San Gabriel fault, southern California*, Am. Assoc. Petrol. Geol., Bull., vol. 36, p. 2026–2035.

Dibblee, Thomas W., Jr., and Hill, Mason L. (1948) *Big Pine fault, California* (Abstract), Geol. Soc. Am., Bull., vol. 59, p. 1369.

Gilluly, James; Waters, Aaron; and Woodford, A. O. (1951) *Principles of Geology*, W. H. Freeman & Co.

Hess, F. L. (1910) *Gold mining in the Randsburg quadrangle, California*, U. S. Geol. Survey, Bull. 430, p. 23–47.

Hill, Mason L. (1947) *Classification of faults*, Am. Assoc. Petrol. Geol, Bull., vol. 31, p. 1669–1673.

Hulin, C. D. (1925) *Geology and ore deposits of the Randsburg quadrangle, California*, Calif. State Min. Bur., Bull. 95, 152 pages.

Jenkins, Olaf P. (1938) *Geologic map of California*, Calif. Div. Mines.

Kennedy, W. Q. (1946) *The Great Glen fault*, Geol. Soc. London, Quart. Jour., vol. 102, p. 41–76.

Kerr, P. F., and Schenck, H. G. (1925) *Active thrust fault in San Benito County, California*, Geol. Soc. Am. Bull., vol. 36, p. 471.

Lawson, A. C., et al. (1908) *The California earthquake of April 18, 1906*, Carnegie Inst. Washington, Pub. 87, 254 pages.

Meade, B. K. (1948) *Earthquake investigation in the vicinity of El Centro, California*, Am. Geophys. Union, Tr., vol. 29, p. 27–31.

Miller, W. J. (1944) *Geology of Palm Springs-Blythe strip, Riverside County, California*, Calif. Jour. Mines Geol, vol. 40, p. 12–72.

Nelson, R. N. (1925) *Geology of the hydrographic basin of the upper Santa Ynez River, California*, Univ. Calif., Dept. Geol. Sci., Bull., vol. 15, p. 327–396.

Noble, L. F. (1925–1926) *The San Andreas rift and some other active faults in the desert region of southeastern California*, Carnegie Inst. Washington, Year Book, no. 25, p. 416–422.

———(1933) *Excursion to the San Andreas fault and Cajon Pass*, 16th Intern. Geol. Cong., Guidebook 15, p. 0–21

Nolan, T. B. (1943) *The Basin and Range province in Utah, Nevada and California*, U. S. Geol. Survey, Prof. Paper 197-D, p. 141–196.

Reed, Ralph D., and Hollister, J. S. (1936) *Structural evolution of southern California*, Am. Assoc. Petrol. Geol., Tulsa, Okla.

Sprotte, E. C. (1949) *The San Andreas and San Jacinto rift zones*, The Compass of Sigma Gamma Epsilon, vol. 126, p. 315–319.

Taliaferro, N. L. (1943) *Geologic history and structure of the central Coast ranges of California*, Calif. Min. Bur, Bull. 118, p. 119–162.

Vickery, F. P. (1925) *The structural dynamics of the Livermore region*, Jour. Geol, vol. 33, p. 608–628.

Wagner, C. M., and Schilling, K. H. (1923) *The San Lorenzo group of the San Emigdio region, California*, Univ. Calif., Dept. Geol. Sci., Bull, vol. 14, p. 235–276.

Wallace, Robert E. (1949) *Structure of a portion of the San Andreas rift in southern California*, Geol. Soc. Am., Bull, vol. 60, p. 781–806.

Weaver, C. E. (1949) *Geology of the Coast Ranges immediately north of the San Francisco Bay region, California*, Geol. Soc. Am., Mem. 35, 242 pages.

Wellman, H. W. (1949) *The Alpine fault, New Zealand* (Unpublished).

Whitten, C. A. (1948) *Horizontal earth movement, vicinity of San Francisco, California*, Am. Geophys. Union, Tr., vol. 29, p. 318–323.

Wiese, John H., and Fine, Spencer F. (1950) *Structural features of western Antelope Valley, California*, Am. Assoc. Petrol. Geol, Bull, vol. 34, p. 1647–1658.

Willis, Bailey (1929) *San Andreas rift, California*, Jour. Geol, vol. 46, p. 793–827.

Wilson, I. F. (1943) *Geology of the San Benito quadrangle, California*, Calif. Jour. Mines Geol., vol. 39, p. 184–267.

RICHFIELD OIL CORPORATION, RICHFIELD BUILDING, LOS ANGELES 17, CALIFORNIA, 111 EAST PEDREGOSA STREET, SANTA BARBARA, CALIFORNIA

MANUSCRIPT RECEIVED BY THE SECRETARY OF THE SOCIETY, MAY 19, 1952

## Influence of West Coast Geology and the GSA Cordilleran Section

In 1984 a neophyte geologist arrived in California from the East Coast with a fresh B.S. degree and rudimentary skills. Fifteen years in this geologically ever-exciting state indelibly molded her scientific perceptions and enriched her life experience. A Stanford dissertation in the Coast Ranges brought opportunities to digest 70 m.y. of California history and to learn from some of the state's luminaries, notably Steve Graham, Jim Ingle, and the irreplaceable Ben Page. A job at San Francisco State University has made it possible to share with others a passion for nature's puzzles and to study classic localities along the San Andreas fault.

Karen Grove, Assoc. Professor
Department of Geology
San Francisco State University

# A lifetime of field work along the San Andreas fault system, California

**Thomas W. Dibblee, Jr.**
*Dibblee Geological Foundation, P.O. Box 60560, Santa Barbara, California 93160;
Department of Geological Sciences, University of California, Santa Barbara, California 93160*

Shortly after the great San Francisco earthquake in 1906, the San Andreas fault was recognized to be a major fault traversing California from Point Delgada southeastward to nearly the Mexican border. A study of the 1906 San Andreas fault rupture (Lawson, 1908; and its review by Prentice, this volume) led to an understanding of the physiographic expression of the fault and its significance as an important element of California's geology. Several subsequent studies of the San Andreas concluded that right-lateral slip on the order of one mile or less prevailed along this fault only in late Quaternary time and that all previous movements were dip slip (Reed, 1933; Reed and Hollister, 1936; Taliaferro, 1943). However, several authors disagreed, believing that the San Andreas fault originated as a strike-slip fault and accumulated considerable right-lateral offset (Noble, 1933; Crowell, 1950, 1952).

Our publication (Hill and Dibblee, 1953) was prepared by Mason L. Hill, a quiet-spoken, inspiring geologist greatly motivated by discoveries of evidence of strike-slip displacement on faults. He occasionally accompanied me in the field from 1936 to 1947, as I undertook an intensive geologic mapping effort along the San Andreas fault in search of oil and gas for Richfield Oil Company. This 10-year effort by Mason Hill and myself culminated in the 1953 article, and in conjunction with contemporary publications by Noble (1954) and Crowell (1950, 1952, 1962), shook up previous understanding of the character of the San Andreas fault, its amount and type of displacement, and its origin and history.

We stated that the San Andreas fault is a vertical fault with a nearly straight trace, continuous for about 965 km across California. The fault zone ranges from a few feet to a few miles in width, along which there is abundant evidence of lateral separation, including offset stream drainages, apparent throw reversals, subsidiary tension cracks, and associated drag folds. We established an orderly, cumulative progression of right-lateral offset on the fault by cross-fault correlations, as follows (from Hill, 1981): Pleistocene facies, 22 km; upper Miocene strata, 104 km; lower Miocene, 280 km; Eocene, 360 km; Upper Cretaceous, 512 km; and offset of basement rock facies, more than 560 km.

In addition, we determined that the San Andreas fault has been active from at least the middle Cretaceous, with a uniform rate of movement of ~0.51 to 0.85 cm per year. We proposed that other faults in California are strike-slip faults and are probably part of the San Andreas fault system, with either parallel strikes and right-lateral offset, or conjugate orientations and left-lateral offset (notably the Big Pine and Garlock faults).

The interpretations of the mechanics of faulting of California may be summarized as follows: active major structures such as the San Andreas fault and associated parallel faults, and the "conjugate" Garlock–Big Pine and similarly oriented strike-slip faults, are first order, deep-seated structures. All other faults (such as thrust faults) and folds are subsidiary structures. The major structures indicate north-south shortening and relative east-west extension. The mechanism for the stress field was interpreted to be a drag mechanism due to flowage of rock material from the Pacific basin to the continent of North America operative at great depth (greater than 16 km).

Additional geologic mapping in California by myself since 1953 to the present has confirmed the interpretations of Hill and Dibblee (1953) with respect to the cumulative right-lateral offsets of rocks on the San Andreas fault system during the Cenozoic Era. It is gratifying that much of my own work has been tested and confirmed by numerous other investigators, with some modifications, as expected. My additional mapping covered the northern or Coast Ranges segment of the San Andreas fault (Dibblee, 1962, 1966, and 1980), the central or Mojave segment through the big bend of the San Andreas fault (Dibblee, 1967, 1968, 1975a, and 1975b), and the southern or Coachella Valley segment (Dibblee, 1954, 1984, 1996, and 1997). In these papers I confirmed large right-lateral offsets in the Coast Ranges

---

Dibblee, T. W., Jr., 1999, A lifetime of field work along the San Andreas fault system, California, *in* Moores, E. M., Sloan, D., and Stout, D. L., eds., Classic Cordilleran Concepts: A View from California: Boulder, Colorado, Geological Society of America Special Paper 338.

(Dibblee, 1962), investigated ambiguous evidence for lateral offset along the Mojave segment through the Transverse Ranges (Dibblee, 1968), and documented no more than 10 km right-lateral offset of late Miocene to Pleistocene rocks in Coachella Valley (Dibblee, 1997). I also mapped and described many physiographic features and subsidiary structures associated with strike-slip faults in many areas of California (e.g., Dibblee, 1977a).

I addressed strike-slip tectonics and the role of the San Andreas fault in Cenozoic basin genesis and development throughout western California (Dibblee, 1977a, 1977b, 1977c). I also related the ideas originally expressed in Hill and Dibblee (1953) to plate tectonic theory (Dibblee, 1989). All of my geologic mapping and interpretations, along with the work of many others, have confirmed the original concept of large right-lateral displacement along the San Andreas fault during the Cenozoic Era.

## UNRESOLVED PROBLEMS

Despite major contributions by many investigators about the San Andreas fault since 1953, a number of important and unresolved questions still remain. Among these problems are those enumerated by Hill (1981, p. 128).

1. Where are the ends of the fault?
2. Why does it have several prominent bends?
3. When and where did it originate, and was there a proto-San Andreas?
4. Where and when will it generate the next great earthquake?
5. How does the San Andreas relate to the rotated and translated miniblocks in southern California indicated by paleomagnetic studies?

The work of Hill and Dibblee (1953) and of our colleagues, with respect to the character, displacement, and history of the San Andreas fault, has stood the test of time. The description and documentation of the San Andreas as a great strike-slip fault focused attention on this major structure and paved the way for other geoscientists to recognize the role of the San Andreas fault as a great plate boundary. I anticipate more discoveries to be made about the San Andreas fault and its surroundings. I am grateful I was able to make a contribution to these efforts through my field mapping and my collaborations with the remarkable geologist, Mason Hill.

## ACKNOWLEDGMENTS

This review article was condensed by Wendy Lou Bartlett, Helmut E. Ehrenspeck, and Eldridge Moores from a longer manuscript of my own in order to conform to the guidelines of this volume. I thank Wendy Lou Bartlett for her dedicated effort in typing and editing this manuscript.

## REFERENCES CITED

Crowell, J. C., 1950, Geology of Hungry Valley area, southern California: American Association of Petroleum Geologists Bulletin, v. 34, p. 1623–1646.

Crowell, J. C., 1952, Probable large lateral displacement on the San Gabriel fault, southern California: American Association of Petroleum Geologists Bulletin, v. 36, p. 2026–2035.

Crowell, J. C., 1962, Displacement along the San Andreas fault, California: Geological Society of America Special Paper 71, 61 p.

Dibblee, T. W., Jr., 1954, Geology of the Imperial Valley region, California, in Jahns, R. H., ed., Geology of southern California: California Division of Mines Bulletin 170, p. 21-28.

Dibblee, T. W., Jr., 1962, Displacements on the San Andreas rift zone and related structures in Carrizo Plain and vicinity, in Hackel, O., chairman, Geology of Carrizo Plains and San Andreas fault: San Joaquin Geological Society and Pacific Section, Society of Economic Paleontologists and Mineralogists, Guidebook, p. 5–12.

Dibblee, T. W., Jr., 1966, Evidence for cumulative offset on the San Andreas fault in central and northern California, in Bailey, E. H., ed., Geology of northern California: California Division of Mines and Geology Bulletin 190, p. 375–384.

Dibblee, T. W., Jr., 1967, Areal geology of the western Mojave Desert, California: U.S. Geological Survey Professional Paper 522, 153 p., map scale 1:125,000.

Dibblee, T. W., Jr., 1968, Displacements on the San Andreas fault system in the San Gabriel, San Bernardino, and San Jacinto Mountains, southern California, in Dickinson, W. R., and Grantz, A., eds., Proceedings of conference on geologic problems of San Andreas fault system: Stanford University Publications in Geological Sciences, v. 11, p. 260–278.

Dibblee, T. W., Jr., 1975a, Tectonics of the western Mojave Desert near the San Andreas fault, in Crowell, J. C., and Ehlig, P. L., eds., The San Andreas fault in southern California: California Division of Mines and Geology Special Report 118, p. 155–161.

Dibblee, T. W., Jr., 1975b, Late Quaternary uplift of the San Bernardino Mountains on the San Andreas and related faults, in Crowell, J. C., and Ehlig, P. L., 1975, The San Andreas fault in southern California: California Division of Mines and Geology Special Report 118, p. 127–135.

Dibblee, T. W., Jr., 1977a, Strike-slip tectonics of the San Andreas fault and its role in Cenozoic basin evolvement, in Nilsen, T. H., ed., Late Mesozoic and Cenozoic sedimentation and tectonics in California: Bakersfield, California, San Joaquin Geological Society Short Course, p. 26–52.

Dibblee, T. W., Jr., 1977b, Sedimentology and diastrophism during Oligocene time relative to the San Andreas fault system, in Nilsen, T. H., ed., Late Mesozoic and Cenozoic sedimentation and tectonics in California: Bakersfield, California, San Joaquin Geological Society Short Course, p. 86–98.

Dibblee, T. W., Jr., 1977c, Relations of hydrocarbon accumulations to strike-slip tectonics of San Andreas fault system, in Nilsen, T. H., ed., 1977, Late Mesozoic and Cenozoic sedimentation and tectonics in California: Bakersfield, California, San Joaquin Geological Society Short Course, p. 135–143.

Dibblee, T. W., Jr., 1980, Geology along the San Andreas fault from Gilroy to Parkfield, in Streitz, R., and Sherburne, R., eds., Studies of the San Andreas fault zone in northern California: California Division of Mines and Geology Special Report 140, p. 3–18.

Dibblee, T. W., Jr., 1984, Stratigraphy and tectonics of the San Felipe Hills, Borrego Badlands, Superstition Hills and vicinity, in Rigsby, C. A., ed., The Imperial basin—Tectonics, sedimentation and thermal aspects: Society of Economic Paleontologists and Mineralogists Special Publication, v. 40, p. 31–44.

Dibblee, T. W., Jr., 1989, The San Andreas fault and major rock terranes of California displaced by it and its tectonics, in Baldwin, E. J., Foster, J. H., Lewis, W. L., and Hardy, J. K., eds., The San Andreas fault; Cajon Pass to

Stanford Geological Survey at north end of Humboldt Range, Nevada in 1932. Standing in front is teaching assistant Ward Smith; Tom Dibblee is on right side of 1st seat. Photograph from the Geohistory Archives, School of Earth Sciences, Stanford University.

Wallace Creek, Volume 1: Santa Ana, California, South Coast Geological Society guidebook 17, p. 223–275.

Dibblee, T. W., Jr., 1996, Stratigraphy and tectonics of the Vallecito–Fish Creek Mountains, Vallecito Badlands, Coyote Mountains and Yuha Desert, southwestern Imperial basin, in Abbott, P. L., and Seymore, D. C., eds., Sturzstroms and detachment faults, Anza-Borrego Desert State Park, California: Santa Ana, California, South Coast Geological Society Annual Field Trip Guidebook 24, p. 59–79.

Dibblee, T. W., Jr., 1997, Geology of the southeastern San Andreas fault zone in the Coachella Valley area, southern California, in Baldwin, J., Lewis, L., Payne, M., and Roquemore, G., eds., Southern San Andreas fault—Whitewater to Bombay Beach, Salton Trough, California: Santa Ana, California, South Coast Geological Society Annual Field Trip Guidebook 25, p. 35–56.

Hill, M. L., 1981, San Andreas fault: History of concepts: Geological Society of America Bulletin, v. 92, p. 112–131.

Hill, M. L., and Dibblee, T. W., Jr., 1953, San Andreas, Garlock, and Big Pine faults, California—A study of the character, history, and tectonic significance of their displacements: Geological Society of America Bulletin, v. 64, p. 443–458.

Lawson, A. C., 1908, The California earthquake of April 18, 1906, Report of the State Earthquake Investigation Commission: Carnegie Institute of Washington Publication 87, 451 p.

Noble, L. F., 1933, Excursion to the San Andreas fault and Cajon Pass: 16th International Geological Congress Guidebook, 15, p. 10–21.

Noble, L. F., 1954, The San Andreas fault zone from Soledad Pass to Cajon Pass, California, in Jahns, R. H., ed., Geology of southern California: California Division of Mines Bulletin 170, p. 37–48.

Reed, R. D., 1933, Geology of California: Tulsa, Oklahoma, American Association of Petroleum Geologists, 355 p.

Reed, R. D., and Hollister, J. S., 1936, Structural evolution of southern California: American Association of Petroleum Geologists Bulletin, v. 20, p. 1529–1704.

Taliaferro, N. L., 1943, Geologic history and structure of the central Coast Ranges of California, in Jenkins, O. P., ed., Geologic formations and their economic development of oil and gas fields of California: California Division of Mines Bulletin 118, p. 119–162.

MANUSCRIPT ACCEPTED BY THE SOCIETY NOVEMBER 23, 1998.

Geological Society of America
Special Paper 338
1999

# Section III
# PLATE TECTONICS—COAST TO MOUNTAINS

# Plate tectonics: Introduction

> Making your way through the mazes of the Coast Range to the summit of any of the inner peaks or passes opposite San Francisco, in the clear springtime, the grandest and most telling of all California landscapes is outspread before you. At your feet lies the great Central Valley...across its eastern margin rises the mighty Sierra...so gloriously colored and so luminous that it seems to be not clothed with light, but wholly composed of it, like the wall of some celestial city.
> —John Muir in 1894 (1991, p. 2)

Since the advent of the plate tectonics revolution, California geology has become classic in its representations of plate tectonics processes. The cross section of central California from the Pacific Coast to the Sierra Nevada has become the classic model for an ancient Andean-style continental margin. The Franciscan Complex in the California Coast Ranges has served as a classic example of a subduction complex, the Great Valley sequence is the classic arc-trench gap deposit, and the Sierra Nevada batholith and its volcanic carapace have assumed the role of the classic continental arc. The view described by John Muir in the preceding quote (see Muir, 1991) encompasses all these features. His description epitomized the romantic impression one gets from such views, still possible after winter rains clear the air, from the Coast Ranges across the Great Valley to the Sierra Nevada. For example, from near Pacheco Pass on California State Route 152, where Muir reported first seeing the Sierra Nevada, one can stand on Franciscan Complex rocks and look eastward over exposures of the Great Valley sequence, the Great Valley itself, and finally at the high Sierra Nevada between Yosemite and Kings Canyon National Parks.

This section consists of five chapters. Two concentrate on Coast Range–Great Valley geology, two on the Sierra Nevada and its southern extension, the Peninsular Ranges, and one considers the general tectonic development of California and the surrounding region.

In the Coast Ranges–Great Valley region, W. G. Ernst's classic paper on Franciscan geology forms the basis of John Wakabayashi's contribution, entitled "Subduction and the rock record; concepts developed in the Franciscan complex, California." R. W. Ojakangas' article on the Great Valley sequence is the focus of Ray Ingersoll's essay entitled "Post-1968 research on the Great Valley Group." Ernst's article was written shortly after the plate tectonic revolution, whereas Ojakangas' work appeared just as the main papers of the revolution were appearing.

In the Sierra Nevada, a paper written before the advent of plate tectonics, by P. Bateman and C. Wahrhaftig, forms the basis of the Jason Saleeby essay, entitled "On some aspects of the geology of the Sierra Nevada." The Bateman and Wahrhaftig paper consisted of two parts, one on the pre-Upper Cretaceous "framework rocks," written chiefly by Bateman, and another part, on the Upper Cretaceous and Cenozoic rocks, as well as the modern topography, written chiefly by Wahrhaftig. Saleeby's article and the reprinted parts of the Bateman and Wahrhaftig article consider only the "foundational rocks" (also called the Subjacent series by the original U.S. Geological Survey folio mappers), consisting of Paleozoic-Mesozoic metasedimentary, metavolcanic, and ophiolitic rocks and Paleozoic-Mesozoic granitic batholithic and related igneous rocks. The rocks above the "great unconformity," also called the Superjacent Series, are not discussed in Saleeby's essay. This younger section contains Late Cretaceous Great Valley rocks along the western edge of the Sierra Nevada, a complex sequence of lower Tertiary gold-bearing gravels (the Auriferous Gravels), and volcanogenic extrusive and related shallow intrusive rocks ranging in age from Oligocene to Pleistocene, that is discontinuously present across the entire western slope of the Sierra Nevada.

Excerpts from a long early paper on the Southern California batholith by E. S. Larsen form the basis of Gordon Gastil's essay "Esper S. Larsen, Jr. on the batholith of Southern California revisited." Larsen's paper was written well before the plate tectonic revolution, at a time when the origin of granite was a hotly debated topic in petrology. Gastil's essay considers only the batholith, as did Larsen's paper. The prebatholithic rocks are not as well developed in the Peninsular Ranges as in the Sierra Nevada, but they are present (e.g., Gastil et al., 1981; Walawender et al., 1990), and they tell a complex tectonic story.

An early plate tectonic interpretation of California and surrounding regions by E. M. Moores forms the basis of the essay by Moores and others, entitled "California terranes."

These articles cover only part of the plate tectonic evolution of California, principally that of central California. In particular there is no explicit discussion of the modern tectonic composition of that part of California north of the Mendocino triple junction, north of lat 40°N, where subduction is still active along the continental margin, and volcanic activity is manifest in the

---

The 100th anniversary of the Cordilleran Section of the GSA is an event to celebrate. My memories go back almost 40 years and I can recall the great change in the focus of section meetings that occurred in the late 1960s with the advent of plate tectonics. Prior to that time papers were dominated by local presentations of geology and the ever-present arguments, sometimes quite raucous, fueled by such uninhibited personalities as Jim Gilluly and Phil King, about their interpretation. There were obvious separations in interest between workers west, east, and north of the Sierra Nevada, and one often wondered if there were not three or four separate meetings going on at the same time within the section. With the introduction of plate tectonics into the mix the section, which should also include the Rocky Mountain section and the Canadian Cordilleran contingent as well, became not only a unified meeting concerning the broad aspects of evolving plate boundary activity from the Pacific to central North America, but also became one of the international focal points for development of plate tectonic processes. Each meeting seemed to contain a wealth of data and ideas to be tested that generated tremendous excitement and interaction between all the diverse subdisciplines in the earth sciences. It required months to digest and assimilate the proceedings.

I recall the intense excitement at section meetings the early 1970s when models of the development of Pacific/Farallon/North American plate interactions pointed the direction to how plate tectonics could be used to explain previously intractable relations between the San Andreas fault and other apparently disparate observations concerning coeval extensional and compressional features, volcanic evolution within the Coast Ranges, and apparently very different offsets on different parts of the fault and other related faults. During meetings in the 1970s, I was also struck by the rapidity with which plate tectonics was used to relate the late Cretaceous to recent volcanic history of the Cordillera in a marvelous synthesis by Christiansen and Lipman; the relations between plate interactions at the western Cordilleran margin and the formation of the Colorado/Wyoming rockies by Armstrong; and the Cordilleran-wide effects of diachronous accretion of all types of oceanic and continental fragments along the western Cordilleran margin by Jones, Coney, and others. The Cordillera became a natural laboratory for concepts that spread globally. Particularly the various concepts of subduction-related phenomena, such as development of paired metamorphic belts by Ernst and colleagues, emplacement of ophiolites by Moores and others, the development of the Mesozoic Sierran arc by the eastward underflow of oceanic plates by Hamilton and others, and the antithetic nature of the Cordilleran fold and thrust belt by Armstrong and others. Perhaps nowhere has the influence of Cordilleran geology been greater than in the concepts of extensional tectonism first developed by Anderson and Proffett. During the 1970s and 1980s you could not miss a section meeting if you wanted to be abreast of recent advances in plate tectonic developments. It was a heady and exciting time. However, one should not forget that were some very perceptive, but often forgotten, Cordilleran predecessors to some of these "modern" concepts, for example, the underflow of the Pacific by Ransome and large magnitude of extension in the Basin and Range by Ransome, Emmons, and Garrey both published after the turn of the century.

It cannot be forgotten that the Cordilleran region of western North America contains some of the finest and most well exposed geology to found anywhere in the world. The field trips associated with section meetings are often as exciting as the meetings themselves and have provided another venue for an inspiring geological experience. I personally thank all the fine geologists, both past and present, who have made my association with the Cordilleran Section one of the truly memorably experiences of my scientific career.

Clark Burchfiel
Massachusetts Institute
of Technology, 1998

southern Cascade volcanic mountains, such as Lassen Peak and Mount Shasta.

Although there has been much progress in the past three decades, much of California's geology remains poorly understood and controversial. Comparison of Saleeby's Figure 3, Ingersoll's Figures 2 and 4, and Moores' 1998 Figure 1 (at the top of Moores' 1970 article) will give some flavor of the disagreement that is still outstanding.

Eldridge M. Moores

# REFERENCES CITED

Gastil, G., Morgan, G. J., and Krummenacher, D., 1981, The tectonic history of Peninsular California and adjacent Mexico, in Ernst, W.G., ed., The geotectonic development of California (Rubey Volume I): Englewood Cliffs, New Jersey, Prentice-Hall, p. 284–305.

Muir, J., 1991, The mountains of California (new and enlarged edition, reprinted from the 1894 original): Berkeley, California, Ten Speed Press, 389 p.

Walawender, M. J., Gastil, R. G., Wardlaw, M., Gunn, S. H., Eastman, B. G., McCormick, W. V., and Smith, B. M., 1990, Geochemical geochronologic, and isotopic constraints on the origin and evolution of the eastern zone of the Peninsular Ranges batholith, in Anderson, J. L., ed., The nature and origin of Cordilleran magmatism: Geological Society of America Memoir 174, p. 1–18.

Zonenshain, L. P., Kuzmin, M. I., Natapor, L. M., and Page, B. M., eds., 1990, Geology of the U.S.S.R.: A plate tectonic synthesis: American Geophysical Union Monograph 21, 242 p.

MANUSCRIPT ACCEPTED BY THE SOCIETY ON NOVEMBER 23, 1998.

Photograph of Benjamin M. Page, long-time Professor of Geology, Stanford University. Ben's long and productive career concentrated mostly on the geology and tectonics of the California Coast Ranges. A consummate gentleman and supportive colleague, he was revered by his many students and collaborators. His dozens of published works included, in addition to his Coast Range work, forays into other regions of the Pacific Rim, especially Taiwan. He also was Editor of the journal Tectonics, and editor of A. G. U. Monograph 21, entitled " Geology of the USSR; a plate-tectonic synthesis, by Lev P. Zonenshain, Michael I. Kuzmin, and Lev M. Natapov. This volume was a landmark contribution to our understanding of that vast region, which in many respects resembles California and neighboring regions.

Chapter 7

# THE FRANCISCAN: CALIFORNIA'S CLASSIC SUBDUCTION COMPLEX

**John Wakabayashi**
*1329 Sheridan Lane*
*Hayward, California 94544*
*E-mail: wako@tdl.com*

# W. GARY ERNST

W. G. Ernst was born in 1931 in St. Louis, Missouri. He received his bachelor of arts degree from Carleton College in 1953, his master of science degree from the University of Minnesota in 1955, and his doctor of philosophy degee from Johns Hopkins University in 1959. He was a professor in the Department of Earth and Space Sciences and Institute of Geophysics and Planetary Physics at the University of California, Los Angeles (UCLA), from 1960 to 1989. At UCLA, he served as chair of the Department of Geology from 1970 to 1974, chair of the Department of Earth and Space Sciences from 1978 to 1982, and director of the Institute of Geophysics and Planetary Physics from 1987 to 1989. Since 1989 he has been a professor in the Department of Geological and Environmental Sciences at Stanford University, and was the dean of the School of Earth Sciences at Stanford from 1989 to 1994. Ernst is a member of the National Academy of Sciences and served as president of the Geological Society of America in 1986.

Ernst made one of the major contributions to the plate tectonics revolution by relating high-pressure–low-temperature metamorphism (especially blueschist facies metamorphism) to subduction (in the "classic paper" reviewed here). In subsequent research, he has continued to advance concepts linking tectonics and metamorphism; he is widely regarded as one of, if not the, most accomplished geoscientists at relating these two subdisciplines of the earth sciences. His research in experimental petrology, some of which formed part of the foundation for the "classic paper," has continued to provide new data constraining key metamorphic reactions and phase stabilities. Although he has had a long research career as an experimental petrologist, Ernst has always valued, and actively participated in, geologic field studies. It is probably this dual background that has made him so effective at linking metamorphic petrology and tectonics. In addition to his research accomplishments, Ernst has supervised numerous outstanding doctoral students, many of whom have gone on to make a significant impact in geologic research.

## JUSTIFICATION FOR CHOOSING ERNST (1970) AS THE CLASSIC PAPER FOR THIS CHAPTER

Ernst's (1970) paper was chosen as the "classic paper" for the Franciscan Complex because it made the connection between the high-pressure–low-temperature (high $P$-$T$) metamorphism and subduction, in particular the explanation given for the low geothermal gradients associated with high $P$-$T$ metamorphism. This paper was one of the key breakthroughs in the plate tectonic revolution, and it helped establish the Franciscan Complex as the type subduction complex. Variations of Figure 3 of Ernst (1970) have become the textbook model of subduction-zone metamorphism. Geologists now regard high $P$-$T$ (including blueschist facies) metamorphism as the strongest evidence of an exhumed subduction complex.

John Wakabayashi

# Tectonic Contact between the Franciscan Mélange and the Great Valley Sequence—Crustal Expression of a Late Mesozoic Benioff Zone

## W. G. ERNST

*Department of Geology and Institute of Geophysics and Planetary Physics
University of California, Los Angeles, California 90024*

Late Mesozoic Franciscan rocks of the California Coast Ranges, chaotically deformed and consisting of graywacke, micrograywacke, shale, chert, and mafic pillow lava, are considered to have been deposited in and adjacent to a northwest-trending oceanic trench. Principally on the east, contemporaneous, well-bedded conglomerate, lithic sandstone, siltstone, and shale of the Great Valley sequence evidently were laid down on a continental shelf, slope, and sea floor environment. The junction between these two sequences, one the ensimatic eugeosynclinal mélange of Franciscan rocks, the other the more orderly Great Valley miogeosynclinal strata overlying chiefly sialic-type basement, is marked by the South Fork Mountain–Stoney Creek–Ortigalita (–Sur-Nacimiento) fault system. This structural break is interpreted as the crustal expression of a Late Mesozoic Benioff zone. Episodic or relatively continuous underthrusting of the trench mélange beneath this former seismic shear zone is held responsible for the contrasting characteristics of the Franciscan rocks and the simultaneously deposited Great Valley sequence, as well as for their ubiquitous tectonic juxaposition. A speculative four-state tectonic model is proposed that involves: (1) rapid Late Jurassic relative northeastward or eastward spreading of the Pacific Ocean floor, coupled with westward or southwestward encroachment by the continental lithospheric plate; (2) mid-Cretaceous buoyant uplift of portions of the eugeosynclinal prism during a period of less intense spreading; (3) accelerated post-Cretaceous convergence between oceanic and continental lithospheric plates; and (4) Miocene-Pleistocene diapiric uplift of the Franciscan mélange related to northwestward sea-floor spreading.

## INTRODUCTION

Two contrasting, largely sedimentary series of Late Mesozoic rocks constitute the chief non-batholithic units of the California Coast Ranges and crop out extensively along the western side of the Sacramento and San Joaquin Valleys. The controversial interrelationships of these rocks, the eugeosynclinal Franciscan group and the miogeosynclinal Knoxville formation and so-called Shasta series and Chico group, have been discussed by geologists for many years. The areal distribution of these units is shown in Figure 1, a geologic index map of western California.

*Taliaferro* [1943, pp. 208-212, 218-219, p. 194] concluded that Mesozoic sedimentary rocks cropping out in the Coast Ranges and in the Great Valley consisted of an older, Late Jurassic (Tithonian) sequence of deformed graywackes and related rocks (the Franciscan) that was thought to give way upward to a more ordered, younger sequence of latest Jurassic Knoxville strata. These conclusions were based in part on geologic mapping under Taliaferro's supervision, principally in central California, and on sparse fossil evidence available at the time. *Taliaferro* [1943, pp. 208, 216-217] further stated that beds of the Lower Cretaceous so-called Shasta series and the Upper Cretaceous so-called Chico group generally rest disconformably on the Knoxville. *Irwin* [1957], basing his conclusions on reconnaissance mapping in northern California and the accumulation of additional paleontological data, recognized the virtual contemporaneity of Franciscan rocks with both the latest Jurassic and the Cretaceous well-bedded strata; he coined the term Sacramento Valley (later usage: Great Valley) sequence to designate these latter units because of excellent exposures along the western side of the Sacramento and San Joaquin Valleys. Summarizing the present state of knowledge, *Bailey et al.*

886

Copyright © 1970 by the American Geophysical Union.

Reprinted by permission from the Journal of Geophysical Research, v. 75, no. 5, p. 886–901, 1970, published by the American Geophysical Union.

Serpentinite matrix melange in Franciscan Complex, Ring Mountain Preserve, Tiburon Peninsula, Marin County, California. Blocks in foreground of eclogite and glaucophane schist weathered out of melange matrix. Downtown San Francisco to left. Golden Gate Bridge towers visible to right. Photo by E. M. Moores.

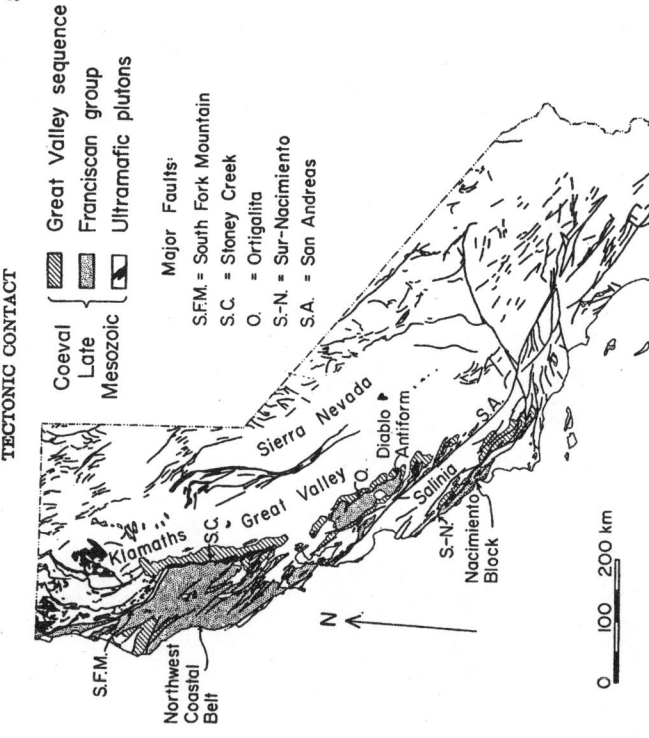

Fig. 1. Generalized geology of western California, including major faults for the entire state, after the 1:2,500,000 scale Geologic Map of California (1966) accompanying California Division of Mines and Geology Bulletin 190. With the exception of the transcurrent (or transform) nature of the San Andreas and related breaks, the major faults specified are all interpreted as thrust or high-angle reverse faults, chiefly dipping to the east or northeast.

[1964, Figure 23] indicated that where the relationships are well understood, Franciscan and some Great Valley strata span approximately the same time interval. Admittedly the oldest Franciscan rocks locally may be slightly older than the associated basal Great Valley units, whereas in other areas the uppermost members of the Great Valley sequence may have depositional ages considerably younger than the latest Franciscan rocks. Junctions between Franciscan and all other Mesozoic sedimentary rocks, once thought to be angular unconformities or gradational sequences, are now regarded as fault contacts [*Bailey et al.*, 1964, pp. 145–146; *Hackel*, 1966, p. 220; *Page*, 1966, p. 272].

The Franciscan terrane consists of a chaotically disturbed mélange of rock types [*Hsü*, 1968]. Graywacke (in part, subgraywacke), siltstone (micrograywacke), and dark shale constitute the major portion of most sections. Somewhat altered pillow basalt (greenstone) and more felsic keratophyre and quartz keratophyre (which are less common) make up much of the remainder, with red siliceous interbedded shale and chert as minor constituents. Limestone is very rare. Although the Franciscan is generally referred to as a eugeosynclinal assemblage of rock types, it is impoverished in volcanics compared to a typical eugeosynclinal series. The entire terrane has been pervasively intruded by partially or completely serpentinized peridotites of the 'cold' or alpine type, but such plutons are particularly abundant along the eastern margin of the California Coast Ranges (see Figure 1). As pointed out by *Bailey et al.* [1964], *Ernst* [1965], and *Ernst et al.* [1970], there is no evidence to suggest that continental basement rocks underlie the Franciscan. On the contrary, judging from the nature of the original sediments, the occurrence of serpentinized peridotites as the only plutons intrusive into the terrane, the style of deformation and unique, relatively low-temperature, high-pressure blueschist-facies metamorphic mineral assemblages developed (e.g., see *Blozam* [1956, 1959, 1960]; *Coleman* [1961, 1965]; *Coleman and Lee* [1962, 1963]; *McKee* [1962a, 1962b]; *Ernst* [1963a, 1963b, 1963c]; *Ghent* [1965]; *Essene and Fyfe* [1967]; *Bailey et al.* [1964, p. 143] tentatively suggested, and *Ernst* [1965, p. 905] concluded that the Franciscan mélange was deposited on oceanic crust, perhaps partially or completely within the confines of a trench system. The thickness of the Franciscan has been estimated by *Taliaferro* [1943, pp. 185–186] as about 8 km (25,000 ft), by *Bailey et al.* [1964, pp. 20, 111] as in excess of 15 km (>50,000 ft), and by *Ernst* [1965, p. 907] as approaching 25–30 km; Taliaferro's estimate was based on measured, relatively undeformed sections, whereas Bailey et al. and Ernst reached their conclusions from the disturbed nature of the rocks themselves and the mineral parageneses. The estimates are not strictly comparable inasmuch as the first involves stratigraphic thickness, whereas the latter two are concerned with tectonic thickness.

Strata of the Great Valley sequence crop out in both the California Coast Ranges and the Great Valley. Compared to the Franciscan, these rocks represent a more ordered miogeosynclinal series. Siltstone and claystone predominate, but arkosic and lithic sandstones are also common [*Hackel*, 1966]; conglomeratic lenses are characteristic of the lower portion of the section, with tuffaceous sandstones and mafic volcanic flows confined to the Late Jurassic Knoxville formation [*Brown and Rich*, 1961; *Brown*, 1964a, 1964b]. Serpentinite intrusives are absent except along some fault contacts, especially those involving juxtaposition against rocks of the Franciscan group. Although customarily considered to be unmetamorphosed, albite, laumontite, and chlorite have developed from detrital plagioclase and biotite in some of the more deeply buried portions of the section, such as those exposed along the western flank of the Sacramento Valley; these strata exhibit transitional stages between authigenesis and zeolite-facies metamorphism [*Dickinson et al.*, 1969]. Exploratory drilling in the eastern half of the Great Valley, and outcrops as well, have demonstrated that the Cretaceous miogeosynclinal series rests on Sierran basement [*Lachenbruch*, 1962; *Safonov*, 1962; *Schilling*, 1962]. However, the Great Valley synclinorium thickens farther to the west, and the nature of the rocks underlying the Knoxville formation has not been clarified as yet (see Figure 2 and further discussion in the section dealing with the envisioned dynamic model). Judging from the well-bedded and only moderately disturbed character of the Great Valley strata and by the presence of some sediments of probable shallow-water deposition (especially in the eastern portions) much of the miogeosynclinal assemblage appears to have been laid down on continental crust, perhaps in a shelf-slope environment [e.g., see *Bailey et al.*, 1964, p. 123]. Aggregate maximum thickness of the Great Valley sequence has been estimated as exceeding 12 km (>40,000 ft) by *Brown* [1964a, p. 7] and by *Bailey et al.* [1964, p. 20]; as approximately 14 km (45,000 ft) by *Dickinson et al.* [1969, p. 519]; and as at least 18 km (60,000 ft) by *Hackel* [1966, p. 217].

The purposes of the present paper are to focus attention on the nature of the junction between the Great Valley sequence and the Franciscan group, and to suggest a tectonic history relating this contact to interaction between western California and the Pacific Ocean lithospheric plates.

### Gross Structure of the Franciscan Terrane

Due to the rarity of fossils, the paucity of distinctive marker beds, and the limited continuity of outcrops of chert and greenstone, straitgraphic control is poor. These effects may be at least partly a consequence of widespread slumping during deposition; however, extensional features, soft sediment deformation, graded bedding, and other current features are scarce in the mélange. *Bailey and Irwin* [1959] and *Bailey et al.* [1964,

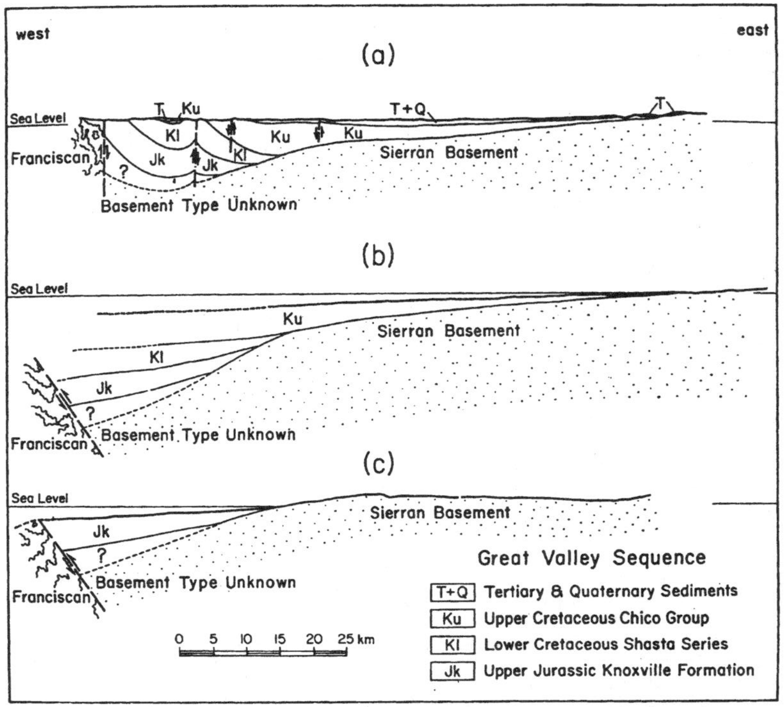

Fig. 2. Structure of the Great Valley of California: (*a*) present configuration in the Sacramento Valley, after *Lachenbruch* [1962, Plate 3]; (*b*) interpretation of same area at the end of Cretaceous time and prior to post-Cretaceous deformation; (*c*) interpretation of same area at the end of Jurassic time. No vertical exaggeration implied.

pp. 139–141] have shown that although K-feldspar is an abundant detrital mineral in mid-Cretaceous rocks of the northwest coastal belt of the Franciscan, this phase is present in only trace amounts in older portions of the terrane; thus, although a regional difference in mineralogy can be discerned, the presence or absence of K-feldspar evidently cannot be employed locally as a stratigraphic tool. A further reason for the absence of stratigraphic control involves the apparently heterogeneous response of the Franciscan to deformation. The terrane to some extent resembles a megabreccia, with coherent, more or less competent blocks, ranging in diameter from several meters to several tens of kilometers in average dimension, floating in a more intensely sheared and granulated matrix [*Hsü*, 1967, 1968; *Hsü and Ohrbom*, 1969].

For all the above reasons, intermediate-scale structures of the order of tens to a thousand or more meters in size, are difficult to recognize in the field. (For a recent attempt to identify such structures, see *Ernst et al.* [1970, Figure 12]). Customarily, small-scale structures observed in a single outcrop can be traced only for short distances. In contrast, large-scale features can be mapped, but this characteristically requires juxaposition of Franciscan rocks against those of broadly contrasting lithologies over the extent of, say, 50 km.



On a local scale, the Franciscan is seen to be broadly folded, with the more competent, boudinaged graywacke layers set in a highly contorted shaly matrix. Individual folds are transected and interrupted by numerous shear zones of diverse orientations. Although linear and planar elements may be systematic within a given outcrop, such features are correlated to nearby exposures with great difficulty, probably because in many cases the adjacent outcrops are representative of different fragments in the megabreccia.

The principal large-scale structural feature that has been recognized to date is the N 35°W trending Diablo antiform (see Figure 1, this paper, and *Bailey et al.* [1964, Figure 29]). The Diablo range is bordered on the east by the San Joaquin Valley and consists of a core of complexly deformed Franciscan, surrounded by predominantly Upper Cretaceous, but also including minor Upper Jurassic and Lower Cretaceous strata of the Great Valley sequence. Serpentinite-invaded piercement domes at Mount Diablo and New Idria mark the northern and southern boundaries, respectively, of the structure. The N 65°W trending Vallecitos syncline and Kettleman hills anticline are superimposed en echelon fashion on the southern portions of the uplift. Broadly antiformal, the contacts between the Franciscan core and the peripheral Great Valley strata are everywhere mapped as steeply dipping faults [*Wilson*, 1943; *Leith*, 1949; *Crittenden*, 1951; *Briggs*, 1953; *Hall*, 1958; *McKee*, 1962*a*; *Enos*, 1963; *Maddock*, 1964; *Ernst*, 1965; *Dickinson*, 1966*a*, *b*]. On the east side of the range, this fault is called the Ortigalita, or Tesla–Ortigalita fault.

The extension of the Diablo antiform north of San Francisco Bay is less distinct. If the problematical mid-Cretaceous, volcanic-poor, K-feldspar-rich graywacke terrane of the northwest coastal belt of Franciscan [*Bailey and Irwin*, 1959; *Bailey et al.*, 1964] can be correlated[1] with coeval miogeosynclinal rocks exposed in the Sacramento Valley, the general trend and structural relations of the Diablo antiform are retained; the exposed core of Upper Jurassic and Lower Cretaceous eugeosynclinal rocks is much more extensive than farther south, however. The San Andreas fault truncates the western portion of this complex. The nature of the junction between the northwest coastal belt and the more typical and older Franciscan mélange is unknown. The eastern margin of the northern California Coast Ranges (the western border of the northern half of the Great Valley) is marked by the Stoney Creek fault zone, which juxtaposes Franciscan rocks on the west against Knoxville and younger strata of the Sacramento Valley on the east. Chiefly a high-angle, east-dipping reverse fault, it locally displays a subhorizontal contact [*Brown*, 1964*a*, 1964*b*]. Westward overthrusting of the Great Valley sequence relative to the Franciscan would explain small belts of miogeosynclinal strata within the northern Coast Ranges as klippen [*Bailey et al.*, 1964, pp. 163–165]. Somewhat similar relations are evident farther north [*Irwin*, 1960, 1964, 1966; *Davis*, 1966], where a series of arcuate, imbricate thrust plates of the Klamath Mountain province, overlain unconformably by Great Valley strata on the east, has ridden westward relatively over the Franciscan terrane; the schists of South Fork Mountain mark the tectonic contact between these two provinces [*Blake et al.*, 1967].

The relationship between the South Fork Mountain and Stoney Creek faults is obscure, but the similarity between Franciscan metamorphic rocks exposed along the northeastern margin of the California Coast Ranges bordering the Sacramento Valley and the South Fork Mountain schists [*Ghent*, 1965; *Blake et al.*, 1967] probably reflects a continuity of structure. Moreover, *Page* [1966, p. 272] has suggested that the faulted margin of the Diablo antiform is correlative with the Stoney Creek fault, its present nearly vertical attitude reflecting a Late Cenozoic diapiric movement of the Franciscan core. Thus it seems possible that the principal sections of Franciscan and Great Valley strata lying to the east of the San Andreas fault owe their present intimate association to a major thrusting event.

The northwest-trending San Andreas fault truncates Coast Range structures and, south-

[1] The Ukiah (1960), Redding (1962), and Santa Rosa (1963) 1:250,000 geologic map sheets of the California Division of Mines and Geology show the northwest coastal belt as undifferentiated Cretaceous and distinguish this unit from the Franciscan; although illustrated by different map patterns, *Bailey et al.* [1964, Plate 1] on the other hand assign both sections to the eugeosynclinal mélange.

## TECTONIC CONTACT

west of the Diablo antiform, has repeated the west-to-east succession of Franciscan group, Great Valley sequence, and the granitic-metamorphic complex along the coast from Monterey Bay to Point Conception [*King*, 1959, pp. 164–173]. Various lines of evidence, including historically observed and measured slip; juxtaposition of contemporaneous sedimentary strata of vastly differing facies and of plutonic igneous rocks against unmetamorphosed, perhaps in part older, sediments or against high-pressure, low-temperature metamorphics; displaced distinctive lithologic contacts and complexes; and offset paleomolluscan provinces all argue convincingly in favor of large-scale Cenozoic right-lateral movement on this profound break [*Hill and Dibblee*, 1953; *Hill*, 1964; *Curtis et al.*, 1958; *Hall*, 1960; *Crowell*, 1960, 1962; *Dibblee*, 1966; *Meade and Small*, 1966; *Dickinson and Grantz*, 1968].

The Franciscan terrane exposed along the Pacific coast south of Monterey Bay is referred to as the South Fork Mountain–Stoney Creek–Ortigalita system, Late Mesozoic eugeosynclinal rocks on the west, presumed to possess an oceanic crust (trench?) basement, are brought into contact with roughly contemporaneous miogeosynclinal strata on the east. The Sur-Nacimiento fault, although typically northeast-dipping, evidently is a complicated series of subparallel breaks, the individual members of which are offset by later cross faults [*Page*, 1970]. Great Valley sequence rocks cropping out on the northeast side of the Sur-Nacimiento fault zone are associated with a granitic and metamorphic basement just as they are in the Klamath Mountains, whereas on the inland side of the San Andreas, Sierran basement is not exposed along the Stoney Creek–Ortigalita fault. Outliers of Upper Jurassic and Cretaceous miogeosynclinal strata cropping out within the Franciscan-cored Nacimiento block and in a narrow strip along the west side of the block (see San Luis Obispo 1959 geologic map sheet of the California Division of Mines and Geology and *Page* [1969])

lend credence to the hypothesis that although it appears to be a high-angle, east- or northeast-dipping reverse fault [*Hall and Corbató*, 1967, p. 577], the Sur-Nacimiento fault zone may locally become a subhorizontal thrust [*Brown*, 1968; *Page*, 1970] analogous to that noted much farther to the northeast. The apparent draping of Great Valley strata over the Franciscan core suggests that the Nacimiento block may be antiformal, at least in a gross way (Figure 1).

### GENERAL GEOLOGY OF THE GREAT VALLEY

The Sacramento and San Joaquin Valleys are the site of a great synclinorium consisting of Upper Mesozoic and Cenozoic miogeosynclinal strata. This downwarp is strongly asymmetric, with the axis of maximum accumulation displaced westward toward the California Coast Ranges [e.g., see *Safonov*, 1962, Figures 3–5]; thin shelf sediments on the east grade laterally into thicker trough deposits on the west. The structural asymmetry reflects the fact that individual series tend to thicken westward and partly results from a systematic eastward transgression of younger Mesozoic beds over the Sierran basement. Individual series show variations in stratigraphic thickness, facies changes, and megafossil assemblages indicative of a northward shoaling of the trough [*Hackel*, 1966, pp. 222, 236], the direction in which such rocks lap up onto the plutonic-metamorphic Klamath Mountain terrane. These depositional relationships are also suggested by south-trending paleocurrent vectors measured in the Upper Jurassic and Cretaceous rocks of the northern portion of the Great Valley by *Ojakangas* [1968].

The Nevadan and pre-Nevadan basement of the Klamath Mountain province ranges in age from at least as old as Devonian (380 m.y.) to approximately Late Jurassic (130–165 m.y.) according to *Lanphere et al.* [1968]. Paleozoic and Lower Mesozoic metamorphics of the Sierra Nevada and White-Inyo ranges have been intruded by composite plutons that span the time interval between about Late Triassic and Latest Jurassic (125–210 m.y.), with a younger igneous event having taken place during the early Late Cretaceous (80–90 m.y.), as reported by *Bateman and Eaton* [1967].

In the Sacramento Valley, a thick, uninter-

rupted sedimentary succession of Knoxville formation, Shasta series, and Chico group strata are exposed, indicating virtually no tectonic interruptions of deposition during Late Jurassic and Cretaceous time. (For an alternative interpretation involving separation of the successive Upper Jurassic and Cretaceous units by unconformities, see *Peterson* [1965, 1967]). Slump deposits [*Crowell*, 1957, 1957, pp. 995–998; *Brown and Rich*, 1960, 1961] may reflect a period of fault movement within the trough, which exposed escarpments of Sierran basement to erosion [*Brown and Rich*, 1967, Figures 3 and 4]. Both Upper Jurassic and Lower Cretaceous rocks also have been recognized along the west side of the San Joaquin Valley, but the lateral extent of these units is small compared with correlative sections of rock cropping out farther north [*Schilling*, 1962; *Bailey et al.*, 1964, Plate 1; *Hackel*, 1966, pp. 220–222]. Greater initial thicknesses of these units may have been decreased locally by erosion during a mid-Cretaceous Diablo uplift [*Repenning*, 1960, Figure 4], prior to the deposition of voluminous Upper Cretaceous sediments [e.g., *Schilling*, 1962] in the southern part of the Great Valley.

General post-Cretaceous and pre-Late Eocene deformation, uplift, and erosion brought about definition of the present structural basin and preceded the subsequent Cenozoic marine and continental deposition [*Hackel*, 1966, pp. 223–227, 236). Vertical movements during earliest Tertiary time led to the formation of two transverse structural highs, the Stockton and Bakersfield arches [*Repenning*, 1960]. The more northerly of these structures divided the Great Valley into two distinct Cenozoic basins of deposition; accumulation of Tertiary marine strata in the San Joaquin far exceeded that in the Sacramento Valley. The period of most extreme deformation commenced in mid-Miocene time and culminated during the mid-Pleistocene [*Hackel*, 1966, p. 236]; this event seems to have been related to upward motion of the Franciscan terrane west of the Stoney Creek–Ortigalita fault system and evidently resulted in the present configuration of the Great Valley synclinorium. The western, thicker and more mobile portion of the synclinorium is characterized by small open folds and faults of small displacement that roughly parallel the large structure. Major cross valley breaks include the Stockton

and White Wolf faults at the northern and southern margins of the San Joaquin Valley, respectively.

A typical geologic cross section of the Sacramento Valley is illustrated in Figure 2a, modified after *Lachenbruch* [1962, Plate 3]. This structure is quite similar to that presented for other portions of the Great Valley [*Safonov*, 1962, Figures 3–5; *Schilling*, 1962; *Brown and Rich*, 1967, Figure 4; *Crowell*, 1968a, Figure 1]. In Figure 2b the post-Cretaceous deposition, folding, and faulting have been removed, and in Figure 2c the Cretaceous system has also been stripped away. Reconstructions presented in Figures 2b and 2c indicate that, similar to the present attitudes of the South Fork Mountain and Sur-Nacimiento faults, the Stoney Creek–Ortigalita fault system initially dipped toward the interior of the continent, that is, toward the Sierran basement, prior to post-Cretaceous deformation. The present nearly vertical attitude of this junction along the west side of the Great Valley apparently reflects rotation during the Late Cenozoic upheaval of the Franciscan-cored Diablo antiform and its northern extension, and perhaps medial portions of the Nacimiento block as well.

### DYNAMIC MODEL

*King* [1959, p. 168] and *Dietz* [1963a] advanced the hypothesis that rocks of the Franciscan group were deposited in deep water on, or at, the foot of the continental slope. Such an environment is compatible with the apparent absence of a continental-type basement complex underlying this eugeosynclinal mélange. The great thickness of this linear belt of chaotically disturbed rocks, intimate association with 'cold' ultramafic plutons, and the development of peculiar, relatively high-pressure, low-temperature metamorphic minerals all favor accumulation in an oceanic trench [*Ernst*, 1965, pp. 905–910; *Ernst et al.*, 1970, Chapter 15; *Bailey et al.*, 1964, pp. 105–112). Some deposition could have taken place on the seaward side of the downwarp, where sea-floor spreading would result in transportation of sediments toward the trench and sediment accumulation within the trench.

But what relationship did this ensimatic prism bear to the well-bedded, largely miogeosynclinal shelf and trough sediments of the

Great Valley sequence? From the conceptual models of *King* [1959, pp. 170-171] and *Dietz* [1963a, Figure 2], one would predict the existence of units reflecting a transition zone in passing from the western eugeosynclinal facies towards the largely epicontinental miogeosynclinal facies. Such relationships have not in fact been recognized, as indicated by the following quotes:

> It is curious that nowhere does the Franciscan clearly grade either laterally or vertically into the Great Valley sequence, despite the fact that the two facies were being deposited contemporaneously through an appreciable segment of geologic time.

> Equally curious is the fact that nowhere has the Franciscan been found to lie depositionally on the Great Valley sequence, even though the youngest Franciscan rocks are clearly younger than some parts of the Great Valley sequence. [*Bailey et al.*, 1964, p. 146]

> It is increasingly doubtful that the Great Valley sequence has ever been observed in depositional contact on the Franciscan. Moreover, it is increasingly doubtful that the base of the Jurassic part of the Great Valley sequence and the Franciscan rocks are in contact, a fault intervenes. [*Page*, 1966, p. 272]

From the general discussion of the Great Valley synclinorium presented in the last section, it is clear that the more easterly strata were deposited on a stable shelf, floored by sialic Klamath-Sierra-Salinia-type basement. The abrupt westward thickening of the units (refer to Figure 2) probably signals deposition on the downsinking continental slope. In the case of an adjacent oceanic trench, the base of this slope, or continental rise [*Menard*, 1964, p. 226] would approximately mark the region of emergence of the Benioff, or seismic shear zone; such a zone of earthquake foci, which is invariably associated with present-day trenches, dips at an angle of 30°-60° beneath the continent or island arc [*Benioff*, 1954; *Richter*, 1959, pp. 413-414; *Sykes*, 1966; *Oliver and Isacks*, 1967; *Isacks et al.*, 1968, Figure 9; *Seyfert*, 1969]. This junction between eugeosynclinal trench and miogeosynclinal shelf-slope deposition would represent the locus of impingement between the spreading and downbuckling oceanic lithosphere and the overriding continental lithospheric plate, in accord with the concept of the new global tectonics [*Hess*, 1962,

1965; *Dietz*, 1963b, 1966; *Isacks et al.*, 1968]. Actually the thickest portion of the Great Valley sequence lies to the east of the Diablo antiform and its northern extension, where no evidence of a continental basement exists. In fact, on the west side of the Sacramento Valley, the Knoxville formation grades down into tuffaceous sandstone and pillow basalt, in turn succeeded by ultramafic material situated along the Stoney Creek fault zone [*Brown*, 1964a, b; *Bailey and Blake*, 1969]. Conceivably the miogeosynclinal strata could be especially thick here because they were deposited on sea floor that locally was situated directly east of the Benioff zone. The Benioff zone everywhere would have marked the junction between the two lithospheric plates, and although in general this contact presumably also would have juxtaposed North American continental and Pacific oceanic crusts, the seismic shear zone could have emerged some tens of kilometers west of the continental rise in the region now occupied by the Great Valley. Such a spatial relationship obviously would account for the complete lack of continental-type basement rocks in the vicinity of the Stoney Creek-Ortigalita fault. Moreover, the greater thickness of miogeosynclinal strata here relative to portions of the Great Valley sequence deposited elsewhere and clearly resting on continental basement might also be a function of the depositional site.

The overall tectonic model envisioned for Late Mesozoic California is presented as Figure 3; in some respects this model is similar to various others proposed by *Hill and Hobson* [1968], *Crowell* [1968b], *Yeats* [1968], *Page* [1969, 1970], *Bailey and Blake* [1969], and *Hamilton* [1969]. The facies change between Franciscan and Great Valley sequence rocks is thought to have taken place chiefly in the vicinity of the continental rise, possibly reflecting hydrographic control. However, with episodic or continuous convergence of the lithospheric plates, the lateral gradation between the trench and the sea floor, slope, and shelf deposits would have been obliterated periodically or continuously by continental overthrusting and deformation of the underthrust Franciscan. Modern Benioff zones penetrate deep within the mantle, as indicated by earthquake hypocenters localized along them; hence, by analogy, shearing could have provided (1) spalled-off portions of the

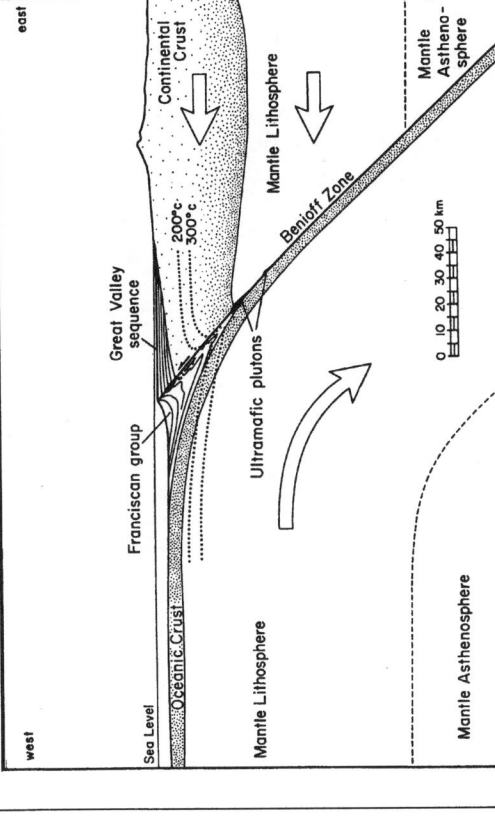

Fig. 3. Tectonic model involving deposition of the Franciscan mélange in a northwest-trending oceanic trench, and the Great Valley sequence principally on the adjacent continental slope and shelf. As illustrated, the Benioff zone in general marks the contact along which the continental lithospheric plate (continental crust plus mantle lithosphere) overrides the oceanic lithospheric plate (oceanic crust plus mantle lithosphere). Hydrated fragments of mantle material rise chiefly along this seismic shear zone, and are in part tectonically mixed into the trench mélange during deformation. No vertical exaggeration implied.

upper mantle peridotite, (2) infusion of aqueous solutions from the deforming trench complex, and (3) ready egress for the buoyant, partly and completely serpentinized products of mantle hydration.

Depression of the isotherms within the deforming, downbuckling eugeosynclinal mélange is also illustrated in Figure 3. Such deflection of the supramantle geotherms is proposed in light of the physical conditions required for the in situ blueschist facies metamorphism [*Ernst*, 1965, p. 909; *Ernst and Seki*, 1967; *Ernst et al.*, 1970, chap. 13; *Taylor and Coleman*, 1968]. It should be pointed out that the downbowing of the isotherms illustrated in Figure 3 is much less pronounced than that computed by *Oxburgh and Turcotte* [1968] for a plate of oceanic lithosphere 100 km thick moving at a spreading rate of 5 cm/year (see also *McKenzie* [1969], Figure 3). The Franciscan metamorphic Pt regime of 150°-300°C and 5 to more than 8 kb inferred from field, experimental, and theoretical phase equilibrium investigations, requires depths of burial in the range of 20-30 km, assuming total pressure is entirely the result of the weight of the overlying material.[2]

---

[2] It must be pointed out that the observed mineral parageneses have been accounted for by alternative schemes; although as controversial, these schemes share the merit of not requiring abnormally great burial of rocks during the Franciscan metamorphism. For instance, the generation of substantial tectonic 'overpressures' accompanying thrust faulting supposedly would allow the formation of blueschist phase compatibilities at less profound depths. Other explanations include the postulated metastable recrystallization of blueschist-facies minerals under the influence of metasomatism at moderate pressures. The issue is by no means settled. A review of the problem is now underway, but the author has stated his convictions previously [*Ernst*, 1963d, 1965, pp. 896-898, 909; *Ernst et al.*, 1970, chap. 13].

Locally the metamorphic event must have been nearly contemporaneous with deposition and tectonic thickening. Such temporal relationships are indicated by radiometric dates of recrystallization: between 130 and 150 m.y. for sections of the Diablo range and its northern extension, which contain Late Jurassic fossils [*Enos*, 1963; *Lee et al.*, 1963; *Ernst et al.*, 1970, chap. 14]; approximately 110–130 m.y. for sparsely fossiliferous Lower Cretaceous rocks of the northern Coast Ranges [*Ghent*, 1963; *Bailey et al.*, 1964, pp. 119–122; *Brown*, 1964a, b; *Suppe*, 1969]; and about 75–80 m.y. for mid-Cretaceous units of the northwest coastal belt of the Franciscan and adjacent San Francisco Bay [*Thalmann*, 1942, 1943; *Schlocker et al.*, 1954; *Hertlein*, 1956; *Keith and Coleman*, 1968].

*Ojakangas* [1968] demonstrated that miogeosynclinal strata of the Sacramento Valley were laid down parallel to an eastern shoreline in a southward-deepening, elongate basin that was open to the west; lithic analysis indicated derivation from the ancestral Sierra Nevada as well as from the Klamath Mountains. Although the rather sparse Franciscan current features have not been investigated systematically, the occurrence of clastic K-feldspar grains in much greater abundance in mélange rocks of known Cretaceous age compared with the Upper Jurassic graywacke [*Bailey and Irwin*, 1959] indicates a similar source area to that from which Great Valley sequence strata were derived.

From the general model presented in Figure 3, the following Late Mesozoic depositional scheme may be suggested. Miogeosynclinal conglomerate, sandstone, siltstone, and shale were deposited successively farther from the strand line on the continental shelf, the slope, and, locally, the adjacent sea floor; sediment transport involved westward (and southward) moving currents, with southward deflection of the bottom flow[3] and longitudinal basin filling near and to the east of an assumed linear topographic prominence. This hydrographic feature is hypothesized to have marked the emergence of the Benioff zone, where largely continental crust presumably was continuously overriding

[3] Although *Ojakangas* [1968] documented this southward current direction for miogeosynclinal strata of the Sacramento Valley, such paleocurrent vectors have not been reported for the San Joaquin.

the oceanic trench system. With a predominantly eastern (ancestral Sierran) source, Franciscan detritus somehow must have been transported westward beyond the site of accumulation of Great Valley sediments during the contemporaneous deposition. Perhaps carried by along-shore currents, moderately coarse clastic debris presumably collected in shallow water at the heads of submarine canyons. It is conjectured that far downslope these submarine canyons (which transected the continental shelf and slope) must have locally breached the buried ridge or surface expression of the seismic shear zone. Evidently, due to oversteepened slopes and continued loading, this clastic material, which formed a temporary accumulation near the eastern margin of the Great Valley sediments, periodically swept down-canyon to the Franciscan trench itself and subsequently filled it longitudinally.

Pillow lava is present in modest amounts within the eugeosynclinal mélange, but except for minor occurrences in the Knoxville formation, volcanics are virtually absent from the miogeosynclinal section. It therefore seems likely that nearby volcanic vents were not disposed on the continental side of the Franciscan trough. Either (1) the magmatic conduits were located within the trench itself, or (2) the primary sources lay to the west. The first alternative is considered improbable, judging from descriptions of the present-day, nonvolcanic nature of trenches [e.g., see *Menard*, 1964, pp. 97–116]. The second alternative would be difficult, if not impossible, to document, granted that Late Mesozoic and Cenozoic sea-floor spreading has occurred; we may conjecture, however, that volcanism associated with a nearby Mesozoic oceanic ridge (possibly the Darwin Rise [*Menard*, 1964, pp. 138–146], although its position relative to the margin of North America in Late Mesozoic time is problematical) provided the pillow basalt and associated deep sea sediments that doubtless would have been incorporated in the trench mélange during lithospheric spreading and downbuckling.

### Schematic Tectonic History of the California Coast Ranges and the Great Valley

Drawing on the information set forth in the first portion of this paper and tentatively ac-

cepting the speculative, dynamic model presented in the last section, a four-stage sequence of tectonic events is advanced to account for geologic relationships observed in western California, as illustrated in Figure 4. This region bears evidence of the complex interaction between converging oceanic and continental lithospheric plates during the past 150 m.y. The Benioff zone evidently played a crucial structural role in this history, as seen from the previous discussion; it was this tectonic junction that appears to have marked the contact between oceanic Franciscan trench deposits and Great Valley sequence strata draped principally on the continental shelf and slope.

During Late Jurassic time (Figure 4a), eugeosynclinal rocks were accumulating in a north-

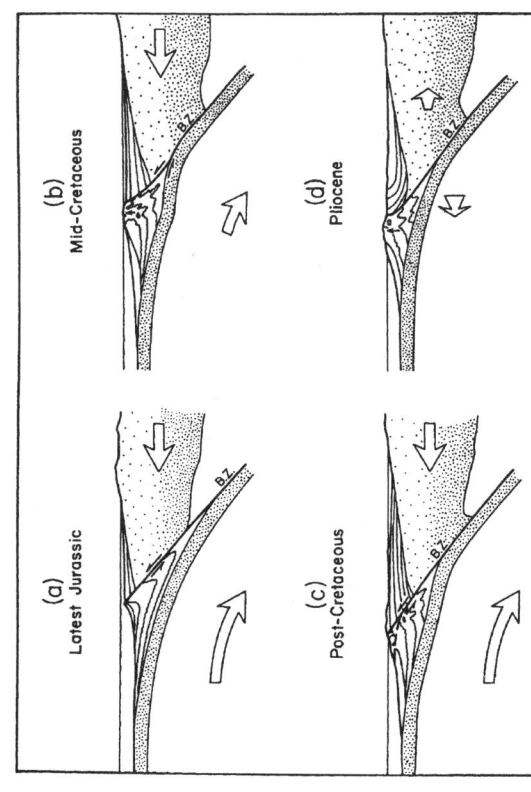

west-trending oceanic deep marginal to an easterly sialic crust; the western periphery of the latter was the chief depositional site of the largely slope facies Knoxville formation. An hypothesized encroachment of the continental lithospheric plate over the descending oceanic plate caused tectonic thickening of the Franciscan mélange; rapid relative northeastward or eastward spreading, associated with a nearby rise, and pressure loading by continental plate overthrusting (i.e., increment of pressure on the underlying Franciscan due to the thrust-over mass, not tectonic 'overpressure') would have allowed for the great downbuckling of buoyant, relatively cold, chaotically deformed eugeosynclinal rocks. The most extensive blueschist-facies metamorphism of the ensimatic prism and the

Fig. 4. Schematic models depicting stages of sea-floor spreading, and impingement of the continental lithosphere on the tectonically thickened and dragged down trench deposits (basically an elaboration on the central portion of Figure 3, same roughly east-west view). Arrow lengths indicate approximate relative convergence velocities; in (d), right-lateral movement is at a low angle to the plane of section; hence arrows are illustrated in perspective. Stages (a)–(c) are related to spreading presumably associated with nearby Late Mesozoic and Early Tertiary rises; in contrast, stage (d) depicts the modern movement picture involving the currently active East Pacific Rise. Geologic times indicated are only approximate. No vertical exaggeration implied.

conversion of basaltic material to eclogite and amphibolite within the oceanic crust seem to have occurred at this time in axial (deepest) portions of the trench.

Significant local uplift of the Franciscan (the Diablan orogeny of *Taliaferro* [1943, pp. 216–217], or Diablo uplift of *Repenning* [1960, Figure 4]) may reflect a diminution in the Pacific spreading rate during approximately mid-Cretaceous time (Figure 4b), or slightly earlier. The great mass of eugeosynclinal material, previously kept at considerable depths due to the dynamics of sea floor spreading, would have tended both to emerge as a positive paleogeographic element locally and to have depressed the underlying oceanic crust and Mohorovicic discontinuity in an attempt to achieve regional isostatic balance. Erosion of exposed portions of the Franciscan mélange could account for the apparently reworked nature of the graywacke which constitutes a large part of the northwest coastal belt; such strata probably were never buried as deeply as portions of the Diablo range and its northern extension, so incipient metamorphism in the northwest Coast Ranges produced zeolites instead of blueschist-facies mineral assemblages [*Bailey et al.*, 1964, pp. 93 and 146, Figure 18].

During latest Cretaceous time in northern California [*Brown*, 1964a, b; *Irwin*, 1964, 1966, p. 31; *Blake et al.*, 1967], and Paleocene farther to the south [*Hackel*, 1966, p. 236; *Page*, 1966, 1969a], a conjectured acceleration of northeastward underflow of the Pacific lithospheric plate may have been responsible for widespread overthrusting along the Benioff zone, that is, the South Fork Mountain–Stoney Creek–Ortigalita (–Sur–Nacimiento) fault system (Figure 4c). Distributary thrust sheets locally could have produced an imbrication of older and younger Franciscan group and Great Valley sequence rocks; such structural complications may be responsible for the observed intimate association of feebly and more thoroughly metamorphosed mélange rocks as well as complex tectonic contacts with roughly coeval miogeosynclinal strata. Subsequent to this period of rapid spreading, which evidently involved relative northeast-southwest compression, activity of the nearby rise apparently waned; its gradual decay may have been related to formation of the Early Tertiary flow regime

of the East Pacific Rise [*Menard*, 1964, p. 137]. Because of its different disposition, Tertiary motion of the Pacific Ocean lithospheric plates related to the East Pacific Rise would have had a rather different tectonic effect on western California than that hypothesized for the Late Mesozoic.

The east-west offset of magnetic anomalies along transform faults in the east Pacific [*Vacquier et al.*, 1961; *Wilson*, 1965] indicates that an east-west spreading regime prevailed during most of Tertiary time. This movement pattern apparently was replaced about 10 m.y. ago by the present northwest-southeast-spreading trend [*Morgan*, 1968, p. 1968; *Heirtzler et al.*, 1968, p. 2132; *Le Pichon*, 1968, Table 2 and p. 3666]. This change in dynamics is also reflected by motion along the San Andreas fault [*Bolt et al.*, 1968] and, possibly, by rifting of the Gulf of California [e.g., see *Larsen et al.*, 1968].

At least as long ago as mid-Miocene, the Diablo antiform began moving upwards, judging from the abundant serpentinite debris shed into the San Joaquin Valley at that time [*Hackel*, 1966, p. 233]. Diapirism continued into Pleistocene time (see Figure 4d), as demonstrated by the steep tilting of Plio-Pleistocene beds [*Wilson*, 1943, p. 264]. The structural configuration illustrated in Figure 4d is most appropriate for northern California. Farther south the Franciscan mélange evidently diapirically intruded relatively farther east into the miogeosynclinal sequence than shown; hence Great Valley strata are also draped along the western flanks of the Diablo antiform and the Nacimiento block. Right-lateral motion along the San Andreas (the latter not shown in Figure 4d) since at least as early as Miocene time [*Hill and Dibblee*, 1953, p. 446; *Hill and Hobson*, 1968; *Hall*, 1960; *Crowell*, 1962, pp. 17–25; *Addicott*, 1968] has paralleled the present spreading direction. This strike slip caused the low-angle truncation of southern Coast Range structures, and it offset the Late Mesozoic Benioff zone of contact between miogeosynclinal strata and their underlying Klamath, Sierran, and Salinian basement on the one hand, and eugeosynclinal trench deposits on the other.

Buoyant rise of the continental margin, continuing to the present, may be responsible for the fact that the depth to the Mohorovicic

discontinuity is approximately the same now under both the California Coast Ranges and the Great Valley; however, the eastern portion of the San Joaquin Valley appears to be underlain by anomalously thin crust according to *Bateman and Eaton* [1967].

### Concluding Remarks

The dynamic model and tectonic history of western California presented in the two preceding sections are, of course, highly speculative, but to some extent they help to explain the nature of the junction between continental shelf-slope and oceanic deep deposits. *Hatherton* [1969] has drawn attention to the fact that the spatial arrangement of Late Paleozoic and Mesozoic eugeosynclinal and miogeosynclinal terranes in New Zealand in many respects is similar to that described for the Franciscan group and Great Valley sequence. Apparently analogous circum-Pacific tectono-sedimentary belts occur in Hokkaido and Sakhalin (*Saito et al.* [1960, pp. 51–54, 65]; *Bogdanov* [1967]; personal communication, *Bogdanov*, 1968) and in southeastern Alaska. [*Burk*, 1965, 1966]. Whether the Benioff zone model here proposed for western California is indeed found to be applicable to these other areas or not, we must eventually account for what appears to have been a profound and widespread, chiefly Late Mesozoic interaction between converging blocks of continental and Pacific Ocean lithosphere.

*Acknowledgments.* M. C. Blake, Jr., D. S. Cowan, J. C. Crowell, W. R. Dickinson, C. A. Hall, Jr., M. L. Hill, and B. M. Page are thanked for criticizing the manuscript. Many of the ideas expressed herein are not new; hopefully they have been acknowledged adequately. However, if concepts and factual data depart from reality, they are, of course, solely the responsibility of the author.

Research reported here has been supported by the University of California, Los Angeles, and by the National Science Foundation through grant GA-11488.

### References

Adicott, W. O., Mid-Tertiary zoogeographic and paleogeographic discontinuities across the San Andreas fault, California, in *Proceedings of Conference on Geologic Problems of San Andreas Fault System, 11*, edited by W. R. Dickinson and A. Grantz, pp. 144–165, Stanford University Publications, Geological Sciences, Stanford, California, 1968.

Bailey, E. H., and W. P. Irwin, K-feldspar content of Jurassic and Cretaceous graywackes of the northern Coast Ranges and Sacramento Valley, California, *Bull. Amer. Ass. Petrol. Geol., 43*, 2797–2809, 1959.

Bailey, E. H., W. P. Irwin, and D. L. Jones, Franciscan and related rocks and their significance in the geology of western California, *Calif. Div. Mines Geol. Bull., 183*, 1964.

Bailey, E. H., and M. C. Blake, Jr., Late Mesozoic sedimentation and deformation in western California, *Geotektonika*, 17–31, 1969.

Bateman, P. C., and J. P. Eaton, Sierra Nevada batholith, *Science, 158*, 1407–1417, 1967.

Benioff, H., Orogenesis and deep crustal structure, *Bull. Geol. Soc. Amer., 65*, 385–400, 1954.

Blake, M. C., Jr., W. P. Irwin, and R. G. Coleman, Upside-down metamorphic zonation, blueschist facies, along a regional thrust in California and Oregon, *U.S. Geol. Surv. Prof. Pap. 575-C*, 1–9, 1967.

Bloxam, T. W., Jadeite-bearing metagraywackes in California, *Amer. Mineral., 41*, 488–496, 1956.

Bloxam, T. W., Glaucophane-schists and associated rocks near Valley Ford, California, *Amer. J. Sci., 257*, 95–112, 1959.

Bloxam, T. W., Jadeite-rock and glaucophane schists from Angel Island, San Francisco Bay, California, *Amer. J. Sci., 258*, 555–573, 1960.

Bogdanov, N. A., Paleozoic geosynclines of the western part of the circum-Pacific belt, *Tectonophysics, 4*, 581, 1967.

Bolt, B. A., C. Lomnitz, and T. V. McEvilly, Seismological evidence on the tectonics of central and northern California and the Mendocino Escarpment, *Bull. Seismol. Soc. Amer., 68*, 1725–1768, 1968.

Briggs, L. I., Geology of the Ortigalita peak quadrangle, California, *Calif. Div. Mines Geol. Bull., 167*, 1953.

Brown, J. A., Jr., Thrust contact between Franciscan group and Great Valley sequence northeast of Santa Maria, California, Ph.D. dissertation, 244 pp., University of Southern California, Los Angeles, 1968.

Brown, R. D., Jr., Thrust-fault relations in the Northern Coast Ranges, California, *U.S. Geol. Surv. Prof. Pap. 475-D*, 7–13, 1964a.

Brown, R. D., Jr., Geologic map of the Stoneyford quadrangle, Glenn, Colusa, and Lake Counties, California, *U.S. Geol. Surv. Miner. Invest. Map MF-279*, 1964b.

Brown, R. D., Jr., and E. I. Rich, Early Cretaceous fossils in submarine slump deposits of Late Cretaceous age, northern Sacramento Valley, California, *U.S. Geol. Surv. Prof. Pap. 400-B*, 318–320, 1960.

Brown, R. D., Jr., and E. I. Rich, Geologic map of Lodoga quadrangle, Glenn and Colusa Counties, California, *U.S. Geol. Surv. Oil and Gas Invest. Map OM-210*, 1961.

Brown, R. D., Jr., and E. I. Rich, Implications of two Cretaceous mass transport deposits, Sacramento Valley, California, Comment on a paper by Gary L. Peterson, *J. Sediment. Petrol*, *37*, 240-248, 1967.

Burk, C. A., Geology of the Alaska Peninsula—Island arc and continental margin, *Geol. Soc. Amer. Mem. 99*, 250 pp., 1965.

Burk, C. A., The Aleutian arc and Alaska continental margin, in *Continental Margins and Island Arcs*, *Paper 66-15*, edited by W. H. Poole, 486 pp., Geological Survey of Canada, 1966.

Coleman, R. G., Jadeite deposits of the Clear Creek area, New Idria district, San Benito County, California, *J. Petrol*, *2*, 209-247, 1961.

Coleman, R. G., Composition of jadeitic pyroxene from the California metagraywackes, *U.S. Geol. Surv. Prof. Pap. 525-C*, 25-34, 1965.

Coleman, R. G., and D. E. Lee, Metamorphic aragonite in the glaucophane schists of Cazadero, California, *Amer. J. Sci.*, *260*, 577-595, 1962.

Coleman, R. G., and D. E. Lee, Glaucophane-bearing metamorphic rock types of the Cazadero area, California, *J. Petrol*, *4*, 260-301, 1963.

Crittenden, M. D., Jr., Geology of the San Jose-Mount Hamilton area, California, *Calif. Div. Mines Geol. Bull*, *157*, 74 pp., 1951.

Crowell, J. C., Origin of pebbly mudstones, *Bull. Geol. Soc. Amer.*, *68*, 993-1010, 1957.

Crowell, J. C., The San Andreas fault in Southern California, *Int. Geol. Congr., 21st*, *18*, 45-52, 1960.

Crowell, J. C., Displacement along the San Andreas fault, California, *Geol. Soc. Amer. Spec. Pap. 71*, 61 pp., 1962.

Crowell, J. C., The California Coast Ranges, *UMR J.*, *1*, 133-156, 1968a.

Crowell, J. C., Movement histories of faults in the Transverse ranges and speculations on the tectonic history of California, in *Proceedings of Conference on Geologic Problems of San Andreas Fault System*, *11*, edited by W. R. Dickinson and A. Grantz, 374 pp., Stanford University Publications, Geological Sciences, Stanford, California, 1968b.

Curtis G. H., J. F. Evernden, and J. I. Lipson, Age determination of some granitic rocks in California by the potassium-argon method, *Calif. Div. Mines Geol. Spec. Rep. 64*, 16 pp., 1958.

Davis, G. A., Metamorphic and granitic history of the Klamath Mountains, in *Geology of Northern California*, edited by E. H. Bailey, *Calif. Div. Mines Geol. Bull.*, *190*, 375-384, 1966.

Dibblee, T. W., Jr., Evidence for cumulative offset on the San Andreas fault in central and northern California, in *Geology of Northern California*, edited by E. H. Bailey, *Calif. Div. Mines Geol. Bull.*, *190*, 217-238, 1966.

Dickinson, W. R., Table Mountain serpentinite extrusion in California Coast Ranges, *Bull. Geol. Soc. Amer.*, *77*, 451-472, 1966a.

Dickinson, W. R., Structural relationships of San Andreas fault system, Cholame Valley, and Castle Mountain Range, California, *Bull. Geol. Soc. Amer.*, *77*, 707-726, 1966b.

Dickinson, W. R., and A. Grantz, Eds., *Proceedings of Conference on Geologic Problems of San Andreas Fault System*, *11*, 374 pp., Stanford University Publications, Geological Sciences, Stanford, California, 1968.

Dickinson, W. R., R. W. Ojakangas, and R. J. Stewart, Burial metamorphism of the Late Mesozoic Great Valley sequence, Cache Creek, California, *Bull. Geol. Soc. Amer.*, *80*, 519-526, 1969.

Dietz, R. S., Alpine serpentinites as oceanic rind fragments, *Bull. Geol. Soc. Amer.*, *74*, 947-952, 1963a.

Dietz, R. S., Collapsing continental rises, an actualistic concept of geosynclines and mountain building, *J. Geol*, *71*, 314-333, 1963b.

Dietz, R. S., Passive continents, spreading sea floors, and collapsing continental rises, *Amer. J. Sci.*, *264*, 177-193, 1966.

Enos, P., Jurassic age of Franciscan formation south of Panoche Pass, California, *Bull. Amer. Assoc. Petrol. Geol.*, *47*, 159-163, 1963.

Ernst, W. G., Petrogenesis of glaucophane schists, *J. Petrol*, *4*, 1-30, 1963a.

Ernst, W. G., Polymorphism in alkali amphiboles, *Amer. Mineral*, *48*, 241-260, 1963b.

Ernst, W. G., Significance of phengitic micas from low-grade schists, *Amer. Mineral*, *48*, 1357-1373, 1963c.

Ernst, W. G., Mineral parageneses in Franciscan metamorphic rocks, Panoche Pass, California, *Bull. Geol. Soc. Amer.*, *76*, 879-914, 1965.

Ernst, W. G., and Y. Seki, Petrologic comparison of the Franciscan and Sanbagawa metamorphic terranes, *Tectonophysics*, *4*, 463-478, 1967.

Ernst, W. G., Y. Seki, H. Onuki, and M. C. Gilbert, Comparative study of low-grade metamorphism in the California Coast Ranges and the Outer Metamorphic belt of Japan, *Geol. Soc. Amer. Mem. 124*, 270 pp., 1970.

Essene, E. J., and W. S. Fyfe, Omphacite in Californian rocks, *Contrib. Mineral. Petrol*, *15*, 1-23, 1967.

Ghent, E. D., Fossil evidence for maximum age of metamorphism in part of the Franciscan formation, Northern Coast Ranges, California, *Calif. Div. Mines Geol. Spec. Rep.*, *82*, 41, 1963.

Ghent, E. D., Glaucophane-schist facies metamorphism in the Black Butte area, Northern Coast Ranges, California, *Amer. J. Sci.*, *263*, 385-400, 1965.

Hackel, O., Summary of the geology of the Great Valley, in *Geology of Northern California*, edited by E. H. Bailey, *Calif. Div. Mines Geol. Bull.*, *190*, 217-238, 1966.

Hall, C. A., Jr., Geology and paleontology of the Pleasanton area, Alameda and Contra Costa Counties, California, *Univ. Calif. Publ. Geol. Sci.*, *34*, 89 pp., 1958.

Hall, C. A., Jr., Displaced Miocene molluscan provinces along the San Andreas fault, California, *Univ. Calif. Publ. Geol. Sci.*, *34*, 281-308, 1960.

Hall, C. A., Jr. and C. E. Corbató, Stratigraphy and structure of Mesozoic and Cenozoic rocks, Nipomo quadrangle, southern Coast Ranges, California, *Bull. Geol. Soc. Amer.*, *78*, 559-582, 1967.

Hamilton, W., Mesozoic California and the underflow of Pacific mantle, *Bull. Geol. Soc. Amer.*, *80*, 2409-2430, 1969.

Hatherton, T., Geophysical anomalies over the eu- and miogeosynclinal systems of California and New Zealand, *Bull. Geol. Soc. Amer.*, *80*, 213-230, 1969.

Heirtzler, J. R., G. O. Dickson, E. M. Herron, W. C. Pitman III, and X. Le Pichon, Marine magnetic anomalies, geomagnetic field reversals, and motions of the ocean floor and continents, *J. Geophys. Res.*, *73*, 2119-2136, 1968.

Hertlein, L. G., Cretaceous ammonite of Franciscan group, Marin County, California, *Bull. Amer. Ass. Petrol. Geol.*, *40*, 1985-1988, 1956.

Hess, H. H., History of ocean basins, in *Petrologic Studies, A Volume in Honor of A. F. Buddington*, edited by A. E. Engel, H. L. James, and B. F. Leonard, pp. 599-620, Geological Society of America, New York, 1962.

Hess, H. H., Mid-oceanic ridges and tectonics of the sea-floor, in *Submarine Geology and Geophysics*, edited by W. F. Whittard and R. Bradshaw, pp. 317-332, Butterworths, London, 1965.

Hill, M. L., Tectonics of faulting in southern California, in *Geology of Southern California*, edited by R. H. Jahns, *Calif. Div. Mines Geol. Bull.*, *170*, 5-13, 1954.

Hill, M. L., and T. W. Dibblee, Jr., San Andreas, Garlock, and Big Pine faults, California, *Bull. Geol. Soc. Amer.*, *64*, 443-458, 1953.

Hill, M. L., and H. D. Hobson, Possible post-Cretaceous slip on the San Andreas fault zone, in *Proceedings of Conference on Geologic Problems of San Andreas Fault System*, *11*, edited by W. R. Dickinson and A. Grantz, pp. 123-128, Stanford University Publications, Geological Sciences, Stanford, California, 1968.

Hsü, K. J., Mesozoic geology of the California Coast Ranges—A new working hypothesis, in *Etages Tectoniques*, edited by J. P. Schaer, pp. 279-296, à la Baconnière, Neuchâtel, Suisse, 1967.

Hsü, K. J., Principles of mélanges and their bearing on the Franciscan-Knoxville paradox, *Bull. Geol. Soc. Amer.*, *79*, 1063-1074, 1968.

Hsü, K. J., and R. Ohrbom, Mélanges of San Francisco Peninsula—Geologic reinterpretation of type Franciscan, *Bull. Amer. Ass. Petrol. Geol.*, *53*, 1348-1367, 1969.

Irwin, W. P., Franciscan group in Coast Ranges and its equivalents in Sacramento Valley, California, *Bull. Amer. Ass. Petrol. Geol.*, *41*, 2284-2297, 1957.

Irwin, W. P., Geologic reconnaissance of the northern Coast Ranges and Klamath Mountains, California, with a summary of the mineral resources, *Calif. Div. Mines Geol. Bull.*, *179*, 80 pp., 1960.

Irwin, W. P., Late Mesozoic orogenies in the ultramafic belts of northwestern California and southwestern Oregon, *U.S. Geol. Surv. Prof. Pap. 501-C*, 1-9, 1964.

Irwin, W. P., Geology of the Klamath Mountains province, in *Geology of Northern California*, edited by E. H. Bailey, *Calif. Div. Mines Geol. Bull.*, *190*, 19-38, 1966.

Isacks, B. J., Oliver, and L. R. Sykes, Seismology and the new global tectonics, *J. Geophys. Res.*, *73*, 5855-5899, 1968.

Keith, T. E. C., and R. G. Coleman, Albite-pyroxene-glaucophane schist from Valley Ford, California, *U.S. Geol. Surv. Prof. Pap. 600-C*, 13-17, 1968.

King, P. B., *The Evolution of North America*, 190 pp., Princeton University Press, Princeton, New Jersey, 1959.

Lachenbruch, M. C., Geology of the west side of the Sacramento Valley, California, in *Geologic Guide to the Gas and Oil Fields of Northern California*, edited by O. E. Bowen, Jr., *Calif. Div. Mines Geol. Bull.*, *181*, 53-66, 1962.

Lanphere, M. A., W. P. Irwin, and P. E. Hotz, Isotopic age of the Nevadan orogeny and older plutonic and metamorphic events in the Klamath Mountains, California, *Bull. Geol. Soc. Amer.*, *79*, 1027-1052, 1968.

Larson, R. L., H. W. Menard, and S. M. Smith, Gulf of California: A result of ocean-floor spreading and transform faulting, *Science*, *161*, 781-783, 1968.

Lee, D. E., H. H. Thomas, R. F. Marvin, and R. G. Coleman, Isotope ages of glaucophane schists from Cazadero, California, *U.S. Geol. Surv. Prof. Pap. 475-D*, 105-107, 1963.

Leith, C. J., Geology of the Quien Sabe quadrangle, California, *Calif. Div. Mines Geol. Bull.*, *147*, 60 pp., 1949.

Le Pichon, X., Sea-floor spreading and continental drift, *J. Geophys. Res.*, *73*, 3661-3697, 1968.

Maddock, M. E., Geology of the Mt. Boardman quadrangle, Santa Clara and Stanislaus Counties, California, *Calif. Div. Mines Geol. Map Sheet 3*, 1964.

McKee, B., Widespread occurrence of jadeite, lawsonite, and glaucophane in central California, *Amer. J. Sci.*, *260*, 596-610, 1962a.

McKee, B., Aragonite in the Franciscan rocks of the Pacheco Pass area, California, *Amer. Mineral*, *47*, 379-387, 1962b.

McKenzie, D. P., Speculations on the consequences and causes of plate motions, *Geophys. J.*, *18*, 1-32, 1969.

Meade, B. K., and J. B. Small, Current and recent movement on the San Andreas fault, in *Geology of Northern California*, edited by E. H. Bailey, *Calif. Div. Mines Geol. Bull.*, *190*, 385-391, 1966.

Menard, H. W., Marine Geology of the Pacific, 271 pp., McGraw-Hill, New York, 1964.

Morgan, W. J., Rises, trenches, great faults, and crustal blocks, J. Geophys. Res., 73, 1959-1982, 1968.

Ojakangas, R. W., Cretaceous sedimentation, Sacramento Valley, California, Bull. Geol. Soc. Amer., 79, 973-1008, 1968.

Oliver, J., and B. Isacks, Deep earthquake zones, anomalous structures in the upper mantle, and the lithosphere, J. Geophys. Res., 72, 4259-4275, 1967.

Oxburgh, E. R., and D. L. Turcotte, Problem of high heat flow and volcanism associated with zones of descending mantle convective flow (abstract), Trans. Amer. Geophys. Union, 49, 318, 1968.

Page, B. M., Geology of the Coast Ranges of California, in Geology of Northern California, edited by E. H. Bailey, Calif. Div. Mines Geol. Bull., 190, 255-276, 1966.

Page, B. M., Relation between ocean-floor spreading and structure of the Santa Lucia Range, California (abstract), Geol. Soc. Amer., Cordilleran Section, Part 3, 51-52, 1969.

Page, B. M., Sur-Nacimiento fault zone of California: Continental margin tectonics, Bull. Geol. Soc. Amer., in press, 1970.

Peterman, Z. E., C. E. Hedge, R. G. Coleman, and P. D. Snavely, $Sr^{87}/Sr^{86}$ ratios in some eugeosynclinal sedimentary rocks and their bearing on the origin of granitic magma in orogenic belts, Earth Planet. Sci. Lett., 2, 433-439, 1967.

Peterson, G. L., Implications of two Cretaceous mass transport deposits, Sacramento Valley, California, J. Sediment. Petrol., 35, 401-407, 1965.

Peterson, G. L., Implications of two Cretaceous mass transport deposits, Sacramento Valley, California: Reply to comment by Brown and Rich, J. Sediment. Petrol., 37, 248-257, 1967.

Reed, R. D., and J. S. Hollister, Structural evolution of southern California, Bull. Amer. Assoc. Petrol. Geol., 20, 1529-1704, 1936.

Repenning, C. A., Geologic summary of the Central Valley of California with reference to disposal of liquid radioactive waste, U.S. Geol. Surv. TEI Rep. 769, 69 pp., 1960.

Richter, C. F., Elementary Seismology, 768 pp., W. H. Freeman, San Francisco, 1959.

Safonov, A., The challenge of the Sacramento Valley, California, in Geologic Guide to the Gas and Oil Fields of Northern California, edited by O. E. Bowen, Jr., Calif. Div. Mines Geol. Bull., 181, 77-98, 1962.

Saito, M., K. Hashimoto, H. Sawata, and Y. Shimazaki, Geology and Mineral Resources of Japan, 2nd ed, 504 pp., Geological Survey of Japan, Tokyo, Japan, 1960.

Schilling, F. A., Jr., Cretaceous geology of the Pacheco Pass area, California, in Geologic Guide to the Gas and Oil Fields of Northern California, edited by O. E. Bowen, Jr., Calif. Div. Mines Geol. Bull., 181, 163-164, 1962.

Schlocker, J., M. G. Bonilla, and R. W. Imlay, Ammonite indicates Cretaceous age for part of Franciscan group in San Francisco Bay area, California, Bull. Amer. Ass. Petrol. Geol., 38, 2372-2381, 1954.

Seyfert, C. K., Undeformed sediments in oceanic trenches with sea-floor spreading, Nature, 222, 70, 1969.

Suppe, J., Times of metamorphism in the Franciscan terrain of the Northern Coast Ranges, California, Bull. Geol. Soc. Amer., 80, 135-142, 1969.

Sykes, L. R., The seismicity and deep structure of island arcs, J. Geophys. Res., 71, 2981-3006, 1966.

Taliaferro, N. L., Franciscan-Knoxville problem, Bull. Amer. Ass. Petrol. Geol., 27, 109-219, 1943.

Taylor, H. P., and R. G. Coleman, $O^{18}/O^{16}$ ratios of coexisting minerals in glaucophane-bearing metamorphic rocks, Bull. Geol. Soc. Amer., 79, 1727-1756, 1968.

Thalmann, H. E., Globotruncana in the Franciscan limestone, Santa Clara County, California (abstract), Bull. Geol. Soc. Amer., 53, 1838, 1942.

Thalmann, H. E., Upper Cretaceous age of the 'Franciscan' limestone near Laytonville, Mendocino County, California (abstract), Bull. Geol. Soc. Amer., 54, 1827, 1943.

Vacquier, V., A. D. Raff, and R. E. Warren, Horizontal displacements in the floor of the northeastern Pacific Ocean, Bull. Geol. Soc. Amer., 72, 1251-1258, 1961.

Wilson, I. F., Geology of the San Benito quadrangle, California, Calif. Div. Mines Geol. Bull., 59, 183-270, 1943.

Wilson, J. T., Transform faults, oceanic ridges, and magnetic anomalies southwest of Vancouver Island, Science, 160, 482-485, 1965.

Yeats, R. S., Southern California structure, sea-floor spreading, and history of the Pacific Basin, Bull. Geol. Soc. Amer., 79, 1693-1702, 1968.

(Received August 11, 1969.)

# Subduction and the rock record: Concepts developed in the Franciscan Complex, California

**John Wakabayashi**
*1329 Sheridan Lane, Hayward, California 94544; e-mail: wako@tdl.com.*

## INTRODUCTION

Ernst's (1970) paper was chosen as the classic paper for the Franciscan Complex because it related high-pressure, low-temperature (high *P-T*) metamorphism to subduction. Perhaps most significantly, the paper explained the association of low geothermal gradients and the metamorphism. The paper also pointed out the difficult tectonic problem of the exhumation of the high *P-T* rocks, a problem still vigorously debated today, and proposed a tectonic model explaining the exhumation of the deeply buried rocks. In addition, the paper explained the tectonic contact of the Great Valley forearc over the Franciscan subduction complex in the context of plate tectonics theory (Hamilton, 1969, gave a similar explanation; see following). Ernst's paper was one of the key advances in the plate tectonics revolution. Variations of Figure 3 of Ernst (1970) have become the textbook model of subduction-zone metamorphism. Geologists now regard high *P-T* (including blueschist facies) metamorphism as the strongest evidence of an exhumed subduction complex. Evaluation of thermal gradients associated with subduction and their connection to metamorphic assemblages, introduced by Ernst (1970), has become an important concept in understanding the evolution of orogenic belts (e.g., Ernst, 1975, 1988). An example of this type of analysis is the premise that Franciscan subduction was continuous, from its inception in the late Mesozoic to conversion to a transform plate boundary in the late Cenozoic, because Franciscan high *P-T* rocks lack thermal overprints (such as late greenschist facies assemblages) which should have resulted from any cessation of subduction (Cloos and Dumitru, 1987; Ernst, 1988).

Ernst's (1970) paper was one of several key papers that related the Franciscan Complex to subduction processes and established the Franciscan as the type subduction complex. Hamilton (1969) equated the Franciscan to a subduction complex, related subduction to arc volcanism in the Sierra Nevada, and pointed out the far-traveled nature of some Franciscan Complex rocks. Hsü (1968, 1971) formalized the concept and principles of melange (Bailey et al., 1964, had recognized the shear-zone character of what were later called melanges). Dickinson (1970) placed the Franciscan in the context of an arc-trench system, with the Franciscan, Great Valley Group, and Sierra Nevada as the subduction complex, forearc basin, magmatic arc (and main terrigenous sediment source), respectively. Bailey et al. (1964) set the stage for these papers by compiling and evaluating an enormous amount of data and presenting ideas that forecast the plate tectonic interpretation of the Franciscan; this is still a useful reference on the Franciscan.

Many major conclusions of these landmark papers have not been significantly challenged since their publication. Subsequent research has continued to provide insight into fundamental processes in subduction zones. Some developments in Franciscan geology since 1970, as well as major controversies, are discussed in the following. The general geology of the Franciscan Complex is shown in Figure 1.

## DEVELOPMENTS IN FRANCISCAN GEOLOGY SINCE 1970

### Reexamination of the tectonic boundary between the Franciscan Complex and the Coast Range ophiolite and/or Great Valley Group

Various studies led to the formulation of a general model for the arc-trench system, in which a forearc basin (the Great Valley Group) and its basement (the Coast Range ophiolite) rode passively and undeformed on the upper plate of the arc-trench system, while the Franciscan Complex was complexly deformed structurally beneath as the subduction complex (e.g., Dickinson and Seely, 1979). Recognition of east-vergent

---

Wakabayashi, J., 1999, Subduction and the rock record: Concepts developed in the Franciscan Complex, California, *in* Moores, E. M., Sloan, D., and Stout, D. L., eds., Classic Cordilleran Concepts: A View from California: Boulder, Colorado, Geological Society of America Special Paper 338.

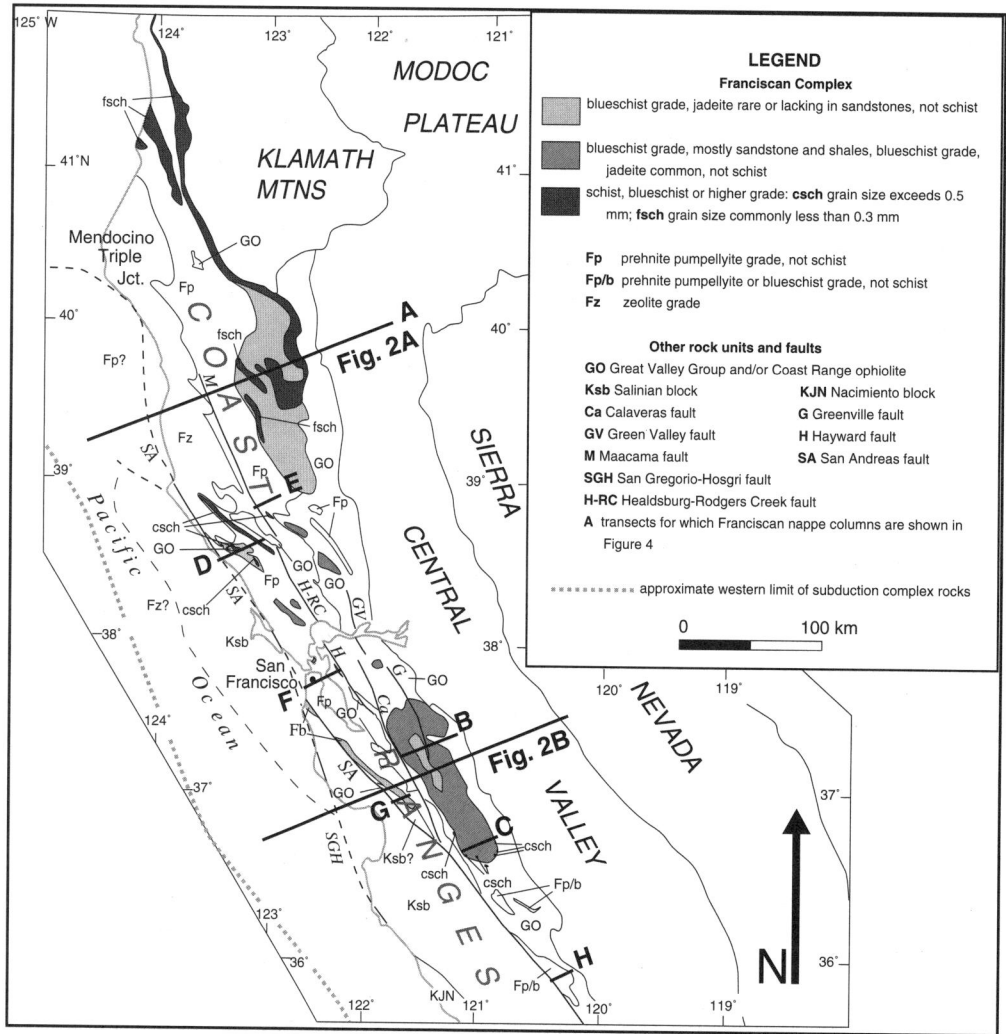

Figure 1. Distribution of Franciscan and other basement rocks of central and northern California, showing Franciscan Complex rocks of different metamorphic grades. Map derived from Jennings (1977) with modifications from Wakabayashi unpublished field data.

structures, comprising part of a tectonic wedge system that faulted the Great Valley Group and Coast Range ophiolite, and locally placed Franciscan Complex rocks *over* Great Valley Group rocks, greatly complicated this model (Wentworth et al., 1984) (Fig. 2).

The tectonic contact of the comparatively unmetamorphosed Great Valley Group and Coast Range ophiolite with the structurally underlying Franciscan Complex was initially called the Coast Range thrust (Bailey et al., 1970). This nomenclature reflected the recognition of the tectonic contact as a subduction boundary (Hamilton, 1969; Ernst, 1970). However, Suppe (1973) noted that the contrast of metamorphic assemblages (low-$P$ rocks on high-$P$ rocks) across this contact was consistent with normal rather than reverse slip. Neogene normal fault movement between the Great Valley Group and Franciscan was proposed by Ernst (1970). Platt (1986) indicated that the metamorphic contrast across the tectonic contact implied normal slip, and noted similar relationships in other high $P$-$T$ belts. Wakabayashi and Unruh (1995) proposed that normal slip took place during one period of time along the ophiolite–Franciscan Complex contact and that tectonic wedging (and periodic thrust faulting along the contact) were operative during other periods (Fig. 2). Godfrey et al. (1997) presented evidence for a buried ophiolitic suture beneath the Great Valley Group that may be a consequence of an east-vergent collision event similar to that originally proposed by Moores (1970; see Moores and others, this volume) (Fig. 2A).

*Metamorphism*

Franciscan metamorphic rocks occur as tectonic blocks in melange and as intact units or thrust sheets (e.g., Bailey et al., 1964). Intact (termed coherent) Franciscan units range from blueschist-greenschist transition assemblages to zeolite facies (e.g., Blake et al., 1988) (Figs. 1 and 2), and grain sizes of meta-

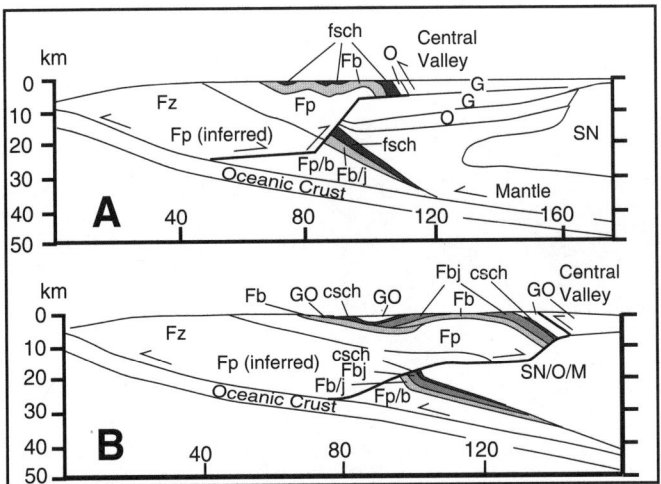

Figure 2. Schematic cross sections along lines shown in Figure 1. San Andreas fault system dextral slip is restored according to Wakabayashi and Hengesh (1995), and subsurface relations are restored to hypothetical geometry immediately prior to conversion to transform margin. Cross-section A is modified from Wakabayashi and Unruh (1995) and Godfrey et al. (1997). Abbreviations as in Figure 1, except: G—Great Valley Group; O—Coast Range ophiolite; SN/O/M—undifferentiated Sierran basement, ophiolitic rocks, and mantle beneath ophiolitic rocks; Fb/j, Fp/b designations reflect higher grade of metamorphism for deeper parts of equivalent unit.

morphic minerals are generally tenths of a millimeter or smaller. Tectonic blocks in melanges range from amphibolite to zeolite grade. Eclogite and amphibolite facies metamorphism is generally restricted to tectonic blocks (Bailey et al., 1964; Coleman and Lanphere, 1971). Amphibolite and eclogite blocks, with metamorphic grain sizes to 1 cm, along with blocks of similar coarse-grained blueschist, have been termed high-grade blocks (Coleman and Lanphere, 1971).

Although there have been many studies of Franciscan metamorphism since 1970, estimates for pressure ($P$) and temperature ($T$) of metamorphism for most coherent Franciscan rocks have not changed significantly. For blueschist and blueschist-greenschist transition facies rocks, these estimates range from 100 to 380 °C and 4 to 9 kbar (Maruyama and Liou, 1988; Blake et al., 1988; Ernst, 1993). The most significant update for $P$-$T$ estimates of coherent rocks has been the characterization of blueschist-greenschist transition assemblages (Brown and Ghent, 1983; Maruyama and Liou, 1988). Direct estimates of $T$ in coherent rocks are relatively scarce and have been based on oxygen isotope studies (Taylor and Coleman, 1968), data from thermal maturity of hydrocarbons, illite crystallinity, fluid-inclusion studies (e.g., Bostick, 1974; Cloos, 1983; Blake et al., 1988; Underwood, 1989; Dalla Torre et al., 1996), and chlorite geothermometry (Bröcker and Day, 1995).

Polymetamorphism was noted in high-grade blocks (Bailey et al., 1964; Dudley, 1972), and quantitative $P$ and $T$ estimates have been made for different stages of metamorphism (e.g., Brown and Bradshaw, 1979; Moore, 1984; Moore and Blake, 1989; Wakabayashi, 1990; Krogh et al., 1994). Petrologic studies show that the $P$-$T$ conditions of these blocks evolved progressively from high to low geothermal gradients, probably in less than 5 m.y. (Wakabayashi, 1990) (Fig. 3). This type of thermal evolution has been suggested to be the result of metamorphism at the inception of Franciscan subduction (Wakabayashi, 1990) (Fig. 3). Alternatively, high-grade blocks may have been metamorphosed prior to Franciscan subduction (e.g., Coleman and Lanphere, 1971; Moore, 1984).

Coherent rocks are part of an inverted metamorphic gradient from zeolite to blueschist-greenschist grade (e.g., Blake et al., 1988; inverted metamorphic gradient first recognized by Blake, 1967). Thrust faults apparently separate the rocks of different metamorphic grade, but the metamorphic gradient may record changing thermal conditions within the subduction zone, and the structurally higher parts of the metamorphic gradient reflect the thermal influence and proximity of the hanging wall of the subduction zone (e.g., Ernst, 1971; Platt, 1975; Suppe and Foland, 1978; Cloos, 1985; Peacock, 1988) (Fig. 3). The $P$-$T$ paths of representative high-grade blocks and coherent metamorphic rocks are shown in Figure 3.

### Melanges

Following studies of Hsü (1968, 1971), research has further clarified details of melanges. Cloos (1982, 1985, and other papers) modeled a return-flow material path in melanges and used this flow pattern to explain the mixture of different blocks, the progressive accretion of melange terranes, and the large-scale thermal patterns noted in the Franciscan Complex. Cowan (1985) examined strain patterns in melanges and concluded that Franciscan melanges formed in several different tectonic settings in the subduction complex. Page (1978) and Aalto (1981) proposed that melanges resulted from a combination of tectonic shearing and earlier olistostromal development. There is evidence for Franciscan or ophiolitic material that had been exhumed and resedimented, supporting an olistostromal component for some melanges (Cowan and Page, 1975; Moore and Liou, 1980; Moore, 1984; Macpherson et al., 1990). Distinctive populations of blocks and structural relationships relative to other units have proved useful for distinguishing melanges as mappable units (e.g., Hsü and Ohrbom, 1969; Maxwell, 1974).

### Major structure: Cross-sectional and along-strike variation

Following the recognition of melanges, units were mapped as chaotic melanges or coherent stratal packages (e.g., Raymond, 1970; Suppe, 1973; Cowan, 1974; Maxwell, 1974). Mapping of fault-bounded units developed into the formalized terrane concept (e.g., Blake et al., 1982; developed elsewhere in California by Irwin, 1972). The terrane concept added new levels of detail to Franciscan mapping and interpretation but did not separate discrete melanges; all melanges were considered to be one terrane.

Imbricate low-angle faults involving coherent units and intervening melanges were recognized in the eastern part of the

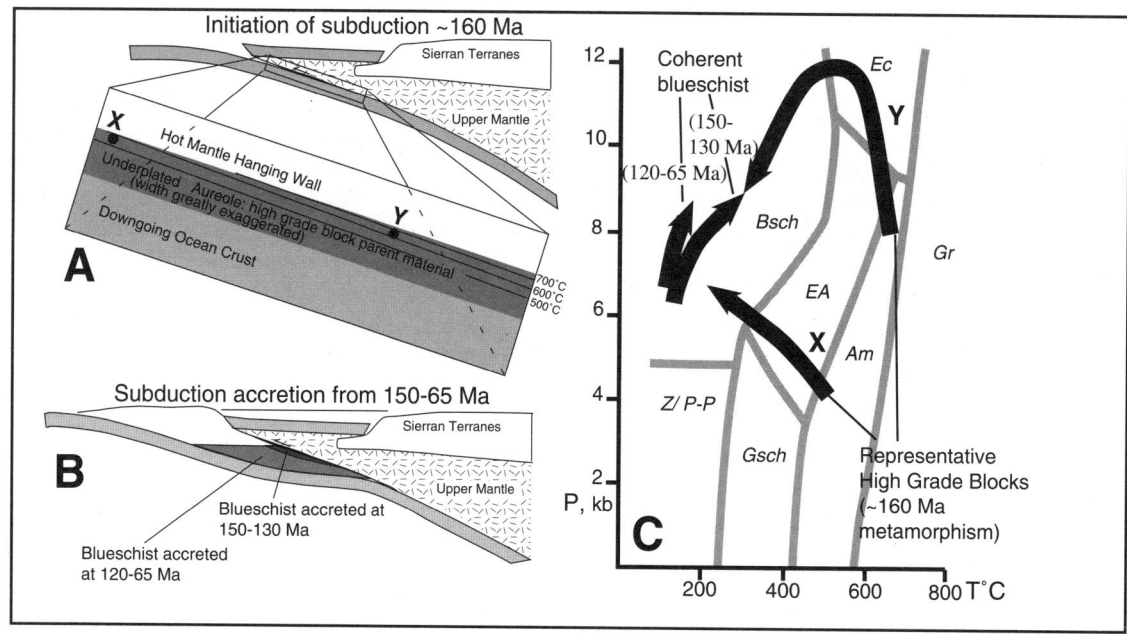

Figure 3. Pressure-temperature (P, T) paths of Franciscan metamorphism and schematic cartoons showing the tectonic setting of metamorphism. Letters X and Y in A correspond to locations of the similarly labeled P-T paths for high-grade blocks shown in C. P-T paths X and Y are from Wakabayashi (1990);150–130 Ma P-T path is from Maruyama and Liou (1988), and 120–65 Ma P-T path is from Maruyama et al., (1985). Tectonic setting diagram is modified from Wakabayashi (1992). Metamorphic facies abbreviations: Am, amphibolite; Bsch, blueschist; Ec, eclogite; EA, epidote amphibolite; Gr, granulite; Gsch, greenschist; Z/P-P, zeolite and prehnite-pumpellyite; Am, amphibolite.

Franciscan in the northern Coast Ranges (Blake, 1967; Suppe, 1973; Worrall, 1981) and the Diablo Range (Cowan, 1974; Page, 1981). Suppe (1978) interpreted low-angle structures across the entire northern Coast Ranges; ultramafic rocks were interpreted as Coast Range ophiolite or klippen thereof. Blake et al. (1984) characterized the coherent units and melanges of the San Francisco Bay region as a set of gently folded thrust nappes, with discrete melange zones separating the coherent sheets. Wahrhaftig (1984) documented internal imbrication within a coherent nappe. Wakabayashi (1992) extended the nappe characterization to the entire Franciscan Complex (Fig. 4), related nappe emplacement to orogenic processes within the Franciscan and the Cordillera, and developed criteria for distinguishing Franciscan from post-Franciscan structures, on the basis of the progressive down-structure decrease in the times at which nappes were incorporated. Ultramafic rocks recently have been interpreted as both remnants of Coast Range ophiolite (structurally above the Franciscan Complex), and units structurally interleaved within the Franciscan (Blake et al., 1984; Wakabayashi, 1992; Coleman, 1996).

The Franciscan Complex exhibits considerable along-strike variation, even if dextral faulting associated with the San Andreas transform system is restored (Wakabayashi, 1992) (Figs. 1 and 2). This variation may be expected in subduction complexes, because offscraped elements, such as packets of trench and pelagic sediment and seamounts, would not be expected to extend the full length of the trench. Along-strike variation makes the "Coastal Belt, Central Belt, Eastern Belt" subdivision of the Franciscan Complex, in common usage since its introduction by Berkland et al. (1972), difficult to apply south of the northern Coast Ranges.

## Geochronology

Analysis of tectonics relies heavily on geochronologic data. Many of the key geochronologic studies in the Franciscan Complex have been conducted since 1970. Geochronologic data from the Franciscan are summarized in Figure 5.

Initial data from sparse macrofossils in clastic sedimentary rocks (e.g., Bailey et al., 1964) have been supplemented by microfossil data from cherts and limestones (Sliter, 1984; Murchey and Jones, 1984). Detailed biostratigraphy from cherts demonstrated that what were once thought to be chert interbeds with basalt or graywacke were instead a consequence of structural imbrication (Murchey, 1984; Isozaki and Blake, 1994). Although clastic rocks of the Franciscan Complex are coeval with the forearc basin strata of the Great Valley Group (ca. 150 Ma and younger; Bailey et al., 1964; Blake et al., 1988), some oceanic rocks are older, such as cherts in the Marin Headlands and Yolla Bolly terranes (175–195 Ma; Murchey, 1984; Isozaki and Blake, 1994). This reflects the far-traveled nature of oceanic and/or pelagic rocks incorporated into the Franciscan Complex (e.g., Hamilton, 1969).

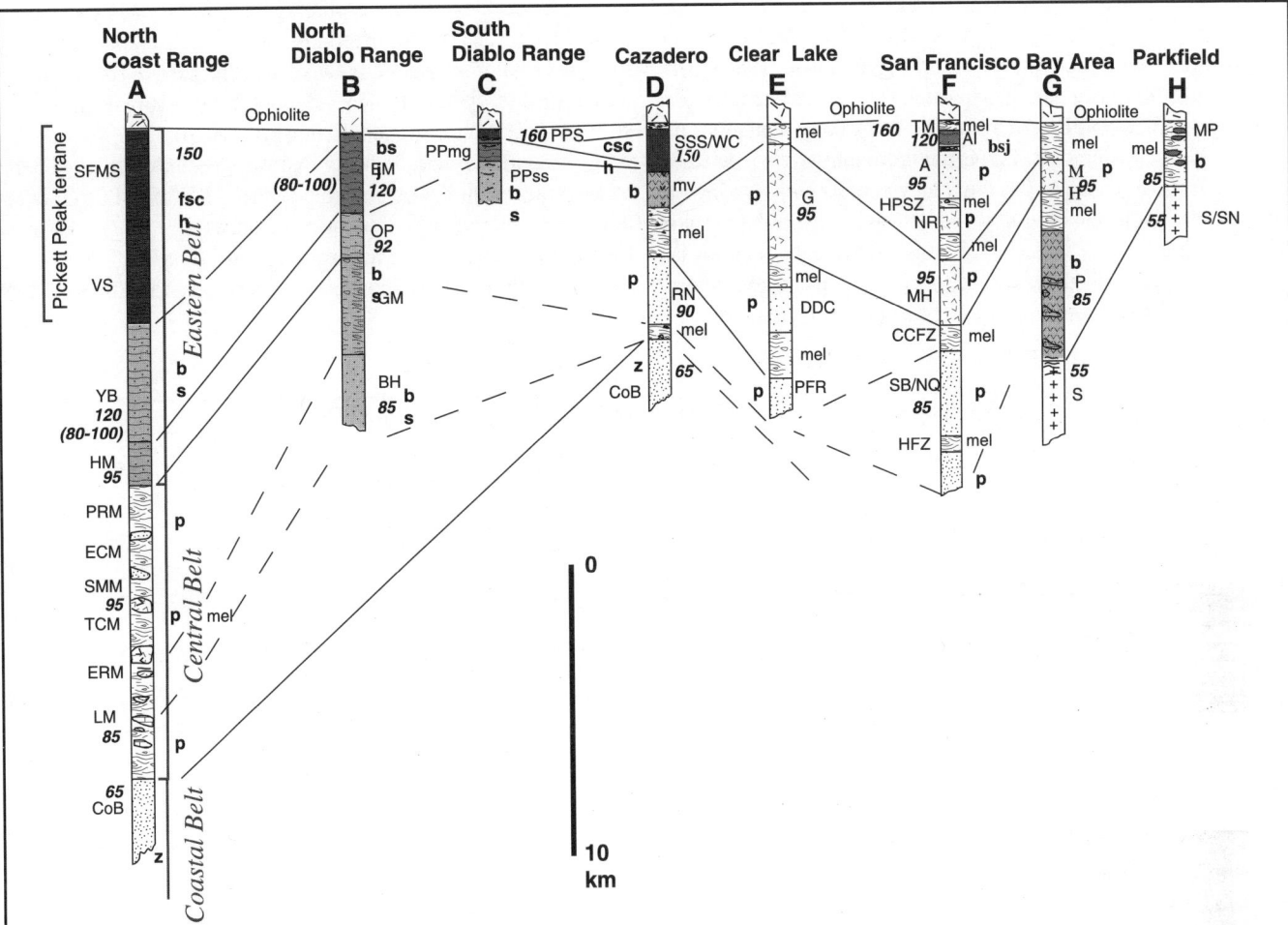

Figure 4. Correlation of Franciscan Complex units as thrust nappes to relative structural position and estimated accretion ages. Blueschist facies units are shaded. Column locations are keyed to Figure 1. All contacts are tectonic; the incorporation age (not depositional age) decreases downward. Scale is approximate. Figure is modified from depository supplement of Wakabayashi (1992); additional data are from P. Renne, T. Dumitru and J. Wakabayashi.

Numerous isotopic dating studies conducted in the Franciscan Complex since 1970 (Coleman and Lanphere, 1971; Suppe and Armstrong, 1972; Lanphere et al., 1978; Mattinson and Echeverria, 1980; McDowell et al., 1984; Mattinson, 1986; Ross and Sharp, 1988) have greatly expanded the preexisting database (e.g., Lee et al., 1964; Peterman et al., 1967). Only metamorphic rocks of blueschist grade or higher, and rare intrusive rocks, have yielded dates. Because of the different methods employed, with different precision and accuracy, and because of the different closure temperatures of isotopic systems, many dates cannot be directly compared (Wakabayashi, 1992). Ages span a continuous range, but the precision of the dating techniques is not sufficient to indicate that they record a continuous or episodic metamorphism. The oldest metamorphic ages, 159–163 Ma (Ross and Sharp, 1986, 1988), have been interpreted to date approximately the inception of subduction (Wakabayashi, 1990). The youngest metamorphic ages obtained are 80–90 Ma (Suppe and Armstrong, 1972; Mattinson and Echeverria, 1980).

Fission-track data indicate that the structurally higher parts of the Franciscan Complex cooled from ~300 to ~100 °C, coinciding with the major exhumation of these high $P$-$T$ rocks from depths of ~30 to ~10 km, from 100 to 70 Ma (Tagami and Dumitru, 1996). Many Franciscan rocks cooled below 90 °C, at 20 to 40 Ma (Dumitru, 1989).

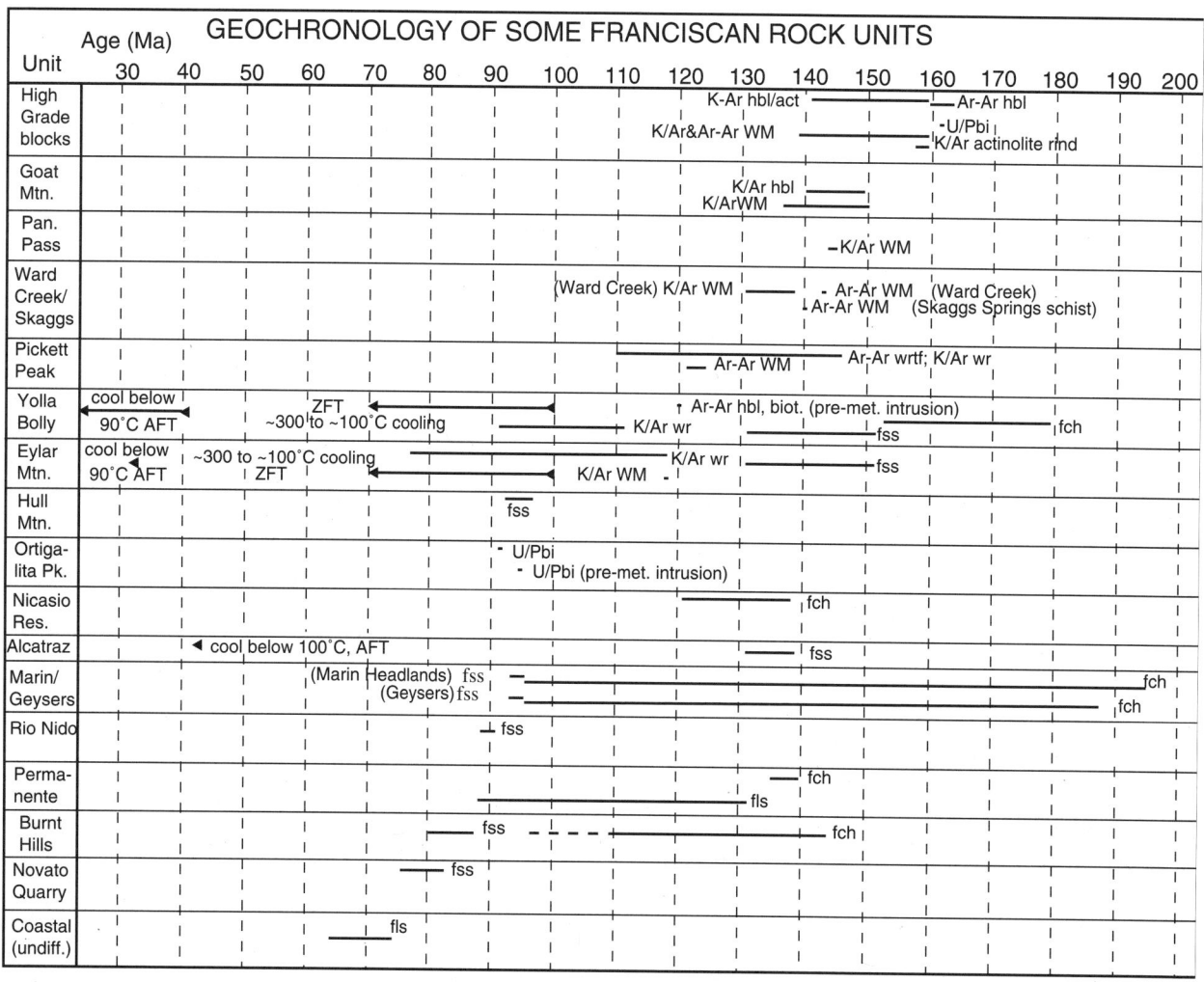

Figure 5. Geochronology of some Franciscan units. Blueschist and higher grade units are in bold. Abbreviations: AFT—apatite fission track; Ar-Ar—$^{40}$Ar/$^{39}$Ar step heating, except wrtf—whole-rock total fusion; fch—fossils from chert; fls—fossils from limestone; fss—fossils from clastic rocks; K/Ar—potassium-argon; U/Pbi—U/Pb isochron method, ZFT—zircon fission track. Mineral abbreviations: act—actinolite, biot—biotite, hbl—hornblende, WM—white mica; wr—whole rock. Sources: Lee et al. (1964), Coleman and Lanphere (1971), Suppe and Armstrong (1972), Lanphere et al. (1978), Suppe and Foland (1978); Mattinson and Echeverria (1980); McDowell et al. (1984); Mattinson (1986); Ross and Sharp (1988); Dumitru (1989); Wakabayashi and Deino (1989); Tagami and Dumitru (1996); Weinrich (1997). Older K/Ar dates recalculated using decay constants of Steiger and Jäger (1977).

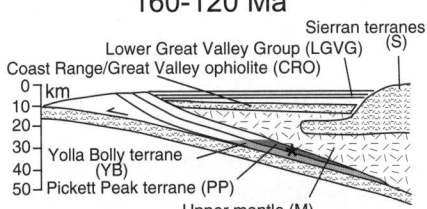

**160-120 Ma**

Initial underplating of coherent high *P/T* rocks (shaded)

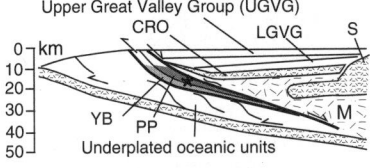

**100-70 Ma**

Underplating drives return flow constrained by mantle backstop; high *P/T* rocks juxtaposed with low P ophiolite

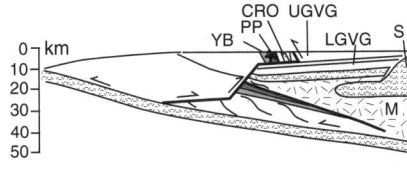

**60 Ma to Present**

Tectonic wedging; high *P/T* rocks, ophiolite, Great Valley Group, exhumed together and exposed.

Figure 6. Cartoons illustrating exhumation of high pressure-temperature (*P-T*) rocks in the Franciscan. Asterisk marks position of high *P-T* rocks exposed today. Modified from Wakabayashi and Unruh (1995).

## CONTROVERSIES AND UNSOLVED PROBLEMS

Some issues fundamental to the history of the Franciscan Complex, and to convergent plate margin tectonics in general, have not been resolved. Still others have not been addressed in a major way. Two major issues are discussed in the following.

### To strike slip or not to strike slip

Plate tectonics theory predicts that pelagic sediments and oceanic volcanic rocks are transported long distances on the oceanic plate as it moves toward the subduction zone. In the Franciscan Complex, paleomagnetic studies of basalts and limestones have indicated southerly latitudes of deposition (e.g., Alvarez et al., 1980; Tarduno et al., 1985), and plate convergence was oblique during much of Franciscan history (Engebretson et al., 1985). However, the presence or absence of large-scale (thousands of kilometers of displacement) strike-slip faults passing through the Franciscan Complex is a subject of debate. Data from conglomerates have been alternatively interpreted to indicate negligible or large synsubduction strike-slip displacement between coeval rocks of the Franciscan and Great Valley Group (e.g., Seiders and Blome, 1988; Jayko and Blake, 1993; respectively). Some researchers have interpreted up to thousands of kilometers of strike-slip faulting within the Franciscan Complex (e.g., McLaughlin et al., 1988; Jayko and Blake, 1993), only a few hundred kilometers of which are attributed to the San Andreas transform fault system, which postdated the Franciscan Complex (e.g., Graham et al., 1989). I (Wakabayashi, 1992) suggested that Franciscan nappe structures were incompatible with major synsubduction strike-slip faulting in any part of the Franciscan Complex accreted prior to 80 Ma, and that most synsubduction strike-slip faulting was partitioned in the vicinity of the Sierra Nevada arc, similar to the position of such faulting in modern obliquely convergent plate margins (e.g., Fitch, 1972; Beck, 1983). The assumption of 100% partitioning of the strike-slip component of oblique subduction into strike-slip faulting has been adopted in some analyses of strike-slip faulting within the Franciscan Complex (e.g., McLaughlin et al., 1988; Jayko and Blake, 1993), but partitioning of the tangential component of active oblique subduction into strike-slip faulting varies from 0% to 100%, depending on the subduction zone (McCaffrey, 1992). Wakabayashi and Hengesh (1995) suggested that the total offset of Franciscan nappes across faults in the central Coast Ranges was identical to the late Cenozoic (transform) fault offsets in that region, indicating negligible synsubduction strike-slip faulting of the Franciscan Complex east of the Salinian block. Synsubduction offset of ~200 km is permissible on the Salinian-Franciscan contact (the proto San Andreas of Page, 1981); additional synsubduction dextral faulting may be permissible west of the Salinian block (e.g., Sedlock and Hamilton, 1991). Better estimates of synsubduction strike-slip faulting within the Franciscan Complex will illuminate processes of obliquely convergent plate margins and may test various models of large-scale latitudinal terrane translation in the Cordillera, such as the Baja British Columbia controversy (e.g., Cowan et al., 1997).

### What went down came up: Exhumation of blueschists and other high P-T rocks

As the type subduction complex, the Franciscan Complex has become a testing ground for models of the exhumation of high *P-T* rocks. Many mechanisms can account for the exhumation of high-grade blocks in melanges, including melange return flow and shale diapirism (e.g., Cloos, 1982), or serpentinite diapirism (Carlson, 1981). Coherent high *P-T* metamorphic rocks present a more difficult problem because they are not thought to be as mobile as blocks in melanges; this problem is reviewed here. The critical part of the exhumation (hereafter called the critical path) is the rise of the high *P-T* rocks from depth (25 km or more) to the same crustal level as the ophiolite and Great Valley Group rocks that are now in contact with them, a difference in burial depth of 15 km or more; exhumation of the shallowly buried ophiolite and Great Valley Group is considered trivial.

Ernst (1970) proposed that high $P$-$T$ rocks were less dense than the mantle they were subducted into and had to rise buoyantly back to the Earth's surface. Platt (1986) dismissed this model, noting that the high $P$-$T$ rocks are more dense than the rocks that are currently found in contact with them. However, while traversing the critical path, the high $P$-$T$ rocks may have been beneath mantle and therefore positively buoyant (Fig. 6). Platt (1986), noting that nearly unmetamorphosed ophiolite tectonically overlies deeply buried Franciscan Complex rocks, concluded that the high $P$-$T$ rocks were exhumed by normal faulting associated with synsubduction extension. Ring and Brandon (1994) advocated exhumation by thrust faulting on the basis of brittle structures along the Franciscan-ophiolite contact. Wakabayashi and Unruh (1995) suggested that late Cenozoic (and active) thrust faulting on the ophiolite–Franciscan Complex contact obliterated earlier kinematic indicators, and that the critical path took place from 100 to 70 Ma, consistent with not only fission-track data of Tagami and Dumitru (1996), but also age constraints on normal slip on the ophiolite–Franciscan Complex contact.

The presence of high $P$-$T$ rocks structurally above lower $P$ metamorphic rocks (Figs. 2, 4, and 6) may not be well explained by the Platt (1986) model and is a common feature of many of the world's high $P$-$T$ belts (Maruyama et al., 1996). These structures suggest that the high $P$-$T$ rocks rose relative to their hanging wall *and* footwall, an extrusion-like process (Maruyama et al., 1996) (Fig. 6). If the faults bounding the high $P$-$T$ rocks were coeval, the material flow pattern in the subduction complex was similar to that of corner flow (e.g., Cowan and Silling, 1978; Cloos, 1982; Pavlis and Bruhn, 1983), and exhumation may have been driven by large-scale underplating of oceanic terranes from 100 to 70 Ma (Ernst, 1971; Wakabayashi and Unruh, 1995) (Fig. 6). However, the timing of movement on the faults bounding the high $P$-$T$ rocks has not been constrained with sufficient precision to determine whether movement on them was indeed coeval. Return flow in the accretionary prisms may be influenced by the shape of the backstop of the subduction complex, greater strength of the backstop (presumably mantle, see Fig. 6) compared to accreted materials, and lower density of the high $P$-$T$ rocks relative to the mantle (e.g., Mancktelow, 1995).

Further detailed structural, geophysical, and geochronologic studies in the Franciscan Complex can help refine and test exhumation models for high $P$-$T$ rocks.

## CONCLUDING STATEMENT

In connecting high $P$-$T$ metamorphism with subduction, Ernst (1970) made a key breakthrough in plate tectonics and helped make the Franciscan Complex one of the most famous rock units in the world. Nearly 30 years later, the Franciscan still poses fundamental problems, the resolution of which promise to advance our understanding of subduction complex processes significantly. It will be interesting to see what the next 30 years of research in the Franciscan Complex will yield.

## ACKNOWLEDGMENTS

Writing this paper partly repays an intellectual debt to Gary Ernst, whose concepts relating metamorphism and tectonics, along with regional tectonic concepts of Eldridge Moores, have most strongly influenced my approach to tectonic problems. My fledgling knowledge of the Franciscan Complex has benefited from discussions with many individuals, especially Clark Blake, Vic Seiders, J. G. Liou, Trevor Dumitru, Gary Ernst, and Mark Cloos, and from guidance from Eldridge Moores and Clyde Wahrhaftig. This manuscript has benefited from constructive reviews by Darrel Cowan, Gary Ernst, and Doris Sloan.

## REFERENCES CITED

Aalto, K. R., 1981, Multistage melange formation in the Franciscan Complex, northernmost California: Geology, v. 9, p. 602–607.

Alvarez, W., Kent, D. V., Silva, I. P., Schweickert, R. A., and Larsen, R. A., 1980, Franciscan Complex limestone deposited at 17° south paleolatitude: Geological Society of America Bulletin, v. 91, p. 476–484.

Bailey, E. H., Irwin, W. P., and Jones, D. L., 1964, The Franciscan and related rocks and their significance in the geology of western California: California Division of Mines and Geology Bulletin 183, 177 p.

Bailey, E. H., Blake, M. C., Jr., and Jones, D. L., 1970, On-land Mesozoic ocean crust in California Coast Ranges: U.S. Geological Survey Professional Paper 700-C, p. 70–81.

Beck, M. E., Jr., 1983, On the mechanism of tectonic transport in zones of oblique subduction: Tectonophysics, v. 93, p. 1–11.

Berkland, J. O., Raymond, L. A., Kramer, J. C., Moores, E. M., and O'Day, M., 1972, What is Franciscan?: American Association of Petroleum Geologists Bulletin, v. 56, p. 2295–2302.

Blake, M. C., Jr., 1967, Upside-down metamorphic zonation, blueschist facies, along a regional thrust in California and Oregon: U.S. Geological Survey Professional Paper 575-C, p. 1–9.

Blake, M. C., Jr., Howell, D. G., and Jones, D. L., 1982, Preliminary tectonostratigraphic terrane map of California: U.S. Geological Survey Open File Report 82–593, scale ~1:730,000.

Blake, M. C., Jr., Howell, D. G., and Jayko, A. S., 1984, Tectonostratigraphic terranes of the San Francisco Bay region, *in* Blake, M. C., Jr., ed., Franciscan geology of northern California: Society of Economic Paleontologists and Mineralogists, v. 43, p. 5–22.

Blake, M. C., Jr., Jayko, A. S., McLaughlin, R. J., and Underwood, M. B., 1988, Metamorphic and tectonic evolution of the Franciscan Complex, northern California, *in* Ernst, W. G., ed., Metamorphism and crustal evolution of the western United States (Rubey Volume VII): Englewood Cliffs, New Jersey, Prentice-Hall, p. 1035–1060.

Bostick, N. H., 1974, Phytoclasts as indicators of thermal metamorphism, Franciscan assemblage and Great Valley sequence, *in* Geological Society of America Special Paper 153, p. 1–17.

Bröcker, M., and Day, H. W., 1995, Low-grade blueschist metamorphism in metagreywackes, Eastern Franciscan belt, northern California: Journal of Metamorphic Geology, v. 13, p. 61–78.

Brown, E. H., and Bradshaw, J. Y., 1979, Phase relations of pyroxene and amphibole in greenstone, blueschist and eclogite of the Franciscan Complex, California: Contributions to Mineralogy and Petrology, v. 71, p. 67–83.

Brown, E. H., and Ghent, E. D., 1983, Mineralogy and phase relations in the blueschist facies of the Black Butte and Ball Rock areas, northern California Coast Ranges: American Mineralogist, v. 68, p. 365–372.

Carlson, C., 1981, Upwardly mobile melanges, serpentinite protrusions, and transport of tectonic blocks in accretionary prisms: Geological Society of America Abstracts with Programs, v. 13, p. 48.

Cloos, M., 1982, Flow melanges: Numerical modelling and geologic constraints on their origin in the Franciscan subduction complex, California: Geological Society America Bulletin, v. 93, p. 330–345.

Cloos, M., 1983, Comparative study of melange matrix and metashales from the Franciscan subduction complex with the basal Great Valley sequence, California: Journal of Geology, v. 91, p. 291–306.

Cloos, M., 1985, Thermal evolution of convergent plate-margins: thermal modelling and re-evaluation of isotopic Ar-ages for blueschists in the Franciscan Complex of California: Tectonics, v. 4, p. 421–433.

Cloos, M., and Dumitru, T. A., 1987, Blueschist terranes in the Franciscan Complex of California: Their future character and implications for past plate interactions: Geological Society of America Abstracts with Programs, v. 19, no. 6, p. 366.

Coleman, R. G., 1996, Prospecting for ophiolites in the California continental margin: Eos (Transactions, American Geophysical Union), v. 77, p. F743.

Coleman, R. G., and Lanphere, M. A., 1971, Distribution and age of high-grade blueschists, associated eclogites, and amphibolites from Oregon and California: Geological Society of America Bulletin, v. 82, p. 2397–2412.

Cowan, D. S., 1974, Deformation and metamorphism of the Franciscan subduction zone complex northwest of Pacheco Pass, California: Geological Society of America Bulletin, v. 85, p. 1623–1634.

Cowan, D. S., 1985, Structural styles in Mesozoic and Cenozoic melanges in the Western Cordillera of North America: Geological Society of America Bulletin, v. 96, p. 451–462.

Cowan, D. S., and Page, B. M., 1975, Recycled Franciscan material in Franciscan melange west of Paso Robles, California: Geological Society of America Bulletin, v. 86, p. 1089–1095.

Cowan, D. S., and Silling, R. M., 1978, A dynamic scaled model of accretion at trenches and its implication for the tectonic evolution of subduction complexes: Journal of Geophysical Research, v. 83, p. 5389–5396.

Cowan, D.S., Brandon, M. T., and Garver, J. I., 1997, Geologic tests of hypotheses for large coastwise displacements—A critique illustrated by the Baja British Columbia controversy: American Journal of Science, v. 297, p. 117–173.

Dalla Torre, M., De Capitani, C., Frey, M., Underwood, M. B., Mullis, J., and Cox, R., 1996, Very low-temperature metamorphism of shales from the Diablo Range, Franciscan Complex, California: New constraints on the exhumation path: Geological Society of America Bulletin, v. 108, p. 578–601.

Dickinson, W. R., 1970, Relations of andesites, granites and derivative sandstones to arc-trench tectonics: Reviews of Geophysics and Space Physics, v. 8, p. 813–860.

Dickinson, W. R., and Seely, D. R., 1979, Structure and stratigraphy of fore-arc regions: American Association of Petroleum Geologists Bulletin, v. 63, p. 2–31.

Dudley, P. P., 1972, Comments on the distribution and age of high-grade blueschists, associated eclogites and amphibolites from the Tiburon Peninsula, California: Geological Society of America Bulletin, v. 83, p. 3497–3500.

Dumitru, T. A., 1989, Constraints on uplift in the Franciscan subduction complex from apatite fission track analysis: Tectonics, v. 8, p. 197–220.

Engebretson, D. C., Cox, A., and Gordon, R. G., 1985, Relative motion between oceanic and continental plates in the Pacific basin: Geological Society of America Special Paper 206, 59 p.

Ernst, W. G., 1970, Tectonic contact between the Franciscan melange and the Great Valley Sequence, crustal expression of a late Mesozoic Benioff Zone: Journal of Geophysical Research, v. 75, p. 886–902.

Ernst, W. G., 1971, Metamorphic zonations on presumably subducted lithospheric plates from Japan, California, and the Alps: Contributions to Mineralogy and Petrology, v. 34, p. 43–59.

Ernst, W. G., 1975, Systematics of large-scale tectonics and age progression in Alpine and circum-Pacific blueschist belts: Tectonophysics, v. 17, p. 255–272.

Ernst, W. G., 1988, Tectonic history of subduction zones inferred from retrograde blueschist P-T paths: Geology, v. 16, p. 1081–1084.

Ernst, W. G., 1993, Metamorphism of Franciscan tectonostratigraphic assemblage, Pacheco Pass area, east-central Diablo Range, California Coast Ranges: Geological Society of America Bulletin, v. 105, p. 618–636.

Fitch, T. J., 1972, Plate convergence, transcurrent faults and internal deformation adjacent to southeast Asia and the western Pacific: Journal of Geophysical Research, v. 74, p. 4432–4461.

Godfrey, N. J., Beaudoin, B. C., Klemperer, S. L., and Mendocino Working Group, 1997, Ophiolitic basement to the Great Valley forearc basin, California, from seismic and gravity data: Implications for crustal growth at the North American continental margin: Geological Society of America Bulletin, v. 108, p. 1536–1562.

Graham, S. A., Stanley, R. G., Bent, J. V., and Carter, J. B., 1989, Oligocene and Miocene paleogeography of central California and displacement along the San Andreas fault: Geological Society of America Bulletin, v. 101, p. 711–730.

Hamilton, W. B., 1969, Mesozoic California and underflow of the Pacific mantle: Geological Society of America Bulletin, v. 80, p. 2409–2430.

Hsü, K. J., 1968, The principles of melanges and their bearing on the Franciscan-Knoxville paradox: Geological Society of America Bulletin, v. 79, p. 1063–1074.

Hsü, K. J., 1971, Franciscan melanges as a model for eugeosynclinal sedimentation and underthrusting tectonics: Journal of Geophysical Research, v. 76, p. 1162–1170.

Hsü, K. J., and Ohrbom, R., 1969, Melanges of the San Francisco Peninsula—A geological reinterpretation of the type Franciscan: American Association of Petroleum Geologists Bulletin, v. 53, p. 1348–1367.

Irwin, W. P., 1972, Terranes of the western Paleozoic and Triassic belt in the southern Klamath Mountains, California: U.S. Geological Survey Professional Paper 800-C, p. C103–C111.

Isozaki, Y., and Blake, M. C., Jr., 1994, Biostratigraphic constraints on the formation and timing of accretion in a subduction complex: An example from the Franciscan Complex of northern California: Journal of Geology, v. 102, p. 283–296.

Jayko, A. S., and Blake, M. C., Jr., 1993, Northward displacements of forearc slivers in the Coast Ranges of California and southwest Oregon during the late Mesozoic and early Cenozoic, in Dunn, G., and McDougall, K., eds., Mesozoic paleogeography of the western United States II: Pacific Section, SEPM (Society for Sedimentary Geology) Book 71, p. 19–36.

Jennings, C., compiler, 1977, Geologic map of California: San Francisco, California Division of Mines and Geology, scale 1:750,000.

Krogh, E. J., Oh, C. W., and Liou, J. G., 1994, Polyphase and anticlockwise P-T evolution for Franciscan eclogites and blueschists from Jenner, California, USA: Journal of Metamorphic Geology, v. 12, p. 121–134.

Lanphere, M. A., Blake, M. C., Jr., and Irwin, W. P., 1978, Early Cretaceous metamorphic age of the South Fork Mountain Schist in the northern Coast Ranges of California: American Journal of Science, v. 278, p. 798–815.

Lee, D. E., Thomas, H. H., Marvin, R. F., and Coleman, R. G., 1964, Isotopic ages of glaucophane schists from the area of Cazadero, California: U.S. Geological Survey Professional Paper 475-D, p. D105–D107.

Macpherson, G. J., Phipps, S. P., and Grossman, J. N., 1990, Diverse sources for igneous blocks in Franciscan melanges, California Coast Ranges: Journal of Geology, v. 98, p. 845–862.

Mancktelow, N. S., 1995, Nonlithostatic pressure during sediment subduction and the development and exhumation of high pressure metamorphic rocks: Journal of Geophysical Research, v. 100, p. 571–583.

Maruyama, S., and Liou, J. G., 1988, Petrology of Franciscan metabasites along the jadeite-glaucophane type facies series, Cazadero, California: Journal of Petrology, v. 29, p. 1–37.

Maruyama, S., Liou, J. G., and Sasakura, Y., 1985, Low-temperature recrystallization of Franciscan graywackes from Pacheco Pass, California: Mineralogical Magazine, v. 49, p. 345–355.

Maruyama, S., Liou, J. G., and Terabayashi, M., 1996, Blueschists and eclogites of the world and their exhumation: International Geology Review, v. 38, p. 485–594.

Mattinson, J. M., 1986, Geochronology of high-pressure–low-temperature Franciscan metabasites: A new approach using the U-Pb system, *in* Evans, B. W., and Brown, E. H., eds., Blueschists and eclogites: Geological Society of America Memoir 164, p. 95–105.

Mattinson, J. M., and Echeverria, L. M., 1980, Ortigalita Peak gabbro, Franciscan complex—U/Pb date of intrusion and high-pressure–low temperature metamorphism: Geology, v. 8, p. 589–593.

Maxwell, J. C., 1974, Anatomy of an orogen: Geological Society of America Bulletin, v. 85, p. 1195–1204.

McCaffrey, R., 1992, Oblique plate convergence, slip vectors, and forearc deformation: Journal of Geophysical Research, v. 97, p. 8905–8915.

McDowell, F., Lehman, D. H., Gucwa, P. R., Fritz, D., and Maxwell, J. C., 1984, Glaucophane schists and ophiolites of the northern California Coast Ranges: Isotopic ages and their tectonic implications: Geological Society of America Bulletin, v. 95, p. 1373–1382.

McLaughlin, R. J., Blake, M. C., Jr., Griscom, A., Blome, C. D., and Murchey, B., 1988, Tectonics of formation, translation and dispersal of the Coast Range ophiolite of California: Tectonics, v. 7, p. 1033–1056.

Moore, D. E., 1984, Metamorphic history of a high-grade blueschist exotic block from the Franciscan Complex, California: Journal of Petrology, v. 25, p. 126–150.

Moore, D. E., and Blake, M. C., Jr., 1989, New evidence for polyphase metamorphism of glaucophane schist and eclogite exotic blocks in the Franciscan Complex, California and Oregon: Journal of Metamorphic Geology, v. 7, p. 211–228.

Moore, D. E., and Liou, J. G., 1980, Detrital blueschist pebbles from Franciscan metaconglomerates of the northeast Diablo Range, California: American Journal of Science, v. 280, p. 249–264.

Moores, E. M., 1970, Ultramafics and orogeny, with models of the U.S. Cordillera and the Tethys: Nature, v. 228, p. 837–842.

Murchey, B. M., 1984, Biostratigraphy and lithostratigraphy of chert in the Franciscan Complex, Marin Headlands block, California, *in* Blake, M. C., Jr., ed., Franciscan geology of northern California: Society of Economic Paleontologists and Mineralogists, Special Publication v. 43, p. 23–30.

Murchey, B. M., and Jones, D. L., 1984, Age and significance of chert in the Franciscan Complex, San Francisco Bay region, *in* Blake, M. C., Jr., ed., Franciscan geology of northern California: Society of Economic Paleontologists and Mineralogists, Special Publication v. 43, p. 51–70.

Page, B. M., 1978, Franciscan melanges compared with olistostromes of Taiwan and Italy: Tectonophysics, v. 47, p. 223–246.

Page, B. M., 1981, The southern Coast Ranges, *in* Ernst, W. G., ed., The geotectonic development of California (Rubey Volume 1): Englewood Cliffs, New Jersey, Prentice-Hall, p. 329–417.

Pavlis, T. L., and Bruhn, R. L., 1983, Deep-seated flow as a mechanism for the uplift of broad fore-arc ridges and its role in the exposure of high P/T metamorphic terranes: Tectonics, v. 2, p. 473–497.

Peacock, S. M., 1988, Inverted metamorphic gradients in the westernmost Cordillera, *in* Ernst, W. G., ed., Metamorphism and crustal evolution, western United States (Rubey Volume VII): Englewood Cliffs, New Jersey, Prentice-Hall, p. 954–975.

Peterman, Z. E., Hedge, C. E., Coleman, R. G., and Snavely, P. D., 1967, $Sr^{87}/Sr^{86}$ ratios in some eugeosynclinal sedimentary rocks and their bearing on the origin of granitic magma in orogenic belts: Earth and Planetary Science Letters, v. 2, p. 433–439.

Platt, J. P., 1975, Metamorphic and deformational processes in the Franciscan Complex, California: Some insights from the Catalina Schist terrane: Geological Society of America Bulletin, v. 86, p. 1337–1347.

Platt, J. P., 1986, Dynamics of orogenic wedges and the uplift of high-pressure metamorphic rocks: Geological Society of America Bulletin, v. 97, p. 1037–1053.

Raymond, L. A., 1970, Cretaceous sedimentation and regional thrusting, northeastern Diablo Range, California: Geological Society of America Bulletin, v. 81, p. 2123–2128.

Ring, U., and Brandon, M. T., 1994, Kinematic data for the Coast Range fault and implications for exhumation of the Franciscan subduction complex: Geology, v. 22, p. 735–738.

Ross, J. A., and Sharp, W. D., 1986, $^{40}Ar/^{39}Ar$ and Sm/Nd dating of garnet amphibolite in the Coast Ranges, California: Eos (Transactions, American Geophysical Union), v. 67, p. 1249.

Ross, J. A., and Sharp, W. D., 1988, The effects of sub-blocking temperature metamorphism on the K/Ar systematics of hornblendes: $^{40}Ar/^{39}Ar$ dating of polymetamorphic garnet amphibolite from the Franciscan Complex, California: Contributions to Mineralogy and Petrology, v. 100, p. 213–221.

Sedlock, R. L., and Hamilton, D. H., 1991, Late Cenozoic tectonic evolution of southwestern California: Journal of Geophysical Research, v. 96, p. 2325–2351.

Seiders, V. M., and Blome, C. D., 1988, Implications of upper Mesozoic conglomerate for suspect terrane in western California and adjacent areas: Geological Society of America Bulletin, v. 100, p. 374–391.

Sliter, R. V., 1984, Foraminifers from Cretaceous limestone of the Franciscan Complex, northern California, *in* Blake, M. C., Jr., ed., Franciscan geology of northern California: Society of Economic Paleontologists and Mineralogists, Special Publication v. 43, p. 149–162.

Steiger, R. H., and Jäger, E., 1977, Subcommission on geochronology: Convention on the use of decay constants in geo- and cosmochronology: Earth and Planetary Science Letters, v. 36, p. 359–362.

Suppe, J., 1973, Geology of the Leech Lake Mountain–Ball Mountain region, California: University of California Publications in the Geological Sciences, v. 107, 82 p.

Suppe, J., 1978, Cross section of southern part of northern Coast Ranges and Sacramento Valley, California: Geological Society of America Map and Chart Series MC-28B, scale 1:250,000.

Suppe, J., and Armstrong, R. L., 1972, Potassium-argon dating of Franciscan metamorphic rocks: American Journal of Science, v. 272, p. 217–233.

Suppe, J., and Foland, K. A., 1978, The Goat Mountain Schists and Pacific Ridge Complex: A redeformed, but still intact, late Mesozoic schuppen complex, *in* Howell, D. G., and McDougall, K. A., eds., Mesozoic paleogeography of the western United States: Pacific Section, Society of Economic Paleontologists and Mineralogists, Pacific Coast Paleogeography Symposium 2, p. 431–451.

Tagami, T., and Dumitru, T. A., 1996, Provenance and thermal history of the Franciscan accretionary complex-constraints from zircon fission track thermochronology: Journal of Geophysical Research, v. 101, p. 11353–11364.

Tarduno, J. A., McWilliams, M. O., Debiche, M. G., Sliter, W. V., and Blake, M. C., Jr., 1985, Franciscan Complex Calera limestones: Accreted remnants of Farallon Plate oceanic plateaus: Nature, v. 317, p. 345–347.

Taylor, H. P., Jr., and Coleman, R. G., 1968, $O^{16}/O^{18}$ ratios of coexisting minerals in glaucophane-bearing metamorphic rocks: Geological Society of America Bulletin, v. 79, p. 1727–1756.

Underwood, M. B., 1989, Temporal changes in geothermal gradients, Franciscan subduction complex, northern California: Journal of Geophysical Research, v. 94, p. 3111–3125.

Wahrhaftig, C., 1984, Structure of the Marin Headlands block, California: A progress report, *in* Blake, M. C., ed., Franciscan geology of northern California: Society of Economic Paleontologists and Mineralogists, Special Publication v. 43, p. 31–50.

Wakabayashi, J., 1990, Counterclockwise P-T-t paths from amphibolites, Franciscan Complex, California: Metamorphism during the early stages of subduction: Journal of Geology, v. 98, p. 657–680.

Wakabayashi, J., 1992, Nappes, tectonics of oblique plate convergence, and metamorphic evolution related to 140 million years of continuous subduction, Franciscan Complex, California: Journal of Geology, v. 100, p. 19–40.

Wakabayashi, J., and Deino, A., 1989, Laser-probe $^{40}Ar/^{39}Ar$ ages from high grade blocks and coherent blueschists, Franciscan Complex, California: Preliminary results and implications for Franciscan tectonics: Geological Society of America Abstracts with Programs, v. 21, no. 6, p. A267.

Wakabayashi, J., and Hengesh, J. V., 1995, Distribution of late Cenozoic displacement on the San Andreas fault system, northern California, *in* Sangines, E. M., Andersen, D. W., and Buising, A. W., eds., Recent geologic studies in the San Francisco Bay Area: Pacific Section, SEPM (Society for Sedimentary Geology) Book 76, p. 19–30.

Wakabayashi, J., and Unruh, J. R., 1995, Tectonic wedging, blueschist metamorphism, and exposure of blueschist: Are they compatible?: Geology, v. 23, p. 85–88.

Weinrich, A., 1997, Geotectonic setting of the Leech Lake Mountain alkaline diabase sills, northern California: Constraints from petrography, geochemistry and $^{40}Ar/^{39}Ar$ dating [diplomarbeit (M.S. thesis)}: Mainz, Germany, Institut für Geowissenschaften, Johannes Gutenberg-Universität Mainz, 119 p.

Wentworth, C. M., Blake, M. C., Jr., Jones, D. L., Walter, A. W., and Zoback, M. D., 1984, Tectonic wedging associated with emplacement of the Franciscan assemblage, California Coast Ranges, in Blake, M. C., ed., Franciscan geology of northern California: Society of Economic Paleontologists and Mineralogists, Special Publication v. 43, p. 163–173.

Worrall, D. M., 1981, Imbricate low-angle faulting in uppermost Franciscan rocks, south Yolla Bolly area, northern California: Geological Society of America Bulletin, v. 92, p. 703–729.

MANUSCRIPT ACCEPTED BY THE SOCIETY NOVEMBER 23, 1998.

Geological Society of America
Special Paper 338
1999

# Chapter 8
# THE GREAT VALLEY GROUP: THE ARC-TRENCH GAP

*Raymond V. Ingersoll*
*Department of Earth Sciences*
*University of California*
*Los Angeles, California 90095-1567*
*E-mail: ringer@ess.ucla.edu*

## RICHARD W. OJAKANGAS

Richard W. Ojakangas grew up in Minnesota and attended the University of Minnesota at Duluth as an undergraduate. He received a masters degree from the University of Missouri before spending a year at the University of Helsinki, Finland, as a Fulbright Fellow. In 1964 he completed his doctoral dissertation at Stanford University and accepted a faculty position at the University of Minnesota at Duluth, where he is currently professor, having previously served as head of the department. He has done field work in Antarctica, India, and Finland, as well as North America. His classic 1968 paper resulted directly from his dissertation research. Most of his subsequent research has involved Precambrian rocks of the Great Lakes area. (All of his research seems to involve something Great.)

## WHY THIS PAPER IS A "CLASSIC"

In his 1968 paper Ojakangas was the first author systematically to study the sedimentology and stratigraphy of upper Mesozoic strata of the Great Valley. His work was completed before the plate tectonics revolution, so that the tectonic setting of the Great Valley was not known in an actualistic sense. Ojakangas worked within a geosynclinal context, as did most geologists of the 1960s (e.g., Kay, 1951). The hot topics of the early 1960s were turbidites and paleocurrents in flysch deposits (e.g., Kuenen, 1953, 1964; Crowell, 1955, 1957; Bouma, 1959, 1962; Dott, 1963; Dzulynski and Smith, 1963; Potter and Pettijohn, 1963; McBride, 1964; Walker, 1967). Ojakangas (1968) interpreted Great Valley strata as having been deposited as turbidites and carefully documented paleocurrent directions. In addition, he studied provenance and diagenesis using sandstone petrology, accessory minerals, conglomerate composition, and clay mineralogy. Ojakangas (1968) integrated sedimentology, stratigraphy, petrology, and regional relations into a basin analysis that culminated in the proposal of three possible models involving turbidites derived from the east and transported primarily to the south, following deflection in a deep-marine environment. Differences among the models involved the degree to which the turbidites built fans or were transported in basin-plain settings, and the degree to which sediment was contributed from the north (Klamath Mountains) versus the east (Sierra Nevada). It is important to keep in mind that these paleogeographic models were proposed without the benefit of subsequent knowledge of forearc or submarine-fan models.

This paper is a "classic" paper because it provided fundamentally important data on paleocurrents, petrology, sedimentary structures, and stratigraphy, which provided the basis for all subsequent work. In addition, it provided insightful paleogeographic models, which led directly to subsequent models based on paradigm shifts in our understanding of submarine-fan facies, sandstone petrofacies and forearc basins. It is impressive that so many insights can still be gained by reading this classic paper, even after 30 years of additional research.

Raymond V. Ingersoll

RICHARD W. OJAKANGAS *University of Minnesota, Duluth, Minnesota*

# Cretaceous Sedimentation, Sacramento Valley, California

**Abstract:** Upper Jurassic and Cretaceous sedimentary rocks more than 35,000 feet thick are exposed along the west side of the Sacramento Valley of California. To ascertain the framework in which the sediments were deposited, a detailed study of these rocks was made in the Cache Creek-Rumsey Hills area, and a general study was made at other localities.

The sedimentary section consists of interbedded sandstones, which are commonly graded, mudstones, and siltstones, with minor conglomerates and bentonitic rocks. Units that are dominantly sandstone alternate with units that are dominantly mudstone. The sequence is sparsely fossiliferous, but approximate stage boundaries were determined.

Nine hundred paleocurrent measurements were made on sole marks, parting lineations, cross-bedding, and other structures. These indicate that paleocurrents in general flowed from north to south, parallel to the regional tectonic trend, the eastern shoreline of the basin of deposition, and the isopachous lines of the sequence. During Cenomanian and early Turonian time, however, currents moving toward the west were prominent.

The mineralogy of the sandstones was determined through the Cache Creek-Rumsey Hills section, and check samples from elsewhere in the belt were found to be similar. The sandstones include feldspathic, arkosic, and lithic wackes and arenites, with the main grain constituents being quartz, feldspar, and volcanic rock fragments. The K-feldspar content is greatest in the Upper Cretaceous rocks. Epidote is the dominant non-opaque, non-micaceous heavy mineral; apatite, sphene, hornblende, and zircon are also common. Six intervals of differing composition were established. In Interval I (Upper Jurassic), andesitic (?) volcanic detritus is dominant in the few samples studied; Interval II (Valanginian) contains abundant quartz and plagioclase; in Interval III (Hauterivian through middle Albian), quartz is dominant; Interval IV (upper Albian and Cenomanian) is characterized by K-feldspar and andesitic (?) volcanic rock fragments; Interval V (Turonian to upper Santonian) contains K-feldspar, plagioclase, quartz, and acidic volcanic detritus; and Interval VI (upper Santonian and Campanian) is characterized by K-feldspar, plagioclase, and quartz. The clayey fractions of sandstones and associated mudstones are similar, with chlorite dominant through much of the Lower Cretaceous, and mica and montmorillonite dominant in the Upper Cretaceous. Major diagenetic changes in the sandstones include widespread carbonate cementation and replacement, recrystallization of matrix, and alteration of plagioclase and biotite. Albitization of plagioclase and chloritization of biotite were found to vary with depth and the content of calcite in the sandstones. These alterations proceeded to a small degree at depths of burial estimated to be as shallow as 10,000 feet, they affected all noncalcareous strata buried from 20 to 30,000 feet, and were extensively developed in even the calcareous rocks that were buried 35,000 feet.

The ancestral Klamath Mountains and Sierra Nevada evidently were the sources of the detritus. Major events in the source areas, such as volcanism, pluton emplacement, and unroofing of plutons, apparently are reflected in the sedimentary rock column. The sequence was deposited below wave base, generally far from shore in an outer neritic-upper bathyal, probably miogeosynclinal, environment. Sedimentary structures and grading suggest deposition by turbidity currents. The sequence strikingly resembles other rock sequences which have been interpreted in the literature as "turbidites."

In an attempt to explain the sedimentation, three models are considered, each consisting of an elongated north-south basin open to the west: (1) a southerly paleoslope trending

Geological Society of America Bulletin, v. 79, p. 973-1008, 11 figs., 5 pls., August 1968

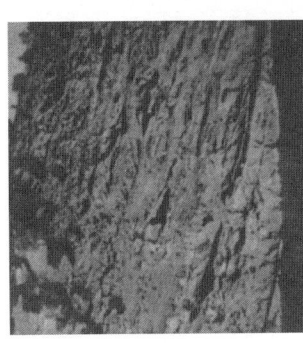

Figure 1. Upper part of Sites Formation, representative of sandier portions of the column.

Figure 2. Member 2 of Formation D, representative of "packet" of sandy beds in dominantly mudstone portion of the column.

Figure 3. Member 2 of Formation D, representative of mudstone portions of the column.

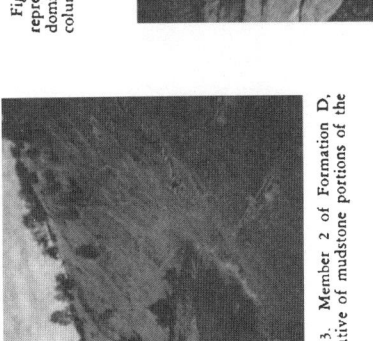

Figure 4. Guinda Formation, representative of sandier portions of the column. Note sharp contacts of bottoms of sandstone beds and underlying mudstones, and gradational contacts of tops of sandstone beds and overlying mudstones. "Ideal" grading.

EXPOSURES ALONG CACHE CREEK-RUMSEY HILLS SECTION

parallel to the long axis and the eastern shoreline of the basin, with currents carrying sediment southward down the paleoslope; (2) a paleoslope to the west from the north-south trending shoreline, with currents moving down the paleoslope and subsequently deflected to the south by deep oceanic currents; and (3) a subsea fan with an asymmetric cross section that caused the deflection of westerly moving turbidity currents toward the south. Modern counterparts of the last model are present off the western coast of North America. This model is tentatively suggested as the one which best fits the data.

## CONTENTS

| | |
|---|---|
| Introduction | 974 |
| Acknowledgments | 976 |
| Methods | 976 |
| Structure and thickness | 976 |
| Stratigraphy, gross lithology, and paleontology | 977 |
| Sedimentary structures | 979 |
| Internal structures | 979 |
| Graded bedding | 979 |
| Horizontal lamination | 979 |
| Convolute lamination | 979 |
| Cross-bedding | 979 |
| Parting lineation | 981 |
| Pebble imbrication | 981 |
| Sole marks | 981 |
| Flute casts | 982 |
| Groove casts | 982 |
| Miscellaneous casts | 982 |
| Frondescent casts | 982 |
| Channel fillings | 982 |
| Other structures | 982 |
| Aligned fossils | 982 |
| Ripple marks | 983 |
| Paleocurrent analysis | 983 |
| Other related paleocurrent data | 985 |
| Petrology | 985 |
| Sandstones | 985 |
| Stratigraphic and areal variation in sandstone composition | 988 |
| Diagenesis | 990 |
| Siltstones | 993 |
| Mudstones | 993 |
| X-ray analysis of mudstones and matrix of sandstones | 993 |
| Bentonic rocks | 993 |
| Conglomerates | 993 |
| Carbonates | 995 |
| Sedimentation | 995 |
| Provenance | 995 |
| Tectonics and sedimentation | 996 |
| Current mechanisms | 998 |
| Volume of sediment and rate of deposition | 999 |
| Environment of deposition | 1000 |
| Basin geometry and sedimentation models | 1002 |
| Conclusions | 1004 |
| References cited | 1005 |

Figure
1. Generalized geologic map of northern and central California ............ 975
2. Location map of Sacramento Valley sequence on west side of Sacramento Valley ............ 976
3. Generalized east-west cross section of Sacramento Valley sequence ............ 977
4. Stratigraphic column, Cache Creek-Rumsey Hills section ............ 978
5. Orientations and abundances of sedimentary structures ............ 984
6. Stratigraphic variations in paleocurrent trends ............ 985
7. Areal and stratigraphic variations in paleocurrent trends ............ 987
8. Variation in paleocurrent trends of certain formations ............ 989
9. Sandstone classification ............ 990
10. Variations in compositions of sandstones in the Cache Creek-Rumsey Hills section ............ 992
11. Sedimentation models for the Sacramento Valley sequence ............ 994

Plate
1. Exposure along Cache Creek-Rumsey Hills section facing ............ 985
2. Grading and sedimentary structures in sandstone beds along Cache Creek following ............ 985
3. Photomicrographs of sandstone beds in Cache Creek-Rumsey Hills section, Intervals I and II ............ 979
4. Photomicrographs of sandstone beds in Cache Creek-Rumsey Hills section, Intervals III and IV ............ 979
5. Photomicrographs of sandstone beds in Cache Creek-Rumsey Hills section, Intervals V and VI ............ 979

Table
1. General bedding characteristics ............ 980
2. Abundance and types of graded beds ............ 981
3. Orientations of associated paleocurrent indicators ............ 982
4. Abundance of primary sedimentary structures ............ 986
5. Heavy mineral data ............ 988
6. K-feldspar content ............ 991

## INTRODUCTION

A thick sequence of Upper Jurassic and Cretaceous sandstones and mudstones, with minor siltstones and conglomerates, crops out in a 150-mile-long belt along the west side of the Sacramento Valley of California (Fig. 1). An unusually well-exposed section of these rocks along Cache Creek and in the adjacent Rumsey Hills was investigated in detail, and other localities along the outcrop belt and elsewhere in northern California were studied less intensively (Fig. 2). The

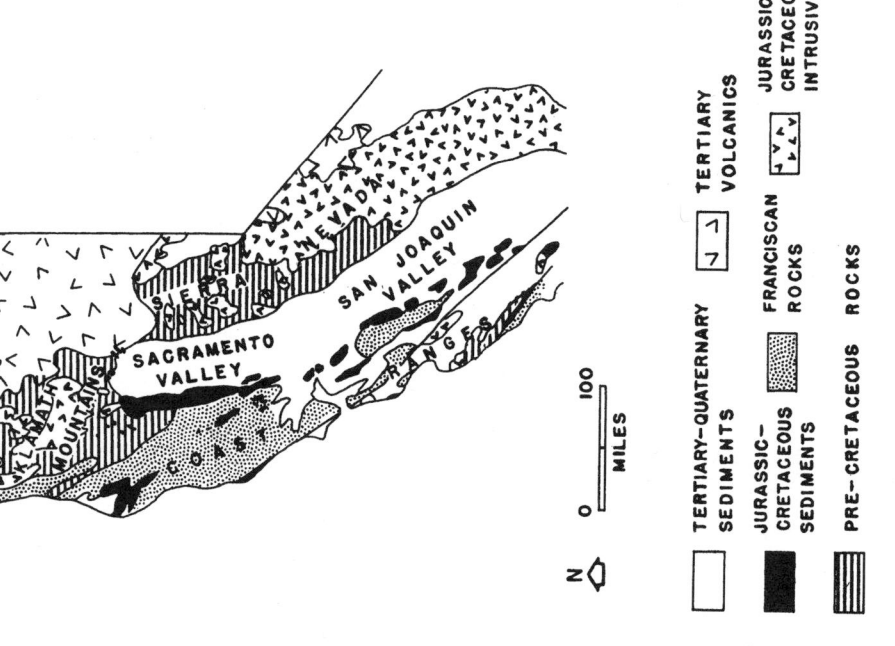

Figure 1. Generalized geologic map of northern and central California (*after* Chuber, 1962). Jurassic-Cretaceous belt on west side of the Sacramento Valley is the Sacramento Valley sequence of this paper.

primary objective of this study was to reconstruct the uppermost Jurassic and Cretaceous depositional framework, including the source areas, paleocurrent patterns, transporting mechanisms, and environments of deposition. Good exposure over a wide area permitted the determination of changes in mineralogy and paleocurrent movement through both space and time.

## ACKNOWLEDGMENTS

This paper is based on a doctoral dissertation completed at Sanford University in the spring of 1964. William R. Dickinson supervised the research, was especially helpful in the study of plagioclase alteration, and read the manuscript. Ben Page, Edgar Bailey, and David L. Jones read preliminary versions of the manuscript and offered numerous suggestions. Arthur O. Beall advised and assisted with the clay mineralogy. Keith D. Berry, of the Standard Oil Company of California, processed and studied microfossil samples collected during this investigation. Numerous other persons, including fellow graduate students, helped in various ways. Financial support was provided by fellowships from the Standard Oil Company of California and the National Science Foundation, by the Shell Fund for Fundamental Research, and by the University of Minnesota Graduate School. To all of the above, I express my sincere thanks.

## METHODS

The orientations of primary current structures on more than 900 sandstone beds were measured and recorded, and for each of these beds the general internal characteristics were also noted. In selected parts of the section in the detailed study area, every bed was examined to ascertain stratigraphic variation in bedding characteristics. For laboratory studies of the sandstones, 175 samples in stratigraphic sequence were collected elsewhere along the belt (see sample traverses on Fig. 2). About 225 thin sections, 200 heavy mineral residues, and 150 stained slabs were prepared and studied. The clay mineralogy of mudstones, sandstone matrix, and bentonitic rocks was determined by X-ray diffraction. Pebble counts were made of the conglomerates. One hundred and eighteen Upper Jurassic and Lower Cretaceous mudstone samples were examined for microfossils.

## STRUCTURE AND THICKNESS

The Upper Jurassic and Cretaceous sedimentary rocks along the west side of the Sacramento Valley form an east-dipping homocline (Fig. 3). Subsidiary folds occur in only a few places. Faults with small displacements, though abundant, do not seriously distort the succession.

The lowest unit of the sequence, the Upper Jurassic "Knoxville" Formation, lies in fault contact with Franciscan rocks (Upper Jurassic and Cretaceous) or a serpentine sheet. The Sacramento Valley sequence may have been thrust westward onto the Franciscan, with the serpentine serving as a lubricant. Thick conglomerates near the base of the Cretaceous column (Fig. 4) may represent a disconformity, but angular discordances were not recognized anywhere within the Sacramento Valley sequence in the Cache Creek-Rumsey Hills area. However, unconformable contacts have been reported in some parts of the belt by Anderson (1938; 1958), Lachenbruch (1962), and Peterson (1964; 1965).

Because of faulting, only a few thousand feet of the Upper Jurassic beds are exposed in the Cache Creek-Rumsey Hills area, but the thickness of the exposed overlying Cretaceous beds is about 31,000 feet. Farther north along the belt, the Upper Jurassic rocks attain a maximum thickness of 20,000 feet, the Lower Cretaceous 22,000 feet, and the exposed Upper Cretaceous about 15,000 feet (Chuber, 1962). Isopachous lines of different portions of the sequence trend north-south, parallel to the outcrop belt (Chuber, 1962). In contrast, correlative Upper Cretaceous rocks at the northern end of the Sacramento Valley and along its eastern side have a maximum thickness of a few thousand feet.

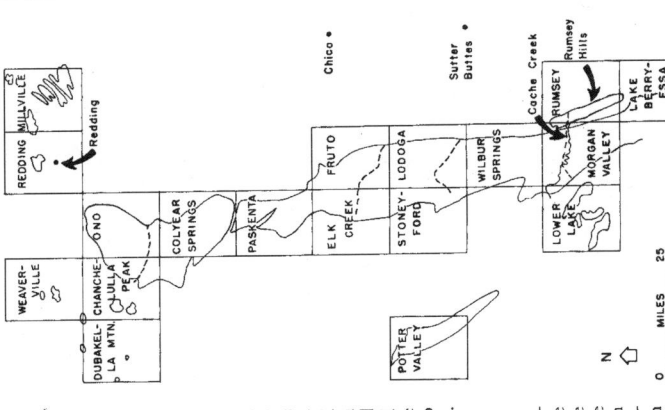

Figure 3. Generalized east-west cross section of Sacramento Valley from west edge of north half of Wilbur Springs quadrangle (see Fig. 2) to a point near Marysville, about 15 miles east of Sutter Butte. Lower part of west half of section is hypothetical (after Safonov, 1962.)

## STRATIGRAPHY, GROSS LITHOLOGY, AND PALEONTOLOGY

The development of the Upper Jurassic and Cretaceous nomenclature and stratigraphic opinion in the Sacramento Valley over the past 75 years has been summarized by Popenoe and others (1960). In the older literature, the Lower Cretaceous beds are called the Shasta "series" (including an older Paskenta "group" and a younger Horsetown "group"), and the Upper Cretaceous beds are referred to as the Chico "series". Formations and members, both formal and informal, have been established in the Cache Creek-Rumsey Hills area on the basis of gross lithology; predominantly sandstone sequences alternate with predominantly mudstone sequences (Fig. 4; Pl. 1). These units form alternating ridges and valleys which persist along strike for about 60 miles. Kirby (1943), working along the west side of the Sacramento Valley, named the Upper Cretaceous Venado, Yolo, Sites, Funks, Guinda, and Forbes Formations, from oldest to youngest. J. E. Lawton, in a Ph.D. thesis (1956, Geology of the north half of the Morgan Valley quadrangle: Stanford University, Palo Alto, California), assigned the following informal formation names, from oldest to youngest, to units in the Cache Creek area: Sulphur Creek ("Knoxville"), Crack Canyon (including the Leesville and Round Mountain Members),

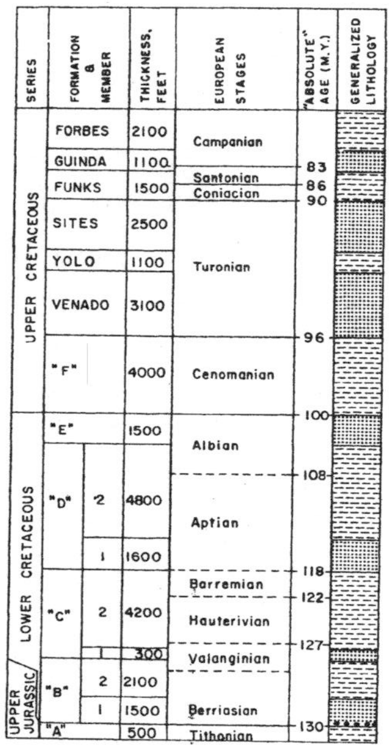

Figure 4. Stratigraphic column, Cache Creek-Rumsey Hills section. Dotted pattern represents lithology which is dominantly sandstone; dashed pattern represents lithology which is dominantly mudstone; large dots at base of Formation C represent conglomerate. Upper Cretaceous stage boundaries from Popenoe and others (1960); most Lower Cretaceous stage boundaries based on micropaleontology by Keith D. Berry (1964, personal commun.); Berriasian-Valanginian boundary based on age determination of *Buchias* collected in the section (Dr. David L. Jones, 1966, personal commun.); 96 m.y. date (Everndon and others, 1961) on biotite from a bentonite bed in upper portion of Formation F (Dr. C. M. Gilbert, 1964, personal commun.); approximate date of 130 m.y. (Curtis and others, 1958) based on relationships of sequence to underlying igneous rocks; other dates are "best guesses" based on isotopes dates and stratigraphic relations elsewhere (Dr. Adolph Knopf, 1964, personal commun.). Only the upper 500 feet of Formation A was studied.

Davis Canyon (including the Bald Mountain and Buck Island Members), Brophy Canyon, and Fiske Creek (the "Antelope shale"). Lawton's formations, because of their informal status, are hereafter referred to as Formations A-F, from oldest to youngest (Fig. 4). These formations are not to be confused with Goudkoff's (1945) microfossil zones which have similar letter designations.

Some of the members and formations contain an estimated 65 to 75 percent sandstone, whereas the intervening units contain only 10 to 30 percent sandstone (Pl. 1). The Cache Creek-Rumsey Hills area, with a 2:1 mudstone to sandstone ratio, contains more sandstone than other areas along the outcrop belt. In general, the Upper Cretaceous Series along the west side of the Sacramento Valley contains more sandstones than does the Lower Cretaceous and Upper Jurassic Series, and the entire sequence shows a decreasing sand content northward from the Cache Creek-Rumsey Hills section. Near the north end of the belt along Dry Creek in the Ono and Chanchellula Peak quadrangles, only 1 to 2 percent of the section is sandstone and conglomerate.

Individual sandstone beds, generally 2 inches to 3 feet thick, with a maximum thickness of 40 feet, commonly occur in "packets" of several beds separated by dark gray to black mudstones which commonly contain thin gray siltstone laminae. Most of the sandstone beds maintain a fairly constant thickness throughout a given exposure, although a few beds pinch out completely. A lack of marker beds makes it impossible to correlate individual strata from outcrop to outcrop.

Minor rock types, together making up only a few percent of the entire thickness, include conglomerates, bentonitic rocks, and carbonate concretions. "Sedimentary serpentine," an unusual rock type, occurs in the Lower Cretaceous of the northern Morgan Valley quadrangle (Emerson and Rich, 1966, p. 473).

Megafossils, other than burrows or feeding tracks, are generally rather scarce in the Cretaceous beds, with a few exceptions. In the northern part of the sequence (Murphy and Rodda, 1960) fossils are fairly common. In the Upper Jurassic and lowermost Cretaceous strata, *Buchia sp.* are abundant. Most of the other fossils in the Lower Cretaceous units are abraded and fragmentary, and occur in conglomerates or coarse basal portions of graded sandstone beds. High in the sequence in the Upper Cretaceous Forbes and Funks Formations, fossils are again more common. Planktonic and benthonic foraminifera are common in the Upper Cretaceous Series.

The study of microfossils permitted the approximate positioning of most Lower Cretaceous stage boundaries in the Cache Creek-Rumsey Hills section (Fig. 4). The Upper Cretaceous stage boundaries had been established earlier (e.g., Popenoe and others, 1960).

## SEDIMENTARY STRUCTURES

### Internal Structures

The internal characteristics of the sandstone beds were studied because they provide evidence regarding the mechanisms of transportation and environments of depositions. The presence or absence of grading, the type of grading, and the internal structure (whether massive, laminated, or both) were recorded for nearly 1500 beds, both in the Cache Creek-Rumsey Hills section and in other parts of the Sacramento Valley area (Tables 1 and 2).

*Graded bedding.* Two-thirds of the studied beds are graded, with grading most common in the more sandy formations and members (Tables 1 and 2). Only about a third of the beds studied in the northernmost three quadrangles (Chanchellula Peak, Ono, and Colyear Springs) of the outcrop belt are graded, whereas from 64 to 84 percent of the studied beds are graded in the quadrangles farther south along the belt. There does not appear to be any major difference in the abundance of grading in the Upper and Lower Cretaceous Series.

The distribution of five types of grading, all of which occur in both massive and laminated beds, is shown in Table 2. The "ideal" type, with a general upward decrease in grain size from sand to a silty-clayey top which tends to merge with the overlying mudstone, (Pl. 1, fig. 4) and a "modified ideal" type without the silty-clayey top (Pl. 2, fig. 1) constitute most of the graded beds. Kuenen (1953) illustrated types closely comparable to those observed in this study.

*Horizontal lamination.* The internal structure of the sandstone beds, irrespective of grading, is either entirely massive, entirely laminated, or massive with an upper laminated portion. Massive or partially massive internal structure is present in slightly more than half of the beds studied (Table 1). In general, massive beds are more abundant in the units with many sandstone beds, and laminated beds are more abundant in the units which contain more mudstones. Although most members and formations are composed predominantly of one type, the other types are usually present.

Kuenen (1953) and Bouma (1962) noted that massive graded beds commonly have laminated tops. Bouma's "complete" bed includes, from the bottom up, a graded massive interval, a lower interval of parallel lamination, an interval of ripple cross-bedding or convolutions or both, an upper interval of parallel lamination, and finally, a pelitic interval. His "derived types" include a "base cut-out sequence" (most common in the rocks he studied) with the lower 1, 2, 3, or 4 of the above intervals absent; a "truncated sequence" with the upper 1, 2, 3, or 4 units absent; and a combination "truncated base cut-out sequence." The beds studied fit well into Bouma's scheme; about half have truncated tops, and the remainder have both upper and lower portions missing. Detailed bedding observations are recorded elsewhere (R. W. Ojakangas, 1964, Petrology and Sedimentation of the Cretaceous Sacramento Valley Sequence, Cache Creek, California, Ph.D. thesis, Stanford University, Palo Alto, California). However, there do not appear to be any changes in the types of bedding in the Sacramento Valley sequence that can be attributed to geographic or stratigraphic position.

*Convolute lamination.* Convolute lamination (Dzulynski and Smith, 1963) is fairly common (Pl. 2, fig. 1). The axes of convolutions exposed on bedding surfaces generally do not show preferred orientations, neither within a single bed nor on associated beds, substantiating ten Haaf's (1956, p. 24) statement that they are not a direct result of gravitational slumping down a slope.

*Cross-bedding.* Small-scale cross-bedding is fairly common in the upper portions of laminated sandstone beds and throughout siltstone beds. In general, a single cross-bedding set is intermittently visible over the exposed length of a bed. Large-scale cross-bedding,

## SEDIMENTARY STRUCTURES

TABLE 2.—ABUNDANCE AND TYPES OF GRADED BEDS, CACHE CREEK-RUMSEY HILLS SECTION

| SERIES | FORMATION & MEMBER | NUMBER OF BEDS | PERCENT OF BEDS GRADED | MASSIVE INTERNAL STRUCTURE (245 Beds) | | | | | LAMINATED INTERNAL STRUCTURE (128 Beds) | | | | |
|---|---|---|---|---|---|---|---|---|---|---|---|---|---|
| | | | | IDEAL | MODIFIED IDEAL | GRADED BASE ONLY | GRADED TOP ONLY | MULTIPLE GRADED | IDEAL | MODIFIED IDEAL | GRADED BASE ONLY | GRADED TOP ONLY | MULTIPLE GRADED |
| UPPER CRETACEOUS | FORBES | 50 | 58 | 7% | 3% | % | 3% | 14% | 59% | % | % | 14% | % |
| | GUINDA | 30 | 100 | 83 | 10 | | 24 | 9 | 7 | | | 2 | 5 |
| | SITES | 100 | 89 | 56 | 32 | 1 | 5 | 44 | | | 5 | 2 | 11 |
| | VENADO | 41 | 100 | 10 | 12 | 2 | 11 | 14 | 43 | 43 | 25 | 10 | 9 |
| | "F" | 52 | 54 | 4 | 8 | | 3 | 5 | 14 | 6 | | 4 | 13 |
| AVERAGES | | 273 | 76 | 38 | 27 | 1 | 11 | 14 | 29 | 23 | 2 | 5 | 4 |
| LOWER CRETACEOUS | "E" | 50 | 62 | 23 | 13 | | 3 | 5 | 17 | 68 | 5 | 10 | 9 |
| | "D" | 2 75 | 29 | | 9 | | 13 | 4 | | 63 | 25 | 4 | |
| | "C" | 1 60 | 40 | 4 | 17 | | 25 | 4 | | 25 | | 13 | |
| | "B" | 2 55 | 15 | | 13 | | | 8 | | 20 | | | |
| | | 2 25 | 48 | 42 | 8 | | 5 | 33 | | 7 | 2 | | 7 |
| | "A" | 30 | 50 | | 53 | | | 6 | 8 | 33 | | 5 | 4 |
| AVERAGES | | 320 | 41 | 10 | 27 | | 5 | 10 | 8 | 17 | 1 | 3 | |
| UPPER JURASSIC | | 47 | 51 | | 79 | | 8% | | 12% | | | | |
| GRAND AVERAGES | | 640 | 58 | 25% | 21% | 1% | | | | 16% | 1% | 3% | 3% |

believed to reflect a near-shore environment, is common in Upper Cretaceous strata on the east side and the north end of the Sacramento Valley and near Hornbrook in northernmost California.

*Parting lineation.* Parting lineation (Stokes, 1947; McBride and Yeakel, 1963), the manifestation of the "streaming" of sand grains in a current and consisting of subparallel grooves and ridges of very low relief on internal bedding planes, is common in the laminated sandstones (Pl. 2, fig. 2). Allen (1963a; 1963b) concluded that parting lineation in well-washed sands forms at higher current velocities than does small-scale cross-stratification. His conclusion seems applicable to these muddy sands as well, assuming a generally decreasing velocity of the currents during deposition; the massive lower portion of a bed is commonly overlain by a laminated portion (with parting linea-

tion) which is in turn overlain by a crossbedded or convoluted portion.

The common parallelism of parting lineation and sole marks of the same bed (Table 3) suggests that the eroding currents and the depositing currents moved in the same direction and probably were, in fact, the same currents.

*Pebble imbrication.* Crude pebble imbrication in conglomeratic beds was seen only locally, and in one case it agreed in orientation with flute and groove casts on the sandy sole of the same bed.

### Sole Marks

Sole (bottom) marks on sandstone beds are readily observed in roadcuts and stream valleys in most of the sequence, and are least abundant in the Upper Jurassic and lowermost Cretaceous strata. Commonly they are co-parallel in a single outcrop, but divergent

rarely do they occur on the same sole with groove casts, and in all particulars they are identical to many described in the literature (Pl. 2, figs. 2 and 3).

*Groove casts.* Groove casts (Kuenen, 1957a), as seen on soles, are linear ridges resulting from sand fillings of long grooves which were probably scribed into mud surfaces by current-propelled objects. However, no scribing tools were observed during this study. The groove casts (Pl. 2, fig. 4) are generally numerous on a given sole, are invariably longer than the exposed portion of the sole (although some are only intermittently developed), are commonly striated or grooved along their length, and are generally narrow with low relief. Two differently oriented sets of groove casts were noted on a few soles (Table 3).

*Miscellaneous casts.* Casts of uncertain origin and type include (a) "flute-cast-like" structures, commonly of very low relief and without a clear sense of current movement, and (b) "groove-cast-like" structures which are more irregularly shaped than groove casts and are less abundant on a given sole. Many of these structures probably are flute or groove casts which have been altered by loading, flattening, organic burrowing, partial filling of the original depressions by the mud host, or possible tectonism. Also noted are cigar- or canoe-shaped casts, numerous types of small-scale discontinuous casts, parallel sets of rather irregular sole marks, and sole "striations". Several structures that have been assigned specific names by other authors are grouped together here.

*Frondescent casts.* Frondescent casts (ten Haaf, 1959) in subradiating patterns were noted on seven soles. They are evidently reliable indicators of current direction if an average value for each sole is used, as they agree fairly well with orientations of associated flute and groove casts (Table 3).

*Channel fillings.* A few elongate "channels" in mudstone or sandstone beds, generally filled with gravelly detritus, were also noted. Their trend is in agreement with other current indicators on adjacent strata.

### Other Structures

*Aligned fossils.* Several occurrences of elongate plant fragments or invertebrate fossils with subparallel alignment were noted

TABLE 3.—ORIENTATIONS OF ASSOCIATED PALEOCURRENT INDICATORS

| | Number of occurrences with same orientation (within 5°) | Number of occurrences with different orientations, and their mean differences |
|---|---|---|
| Sole Marks and Parting Lineation | 46 | 8 – 15° |
| Sole Marks and Cross-bedding | 30 | 5 – 40° |
| Sole Marks and Plant Fragments | 6 | |
| Cross-bedding and Plant Fragments | 1 | |
| Cross-bedding and Parting Lineation | 3 | 1 – 45° |
| Plant Fragments and Parting Lineation | 4 | |
| Two Sets of Groove Casts | | 6 – 15° |
| Two Sets of Frondescent Casts | | 1 – 80° |
| Two or More Sets of Parting Lineation | More than 100 | 6 – 40° |

sole marks also occur. A few soles are flat and unmarred by either inorganic or organic markings, but most are irregular surfaces covered with miscellaneous nondirectional casts.

The literature contains numerous descriptions and discussions of sole marks (e.g., Kuenen, 1957a), and hence the following descriptions are brief. Most workers (e.g., Dzulynski and Sanders, 1962) attribute "scour marks" (notably flute casts) to the effect of sediment-laden turbulent eddies, and "tool marks" (notably groove casts) to the effect of traction of larger objects which acted as scribing instruments. McBride (1962, p. 59) detected preferred associations of flute casts with beds more than 2 feet thick and groove casts with thinner beds, but no such preference was noted in the Sacramento Valley sequence where both structures generally occur on beds anywhere between 2 inches to 4 feet thick.

*Flute casts.* Flute casts (Crowell, 1955, p. 1350) are elongate, sand-filled counterparts of current-scoured depressions in the underlying mud, with a narrow upcurrent end of higher relief than the wider downcurrent end which gradually merges with the sole. Those observed in the present study occur either in bunches or singly. They range from 1 inch to 8 feet in length, though large ones are rare and are confined to thick beds. Only

on bedding planes. The general orientation of the fragments commonly agrees with the orientation of other directional structures (Table 3). On the sole of one bed, numerous belemnite guards were aligned parallel to flute casts, with the pointed ends toward the upcurrent direction.

*Ripple marks.* Ripple marks were observed on only 20 to 30 beds in the entire outcrop belt, and only a few were of value for paleocurrent analysis.

## PALEOCURRENT ANALYSIS

A device consisting of a protractor and a hinged arm mounted near one end of a 7-inch wooden level was used to measure the angle between sole marks and the strike of the bed on which the marks occurred. This device was held against the tilted sole of the bed and leveled to a horizontal position; then the hinged arm was aligned parallel to the sole mark, and the angle between the sole mark and the horizontal (in the plane of the bed) was read off the protractor. A Brunton compass was then used to measure the strike. The direction of the original, prefolding, sedimentary current indicator was then recorded to the nearest 5° at the outcrop. Little, if any, correction of the paleocurrent measurements for plunge is required by the generally simple homoclinical folds of most of the sequence. Perhaps the Upper Jurassic and Lower Cretaceous section along Cache Creek on the flank of a small syncline might require correction, although this depends on assumptions regarding the style of folding. The maximum correction here, however, would be a shift of 15° on the compass in a clockwise direction and would not significantly alter the results of this study.

Paleocurrent data are summarized in Figures 5 through 8 and in Table 4. Simple arithmetic means, rather than vectorial means, were used because of the close grouping of the directional data and because the readings in general did not fall near and on both sides of north. In the calculations, each nonsensical indicator (Table 4) was assigned the directional value closest to those of the sensical indicators in the same member of formation. Forty-five percent of the 902 measurements are from the Cache Creek-Rumsey Hills area, and the remainder scattered over 19 fifteen-minute quadrangles (Fig. 2).

In the Cache Creek section, the sedimentary structures in the Upper Jurassic and Lower Cretaceous formations indicate paleocurrent movement from the north to south with some variation, and readings from the same formations elsewhere along the belt are in good agreement (Figs. 5 through 7). The trend in most of the Upper Cretaceous units is also southerly. The most prominent divergence is in the lowermost Upper Cretaceous units (Formation F of Cenomanian age and the Venado Formation of Turonian age) where many readings indicate currents flowing toward the southwest, west, and northwest.

The paleocurrent pattern over the entire study area is shown in Figure 7. A general current movement from north to south is again indicated for the Upper Jurassic beds. A similar trend is evident for the Lower Cretaceous with the only apparent major deviation occurring at the south end of the Sacramento Valley where a small number of readings indicate a shift in flow toward the west. The Upper Cretaceous rocks have twice as many "transverse" or east-to-west indicators as the Lower Cretaceous, with the greatest number in Formation F and the Venado Formation, but the dominant trend of the currents is still from north to south (Figs. 5 and 7). The variation evident in these two formations is localized to a few quadrangles, as shown by Figure 8.

Orientations of primary sedimentary structures were also obtained in several areas of Cretaceous outcrops which are geographically isolated from the Sacramento Valley sequence (Figs. 5 and 7). In the Lower Cretaceous strata of the Potter Valley and Lower Lake quadrangles west of the main outcrop belt, paleocurrent trends are toward the south. Measurements in small patches of probable Lower Cretaceous rocks located northwest of the main outcrop belt (in the Weaverville, Dubakella Mountain, and Chanchelulla Peak quadrangles) indicate a paleocurrent trend toward the south-southwest. Large-scale cross-bedding in Upper Cretaceous beds near Redding, northeast of the main outcrop belt, shows that the currents in general moved southward.

In Upper Cretaceous beds near Hornbrook in northernmost California, about 75 miles north of the northern end of the Sacramento Valley, the currents moved toward the east and north. This is opposite to the paleocurrent trend of the Sacramento Valley sequence, and may indicate deposition either in a different part of the basin, or perhaps more probably, in a separate basin.

### Other Related Paleocurrent Data

Paleocurrent data collected by other workers from individual localities in the Sacramento Valley sequence are in good agreement with this study (Gilbert, 1955; Crowell, 1957; Chuber, 1962; as well as some unpublished data). I. P. Colburn, in a Ph.D. thesis (1961, The tectonic history of Mt. Diablo, California: Stanford University, Palo Alto, California), reported paleocurrent trends to the northwest and south in the Cretaceous rocks of the Mt. Diablo area east of San Francisco Bay, and C. A. Hall, Jr., in a Ph.D. thesis (1958, Geology and paleontology of Pleasanton area, Alameda and Contra Costa Counties, California: Stanford University, Palo Alto, California), noted currents to the southwest in Upper Cretaceous units of the same region.

## PETROLOGY

### Sandstones

More than 250 samples were collected for petrologic and mineralogic study. Detailed mineralogical descriptions and quantitative data are recorded elsewhere (R. W. Ojakangas, 1964, Petrology and sedimentation of the Cretaceous Sacramento Valley sequence, Cache Creek, California: Ph.D. thesis, Stanford University, Palo Alto, California). Most samples are from the basal portions of beds, many of which are graded, and most have an average grain size of about 0.25 mm or less. In general, the thicker beds contain the coarsest material. The grains are generally subangular to angular and poorly sorted. Unabraded mudstone fragments, probably derived from the mudstone bed underlying the containing sandstone bed, are relatively common, usually occurring in zones near the tops of sandstone beds. Glauconite grains and abraded fossils occur in the basal parts of several graded beds.

Heavy minerals from 150 of the Cache Creek-Rumsey Hills samples were studied, and grain counts were made on 30 representative

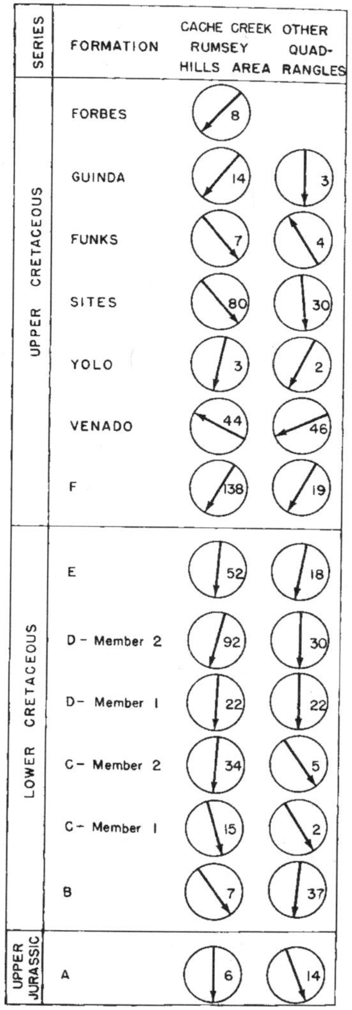

Figure 6. Stratigraphic variations in paleocurrent trends. (Arrows represent azimuthal arithmetic means, and digits represent number of paleocurrent indicators measured.)

tative, unfractionated mounts (Table 5). Epidote minerals, apatite, sphene, hornblende, and zircon dominate; also present are tourmaline, garnet, allanite, rutile, fluorite, and chromian spinel. Augite, oxyhornblende, diopside, hypersthene, glaucophane, and pumpellyite are rare. Sixty-four additional heavy mineral mounts were prepared from samples collected on the four additional traverses (Fig. 2), from Chico on the east side of the Sacramento Valley, and from Sutter's Butte in the center of the Valley. These were found to be quite comparable to the counted samples from the Cache Creek-Rumsey Hills section. Two hundred grains were first counted on each mount to determine the content of epidote minerals, which overwhelm all other heavy minerals, and then the count of non-epidote minerals was expanded to 300 grains where possible. Opaque minerals, biotite, chlorite, and composite grains were not counted because of their great abundance and because they are generally not diagnostic of specific source rock lithologies.

The K-feldspar contents of 148 stained slabs from the Cache Creek-Rumsey Hills section were determined by counting 400 points on each slab (Table 6). The Upper Cretaceous rocks contain the most K-feldspar, as reported by Bailey and Irwin (1959) and Bailey and others (1964). The data suggest that the vertical change in K-feldspar content within the Lower Cretaceous is rather abrupt and occurs approximately at the base of Formation E.

The mineral compositions of 125 thin-sectioned samples from the Cache Creek-Rumsey Hills area were obtained by counting along six random 100-point traverses perpendicular to bedding. The sandstones are primarily composed of quartz, plagioclase, K-feldspar, volcanic rock fragments, clayey matrix, and carbonate cement (Figs. 9 and 10; Pls. 3, 4, and 5). Minor light minerals include biotite, chlorite, muscovite, and glauconite; minor rock fragments include chert, shale, mudstone-siltstone, phyllite, schist, quartzite, greenstone, serpentine, and carbonates. Carbonaceous material is common, especially in the upper parts of graded beds. Forty additional thin sections from samples collected on traverses across the belt (Fig. 2), from single samples from Redding at the north end of the Sacramento Valley, and from Sutter's Butte, are similar, unit by unit, to the point-counted samples.

Using Gilbert's classification (in Williams and others, 1955), 68 of the 125 point-counted samples are wackes with more than 10 percent clayey matrix, and the remainder are arenites with less than 10 percent clayey matrix (Fig. 9). Following Pettijohn (1957), 27 of the samples are feldspathic graywackes, 18 are lithic graywackes, 40 are subgraywackes, and 39 are arkoses.

A survey of the literature shows that the general compositions of related Cretaceous sandstones to the south along the west side of the San Joaquin Valley are generally similar to those of this study. Nearly all workers report abundant clayey matrix and rock fragments, and most report volcanic fragments and chert. The vertical increase in K-feldspar in the section has been noted in other studies. The heavy mineral suites are similar, with epidote the common mineral. The sandstones are also similar in composition to the sandstones of the Franciscan assemblage described by Bailey and others (1964) and Soliman (1965).

*Stratigraphic and areal variation in sandstone composition.* The foregoing data permit the 15 members and formations of the Cache Creek-Rumsey Hills section to be grouped into six compositional intervals, herein labeled I-VI from the oldest to the youngest, with each interval having a distinctive composition as shown in Figure 10. Photomicrographs of a sample from each of the intervals are shown in Plates 3, 4, and 5.

Interval I, comprised of Formation A, is characterized by the highest volcanic rock fragment content (commonly with a "felty" texture due to plagioclase laths), low feldspar and quartz contents, and the presence of mudstone-siltstone fragments. Epidote, apatite, zircon, and chromian spinel are abundant heavy minerals; also distinctive are tourmaline, "dusky" apatite, and well-rounded (multicycle?) grains of tourmaline, zircon, and apatite. An andesitic (?) volcanic terrain with minor ultrabasic rocks (sources for the spinel), acidic plutonic rocks, and sandstones evidently contributed the sand. However, this interval is based on only three samples from the uppermost Jurassic beds and may not be representative of the Upper Jurassic Series.

Interval II, comprised of Formation B, is high in plagioclase throughout. Quartz increases and volcanic rock fragments decrease from the bottom of the interval to the top. The heavy mineral suite is rich in epidote, apatite, zircon, and hornblende. Plutonic rocks, probably sodic, were contributing detritus in volumes about equal to the volcanic terrain during deposition of the lower part of the formation, and became dominant near the end of the interval.

Interval III, comprised of Formations C and D, has the highest quartz content, a moderate plagioclase content that decreases upward, a low K-feldspar content that shows a slight increase upward, and the lowest amount of volcanic rock fragments. Zircon is at its maximum and epidote, apatite, and sphene are also common. These sandstones are mineralogically the most mature rocks on the sequence, yet neither the light nor heavy minerals show evidence of long abrasion. However, derivation from older sedimentary terranes is more likely for this interval than for the others. Plausibly the maturity reflects both source-rock composition and weathering in the source area. The quartz to feldspar ratio (Fig. 10) increases

---

> From [the western Sierran foothills], the Great Central Valley of California, the immense flatland runs so far off the curve of the earth that its western horizon makes a simple line to the extremes of peripheral vision. In California's exceptional topography—with its crowd-gathering glacial excavations, its High Sierran hanging wall, its itinerant Salinian coast—nothing seems more singular to me than the Great Central Valley. It is far more planar than the plainest of the plains. With respect to its surroundings, it arrived first. At its edges are mountains that were set up around it like portable screens.
>
> John McPhee, *Assembling California*, p. 172

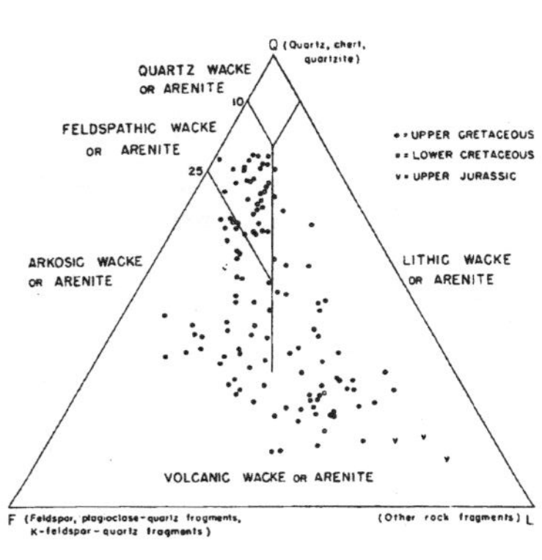

plagioclase. Although these components are not volumetrically important, their presence reflects a change in the source area: the volcanic rocks were more acidic, and some were rhyolitic. Granitic rocks remained dominant.

Interval VI, comprised of the Guinda and Forbes Formations, is the most heterogeneous interval. It is high in K-feldspar and mica (biotite), contains both high and low plagioclase contents, has quite high quartz content, and is low in volcanic rock fragments. The minor light minerals present in Interval V also occur here, especially in the Guinda Formation. Epidote, apatite, sphene, and hornblende are the common heavy minerals, and allanite is also present.

The samples from the other localities have light and heavy mineral compositions quite similar to those of the Cache Creek-Rumsey Hills section. Most contain volcanic rock fragments, feldspar, and quartz in about the same relative abundance and variety as correlative samples from the Cache Creek area. Epidote and apatite are abundant in nearly all samples. The Upper Jurassic samples, and locally those from the Lower Cretaceous, are characterized by chromian spinel; the Lower Cretaceous samples by garnet and zircon; and the Upper Cretaceous samples by hornblende, sphene, and in some, tourmaline.

*Diagenesis.* Diagenetic changes in the sandstones include the following: (1) Soft grains, especially mica and pelitic fragments, are commonly bent and deformed by the harder grain constituents, indicating compaction.

(2) Large amounts of introduced carbonate (up to 72 percent) are present in about half of the samples. It evidently has replaced several constituents, especially plagioclase and clayey matrix. Although some of the samples may have originally been relatively clean-washed sands, much of the carbonate appears "dirty" and "cloudy" with many small inclusions, suggestive of remnants of an almost completely replaced clayey matrix. It seems likely that many of the arenites in the column were once wackes (see Pl. 4, fig. 1). Small amounts of silica cement occur in some Lower Cretaceous samples, and a K-feldspar(?) cement was found in one sample from the Venado Formation.

(3) Intrastratal solution of green hornblende grains is shown by the common occurrence of cockscomb structures. A calcareous sandstone concretion in a massive sandstone bed (lower Sites Formation) and the adjacent noncalcareous sandstone at the same level have markedly different amounts of hornblende in the heavy mineral suites (83 percent vs 3 percent); the carbonate evidently prevented the intrastratal solution of the hornblende. The inevitable conclusion is that hornblende was present in larger quantities prior to diagenesis; perhaps some other heavy minerals have also been dissolved (Table 5).

(4) Recrystallization of matrix, chloritization of biotite, and albitization of plagioclase have occurred.

The matrix has assumed a stringy appearance, and aggregates of small crystals in subradiating patterns occur in voids, apparently as authigenic cement. X-ray diffraction of one such cemented sample showed the cement-matrix material to be dominantly chlorite. The amount of recrystallization seems to increase with depth (see *X-ray Analysis of Mudstones and Matrix of Sandstones*).

Altered biotite has fuzzy borders, contains sheaves and stringers of chlorite, and has a greenish color due to the chlorite content. This greenish color is opposed to a deep brown color where fresh, and a yellow-brown color where weathered. The biotite is freshest in calcareous samples.

Albitization has given the plagioclase grains a mottled appearance, with different portions of a single grain having different indices of refraction. Alteration of some grains is most advanced near the edges. Small crystals or patches of fine-grained greenish or colorless alteration products, probably pumpellyite, epidote-clinozoisite, and possibly prehnite, are common in many grains. In Interval VI, with an estimated depth of burial of only 5000 to 8000 feet, the plagioclase is predominantly fresh and affords no clear evidence of alteration in place (see Pl. 5, fig. 2). However, in underlying intervals it is altered, except in rocks with carbonate cement. As high as in the Sites Formation, with an estimated depth of burial of 10,000 to 12,500 feet, some of the noncalcareous rocks display clear-cut evidence of at least partial alteration of plagioclase, but in other noncalcareous rocks in the same interval, it is fresh. Lower in Interval V and in Interval IV, the amount of alteration appears to have increased (see Pl. 5, fig. 1; Pl. 4, fig. 2). In Interval III, having an estimated depth of burial of 20,000 to 30,000 feet, the plagioclase in all noncalcareous rocks examined shows at least partial alteration, and some shows a rather thorough albitization (see Pl. 4, fig. 1). However, plagioclase, in a majority of the calcareous rocks in the same interval, is unaltered. Interval II displays alteration similar to that in III, but in Interval I, with an estimated depth of burial of 35,000 feet, even the plagioclase in the calcareous samples is thoroughly altered (see Pl. 3).

The presence of carbonate cement evidently inhibited the alteration of both biotite and plagioclase. Although the albitization in possible source rocks of the Klamath Mountain igneous and metamorphic complex near Redding is commonly albitized (Kinkel and

TABLE 6. K-FELDSPAR CONTENT*
Cache Creek-Rumsey Hills Section

|  | 1 | 2 | 3 | 4 | 5 | 6 | | |
|---|---|---|---|---|---|---|---|---|
|  | Upper Cretaceous | Lower Cretaceous | Upper Jurassic | Upper Cretaceous plus Fm. E | Lower Cretaceous minus Fm. E | Bailey and Irwin (1959) | | |
|  |  |  |  |  |  | Upper Cretaceous | Lower Cretaceous | Upper Jurassic |
| Number of Samples | 74 | 71 | 3 | 83 | 62 | 22 | 45 | 24 |
| Mean % | 9.2 | 4.7 | 2.8 | 9.6 | 3.5 | 10.6 | 2.8 | .5 |
| Median % | 9.0 | 4.0 |  | 9.0 | 3.0 | 6.0 | 1.0 | .5 |
| Mode % | 6.0 | 3.0 and 6.0 |  | 4.0 | 3.0 |  |  |  |
| Range % | 1-24 | trace-24 | 1-4 | 1-24 | trace-10 | .5-35 | trace-25 | 0-4 |

*In columns 4 and 5, Formation E of Lower Cretaceous age is added to the Upper Cretaceous column and removed from the Lower Cretaceous column because of its higher K-feldspar content and to emphasize the rather abrupt difference in K-feldspar content between the upper and lower parts of the sedimentary column.

Figure 9. Sandstone classification. Triangle after Gilbert in Williams and others (1955).

through Intervals II and III. Most of the sediment was probably derived from quartz-plagioclase plutonic rocks or older sediment or both.

Interval IV, comprised of Formations E and F, spans the Lower to Upper Cretaceous boundary. It has a high K-feldspar content, medium amounts of andestic(?) volcanic rock fragments and quartz, and a low plagioclase content. Apatite, tourmaline, sphene, and zircon are abundant, and epidote is common. K-feldspar-bearing granitic rocks were consistently important source materials, evidently for the first time.

Interval V, comprised of the Venado, Yolo, Sites, and Funks Formations, is high in K-feldspar and volcanic rock fragments, and contains fair amounts of plagioclase. Epidote, apatite, sphene, and hornblende are abundant, and allanite is present. Zoned plagioclase, volcanic quartz, hornblende, and perthite occur in the upper part of the Sites Formation, as noted by R. K. Rozelle (1952, Feldspar content, current lineations, and bedding characteristics of the Sites Formation, northern Yolo County, California: M.A. thesis, University of California, Berkeley, California), and in the Funks Formation. Throughout the interval are small amounts of microcline, sanidine, muscovite, and K-feldspar-quartz and plagioclase-quartz fragments. Volcanic rock fragments contain phenocrysts of quartz, biotite, sanidine, and

gray or yellow color, "popcorn" texture on some weathered and dried surfaces, and a stickiness in fresh samples. These latter beds are commonly 1 to 2 inches thick, but one bed in Formation F has a thickness of more than 6 feet.

Heavy mineral suites of two bentonitic samples are mainly composed of euhedral apatite, including a grayish "dusky" type, euhedral zircon, traces of fluorite, and in one sample, hexagonal flakes of biotite. The simple nature of this suite, compared to the more varied suites of the associated sandstone beds, supports a volcanic origin. The dominance of apatite and zircon agrees with Weaver's data (1963) on heavy minerals in bentonites.

X-ray diffraction of the clay in three suspected bentonites and three suspected bentonitic sandstones showed that it is largely montmorillonite, supporting a volcanic origin for these beds. In addition, thin section study indicated the possible presence of relict volcanic shards.

### Conglomerates

Minor conglomerates were observed at seven horizons in the Cache Creek-Rumsey Hills section. All except one are less than 10 feet thick, and probably lensoid in shape. The commonly well-rounded pebbles are generally suspended in a sand matrix, suggesting a viscous medium of transport such as a submarine slump, with the material derived from shelf deposits. At least two Lower Cretaceous conglomerates are the basal beds of sandstone "packets."

From randomly selected, carefully disaggregated pieces of the conglomeratic beds, clasts varying in size from 4 to 80 mm were separated, identified, and counted. The clasts from the different lenses have similar lithologies, with an abundance of rock types such as limestone concretions, calcareous sandstones, and mudstones, all of which were probably derived through cannibalism within the depositional basin. Clasts containing *Buchia piochii* and reworked belemnites occur in the lowermost Cretaceous units. The smaller amounts of material derived from outside of the basin consist of volcanic rocks, cherts which are important in some Upper Cretaceous units, and plutonic rocks which are generally uncommon except in the lowest unit.

### Siltstones

Lensoid, light gray siltstone laminae and beds to 2 inches thick are interbedded with mudstone in much of the sequence. These are commonly cross-bedded, with the current sense the same as that indicated by structures on interbedded sandstones. Thin section study indicated that they have compositions similar to the sandstones.

### Mudstones

Mudstones in the sequence are blocky, dark gray to black, and commonly laminated. Thin sections of samples from five formations are similar in that they contain abundant angular silt-sized grains of quartz, mica, organic material, and opaque minerals. Also common are small, elongated, saucer-shaped, darker colored fragments of slightly finer-grained mudstone, evidently derived from nearly coeval deposits.

### X-ray Analysis of Mudstones and Matrix of Sandstones

The clay minerals in 20 mudstones and 15 sandstones, both selected from each of the 15 members and formations in the Cache Creek-Rumsey Hills area, were studied by X-ray diffraction. The matrix of the sandstones has the same general composition as associated mudstones. In the Upper Jurassic and lowermost Cretaceous beds, chlorite is as abundant as montmorillonite plus fine-grained mica. Through the remainder of the Lower Cretaceous section and in the lowermost Upper Cretaceous beds, chlorite is more abundant than montmorillonite plus fine-grained mica. Through the rest of the Upper Cretaceous section, montmorillonite is dominant, mica is common, and chlorite is unimportant.

### Bentonitic Rocks

A few bentonitic beds were noted in the uppermost Jurassic, Aptian, Albian, and Cenomanian rocks. Bentonitic sandstones up to 11 feet thick are characterized by a dark green color, very poor lithification, and the presence of a great amount of muddy matrix. Purer bentonites are characterized by a light

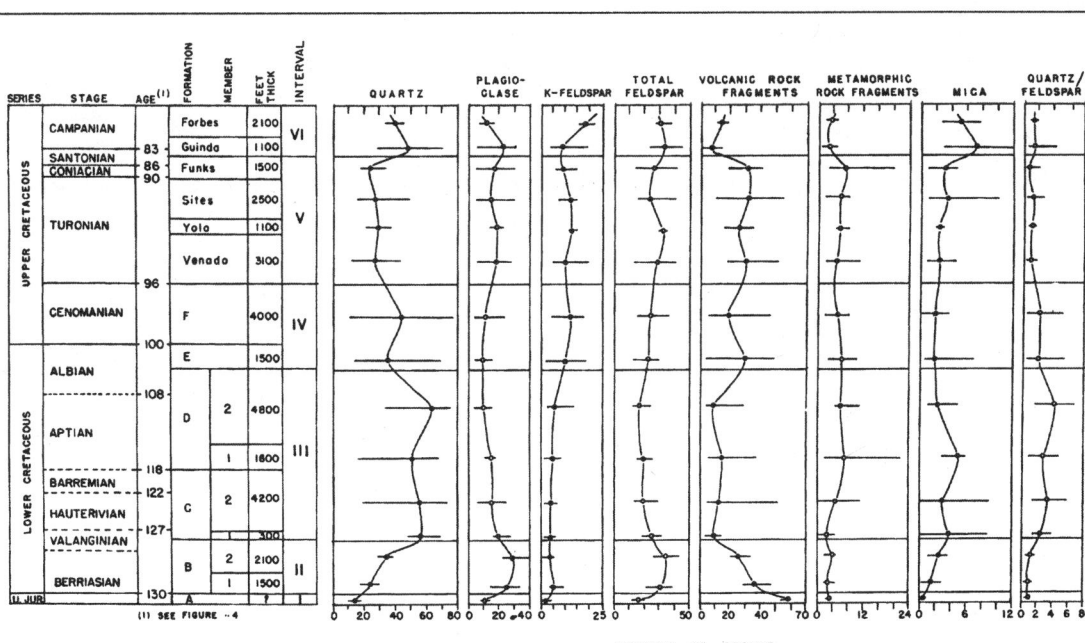

Figure 10. Variations in compositions of sandstones in the Cache Creek-Rumsey Hills section, based on averages of modal analyses of thin sections. Number of counted thin sections per unit is as follows, from Formation A through the Forbes Formation: 3, 5, 4, 6, 14, 5, 15, 8, 14, 11, 2, 15, 9, 10, and 4. Grains percentages in original point counts were recalculated to 100 percent; miscellaneous components (commonly opaques or sedimentary rock fragments) are not shown on the graph. Horizontal bars show ranges in compositions.

The lowest conglomeratic unit, at the base of Formation B, is 75 feet thick, and is composed of several conglomeratic beds separated by sandstones or mudstones. It contains clasts to boulder size, and a few of the beds are well-graded from 6-inch boulders to coarse sand. This unit and an associated pebbly mudstone contain much less intrabasinal material than the other conglomeratic rocks. In contrast, volcanic clasts are abundant, and plutonic rocks, commonly quartz-plagioclase rocks, are present in appreciable quantities. Also present are lesser amounts of a great variety of other rock types.

In summary, the counts of pebbles in the conglomerates showed that (1) much of the material is intrabasinal in origin, (2) volcanic rocks supplied the majority of the (3) quartz-plagioclase plutonic rocks were clasts derived from outside the basin, and present in the source areas in earliest Cretaceous time.

### Carbonates

Dark gray carbonate concretions are common in the mudstones. Although they occur in such profusion at certain horizons that they coalesce and form irregular carbonate lenses, true carbonate beds more than 2 to 3 inches thick are rare. X-ray studies of four Lower Cretaceous concretions showed that three were impure calcite and one contained much siderite.

### SEDIMENTATION

### Provenance

The common parallelism of parting lineation and sole marks of the same beds (Table 3) indicates that many of the eroding (sole mark scribing) currents and depositing currents which existed during Cretaceous time in the study area were moving in the same direction. Therefore, at least part of the paleocurrent indicators seem to point back to the source areas. The paleocurrent patterns thus suggest that the sediment moved into the basin from the ancestral Klamath Mountains to the north and the ancestral Sierra Nevada to the east. The general absence of paleocurrent indicators showing movement from south to north eliminates the possibility of a southern source area. There is virtually no paleocurrent evidence that landmasses lay to the west, although several workers proposed their existence (e.g., Reed, 1933; Anderson, 1933, p. 1255; 1938, p. 25; 1958, p. 3; Taliaferro, 1943, p. 110; 1944). Curtis and others (1958) discussed several objections to a western source for correlative Cretaceous strata to the south.

The Klamath Mountains contain rocks of Ordovician to Late Jurassic age, including graywackes, mudstones, greenstones, cherts, limestones, schists, phyllites, various granitic rocks (most dating from 146 to 127 m.y., but with some older ones), peridotites (Late Jurassic), and serpentinized peridotites (Davis, 1966; Irwin, 1966). Detailed rock descriptions can be found in Hinds (1933), Kinkel and others (1956), and Lipman (1964). Two province-wide periods of regional metamorphism, deformation, and igneous intrusion are dated as (1) Late Pennsylvanian to Early or Middle Triassic and (2) Late Jurassic (Davis, 1966). According to Davis (1963), the Klamath plutons seem to contain relatively little K-feldspar, in contrast to those in the central Sierra Nevada as reported by Curtis and others (1958).

Most of the southern half of the Sierra Nevada and the eastern part of the northern half are granitic rocks of Mesozoic age. The western part of the northern half is comprised of a variety of Paleozoic and Mesozoic metasediments and metavolcanics, and to the south in the western foothills and along the crest, remnants of these metamorphic rocks are found as roof pendants in the batholith (Bateman and Wahrhaftig, 1966). Kistler and others (1965) dated plutons in the central Sierra Nevada and the Inyo Mountains to the east, and concluded that they represent two or three pulses of emplacement: 80 to 90 m.y., 124 to 136 m.y., and 150 to 180 m.y. ago. Bateman and Wahrhaftig (1966, p. 123) summarized the isotopic ages of magmatism in the Sierra Nevada as follows:

The oldest epoch is represented by granitic rocks that lie along the east side of the Sierra Nevada and have isotopic ages of $200 \pm 20$ m.y. (Late Triassic or Early Jurassic). They are chiefly quartz monzonite and granodiorite and are not as mafic as the next younger granitic rocks of the western foothills, which are chiefly granodiorite and quartz diorite and have isotopic ages of $130 \pm 10$ m.y. (Late Jurassic). The youngest plutonic rocks are those along the crest of the range, which have isotopic ages of 80-90 m.y. (early Late Cretaceous) and are more felsic than the rocks to the west.

Detailed rock descriptions of certain areas in the Sierra have been made by Hietanen

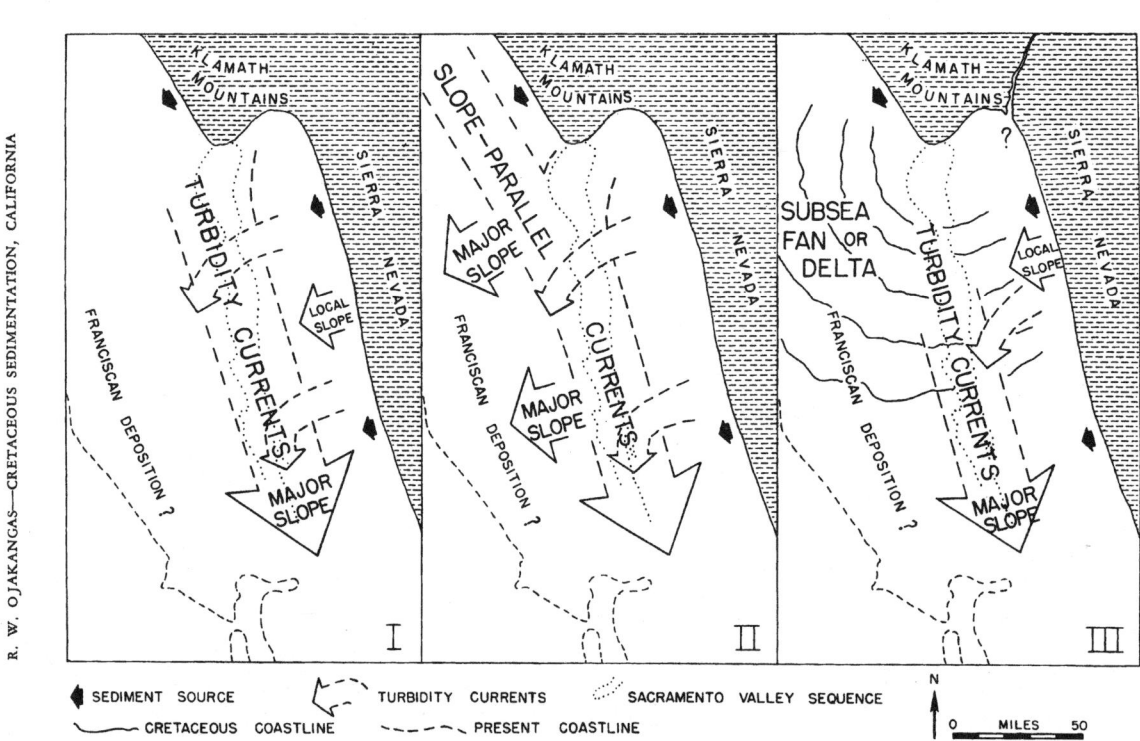

Figure 11. Sedimentation models for the Sacramento Valley sequence.

(1951) and Compton (1955), and Clark (1960) mentioned several rock types, including serpentine, in the western foothills. Pabst (1938) and F. C. Dodge (1963, A mineralogic study of the intrusive rocks of the Yosemite Valley area, California: Ph.D. thesis, Stanford University, Palo Alto, California) studied accessory minerals in the granitic rocks of the Yosemite Valley area; hornblende, and biotite dominate, and sphene, epidote, apatite, zircon, and allanite comprise the remainder of the suite. All occur in the Sacramento Valley sequence, especially in the Upper Cretaceous strata where K-feldspar is abundant.

The primary objective of the petrologic and mineralogic studies of the sandstones of the sequence was to find components, such as varieties of heavy minerals, which could be traced back to specific source areas. This objective was not realized, in large part because the source rocks are found in both of these areas even at the present time. It is possible that some volcanic detritus was derived from within the basin; Brown (1964) mapped volcanic rocks which are probably a part of the Upper Jurassic portion of the Sacramento Valley sequence in the Stonyford 15-minute quadrangle. Furthermore, the Franciscan rocks to the west contain much mafic volcanic material, and submarine erosion of these units could have occurred. However, it seems unlikely that much of the volcanic detritus, expecially that containing quartz and K-feldspar phenocrysts, was derived from within the geosyncline where the volcanic rocks are, theoretically at least, more mafic. Bailey and others (1964, p. 44) reported that the Franciscan volcanics, which presumably were being extruded in the geosyncline to the west during sedimentation of the Sacramento Valley sequence, are basaltic in composition. Other workers have also proposed that the Sierra Nevada or the Klamath Mountains, or both, were the source areas for the Sacramento Valley sequence and related rocks to the south (e.g., Briggs, 1953; Chuber, 1962), and the Franciscan sediments (Bailey and others, 1964; Soliman, 1965).

Weathering occurred in the source areas and affected most of the K-feldspar, but the generally sodic plagioclase in all formations except A and B is commonly fresh with minor weathered grains. Unstable rock fragments and heavy minerals are abundant, and the clay in the sandstones and mudstones does not include kaolinite, an indicator of rigorous or long weathering. Nevertheless, the great quantities of clay in the sequence show that much chemical weathering did occur, thereby supporting other evidence indicating a warm or tropical Cretaceous climate (Durham, 1962; Schwarzbach, 1963). Rapid erosion probably limited the time during which some of the source rocks were exposed to chemical agents, and the association of fresh and minor weathered sodic plagioclase in the same samples indicates that fresh and weathered rocks were being eroded simultaneously.

*Tectonics and Sedimentation*

Based on compositions of the sandstones in the sequence, the source rocks which contributed sediment from latest Jurassic to Late Campanian time may be summarized as follows. In the latest Jurassic, andesitic (?) volcanic sources may have been dominant, and plutonic rocks (including ultrabasics) and sandstones were minor. In the earliest Cretaceous, quartz-plagioclase plutonic rocks became dominant. In middle Early Cretaceous (Hauterivian, Barremian, and Aptian), quartz-plagioclase plutonic sources still dominated, and volcanic and metamorphic sources were minor. During late Albian and Cenomanian time, andesitic (?) volcanic rocks were again important and were accompanied for the first time by volumetrically important K-feldspar-bearing granitic rocks. From early Turonian into early Campanian time, granitic and volcanic rocks were important, and from late Turonian into early Campanian, a prominent part of the volcanic rocks were acidic. In late Campanian time, granitic rocks were dominant with volcanic rocks less important. Throughout the deposition of the entire sequence, low-grade metamorphic rocks, pre-existing sandstones, and ultrabasic rocks contributed small amounts of sediment. Some of the volcanic detritus, especially in Formation B and the Venado Formation, contains chlorite and may have been metamorphosed in the source areas.

It is theoretically possible that all or part of the sediment may have come from within the basin because of volcanism or reworking of older basin deposits. However, the lithology, mineralogy, and geometry of the sequence (Figs. 4 and 10) could also, in a general way, reflect major events in the source areas. Certainly individual mudstone and sandstone beds are simply the result of minor alternations of normal hemipelagic sedimentation of clay and silt, and sporadic releases of reworked clay, silt, and sand, probably from the shelf to the east. The thick intervals of abundant sandstone beds, although lenticular on a large scale within the basin, perhaps because of some control on the paths taken by turbidity currents, may be related to periods of intrusion and uplift in the source areas. The thick mudstone intervals probably formed during major periods of quiescence.

The entire Upper Jurassic and Cretaceous sequence thickens westward from the Sierra Nevada, the Lower Cretaceous Series generally thickens to the north as well, along the outcrop belt, and the studied portion of the Upper Cretaceous Series generally thickens to the south along the belt (Chuber, 1962). This may be related to the amount of intrusion and uplift in different parts of the source areas, with Early Cretaceous uplift dominant in the Klamath Mountains and northern Sierra Nevada, and Late Cretaceous uplift more important farther to the south in the central Sierra Nevada. The paleocurrent trends are compatible with this suggestion; north to south currents dominated through most of Cretaceous time, whereas east to west currents were important during Cenomanian and Turonian time in the southern portion of the study area. Alternatively, the thickening may just be a result of differential subsidence within the basin, but such a hypothesis is hard to evalute where relatively deep-water sediments are involved.

The major events in the source areas, arranged chronologically and assigned tentative dates based on source rock radiometric dates as well as on dates for the Cretaceous stage boundaries from California and elsewhere in the world (Fig. 4), are suggested to have occurred as follows.

(1) During latest Jurassic time, the source areas seem to have supplied predominantly volcanic detritus to the sea. If this sediment came from the Klamath and Sierra Nevada regions, it means that plutonic rocks, although evidently already emplaced (Curtis and others, 1958; Kistler and others, 1965), were not yet markedly unroofed. However, at least a partial derivation from within the basin is also possible, as volcanics occur in the Upper Jurassic portion of the sequence (Brown, 1964).

(2) Quartz-plagioclase plutonic detritus appears abruptly in lowermost Cretaceous conglomerates and sandstones, and may reflect an initial unroofing of plutons about 130 m.y. ago. Biotite and hornblende from two boulders at this horizon were dated at 138 to 152 m.y. (Irwin, 1966). Alternatively, much of this sediment may have been derived from older sediments. A sympathetic decrease in the area of exposed volcanic terrain due to erosion, or cessation of nearby intrabasinal volcanism, is indicated by a decrease in volcanic detritus in Formation B and its minor importance through most of the younger Lower Cretaceous units.

(3) Minor volcanism is indicated by bentonitic sandstones in the lower part of member 2 of Formation D, and possibly also by abundant volcanic rock fragments in a few samples high in Formation C and low in Formation D. This activity is estimated to have occurred approximately 110 to 120 m.y. ago.

(4) Bentonitic sandstones occur in the Upper Albian beds (Formation E), and volcanic rock fragments and K-feldspar rather abruptly become important at the same horizon. (K-feldspar increases from about 7 to 11½ percent, and volcanic rock fragments increase from 9½ to 29 percent.) Bailey and Irwin (1959) and Bailey and others (1964) suggested that the increase in K-feldspar content through the Cretaceous column may indicate the progressive unroofing of the Late Jurassic to Mid-Cretaceous Klamath Mountain and Sierra Nevada batholiths. Evidently in late Albian time, an estimated 105 m.y. ago, major K-feldspar-bearing plutons and

volcanic rocks that were associated with the plutons began shedding detritus. This date, however, is too old to be correlated with the 80 to 90 m.y. pulse in the source areas. If it corresponds to the erosion of the 124 to 136 m.y. emplacements, 20 to 30 m.y. elapsed before the mixed K-feldspar and volcanic detritus appeared in the basin of deposition in the study area. If unroofing took this long, the volcanic detritus should have appeared long before the K-feldspar, especially if the plutons were emplaced under a roof several miles thick. There is still another possibility; perhaps a major pulse of igneous activity, for which there are at present no radiometric age dates, did occur in the Sierra Nevada or the Klamath Mountains, or both, about 105 to 110 m.y. ago or 110 to 120 m.y. ago, as also suggested in paragraph (3) above. The mixed K-feldspar and volcanic detritus suggests that the plutons may have been intruded at a shallow depth and were unroofed rather rapidly, perhaps in a manner similar to that demonstrated by Fiske and others (1963) for the Mt. Ranier area where Pliocene plutons intruded associated volcanic rocks, with subsequent unroofing and deposition of mixed volcanic and plutonic materials. Admittedly the evidence based on the sedimentary column is indirect, but the latter alternative must at least be considered a possibility.

(5) Sporadic volcanism in the region an estimated 95 to 100 m.y. ago is evidenced by a few bentonites and bentonitic sandstones in Formation F of Cenomanian age.

(6) Slumps containing large quartz diorite blocks and abundant Albian fossils occur near the base of the sandy Venado Formation which was deposited in early Turonian time (Brown and Rich, 1960), and slumps at the same horizon occur elsewhere in the area (Peterson, 1965). On the basis of paleocurrent indicators in the sandstones of the area, the lensoid north to south cross section of the slumps, the abundance of fossils (some probably indicative of a shallow nearshore environment), and the quartz diorite composition of the blocks, both the slumped material and the sandstones seem to have been derived from the east (near the Sierra Nevada?), and may correspond to early uplift and erosion accompanying the emplacement of plutons, as suggested by Chuber (1962). Brown and Rich (1967) ascribe the slumps to derivation from uplifted fault blocks which formed within the basin to the east. As the sea was evidently advancing eastward onto the Sierra Nevada block during Turonian time (an onlap relationship is indicated), the uplift evidently did not cause any great changes within the basin of deposition, other than local erosion and slumping (probably submarine) of Albian beds.

(7) The distinct change in the nature of the volcanic detritus high in the Turonian column reflects erosion of acidic volcanics which were possibly associated with intrusions in the Sierra Nevada an estimated 90 to 95 m.y. ago.

(8) By Upper Santonian and Campanian time, the volcanic rocks had been essentially stripped off the source areas and K-feldspar-bearing plutonic rocks again predominated. The "absolute" dates used in the above analysis, both those in the Cretaceous sedimentary column and those in the source areas, are tentative dates, and hence the conclusions based on these dates are certainly of a preliminary nature. However, it was thought that emphasizing the possible correlation of igneous events in the source areas and the resultant sedimentation in the Sacramento Valley sequence might help initiate further work in this area. More precise radiometric dates from the Klamath Mountains and the Sierra Nevada, and more precise dating of the Cretaceous stage boundaries here and elsewhere in the world, should make possible more detailed and more certain correlations of igneous activity and subsequent sedimentation.

## Current Mechanisms

Turbidity currents have been widely accepted as the mechanisms which deposited the sandstones in sequences of alternating mudstones and graded sandstones. A few workers (e.g., Kuenen, 1957a, p. 233; Murphy and Schlanger, 1962; Dott and Howard, 1962) cautioned that a turbidity current origin cannot be indiscriminately applied to such deposits just because the sandstone beds exhibit the sole marks, convolutions, and other structures commonly attributed to this mechanism. Dott (1963) concluded that repeated graded bedding and displaced faunas are the most valid criteria of turbidity current deposits.

The alternating mudstone-sandstone lithology, the sedimentary structures, the graded beds, the unabraded mudstone fragments, the reworked fossil detritus and glauconite in the basal portions of some graded sandstone beds, the absence of large-scale cross-bedding, and many other features in the Sacramento Valley sequence are seemingly typical of alleged turbidite sequences and deep-sea sands, as summarized by Kuenen (1964). Furthermore, the general upward transition in the internal characteristics of individual beds (from massive to laminated to cross-bedded), as well as the abundance of graded beds, indicates currents of decreasing velocity (McBride, 1962), such as turbidity currents.

Several workers think that "normal" bottom currents are important on the modern deep-sea floor. Heezen and Hollister (1964) summarized evidence for weak bottom currents. Hubert (1964) stated that bottom currents were an important mechanism in the deposition of deep-sea sands of the western North Atlantic, and Von Rad and others (1965) suggested that normal bottom currents transported or reworked sands at a depth of 3600 feet off LaJolla, California. Klein (1967) summarized and supported evidence for reworking of deep-sea sands by normal currents, but did not present evidence that these currents are capable of depositing graded beds.

Kuenen (1964, p. 9) brought out a pertinent and important point concerning the role of such currents on the modern sea floor. Normal deep-sea currents are of a permanent nature and do not come and go at random like surface currents can do. Therefore, deep-sea sands interbedded with clay-rich hemipelagic strata cannot be attributed to normal currents, as the hemipelagic clays reflect a slow, normal "raining down" of fine sediment from waters near the surface of the sea, and hence the absence or near-absence of currents at the site of deposition.

It is suggested here that turbidity currents were important in the deposition of the Sacramento Valley sequence. "Normal" bottom currents may have been present, but it does not seem likely that they played a major role in the movement of sediment.

A few of the conglomerate beds of the sequence are graded, suggestive of turbidity current deposition. Most, however, are un-graded and have "disrupted" frameworks (Pettijohn, 1957) with sand and mud between the larger clasts. Submarine slides probably deposited these beds.

## Volume of Sediment and Rate of Deposition

The volume of sediment in the sequence was crudely estimated, assuming an average total thickness along the west side of the Sacramento Valley of 5 miles for the Upper Jurassic and Lower Cretaceous Series (together) and 3 miles for the exposed Upper Cretaceous Series, and assuming a uniform wedging out to nothing toward the east beneath the Sacramento Valley (Fig. 3). The length of the belt under study is 150 miles, the width of the Upper Jurassic-Lower Cretaceous wedge which extends eastward to approximately the center of the Valley (Chuber, 1962) is about 15 miles, and the width of the Upper Cretaceous wedge from the west edge of the known outcrops to the Sierran foothills is about 35 miles. The volume thus estimated is about 5500 mi³ for the Upper Jurassic-Lower Cretaceous units and 8000 mi³ for the Upper Cretaceous units. The total volume is about equal to a wedge-shaped mountain belt 150 miles long, 60 miles wide, and rising to a height of 3 miles at its crest; this roughly corresponds to the volume of that portion of the modern Sierra Nevada located east of the Sacramento Valley. The volume of sediment in the adjacent Franciscan Group, at least partially correlative with the Sacramento Valley sequence, is very probably much greater (Bailey and others, 1964). Much larger estimates of sediment volume have been made. Bateman and Wahrhaftig (1966) stated that at least a 9-mile thickness of rock was eroded from the Sierra Nevada since Late Jurassic time, and that this erosion was probably completed along the western Sierra Nevada by about 85 m.y. ago.

The rate of deposition of the exposed Cretaceous sequence in the Cache Creek-Rumsey Hills area was also crudely estimated, assuming, as in the turbidity current hypothesis, that individual sandstone beds were deposited almost instantaneously relative to the interbedded sandstones. About two-thirds of the 30,000-foot section is mudstone which was deposited over about a 50 m.y. span (Fig. 4); this is an average rate of 4.8 inches per 1000 years. Even if the estimate of 50 m.y. for the deposition of the sequence is in great error, the calculated sedimentation rates will not be very different; rates of only a few inches per 1000 years are a certainty. The rate for the exposed Upper Cretaceous sequence (9000 feet of mudstone in 20 m.y.) is 5.4 inches per 1000 years, and for the muddier Lower Cretaceous Series (10,500 feet of mudstone in 30 m.y.), it is 4.2 inches per 1000 years. The rates are much higher than the 1 inch per 1000 years calculated by Sujkowski (1957) for the Carpathian flysch, but corresponds well with observed sedimentation rates of mud in the center of the San Diego Trough (Emery, 1960, p. 250; Hand and Emery, 1964). Previously published estimates on the rate of deposition of the Sacramento Valley Cretaceous belt (7 to 9 inches per 1000 years) were based on the total thickness (mudstone, sandstone, and miscellaneous rock types) of the sequence (Kay, 1955, p. 674). Bateman and Wahrhaftig (1966) estimated the erosion rate of the Sierra Nevada during this time at 6 to 18 inches per 1000 years; allowing for deposition of the great volume of Franciscan sediments as well, the erosional and depositional rates seem quite compatible.

Based on the above calculated rates of deposition for the interbedded hemipelagic mudstones, the average time spans between sand-depositing turbidity currents can be calculated if the thicknesses of the interbedded mudstone beds are known. These times, however, may actually be much longer, as the amount of erosion of individual mudstone beds by each subsequent current is not known. The average time spans between turbidity currents in measured portions of the sequence vary from about 800 years in the Sites Formation to about 10,000 years in Formations D (member 2) and F. Strikingly similar figures have been obtained for modern calcareous "turbidites" by radiocarbon dating (in Kuenen, 1964, p. 13). Mudstone "beds" up to 50 feet thick are present in the section, and may represent spans of up to 150,000 years without any turbidity current deposition at those localities. In other portions of the section where individual sandstone beds are composed of several graded units ("multiple grading"), turbidity currents may have followed each other in more rapid succession, although it is just as probable that hemipelagic muds were deposited after each graded unit but were completely removed by the succeeding turbidity currents. The sandier sequences with thinner mudstones thus suggest more frequent turbidity currents, and are likely related to a greater supply of coarse clastics in the source areas.

## Environment of Deposition

The Early Cretaceous sea, at the present north end of the outcrop belt, transgressed northward onto the flank of the ancestral Klamath Mountains, depositing fossiliferous mudstones, sandstones, and conglomerates (e.g., Diller and Stanton, 1894; Murphy, 1956). Wells have not penetrated through the Lower Cretaceous of the Sacramento Valley, but data from wells near the middle of the Valley (Kirby, 1943; Chuber, 1962) reveal that the Upper Jurassic and Lower Cretaceous deposits do not extend very far east of the outcrop belt (Fig. 3). The Upper Jurassic and Lower Cretaceous foraminifera from the Cache Creek section consist predominantly of robust arenaceous forms and minor calcareous species that are probably indicative of a lower neritic to upper bathyal environment (K. D. Berry, 1964, personal commun.). The clayey tops of six graded sandstone beds were processed in vain for microfossils in an attempt to find shallow water forms interbedded with the deeper water forms of the interbedded mudstones, as did Natland and Kuenen (1951).

Transgression of Late Cretaceous seas eastward onto the ancestral Sierra Nevada is well documented. For example, Diller and Stanton (1894) and Kirby (1943) illustrated the west-to-east onlapping of successively younger Upper Cretaceous units beneath the Sacramento Valley. Goudkoff (1945, p. 1003) found deeper water microfossils near the outcrop belt of the Forbes Formation, and shallower forms to the east. He also stated that the western parts of the Sacramento and San Joaquin Valleys were covered by fairly deep water during deposition of the Upper Cretaceous units studied in this investigation. The massive, cross-bedded, fossiliferous sandstones near Chico on the east side of the Valley are nearshore equivalents of the upper Funks, Guinda, and lower Forbes Formations, and similar sandstones near Redding at the north end of the Valley are equivalent to the upper Venado, Yolo, Sites, and Funks Formations (Popenoe and others, 1960). The sedimentary rocks now exposed in the outcrop belt of the Venado Formation were deposited 25 to 30 miles from the eastern shoreline of the basin, whereas the rocks now exposed in the outcrop belt of the Forbes Formation were deposited about 40 miles offshore (based on Chuber, 1962). This is compatible with the concept of the sea transgressing eastward throughout Cretaceous time.

Jones (1960) suggested that the Late Cretaceous seas transgressed northeastward in the vicinity of Redding, and southward (possibly in a more northerly basin) in the vicinity of Hornbrook near the California-Oregon border. McKnight (1964) also reported that paleocurrents moved toward the north during Albian deposition in central Oregon. Paleocurrent measurements made during the present investigation further substantiate the existence of a Late Cretaceous landmass in the Klamath Mountain area between these two areas, either separating two basins of deposition or two parts of the same basin.

The lamination and small-scale cross-bedding in the sandstones and siltstones, the lamination in the mudstones, and the lack of large-scale cross-bedding, ripple marks, and shallow water fossils all indicate that most of the Sacramento Valley sequence now exposed on the western edge of the Valley was deposited below wave base. Excluding the shallow water deposits at the northern end of the belt in the northern half of the Ono quadrangle, three time-successive environments are suggested in the studied sequence along most of the west side of the Sacramento Valley.

(1) In Late Jurassic and earliest Cretaceous time, a shallow sea in which *Buchias* locally lived in profusion probably prevailed (Imlay, 1959). Sole marks seem to be less abundant than in younger rocks.

(2) Through much of Cretaceous time, a deeper sea covered the basin in the vicinity of the present outcrop belt. More than 23,000 feet of generally unfossiliferous sediments were deposited during the 40 m.y. or so between the end of Valanginian time and the end of Coniacian time. The scarcity of fossils may be the result of water depths too great to permit shelly molluscan faunas to exist. The shelly debris which occurs in the basal portions of some graded beds probably originated in shallower water. Sedimentary structures indicate deposition below wave base.

(3) A somewhat shallower sea may have prevailed during Campanian time. Fossils are more common in the Forbes Formation and locally in the Funks Formation, and might indicate a shallowing of the sea, perhaps owing to a gradual filling of the basin. Sedimentary structures, however, still indicate deposition below wave base.

Kuenen (1953) and ten Haaf (1959, p. 61) suggested that bedding characteristics can be used as evidence of deposition on steep slopes versus gentle slopes or level bottoms. The abundance of finely laminated mudstones and fine tops on graded, regularly bedded sandstones are indicative of deposition on gentle slopes or level bottoms, whereas the presence of slumps, channel fills, and conglomeratic deposits suggest the proximity of at least moderate slopes. The former characteristics are most common in the Sacramento Valley sequence. Ten Haaf (1959, p. 61) also suggested that flute casts form nearer the sources of currents and higher on the slopes than do groove casts. No systematic regional variation in the flute to groove cast ratio was noted in this study; however, Formations D, E, and F along Cache Creek have the highest ratios and hence may have been deposited on slopes which were steeper and nearer the sources of the turbidity currents than were the other formations.

Bouma (1959) suggested that beds with the lower massive interval present represent deposition closer to the source of the current (proximal) than do beds without the lower massive unit which represent deposition farther from the source of the current (distal). On this basis, Formations A through C-1 and the Venado, Sites, and Guinda were deposited in proximal environments nearer to the current sources than were Formations C-2, D, E, F, and the Forbes. Thicker beds, amalgamation of individual sandstones to form thick beds, and high sand to mud ratios, all indicators of a proximal environment according to a summary by Walker (1967), are generally characteristic of the above "proximal" units. However, on the basis of other features named by Walker as representative of proximal environments (a prevalence of scour marks over tool marks, the absence of grading or the presence of crude grading, and an abundance of complete "Bouma beds"), more than half of the above proximal versus distal determinations are reversed.

Subsurface data from the Sacramento Valley (Chuber, 1962) indicates that a southwesterly-sloping shelf existed during deposition of the Upper Cretaceous rocks which are found beneath the Valley (Fig. 3). The considerable thickness of the Upper Cretaceous sediments (up to 10,000 feet, with isopach lines trending north-south) indicates that the shelf was subsiding during deposition.

If the entire width of the Sacramento Valley sequence had been deposited on a shelf, then water depths at the outer limits of deposition would have been considerable, on the order of a few thousand feet for a one-degree slope. However, the much greater thickness of the sequence along the outcrop belt on the west side of the Valley, as compared to the thickness beneath the Valley, strongly suggests that deposition occurred beyond the shelf, either on a down-faulted deeper shelf, on the continental slope, or beyond the slope. The shelf may have been narrow, like most modern California shelves, and may have extended from the present Sierra foothills to about the middle of the present Sacramento Valley, a distance of about 25 miles. Bailey and others (1964) suggested that a possible marked change in basement lithology, from granitic on the east to basalt or peridotite on the west, coincides with the magnetic high which runs along the center of, and extends the length of, the Sacramento and San Joaquin Valleys. Perhaps this feature is somehow related to the location of the continental slope during Cretaceous time.

Kuenen (1964) compared features in ancient turbidites and modern deep-sea sands; features of the Sacramento Valley sequence are remarkably similar to those he discussed. Dietz and Holden (1966) concluded that most modern turbidites occur at depths greater than 2000 meters.

## Basin Geometry and Sedimentation Models

Irwin (1957, 1960) and Bailey and others (1964, p. 124) proposed that the Franciscan graywacke and volcanic complex represents a eugeosynclinal assemblage, with the Sacramento Valley sequence representing the miogeosynclinal deposits. Although thrusting and strike-slip faulting obscure the original relationships of these rocks, paleogeographic and paleocurrent evidence indicate that the basin of deposition during the Late Jurassic and Cretaceous was elongate, relatively wide, on the continental margin, and bordered on the west by open ocean. The dominant north to south paleocurrent trend is parallel to the shoreline of the basin, to the regional tectonic trend, and to isopachous lines (Chuber, 1962) of the sedimentary strata involved. Kuenen (1957b) advanced the concept of longitudinal basin fill from one end to explain such paleocurrent trends, and this interpretation has been applied to several basins (Potter and Pettijohn, 1963). Knill (1959) suggested that marginal facies of such basins are generally derived from the sides, however, Scott (1966) concluded that the sediment in a similar basin in Chile was derived from the sides, and then reworked and redeposited by currents moving parallel to the slope.

The fact that Upper Jurassic intrusion occurred in both the Klamath Mountains and the Sierra Nevada suggests that both areas were elevated and that they both shed sediment throughout the Cretaceous. However, as Lower Cretaceous paleocurrent indicators show little current movement from east to west and Upper Cretaceous indicators show east to west movement only in the Cenomanian and Turonian, any currents which initially moved to the west were evidently turned toward the south. Analysis of the different types of information obtained in this study indicates that several suitable sedimentation models, each satisfying the data but each with weaknesses, can be constructed. The three most basic models are discussed below and are illustrated by Figure 11.

(1) A southerly paleoslope may have existed along the studied 150-mile length of the basin; however, even a one-degree slope would have resulted in a water depth of 13,000 feet at the southern end of the study area. Possibly a shorter, steeper paleoslope at the north end of the basin could have given turbidity currents the momentum necessary to flow great distances to the south over a nearly flat sea floor, as such currents can flow on slopes of 0.001, or about 5 feet per mile (Menard, 1955; 1964, p. 194). Turbidity currents moving westward, down the side of the basin, may have been turned southward by the major paleoslope. However, a north-south bathymetric axis, which trended parallel to the Sierra Nevada and in which the Sacramento Valley sequence could have been deposited, seems somewhat unlikely beneath a body of water with no apparent western shore, although perhaps a rise within the geosyncline or a tectonic subsea ridge could have controlled current movement. For example, Cummins (1959) proposed that tectonic troughs in Cambrian seas of Wales governed paleocurrent patterns.

(2) Deep, strong ocean-bottom currents moving southward, parallel to the shoreline and the strike of the slope, could have deflected to the south turbidity currents which originally flowed westward, perpendicular to the strike of the slope. Menard (1955, p. 251) considered this possibility in the modern Pacific and concluded that the southerly moving currents present off the California coast are very weak below a depth of 1500 feet, do not reach the bottom in deeper waters, and would not be a likely mechanism for deflecting turbidity currents (see *Current Mechanisms* above for additional discussion of "normal" bottom currents). Even if some strong currents did exist at depth, they would have to have been unusually strong to deflect turbidity currents, and it is certainly doubtful whether they would have been capable of reworking turbidity current deposits and redepositing them as graded beds.

Murphy and Schlanger (1962), working in Brazil, and Scott (1966), in Chile, cited the parallelism of the axes of slump folds (slope indicators?) and paleocurrent indicators as evidence of current movement parallel to the strike of the slope. Cummins (1959), with similar evidence, proposed that turbidity currents flowed down the plunge of tectonic troughs and that the slumped beds moved down the steep sides. Others (e.g., Rich, 1950) have noted axes of crumpled slump zones trending perpendicular to the paleocurrent directions, indicating that the paleocurrents were moving down the paleoslope. It was realized that the measurement of many slump axes in the Sacramento Valley sequence would help resolve some fundamental paleogeographical problems. However, a concerted search for slumps with measurable axes revealed only two slump folds and two consistently convoluted layers that could be measured. They are, interestingly, approximately parallel to the dominant southerly paleocurrent trend, and if more could be located with similar trends, they would be important evidence in support of currents moving parallel to the strike of the slope.

(3) Menard (1955) noted that most deep-sea channels on numerous deep-sea fans off western North America hook to the left (south). He proposed that the fans were built by large turbidity currents which flowed out of submarine canyons and over levees of the deep-sea channels, spreading sediment out in a semicircular sheet. The levee deposits would be built up higher on the right or north side of the channels due to the Coriolis effect on the turbidity currents; hence an asymmetrical channel cross section would develop and younger turbidity currents would migrate toward the left (south) where the levees would be lower. As an example, Menard (1964, p. 229 and map) showed that sediment originating at the mouth of the Columbia River is being deposited as a deep-sea fan at the base of the continental slope by southward-flowing turbidity currents.

A similar shift of currents to the south side of a large subsea fan or delta located on or beyond the continental slope of that time (with or without a submarine canyon), or perhaps even located on the shelf off the mouth of a major river system, could explain the predominantly southerly paleocurrent trend. Such a structure would have provided an initial slope down which currents originating at the north end of the present Valley might have moved, and could have extended far to the south as well, possibly turning most of the currents which moved westward down the side of the basin. The general increase in the sandstone: mudstone ratio from north to south along the outcrop belt suggests that the coarser material was being deposited nearer the seaward extremities of the fan, a relationship which seems plausible if turbidity currents were involved in moving sediment across the fan. Finer sediment may have been left behind as it spilled over the channel levees, as noted by Hand and Emery (1964) in studies on the modern San Diego Trough. The gross geography of the modern west coast of North America might not be very different from that which probably existed in Cretaceous time. The size of such fans is no problem; the Monterey and Delgada fans off the coast of central California, for example, are each larger than a hypothetical fan which could have encompassed both the Sacramento Valley sequence and the Franciscan Group deposition.

## CONCLUSIONS

This investigation has provided new paleocurrent, petrographic, and mineralogic information on a Late Jurassic-Cretaceous depositional basin. The remarkably consistent north to south paleocurrent trend through most of the thick Cretaceous sequence illustrates a gross uniformity of conditions during a long span of time. Detailed mineralogic and petrographic studies through one complete stratigraphic section revealed distinct changes in source rocks through time; a generalized sampling of the sequence along the remainder of the outcrop belt indicated that the compositions of the sandstones, unit by unit, are quite similar to those in the detailed section.

Stratigraphic and paleogeographic evidence indicate that much of the studied sequence was deposited a few tens of miles offshore in at least moderately deep water. The great thickness of the sequence, and other features, indicate a geosynclinal environment, and the general lack of volcanic rocks suggest a miogeosyncline rather than a eugeosyncline. Bedding characteristics, especially grading, are indicative of deposition below wave base by turbidity currents.

An attempt was made to integrate all these data with radiometric age dates and other paleogeographic information on the ancestral Klamath Mountains and the ancestral Sierra Nevada, the logical source areas. Major events in the source areas, including volcanism, intrusion, and unroofing of intrusions, are recorded in the sedimentary column.

Basins with similar sedimentary rocks and sedimentary structures have been studied by other workers. In these investigations, a basic problem has been the interpretation of the significance of the paleocurrent indicators. Do they point back toward the source areas, or have the currents which deposited these beds been turned or deflected prior to depositing their loads? Longitudinal basin fill, marginal fill, and longitudinal plus marginal fill have been offered as solutions by these workers. In the present study, the same problem exists. Considered alone, the paleocurrent data might suggest that the basin may have been filled from the north end, with the sediment derived from the Klamath Mountains. However, the petrographic, mineralogic, radiometric, and paleogeographic data indicate that, in spite of the predominant "longitudinal" paleocurrent trend, much material must have been supplied by the Sierra Nevada along the east side of the basin. A synthesis of numerous types of data is necessary if the sedimentational history of such basins is to be solved.

Three sedimentation models have been presented as possible solutions. One involves a southerly paleoslope trending parallel to the length of the basin; another is based on southerly-moving deep oceanic currents which deflected to the south turbidity currents moving westward down the side of the basin; and the third suggests a subsea fan or delta which turned westerly moving turbidity currents toward the south. It is suggested that the latter model best fits the data; modern seafans with sedimentation patterns similar to that of the Sacramento Valley sequence are present at the base of the slope off the west coast of North America.

Oceanographic investigations (e.g., Kuenen, 1964) have shown that modern deepsea sands at the foot of the slopes strongly resemble the graded graywacke "turbidite" or "turbidite-like" sequences of the rock column. Information obtained during this study suggests that a major part of the Sacramento Valley sequence may have been deposited in such an environment. Dietz (1963) suggested that the prisms of turbidity-current deposits at the base of the continental slopes may be modern eugeosynclinal accumulations, with the material at the edge of the shelves (continental terraces) representing the miogeosynclinal deposits. Dietz and Holden (1966) further suggested that turbidites, such as comprise the continental rise prisms at the bases of the slopes, are confined to the eugeosynclinal facies. The evidence from this investigation suggests that miogeosynclinal deposits, which do not contain the volcanics and radiolarian cherts typical of the eugeosynclines, can also be composed of "hemipelagic turbidite" beds. McBride (1964) noted that deeper waters may have been common in those geosynclines which contain "turbidites".

The interpretations are necessarily conjectural and tentative. Oceanographic studies are providing information which is essential to the solution of problems such as these. In addition, further investigation: (1) of the Sacramento Valley sequence, including more work on the Upper Jurassic strata; (2) of the correlative sequence in the San Joaquin Valley to the south, which may have been part of the same depositional basin; (3) of the closely related Franciscan Group; (4) of the possible source rocks in the Klamath Mountains and the Sierra Nevada; and (5) of the regional tectonic history will be necessary before the Upper Jurassic and Cretaceous depositional framework in California can be fully understood.

## REFERENCES CITED

Allen, J. R. L., 1963a, Henry Clifton Sorby and the sedimentation structures of sands and sandstones in relation to flow conditions: Geologie en Mijnbouw, v. 42, p. 223–228.
—— 1963b, Internal sedimentation structures of well-washed sands and sandstones in relation to flow conditions: Nature, Oct. 26, p. 326–327.
Anderson, F. M., 1933, Knoxville-Shasta succession in California: Geol. Soc. America Bull., v. 44, p. 1237–1270.
—— 1938, Lower Cretaceous deposits in California and Oregon: Geol. Soc. America Special Paper 16, 339 p.
—— 1958, Upper Cretaceous of the Pacific Coast: Geol. Soc. America Mem. 71, 378 p.
Bailey, E. H., and Irwin, W. P., 1959, K-feldspar content of Jurassic and Cretaceous graywackes of northern Coast Ranges and Sacramento Valley, California: Am. Assoc. Petroleum Geologists Bull., v. 43, p. 2797–2809.
Bailey, E. H., Irwin, W. P., and Jones, D. L., 1964, Franciscan and related rocks, and their significance in the geology of western California: California Div. Mines and Geology Bull. 183, 177 p.
Bateman, P. C., and Wahrhaftig, Clyde, 1966, Geology of the Sierra Nevada: Bull. 190, Geology of Northern California Div. Mines and Geology. p. 107–172.
Bouma, A. H., 1962, Sedimentology of some flysch deposits: New York, Elsevier Publishing Co., 168 p.
Briggs, L. I., 1953, Some data on turbidites from the Alpes Maritimes France: Geologie en Mijnbouw, v. 21, p. 223–227.
Brown, R. D., 1964, Geologic map of the Stonyford quadrangle, Glenn, Colusa, and Lake Counties, California: U.S. Geol. Survey Map MF-279.
—— and Rich, E. I., 1960, Early Cretaceous fossils in submarine slump deposits of Late Cretaceous age, Northern Sacramento Valley, California: U.S. Geol. Survey Prof. Paper 400-B, p. 318B–320B.
—— 1967, Implications of two Cretaceous mass transport deposits, Sacramento Valley, California: Comment on a paper by Gary L. Peterson: Jour. Sed. Petrology, v. 37, p. 240–248.
Chuber, Stewart, 1962, Late Mesozoic stratigraphy of the Sacramento Valley: San Joaquin Geol. Soc., v. 1, p. 3–16.
Clark, L. D., 1960, Foothills fault system, western Sierra Nevada, California: Geol. Soc. America Bull., v. 71, p. 483–496.
Compton, R. R., 1955, Trondhjemite batholith near Bidwell Bar, California: Geol. Soc. America Bull., v. 66, p. 1351–1384.
Crowell, J. C., 1955, Directional current features from the Pre-alpine flysch, Switzerland: Geol. Soc. America Bull., v. 68, p. 993–1010.
Cummins, W. A., 1959, The Lower Ludlow Grits in Wales: Liverpool and Manchester Geol. Jour., v. 2, pt. 2, p. 168–179.
Curtis, G. H., Evernden, J. F., and Lipson, J., 1958, Age determination of some granitic

# REFERENCES CITED

Davis, G. A., 1963, Structure and mode of emplacement of Caribou Mountain Pluton, Klamath Mountains, California: Geol. Soc. America Bull., v. 74, p. 331–348.
—— 1966, Metamorphic and granitic history of the Klamath Mountains: Bull. 190, Geology of Northern California, California Div. Mines and Geology, p. 39–50.
Dietz, R. S., 1963, Alpine serpentines as oceanic rind fragments: Geol. Soc. America Bull., v. 74, p. 947–952.
Dietz, R. S., and Holden, J. C., 1966, Deep-sea deposits in but not on the continents: Am. Assoc. Petroleum Geologists Bull., v. 50, p. 351–362.
Diller, J. S., and Stanton, T. W., 1894, The Shasta-Chico Series: Geol. Soc. America Bull., v. 5, p. 435–464.
Dott, R. H., Jr., 1963, Dynamics of subaqueous gravity depositional processes: Am. Assoc. Petroleum Geologists Bull., v. 47, p. 104–128.
—— and Howard, J. K., 1962, Convolute lamination in non-graded sequences: Jour. Geology, v. 70, p. 114–121.
Durham, J. W., 1962, The Late Mesozoic of Central California: California Div. Mines and Geology Bull. 181, p. 31–38.
Dzulynski, Stanislaw, and Sanders, J. E., 1962, Current marks on firm bottoms: Trans. Connecticut Acad. Arts and Sciences, v. 42, p. 57–96.
Dzulynski, Stanislaw, and Smith, A. J., 1963, Convolute lamination, its origin, preservation, and directional significance: Jour. Sed. Petrology, v. 33, p. 616–627.
Emerson, D. O., and Rich, E. I., 1966, Guide to Sacramento Valley and Northern Coast Ranges, Field Trip E: Bull. 190, Geology of Northern California, California Div. Mines and Geology, p. 473–485.
Emery, K. O., 1960, The sea off southern California—a modern habitat of petroleum: New York, John Wiley, 366 p.
Evernden, J. F., Curtis, G. H., Obradovich, J., and Kistler, R. W., 1961, On the evaluation of glauconite and illite for dating sedimentary rocks by the Potassium-Argon method: Geochim. et Cosmochim. Acta, v. 23, p. 78–99.
Fiske, R. S., Hopson, C. A., and Waters, A. C., 1963, Geology of Mount Rainier National Park, Washington: U.S. Geol. Survey Prof. Paper 444, 93 p.
Gilbert, C. M., 1955, "Flow" casts on sandstone beds (Abstract): Geol. Soc. America Bull., v. 66, p. 1650.
Goudkoff, P. P., 1945, Stratigraphic relations of Upper Cretaceous in Great Valley, California: Am. Assoc. Petroleum Geologists Bull., v. 29, p. 956–1007.
Hand, B. M., and Emery, K. O., 1964, Turbidites and topography of the north end of the San Diego Trough, California: Jour. Geology, v. 72, p. 526–542.
Heezen, B. C., and Hollister, Charles, 1964, Deep-sea current evidence from abyssal sediments: Marine Geol., v. 1, p. 141–174.
Hietanen, Anna 1951, Metamorphic and igneous rocks of the Merrimac area, Plumas National Forest, California: Geol. Soc. America Bull., v. 62, p. 565–608.
Hinds, N. E., 1933, Geologic formations of the Redding-Weaverville districts, northern California: California Jour. Mines and Geology Bull., v. 29, p. 77–122.
Hubert, J. F., 1964, Textural evidence for deposition of many western North Atlantic deepsea sands by ocean-bottom currents rather than turbidity currents: Jour. Geology, v. 72, p. 757–785.
Imlay, R. W., 1959, Succession and speciation of the pelecypod *Aucella* and its equivalents in Sacramento Valley, California: U.S. Geol. Survey Prof. Paper 314-G, p. 155–167.
Irwin, W. P., 1957, Franciscan group in Coast Ranges and its equivalents in Sacramento Valley, California: Am. Assoc. Petroleum Geologists Bull., v. 43, p. 2770–2785.
—— 1960, Geologic reconnaissance of the northern Coast Ranges and Klamath Mountains, California: California Div. Mines and Geology Bull. 179, 80 p.
—— 1966, Geology of the Klamath Mountains Province: Bull. 190, Geology of Northern California, Calif. Div. Mines and Geology, p. 19–38.
Jones, D. L., 1960, Cretaceous stratigraphy of northern California and southern Oregon: Pacific Petroleum Geol., v. 14, no. 5, p. 4.
Kay, Marshall, 1955, Sediments and subsidence through time: Geol. Soc. America Spec. Paper 62, p. 665–684.
Kinkel, A. R., Hall, W. E., and Albers, J. P., 1956, Geology and base-metal deposits of West Shasta copper-zinc district, Shasta County, California: U.S. Geol. Survey Prof. Paper 285, 156 p.
Kirby, J. M., 1945, Upper Cretaceous stratigraphy of west side of Sacramento Valley south of Willows, Glenn County, California: Am. Assoc. Petroleum Geologists Bull., v. 27, p. 279–305.
Kistler, R. W., Bateman, P. C., and Brannock, W. W., 1965, Isotopic ages of minerals from granitic rocks of the central Sierra Nevada and Inyo Mountains, California: Geol. Soc. America Bull., v. 76, p. 155–164.
Klein, G. de V., 1967, Paleocurrent analysis in relation to modern marine sediment dispersal patterns: Am. Assoc. Petroleum Geologists Bull., v. 51, p. 366–382.
Knill, J. L., 1959, Axial and marginal sedimentation in geosynclinal basins: Jour. Sed. Petrology, v. 29, p. 317–325.
Kuenen, Ph. H., 1953, Significant features of graded bedding: Am. Assoc. Petroleum Geologists Bull., v. 37, p. 1044–1066.
—— 1957a, Sole markings of graywacke beds: Jour. Geology, v. 65, p. 231–258.
—— 1957b, Longitudinal filling of oblong sedimentary basins: Verhand. Konin. Neder. Geol. Mijnbou. Genootschap Geol. Series, v. 18, p. 189–195.
—— 1964, Deep-sea sands and turbidites, *in* Bouma, A. H., *Editor*, Developments in Sedimentology: Turbidites, v. 3, p. 3–3. New York, Elsevier Publishing Co.
Lachenbruch, M. C., 1962, Geology of the west side of the Sacramento Valley: California Div. Mines and Geology Bull. 181, p. 53–66.
Lipman, P. W., 1964, Structure and origin of an ultramafic pluton in the Klamath Mountains, California: Am. Jour. Sci., v. 262, p. 199–222.
McBride, E. F., 1962, Flysch and associated beds of the Martinsburg Formation (Ordovician), central Appalachians: Jour. Sed. Petrology, v. 32, p. 39–91.
—— 1964, Review of turbidite studies in the United States, *in* Bouma, A. H., *Editor*, Developments in Sedimentology: Turbidites, v. 3, p. 93–105. New York, Elsevier Publishing Co.
—— and Yeakel, L. S., 1963, Relationship between parting lineation and rock fabric: Jour. Sed. Petrology, v. 33, p. 779–782.
McKnight, B. K., 1964, Stratigraphic study of Cretaceous rocks near Mitchell, Oregon (Abstract): Geol. Soc. America Spec. Paper 82, p. 264.
Menard, H. W., Jr., 1955, Deep-sea channels, topography, and sedimentation: Am. Assoc. Petroleum Geologists Bull., v. 39, p. 236–255.
—— 1964, Marine geology of the Pacific: New York, McGraw-Hill, 271 p.
Murphy, M. A., 1956, Lower Cretaceous stratigraphic units of northern California: Am. Assoc. Petroleum Geologists Bull., v. 40, p. 2098–2119.
Murphy, M. A., and Rodda, P. U., 1960, Mollusca of Cretaceous Bald Hills Formation of California: Jour. Paleontology, v. 34, p. 835–858.
Murphy, M. A., and Schlanger, S. O., 1962, Sedimentary structures in Ilhas and Sao Sebastiao Formations (Cretaceous), Reconavo Basin, Brazil: Am. Assoc. Petroleum Geologists Bull., v. 46, p. 457–477.
Natland, M. L., and Kuenen, Ph. H., 1951, Sedimentary history of the Ventura Basin, California, and the action of turbidity currents: Soc. Econ. Paleontologists and Mineralogists Spec. Pub. 2, p. 76–107.
Pabst, Adolph, 1938, Heavy minerals in the granitic rocks of the Yosemite region: Am. Mineralogist, v. 23, p. 46–53.
Peterson, G. L., 1964, Late Mesozoic stratigraphic framework, Sacramento Valley, California (Abstract): Geol. Soc. America Spec. Paper 82, p. 270.
—— 1965, Implications of two Cretaceous mass transport deposits, Sacramento Valley, California: Jour. Sed. Petrology, v. 35, p. 401–407.
Pettijohn, F. J., 1957, Sedimentary rocks: 2nd ed., New York, Harper, 718 p.
Popenoe, W. P., Imlay, R. W., and Murphy, M. A., 1960, Correlation of the Cretaceous Formations of the Pacific Coast (United States and northwestern Mexico): Geol. Soc. America Bull., v. 71, p. 1491–1540.
Potter, P. E., and Pettijohn, F. J., 1963, Paleocurrents and basin analysis: New York, Academic Press, Inc, 296 p.
Reed, R. D., 1933, Geology of California: Tulsa (Oklahoma), Am. Assoc. Petroleum Geologists, 355 p.
Rich, J. L., 1950, Flow markings, groovings, and intrastratal crumplings as criteria for recognition of slope deposits, with illustrations from Silurian rocks of Wales: Am. Assoc. Petroleum Geologists Bull., v. 34, p. 717–741.

Safonov, Anatole, 1962, The challenge of the Sacramento Valley, California: California Div. Mines and Geology Bull. 181, p. 77–97.
Schwarzbach, Martine, 1963, Climates of the past: New York, D. Van Nostrand Co., 328 p.
Scott, K. M., 1966, Sedimentology and dispersal pattern of a Cretaceous flysch sequence, Patagonian Andes, southern Chile: Am. Assoc. Petroleum Geologists Bull., v. 50, p. 72–107.
Soliman, S. M., 1965, Geology of the east half of the Mount Hamilton Quadrangle, California: California Div. Mines and Geology Bull. 185, 32 p.
Stokes, W. L., 1947, Primary lineation in fluvial sandstones: a criterion of current direction: Jour. Geology, v. 55, p. 52–54.
Sujkowski, Z. L., 1957, Flysch sedimentation: Geol. Soc. America Bull., v. 68, p. 543–544.
Taliaferro, N. L., 1943, Franciscan-Knoxville problem: Am. Assoc. Petroleum Geologists Bull., v. 27, p. 109–219.
—— 1944, Cretaceous and Paleocene of Santa Lucia Range, California: Am. Assoc. Petroleum Geologists Bull., v. 28, p. 449–521.
ten Haaf, Ernst, 1956, Significance of convolute lamination: Geologie en Mijnbouw, n. ser., v. 18, p. 188–194.
—— 1959, Graded beds of the Northern Apennines: Ph.D. thesis, Univ. Groningen, Amsterdam, The Netherlands, 102 p.
Von Rad, Ulrich, Shephard, F. P., Rosfelder, A., and Dill, R. F., 1965, Origin of deepwater sands off La Jolla, California (Abstract): Geol. Soc. America Spec. Paper 87, p. 180.
Walker, R. G., 1967, Turbidite sedimentary structures and their relationship to proximal and distal depositional environments: Jour. Sed. Petrology, v. 37, p. 25–43.
Weaver, C. E., 1963, Interpretive value of heavy minerals from bentonites: Jour. Sed. Petrology, v. 33, p. 343–349.
Williams, Howell, Turner, F. J., and Gilbert, C. M., 1955, Petrography: San Francisco, W. H. Freeman and Co., 406 p.

MANUSCRIPT RECEIVED BY THE SOCIETY JUNE 17, 1966
REVISED MANUSCRIPT RECEIVED FEBRUARY 26, 1968

# Post-1968 research on the Great Valley Group

**Raymond V. Ingersoll**
*Department of Earth and Space Sciences, University of California, Los Angeles, California 90095; e-mail: ringer@ess.ucla.edu*

## INTRODUCTION

Ojakangas (1964, 1968) documented paleocurrent directions and associations of sedimentary structures, stratigraphic changes in sandstone composition primarily reflecting changes in provenance, and overall stratigraphic-structural relations of the Sacramento Valley. These three components (sedimentology, petrology, and structure) formed the basis for Ingersoll's (1976) dissertation (both Ojakangas and Ingersoll were supervised by W. R. Dickinson at Stanford University). Fundamental breakthroughs in understanding occurred between the times of Ojakangas' and Ingersoll's dissertations, which allowed the latter to integrate these fields in a way not previously possible.

## SEDIMENTOLOGY

Following Ojakangas' (1968) suggestion that turbidite sedimentation produced the upper Mesozoic strata of the Great Valley, Swe and Dickinson (1970), Mansfield (1971, 1979), and Lowe (1972) confirmed this general conclusion with detailed analyses of local deep-marine components of the Great Valley basin. Meanwhile, researchers of both modern and ancient deep-marine sedimentation were developing models for the four-dimensional evolution of submarine fans (e.g., Normark, 1970, 1974; Walker, 1970; Haner, 1971; Nelson and Kulm, 1973; Walker and Mutti, 1973; Mutti, 1974; Nelson and Nilsen, 1974), culminating in the seminal model for ancient submarine fans of Mutti and Ricci-Lucchi (1972).

Ingersoll (1978b) utilized the model of Mutti and Ricci-Lucchi (1972) to delineate horizontal and vertical trends in submarine-fan facies of the Upper Cretaceous part of the Great Valley Group. This facies analysis was combined with a synthesis of all available paleocurrent and paleoecological data to outline the depositional history of the Great Valley basin (Ingersoll, 1979).

Subsequent sedimentological research has refined interpretations of processes and products of submarine deposition, paleoecology, and paleogeography (e.g., Garcia, 1981; Cherven, 1983; Suchecki, 1984; Ingersoll and Nilsen, 1990; Graham and Lowe, 1993; Williams et al., 1998).

## PETROLOGY AND PROVENANCE

Bailey and Irwin (1959) first recognized systematic stratigraphic changes in sandstone composition in the Sacramento Valley, and Brown and Rich (1961) utilized these petrologic intervals for mapping. Following Ojakangas' (1968) refinement of these petrologic intervals, Rich (1971) and Dickinson and Rich (1972) named five petrofacies along the west side of the Sacramento Valley. At the same time, Mansfield (1971, 1979) studied comparable petrofacies of part of the San Joaquin Valley. This petrofacies work overlapped publication of the definitive method of petrographic analysis of graywacke and arkose by Dickinson (1970a). Dickinson et al. (1969) also documented in greater detail diagenetic changes in Great Valley strata, as previously discussed by Ojakangas (1968).

Ingersoll (1978a, 1983) refined the Sacramento and San Joaquin petrofacies and designated petrostratigraphic units, which are mappable over the entire Great Valley. Ojakangas (1968), Dickinson and Rich (1972), Ingersoll (1978a, 1983) and Mansfield (1979) all related changes in petrofacies to provenance changes in the Sierra Nevada and Klamath Mountains (Fig. 1). Additional work by Bertucci (1983), Seiders (1983, 1989), Suchecki (1984), Seiders and Blome (1988), and Short and Ingersoll (1990) refined provenance interpretations through combined study of sandstone and conglomerate petrofacies. Additional detailed insight regarding evolution of the Sierra Nevada magmatic arc was provided by analysis of radiogenic isotopes of Great Valley petrofacies (Linn et al., 1991, 1992).

Ingersoll, R. V., 1999, Post-1968 research on the Great Valley Group, *in* Moores, E. M., Sloan, D., and Stout, D. L., eds., Classic Cordilleran Concepts: A View from California: Boulder, Colorado, Geological Society of America Special Paper 338.

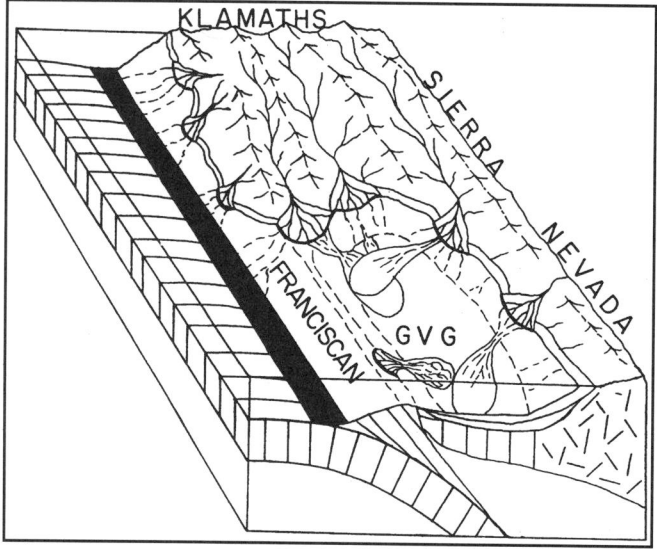

Figure 1. Schematic paleogeographic block diagram looking northeast, showing north end of Sacramento forearc basin during Early Cretaceous (early Hauterivian). Black trough represents Franciscan trench, where oceanic lithosphere is being subducted beneath Klamath terranes to north, and Coast Range ophiolite underlying Great Valley Group (GVG) to south. Sierra Nevada magmatic arc is east of forearc basin. Diagram illustrates relations prior to extensive transgression of Klamaths. Note mixing of Klamath-derived (Platina petrofacies) and northern Sierra-derived (Stony Creek petrofacies) detritus in northwest part of forearc basin (from Short and Ingersoll, 1990).

## STRATIGRAPHY

Stratigraphic nomenclature of the Great Valley has a complex history (discussed in Ingersoll, 1990). The enormous thickness, lithologic homogeneity, and lateral stratal lenticularity have resulted in proliferation of lithostratigraphic names of local utility, and alternatively, designation of individual units several kilometers thick, with little structural or stratigraphic utility. Following the mapping of Brown and Rich (1961), Rich (1971), and Mansfield (1971, 1979), Ingersoll et al. (1977) and Ingersoll and Dickinson (1981) proposed using petrofacies to define regionally extensive petrostratigraphic formations; these formations constitute the Great Valley Group, as formalized in Ingersoll (1990).

Chronostratigraphic units have been refined since 1968, based on megafossils, microfossils, geochronology, and magnetostratigraphy (e.g., Douglas, 1969; Imlay and Jones, 1970; Pessagno, 1976, 1977; Ward et al., 1983; Haggart and Ward, 1984; Almgren, 1986; Verosub et al., 1989; Bralower, 1990). Moxon (1988, 1990) reinterpreted the Great Valley Group in terms of sequence stratigraphy, as summarized by Williams (1993, 1997).

## BASIN ANALYSIS AND TECTONICS

During the plate tectonics revolution of the late 1960s and early 1970s, major aspects of California geology were reinterpreted. Hamilton (1969) suggested that Pacific oceanic plate "underflowed" California throughout the Mesozoic and created the Franciscan subduction complex, as well as the Sierra Nevada batholith, by generating melts at depth. Ernst (1970) proposed that the fault contact between the Franciscan Complex and the Coast Range ophiolite was the crustal remnant of the late Mesozoic subduction zone. At the same time, radiometric dating of plutons in the Sierra Nevada was delineating migrating patterns of magmatism (e.g., Evernden and Kistler, 1970) (Fig. 2). Dickinson (1970b, 1971, 1973, 1974a, 1974b) discussed how forearc basins such as the Great Valley formed between growing subduction complexes and active magmatic arcs. In fact, the Great Valley forearc basin has served as the type forearc basin in subsequent discussions (e.g., Dickinson and Seely, 1979; Ingersoll, 1982; Dickinson, 1995).

The most enigmatic aspect of Great Valley geology is the origin of the Great Valley ophiolite, which in places depositionally underlies the western Great Valley Group. Bailey et al. (1970) and Moores (1970) proposed that the ophiolite represented oceanic crust accreted to the continental margin prior to initiation of the Great Valley forearc basin. Schweickert and Cowan (1975) proposed a model involving the collision of the west-facing continental-margin arc with an east-facing intraoceanic arc with backarc spreading; collision created the Nevadan orogeny, which immediately preceded and overlapped with initiation of Franciscan subduction in the latest Jurassic (Fig. 3). Ingersoll and Schweickert (1986) integrated this model with the contrasting Nevadan tectonic history of the Klamath area (e.g., Harper and Wright, 1984) and showed how a wide oceanic forearc basin could have formed in the Great Valley area at the same time that no such forearc basin formed in the Klamath area (Fig. 4). Dickinson et al. (1996) reviewed ongoing controversy concerning origin of the Great Valley ophiolite. Godfrey et al. (1997) provided seismic and gravity data consistent with thrust emplacement of the Great Valley ophiolite over Sierran basement during the Nevadan orogeny, as predicted by Schweickert and Cowan's (1975), Moores and Day's (1984), and Ingersoll and Schweickert's (1986) models.

Additional insights regarding the Great Valley forearc basin were provided by subsidence and thermochronologic analyses. Dickinson et al. (1987) suggested that Cretaceous subsidence was primarily due to isostatic sediment loading on top of the residual deep oceanic crust, followed by rapid shallowing during flat-slab subduction. Moxon and Graham (1987) documented Late Cretaceous thermal subsidence along the east side of the basin, corresponding to eastward migration of the magmatic arc (i.e., Ingersoll, 1979) (Fig. 3), and latest Cretaceous uplift along the west side of the basin, corresponding to initiation of Laramide flat-slab subduction. Bostick (1974) and Dumitru (1988) documented low geothermal gradients within the Great Valley forearc, as predicted by the forearc model.

By the end of the Cretaceous, most of the Sacramento forearc basin had been filled nearly to sea level, to form a broad shelf (Dickinson et al., 1979; Ingersoll, 1982). Diverse depositional environments, ranging from nonmarine to coastal and deltaic to

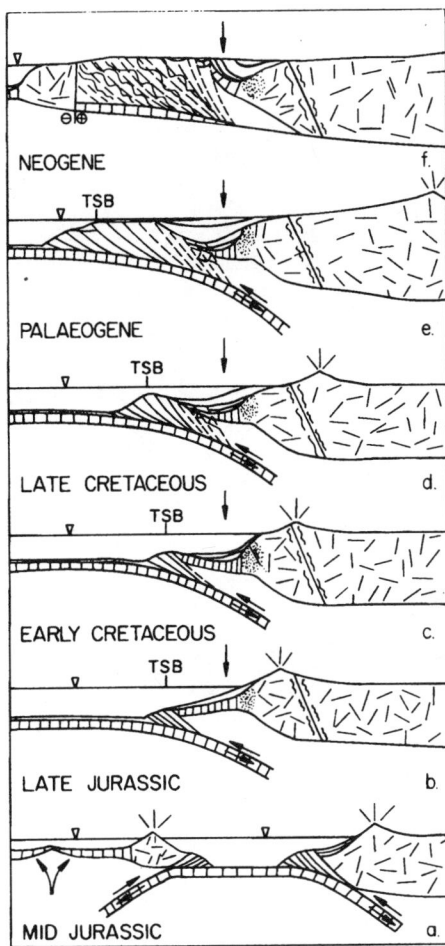

Figure 2. Schematic cross sections of northern California from formation of Coast Range ophiolite behind an east-facing intraoceanic arc (bottom) to termination of Great Valley forearc by conversion to transform margin (top). Vertical arrows are reference points for each cross section and indicate location of present outcrop of Great Valley Group along west side of Sacramento Valley. Stippled pattern indicates upper-slope discontinuity. TSB is trench-slope break. Foothill suture shown by diagonal and wiggly lines (from Ingersoll, 1982).

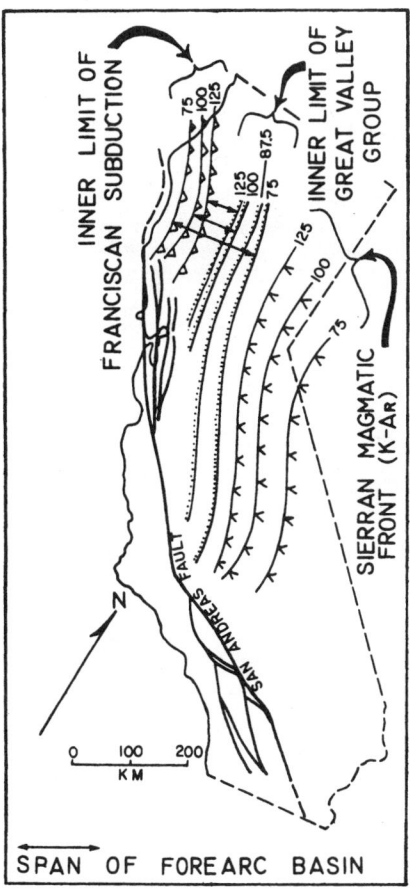

Figure 3. Schematic map illustrating increase in width of Great Valley forearc basin. Approximate positions of migratory boundaries are shown for the Early Cretaceous (125 Ma), the mid-Cretaceous (100 Ma) and the Late Cretaceous (75 Ma). Positions of western boundary at migratory trench-slope break marking inner limit of active subduction are inferred from easternmost extent of successively younger strata within Franciscan Complex. Positions of eastern boundary of Great Valley Group primarily from subsurface information. Positions of magmatic front from western limits of radiometric dates for Sierra Nevada plutons (Evernden and Kistler, 1970) (from Ingersoll, 1982).

slope to basin plain and submarine fan, during the latest Cretaceous to early Paleogene, are well documented in the subsurface, primarily based on oil and gas wells and seismic studies (e.g., Garcia, 1981; Ingersoll, 1982; Cherven, 1983; Nilsen, 1990). Almgren (1978) and Dickinson et al. (1979) summarized the Paleogene history of northern California, including the cutting and filling of submarine canyons in the forearc shelf.

Structural studies have demonstrated two important characteristics of the west side of the Great Valley. In many localities, the Coast Range fault (formerly considered only a thrust representing subduction displacement; e.g., Ernst, 1970) has demonstrable normal displacement that overprints older thrust displacement (Jayko et al., 1987; Krueger and Jones, 1989; Harms et al., 1992). Cenozoic and possibly Cretaceous uplift of the Coast Ranges was accomplished in part by extension along the Coast Range and related faults. Contractional deformation, in places of demonstrable Quaternary age, has overprinted this extension along the east side of the Franciscan Complex (e.g., Wentworth et al., 1984; Namson and Davis, 1988; Wentworth and Zoback, 1989; Unruh and Moores, 1992; Unruh et al., 1995).

## CONCLUSIONS

Ojakangas (1968) set the stage for our present understanding of the Great Valley forearc basin by documenting the sedimentology, stratigraphy, petrology, and paleoenvironments of the Great Valley Group. Ingersoll (1978a, 1978b, 1978c, 1979, 1982, 1983), Dickinson and Seely (1979), and Ingersoll and Dickinson (1981) applied new concepts of submarine-fan facies analysis, petrostratigraphy, provenance analysis, and forearc models to outline the late Mesozoic evolution of the Great Valley forearc basin. Schweickert and Cowan (1975) and Ingersoll and Schweickert (1986)

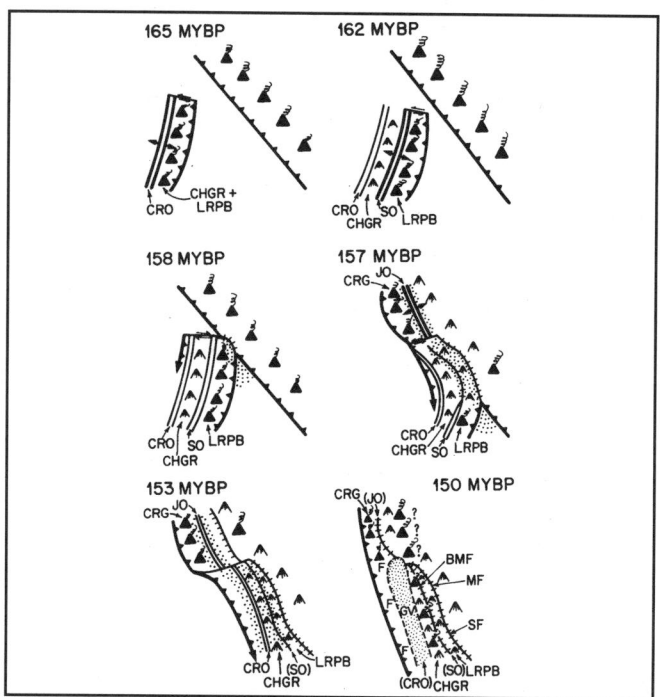

Figure 4. Sequential paleotectonic diagrams for Middle to Late Jurassic in northern California. Active magmatic arcs are shown with smoke, inactive without. Active subduction zones are shown by barbed symbols; suture zones are shown by suture pattern. Rifted continental margin is shown by hachured line. Active spreading centers are shown by divergent arrows on double lines, without implication of exact spreading orientation; inactive spreading centers are shown by double lines without arrows. Transforms are shown by thin arrows. Southward propagating trench is shown by large arrow. Stippled pattern shows sites of deposition of the Mariposa and Galice formations, and the Great Valley (GV) forearc basin. Abbreviations: CRO—Coast Range (Great Valley) ophiolite; SO—Smartville ophiolite; JO—Josephine ophiolite; CRG—Chetco, Rogue, Galice arc complex; F—Franciscan Complex; LRPB—Logtown Ridge arc complex and 200my-old Peñon Blanco arc complex; CHGR—Copper Hill, Gopher Ridge arc complex; BMF—Bear Mountain fault; MF—Melones fault; SF—Sonora fault. Parentheses around ophiolite names indicate partial preservation within fault zones (from Ingersoll and Schweickert, 1986).

provided viable models to explain the initiation of the Great Valley forearc basin following the Nevadan orogeny. Subsequent work has added to the data base and modified existing models, but has not challenged the fundamental models.

## ACKNOWLEDGMENTS

I thank Eldridge Moores, Doris Sloan, and Dottie Stout for inviting me to write this review. I also thank Bill Dickinson and Steve Graham for reviewing the manuscript and for ongoing advice and friendship. And, of course, I thank Dick Ojakangas for laying the framework upon which I worked, for providing additional data during my research, and for providing a picture and additional insights while writing this manuscript.

## REFERENCES CITED

Almgren, A. A., 1978, Timing of Tertiary submarine canyons and marine cycles of deposition in the southern Sacramento Valley, California, in Stanley, D. J., and Kelling, G., eds., Sedimentation in submarine canyons, fans and trenches: Stroudsburg, Dowden, Hutchinson and Ross, p. 276–291.

Almgren, A. A., 1986, Benthic foraminiferal zonation and correlations of Upper Cretaceous strata of the Great Valley of California—A modification, in Abbott, P. L., ed., Cretaceous stratigraphy western North America: Pacific Section, Society of Economic Paleontologists and Mineralogists, Book 46, p. 137–152.

Bailey, E. H., and Irwin, W. P., 1959, K-feldspar content of Jurassic and Cretaceous graywackes of northern Coast Ranges and Sacramento Valley, California: American Association of Petroleum Geologists Bulletin, v. 43, p. 2797–2809.

Bailey, E. H., Blake, M. C., Jr., and Jones, D. L., 1970, On-land Mesozoic oceanic crust in California Coast Ranges: U.S. Geological Survey Professional Paper 700-C, p. C70–C81.

Bertucci, P. F., 1983, Petrology and provenance of the Stony Creek Formation, northwestern Sacramento Valley, California, in Bertucci, P. F., and Ingersoll, R. V., eds., Guidebook to the Stony Creek Formation, Great Valley Group, Sacramento Valley, California (Annual Meeting, Pacific Section SEPM): Pacific Section, Society of Economic Paleontologists and Mineralogists, p. 1–16.

Bostick, N. H., 1974, Phytoclasts as indicators of thermal metamorphism, Franciscan Assemblage and Great Valley sequence (upper Mesozoic), California, in Dutcher, R. R., Hacquebard, P. A., Schopf, J. M., and Simon, J. A., eds., Carbonaceous Materials as Indicators of Metamorphism: Geological Society of America Special Paper 153, p. 1–17.

Bralower, T. J., 1990, A field guide to Lower Cretaceous calcareous nannofossil biostratigraphy of the Great Valley Group, Sacramento Valley, California, in Ingersoll, R. V., and Nilsen, T. H., eds., Sacramento Valley symposium and guidebook: Pacific Section, Society of Economic Paleontologists and Mineralogists, Book 65, p. 31–37.

Brown, R. D., Jr., and Rich, E. I., 1961, Geologic map of the Lodoga quadrangle, Glenn and Colusa counties, California: U.S. Geological Survey Oil and Gas Investigations Map OM-210, scale 1:48,000.

Cherven, V. B., 1983, A delta-slope-submarine fan model for Maestrichtian part of Great Valley sequence, Sacramento and San Joaquin basins, California: American Association of Petroleum Geologists Bulletin, v. 67, p. 772–816.

Crowell, J. C., 1955, Directional-current structures from the Prealpine flysch, Switzerland: Geological Society of America Bulletin, v. 66, p. 1351–1384.

Crowell, J. C., 1957, Origin of pebbly mudstones: Geological Society of America Bulletin, v. 68, p. 993–1009.

Dickinson, W. R., 1970a, Interpreting detrital modes of graywacke and arkose: Journal of Sedimentary Petrology, v. 40, p. 695–707.

Dickinson, W. R., 1970b, Relations of andesites, granites, and derivative sandstones to arc-trench tectonics: Reviews of Geophysics and Space Physics, v. 8, p. 813–860.

Dickinson, W. R., 1971, Clastic sedimentary sequences deposited in shelf, slope, and trough settings between magmatic arcs and associated trenches: Pacific Geology, v. 3, p. 15–30.

Dickinson, W. R., 1973, Widths of modern arc-trench gaps proportional to past duration of igneous activity in associated magmatic arcs: Journal of Geophysical Research, v. 78, p. 3376–3389.

Dickinson, W. R., 1974a, Sedimentation within and beside ancient and modern magmatic arcs: Society of Economic Paleontologists and Mineralogists Special Publication 19, p. 230–239.

Dickinson, W. R., 1974b, Plate tectonics and sedimentation, in Dickinson, W. R., ed., Tectonics and sedimentation: Society of Economic Paleontologists and Mineralogists Special Publication 22, p. 1–27.

Dickinson, W. R., 1995, Forearc basins, in Busby, C. J., and Ingersoll, R. V., eds., Tectonics of sedimentary basins: Cambridge, Blackwell, p. 221–261.

Dickinson, W. R., and Rich, E. I., 1972, Petrologic intervals and petrofacies in the Great Valley sequence, Sacramento Valley, California: Geological Society of America Bulletin, v. 83, p. 3007–3024.

Dickinson, W. R., and Seely, D. R., 1979, Structure and stratigraphy of forearc regions: American Association of Petroleum Geologists Bulletin, v. 63, p. 2–31.

Dickinson, W. R., Ojakangas, R. W., and Stewart, R. J., 1969, Burial metamorphism of the late Mesozoic Great Valley Sequence, Cache Creek, California: Geological Society of America Bulletin, v. 80, p. 519–526.

Dickinson, W. R., Ingersoll, R. V., and Graham, S. A., 1979, Paleogene sediment dispersal and paleotectonics in northern California: Geological Society of America Bulletin, v. 90, part I, p. 897–898, part II, p. 1458–1528.

Dickinson, W. R., Armin, R. A., Beckvar, N., Goodlin, T. C., Janecke, S. U., Mark, R. A., Norris, R. D., Radel, G., and Wortman, A. A.,1987, Geohistory analysis of rates of sediment accumulation and subsidence for selected California basins, in Ingersoll, R. V., and Ernst, W. G., eds., Cenozoic basin development of coastal California (Rubey Volume VI): Englewood Cliffs, New Jersey, Prentice-Hall, p. 1–23.

Dickinson, W. R., Hopson, C. A., and Saleeby, J. B., 1996, Alternate origins of the Coast Range ophiolite (California): Introduction and implications: GSA Today, v. 6, p. 1–2.

Dott, R. H., Jr., 1963, Dynamics of subaqueous gravity depositional processes: American Association of Petroleum Geologists Bulletin, v. 47, p. 104–128.

Douglas, R. G., 1969, Upper Cretaceous planktonic foraminifera in northern California, Part 1—Systematics: Micropaleontology, v. 15, p. 151–209.

Dumitru, T. A., 1988, Subnormal geothermal gradients in the Great Valley forearc basin, California, during Franciscan subduction: A fission track study: Tectonics, v. 7, p. 1201–1221.

Ernst, W. G., 1970, Tectonic contact between the Franciscan melange and the Great Valley sequence, crustal expression of a late Mesozoic Benioff zone: Journal of Geophysical Research, v. 75, p. 886–902.

Evernden, J. F., and Kistler, R. W., 1970, Chronology of emplacement of Mesozoic batholithic complexes in California and western Nevada: U.S. Geological Survey Professional Paper 623, 42 p.

Garcia, R., 1981, Depositional systems and their relation to gas accumulation in Sacramento Valley, California: American Association of Petroleum Geologists Bulletin, v. 65, p. 653–673.

Godfrey, N. J., Beaudoin, B. C., Klemperer, S. L., and Mendocino Working Group, 1997, Ophiolitic basement to the Great Valley forearc basin, California, from seismic and gravity data: implications for crustal growth at the North American continental margin: Geological Society of America Bulletin, v. 108, p. 1536–1562.

Graham, S. A., and Lowe, D. R., eds., 1993, Advances in the sedimentary geology of the Great Valley Group, Sacramento Valley, California: Pacific Section, Society of Economic Paleontologists and Mineralogists, Book 73, 65 p.

Haggart, J. W., and Ward, P. D., 1984, Late Cretaceous (Santonian-Campanian) stratigraphy of the northern Sacramento Valley, California: Geological Society of America Bulletin, v. 95, p. 618–627.

Hamilton, W., 1969, Mesozoic California and the underflow of Pacific mantle: Geological Society of America Bulletin, v. 80, p. 2409–2430.

Haner, B. E., 1971, Morphology and sediments of Redondo submarine fan, southern California: Geological Society of America Bulletin, v. 82, p. 2413–2432.

Harms, T. A., Jayko, A. S., and Blake, M. C., Jr., 1992, Kinematic evidence for extensional unroofing of the Franciscan Complex along the Coast Range fault, northern Diablo Range, California: Tectonics, v. 11, p. 228–241.

Harper, G. D., and Wright, J. E., 1984, Middle to Late Jurassic tectonic evolution of the Klamath Mountains, California-Oregon: Tectonics, v. 3, p. 759–772.

Imlay, R. W., and Jones, D. L., 1970, Ammonites from the Buchia zones in northwestern California and southwestern Oregon: U.S. Geological Survey Professional Paper 647B, 29 p.

Ingersoll, R. V., 1976, Evolution of the Late Cretaceous fore-arc basin of northern and central California [Ph.D. thesis]: Stanford, California, Stanford University, 200 p.

Ingersoll, R. V., 1978a, Petrofacies and petrologic evolution of the Late Cretaceous forearc basin, northern and central California: Journal of Geology, v. 86, p. 335–352.

Ingersoll, R. V., 1978b, Submarine fan facies of the Upper Cretaceous Great Valley Sequence, northern and central California: Sedimentary Geology, v. 21, p. 205–230.

Ingersoll, R. V., 1978c, Paleogeography and paleotectonics of the late Mesozoic forearc basin of northern and central California, in Howell, D. G., and McDougall, K. A., eds., Mesozoic paleogeography of the western United States (Pacific Coast Paleogeography Symposium 2): Pacific Section, Society of Economic Paleontologists and Mineralogists, p. 471–482.

Ingersoll, R. V., 1979, Evolution of the Late Cretaceous forearc basin, northern and central California: Geological Society of America Bulletin, v. 90, part I, p. 813–826.

Ingersoll, R. V., 1982, Initiation and evolution of the Great Valley forearc basin of northern and central California, in Leggett, J. K., ed., Trench-forearc geology: Sedimentation and tectonics on modern and ancient active plate margins: Geological Society of London Special Publication 10, p. 459–467.

Ingersoll, R. V., 1983, Petrofacies and provenance of late Mesozoic forearc basin, northern and central California: American Association of Petroleum Geologists Bulletin, v. 67, p. 1125–1142.

Ingersoll, R. V., 1990, Nomenclature of upper Mesozoic strata of the Sacramento Valley of California: review and recommendations, in Ingersoll, R. V., and Nilsen, T. H., eds., Sacramento Valley symposium and guidebook: Pacific Section, Society of Economic Paleontologists and Mineralogists, Book 65, p. 1–3.

Ingersoll, R. V., and Dickinson, W. R., 1981, Great Valley Group (sequence), Sacramento Valley, California, in Frizzell, V., ed., Upper Mesozoic Franciscan rocks and Great Valley sequence, central Coast ranges, California (Annual Meeting, Pacific Section SEPM field trips 1 and 4): Pacific Section, Society of Economic Paleontologists and Mineralogists, p. 1–33.

Ingersoll, R. V., and Nilsen, T. H., eds., 1990, Sacramento Valley symposium and guidebook: Pacific Section, Society of Economic Paleontologists and Mineralogists, Book 65, 215 p.

Ingersoll, R. V., and Schweickert, R. A., 1986, A plate-tectonic model for Late Jurassic ophiolite genesis, Nevadan orogeny and forearc initiation, northern California: Tectonics, v. 5, p. 901–912.

Ingersoll, R. V., Rich, E. I., and Dickinson, W. R., 1977, Great Valley sequence, Sacramento Valley: Cordilleran Section, Geological Society of America, Fieldtrip guidebook, 72 p.

Jayko, A. S., Blake, M. C., Jr., and Harms, T., 1987, Attenuation of the Coast Range ophiolite by extensional faulting, and nature of the Coast Range "thrust," California: Tectonics, v. 6, p. 475–488.

Krueger, S. W., and Jones, D. L., 1989, Extensional fault uplift of regional Franciscan blueschists due to subduction shallowing during the Laramide orogeny: Geology, v. 17, p. 1157–1159.

Linn, A. M., DePaolo, D. J., and Ingersoll, R. V., 1991, Nd-Sr isotopic provenance analysis of Upper Cretaceous Great Valley fore-arc sandstones: Geology, v. 19, p. 803–806.

Linn, A. M., DePaolo, D. J., and Ingersoll, R. V., 1992, Nd-Sr isotopic, geochemical, and petrographic stratigraphy and paleotectonic analysis: Mesozoic Great Valley forearc sedimentary rocks of California: Geological Society of America Bulletin, v. 104, p. 1264–1279.

Lowe, D. R., 1972, Implications of three submarine mass-movement deposits, Cretaceous, Sacramento Valley, California: Journal of Sedimentary Petrology, v. 42, p. 89–101.

Mansfield, C. F., 1971, Petrology and sedimentology of the late Mesozoic Great Valley sequence, near Coalinga, California [Ph.D. thesis]: Stanford, California, Stanford University, 71 p.

Mansfield, C. F., 1979, Upper Mesozoic subsea fan deposits in the southern Diablo Range, California: Record of the Sierra Nevada magmatic arc: Geological Society of America Bulletin, v. 90, part I, p. 1025–1046.

Moores, E., 1970, Ultramafics and orogeny, with models of the US Cordillera and the Tethys: Nature, v. 228, p. 837–842.

Moores, E. M., and Day, H. W., 1984, Overthrust model for the Sierra Nevada: Geology, v. 12, p. 416–419.

Moxon, I. W., 1988, Sequence stratigraphy of the Great Valley basin in the context of convergent margin tectonics, in Graham, S. A., and Olson, H. C., eds., Studies of the geology of the San Joaquin basin: Pacific Section, Society of Economic Paleontologists and Mineralogists, Book 60, p. 3–28.

Moxon, I. W., 1990, Stratigraphy and structure of Upper Jurassic–Lower Cretaceous strata, Sacramento Valley, in Ingersoll, R. V., and Nilsen, T. H., eds., Sacramento Valley symposium and guidebook: Pacific Section, Society of Economic Paleontologists and Mineralogists, Book 65, p. 5–29.

Moxon, I. W., and Graham, S. A., 1987, History and controls of subsidence in the Late Cretaceous–Tertiary Great Valley forearc basin, California: Geology, v. 15, p. 626–629.

Mutti, E., 1974, Examples of ancient deep-sea fan deposits from circum-Mediterranean geosynclines, in Dott, R. H., Jr., and Shaver R. H., eds., Modern and ancient geosynclinal sedimentation: Society of Economic Paleontologists and Mineralogists Special Publication 19, p. 92–105.

Mutti, E., and Ricci-Lucchi, F., 1972, Le torbiditi dell' Appennino settentrionale: introduzione all' analisi di facies: Memorie della Societa Geologica Italiana, v. 11, p. 161–199.

Namson, J. S., and Davis, T. L., 1988, Seismically active fold and thrust belt in the San Joaquin Valley, central California: Geological Society of America Bulletin, v. 100, p. 257–273.

Nelson, C. H., and Kulm, L. D., 1973, Submarine fans and deep-sea channels, in Turbidites and deep-water sedimentation, short course: Los Angeles, Pacific Section, Society of Economic Paleontologists and Mineralogists, p. 39–78.

Nelson, C. H., and Nilsen, T. H., 1974, Depositional trends of modern and ancient deep-sea fans in Dott, R. H., Jr., and Shaver, R. H., eds., Modern and ancient geosynclinal sedimentation: Society of Economic Paleontologists and Mineralogists Special Publication 19, p. 69–91.

Nilsen, T. H., 1990, Santonian, Campanian, and Maestrichtian depositional systems, Sacramento basin, California, in Ingersoll, R. V., and Nilsen, T. H., eds., Sacramento Valley symposium and guidebook: Pacific Section, Society of Economic Paleontologists and Mineralogists, Book 65, p. 95–132.

Normark, W. R., 1970, Growth patterns of deep-sea fans: American Association of Petroleum Geologists Bulletin, v. 54, p. 2170–2195.

Normark, W. R., 1974, Submarine canyons and fan valleys: Factors affecting growth patterns of deep-sea fans in Dott, R. H., Jr., and Shaver, R. H., eds., Modern and ancient geosynclinal sedimentation: Society of Economic Paleontologists and Mineralogists Special Publication 19, p. 56–68.

Ojakangas, R. W., 1964, Petrology and sedimentation of the Cretaceous Sacramento Valley sequence, Cache Creek [Ph.D. thesis]: Stanford, California, Stanford University, 190 p.

Ojakangas, R. W., 1968, Cretaceous sedimentation, Sacramento Valley, California: Geological Society of America Bulletin, v. 79, p. 973–1008.

Pessagno, E. A., Jr., 1976, Radiolarian zonation and startigraphy of the Upper Cretaceous portion of the Great Valley sequence, California Coast Ranges: Micropaleontology, Special Publication 2, 67 p.

Pessagno, E. A., Jr., 1977, Upper Jurassic radiolaria and radiolarian biostratigraphy of the California Coast Ranges: Micropaleontology, v. 23, p. 56–113.

Rich, E. I., 1971, Geologic map of the Wilbur Springs quadrangle, Colusa and Lake counties, California: U.S. Geological Survey Miscellaneous Geologic Investigations Map I-538, scale 1:48,000.

Schweickert, R. A., and Cowan, D. S., 1975, Early Mesozoic tectonic evolution of the western Sierra Nevada, California: Geological Society of America Bulletin, v. 86, p. 1329–1336.

Seiders, V. M., 1983, Correlation and provenance of upper Mesozoic chert-rich conglomerate of California: Geological Society of America Bulletin, v. 94, p. 875–888.

Seiders, V. M., 1989, Conglomerate clast compositions and sediment distribution in the upper Mesozoic Franciscan assemblage and Great Valley sequence, northern California, in Colburn, I. P., Abbott, P. L., and Minch, J., eds., 1989, Conglomerates in basin analysis: a symposium dedicated to A. O. Woodford: Pacific Section, Society of Economic Paleontologists and Mineralogists, Book 62, p. 161–168.

Seiders, V. M., and Blome, C. D., 1988, Implications of upper Mesozoic conglomerate for suspect terrane in western California and adjacent areas: Geological Society of America Bulletin, v. 100, p. 374–391.

Short, P. F., and Ingersoll, R. V., 1990, Petrofacies and provenance of the Great Valley Group, southern Klamath Mountains and northern Sacramento Valley, in Ingersoll, R. V., and Nilsen, T. H., eds., Sacramento Valley symposium and guidebook: Pacific Section, Society of Economic Paleontologists and Mineralogists, Book 65, p. 39–52.

Suchecki, R. K., 1984, Facies history of the Upper Jurassic–Lower Cretaceous Great Valley sequence: Response to structural development of an outer-arc basin: Journal of Sedimentary Petrology, v. 54, p. 170–191.

Swe, W., and Dickinson, W. R., 1970, Sedimentation and thrusting of late Mesozoic rocks in the Coast Ranges near Clear Lake, California: Geological Society of America Bulletin, v. 81, p. 165–188.

Unruh, J. R., and Moores, E. M., 1992, Quaternary blind thrusting in the southwestern Sacramento Valley, California: Tectonics, v. 11, p. 192–203.

Unruh, J. R., Loewen, B. A., and Moores, E. M., 1995, Progressive arcward contraction of a Mesozoic-Tertiary fore-arc basin, southwestern Sacramento Valley, California: Geological Society of America Bulletin, v. 107, p. 38–53.

Verosub, K. L., Haggart, J. W., and Ward, P. D., 1989, Magnetostratigraphy of Upper Cretaceous strata of the Sacramento Valley, California: Geological Society of America Bulletin, v. 101, p. 521–533.

Walker, R. G., 1970, Review of the geometry and facies organization of turbidites and turbidite-bearing basins, in Lajoie, J., ed., Flysch sedimentation of North America: Geological Association of Canada Special Paper 7, p. 219–251.

Walker, R. G., and Mutti, E., 1973, Turbidite facies and facies associations, in Turbidites and deep-water sedimentation, short course: Los Angeles, Pacific Section, Society of Economic Paleontologists and Mineralogists, p. 119–157.

Ward, P. D., Verosub, K. L., and Haggart, J. W., 1983, Marine magnetic anomaly 33-34 identified in the Upper Cretaceous of the Great Valley sequence of California: Geology, v. 11, p. 90–93.

Wentworth, C. M., and Zoback, M. D., 1989, The style of late Cenozoic deformation at the eastern front of the California Coast Ranges: Tectonics, v. 8, p. 237–246.

Wentworth, C. M., Blake, M. C., Jones, D. L., Walter, A. W., and Zoback, M. D., 1984, Tectonic wedging associated with emplacement of the Franciscan assemblage, California Coast Ranges, in Blake, M. C., Jr., ed., Franciscan geology of northern California: Pacific Section, Society of Economic Paleontologists and Mineralogists, Book 43, p. 163–173.

Williams, T. A., 1993, Current views on the deposition and sequence stratigraphy of the Great Valley Group, Sacramento basin, in Graham, S. A., and Lowe, D. R., eds., Advances in the sedimentary geology of the Great Valley Group, Sacramento Valley, California: Pacific Section, Society of Economic Paleontologists and Mineralogists, Book 73, p. 5–17.

Williams, T. A., 1997, Basin-fill architecture and forearc tectonics Cretaceous Great Valley Group, Sacramento basin, northern California [Ph.D. thesis]: Stanford, California, Stanford University, 418 p.

Williams, T. A., Graham, S. A., and Constenius, K. N., 1998, Recognition of a Santonian submarine canyon, Great Valley Group, Sacramento basin, California: Implications for petroleum exploration and sequence stratigraphy of deep-marine strata: American Association of Petroleum Geologists Bulletin, v. 82, p. 1575–1595.

MANUSCRIPT ACCEPTED BY THE SOCIETY ON NOVEMBER 23, 1998.

Printed in U.S.A.

Geological Society of America
Special Paper 338
1999

# Chapter 9
# THE SIERRA NEVADA: CENTRAL CALIFORNIA'S ARC

**Jason Saleeby**
*California Institute of Technology*
*Division of Geological and Planetary Sciences*
*Pasedena, California 91125*
*E-mail: jason@gps.caltech.edu*

## PAUL C. BATEMAN

Paul Bateman in 1953.

Paul C. Bateman is a leading field geologist and igneous petrologist, best known for his studies of batholiths, especially the Sierra Nevada batholith. In addition, he has been a pioneer in fostering relations with geologists from nations around the Pacific Rim. After receiving his bachelor of science degree from the University of California, Los Angeles (UCLA) in 1936, he worked first in commercial geophysics in California, New Mexico, Texas, and Kansas, then as a geologist and supervisor of exploration by Du Pont for ilmenite in the San Gabriel Mountains of southern California. In 1940, Bateman enrolled in the doctorate program at UCLA to study under James Gilluly, but he continued part-time work with Du Pont's ilmenite exploration. In 1942, Bateman joined the U.S. Geological Survey's (USGS) strategic minerals program. His investigation of tungsten deposits led to an interest in contact-metamorphic deposits. At the end of World War II, he began a study of the contact-metamorphic tungsten deposits in the eastern Sierra Nevada near Bishop, California, in the Owens Valley, just east of the Sierra Nevada. Thus developed his life-long interest in granites and batholiths. After completion of the tungsten study, he began an extensive years-long regional mapping program westward across the Sierra Nevada batholith.

Bateman served three two-year administrative stints in the USGS—Western District Supervisor for Mineral Deposits (1957–1958), Regional Geologist in Menlo Park (1958–1960), and Chief of the Field Geochemistry and Petrology Branch (1966–1968). In 1969, Bateman was one of six U. S. delegates at the organizational meeting of the International Geological Correlation Program (IGCP) in Budapest, Hungary. This experience led to his organizing IGCP Project 30, "Circum-Pacific Plutonism," which held annual meetings in different Pacific Rim nations between 1972 and 1981. More than 100 publications resulted from this project, including GSA Memoir 1519, *Circum-Pacific Plutonic Terranes* (edited by J. A. Roddick, 1983). Bateman is the author or co-author of more than 100 published works, culminating in 1992 with USGS Professional Paper 1483, entitled "Plutonism in the central part of the Sierra Nevada Batholith."

Eldridge M. Moores

Trees at Dorothy Lake. Sketch by Clyde Wahrhaftig

# CLYDE WAHRHAFTIG

Clyde Wahrhaftig was born and raised in Fresno, California, in 1919 and died in 1994. He earned a bachelor's degree in geology from the California Institute of Technology in 1941 and a doctorate in geology from Harvard University in 1953. He worked for the U.S. Geological Survey full time from 1941 until 1960, when he joined the Department of Geology and Geophysics at the University of California at Berkeley (UC). From then until his retirement in 1982 he worked part time for the U.S.G.S., returning to the survey full-time until his death.

Most of his work with the survey was in Alaska, and during his years at UC Berkeley Wahrhaftig worked in the Sierra Nevada. His studies there led to the paper for which he was awarded GSA's Kirk Bryan Award in 1965, "Stepped Topography of the Southern Sierra Nevada." In 1989 he was awarded the Distinguished Career Award of GSA's Quaternary Geology and Geomorphology division.

Wahrhaftig was a consummate field geologist, whose field work and maps are legendary. The Tower Peak quadrangle was meticulously mapped with plane table during summer vacations. He kept sketchbooks wherever he went, and some of his drawings illustrate this Centennial volume.

In addition to his work at the survey and UC Berkeley, Wahrhaftig was actively involved in public education in geology, preparing a large number of field guides to local areas of geological interest for students and the public. In the 1960s he became a strong supporter of bringing women and minorities into geology and was the first chairman of GSA's Committee on Minorities and Women in the Geosciences.

Doris Sloan

# GEOLOGY OF THE SIERRA NEVADA

By Paul C. Bateman and Clyde Wahrhaftig [*]
U.S. Geological Survey, Menlo Park, California;
U.S.G.S. and University of California, Berkeley

The Sierra Nevada is a strongly asymmetric mountain range with a long gentle western slope and a high and steep eastern escarpment. It is 50 to 80 miles wide, and it runs west of north through eastern California for more than 400 miles—from the Mojave Desert on the south to the Cascade Range and the Modoc Plateau on the north (pl. 1). Mount Whitney, in the southeastern part of the range, attains a height of 14,495 feet and is the highest point in the conterminous United States. The "High Sierra," a spectacular span of the crestal region, which extends north from Mount Whitney for about a hundred miles, is a glaciated region characterized by numerous lakes and a procession of 13,000- and 14,000-foot peaks.

The range is a tremendous physical barrier to the passage of moisture eastward from the Pacific. Polar front cyclones expand adiabatically as they pass over the Sierra Nevada during the winter and cool well below their dewpoint. Most of the moisture that was obtained during the passage of warm air masses across the Pacific is precipitated as snow, which is preserved as a heavy snow pack at high altitudes and in the shade of the forests at lower elevations until late spring or summer. On the east side of the Sierra Nevada the descending air is warmed adiabatically and can hold more moisture than it contains. Hence the arid valleys of the Great Basin are "lands of little rain."

But the Sierra Nevada is more than a physical and climatic barrier; until recently it has been a remarkably effective barrier to geologic thought. Its towering eastern escarpment has been a boundary for thinking about problems in the Great Basin; and geologists working in the Great Valley, the Coast Ranges, or even along the west slope of the Sierra Nevada itself, have seldom looked eastward for correlations. Even now, we are only on the threshold of understanding the tremendous role the Sierra Nevada has played in the geologic history of the West.

## GENERAL GEOLOGIC RELATIONS

The Sierra Nevada is a huge block of the earth's crust that has broken free on the east along the Sierra Nevada fault system and been tilted westward. It is overlapped on the west by sedimentary rocks of the Great Valley and on the north by volcanic sheets extending south from the Cascade Range. A blanket of volcanic material caps large areas in the north part of the range.

Most of the south half of the Sierra Nevada and the eastern part of the north half are composed of plutonic (chiefly granitic) rocks of Mesozoic age. These rocks constitute the Sierra Nevada batholith, which is part of a more or less continuous belt of plutonic rocks that extends from Baja California northward through the Peninsular Ranges and the Mojave Desert, through the Sierra Nevada at an acute angle to the long axis of the range, and into western Nevada; it may continue at depth beneath the volcanic rocks of the Snake River Plains and connect with the Idaho batholith.

In the north half of the range the batholith is flanked on the west by the western metamorphic belt, a terrane of strongly deformed, but weakly metamorphosed sedimentary and volcanic rocks of Paleozoic and Mesozoic age. The famed Mother Lode passes through the heart of this belt. Farther south, scattered remnants of metamorphic rock are found within the batholith, especially in the western foothills and along the crest in the east-central Sierra Nevada. The batholith extends eastward to the east edge of the range, but in the south half of the range one can look eastward across Owens Valley to the wall rocks on the east side of the batholith making up the White and Inyo Mountains.

The story of the Sierra Nevada is in four overlapping parts: (1) a long period in the Paleozoic when the area was mostly under the sea receiving sediments; (2) a shorter period during the Mesozoic when the Paleozoic strata were downwarped into a gigantic complexly faulted synclinorium which was filled with contemporaneous volcanic and sedimentary detritus, strongly deformed, intruded repeatedly by granitic masses, and eroded to a depth of at least 5 miles; (3) a period of relative stability in the early Cenozoic; and (4) a period of uplift, tilting, and faulting, preceded and accompanied by volcanic activity, in the late Cenozoic.

\* \* \* \* \*

---

[*] Publication authorized by the Director, U.S. Geological Survey. Bateman prepared the parts of the report that deal with the bedrock geology and Wahrhaftig the parts that deal with the Cenozoic geology.

Stubblefield Canyon, 1962 — Sketch by Clyde Wahrhaftig

### Structure of the "Framework" Rocks

The Paleozoic and Mesozoic strata of the Sierra Nevada have been complexly folded and faulted, and beds, cleavage, and lineations, including fold axes, are commonly steep or vertical (photo 1). A predominance of opposing, inward-facing top directions in the strata on the two sides of the range define a complexly faulted synclinorium. This synclinorium is not readily apparent in the patterns of geologic maps chiefly because strike faults of large displacement interrupt the sequence of strata in the western metamorphic belt. Apparently the axis of the synclinorium lies between miogeosynclinal Paleozoic strata on the east and eugeosynclinal strata on the west. The axial part of the synclinorium is occupied by the granitic rocks of the Sierra Nevada batholith. It trends N. 40° W. in the central Sierra Nevada, but probably bends to northward in the northern Sierra Nevada. The east side of the batholith follows approximately the east side of the volcanogenic and epiclastic Triassic and Jurassic strata. The eastern limit of the synclinorium is marked by a belt of Precambrian and Cambrian rocks that extends from the White Mountains southeastward into the Death Valley region and beyond. The western limit presumably lies beneath the Cretaceous and Tertiary strata of the Great Valley. An interesting speculation is that the sharply arcuate pattern of outcrops of Precambrian rock along the east and south sides of the Mojave Desert (pl. 1) may result from a southeastward continuation of the synclinorium.

From the belt of older rocks that extends between the White Mountains and Death Valley, the strata on the east side of the batholith are progressively younger westward. The range-front faults that bound Owens Valley and the east side of the central Sierra Nevada strike obliquely across the major structures in these Paleozoic and Mesozoic strata. The strata east of the White and Inyo Mountains, and in many remnants within the Sierra Nevada batholith, are strongly folded and faulted, causing repetitions of formations, but in the Mount Morrison and Ritter Range pendants of the eastern Sierra Nevada the gross structure is homoclinal, and bedding tops face west across more than 50,000 feet of vertical or steeply dipping strata ranging in age from Ordovician to Jurassic. Folds in the western part of the Ritter Range pendant may be related to the axial region of the synclinorium.

In the western metamorphic belt the gross distribution of strata resulting from the development of the synclinorium has been reversed by movement along steeply dipping fault zones of large displacement, and the Paleozoic strata lie between two belts of Mesozoic strata (see pl. 1). The internal structure of individual fault blocks is, in general, homoclinal, and most bedding tops face east; the dip of the beds is generally more than 60° eastward (Clark, 1964, p. 44). The homoclinal structure is interrupted in parts of the belt by both isoclinal and open folds, but the east limbs of anticlines commonly are longer than the west limbs, and the older strata in a fault block generally are exposed near its west side and the younger strata near its east side.

Unconformities have been recognized both within and between the Paleozoic and Mesozoic units, and they indicate repeated movement since middle Paleozoic time. The geometry of the structures of the "framework" rocks also indicates that the strata have been deformed either during several different episodes or during a single complex episode. Minor folds with steeply dipping axes are common and represent either refolding of earlier folds that were initially formed with subhorizontal axes or else folds that were formed in strata that had been previously so folded as to have steep dips. Some terranes contain two or more axial surfaces of systematically different orientation.

The Sierra Nevada lies within the Cordilleran mobile belt and its rocks reflect part of the deformation that has taken place there since mid-Paleozoic time. Probably the faulted synclinorium began to take form in Permian or Triassic time, and intermittent disturbances occurred through the Jurassic. The very severe disturbance that took place near the close of the Jurassic and caused the principal folds in the Upper Jurassic strata of the western metamorphic belt is referred to as the Nevadan orogeny, but both earlier and later disturbances are known to have occurred. Unconformities in the Taylorsville region indicate disturbances between the Silurian and Mississippian, and at the end of the Permian, Triassic, and Jurassic (McMath, this bulletin); and, in the eastern and central parts of the range, folds of two periods of deforma-

Photo 2. Contact between porphyritic and equigranular quartz monzonite.

tion antedate a third set that appears to have been formed during the Late Jurassic Nevadan orogeny. Kistler (in press) believes the earlier deformations occurred during the Late Permian and in the Early or Middle Triassic. In the western metamorphic belt, Clark (1964, p. 44) has recognized a stage of deformation that occurred after the principal folding of the Upper Jurassic strata. This deformation is characterized by the development of slip cleavage, steeply dipping minor folds, and steeply dipping lineations. Clark believes the large faults in the western metamorphic belt formed during this deformation, probably by strike-slip movement. The presence of ultramafic rocks, especially serpentine, along these faults suggests deep penetration, possibly penetration into the upper mantle.

The major deformation of the stratified rocks took place in parts of the range before the emplacement of the adjacent plutonic rocks, but locally plutons have been affected by regional deformation, indicating an overlap of the period of regional deformation with that of magma emplacement. For example, in the Goddard roof pendant of the east-central Sierra Nevada (fig. 1), a second deformation of the metamorphic rocks has affected plutons that were intruded after the first deformation. Farther west the eastern margin of the granodiorite of "Dinkey Creek" type, one of the largest plutons in the western Sierra Nevada, was intensely sheared and lineated before the emplacement of the Mount Givens Granodiorite which is the largest pluton in the central part of the range. Some of the plutons within the western metamorphic belt also were sheared during the second deformation.

The strata in the western metamorphic belt, except where adjacent to the batholith or smaller intrusive bodies, exhibit greenschist facies regional metamorphism. East of the batholith, in the White and Inyo Mountains, the only evidence of regional metamorphism is the presence of slaty cleavage in some pelitic and calcareous rocks. The strata adjacent to intrusive bodies and in remnants within the batholith are chiefly in the hornblende hornfels facies of contact metamorphism, although in the inner aureoles of some plutons that were intruded at unusually high temperatures the minerals sillimanite and brucite indicate the next higher pyroxene hornfels facies.

### THE BATHOLITH

The batholith has been studied most intensively across the central part, between the 37th and 38th parallels of latitude (fig. 1), and much of the following discussion pertains specifically to that area (Bateman and others, 1963). However, the rest of the batholith appears to be very similar. The batholith is composed chiefly of quartz-bearing granitic rocks ranging in composition from quartz diorite to alaskite, but includes scattered smaller masses of darker and older plutonic rocks and remnants of metamorphosed sedimentary and volcanic rocks. Rocks in the compositional range of quartz monzonite and granodiorite predominate and are about equally abundant.

#### Mafic Rocks

The oldest and most mafic rocks of the batholith are small bodies of diorite, quartz diorite, and hornblende gabbro, which have been aptly called by Mayo (1941, p. 1010) "basic forerunners" or simply "forerunners." Their distribution is reminiscent of the metamorphic rocks, for they occur as inclusions or small roof pendants within individual plutons of more silicic rock or as septa between plutons. Commonly they are associated with metamorphic rocks, and many are crowded with metamorphic inclusions. This intimate association with the metamorphic rocks probably results from the mafic plutonic rocks being the first to be emplaced, and consequently coming in contact with the metamorphic rocks on all sides. The original sizes and shapes of most masses were destroyed by later granitic intrusives, which tore them apart and recrystallized, granitized, and assimilated their fragments. Partly as a result of original differences, and partly because of subsequent modification, the mafic plutonic rocks are heterogeneous in composition and texture. Very likely they include rocks of diverse origin, some having been mafic volcanic or calcareous sedimentary rocks. Many bodies of dark granodiorite and some bodies of quartz diorite may be hybrids of more silicic granitic rocks and diorite, hornblende gabbro, or amphibolite. The fabric of these suspected hybrid rocks is generally highly irregular; some rocks are very coarse grained and in places contain poikilitic hornblende crystals an inch or more long.

### Larger Features of the Granitic Rocks

The more leucocratic granitic rocks, which make up the bulk of the batholith, are in discrete masses or plutons, which are in sharp contact with one another or are separated by thin septa of metamorphic or mafic igneous rock. Individual plutons vary greatly in size; their outcrop area ranges from less than a square mile to more than 500 sq mi. The limits of many of the large plutons have not yet been delineated. On the whole the batholith appears to consist of a few large plutons and a great many smaller ones, which are grouped between the large plutons (fig. 1). All of the large plutons, and some of the smaller ones, are elongate in a northwesterly direction, parallel with the long direction of the batholith, but many of the small plutons are elongate in other directions or are rounded or irregularly shaped.

The major plutons in the western part of the batholith are generally older than those along the crest of the range, and in the Yosemite region Calkins (1930) has mapped two series of granitic formations in which the plutons are successively younger toward the east. Nevertheless, the pattern of intrusion is more complicated than a simple west-to-east sequence of emplacement.

\* \* \* \* \*

Bateman and others (1963, p. D38) have listed ten relations, which taken together indicate that the major part of the granitic rocks crystallized from melts. These include such relations as sharp contacts of plutons with wall rocks and with one another, dikes and inclusions along contacts between plutons from which the relative ages of the plutons can be determined with consistent results, finer grain size in apophyses and in the margins of some plutons, wall-rock geometry that suggests dislocation by the emplacement of plutons, and dilated walls of aschistic dikes. The plutons are pictured as having moved upward from a deeper source region, much like salt domes, because molten granitic magma has a significantly lower specific gravity than rock of the same composition. As the plutons rose, the country rock is believed to have settled downward around the magma, thus providing room for its continued upward migration. Conceivably, some plutons may have become entirely detached from their source region and be underlain by country rock, but no field evidence for downward bottoming of plutons has been found.

The relative importance of several processes in the emplacement of the plutons has been only incompletely evaluated. Wall-rock deformation indicates that rising magma squeezed the wall and roof rocks aside and upward. Stoping appears to have been important locally, but there is little evidence in support of stoping as the principal mechanism of emplacement. Processes of granitization and assimilation have operated on a small scale where the wall and roof rocks were amphibolites or other mafic rock, but these processes are of

Photo 3. Intrusive breccia. A (left)—Fragmentation of metamorphic rock by granitic magma. B (right)—Fragmentation of early quartz diorite by later quartz monzonite.

Photo 4. Felsic dikes in hornblende gabbro. Parent to dikes is quartz monzonite in foreground.

possible quantitative importance only in terranes of mafic volcanic rocks. Melting and assimilation of pelitic rocks has not been proved, but very likely has taken place and may have been of considerable importance.

Broad chemical and mineralogic changes take place across the batholith. In general, the granitic rocks are more mafic toward the west and more silicic toward the east, but this is a gross trend, and some silicic plutons occur in the western half of the range and some mafic plutons are within the eastern half. The simple explanation that the more felsic rocks in the east half of the range are differentiates of the more mafic rocks in the west half does not hold for several reasons: (1) Lengthy time gaps probably exist between the different age groups. (2) Large granodiorite and quartz diorite plutons in the west half of the range are accompanied by younger and more felsic plutons, which are older than the large granodiorite and quartz monzonite plutons in the east side of the range. (3) The limited analytic data now available indicate systematically lower $K_2O/Na_2O$ ratios in the Jurassic granitic rocks of the western part of the range than in either the Cretaceous or the Early Jurassic or Late Triassic rocks farther east.

Photo 5. An exceptional abundance of K-feldspar phenocrysts in the marginal part of a pluton of prophyritic quartz monzonite.

## SPECULATIONS ON THE ORIGIN OF THE BATHOLITH

The localization of the batholith in the axial region of a synclinorium of great size and depth compels serious consideration of the hypothesis that the granitic magmas were generated by the melting of sialic rocks of the upper crust as a result of their being depressed into deeper regions of high temperature. Experimental studies show that at atmospheric pressure sialic rocks begin to melt fractionally at 960°C, but that an increase in water-vapor pressure causes the melting temperature to drop spectacularly (Tuttle and Bowen, 1958). At pressures above 1 kilobar and in the presence of enough water to saturate any amount of magma that may be generated, melting begins at temperatures between 600° and 700°C, the temperature varying inversely with the pressure. If certain other substances, such as fluorine or chlorine, are present, the temperature at which melting begins is lowered further.

The first melt is composed chiefly of normative quartz, orthoclase, and albite, in proportions that are different at different pressures of water vapor; this melt probably is saturated or nearly saturated with water. To form more calcic magma more of the source rock must be melted, which requires higher temperatures. Early emplaced plutons of quartz diorite or granodiorite, which represent the least differentiated magmas, commonly contain zoned plagioclase crystals in which the composition of the most calcic zone is about $An_{50}$. Plagioclase of $An_{50}$ composition crystallizes from melt with plagioclase of $An_{15}$ composition

The amount of water soluble in magma is quite high (about 17 percent by weight at 10 kilobars); consequently undifferentiated parent magmas are probably undersaturated. Magma containing about 2 percent water and which can crystallize plagioclase of $An_{50}$ composition can form by fractional melting at a temperature of about 900°C.

The depth at which sialic rocks can be expected to melt to granitic magma is difficult to evaluate because our present knowledge of the generation and distribution of heat in the earth is still in a primitive state. In stable parts of the crust where all the heat is carried to the surface by conduction, temperatures of 600° to 700°C may be attained at depths of 30 to 50 km, but a temperature of 900°C may not be reached at depths twice as great. However, the situation in a downfolding synclinorium is not ordinary because the crustal rock is greatly thickened in the downfold. Sialic rocks generally produce more heat than basaltic rocks, and the heat production in a downfolded sialic layer thickened two or three times may be of sufficient magnitude to cause melting to occur in the lower part of the downfold (A. H. Lachenbruch, oral communication, 1965). Measurements of the heat being produced in the exposed framework rocks by radioactivity would make possible some limiting calculations, but have not yet been made. Another possible source for heat is by upward movement of hotter material from

Forsythe Peak, 1964

Sketch by Clyde Wahrhaftig

deeper levels either by intrusion or by convection within the mantle. Convection currents in the mantle have been postulated by Vening Meinesz (1934), Griggs (1939), and many others in connection with the formation of deep geosynclinal troughs.

The model that emerges from these considerations is a magma zone that begins at a depth of 30 to 50 km and extends downward to the lowest level of depressed sial. At the top of the zone of melting, the ratio of hydrous felsic magma to solid rock is probably very small. If no migration of material occurs within the zone of melting, the ratio of magma to rock increases downward, and the melt is progressively less hydrous and richer in calcium, iron, and magnesium. Very likely, however, this model would be modified by diffusion of water from the hydrous upper part downward into the less hydrous lower part, and possibly also by convective circulation in the more completely fluid lower part, which would cause homogenization of the magma.

In terms of a developing synclinorium, the first magma would form when the lowest sialic rocks were depressed into the zone of melting. Further depression of these same rocks would cause additional melting to form a magma richer in calcium, iron, and magnesium, and poorer in water. At the same time, new magma would form at the top of the melting zone in higher rocks. Because the density of granitic magma is significantly less than granitic rock of the same composition, the magma would tend to work upward in the manner of a salt intrusion, exploiting lines of structural weakness wherever possible. Some magma would perhaps break through to the surface in volcanic eruptions, but much of it would crystallize at depth as plutons—which is what happened. Closely related sequences of granitic rocks that are successively more felsic with age and concentrically zoned plutons that are progressively more felsic toward the center indicate magmatic differentiation during emplacement from parent magma of granodiorite or quartz diorite composition. However, some compositional differences in genetically related sequences, especially among the volcanic rocks, may reflect differences in parent magmas formed at different depths in the zone of melting.

In the Sierra Nevada, the rocks that were melted doubtless included upper Precambrian, Paleozoic, and Mesozoic strata, but a major source of material probably was the underlying Precambrian. In the Death Valley region the Precambrian terrane is composed of quartzose and micaceous schist, amphibolite, and granitic gneiss. In addition to these sialic rocks, femic material from the lower crust and perhaps also the mantle must have been incorporated in the parent granitic magmas to supply the required calcium, iron, and magnesium, although upper Precambrian and Paleozoic carbonate formations like those exposed in the White and Inyo Mountains and in roof pendants of the eastern Sierra Nevada doubtless supplied some calcium and magnesium. Ultramafic and mafic intrusions in the western metamorphic belt, and basalt flows and hypabyssal intrusives, indicate that mafic material was introduced into the upper crust, and some of this mafic material must have been depressed with the enclosing rocks into the zone of melting; mixing of sialic and femic material could also have occurred at the base of the sial, in the zone of melting. Hurley and others (1963, p. 172) have determined the initial $Sr^{87}/Sr^{86}$ ratio of the granitic rocks of the central Sierra Nevada to have been in the range of $0.7073 \pm 0.0010$; they deduced from this ratio that if the granitic magmas originated from a mixture of sialic and basaltic materials the ratio was one-third basalt and two-thirds sial.

The fact that the plutonic rocks in the east side of the range are both younger and more felsic than those in the west side seems compatible with differentiation processes within a single parent magma. However, Moore (1959, p. 206) has inferred from systematic difference in $K_2O/Na_2O$ ratios that the rocks on the two sides of his quartz diorite line, which runs through the Sierra Nevada parallel with the long axis of the range, are derived from different parent magmas and are not differentiates of a single parent magma. This interpretation is supported by isotopic dates that suggest gaps of many millions of years between the time of emplacement of some of the largest plutons on the two sides of the range (Kistler and others, 1965). Furthermore, the Mesozoic volcanic rocks also are more felsic toward the east even though some, and perhaps most, of the eastern volcanic rocks are older than the western ones.

Moore (1959, p. 206) suggested that his quartz diorite line "is probably parallel to and not far distant from the edge of the granitic (sialic) layer in the continental crust," and that "granitic rocks emplaced east of the line are generated within a thick sialic layer, whereas those emplaced west of the line are developed within the sima or a thinner sialic layer with great thickness of associated geosynclinal sediments and volcanic rocks." Westward thinning of the sialic layer is in accord with interpretations of Bailey and others (1964, p. 7, p. 143, fig. 35 on p. 164) that the Franciscan Formation was deposited on an oceanic rather than a continental crust.

A reasonable interpretation of the compositional, geographic, and temporal relations of the granitic rocks is that magma was generated several times, each time in a somewhat different area as a result of shift in the locus of downfolding (fig. 2). The isotopic ages indicate three epochs of magmatism but do not preclude other epochs. The oldest epoch is represented by granitic rocks that lie along the east side of the Sierra Nevada and have isotopic ages of $200 \pm 20$ m.y. (Late Triassic or Early Jurassic). They are chiefly quartz monzonite and granodiorite and are not as mafic as the next younger granitic rocks of the western foothills, which are chiefly granodiorite and quartz diorite and have isotopic ages of 130 ± 10 m.y. (Late Jurassic). The youngest plutonic rocks are those along the crest of the range, which have isotopic ages of 80 to 90 m.y. (early Late Cretaceous), and are more felsic than the rocks to the west. These young rocks also may be more felsic than the Late Triassic or Early Jurassic rocks to the east, but before meaningful comparison can be made of these two groups more analytic data than are now available will be required, and the age assignments of plutons will have to be made more precise.

According to Eaton and Healy (1963) the crust beneath the high central Sierra Nevada is at least 50 kilometers thick, about twice as thick as the crust west of the range and 15 to 20 kilometers thicker than the crust beneath the Basin and Range province east of the range. Not only is this geophysical root centered beneath the highest peaks, at least in the central Sierra Nevada, but it also lies beneath the belt of Cretaceous plutons. These coincidences suggest that the root was formed by downfolding of the crust at the time of the early Late Cretaceous magmatism, and that uplift of the Sierra Nevada during the Late Cretaceous and Cenozoic was caused by a mass deficiency of the root. Negative isostatic anomalies (Oliver, 1960) and analysis of first-order leveling by the U.S. Coast and Geodetic Survey by Oliver (written communication, 1965) that indicates the central Sierran crest has risen during a 60-year period at the rate of about a centimeter per year support the concept of a net mass deficiency extending from the surface downward to some depth of compensation. A serious difficulty with this hypothesis is that an expectable consequence is continuous uplift, though at a steadily diminishing rate, until the root has disappeared. The geologic record indicates, however, that little or no uplift took place during an extended period of 25 to 30 million years that included much of Eocene and Oligocene time. Nevertheless, the Sierra Nevada has certainly been uplifted and deeply eroded, and a better explanation for the uplift than mass deficiency has not been devised.

#### DEPTH OF EROSION SINCE MESOZOIC PLUTONISM

In 1963 Bateman and others published a speculation that 11 miles (about 17 km) of rock has been stripped off the Sierra Nevada since the batholith was emplaced. To arrive at this speculation they inferred that the water-vapor pressure ($P_{H_2O}$) in the most highly fractionated magmas and in the thermally metamorphosed wall rocks was about 5 kilobars, and assumed that $P_{H_2O} = P_{load}$ (pressure equivalent to the weight of the overlying rocks). Since then, Putman and Alfors (1965) have inferred that the $P_{H_2O}$ during crystallization of the Late Jurassic Rocky Hill stock in the western foothills was about 1.5 kilobars, and Sylvester and Nelson (1965) have inferred that the $P_{H_2O}$ during crystallization of the Cretaceous Birch Creek pluton in the White Mountains was 2 to 3 kilobars. In both reports $P_{H_2O}$ was considered to equal $P_{load}$.

Since Bateman and others' speculations were published much additional information has become available, some of which casts serious doubt on the validity of the assumption that $P_{H_2O} = P_{load}$. In view of this doubt, arguments that require making the assumption that $P_{H_2O} = P_{load}$ are excluded from the following discussion.

Unfortunately, there is no simple way to measure the amount of rock that has been stripped away, and the problem can be attacked only indirectly. In the following discussion, two approaches are pursued: One is an evaluation of the volume of sediment of latest Jurassic, Cretaceous, and Cenozoic age that had its source in the Sierra Nevada, and the other is consideration of the significance in terms of load pressure of andalusite and sillimanite in thermally metamorphosed wall and roof rocks.

\* \* \* \* \*

Thus both lines of evidence followed here suggest erosion from the Sierra Nevada of the same order of magnitude as was suggested, perhaps fortuitously, by Bateman and others in 1963. The time span of this amount of erosion is the time since the Nevadan orogeny and intrusion of the Late Jurassic granitic rocks. The exposed early Late Cretaceous granitic rocks must have consolidated under less cover than the exposed Late Jurassic granitic rocks, and it may not be accidental that all occurrences of sillimanite thus far reported, with the possible exception of the sillimanite reported by Rose (1957), are along contacts with the older granitic rocks.

# On some aspects of the geology of the Sierra Nevada

**Jason B. Saleeby**
*California Institute of Technology, Division of Geological and Planetary Sciences, Pasadena, California 91125*

## INTRODUCTION

The Sierra Nevada is recognized globally for its spectacular exposures of a Phanerozoic composite batholithic belt. In their "classic paper," "The Geology of the Sierra Nevada," Bateman and Wahrhaftig (1966) focused a great deal on the nature and petrogenesis of the batholith, but they also related the batholith to the structure of its metamorphic framework rocks, as well as to the modern physiography of the range. This classic work was the first effort to synthesize such a broad range of topics on Sierra Nevada geology. Paul Bateman was then spearheading a major effort by the U.S. Geological Survey (USGS) to map systematically a large swath across the central part of the Sierra. Even though Bateman's emphasis was on the batholith, he made major contributions in the exploration of the metamorphic framework for the batholith and also vigorously pursued geomorphic aspects of the range. Clyde Wahrhaftig was part of the USGS team that systematically worked across the central part of the range. He was a major driving force in ideas concerning geomorphology of the range, and he also made significant contributions in the study of batholithic and framework rocks. My initial research in the Sierra Nevada focused primarily on the framework rocks, but over the past decade has shifted to issues in broad batholithic structure and petrogenesis, as well as the range's physiography. For this reason, Bateman and Wahrhaftig (1966) is of great interest to me in respect to identifying classic works in Cordilleran geology.

In the past three decades, a tremendous amount of research has been done on the batholith and its metamorphic framework, and to a lesser extent on the physiographic evolution of the range. The gross structure of the metamorphic framework, the structure and petrogenesis of the batholith, and relations between the batholith and modern physiography of the range are the topics that will be covered in this paper. We begin with a general discussion of the framework rocks.

## STRUCTURE OF THE FRAMEWORK ROCKS

Metamorphic framework rocks of the Sierra Nevada batholith occur in three regional structural domains (Fig. 1). The first domain consists of the eastern wall, which is primarily to the east of the modern Sierra Nevada range. Strata of this domain may be characterized as being the southwest limit of the Cordilleran miogeocline (Moore and Foster, 1980), or as having depositional and/or structural basement of the miogeocline and its Proterozoic crystalline underpinnings. Intruded strata above the miogeocline include Paleozoic–early Mesozoic allochthons related to the Antler and Sonoma orogenies, and Mesozoic overlap and arc-related deposits. The second domain consists of the contiguous western Foothills metamorphic belt. These rocks consist of Eocambrian to Late Jurassic ensimatic assemblages that were tectonically accreted to the North American plate edge in Paleozoic to Late Jurassic time. The third domain consists of pendants engulfed in the composite batholith. These were derived primarily from rocks of the other two domains.

The composite batholith is about 90% Cretaceous in age. The remaining 10% consists of Triassic and Jurassic plutons. Regional patterns in plutonism changed dramatically at the end of the Jurassic. Thus, for our purposes, we will consider the older plutons as part of the metamorphic framework. Herein, "batholith" refers to the Cretaceous composite batholith.

### Comments on "Structure of the 'Framework' Rocks"

The "predominance of opposing, inward-facing top directions in the strata of the two sides of the range" (Bateman and Wahrhaftig, 1966, p. 115) was the main observational basis for the concept of a regional synclinorium having characterized the prebatholithic framework. Implicit in the original description of the synclinorium was the fact that conventional stratigraphic

---

Saleeby, J. B., 1999, On some aspects of the geology of the Sierra Nevada, *in* Moores, E. M., Sloan, D., and Stout, D. L., eds., Classic Cordilleran Concepts: A View from California: Boulder, Colorado, Geological Society of America Special Paper 338.

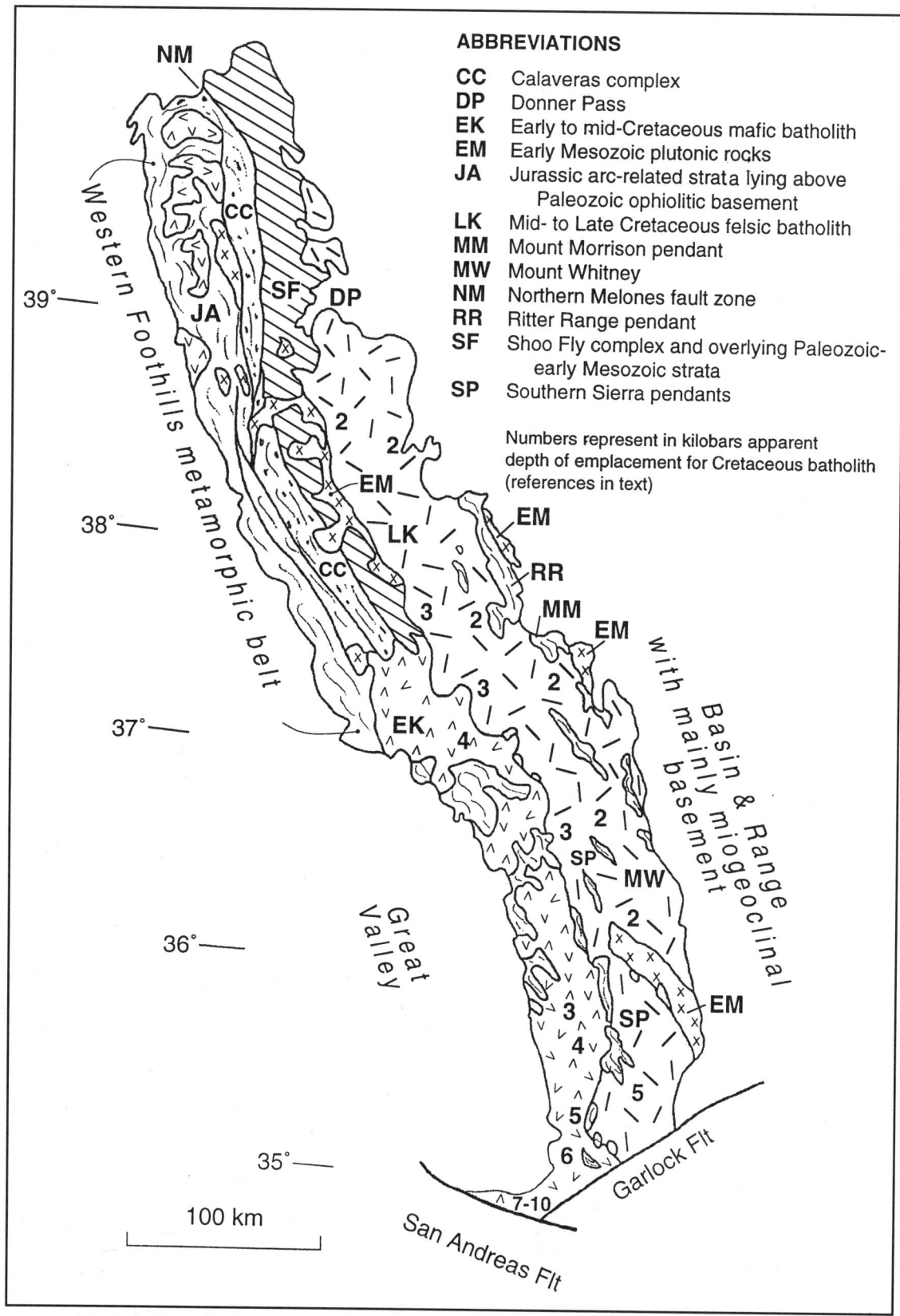

Figure 1. Generalized map showing major batholithic domains and prebatholithic framework rocks, and localities referred to in text.

correlations could not be made from opposing limb domains: "the axis of the synclinorium lies between miogeosynclinal Paleozoic strata on the east and eugeosynclinal strata on the west" (Bateman and Wahrhaftig, 1966, p. 115).

Subsequent work has brought the synclinorium concept into question. Bateman et al. (1985) presented data indicating that the Calaveras Complex of the western "synclinorium" limb (Fig. 1) is progressively younger westward. Saleeby et al. (1986) attributed this younging pattern to an accretionary prism origin for the Calaveras; this younging pattern continues to the western edge of the Foothills belt, where the youngest (Late Jurassic) prebatholithic framework rocks occur. Furthermore, Merguerian and Schweickert (1987) and Sharp (1988) demonstrated that the oldest (lower Paleozoic) eugeoclinal rocks of the western Sierra region, the Shoo Fly Complex, occur east of the Calaveras Complex. This younging sense in the west is precisely opposite to what is required for the synclinorium concept. This pattern can be explained by ensimatic terrane-accretion processes that are discussed by Moores (this volume).

The synclinorium concept is further questioned by a number of lines of evidence suggesting that major prebatholithic structural breaks extend along the axis of the batholith. Hamilton and Myers (1966) suggested that the southwest-trending Cordilleran miogeocline was truncated and displaced in late Paleozoic time along a northwest trend within or adjacent to the Sierran metamorphic framework, whereupon a new active margin-batholithic belt formed. The accretion locus of the western Sierra eugeoclinal terranes appears to have been controlled by the truncation zone. Kistler (1990, 1993), using mainly initial Sr isotope data on the batholith, discussed possible sinistral and dextral offset phases along the truncation locus. Lahren et al. (1990) proposed stratigraphic correlations and piercing point relations that suggest a 400–500 km phase of dextral offset in Early to mid-Cretaceous time. Saleeby (1992) and Saleeby et al. (1992) further suggested that the Sierra Nevada metamorphic framework has undergone multiple phases of sinistral and dextral translations and/or truncations. The net result is that the opposing limb domains of the hypothetical synclinorium cannot be directly related to one another in a paleogeographic sense.

## Local structural patterns

We turn attention now to some of the more local structural patterns noted by Bateman and Wahrhaftig (1966) for the metamorphic framework rocks.

"Steeply-plunging to vertical fold axes that are so common" (Bateman and Wahrhaftig, 1966, p. 115) have been shown to have a number of different ages and origins. In the western Foothills, multiple generations of structures with folds of such orientation are interpreted as having developed in response to strike-slip displacements, oblique convergence, and downdip transposition along thrust faults (Saleeby, 1978; Schweickert et al., 1984; Saleeby et al., 1986; Wolf and Saleeby, 1995). These structures developed in at least three episodes, including mid-Paleozoic, Early to Middle Jurassic, and Late Jurassic. In pendants of the axial central to southern Sierra and of the southwestern Foothills, steeply plunging fold axes are parallel to high finite strain stretching lineations that developed in conjunction with batholith emplacement and related high-grade metamorphism (Saleeby, 1990; Saleeby and Busby, 1993). We return to this Cretaceous deformation pattern in our discussion of the batholith.

Sequences referred to in Bateman and Wahrhaftig (1966) as "homoclinal" in the western Foothills and in the Mount Morrison–Ritter Range pendants have subsequently been shown to have complex internal structures. In the eastern Sierra, the apparent homoclinal sequences of the Ritter Range and neighboring pendants are now shown to be internally imbricated by fault systems that are currently subparallel to bedding (Tobisch et al., 1986, 1999; Schweickert and Lahren, 1987). Rigid rotation and transposition during batholith-related strain of thrust-fault and possibly normal fault systems have been suggested by these studies. The rotation and strain have obscured the structurally imbricated relations and have rendered them homoclinal in appearance. In the western Foothills, the origin of a number of large ultramafic rock bodies is critical to the question of apparent regional-scale homoclinal sections. "Deep penetration...possibly...into the upper mantle" (Bateman and Wahrhaftig, 1966, p. 116) was the commonly held view for introduction of ultramafic belts into crustal level sequences. Subsequent work by Saleeby (1979, 1982) showed that a number of the large ultramafic bodies represent deformed and eroded remnants of a regional ophiolite belt that formed depositional basement for Jurassic sequences of the western Foothills. These basement rocks are exposed in core areas of faulted regional anticlines that are structurally packaged into apparent homoclinal sequences. The large ultramafic belt exposed along the northern Melones fault zone (Fig. 1) is a correlative fault slice of this regional basement terrane (Saleeby et al., 1989). The western Foothills may be more correctly interpreted as a tightly appressed fold-thrust belt.

Bateman and Wahrhaftig (1966) recognized "several different deformation episodes" in the framework rocks. Subsequent studies have refined this view considerably. Small remnants of the Roberts Mountains and Golconda allochthons are recognized in pendants of the east-central Sierra (Schweickert and Lahren, 1987), indicating that the mid-Paleozoic and Permian-Triassic Antler and Sonoma orogenies affected the eastern framework domain of the Sierra Nevada. A major unconformity between polydeformed Shoo Fly Complex and mid-Paleozoic to lower Mesozoic strata of the northern Sierra (McMath, 1966), as well as structural relations within the Shoo Fly Complex (Merguerian and Schweickert, 1987; Sharp, 1988), record major early to mid-Paleozoic deformation(s) in the western framework domain. Mid-Paleozoic deformation within the western Foothills ophiolite belt was resolved by Saleeby (1978, 1979, 1982) and attributed to both intraoceanic tectonism and continent-edge emplacement during the aforementioned truncation event. Saleeby et al. (1992) suggested that the initial truncation event was Sonoman in age and involved sinistral transform tectonics through the Sierran framework in conjunction with the southeast emplacement of the Golconda allochthon into the eastern framework domain. Post-truncation early Mesozoic pre-batholithic deformations are numerous and may be interpreted as semicontinuous or resulting from at least three distinct pulses between Late Triassic and Late Jurassic time. The literature is nearly as complex as the geology; intraplate contractile deformations, collisional or terrane-accretion deformations, regional extension, and dextral and sinistral lateral translations are evoked (Saleeby and Sharp, 1978; Nokleberg and Kistler, 1980; Saleeby, 1982; Schweickert et al., 1984; Saleeby et al., 1986; Tobisch et al., 1986, 1989, 1999; Schweickert and Lahren, 1987; Sharp, 1988; Saleeby and Busby, 1992; Wolf and Saleeby, 1995). Each of these deformational regimes is related to a complex convergent or oblique convergent margin having formed along the Sierran metamorphic framework in early Mesozoic time. The literature is packed with highly charged discussions evoking a wide spectrum of plate tectonic scenarios.

Bateman and Wahrhaftig (1966) recognized overlapping periods of deformation and plutonism. This is most clearly the case for Triassic and Jurassic plutons herein treated as part of the metamorphic framework. As discussed in the following, considerable regional deformation occurred during Cretaceous batholith emplacement as well.

## THE BATHOLITH

Prior to Bateman and Wahrhaftig (1966), the Sierra Nevada batholith was understood in only a very broad context as being a composite of many plutons having great variations in length scales. The batholith was at that time also recognized to have considerable compositional variation and broad geographic patterns in compositional variation (Moore, 1959). Extremes in compositional variation ranging from ultramafic to Alaskitic plutons were particularly well documented in a classic field study of the Sierra Nevada batholith by Moore (1963). From my perspective, such compositional variations, particularly viewed in the context of framework structure and tectonics, place important geological constraints on the petrogenesis of the batholith.

### Comments on "The Batholith"

A tremendous amount of field mapping and geochronological work has been done subsequent to the classic description of the batholith (Bateman and Wahrhaftig, 1966, p. 115–119). Major age patterns and the map distribution of the principal batholithic units have been filled in for much of the range south of lat 38°N (Evernden and Kistler, 1970; Bateman and Nokleberg, 1978; Bateman and Chappell, 1979; Saleeby and Sharp, 1980; Stern et al., 1981; Chen and Moore, 1982; Ross, 1990; Saleeby et al., 1987; Moore and Sisson, 1987). These and other works have shown Bateman and Wahrhaftig's (1966) discussion to be accurate to a first order. One of the spectacular features resolved by these later studies is the presence of a number of very large individual batholith-scale plutons, a number of which are more than 1000 km² in outcrop area.

An important point that was not stressed in the Bateman and Wahrhaftig (1966) description of the batholith, but that was treated in the petrogenesis section, was the recognition by Moore (1959) of a general compositional gradient across the batholith with more mafic "quartz dioritic" (tonalitic) plutons to the west and more granitic plutons to the east. This pattern was further substantiated by regional bulk chemical data (Bateman and Dodge, 1970). These important early observations have, to a first order, withstood the test of time and, as discussed in the following, have been corroborated by radiogenic isotopic studies.

Geochronological studies that are cited here have likewise defined a transverse gradient in emplacement ages for the Cretaceous batholithic plutons. Voluminous plutonism began in the western, more mafic phases of the batholith ca. 125 Ma. There was a general eastward migration of plutonism into Late Cretaceous time, and a number of larger culminating plutons were emplaced ca. 85 Ma in what was subsequently to become the high eastern Sierra region.

## SPECULATIONS ON THE ORIGIN OF THE BATHOLITH

The Bateman and Wahrhaftig (1966) discussion on the petrogenesis of the batholith also leads into the topics of modern crustal structure and the depth of erosion for the batholith. These are areas for which there has been substantial new work.

### Comments on batholith petrogenesis

Bateman and Wahrhaftig (1966) envisaged generation of batholithic magmas from water-saturated melting of sialic material in the "downfolding synclinorium" at a depth of 30 to 50 km. They further recognized that "mixing of sialic and femic material could have also occurred at the base of the sial, in the zone of melting" (Bateman and Wahrhaftig, 1966, p. 123). A "possible source for heat...by upward movement of hotter material from deeper levels by intrusion or convection" (Bateman and Wahrhaftig, 1966, p. 122–123) was also recognized. Their classic model for batholith genesis and subsequent uplift is presented graphically with the original artwork in Figure 2.

The focus of Bateman and Wahrhaftig's (1966) discussion on the genesis of the batholith resulting from the downfolding of the regional synclinorium clearly needs modification in the light of the subsequent work discussed here. Their discussion predated the broad application of plate tectonics theory to the origin of continental margin orogenic zones. The first attempt to apply the emerging plate tectonics paradigm to the Sierra Nevada batholith was Hamilton's (1969) classic paper, "Mesozoic California and the underflow of Pacific mantle." This paper, more than any other individual paper, changed our thoughts on the relationships between tectonics and California

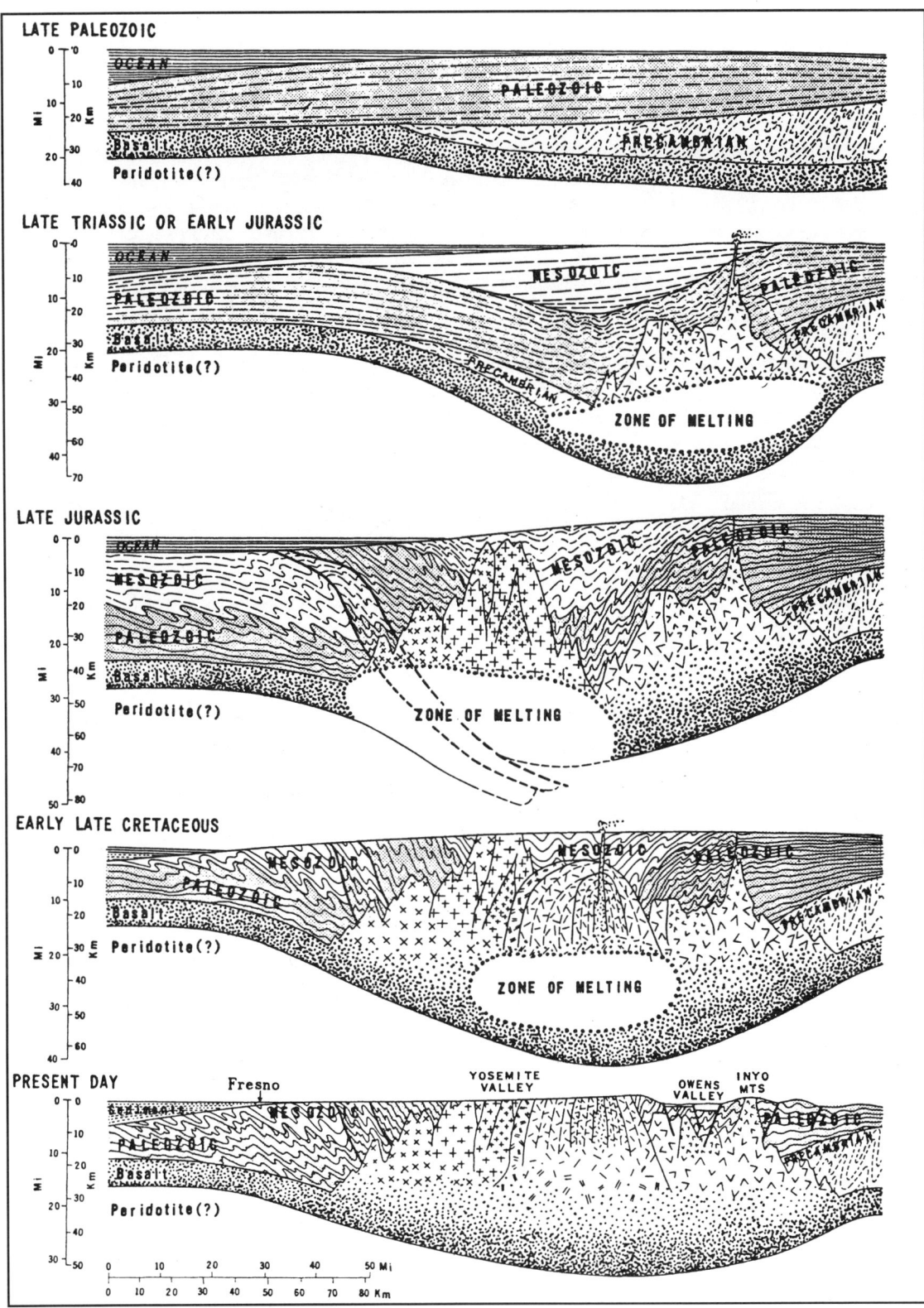

Figure 2. Model for crustal evolution of Sierra Nevada taken directly from Bateman and Wahrhaftig (1966).

crustal evolution. Hamilton's paper focused primarily on subduction-zone magmatism as the prime source of the batholithic magmas and called into question the synclinorium model. In the discussion that follows, we will find that perhaps a view intermediate between these two classic views best satisfies the current data.

A prime consideration in the genesis of the batholith is the nature of its source materials. A tremendous amount of isotopic work has been done over the past 25 years and has shed considerable light on this problem. Another important consideration is the well-documented tectonic boundary between oceanic wall rocks in the western domain and North American wall rocks in the eastern domain of the prebatholithic framework. This boundary, at least at shallow to mid-crustal levels, presents an opportunity to relate batholithic petrochemistry to possible crustal assimilation regimes. Bateman and Wahrhaftig (1966) recognized this important possibility, but only in passing by mentioning Moore's (1959) interpretation of his "quartz diorite boundary line" within the batholith. We focus here on the evolution of thought for the batholithic source and the broader implications for the relations between petrogenesis and tectonics.

Isotopic variations in Cordilleran-type batholiths in general have been interpreted as either reflecting the nature of deep-crustal and upper mantle source regimes (Kistler and Peterman, 1973, 1978; Silver and Chappel, 1988; Hildreth and Moorbath, 1988), or as reflecting the net results of higher level assimilation and fractional crystallization (DePaolo, 1981; Saleeby, 1990; Clemens-Knott, 1992). These end-member interpretations are based on regional isotopic data sets from the batholiths, and from sparser data sets and assumptions on the isotopic character of the proposed source components. Kistler and Peterman (1973, 1978), using mainly initial Sr data, identified contrasting deep-crust/upper mantle source regimes for the Sierra Nevada batholith whereby the eastern domain of the batholith acquired a Proterozoic sialic basement component, while the western, more mafic domain lacked such a component and by implication reflected a predominance of subducted slab and/or mantle wedge components. This made geologic sense with the existence of ophiolitic basement for the wall rocks along the western margin of the batholith and known or presumed miogeoclinal basement for wall rocks to the east. Subsequently, Saleeby and Chen (1978), Saleeby and Sharp (1980), and Knott et al. (1990) identified depleted-mantle source end-member intrusions in the extreme western foothills of the southern Sierra Nevada batholith within solely ophiolitic wall rocks. Isotopically and petrographically, these extreme western intrusions resemble intraoceanic arc magma systems, and thus by analogy, they are presumed to reflect Hamilton's (1969) view of subduction-zone magma systems.

In a regional study of west coast batholithic Nd and Sr isotopes, DePaolo (1981) noted that most samples scatter along a trajectory extending from the mantle array to a low $\varepsilon_{Nd}$ high $\varepsilon_{Sr}$ end-member typical of North American craton basement, but with an added correlation of higher $\delta^{18}O$ toward the basement-contaminated samples. DePaolo (1981) interpreted these data to suggest that assimilated craton-derived metasedimentary material constituted the main continental component within well-mixed and fractionated magma systems. This view was seen generally to be consistent with later studies of Sr, Nd, and O isotopes in the deep-level batholithic rocks of the southernmost Sierra (Saleeby et al., 1987; Pickett and Saleeby, 1994), although a high proportion of the assimilation and fractionation was required to have occurred at depths greater than the ~30 km level of exposure. These results also appeared to be consistent with batholith mineral chemistry studies conducted by Ague and Brimhall (1988a, 1988b), wherein regional zones of deep-level assimilation were resolved as well as local higher level occurrences of highly contaminated plutons.

More recent studies have again brought into focus the possible involvement of Proterozoic basement as a significant source component for the eastern domain of the Sierra Nevada batholith. Regional variation patterns in initial Sr and initial $^{206}Pb/^{204}Pb$ versus initial $^{208}Pb/^{204}Pb$ on batholithic samples were interpreted in this light by Chen and Tilton (1991). In an extensive study of late Cenozoic volcanic rock-hosted lower crustal/upper mantle xenoliths, Ducea et al. (1995), Ducea and Saleeby (1996a, 1996b, 1998a, 1998b, 1998c) resolved a thick (~50 km) mafic keel for the batholith, the lower levels of which equilibrated in eclogite facies in the Late Cretaceous. Isotopic, trace element, and Sm/Nd–Rb/Sr isochron data show this keel to be petrogenetically linked to the overlying batholith as its cumulates and/or residues. Initial Nd and Sr data on these deep mafic rocks encompass DePaolo's (1981) two-component mixing trajectory, but also disperse off this array toward a distinct low $\varepsilon_{Nd}$–low $\varepsilon_{Sr}$ end member that reflects a Proterozoic lower continental basement component; much like the $^{206}Pb/^{208}Pb$ relations discussed in Chen and Tilton (1991). Furthermore, the deep-crustal/upper mantle xenolith data show elevated $\delta^{18}O$ values relative to mantle values (Ducea et al., 1999) indicating the presence of a supracrustal component within the deepest magmatic systems of the batholith.

In summary, it appears that the regional isotopic data clearly identify deep Proterozoic sialic basement, depleted mantle, and supracrustal components in the batholithic source regime. The regional variations in the depleted mantle versus Proterozoic basement components can be readily interpreted in terms of regional patterns in wall rock geology and tectonics (Kistler, 1990). The supracrustal component, however, is poorly understood, both in terms of its origin, and how it was transmitted to such great depths. Two possible mechanisms for the introduction of these supracrustal components into the deep magmatic systems appear to be viable; both in their own ways resemble some aspects of the synclinorium model. The first of these results directly from the prebatholithic framework structure. By the end of the Jurassic, there had been substantial crustal thickening by a long history of crustal shortening and terrane accretion. It seems likely that supracrustal components were introduced deeply into the prebatholithic framework by this process and thereby delivered to deep crustal melting levels for the batholithic magma systems. Direct observations are not available for the testing of this mechanism, although intuitively some contribution from this mechanism seems inevitable. The second mechanism for the introduction of supracrustal components is derived from recent findings on the current level of surface exposure for the batholith. We now turn our attention to this topic.

## COMMENTS ON "DEPTH OF EROSION SINCE MESOZOIC PLUTONISM"

In Bateman and Wahrhaftig's (1966) discussion on the depth of erosion of the batholith, consideration was given to assumptions regarding $P_{H2O}$ and $P_{load}$ ($P$ = pressure) conditions of the batholithic magma system, phase relations in aluminosilicates in metapelites, and Cretaceous-Cenozoic sedimentary sections that presumably had their source in the Sierra Nevada region. These lines of reasoning all substantially predate the advent of modern thermobarometry, and a new paradigm on how continental basement rocks may be unroofed. Bateman and Wahrhaftig's (1966) reasoning led to the conclusion that the batholith had been unroofed to a depth of ~17 km. This value appears fortuitously to be roughly correct if our current values for the entire batholith south of lat 39°N are integrated. However, there is tremendous variation within this field.

First, within the area of greatest focus of Bateman and Wahrhaftig (1966), the shallowest level of the batholith is exposed. Along the western margin of the Ritter Range pendant (Fig. 1) are the remnants of a mid-Cretaceous caldera complex, including a subcaldera batholithic pluton (Fiske and Tobisch, 1994). The preservation of near-surface levels for the batholith is clearly implied here. Igneous barometric data from plutons adjacent to the caldera indicate shallow exposure levels of <5 km (Ague and Brimhall, 1988b). Similar shallow exposure levels are resolved in the Sierra crest southward to the Mount Whitney region, and again farther north in the Donner Pass region (Fig. 1). To the west and southwest of the Sierra crest shallow domains, Ague and Brimhall's (1988b) data show a general southward deepening. Igneous and metamorphic thermobarometric work in the southernmost Sierra (Pickett and Saleeby, 1993; Ague, 1997) shows a continued southward deepening to depths of ~30 km in the contiguous batholith. Thus, an integrated value of ~15 km depth of erosion is reasonable for much of the batholith, but with tremendous local variation.

The net result of the southward deepening in level of exposure is that the southern Sierra Nevada exposes an oblique section through the batholith from near-surface (volcanic) levels to deep, ~30 km depths (Saleeby, 1990). Structural relations of metamorphic framework rocks along this oblique section offer some insight into a possible second mechanism for the introduction of supracrustal components into the batholithic magma systems. This discussion continues in part from the framework structure discussion on pendants of the central to southern axial Sierra and of the southwestern Foothills. North of lat 35.7°N, these pendants are for the most part andalusite bearing in their pelite units, and according to the regional igneous barometric data of Ague and Brimhall (1988b), were intruded by the batholith mainly at 2 to 4 kbar levels. These pendants are virtually devoid of features suggestive of any partial melting. In contrast, pendants to the south that are within the contiguous batholith were intruded and metamorphosed at progressively greater depths southward, and correspondingly show transitions into sillimanite and kyanite paragenesis, as well as various degrees of partial melting in their pelitic and psammitic units. What remains of the original stratigraphy in the deepest-level exposures to the south are lenses of the refractory marble and quartzite units and local restitic "ghosts" of the pelite-psammite units dispersed within the batholith, most commonly adjacent to the refractory lenses.

This entire domain of pendants is characterized by steep linear deformation fabrics that developed during peak thermal metamorphism and pluton emplacement (Saleeby, 1987; Saleeby and Busby, 1993). At several locations along this belt of pendants, mid-Cretaceous metamorphosed silicic volcanic-hypabyssal units are "tucked-in" along pendant-batholithic contacts, and share the same steep linear deformation fabrics that are present in the adjacent metasedimentary rocks (Saleeby et al., 1990; Saleeby and Busby, 1993). These volcanic units are characteristically 2 to 5 m.y. older than the adjacent batholithic units: they are batholithic ejecta. Transposed unconformities are locally preserved between the mid-Cretaceous volcanic units and the adjacent prebatholithic framework rocks (Saleeby and Busby, 1993). The age patterns and structural settings of these metavolcanic sections dictate that they, and their adjacent framework sections, moved substantially downward at high rates (to 5 mm/yr) during batholith emplacement. The steep linear fabrics and associated passive folds within the framework sequences are believed to be a finite strain expression of this downward flow pattern. This view for the synbatholithic transport pattern of the group of pendants in question is contrary to the view expressed in Bateman and Wahrhaftig (1966, p. 118): "Wall-rock deformation indicates that rising magma squeezed the wall and roof aside and upward." However, in another passage the authors state in passing that the wall rocks settled to make room for rising magma (Bateman and Wahrhaftig, 1966, p. 118).

The downward flow and displacement pattern of pendant rocks and their adjacent silicic metavolcanic units is suggested to represent return flow dynamically linked to voluminous magma ascent (Saleeby, 1990). The result of this downward flow pattern is that fertile pelite-psammite intervals of the framework sequences were transported to great depth and thereby partially to completely melted, and presumably assimilated into the deeper levels of the batholithic magma systems. Remnants of restitic ghost trains have been recognized in the deepest exposed (~30 km) levels of the batholith, and conceivably they were transported to even deeper levels. This view is consistent with the interpretation set forth in the isotopic studies of the deep rocks whereby contamination by supracrustal components, widespread mixing, and fractionation was seen as having occurred at a deeper level than the current level of exposure (Saleeby et al., 1987; Pickett and Saleeby, 1994).

The synbatholithic downflow mechanism outlined here, as well as deep thrust imbrication of supracrustal components into the prebatholithic framework discussed earlier, are not that different than the classic view of downward buckling of supracrustal rocks into the zone of batholith melting. The view adopted here, though, is that subduction-zone magmatism, as expressed in the westernmost mafic batholithic plutons, was the primary driving force of voluminous magmatism, and that at a depth of >30 km there was large-scale interaction with crustal materials. We return to this in the following.

The oblique-section view of the southern Sierra Nevada batholith offers additional insights regarding classic treatments of the batholith. Hamilton and Myers (1967) objected to Bateman and Wahrhaftig's (1966) view on the shapes of the batholithic magma chambers and the resulting plutons. As seen in Figure 2, Bateman and Wahrhaftig (1966) envisaged very deep seated magma systems that became progressively more mafic with depth, all the way to the zone of melting. In contrast, Hamilton and Myers (1967) contended that the plutons detached as diapirs from their source, ascended, and then spread and coalesced into an upper crustal level batholithic layer. At least to the ~30 km level of exposure observed in the southernmost Sierra Nevada, there is no sign of this mechanism having operated. The batholith gets progressively more mafic by typically no more than 10 modal percent, as suggested in Bateman and Wahrhaftig (1966). Some of the more mafic Cretaceous plutons of the western and southern Sierra, as well as a number of early Mesozoic plutons, have outcrop patterns suggestive of detached diapirs, but if the larger Cretaceous plutons have such shapes, their detached bases appear to have been at a depth substantially greater than 30 km. Bateman and Wahrhaftig (1966, p. 118) stated that there is "no field evidence for downward bottoming of plutons" although the central Sierra region of their focus was not the best area in the Sierra to pursue this problem.

A final important finding from the southernmost Sierra addresses Bateman and Wahrhaftig's (1966) discussion on depth of erosion and the mechanism of unroofing. Not shown in the classic text excerpts is a lengthy discussion on the current location of the sediment that was stripped off the Sierra Nevada batholith. Such a treatment of this problem is only applicable for the northern and central Sierra. Structural and thermochronologic studies in the southernmost Sierra, coupled with field studies in the adjacent Mojave Desert region, indicate that the deeper levels of the Sierra Nevada batholith south of lat 35.5°N were denuded very rapidly in latest Cretaceous–Paleocene time by a series of low-angle detachment faults (Wood and Saleeby, 1997). Some of the higher level sheets and their detachment surfaces are still exposed in the southernmost Sierra adjacent to the Garlock fault. Wood and Saleeby (1997) believe that other displaced remnants occur in ranges of the northern Mojave Desert, and in the Salinian block of the central California Coast Ranges. Initial exposure of the deepest-level batholithic rocks resulted from this important and newly recognized tectonic event. Mass tectonic transport of upper crustal detachment sheets accounts for missing sedimentary sections, which could conceivably correspond to the unroofing of the deep-level rocks of the southern Sierra.

## TECTONICS AND CRUSTAL EVOLUTION

Figure 3 is an updated version of the model (Fig. 2) presented in Bateman and Wahrhaftig (1966) for the crustal evolution and batholith petrogenesis of the Sierra Nevada. The time frames have been modified slightly in order to emphasize major developmental phases in crustal evolution. The frames are highly generalized and are intended generally to be parallel in representation to Bateman and Wahrhaftig (1966).

Late Paleozoic time is characterized by transform truncation of the Cordilleran miogeocline along the axial region of the Sierra Nevada and the initial accretion of ensimatic or Panthalassan crust/mantle sequences to the western wall rock domain. Also shown diagrammatically are oblique sections through thrust sequences within and above the miogeocline; these are related to the Antler and Sonoma orogenies. The first-order consequence of this phase of geologic history was the establishment of an oceanic-continental lithospheric boundary in what was to become the axial region of the prebatholithic framework.

Early Mesozoic time is characterized by arc magmatism and crustal imbrication. This time frame is highly generalized relative to the wealth of geologic information available. Phases of regional extension and ignimbrite eruption were interspersed and conceivably predominated in time duration over the crustal shortening phases. Likewise, phases of sinistral and dextral shear appear to have operated along the earlier truncation boundary, and within the actively accreting ensimatic terrane complex. A two-sided orogen is depicted with west-verging structures along the west side of the Sierra and east-verging structures along the east side. Such a structural geometry may only be applicable for the central and southern Sierra; in the northern Sierra, east-verging structures as well as west-verging structures are present (Moores, this volume). Likewise, only one subduction zone is depicted for this time frame; there may have been multiple active subduction zones (Moore et al., this volume).

During the early Mesozoic time frame, marine conditions persisted throughout the Sierran region until the later part of the Late Jurassic, when the culminating phase of crustal shortening (Nevadan orogeny) marked the last phases of ensimatic terrane accretion. The first-order consequences of this phase of geologic history appear to have been crustal thickening and the introduction of labile components to deep crustal levels.

Mid-Cretaceous time is characterized by voluminous batholithic magmatism. Such magmatism commenced in the western Foothills and adjacent Great Valley basement ca.125 Ma and was expressed primarily by subduction-zone magmas that, for

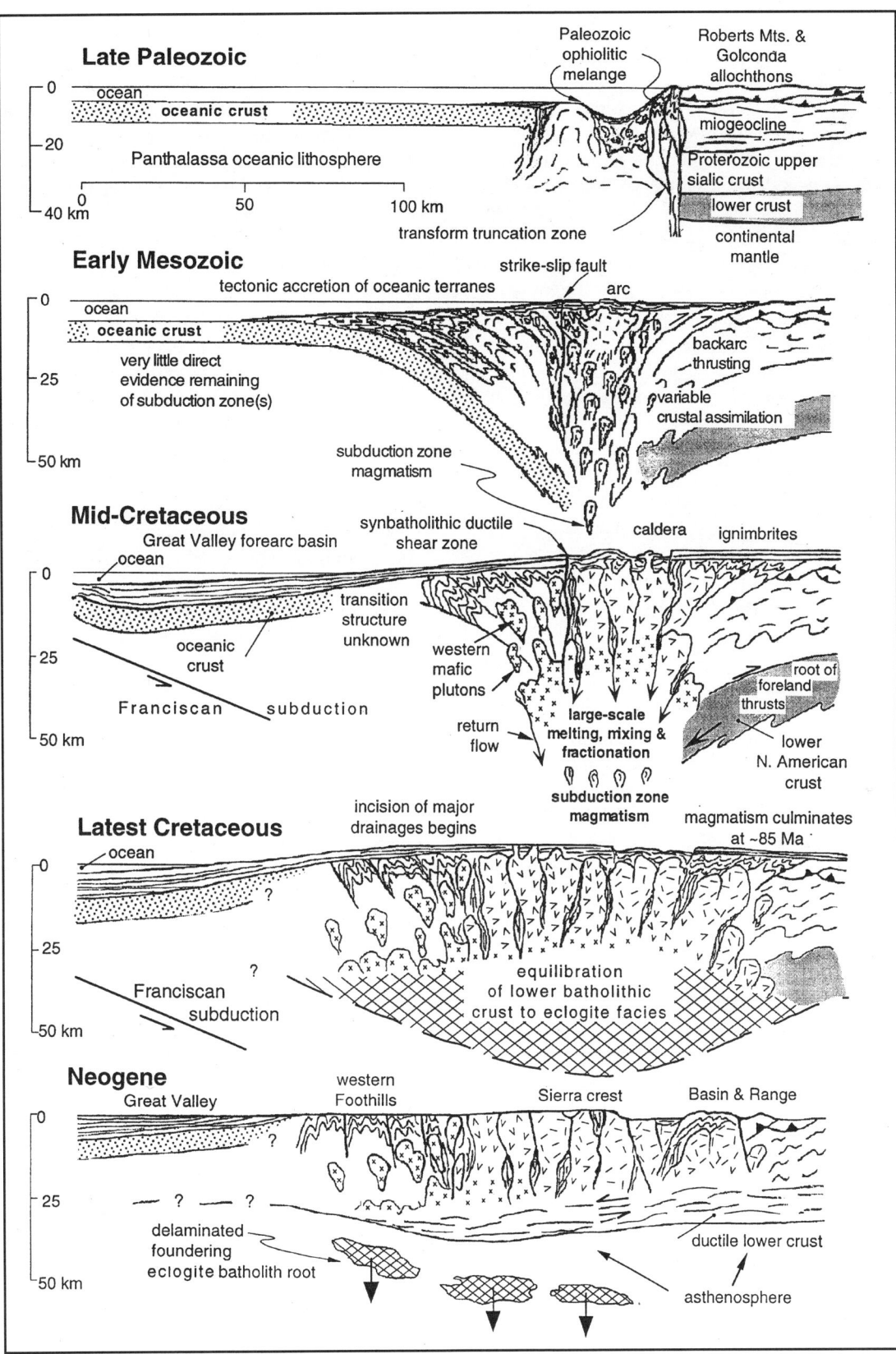

Figure 3. Updated highly generalized model for crustal evolution of Sierra Nevada based on the discussion of Bateman and Wahrhaftig (1966) and the Figure 2 model.

the most part, lacked continental crustal components. As magmatism migrated eastward in time, progressively more continental crustal components were encountered by the ascending mafic magmas, resulting in melting of the crustal components and the commingling and mixing of the ascending mafic magmas and the crustal melt products. Very large mixing and fractionation domains formed at great depth. Isotopic studies of deep-level exposures of the southernmost Sierra (Saleeby et al., 1987; Pickett and Saleeby, 1994) coupled with Neogene volcanic-hosted deep-crust and mantle xenolith studies (Ducea et al., 1995; Ducea and Saleeby, 1996a, 1996b, 1998a, 1998b, 1998c) constrain the depth of these domains to between 30 and 80 km. Mixing in these deep domains appears to have been fairly efficient, as indicated by a continuity of isotopic and trace element patterns throughout the entire ~80 km section of the silicic batholith and its mafic keel.

In the crustal evolution model of Bateman and Wahrhaftig (1966) (Fig. 2), only local volcanism was depicted as occurring above the evolving batholith. The model shown in Figure 3 envisages large calderas and large-volume ignimbrite eruptions closely related to the batholith. This view was suggested in Hamilton and Myers (1967) and Hamilton (1969) and has been corroborated by the subsequent discovery of widespread mid-Cretaceous silicic volcanic units as cited herein.

The axial region of the actively forming batholith was also the locus of steep ductile shearing throughout middle to Late Cretaceous time. Shearing was concentrated along discrete zones, commonly the margins of large plutons at or close to solidus conditions. Mid-Cretaceous shearing was predominantly dip slip (Tobisch et al., 1995), whereas Late Cretaceous time was characterized by dextral shearing (Busby-Spera and Saleeby, 1990; Saleeby and Busby, 1993; Tikoff and de Saint Blanquat, 1997).

Latest Cretaceous time was characterized by the cessation of batholithic magmatism, the cooling of the crustal column, and the equilibration of the batholithic mafic keel in the eclogite facies (Ducea and Saleeby, 1998c). An additional ≥75 km column of mantle lithosphere garnet peridotite was beneath the eclogite facies mafic keel (Ducea and Saleeby, 1996, a and b). By analogy with the modern Andes, one would presume that throughout much of the Cretaceous the Sierra Nevada was a high-standing plateau. Recent low-temperature thermochronologic studies of the central Sierra Nevada batholith suggest deep stream incision of the axial part of the range in Late Cretaceous time with an overall decrease in relief and mean elevation having occurred throughout Cenozoic time (House et al., 1998). A significant factor in the relief and mean elevation loss could be broad flexural subsidence related to the dense eclogite facies keel. Thus the root (keel) of the Sierra Nevada batholith as resolved by the volcanic-hosted xenolith studies may have had the inverse effect of the root as envisaged by Bateman and Wahrhaftig (1966, p. 125): "uplift of the Sierra Nevada...was caused by a mass deficiency of the root."

Recent seismic refraction studies indicate that there is no low-velocity crustal root beneath the Sierra Nevada and that there is only minimal difference in the depth to the Moho between the high Sierra crest and the adjacent extensional province of the Basin and Range (Fliedner et al., 1996). This leads to a paradox in terms of a buoyancy source for the Sierra Nevada within, or immediately adjacent to, a region of high-magnitude crustal thinning. Studies of Neogene volcanic-hosted mantle xenoliths of the Sierra Nevada–Owens Valley region have revealed a time dependency on mantle composition and metamorphic facies condition (Ducea and Saleeby, 1996a, 1998, a and b). Mid-Miocene mantle assemblages contain garnet peridotite and eclogite facies mafic batholithic root rocks. Pliocene-Pleistocene assemblages lack these components and contain primarily spinel peridotites with glass (quenched melt) inclusions. It follows that between mid-Miocene and Pliocene-Pleistocene time the mafic root for the batholith and its lithospheric mantle underpinnings were replaced by asthenospheric mantle. This corresponds in time with large-scale extension in the adjacent Basin and Range province (Wernicke and Snow, 1998).

The Neogene time frame of Figure 3 shows the loss of the subbatholithic mafic root and lithosphere and its replacement by asthenosphere, as envisaged by Ducea and Saleeby (1998a, 1998b). Regional extension and presumably the related ascent of asthenospheric mantle are thought to have instigated the delamination of the eclogite facies mafic root, and its foundering into the deeper mantle. The last, or perhaps one of the last, fragments of the foundering mafic root may be represented by the high-velocity anomaly resolved by passive seismic experiments centered at about lat 36°N in the southern Sierra (Jones et al., 1994). The replacement of the eclogitic root and its garnet peridotite lithospheric underpinnings by spinel peridotite at or near solidus conditions provides an adequate buoyancy source for the support of the modern Sierra Nevada within or adjacent to the Basin and Range extensional province (Ducea and Saleeby, 1996, a and b). Thus the root (its loss) of the Sierra Nevada batholith appears to have played a previously unsuspected role in the generation of the modern physiography. Whether the Sierra Nevada batholith region subsided in early Cenozoic time to a region of lower elevation and relief relative to modern and past times and was again uplifted in late Neogene time, or whether it has remained intermediately high in elevation and in relief throughout Cenozoic time, is an area of current interest (House et al., 1998).

## CLOSING REMARKS

Bateman and Wahrhaftig (1966) presented a unique synthesis of Sierra Nevada geology in that they related descriptive aspects of metamorphic framework, batholith, physiographic geology, and batholith petrogenesis. Their paper was based primarily on sound field observations and descriptions. It was published just prior to the inception of the plate tectonic paradigm, but most of the efforts set forth in their 1966 synthesis survived this paradigm shift in terms of the overall usefulness to subsequent workers. Perhaps one of the unique aspects of the science of geology is the time transcendence of sound field observations and descriptions regardless of the paradigm in genetic lines of thought. The work of Bateman and Wahrhaftig (1966) embodies this principal.

## ACKNOWLEDGMENTS

I had the pleasure of interacting extensively with Paul Bateman and Clyde Wahrhaftig earlier in my career. Other important individuals who influenced my Sierran work considerably include Cliff Hopson, Ron Kistler, Jim Moore, Eldridge Moores, Don Ross, Richard Schweickert, Leon Silver, and Howel Williams. Recent support for my Sierran research was provided by National Science Foundation grant EAR-9526859.

## REFERENCES CITED

Ague, J. J., 1997, Thermodynamic calculation of emplacement pressures for batholithic rocks, California: Implications for the aluminum-in-hornblende barometer: Geology, v. 25, p. 563–566.

Ague, J. J., and Brimhall, G. H., 1988a, Regional variations in bulk chemistry, mineralogy, and the compositions of mafic and accessory minerals in the batholiths of California: Geological Society of America Bulletin, v. 100, p. 891–911.

Ague, J. J., and Brimhall, G. H., 1988b, Magmatic arc asymmetry and distribution of anomalous plutonic belts in the batholiths of California: Effects of assimilation, crustal thickness, and depth of crystallization: Geological Society of America Bulletin, v. 100, p. 912–927.

Bateman, P. C., and Chappell, B. W., 1979, Crystalllization, fractionation and solidification of the Tuolumne intrusive series, Yosemite National Park, California: Geological Society of America Bulletin, v. 90, p. 465–482.

Bateman, P. C., and Dodge, F. C. W., 1970, Variations of major chemical constituents across the central Sierra Nevada batholith: Geological Society of America Bulletin, v. 81, p. 409–420.

Bateman, P. C., and Nokleberg, W. J., 1978, Solidification of the Mount Givens Granodiorite, Sierra Nevada, California: Journal of Geology, v. 86, p. 59–75.

Bateman, P. C., and Wahrhaftig, C., 1966, Geology of the Sierra Nevada, in Bailey, E. A., ed., Geology of northern California: California Division of Mines and Geology Bulletin 190, p. 107–172.

Bateman, P. C., Harris, A. G., Kistler, R. W., and Krauskopf, K. B., 1985, Calaveras reversed—Westward younging is indicated: Geology, v. 13, p. 338–341.

Busby-Spera, C., and Saleeby, J. B., 1990, Intra-arc strike-slip fault exposed at batholithic levels in southern Sierra Nevada, California: Geology, v. 18, p. 255–259.

Chen, J. H., and Moore, J. G., 1982, Uranium-lead isotopic ages from the Sierra Nevada batholith, California: Journal of Geophysical Research, v. 87, p. 4761–4784.

Chen, J. H., and Tilton, G. R., 1991, Applications of lead and strontium isotopic relationships to the petrogenesis of granitoid rocks, central SIierra Nevada batholith, California: Geological Society of America Bulletin, v. 103, p. 439–447.

Clemens-Knott, D., 1992, Geologic and isotopic investigations of the Early Cretaceous Sierra Nevada batholith, Tulare, CA, and the Ivrea zone, NW Italian Alps: Example of interaction between mantle-derived magma and continental crust [Ph.D. thesis]: Pasadena, California Institute of Technology, 345 p.

DePaolo, D. J., 1981, A neodymium and strontium isotopic study of the Mesozoic calc-alkaline granitic batholiths of the Sierra Nevada and Peninsular Ranges, California: Journal of Geophysical Research, v. 86, p. 10,470–10,488.

Ducea, M. N., and Saleeby, J. B., 1996a, Buoyancy sources for a large, unrooted mountain range, the Sierra Nevada, California; evidence from xenolith thermobarometry: Journal of Geophysical Research, v. 101, p. 8229–8244.

Ducea, M. N., and Saleeby, J. B., 1996b, Rb-Sr and Sm-Nd mineral ages of some Sierra Nevada xenoliths; implications for crustal growth and thermal evolution: Eos (Transactions, American Geophysical Union), v. 77, p. 780.

Ducea, M. N., and Saleeby, J. B., 1998a, Crustal recycling beneath continental arcs: Silica-rich glass inclusions in ultramafic xenoliths from the Sierra Nevada, California: Earth and Planetary Letters, v. 156, p. 101–116.

Ducea, M. N., and Saleeby, J. B., 1998b, A case for delamination of the deep batholithic crust beneath the Sierra Nevada, California: International Geology Review, v. 40, p. 78–93.

Ducea, M. N., and Saleeby, J. B., 1998c, The age and origin of a thick mafic-ultramafic root from beneath the Sierra Nevada batholith; Part 1: Geochronology: Contributions to Mineralogy and Petrology.

Ducea, M. N., Kistler, R. W., and Saleeby, J. B., 1995, Testing petrogenetic models of the deep Sierra Nevada crust and upper mantle using REE data on xenolith assemblages: Geological Society of America Abstracts with Programs, v. 27, p. 15.

Ducea, M.N., Saleeby, J.B., Taylor, H.P., and Clemens-Knott, D., 1999, Isotopic heterogenetics in the vertical dimension of the central Sierra Nevada batholith: Petrologic and tectonic implications: Geology, in press.

Evernden, J. F., and Kistler, R. W., 1970, Chronology of emplacement of Mesozoic batholithic complexes in California and western Nevada: U.S. Geological Survey Professional Paper 623, 42 p.

Fiske, R. S., and Tobisch, O. T., 1994, Middle Cretaceous ash-flow tuff and caldera-collapse deposit in the Minarets Caldera, east-central Sierra Nevada, California: Geological Society of America Bulletin, v. 106, p. 582–593.

Fliedner, M. M., Ruppert, S., and the Southern Sierra Nevada Continental Dynamics Working Group, 1996, Three-dimensional crustal structure of the southern Sierra Nevada from seismic fan profiles and gravity modeling: Geology, v. 24, p. 367–370.

Hamilton, W., 1969, Mesozoic California and the underflow of the Pacific mantle: Geological Society of America Bulletin, v. 80, p. 2409–2430.

Hamilton, W., and Myers, W. B., 1966, Cenozoic tectonics of the western United States: Reviews of Geophysics, v. 4, p. 509–549.

Hamilton, W., and Myers, W. B., 1967, The nature of batholiths: U.S. Geological Survey Professional Paper 554-C, p. C1–C30.

Hildreth, W., and Moorbath, S., 1988, Crustal contributions to arc magmatism in the Andes of central Chile: Contributions to Mineralogy and Petrology, v. 98, p. 455–489.

House, M. A., Wernicke, B. P., and Farley, K. A., 1998, Dating topography of the Sierra Nevada, California, using apatite (U-Th)/He ages: Nature, v. 396, no. 6706, p. 66–69.

Jones, C. H., Kanamori, H., and Roecker, S. W., 1994, Missing roots and mantle "drips": Regional Pn and teleseismic arrival times in the southern Sierra Nevada and vicinity, California: Journal of Geophysical Research, v. 99, p. 4567–4601.

Kistler, R. W., 1990, Two different lithosphere types in the Sierra Nevada, California, in Anderson, J. L., ed., The nature and origin of Cordilleran magmatism: Geological Society of America Memoir 174, p. 271–281.

Kistler, R. W., 1993, Mesozoic intrabatholithic faulting, Sierra Nevada, California, in Dunne, G., and McDougall, K., eds., Mesozoic paleogeography of the western United States II: Pacific Section, SEPM (Society for Sedimentary Geology), Book 71, p. 247–262.

Kistler, R. W., and Peterman, Z. E., 1973, Variations in Sr, Rb, K, Na, and initial $Sr_{87}/Sr_{86}$ in Mesozoic granitic rocks and intruded wall rocks in Central California: Geological Society of America Bulletin, v. 84, p. 3489–3512.

Kistler, R. W., and Peterman, Z. E., 1978, A study of regional variation of initial strontium isotopic composition of Mesozoic granitic rocks in California: U.S. Geological Survey Professional Paper 1071, 17 p.

Knott, D. C., Saleeby, J. B., and Taylor, H. P., Jr., 1990, Petrology of the Early Cretaceous Sierra Nevada batholith: The Stokes Mountain region, California: Eos (Transactions, American Geophysical Union), v. 71, p. 1576.

Lahren, M. M., Schweickert, R. A., Mattison, J. M., and Walker, J. D., 1990, Evidence of uppermost Proterozoic to Lower Cambrian miogeoclinal rocks and the Mojave–Snow Lake fault: Snow Lake pendant, central Sierra Nevada, California: Tectonics, v. 9, p. 1585–1608.

McMath, V. E., 1966, Geology of the Taylorsville area, northern Sierra Nevada, California, in Bailey, E. H., Geology of Northern California: California Division of Mines and Geology Bulletin 190, p. 173–183.

Merguerian, C., and Schweickert, R. A., 1987, Paleozoic gneissic granitoids in the Shoo Fly complex, central Sierra Nevada, California: Geological Society of America Bulletin, v. 99, p. 699–717.

Moore, J. G., 1959, The quartz diorite boundary line in the western United States: Journal of Geology, v. 67, p. 197–210.

Moore, J. G., 1963, Geology of the Mount Pinchot quadrangle, southern Sierra Nevada, California: U.S. Geological Survey Bulletin 1130, 152 p.

Moore, J. G., and Sisson, T., 1987, Preliminary Geologic Map of Sequoia–Kings Canyon National Park, California: U.S. Geological Survey Open File Report 87–651, scale 1:125,000.

Moore, J. N., and Foster, C. T., Jr., 1980, Lower Paleozoic metasedimentary rocks in the east-central Sierra Nevada, California: Correlation with Great Basin formations: Geological Society of America Bulletin, v. 91, p. 37–43.

Nokleberg, W. J., and Kistler, R. W., 1980, Paleozoic and Mesozoic deformations in the central Sierra Nevada, California: U.S. Geological Survey Professional Paper 1145, 24 p.

Pickett, D. A., and Saleeby, J. B., 1993, Thermobarometry of Cretaceous rocks of the Tehachapi Mountains, California: Plutonism and metamorphism in deep levels of the Sierra Nevada batholith: Journal of Geophysical Research, v. 98, p. 609–629.

Pickett, D. A., and Saleeby, J. B., 1994, Nd, Sr and Pb isotopic characteristics of Cretaceous intrusive rocks from deep levels of the Sierra Nevada batholith, Tehachapi Mountains, California: Contributions to Mineralogy and Petrology, v. 118, p. 198–215.

Ross, D. C., 1990, Reconnaissance geologic map of the southern Sierra Nevada, Kern, Tulare, and Inyo counties, California: U.S. Geological Survey Open-File Report 90-337, 163 p.

Saleeby, J. B., 1978, Kings River Ophiolite, southwest Sierra Nevada Foothills, California: Geological Society of America Bulletin, v. 89, p. 617–636.

Saleeby, J. B., 1979, Kaweah serpentine melange, southwest Sierra Nevada Foothills, California: Geological Society of America Bulletin, v. 90, p. 29–46.

Saleeby, J. B., 1982, Polygenetic ophiolite belt of the California Sierra Nevada, geochronological and tectonostratigraphic development: Journal of Geophysical Research, v. 87, p. 1802–1824.

Saleeby, J. B., 1990, Progress in tectonic and petrogenetic studies in an exposed cross-section of young (~100 Ma) continental crust, southern Sierra Nevada, California, in Salisbury, M. H., ed., Exposed cross sections of the continental crust: Dordrecht, D. Reidel Publishing Co., p. 137–158.

Saleeby, J. B., 1992, Petrotectonic and paleogeographic settings of U.S. Cordilleran ophiolites, in Burchfiel, B. C, Lipman, P. W., and Zoback, M. L., eds., The Cordilleran orogen: Conterminous U.S.: Boulder, Colorado, Geological Society of America, Geology of North America, v. G-3, p. 653–682.

Saleeby, J. B., and Busby, C., 1993, Paleogeographic and tectonic setting of axial and western metamorphic framework rocks of the southern Sierra Nevada, California, in Dunne, G., and McDougall, K., eds., Mesozoic paleogeography of the western United States—II: Pacific Section, Society of Economic Paleontologists and Mineralogists, Book 71, p. 197–226.

Saleeby, J. B. and Chen, J. H., 1978, Preliminary report on initial lead and strontium isotopes from ophiolitic and batholithic rocks, southwestern foothills, Sierra Nevada, California: U.S. Geological Survey Open-File Report 78-701, p. 375–376.

Saleeby, J. B. and Sharp, W. D., 1978, Preliminary report on the behavior of U-Pb zircon and K/Ar systems in polymetamorphosed ophiolitic and batholithic rocks, southwestern Sierra Nevada Foothills, California: U.S. Geological Survey Open-File Report 78-701, p. 376–378.

Saleeby, J. B. and Sharp, W. D., 1980, Chronology of the structural and petrologic development of the southwest Sierra Nevada Foothills, California: Geological Society of America *Bulletin*, v. 91, part II, p. 1416–1535.

Saleeby, J. B., Speed, R. C., Blake, M. C., Allmendinger, R. W., Gans, P. B., Kistler, R. W., Ross, D. C., Stauber, D. A., Zobak, M. L., Griscom, A., McCulloch, D. S., Lachenbruch, A. H., Smith, R. B., and Hill, D. P., 1986, Continent-ocean transect, Corridor C2, Monterey Bay offshore to the Colorado Plateau: Geological Society of America Map and Chart Series TRA C2, 2 sheets, scale 1:500,000, 87 p.

Saleeby, J. B., Sams, D. B. and Kistler, R. W., 1987, U/Pb zircon, strontium, and oxygen isotopic and geochronological study of the southernmost Sierra Nevada batholith, California: Journal of Geophysical Research, v. 92, p. 10,443–10,446.

Saleeby, J. B., Shaw, H. F., Niemeyer, S., Edelman, S. H., and Moores, E. M., 1989, U/Pb, Sm/Nd and Rb/Sr geochronological and isotopic study of northern Sierra Nevada ophiolitic assemblages, California: Contributions to Mineralogy and Petrology, v. 102, p. 205–220.

Saleeby, J. B., Busby-Spera, C., Oldow, J. S., Dunne, G. C., Wright, J. E., Cowan, D. S., Walker, N. W., and Allmendinger, R. W., 1992, Early Mesozoic tectonic evolution of the western U.S. Cordillera, in Burchfiel, B. C, Lipman, P. W., and Zoback, M. L., eds., The Cordilleran orogen: Conterminous U.S.: Boulder, Colorado, Geological Society of America, Geology of North America, v. G-3, p. 107–168.

Schweickert, R. A., and Lahren, M. M., 1987, Continuation of Antler and Sonoma orogenic belts to the eastern Sierra Nevada, California, and Late Triassic thrusting in a compressional arc: Geology, v. 15, p. 270–273.

Schweickert, R. A., Bogen, N. L., Girty, G. H., Hanson, R. E., and Merguerian, C., 1984, Timing and structural expression of the Nevadan orogeny, Sierra Nevada, California: Geological Society of America *Bulletin*, v. 95, p. 967–979.

Sharp, W. D., 1988, Pre-Cretaceous crustal evolution in the Sierra Nevada region, California, in Ernst, W. G., ed., Metamorphism and crustal evolution of the western United States (Rubey Volume VII): Englewood Cliffs, New Jersey, Prentice-Hall, p. 824–864.

Silver, L. T., and Chappell, B. W., 1988, The Peninsular Ranges batholith; an insight into the evolution of the Cordilleran batholith of southwestern North America: Royal Society of Edinburgh Transactions, Earth Sciences, v. 79, p. 105–121.

Stern, T. W., Bateman, P. C., Morgan, B. A., Newell, M. F., and Peck, D. L., 1981, Isotopic U-Pb ages of zircon from granitoids of the central Sierra Nevada, California: U.S. Geological Survey Professional Paper 1185, 17 p.

Tikoff, B., and de Saint Blanquat, M., 1997, Transpressional shearing and strike-slip partitioning in the Late Cretaceous Sierra Nevada magmatic arc, California: Tectonics, v. 16, p. 442–459.

Tobisch, O. T., Saleeby, J. B., and Fiske, R. S., 1986, Structural history of continental volcanic arc rocks along part of the eastern Sierra Nevada, California: A case for extensional tectonics: Tectonics, v. 5, p. 65–94.

Tobisch, O. T., Patterson, S., Saleeby, J. B., and Geary, E. E., 1989, Nature and timing of deformation in the Foothills Terrane, central Sierra Nevada, California: Its bearing on orogenesis: Geological Society of America Bulletin, v. 101, p. 401–413.

Tobisch, O. T., Saleeby, J. B., Renne, P. R., McNulty, B., and Tong, W., 1995, Variations in deformation fields during development of a large volume magmatic arc, central Sierra Nevada, California: Geological Society of America Bulletin, v. 107, p. 148–166.

Tobisch, O. T., Saleeby, J. B., Fiske, R. S., and Holt, L., 1999, U/Pb zircon geochronology and structural sequence of the Ritter Range roof pendant, Sierra Nevada, California: Geological Society of America Bulletin (in press).

Wernicke, B., and Snow, J. K., 1998, Cenozoic tectonism in the Central Basin and Range: Motion of the Sierran–Great Valley block: International Geology Review, v. 40, p. 403–410.

Wolf, M. B., and Saleeby, J. B., 1995, Late Jurassic dike swarms in the southwestern Sierra Nevada Foothills terrane, California: Implications for the Nevadan orogeny and North American plate motion, in Miller, D. M., and Busby, C., eds., Jurassic magmatism and tectonics of the North American Cordillera: Geological Society of America Special Paper 299, p. 203–228.

Wood, D. J. and Saleeby, J. B., 1997, Late Cretaceous–Paleocene extensional collapse and disaggregation of the southernmost Sierra Nevada batholith: International Geology Review, v. 39, p. 973–1009.

MANUSCRIPT ACCEPTED BY THE SOCIETY NOVEMBER 23, 1998

Printed in U.S.A.

Chapter 10

# GEOCHRONOLOGY OF CALIFORNIA'S ARC

**Warren D. Sharp and Paul R. Renne**
*Berkeley Geochronology Center*
*2455 Ridge Road*
*Berkeley, California 94709*
*E-mail: wsharp@bgc.org; prenne@bgc.org*

## JACK FOORD EVERNDEN

Jack Foord Evernden was born March 12, 1922, in Okeechobee, Florida. After undergraduate matriculation in the College of Mining, he earned a doctoral degree in geophysics at the University of California Berkeley (UCB) in 1951. Evernden joined the UCB faculty in 1953 and served there until 1965, when he left to resume the career in seismology that was temporarily diverted by an extraordinarily productive, yet surprisingly brief, diversion into K-Ar geochronology.

This diversion was occasioned by the move of John H. Reynolds from the University of Chicago to the UCB Physics Department. Reynolds' specific mission was to use K-Ar dating and noble gas isotope geochemistry to clarify the evolution of our solar system, and he sought collaboration with Berkeley earth scientists Evernden and Garniss Curtis to this end. Evernden's active involvement in geochronology lasted scarcely more than a decade, yet his was among the most influential work in transforming the K-Ar isotopic system into a routine dating method of unusual scope.

Evernden's return to seismology in 1965 was focused on seismic monitoring of nuclear testing, a field that he largely shaped. After successful implementation of the program, Evernden turned his energy to the prediction of seismic intensity and expected loss from major earthquakes, interests he continues to hold today. Evernden received the Holcombe-Cleveland Prize of the American Association for the Advancement of Science in 1962, the Outstanding Civilian Service Medal of the U.S. Air Force in 1968, and is a Fellow of the American Geophysical Union.

Jack Evernden seated at the controls of one of the Reynolds-type, rare gas mass spectrometers in the University of California Berkeley Department of Geology and Geophysics, ca. 1960. This instrument was one of those used to make Ar measurements for Evernden and Kistler (1970). Photo by Joachim Hampel.

Kern Canyon, 1963. Sketch by Clyde Wahrhaftig

## RONALD WAYNE KISTLER

Ron Kistler working with his newly constructed Ar extraction line at the U.S. Geological Survey in Menlo Park in 1960. Ar was released from samples using the NaOH flux method to digest the minerals, a technique that did not require an expensive induction heater. Ar was collected and then measured on the Berkeley spectrometer. Ages of Sierran granites determined with this equipment were included in Evernden and Kistler (1970). Photo by Norman Prime.

Ronald Wayne Kistler was born May 18, 1931, in Chicago. He earned his bachelor of science degree in geology from Johns Hopkins University in 1953, and his doctoral degree in geology at the University of California Berkeley in 1960. His longstanding involvement in geology of the Sierra Nevada began with his doctoral thesis, entitled "The geology of the Mono Craters Quadrangle, California."

Moving to the U.S. Geological Survey (USGS) in 1960, Kistler was responsible for establishing a geochronology laboratory in Menlo Park. He was the vanguard of a stream of young Berkeley geochronologists who joined the U.S. Geological Survey in the 1960s and 1970s. He has remained at the USGS (Menlo Park) as a geologist for the rest of his career, with a stint as a visiting professor in 1971 at Northwestern University.

Building upon his roots in the geology of the Sierra Nevada, Kistler continued his earlier work with many important isotopic studies bearing on the petrological and structural history of the Sierra Nevada range as well as other components of the western North American Cordillera. Among his most notable contributions was a study (Kistler and Peterman, 1973) demonstrating the remarkable congruence between patterns of geochemistry and strontium isotopic composition in igneous rocks and the location of the western margin of cratonic North America. This insightful study was one of the earliest applications of radiogenic isotope tracers to the analysis of large-scale crustal evolution, and it continues to be highly influential in Cordilleran geologic thought.

**Why Evernden and Kistler (1970) is a classic paper**

Evernden and Kistler's (1970) extraordinary volume of painstakingly acquired data enabled them to place many constraints on the nature and evolution of California's Mesozoic batholiths. Their work showed, for the first time on a regional scale in the Cordillera, the highly organized pattern of pluton emplacement in space, depth, and time. One measure of the utility of this work is its nearly 200 citations in the scientific literature, more than 30 of which occurred 20 years or more after publication. Subsequent developments in geochronology have modified some of their conclusions, but many remain conceptually intact after nearly 30 years. These conclusions challenged the thinking of the time and contributed significantly to the reinterpretation of batholith genesis and other aspects of Jurassic to Cretaceous California geology in light of plate tectonics theory, which was just emerging at the time of their work.

Paul R. Renne and Warren D. Sharp

# Chronology of Emplacement of Mesozoic Batholithic Complexes In California and Western Nevada

*By* J. F. EVERNDEN *and* R. W. KISTLER

GEOLOGICAL SURVEY PROFESSIONAL PAPER 623

*Prepared in cooperation with the*
*University of California (Berkeley)*
*Department of Geological Sciences*

UNITED STATES GOVERNMENT PRINTING OFFICE, WASHINGTON : 1970

# CHRONOLOGY OF EMPLACEMENT OF MESOZOIC BATHOLITHIC COMPLEXES IN CALIFORNIA AND WESTERN NEVADA

By J. F. Evernden[1] and R. W. Kistler

## ABSTRACT

$Ar^{40}/K^{40}$ was determined for minerals from specimens of Mesozoic granitic rocks collected at 250 localities in California and western Nevada. Analytical reproducibility of $Ar^{40}/K^{40}$ in common minerals from granitic rocks was tested in the laboratories of the U.S. Geological Survey and University of California (Berkeley), and the results were studied for interpretation of the genetic significance of potassium-argon ages derived from the ratios.

The potassium-argon ages of minerals from Mesozoic granitic rocks in California have a continuity from 210 to 80 million years ago. However, a distinct periodicity of magma generation and intrusion is shown when the ages are related to the distribution of genetic groups of plutons based on geologic mapping. Five epochs of magma generation and emplacement that took from 10 to 15 million years to complete were initiated at approximately 30-million year intervals. Each intrusive epoch was preceded by, or was in part contemporaneous with, a period of regional deformation in California or western Nevada. These intrusive epochs and their ages are:

| Maximum to minimum age (m.y.) | Geologic age | Intrusive epoch |
|---|---|---|
| 90–79 | Late Cretaceous | Cathedral Range. |
| 121–104 | Early Cretaceous | Huntington Lake. |
| 148–132 | Late Jurassic | Yosemite. |
| 180–160 | Early and Middle Jurassic | Inyo Mountains. |
| 210–195 | Middle and Late Triassic | Lee Vining. |

## INTRODUCTION

The motivations for this study derive both from the desire to investigate and document further the influence of various factors on the precision and accuracy attainable in the measurement of potassium-argon ratios in minerals of granitic rocks and from an interest in the problem of genesis and emplacement of pluton complexes such as those of California and western Nevada. Study emphasis is on the well-exposed Sierra Nevada batholith in California.

It must be made clear that precision and accuracy as used here refer only to problems of potassium and argon measurement and not to the meaning of "apparent ages." Accepting this as the only possible meaning to be signified by plus or minus estimates on determined "ages," there is very little documentation of factors influencing such plus or minus estimates. Moreover, highly variable statements are still made by different investigators with little or no supporting data. Some investigators report ages with "probable errors of about 5 percent," while others imply precision of 1 percent. The data of the present paper clearly show the influence of diverse factors on precision and accuracy of potassium-argon ages. The investigation of determined potassium-argon ratios as a function of mineral alteration or other factors has indicated the necessity of considering such factors in the interpretation of derived "ages."

Data previously presented (Evernden and Richards, 1962) on the Paleozoic batholiths of eastern Australia strongly suggested that there had been continuity of granite generation and emplacement over a period of some 200 million years. The locus of batholithic intrusion moved more than 300 km, but the "ages" obtained implied the virtual lack of any cessation of emplacement of granitic magma during that long interval. On the other hand, previous interpretations of ages of granitic rocks from the Sierra Nevada of California have implied at least two distinct periods of granite emplacement (Curtis and others, 1958). Kistler, Bateman, and Brannock (1965) have suggested three possible age groups.

We show that the patterns of ages of granitic rocks in the Sierra Nevada, taken altogether, have a continuity similar to that of the ages of granitic rocks in eastern Australia. However, a distinct periodicity of magma generation and intrusion is shown when potassium-argon and rubidium-strontium ages are related to the distribution of genetic groups of plutons based on geologic mapping. We also show that the thermal history of individual plutons is reflected in their mineral ages and that, with a few assumptions, the data permit estimates of the crustal level of emplacement of the plutons.

An essential assumption in the discussion that follows

[1] Present affiliation, Advanced Research Projects Agency, Washington, D.C.

is that the "ages" obtained by potassium-argon analysis result primarily from a more or less complex cooling history for each pluton and mineral. Virtually all investigators have concluded that the plutons of the Sierra Nevada are the product of magmatic emplacement. In fact, the pattern of ages obtained seems to be inconsistent with any other interpretation. Assuming this and assuming no later reheating of the rock, then at best the potassium-argon age of a mineral from the rock will be a determination of the time when this magma mass had cooled sufficiently to allow retention of argon in the crystal lattice. Because of pronounced difference in diffusion rate in relation to temperature for argon in biotite and hornblende (Hart, 1964), the amount of discordance between biotite and hornblende ages from rocks will be, in some sense, a measure of the cooling rate of the pluton. Thus, concordance of these two ages implies rapid cooling and a near approach of the determined age to the time of initiation of cooling of the pluton. On the other hand, reheating of the sample by later intrusion will result in partial or complete loss of previously generated argon and will often give rise to marked discordance in biotite and hornblende ages. If the pattern of discordance is accurately delineated, conclusions about the pattern of growth of the pluton complexes become possible.

Many of the samples collected for this study are from areas that are virtually unmapped in terms of delineation of individual plutons. This is unfortunate, but we point out that the great span of time, Late Triassic to Late Cretaceous, represented by the Mesozoic intrusive rocks of California, is demonstrated mainly by the data of geochronometry and not by stratigraphic correlation. In fact, even in areas of detailed mapping, the ages of intruded rocks are so poorly known, or stratigraphic columns have such large gaps, that geologic assignment of ages of intrusive rocks is usually possible only within the limits of two or more geologic periods. It was decided, therefore, that questions of sufficient interest for this study could be answered without further detailed field control, while such a pattern of dates as here demonstrated may act as a stimulus to field investigations.

Prior to the writing of the paper, we had been dating rocks independently in different areas of the Sierra Nevada and adjacent batholiths of California and western Nevada as part of work at the University of California (Berkeley) and the Geological Survey. In order to present as comprehensive a report as is possible at present and to facilitate the publication of the large mass of data gathered, we combined our efforts for this paper.

The discussions separate logically into two parts. The first part deals with analytical problems such as the precision and accuracy of potassium measurements (a similar treatment for argon measurements is to be found in Evernden and Curtis, 1965) and the influence of chemical and mineralogic factors on apparent potassium-argon ages. The second part is the geologic interpretation of the total body of the absolute age data. The two parts can be read as nearly separate papers.

El Capitan (left) and Half Dome (center), Yosemite National Park. The age of the El Capitan Granite is about 108 my and that of the Half Dome Granodiorite, the youngest plutonic rock in Yosemite Valley, about 87 my (Huber, K., 1987, The Geologic Story of Yosemite National Park, U.S. Geological Survey Bulletin 1595). Photo by Doris Sloan.

# Commentary on "Chronology of emplacement of Mesozoic batholithic complexes in California and western Nevada" by J. F. Evernden and R. W. Kistler

**Warren D. Sharp and Paul R. Renne**
*Berkeley Geochronology Center, 2455 Ridge Road, Berkeley, California 94709; e-mail: wsharp@bgc.org; prenne@bgc.org*

## BACKGROUND

Researchers and laboratories in California have played a crucial role in the development and applications of radioisotopic geochronology. The international prominence of scientists and laboratories at the California Institute of Technology, the University of California at Berkeley and at Santa Barbara, and the U.S. Geological Survey (USGS) in Menlo Park began in the 1950s, led by scientists from the University of Chicago who had worked with Harold Urey and Harrison Brown, and who became the first generation of isotope geochemists.

The decade of the 1960s was a time of great ferment in the field of geochronology, and several laboratories in California were pioneering new developments in various areas. The worldwide production of geochronological data was increasing dramatically. Calibration points for the geologic time scale burgeoned from only five uranium-thorium-lead ages in 1947 (Holmes, 1947) to 337 in 1964 (Harland et al., 1964), of which some 286 were based on the potassium-argon (K-Ar) method. K-Ar was being applied vigorously to a wide variety of geologic problems, ranging from Precambrian geology of the southwestern United States (e.g., Aldrich et al., 1957; Wasserburg and Lanphere, 1965) to establishing a then-surprising antiquity for the Pleistocene (Evernden et al., 1957) and the time span of human evolution (Evernden and Curtis, 1965). K-Ar dating by G. Brent Dalrymple (USGS, Menlo Park) and Ian McDougall (Australian National University) was instrumental in revealing geomagnetic polarity reversals that were key to the nascent theory of plate tectonics (e.g., as recounted by Glen, 1982). Pioneering efforts were also underway in uranium-lead (U-Pb) and rubidium-strontium (Rb-Sr) dating. In particular, the superiority of U-Pb methods applied to highly retentive minerals like zircon, for determining the age of crystallization rather than cooling of plutonic and metamorphic rocks, was becoming appreciated (e.g., Silver, 1967). California laboratories in the 1960s were launching a pattern of leadership in geochronology research that was to flourish for the subsequent decades.

Much of the geochronology taking place in California laboratories in the 1950s and 1960s was directed at geologic problems of general scope, including extraterrestrial environs through studies of meteorites. The early California geochronologists did not overlook problems specific to Cordilleran geology, but major effort in this area was largely deferred until later. A notable exception was California's granitoid batholiths, the age and origin of which had attracted geologists' curiosity since the nineteenth century.

In 1958, only a decade after documentation that $^{40}$K decayed over geologic time to $^{40}$Ar (Aldrich and Nier, 1948), and eight years after the first published account of the quantitative use of this tool for geochronology (Smits and Gentner, 1950), radioisotopic dating was first applied to California's batholiths. Curtis et al. (1958) used the nascent K-Ar method to date granitoids from the Sierra Nevada, Klamath Mountains, and what is now recognized as the Salinian block or terrane in the southern California Coast Ranges.

A relative chronology of intrusion in the Yosemite region based on field relations had been established by Calkins (1930). The K-Ar dates from Curtis et al. (1958), determined solely on micas, were remarkably consistent with Calkins' (1930) chronology, as well as with constraints imposed by the existing geologic time scale. The dates suggested that plutonism was episodic in nature, occurring in pulses at 210–183 Ma, 136–124 Ma, and 90–80 Ma. Some aspects of this study soon stirred controversy. Questions were raised as to whether differences among the K-Ar dates for the Yosemite region were statistically significant, and whether the pattern of K-Ar data reflected the sequence of intrusion or the effects of subsequent cooling histories (e.g., see Dalrymple and Lanphere, 1969, p. 216–219). These questions and

---

Sharp, W. D., and Renne, P. R., 1999, Commentary on "Chronology of emplacement of Mesozoic batholithic complexes in California and western Nevada" by J. F. Evernden and R. W. Kistler, *in* Moores, E. M., Sloan, D., and Stout, D. L., eds., Classic Cordilleran Concepts: A View from California: Boulder, Colorado, Geological Society of America Special Paper 338.

others fueled more than a decade of intensive K-Ar studies by both the UC Berkeley and USGS Menlo Park laboratories.

## STATE OF THE ART IN THE 1960s: K-AR DATING AND THE AGE AND ORIGIN OF THE SIERRA NEVADA BATHOLITH

In the 1960s, when Evernden and Kistler's effort to date the Sierra Nevada batholith and other granitoids was ongoing, the prevailing explanation for the origin of batholiths was downwarping of a thick sedimentary pile, leading to melting and mixing with mantle material (e.g., Bateman and Wahrhaftig, 1966). The classic paper by Hamilton (1969a), which provided a link between California's batholiths and the newly recognized process of subduction, was probably never seen by Evernden and Kistler before they submitted their paper for publication, although a prescient paper by Robert Coats had been published several years earlier. Coats (1962) related andesitic arc magmatism in the Aleutian Islands to partial melting of basalt and sediment thrust down the Benioff zone (Benioff, 1954), the plane of deep earthquake foci dipping beneath the Aleutian volcanic arc. He described in some detail many aspects of convergent margin magmatism, including the location of the volcanic front above the 100-km-depth contour of the Benioff zone, and thermo-chemical interaction of melts with peridotite in the mantle wedge overlying the Benioff zone. Nevertheless, the dating reported in Evernden and Kistler (1970) was conceived and executed before the implications of plate tectonics for continental margin orogenesis were widely appreciated.

Evernden and Kistler (1970) sought to test earlier conclusions, based on more limited data (e.g., Curtis et al., 1958; Kistler et al., 1965; Lanphere et al., 1968), that granitoid emplacement in California's batholiths had occurred episodically with pulses of intrusion separated by intervening periods during which no plutons were intruded. One of their basic goals was to extend the previous work and test the possible episodicity with a more comprehensive data set, especially one including data for the more retentive mineral hornblende. Other fundamental questions were the geographic distribution of contemporaneous plutons and the depth of emplacement of plutons now exposed at the surface. They also realized that by establishing the age span of granitoid intrusion it would be possible to correlate the development of the batholith with the evolution of adjacent sedimentary basins, such as those underlying the Sacramento–San Joaquin Valley. Such age data were essential in order to evaluate the geosynclinal theory and other possible origins of the granitoid batholiths.

Evernden and Kistler (1970) reported K-Ar data for minerals separated from samples collected at 250 localities in granitic rocks of California and western Nevada. Their sampling emphasized the volumetrically dominant Sierra Nevada batholith, spanning the geographic and compositional range of its granitoids. They also analyzed granitoid samples from the Peninsular Ranges, Transverse Ranges, Coast Ranges, and Klamath Mountains, providing the first comprehensive basis for comparing the granitoid emplacement histories of California's diverse geomorphic provinces.

The Evernden and Kistler (1970) paper represented a culmination of effort that spanned more than a decade and was conducted in two laboratories (UC Berkeley and USGS Menlo Park). The sheer number of potassium and argon analyses reported in this paper was unprecedented and has probably never been surpassed in a single publication. Using mass spectrometers that were relatively insensitive by modern standards, Evernden and Kistler analyzed samples consisting of hundreds to thousands of mineral grains, most of which were hand-picked for purity. Each sample required many hours of laborious preparation before analysis, and each analysis required four to eight hours in the laboratory.

Because the ages were determined in two different laboratories using slightly different methods, Evernden and Kistler (1970) conducted extensive comparisons to satisfy themselves that their results were free of interlaboratory bias. They also performed many replicate analyses to establish the reproducibility of measurements. Their duplicate K-Ar analyses had a mean difference of less than 0.5%. This is remarkable for a time in which Ar isotopic ratios were determined by reading strip chart records with a ruler and magnifying glass!

Evernden and Kistler were concerned about the effects of sample heterogeneities on the mineral scale. They did a number of experiments aimed at clarifying the effects of chloritization in biotites. They found a discrepancy between the reproducibility of K analyses of granitic biotites (0.3%, 1$\sigma$) versus those of a leucite standard (0.13%) that they attributed to heterogeneity in the variably chloritized granitic biotites. The effect of chloritization on the ages of biotites was also systematically investigated with a surprising outcome, namely that chloritization by weathering resulted in anomalously old ages, meaning that the K was preferentially lost with respect to radiogenic Ar. Fortunately, this problem could be assessed with petrographic screening of samples.

Another issue at the time was the suitability of potassium feldspars for K-Ar dating. The pendulum of thinking on this topic had swung rapidly in the preceding years. Initially, it was thought that the framework structure of feldspars would make them retentive to radiogenic $^{40}$Ar, but these hopes were dashed by the earliest comparisons of K-Ar ages for pegmatitic potassium feldspar to ages for the same rocks determined from other systems (Wetherill et al., 1956). Subsequently, the volcanic potassium feldspar, sanidine, was shown to be excellent at retaining Ar (e.g., Evernden et al., 1960; Evernden and Curtis, 1965). Evernden and Kistler (1970) showed that granitic potassium feldspars with simple internal structures gave ages that agreed with coexisting biotites, while other potassium feldspars were younger than coexisting biotite. This focused attention on the role of intragranular discontinuities in the feldspar structure, such as exsolution lamellae, as the features that control diffusive loss of Ar. This was an early step on a path that has since been explored in great detail using sophisticated modeling of the effects of the size of intragranular discontinuities on thermally controlled Ar diffusion (e.g., Lovera et al., 1989).

By the time their study began, Evernden and Kistler appreciated two important potential limitations on the accuracy of ages determined by the K-Ar method: (1) the postcrystallization loss of radiogenic $^{40}$Ar ($^{40}$Ar*) from minerals at high temperatures, resulting in apparent ages younger than the true age, and (2) the presence of "excess" $^{40}$Ar trapped in minerals crystallizing at depth, which could result in spuriously old apparent ages.

Evernden and Kistler knew from the work of Hart (1964), who had characterized the resetting of K-Ar ages (and those of other systems) in Proterozoic wall rocks around the Tertiary Eldora stock of the Colorado Front Range, that Ar diffusion and loss is thermally controlled in the plutonic environment. Hart had also shown that at elevated temperatures, hornblende retains $^{40}$Ar* better than biotite, and biotite retains $^{40}$Ar* better than feldspars. Thus, Evernden and Kistler devoted considerable effort to analyzing multiple minerals from a given sample, emphasizing biotite-hornblende pairs. The disparity in Ar retention between biotite and hornblende was exploited as a criterion to recognize intrusive ages. Hornblende-biotite pairs that differed by no more than 5 m.y. were termed concordant. Agreement between apparent ages from these two minerals, they realized, was a powerful basis for inferring an emplacement age for the granitoid. Similarly, they recognized that disagreement between apparent ages from these two minerals, termed discordance, usually with biotite younger than hornblende, likely betokened slow cooling, moderate reheating by younger intrusions, or other failures of the assumption of a closed K-Ar system.

Evernden and Kistler were also concerned about the possible presence of "excess" $^{40}$Ar. In K-Ar dating, a correction is made for $^{40}$Ar from the atmosphere that is trapped in, or adsorbed onto, all samples. However, in some cases, additional $^{40}$Ar not produced by in situ decay of potassium initially may be trapped in, or diffused into, a sample, and this is termed excess $^{40}$Ar. A previous study (Kistler and Dodge, 1966) showed no evidence of excess $^{40}$Ar in Sierran pyroxenes, whose essentially potassium-free composition would be likely to manifest the problem. Nonetheless, Evernden and Kistler (1970) reasoned that analyzing late-crystallizing quartz, with its possible gaseous inclusions, would be another sensitive test for excess $^{40}$Ar. They analyzed five quartz samples for which they also had data from coexisting biotite and showed that quartz contained no significant excess $^{40}$Ar.

Subsequent to the publication of Evernden and Kistler (1970), $^{40}$Ar* loss at elevated temperatures has taken on a new significance. As geochronological systems that are robust at higher temperatures have become more widely applied (e.g., U-Pb in zircon and other minerals, and Sm-Nd dating), the temperature-dependent loss of $^{40}$Ar* by hornblende, muscovite, biotite, and potassium feldspar is no longer viewed as a barrier to determining crystallization ages. Rather, temperature-dependent $^{40}$Ar* loss may be exploited to determine the time-temperature histories of plutonic and metamorphic rocks. This is the basis of thermochronology (e.g., MacDougall and Harrison, 1988). The ability to circumvent the presence of excess $^{40}$Ar has also improved dramatically. The $^{40}$Ar/$^{39}$Ar dating and the Ar-isotope correlation diagram, first suggested by Merrihue and Turner (1966), and the $^{40}$Ar/$^{39}$Ar incremental heating technique (Turner et al., 1966; Turner, 1969) have come into widespread use. These approaches allow the possible presence of excess $^{40}$Ar to be evaluated directly from the analytical data (e.g., Lanphere and Dalrymple, 1971), and in favorable circumstances a quantitative correction for excess $^{40}$Ar may be applied to the age.

## AGE AND DEPTH PATTERNS IN THE CALIFORNIA BATHOLITHS

The centerpiece of Evernden and Kistler (1970) was to produce K-Ar ages for biotite and hornblende for a comprehensive sampling of the Sierra Nevada batholith. Armed with this extensive data set and criteria for recognizing intrusive ages and rapid versus slow cooling, Evernden and Kistler were in a position to address the questions of the age span of batholith emplacement, the geographic distribution of contemporaneous plutons, whether emplacement had occurred in multiple episodes or formed a continuum, and to estimate depths of emplacement.

The age span of the Sierra Nevada batholith was not previously known. Stratigraphic relations provided few constraints, and previous geochronological studies were of insufficient breadth to characterize the age range. Earlier studies (Curtis et al., 1958; Kistler et al., 1965) had revealed several distinct periods of batholith emplacement, but these were separated by intervals of several tens of millions of years for which granitoid ages were lacking. In contrast, Evernden and Kistler's (1970, p. 16) data showed that:

> ...granite emplacement in the environs of the Sierra Nevada was an essentially continuous process covering the period from 210 m.y. ago to approximately 80 m.y. ago.

Nevertheless, Evernden and Kistler inferred that granitoid emplacement was clustered in several pulses, which they termed intrusive epochs. Supplementing the K-Ar data with a few Rb-Sr whole-rock ages and utilizing field- and petrological-based groupings, they defined five such epochs, beginning at 210, 180, 148, 121, and 90 Ma, and each lasting 10 to 20 m.y. A surprisingly simple pattern of ages for the Sierra Nevada batholith resulted from consideration of their data. Intrusions of the three youngest epochs, termed Yosemite, Huntington Lake, and Cathedral Range, are of Early to Late Cretaceous age and form subparallel, northwest-trending belts that become progressively younger eastward. The oldest of these belts crops out in the western foothills, and the youngest crops out along the crest of the Sierra. A belt of Early to Middle Jurassic intrusions, the Inyo Mountains epoch, is split by the Cretaceous batholith, cropping out east of the Sierran crest in the southern Sierra and White-Inyo Mountains, and west of the Cretaceous granitoid belts in the foothills of the central and eastern Mother Lode belt (Fig. 1). The oldest intrusions, of the Triassic-Jurassic Lee Vining epoch, form a smaller, north-trending belt extending about 70 km on either side of Mono Lake. Nearly 30 years later, this basic age pattern for the Sierra Nevada batholith remains valid, although it has been modified and enhanced by many additional geochronological studies (e.g., Chen and Moore, 1982; Saleeby and Sharp, 1980; Stern et al., 1981; Tobisch et al., 1993, 1995).

The number of intrusive epochs proposed by Evernden and Kistler and their apparent 30 m.y. periodicity were soon questioned (Armstrong and Suppe, 1973; Lanphere and Reed, 1973). The epochs are not implicit in Evernden and Kistler's geochronological data. For example, concordant-age pairs populate the interval between the Huntington Lake and Cathedral Peak epochs. Rather, their designation of the epochs apparently rested on more interpretive grounds, such as field- and petrological-based groupings and volume estimates. Additional geochronological data, including many multiple-fraction U-Pb zircon ages, show that Cretaceous granitoids form a continuum of ages from about 125 to 80 Ma, with a steady, eastward migration of the locus of intrusion of 2.7 mm/yr (Chen and Moore, 1982). The apparent lull between Evernden and Kistler's Early to Middle Jurassic Inyo Mountains and Late Jurassic Yosemite epochs has also been filled by reliable emplacement ages. Together, these changes indicate that the Sierra Nevada batholith is principally composed of two great magmatic arcs, one Cretaceous and one Jurassic. The oldest Mesozoic granitoids of the Sierra Nevada, those of the Late Triassic–Early Jurassic Lee Vining epoch, are apparently the more restricted remnant of a third, older magmatic arc.

The voluminous Late Cretaceous granites of the Cathedral Range intrusive epoch that are found along the Sierran crest are characterized by concordant hornblende-biotite ages. They are the youngest group of intrusions in the batholith, so they have not been reheated. Rapid cooling, however, was also necessary to pass quickly through what today would be termed the closure temperatures of hornblende and biotite (about 500 °C and 300 °C, respectively). Evernden and Kistler reasoned that rapid

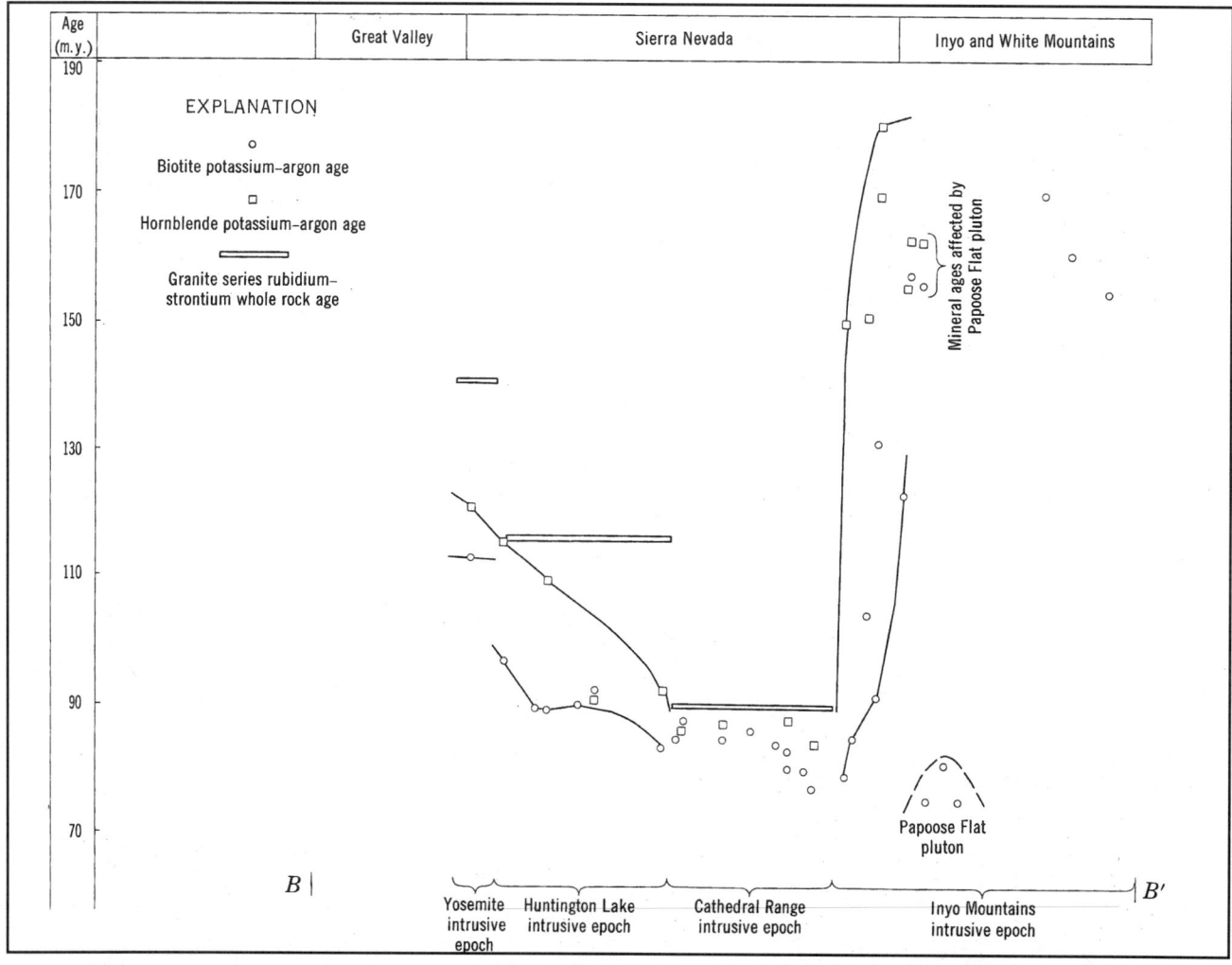

Figure 1. Reproduction of Evernden and Kistler's (1970) Figure 8. K-Ar mineral ages and Rb-Sr whole-rock ages for granitic rocks projected onto a section across the central part of the batholith at about the latitudes of Fresno and Bishop, California. Note, from west to east, the progression towards younger ages in the Cretaceous granitoids of the Yosemite, Huntington Lake and Cathedral Range epochs, followed by the sharp step to Jurassic ages in granitoids in the Inyo Mountains epoch.

cooling resulted from emplacement of granitoids into cold wall rocks at shallow crustal levels of 12 km or less. This conclusion has been confirmed for these intrusions by the recognition of the Ritter Range caldera, which is intruded by plutons of the Tuolumne Intrusive Suite of the Cathedral Range epoch (Fiske and Tobisch, 1978), as well as by study of metamorphic assemblages (Hanson et al., 1993), and geobarometry of the granitoids (Ague and Brimhall, 1988). In addition, cooling of intrusions of the Cathedral Range epoch along the western edge of the belt was enhanced by extensional unroofing (Renne et al., 1993).

Some appreciation of how far our knowledge has advanced in the area of temperature-depth conditions and thermal histories can be gleaned from the opening sentences of the discussion of depth of emplacement. Evernden and Kistler (1970, p. 12) wrote:

> Such an approach must be based upon a heat model of granite emplacement and cooling. We are at once on uncertain ground. But if the tenuousness of the fabric being woven is fully appreciated, no harm results from such discussions, and possibly some useful ideas may result.

Evernden and Kistler also identified deeper crustal levels of batholithic rocks on the basis of strong hornblende-biotite discordances in the western and southern Sierra Nevada batholith, the Santa Lucia region of the central Coast Ranges, and the San Gabriel Mountains of the Transverse Ranges. Thermobarometry has borne out this conclusion for the southernmost Sierra Nevada batholith (Pickett and Saleeby, 1993).

## CALIFORNIA BATHOLITHS AND REGIONAL RELATIONS

The scope of the new geochronological data provided the opportunity to examine the relations between batholith development and sedimentation to the east and west in more detail than had previously been possible. In the Sacramento–San Joaquin Valley, eastward transgression of marine sandstones throughout Cretaceous time coincided with batholith emplacement, but was followed by oscillatory regression after emplacement of granitoids ceased (Fig. 2). In the epicontinental seas to the east of the batholith, widespread transgression also coincided with the Jurassic and Cretaceous age span of the Sierra Nevada batholith. Changes to the geologic time scale and in the $^{40}K$ decay constants ($^{40}K$ decays to both $^{40}Ar$ and $^{40}Ca$) have shifted our current view of these relations somewhat. Nevertheless, these continent-scale correlations argued for a common mechanism that could control regional patterns of sedimentation in concert with batholith emplacement. The authors concluded correctly that the origin of batholiths envisioned in geosynclinal theory was inadequate to explain their observations, and they wrote (Evernden and Kistler, 1970, p. 26):

It is obvious that the fundamental explanation for the mobilization of the magmas subsequently emplaced in eastern California cannot be found in localized down-warping of a deep sedimentary basin. Mechanisms much more profound than this must be found to explain the detailed correspondence of the events here discussed. Present geological thought would probably attempt to find a driving mechanism for these events, either in complimentary phase changes or by horizontal convective transport of vast quantities of material.

Appreciation of the implications of plate tectonic theory for convergent plate margin orogenesis, especially for Mesozoic magmatism, metamorphism, and sedimentation in California, virtually exploded in the late 1960s and early 1970s. Evernden and Kistler's age data for the Sierra Nevada batholith were essential for developing our present view of the batholith's origins, as well as its relations to other regional and global phenomena. Newly understood properties of the batholith were incompatible with formation by melting of sediments in a crustal downwarp. These included the narrow, linear distributions of contemporaneous plutons, the shallow emplacement levels of some Sierran intrusions, and the relatively primitive (mantle-like) initial Sr isotope ratios of the granitoids documented by Evernden and Kistler (1970). Voluminous volcanic-plutonic-derived sandstones in the Great Valley sequence (e.g., Ojakangas, 1968) are broadly synchronous with granitoid emplacement and unconformably overlie the western flank of the batholith. The relations indicate that during magmatism the batholith was capped by high-standing volcanic and volcaniclastic rocks, and was a source of sediment, not a sink. The Sierra Nevada batholith is the root of one or more volcanic-plutonic arcs that was fed by mantle-derived magmas rising from subducting oceanic lithosphere (Hamilton, 1969b; Dickinson, 1970).

The lithologic, metamorphic, and structural characteristics of the Franciscan assemblage (now the Franciscan Complex) led to its interpretation as the product of subduction of oceanic lithosphere (e.g., Ernst, 1965, 1970; Hsü, 1968; see Wakabayashi, this volume). Age data for high-pressure–low-temperature metamorphism in the Franciscan Complex (Lee et al., 1964; Suppe, 1969; Suppe and Armstrong, 1972), and ages for magmatism and metamorphism in the Klamath Mountains, the Mojave Desert, and southern California (e.g., Lanphere et al., 1968; Armstrong and Suppe, 1973), combined with Evernden and Kistler's data, supported interpretation of the Franciscan Complex and the Sierra Nevada batholith and its wall rocks as paired metamorphic belts, analogous to those of the circum-Pacific region (e.g., Miyashiro, 1961). Controversies that raged in the 1970s and 1980s over the extent, number, and original location of arc-derived terranes represented in the Mesozoic geology of California (e.g., Coney et al., 1980; Wright and Fahan, 1988) remain to be resolved. The possible role of large-scale, orogen-parallel, strike-slip displacements is also an area of ongoing research (e.g., Beck, 1980; Whidden et al., 1998). The insights discussed here, however, which relied heavily on the ages of Evernden and

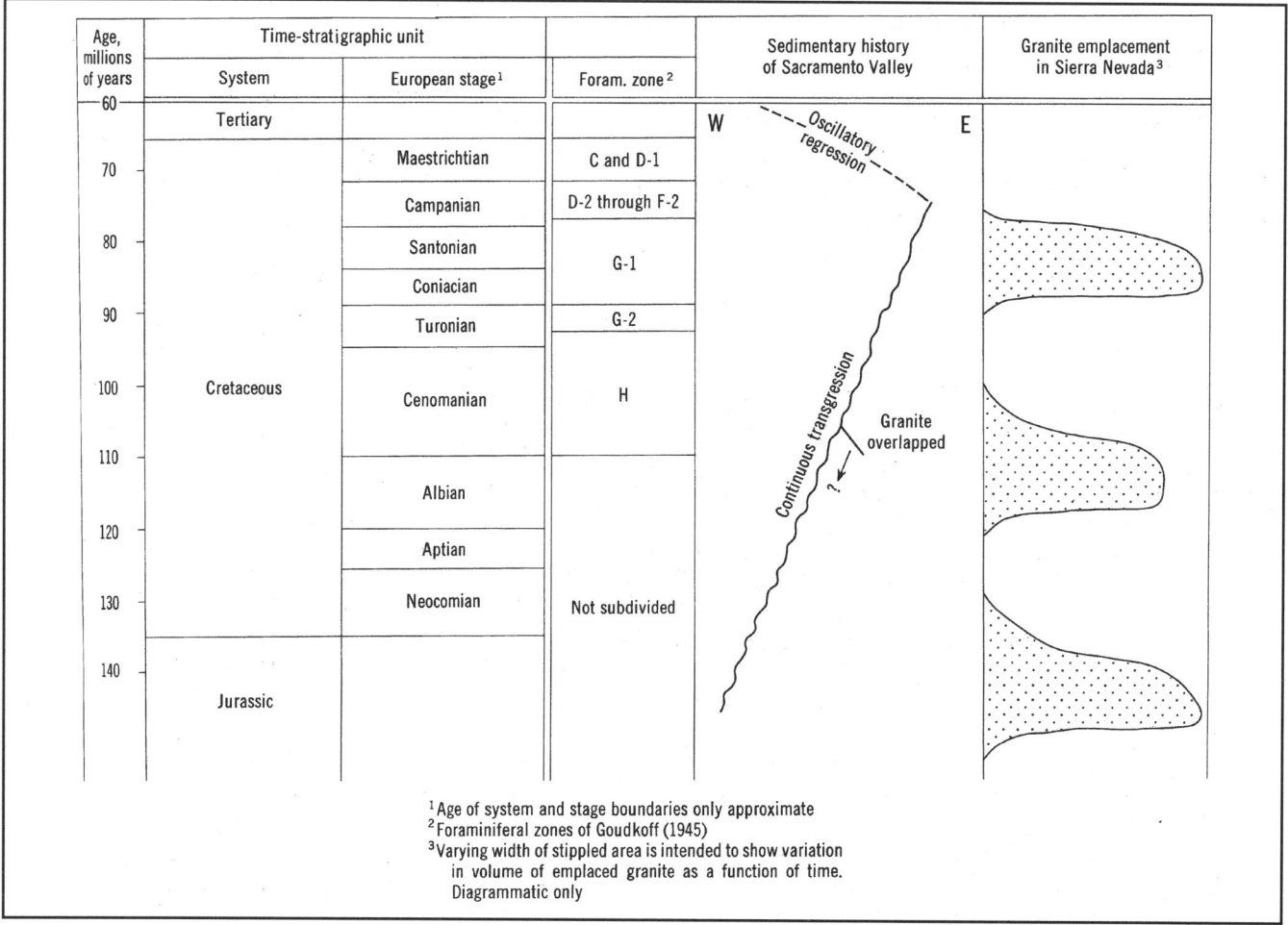

Figure 2. Reproduction of Evernden and Kistler's (1970) Figure 9: Their caption is: "Relation of sedimentation in the Sacramento Valley to granite emplacement in the Sierra Nevada area...." The dotted areas in the right column give a qualitative representation of relative volumes of granitoids emplaced in their three youngest intrusive epochs, Yosemite, Huntington Lake, and Cathedral Peak, from oldest to youngest. The note "Granite overlapped" refers to the overlap of Cenomanian sandstones on Sierran-type granites in the subsurface of the Sacramento Valley.

Kistler (1970), form the basis of modern interpretations of California's geologic development in the Mesozoic. These same insights were also an early example of the application of plate tectonic theory to interpretation of ancient convergent margin orogens that had worldwide influence.

## ACKNOWLEDGMENTS

We thank Paul Bateman and Marvin Lanphere for reviewing an earlier draft of the manuscript, and Garniss Curtis, Bill Glen, and Ken Ludwig for helpful discussions.

We dedicate this commentary to the late Joachim Hampel. Beginning in 1961, Joachim was a staff member in the Department of Geology and Geophysics at UC Berkeley for more than 30 years. He was a contributor to Evernden and Kistler (1970), carrying out mineral separations and potassium analyses, and provided the portrait of Jack Evernden published here. Over the years, he produced thousands of potassium analyses for K-Ar dating and studies in igneous petrology and geochemistry. In later years he also ran the department's X-ray fluorescence laboratory. Hampel was a highly accomplished photographer, and he took great pride in his portraits and photomicrographs, which were widely used in journal articles and departmental publications and displays. He was also a steam train enthusiast, and over the years built an extensive narrow-gauge model railroad layout in his home. He was a friend and colleague to faculty, fellow staff members, and generations of students, remaining in close contact after his retirement in 1993 until his death in August, 1998.

## REFERENCES CITED

Ague, J. J., and Brimhall, G. H., 1988, Regional variations in bulk chemistry, mineralogy, and the compositions of mafic and accessory minerals in the batholiths of California: Geological Society of America Bulletin, v. 100, p. 891–911.

Aldrich, L. T., and Nier, A. O., 1948, Argon 40 in potassium minerals: Physical Reviews, v. 74, p. 876–877.
Aldrich, L. T., Wetherill, G. W., and Davis, G. L., 1957, Occurrence of 1350 million-year-old granitic rocks in western United States: Geological Society of America Bulletin, v. 68, p. 655–656.
Armstrong, R. L., and Suppe, J., 1973, Potassium-argon geochronology of Mesozoic igneous rocks in Nevada, Utah, and southern California: Geological Society of America Bulletin, v. 84, p. 1373–1391.
Bateman, P. C., and Wahrhaftig, C., 1966, Geology of the Sierra Nevada, in Bailey, E. H., ed., Geology of northern California: California Division of Mines and Geology Bulletin 190, p. 107–172.
Beck, M. E., Jr., 1980, Paleomagnetic record of plate-margin tectonic processes along the western edge of North America: Journal of Geophysical Research, v. 85, p. 7115–7131.
Benioff, H., 1954, Orogenesis and deep crustal structure: Additional evidence from seismology: Geological Society of America Bulletin, v. 65, p. 385–400.
Calkins, F. C., 1930, The granitic rocks of the Yosemite region, in Mathers, F. E., ed., Geologic history of the Yosemite Valley: U.S. Geological Survey Professional Paper 160, 137 p.
Chen, J. H., and Moore, J. G., 1982, Uranium-lead isotopic ages from the Sierra Nevada batholith, California: Journal of Geophysical Research, v. 87, no. B6, p. 4761–4784.
Coats, R. R., 1962, Magma type and crustal structure in the Aleutian Arc, in MacDonald, G. A., and Kuno, H., eds., The crust of the Pacific basin: American Geophysical Union Geophysical Monograph 6, p. 92–109.
Coney, P. J., Jones, D. L., and Monger, J. W. H., 1980, Cordilleran suspect terranes: Nature, v. 288, p. 329–333.
Curtis, G. H., Evernden, J. F., and Lipson, J., 1958, Age determination of some granitic rocks in California by the potassium-argon method: California Division of Mines and Geology Special Report, v. 54, 16 p.
Dalrymple, G. B., and Lanphere, M. A., 1969, Potassium-argon dating: San Francisco, W. H. Freeman and Co., 258 p.
Dickinson, W. R., 1970, Relations of andesites, granites, and derivative sandstones to arc-trench tectonics: Reviews of Geophysics and Space Physics, v. 8, p. 813–860.
Ernst, W. G., 1965, Mineral parageneses in Franciscan metamorphic rocks, Panoche Pass, California: Geological Society of America Bulletin, v. 76, p. 879–914.
Ernst, W. G., 1970, Tectonic contact between the Franciscan melange and the Great Valley sequence, crustal expression of a late Mesozoic Benioff zone: Journal of Geophysical Research, v. 75, p. 886–902.
Evernden, J. F., and Curtis, G. H., 1965, The potassium-argon dating of late Cenozoic rocks in East Africa and Italy: Current Anthropology, v. 6, p. 343–385.
Evernden, J. F., and Kistler, R. W., 1970, Chronology of emplacement of Mesozoic batholithic complexes in California and western Nevada: U.S. Geological Survey Professional Paper 623, 28 p.
Evernden, J. F., Curtis, G. H., and Kistler, R. W., 1957, Potassium-argon dating of Pleistocene volcanics: Quaternaria, v. IV, p. 1–5.
Evernden, J. F., Curtis, G. H., Kistler, R. W., and Obradovich, J., 1960, Argon diffusion in glauconite, microcline, sanidine, leucite, and phlogopite: American Journal of Science, v., 258, p. 583–604.
Fiske, R. S., and Tobisch, O. T., 1978, Paleogeographic significance of volcanic rocks of the Ritter Range pendant, central Sierra Nevada, California, in Howell, D. G., and McDougall, K. A., eds., Mesozoic paleogeography of the western United States: Pacific Section, Society of Economic Paleontologists and Mineralogists Pacific Coast Paleogeography Symposium 2, p. 209–221.
Glen, W., 1982, The road to Jaramillo: Stanford, California, Stanford University Press, 459 p.
Hamilton, W. O., 1969a, Mesozoic California and the underflow of Pacific mantle: Geological Society of America Bulletin, v. 80, p. 2409–2430.
Hamilton, W. O., 1969b, The volcanic central Andes, a modern model for the Cretaceous batholiths and tectonics of the western North America: Oregon Department of Geology and Mineral Industries Bulletin, v. 65, p. 175–184.
Hanson, R. B., Sorensen, S. S., Barton, M. D., and Fiske, R. S., 1993, Long-term evolution of fluid-rock interactions in magmatic arcs: Evidence from the Ritter Range pendant, Sierra Nevada, California, and numerical modeling: Journal of Petrology, v. 34, p. 23–62.
Harland, W. B., Smith, A. G., and Wilcock, B., eds., 1964, The Phanerozoic time scale: Geological Society of London Quarterly Journal, v. 120, 458 p.
Hart, S. R., 1964, The petrology and isotopic-mineral age relations of a contact zone in the Front Range, Colorado: Journal of Geology, v. 72, p. 493–525.
Holmes, A., 1947, The construction of a geological time-scale: Geological Society of Glasgow Transactions, v. 21, p. 117–152.
Hsü, K. J., 1968, Principles of melanges and their bearing on the Franciscan-Knoxville paradox: Geological Society of America Bulletin, v. 79, p. 1063–1074.
Kistler, R. W., and Dodge, F. C. W., 1966, Potassium-argon ages of coexisting minerals from pyroxene-bearing granitic rocks of the central Sierra Nevada, California: Journal of Geophysical Research, v. 71, p. 2157–2161.
Kistler, R. W., and Peterman, Z. E., 1973, Variations in Sr, Rb, K, Na, and initial $Sr^{87}/Sr^{86}$ in Mesozoic granitic rocks and intruded wall rocks in central California: Geological Society of America Bulletin, v. 84, p. 3489–3512.
Kistler, R. W., Bateman, P. C., and Brannock, W. W., 1965, Isotopic ages of minerals from granitic rocks of the central Sierra Nevada and Inyo Mountains, California: Geological Society of America Bulletin, v. 76, p. 155–164.
Lanphere, M. A., and Dalrymple, G. B., 1971, A test of the $^{40}Ar/^{39}Ar$ age spectrum technique on some terrestrial materials: Earth and Planetary Science Letters, v. 12, p. 359–372.
Lanphere, M. A., and Reed, B. L., 1973, Timing of Mesozoic and Cenozoic plutonic events in circum-Pacific North America: Geological Society of America Bulletin, v. 84, p. 3773–3782.
Lanphere, M. A., Irwin, W. P., and Hotz, P. E., 1968, Isotopic age of the Nevadan orogeny and older plutonic and metamorphic events in the Klamath Mountains, California: Geological Society of America Bulletin, v. 79, p. 1027–1052.
Lee, D. E., Thomas, H. H., Marvin, R. F., and Coleman, R. G., 1964, Isotopic ages of glaucophane schists from the Cazadero area: U.S. Geological Survey Professional Paper 475-D, p. 105–107.
Lovera, O. M., Richter, F. M., and Harrison, T. M., 1989, The $^{40}Ar/^{39}Ar$ thermochronometry for slowly cooled samples having a distribution of diffusion domain sizes: Journal of Geophysical Research, v. 94, no. 12, p. 17,917–17,935.
MacDougall, I., and Harrison, T. M., 1988, Geochronology and thermochronology by the $^{40}Ar/^{39}Ar$ method: New York, Oxford University Press, 212 p.
Merrihue, C., and Turner, G., 1966, Potassium-argon dating by activation with fast neutrons: Journal of Geophysical Research, v. 71, p. 2852–2857.
Miyashiro, A., 1961, Evolution of metamorphic belts: Journal of Petrology, v. 2, p. 277–311.
Ojakangas, R. W., 1968, Cretaceous sedimentation, Sacramento Valley, California: Geological Society of America Bulletin, v. 79, p. 973–1008.
Pickett, D. A., and Saleeby, J. B., 1993, Thermobarometric straints on the depth of exposure and conditions of plutonism and metamorphism at deep levels of the Sierra Nevada batholith, Tehachapi Mountains, California: Journal of Geophysical Research, v. 98, p. 609–629.
Renne, P. R., Tobisch, O. T., and Saleeby, J. B., 1993, Thermochronologic record of pluton emplacement, deformation and exhumation at Courtright shear zone: Geology, v. 21, p. 331–334.
Saleeby, J. B., and Sharp, W. D., 1980, Chronology of the structural and petrologic development of the southwest Sierra Nevada Foothills, California: Geological Society of America Bulletin, v. 91, part I, p. 317–320, part II, p. 1416–1535.
Silver, L. T., 1967, Apparent age relations in the older Precambrian stratigraphy of Arizona [extended abs.] in Burwash, R. A., and Morton, R. D., eds., Proceedings Conference on Geochronology of Precambrian Stratified Rocks: Edmonton, Canada, University of Alberta, p. 87.

Smits, F., and Gentner, W., 1950, Argonbestimmungen an Kalium-Mineralen, I. Bestimmungen an Tertiären Kalisalzen: Geochimica et Cosmochimica Acta, v. 1, p. 22–27.

Stern, T. W., Bateman, P. C., Morgan, B. A., Newell, M. F., and Peck, D. L., 1981, Isotopic ages of zircon from the granitoids of the central Sierra Nevada, California: U.S. Geological Survey Professional Paper 1185, 17 p.

Suppe, J., 1969, Times of metamorphism in the Franciscan terrain of the northern coast ranges of California: Geological Society of America Bulletin, v. 80, p. 135–142.

Suppe, J., and Armstrong, R. L., 1972, Potassium-argon dating of Franciscan metamorphic rocks: American Journal of Science, v. 272, p. 217–233.

Tobisch, O. T., Renne, P. R., and Saleeby, J. B., 1993, Deformation resulting from regional extension during pluton ascent and emplacement, central Sierra Nevada, California: Journal of Structural Geology, v. 15, p. 609–628.

Tobisch, O. T., Saleeby, J. B., Renne, P. R., McNulty, B., and Tong, W., 1995, Variations in deformation fields during development of a large-volume magmatic arc, central Sierra Nevada, California: Geological Society of America Bulletin, v. 107, p. 148–166.

Turner, G., 1969, Thermal histories of meteorites by the $^{39}$Ar-$^{40}$Ar method, *in* Millman, P. M., ed., Meteorite research: Dordrecht, Reidel Publishing Company, p. 407–417.

Turner, G., Miller, J. A., and Grasty, R. L., 1966, The thermal history of the Bruderheim meteorite: Earth and Planetary Science Letters, v. 1, p. 155–157.

Wasserburg, G. J., and Lanphere, M. A., 1965, Age determinations in the Precambrian of Arizona and Nevada: Geological Society of America Bulletin, v. 76, p. 735–758.

Wetherill, G. W., Wasserburg, G. J., Aldrich, L. T., Tilton, G. R., and Hayden, R. J., 1956, Decay constants of $^{40}$K as determined by the radiogenic argon content of potassium minerals: Physical Review, v. 103, p. 987–989.

Whidden, K. J., Lund, S. P., Bottjer, D. J., Champion, D., and Howell, D. G., 1998, Paleomagnetic evidence that the central block of Salinia (California) is not a far-travelled terrane: Tectonics, v. 17, p. 329–343.

Wright, J. E., and Fahan, M. R., 1988, An expanded view of Jurassic orogenesis in the western United States Cordillera; Middle Jurassic (pre-Nevadan) regional metamorphism and thrust faulting within an active arc environment, Klamath Mountains, California: Geological Society of America Bulletin, v. 100, p. 859–876.

MANUSCRIPT ACCEPTED BY THE SOCIETY ON NOVEMBER 23, 1998.

Chapter 11

# THE SOUTHERN CALIFORNIA BATHOLITH

**Gordon R. Gastil**
*9435 Alto Drive*
*La Mesa, California 91941*

Photo courtesy of Bancroft Library, University of California, Berkeley.

## ESPER SIGNIUS LARSEN, JR.

Esper Signius Larsen, Jr., was born in Oregon in 1879. He received a bachelor of science degree in 1906 and his doctor of philosophy degree in 1918, both from the University of California (Berkeley). Between these degrees he spent two years at the Carnegie Institution and then went to the U.S. Geological Survey in 1909, where he became chief of the Petrology Section and remained until 1923, when he became professor of petrography at Harvard University. Following retirement, he returned to the survey to continue his work in chemical mineralogy. His primary interest was field work, and during his entire career he spent virtually every summer in the field. These studies led to the publication of a number of major regional papers, such as that on the southern California batholith excerpted here.

Larsen was president of the Mineralogical Society of America in 1928 and in 1941 was awarded its Roebling Medal. He received the Penrose Medal of the Geological Society of America in 1953.

Gordon R. Gastil

*The Geological Society of America*
Memoir 29

# BATHOLITH AND ASSOCIATED ROCKS OF CORONA, ELSINORE, AND SAN LUIS REY QUADRANGLES SOUTHERN CALIFORNIA

BY

ESPER S. LARSEN, JR.
Harvard University
Cambridge, Mass.

June 21, 1948

## ABSTRACT

The batholith of Southern and Lower California is exposed continuously from near Riverside, California, southward for a distance of about 350 miles. In central Lower California it is covered in part by younger rocks, but discontinuous bodies extend to the southern end of Lower California, and hence the batholith is probably over 1000 miles long. Its width is about 60 miles. A strip across the northern part of the batholith about 70 miles wide has been studied; the western half was mapped in detail, and the eastern half was covered in rapid reconnaissance.

In the area studied the batholith intrudes Triassic sediments and Jurassic(?) volcanic rocks along its western border and Paleozoic sediments along its eastern border. Screens and roof pendants are common within the batholith. The Triassic rocks are mildly metamorphosed in the western part of the area but become progressively more coarsely crystalline toward the east. The Paleozoic rocks are rather coarsely crystalline. The metamorphism in large part preceded the intrusion of the batholith, and only locally was there appreciable contact metamorphism. The batholith and older rocks are overlain by Upper Cretaceous and younger sediments. Small bodies of andesite and basalt are associated with the Tertiary sediments, and small bodies of nepheline basalt of Quaternary age are present in the area. The batholith was intruded in early Upper Cretaceous time.

The batholith in the area studied was emplaced by over 20 separate injections. Most of the resulting rock types are found in only one or a few small bodies which are confined to a small area. In the area studied in detail (Pl. 1) five types are present in many large, widely separated bodies, making up about 88 per cent of the area underlain by the batholith. In the eastern half of the batholith three more widespread types are present. In the western half of the body the rocks range fro n gabbro to granite, but in the eastern half several tonalites constitute nearly the whole of the mass. The gabbro is composed of many related rocks. Some have hornblende, some pyroxene; in some the plagioclase is anorthite, in others it is as sodic as andesine-labradorite. Some of the tonalites contain abundant inclusions that have been almost completely reworked by the magma and have been softened and stretched into thin discs. These inclusions are well oriented and near the contacts with older rocks they parallel the contacts, but elsewhere they strike about N. 30° W. and d:p steeply to the east. One tonalite, whose feldspar is andesine, has scattered crystals with cores of bytownite, and has well-crystallized hornblende with cores of pale uralitic hornblende and remnants of augite. Hornblende and biotite are the predominant mafic minerals of the tonalites and granodiorites. The iron content of the mafic minerals of the gabbros is moderate, and it increases as the rocks become richer in silica. The norms and the modes are shown on a variation diagram (Figs. 11, 12). The chemical analyses of the rocks fall near smooth variation curves (Fig. 4).

The general strike of the structures of the area have been about N. 30° W. from Paleozoic to the present time. The Paleozoic and Triassic sediments, the orientation of the inclusions and other structures of the batholith, the elongation of the batholith and the mountain ranges, and the strike of the major faults are in about the same direction. In the batholith and the older sediments the dips are steep to the east.

The batholith must have been emplaced by stoping and not by forceful injection. Calculations show that the cooling of a large batholith is chiefly through the roof and not through the walls. Crystallization to a depth of 3 kilometers takes place in about half a million years. The different rocks of the batholith were formed from the intermediate gabbro by crystal differentiation and assimilation in depth.

In early Upper Cretaceous time diastrophism folded the older rocks and formed, in depth, a strip of gabbroic magma about 1000 miles long. A small amount of this

# INTRODUCTION

## LOCATION

The area described in this report is in southern California and includes a part of the group of mountains known as the Peninsular Ranges (Fig. 1). It lies southeast of Los Angeles and north of San Diego. It is bounded on the north by latitude 34°, on the south by latitude 33°, on the west by the Pacific Ocean, and on the east by longitude 117°. It is mapped on the topographic sheets (scale 1:125,000) of the Corona, Elsinore, Capistrano, and San Luis Rey quadrangles....

The area is about 70 miles from north to south and somewhat less from east to west. It includes parts of Orange, Riverside, and San Diego counties. The two quadrangles east of the area mapped in detail have been covered in reconnaissance.

## SAN MARCOS GABBRO

* * * * * * *

### OCCURRENCE AND FIELD RELATIONS

The San Marcos gabbro occurs in the San Luis Rey, Elsinore, and Corona quadrangles in about a hundred separate mappable bodies. Many of these are very small. The largest is southeast of Lake Elsinore on the Elsinore quadrangle and underlies an area of about 25 square miles....

From a reconnaissance study of the area east of the Elsinore and San Luis Rey quadrangles and from the map of the San Jacinto quadrangle by Fraser (1931, p. 508-509), no bodies of San Marcos gabbro are known in the eastern part of the mountains, as far east as the Colorado Desert. It is confined to the area west of the line running from a point a mile east of the southeastern corner of the Elsinore quadrangle, thence to a point a few miles east of the northwestern corner of the Elsinore quadrangle.

The San Marcos gabbro intrudes the Bedford Canyon formation of Triassic age and rocks of the overlying Santiago Peak volcanics. It is older than the Green Valley tonalite and the Lakeview Mountain tonalite and appears to be the oldest rock of the Cretaceous (?) batholith.

* * * * * * *

## BATHOLITH OF SOUTHERN CALIFORNIA

magma was intruded nearly to the surface. The deep magma differentiated quietly until its upper part attained the composition of a tonalite. Earth movements then occurred at least five times in rapid succession and caused the injection of the different tonalites. Some of these carry abundant inclusions, indicating a widespread shattering of the wall rock shortly before final emplacement. From time to time local movements caused the injections of the different granodiorites. When the deep-seated magma reached the composition of a light-colored granodiorite, widespread diastrophism moved the main granodiorite upward. Further local movement caused the emplacement of the many local granodiorites and granites.

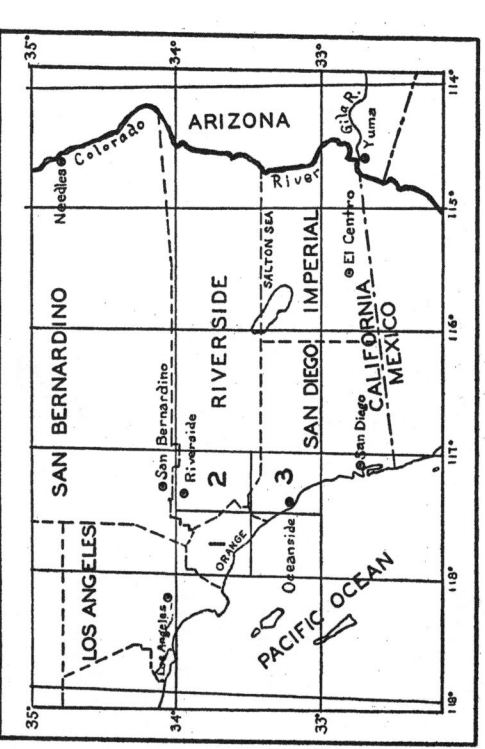

FIGURE 1.—*Index map of Southern California*
Showing area described in this report
1. Corona quadrangle. 2. Elsinore quadrangle
3. San Luis Rey quadrangle

## GENERAL DESCRIPTION

The San Marcos gabbro of the San Luis Rey quadrangle was described in detail by Miller (1937, 1938), and the present description is largely taken from Miller.

In the field one of the most noticeable features of the rock is its variability. In places good-sized bodies of one kind of rock are found, but in others the rock varieties are a few tens of feet or less across. Nearly every large body of the rock shows this variability...

The most widespread and marked type of variation results from abrupt changes in the amount or character of the hornblende. At other places the rock appears to be made of two distinct phases, with one intimately penetrating the other.

## PETROGRAPHY

*...Olivine Rocks.—*...The two outstanding features of the olivine rocks are the calcic plagioclase and the mosaic texture, characterized by nearly equant anhedral grains. The typical calcic feldspar ($An_{93}$) shows no zoning, and in rocks with slight zoning the feldspar is somewhat less calcic ($An_{87}$-$An_{90}$). The mosaic texture is less well developed, and hornblende is likely to be present.

The olivine contains 25 to 31 percent fayalite... The hornblende is pale green, and the hypersthene contains 21 to 25 percent $FeSiO_3$. The hornblende and pyroxene tend to be poikilitic or interstitial to the plagioclase and olivine.

*Norites and Hypersthene Gabbros.—*Miller (1937, p. 1404-1407) gives the following description of this rock in the San Luis Rey quadrangle.

"The most widespread hornblende-poor rock is a typical norite. This is a medium-grained plutonic rock composed essentially of plagioclase and hypersthene with minor augite. Through a local increase in augite relative to hypersthene, the rock grades to hypersthene gabbro without other change in appearance or composition. The norites are notably more uniform in mineral content and grain size than the olivine rocks. The average grain size is 1 to 2 mm., and the principal deviation from this arises through the appearance of a matrix of abundant, much smaller grains (0.05 to 0.4 mm.) in the midst of grains of the ordinary size. This produces seriate and porphyritic textures in some of the norite bodies. Yet on the whole the norites tend to maintain a given textural character over a much greater area than do the olivine rocks. The rocks with porphyritic texture tend to be richer in augite than the other norites, and so include a considerable proportion of the hypersthene gabbros.

"The plagioclase of the typical norites is labradorite with an average composition of $An_{65}$. It does not display the mosaic texture but occurs in well-developed tabular grains, ordinarily showing moderate zoning and sometimes a subparallel arrangement. An unusual feature is the occasional appearance of a core of bytownite-anorthite within some of the larger labradorite grains. The core may have a regular polygonal shape conformable to the crystal structure or consist merely of irregular patches or shreds or even a hollow shell. The core material is in the composition range $An_{80}$-$An_{92}$, but chiefly $An_{87}$-$An_{90}$. Even the inner zones of the enclosing plagioclase have an anorthite content of only $An_{60}$-$An_{70}$. Thus there is a sharp composition break between the calcic cores and the adjacent labradorite, which represents a difference of more than 20 per cent in anorthite content.

"The hypersthene and augite of the norites show a much stronger tendency to attain their characteristic prismatic form than those of the olivine rocks. The hypersthene has definitely higher refractive indices, stronger pleochroism, and a higher Fe/Mg ratio (28-32 per cent iron metasilicate) than the hypersthene of the olivine rocks.

"Most of the hornblende of the norites is a compact deep-green hornblende derived largely from a reaction with the pyroxene. All stages are found in the centripetal replacement of hypersthene and augite by this reaction hornblende up to a point where only small remnants of pyroxene are enclosed poikilitically in large hornblende individuals. Ordinarily the replacement of augite by hornblende is further advanced than that of hypersthene. Coarse, poikilitic, green-brown hornblende, apparently identical with that found in some of the olivine rocks, is also present in a number of norite specimens. Moreover, in three of four specimens, the plagioclase poikilitically included within the large green-brown hornblendes is anorthite, although the main plagioclase of the rock outside of the hornblendes is the usual labradorite. The inclusions are not merely tiny rounded blebs, but well-formed subtabular grains.

"The norites contain more iron ore than the olivine rocks. The average is about 1 per cent by volume, but the porphyritic and finer-grained phases carry as much as 3 per cent, occasionally 5 per cent. In these ore-rich rocks augite is more prominent in comparison with hypersthene than in the typical norites, just as it is in those olivine rocks carrying appreciable iron ore.

"Most of the norites contain neither quartz nor biotite...

## QUARTZ-BIOTITE NORITES

"The norites containing appreciable quartz and biotite are worthy of special mention because of their more sodic plagioclase. Quartz and biotite both begin to appear in rocks with a plagioclase more sodic than $An_{56}$-$An_{58}$. Norites with no distinctive mineralogical or textural features, other than the presence of 0.5 to 8 per cent each of quartz and biotite, have a sodic labradorite plagioclase with an average composition in the range $An_{45}$-$An_{60}$, commonly $An_{50}$-$An_{56}$. In the same rocks the calcic cores in the plagioclase are more numerous and better developed than in the other norites. The absence of hornblende is as great a proportion of the quartz-biotite rocks as of the typical norites shows that these characteristics are not dependent upon the presence of hornblende. The combination of the more sodic plagioclase and the appearance of quartz and biotite marks these rocks as one step farther away from the olivine rocks than the ordinary norites, and this is confirmed by the chemical analysis presented."

* * * * *

*Hornblende Gabbros.*—Miller (1937, p. 1407-1408) gives the following description of the hornblende gabbro of the San Luis Rey quadrangle.

"The hornblende gabbros as a group are the most complex of the San Marcos rocks. They are all composed essentially of plagioclase and hornblende, with minor amounts of pyroxene, but the plagioclases cover such a wide range of composition, and the hornblende such a wide range of habit and color as to produce a confusing assemblage of rock types. . ."

* * * * *

## GREEN VALLEY TONALITE
### NAME AND DISTRIBUTION

A number of bodies of tonalite of rather uniform character form a discontinuous strip just east of the main mass of the Triassic sedimentary and Jurassic volcanic rocks in the San Luis Rey quadrangle. This tonalite is not very different chemically and mineralogically from the Bonsall tonalite but it lacks the abundant streaked inclusions of that tonalite, and it everywhere has some characteristic features, both in the hand specimens and the thin-sections. The name Green Valley tonalite has been used by Miller for this rock from its characteristic development at Green Valley in the extreme southeastern part of the quadrangle.

* * * * *

### PETROGRAPHY

The rock is gray, medium-grained, and rather uniform in character. The minerals, especially the dark minerals, do not show a sharp granularity in the hand specimens. The plagioclase crystals are gray from inclusions. Under the microscope, the rocks show a variable grain size in the average rock, ranging from 1/2 to 3 mm. Some are . . . prominently seriate in texture, with grains varying from 0.1 to 3 mm. The plagioclase forms the largest crystals, and the quartz, dark minerals, the microcline, and some plagioclase form the smaller grains.

The plagioclase is somewhat zoned and averages andesine ($An_{40}$). . . In the dark rocks, the larger plagioclase grains are clouded with submicroscopic, black inclusions which tend to be arranged zonally in the crystals. As many as 12 of these dust zones were counted in some feldspar crystals. In nearly all the rocks some of the large plagioclase crystals have in the interior of the crystals irregular remnants of calcic bytownite or anorthite (Pl. 3).

The microcline is interstitial and intergrown with quartz only in the rocks richer in microcline than usual. In part it forms small, irregular patches in the plagioclase crystals. It carries little intergrown plagioclase. In most of the rocks, it is in small amounts and in films between the plagioclase and other crystals as if the grains were glued together with about the least possible amount of microcline. In some specimens it fills all of the cracks and fractures, borders most of the grains, and penetrates all the minerals but the hornblende. It must be a very late mineral and is probably deuteric and crystallized after some fracturing of the quartz and plagioclase.

The quartz is also interstitial and some grains are nearly as large as those of the plagioclase, but in general they are much smaller. The hornblende and biotite are in small, irregular grains and tend to collect in nests and streaks. The amphibole grains commonly have a core which is light colored and fibrous, and a main, broad border that is darker green and extinguishes as a unit. In some specimens ragged remnants of a pyroxene take the place of the fibrous core of the amphibole (Pl. 3, fig. 2). . . Both the biotite and the amphibole were no doubt formed by reaction on pyroxene. The accessory minerals are magnetite and ilmenite bordered by sphene, zircon, apatite, and occasional allanite. Secondary epidote, chlorite, sericite, uralite, and a clay mineral are present.

* * * * *

## BONSALL TONALITE
### NAME

The name Bonsall tonalite was applied by Hurlbut (1935, p. 611) to the tonalite of the San Luis Rey quadrangle that is characterized by abundant inclusions..

* * * * *

The rock differs greatly in the proportion and distribution of the inclusions, in the extent of the streaking and schistosity, in the proportion of the minerals present, and in the average grain size. It represents more than one injection, probably with little difference in age. The dominant rock varies within moderate limits. Many of the streaked, fine-textured, and gneissoid rocks are near contacts...

## AGE AND CONTACTS

The Bonsall tonalite is younger than the San Marcos gabbro, as it carries abundant inclusions of the gabbro, which in places are drawn out and reacted on almost beyond recognition but in others retain their characteristic features. The flow lines of the tonalite bend around inclusions of gabbro, and the tonalite sends apophyses into the gabbro in many places.

In Moosa Canyon, southwest of Mount Ararat, large, angular xenoliths of gabbro form a breccia, cemented by the Bonsall tonalite. Here the contacts are sharp and the gabbro must have been nearly or completely solidified at the time of the tonalite intrusion...

The Bonsall tonalite is younger than the Lakeview Mountain tonalite. Southeast of Lakeview, where the contact is well shown, for about 50 feet next the contact, the Bonsall is fine-grained and has contorted banding. Along the whole contact in the Lakeview Mountains area the structure of the Bonsall parallels the contact, and the rock near the contact tends to be finer grained than normally and to be irregularly layered.

The Bonsall tonalite is intruded by the Woodson Mountain granodiorite and the Indian Mountain granodiorite, as the latter rocks carry inclusions of the Bonsall and send apophyses into it in many places. The fluidal structure of the granodiorites parallels the contacts with the tonalite, while those of the tonalite are truncated at the contacts.

The relative age of the Bonsall and Green Valley tonalites is not known, but they are probably not very different in age, as both rocks intruded the San Marcos gabbro before the latter was completely consolidated...

\*  \*  \*  \*  \*

## PETROGRAPHY

The Bonsall tonalite is a medium-grained rock ranging from light to dark gray, depending on the proportion of dark minerals present. Hornblende is the chief dark mineral, and it is present in elongated grains up to a centimeter in length or in large patches, poikilitically enclosing the feldspar. This poikilitic type is the main rock of the eastern part of Bear Valley and of the large mass east of the San Luis Rey quadrangle (Table 5, analysis 2). Biotite is present in moderate amount and locally it is more abundant than hornblende. White to gray plagioclase and quartz are easily seen in hand specimen.

The typical Bonsall tonalite (Table 8, analyses 3 and 4) is an inequigranular rock with grains varying from 1/2 to 3 mm; rarely the grains are as large as 10 mm. Plagioclase makes up from 55 to 60 per cent of the rock and it differs a little in different specimens, averaging about $An_{40}$. It is zoned, and in most specimens the difference in composition between the zones is less than 10 per cent, but rarely it is 30 per cent. In mixed rock near the gabbro contact the core may be $An_{90}$ and the border andesine. Orthoclase makes up only a few per cent of the tonalite, but rarely it amounts to 15 per cent. Quartz averages about 20 to 25 per cent of most of the tonalite; rarely it makes up only 10 per cent or as much as 40 per cent.

Hornblende averages about 10 per cent of the tonalite, but locally it is either in much smaller or in much greater amount. The hornblende is green and not uralitic. Its indices of refraction vary somewhat, as shown in the section on mineralogy. Biotite makes up 5 to 15 per cent of the rock... Pyroxenes are almost entirely absent except in the mixed rock near gabbro where both augite and hypersthene are present. The accessory minerals are sphene, zircon, magnetite, apatite, and rarely garnet. Wilson (1937) reports rare tourmaline, monazite, anatase, allanite, thulite, pyrite, and undetermined minerals...

In the northern part of the Elsinore quadrangle the rock is a little coarser in grain than to the south. In this area the most abundant rock, forming most of the outcrops west of the Perris-Riverside highway and present in most of the other larger bodies, is much like the Lakeview Mountain tonalite but has somewhat less mafic minerals and more quartz than that rock. It is nearly white tonalite with black grains of biotite and hornblende. Most of the grains are between 1-5 mm in length. An analysis, norm, mode, and heavy-liquid analysis of a typical specimen are shown in Column 3 of Table 8. A few of the rocks show some allanite. This rock rarely shows a linear orientation...

The fine-grained varieties of the Bonsall are in part much like the main rock except for texture, but some have less dark minerals and some have as much as 20 per cent of orthoclase. A few have biotite as the only dark mineral. Myrmekite is common in these rocks, and some have abundant large sphere crystals. They tend to be porphyritic in texture and are commonly layered or gneissoid....

* * * * *

## STREAKING AND INCLUSIONS

The Lakewood Mountain tonalite contains only scattered inclusions which are not well oriented. In many places this tonalite shows a streaking, with dark layers as much as a foot across; in others it shows some orientation of the dark minerals, and locally the rock has a faint schistosity. In a few places the hornblende has a good linear orientation that is mostly along the dip of the other structures, but in a few places appears to be at an angle to the dip.

Nearly every good outcrop of the Bonsall tonalite and the associated granodiorite west of Lakeview shows inclusions. In most places the inclusions make up only a few per cent of the tonalite, but locally they make up as much as 60 per cent. The inclusions are concentrated in streaks parallel to the structure, and streaks rich in inclusions run across the hills like dikes of dark rocks. In a few outcrops the inclusions are spherical or angular, but for the most part they are much-flattened ellipsoids of revolution, and in some places they are streaked out into thin dark layers in the host. Where the flattened inclusions are observed on a surface cutting their long dimensions, they appear as long, narrow, parallel strips ranging from half an inch to 2 feet in length. Locally, as in the hills east of Reche Canyon in the northeastern part of the Elsinore quadrangle, they are hundreds of feet in length. On a surface parallel to their flat surface they appear as large round discs. That the inclusions were plastic during the later movement of the tonalite is shown by their shape and orientation and by the fact that they bend around quartzite xenoliths. All stages in the streaking of the inclusions are present including the final stage, where they appear as dark parallel streaks in the tonalite.

Plate 4 shows photographs of inclusions in the granodiorite associated with the Bonsall tonalite on the west side of the hill which is about 31/2 miles west of Lakeview on the Elsinore quadrangle. Figure 1 of this plate shows more abundant and more nearly equant inclusions than are usual....

... Hornblende is much more abundant in the inclusions than in the host and biotite is somewhat more abundant. Pyroxenes are locally present in the inclusions, and quartz and orthoclase are in small amount. The biotite and hornblende have higher iron content in the inclusions than in the host. Some of the plagioclase crystals of the inclusions have cores of calcic bytownite or anorthite with borders of calcic andesine. At least a few of these calcic cores are present in nearly every thin section of an inclusion. The mineral composition of the inclusions with pyroxene and very calcic plagioclase, and the gradation of the drawn-out inclusions to typical inclusions of gabbro show that many of them represent fragments of the San Marcos gabbro that have been much reacted on by the tonalite magma. They are finer grained than the gabbro from which they were derived, probably due to recrystallization rather than mechanical granulation. In the Elsinore quadrangle many of the inclusions have been derived from schists and gneisses, and all gradations between schist and typical inclusions are present. Where the reaction has gone far enough it is not possible to determine the origin of the inclusions.

The tabular or lens-like shape of the inclusions is in part due to resorption, reaction, and drawing out of the resulting viscous rock and in part to subparallel fracturing of the rocks. In places the larger inclusions are cut by many subparallel veinlets of magma in the process of being broken up.

The inclusions in the Bonsall tonalite in the San Luis Rey quadrangle have been described in detail and illustrated by Hurlbut (1935).

The granodiorite west of Lakeview has very abundant flat, oriented inclusions and the orthoclase tablets are well oriented.

In the Bonsall tonalite and associated rocks the orientation of the inclusions and mineral crystals results in a regional structure. Near the contacts with older rocks the structure is parallel to the contact. In most places the contacts parallel the general structure of the area, but locally this is not true. Where the Bonsall tonalite circles the base of Lakeview Mountains, the structure turns to parallel the contacts. It is rather unexpected to find that this holds for the well-developed structure of the Bonsall tonalite and for the poorly developed structure of the older Lakeview Mountain tonalite.

In many places the dark minerals are oriented and concentrated in layers, and in some places the rock has a faint gneissoid structure. This is more common in the rocks in which biotite is the chief mafic mineral, as in the rocks east of the Riverside-Perris road. In some places the hornblende and other minerals have a linear orientation. The rocks near contacts and the fine-grained varieties commonly have a pronounced streaking with thin layers of different minerals or textures, even when fine-textured rocks are not associated with known contacts. In places the rock near the contacts is a good gneiss with abundant biotite in layers.

Hurlbut (1935, p. 622-625) found that the plagioclase and hornblende of the streaks and of the enclosing Bonsall tonalite have a good dimensional and crystalographic orientation. In both the inclusions and the tonalite, the plagioclase has the flatface (010) parallel to the streaking and the long dimension (crystal axis *a*) nearly parallel to the dip. The hornblendes likewise have their *c* crystal axes nearly parallel to the dip but are not oriented in the prism zone. The quartz shows no orientation. Osborne has made a detailed study of the orientation of the minerals of the inclusions, the tonalite, and the adjoining schists, and he furnished several petrofabric diagrams. He found that the (101) face of the plagioclase and the (001) of the biotite show more than one maximum and tend to be parallel to the streaking; the *a* axis of the plagioclase is along the dip. The quartz shows girdles around the dip with strong maxima.

In most places the flat inclusions appear to be disclike, but in some they are more or less elongated in or near the dip. Where the hornblende is oriented, it, too, tends to be elongated in the dip. In a few places, the elongation of the hornblende is at a considerable angle to the dip, and in places it is not even in the strike of the plane structure of the flat inclusions. In most places the orientation of the inclusions and the faint schistosity due to the mica are parallel.

The pronounced flatness of the inclusions and the nearly circular form in the plane of the dip is difficult to explain on any theory of flow. The orientation of the hornblende and plagioclase where seen is nearly in the dip, indicating an upward movement of the magma, or possibly a horizontal movement parallel to the strike of the streaking.

The orientation of the inclusions and of the minerals is believed to have taken place before the magma was completely crystalline. Some of the streaking of the mica along nearly plane surfaces to give the rock a rude schistosity may have taken place after the rock was sufficiently crystalline to fracture. In general, the texture is typically igneous, and there is little fracturing or straining of the crystals, indicating that most of the orientation was brought about before the crystal mesh was strong enough to fracture. The orientation is thus essentially a flow phenomenon, and the crystals and inclusions were oriented according to shape. The actual movement during this stage must have been small, as otherwise the thin plastic inclusions would have been streaked out still more, torn apart, and elongated.

The quartz is interstitial and crystallized very late. Its orientation is probably not due to a shape orientation. It is very improbable that the quartz was oriented by fracturing into needles and later recrystallizing. In general, the *c* axis of quartz is normal to the *a* axis of, and tends to be nearly parallel to, the *b* axis of the plagioclase. It is possible that the residual quartz crystallized around the feldspar as a nucleus, and that the feldspar was able to orient the quartz.

The nearly circular form of the inclusions in the plane of the structure is difficult to explain on any theory of flow. The orientation of the flat dimensions of the inclusions and the minerals parallel to the contacts and in the plane of lamellar flow is common in igneous rocks. Prismatic minerals are also oriented with their long dimensions in the plane of lamellar flow, but we do not know with certainty the orientation of their elongation with respect to the direction of flow. . . .

\* \* \* \* \* \* \*

## WOODSON MOUNTAIN GRANODIORITE

### NAME AND DISTRIBUTION

A light-colored, rather coarse-grained granodiorite that contains biotite and a little hornblende and underlies a number of rather large areas in the three mapped quadrangles was called the Woodson Mountain granodiorite by Miller (1937, p. 1399) from its characteristic outcrops on Woodson Mountain, which is a few miles northeast of the southeastern corner of the San Luis Rey quadrangle and a few miles southwest of Ramona. The rocks called by Dudley (1935, p. 502-503) Steele Valley granodiorite and Cajalco quartz monzonite belong to this formation.

This granodiorite is present in many large masses in the western part of the Peninsular range. The largest mass in the mapped quadrangles is in the northwestern part of the San Luis Rey quadrangle and the southwestern part of the Elsinore quadrangle, and underlies an area of about 80 square miles. In the three mapped quadrangles, the Woodson Mountain granodiorite underlies an are of about 220 square miles, and among the rocks of the batholith it is therefore exceeded in area of outcrop only by the Bonsall tonalite. In the eastern part of the batholith typical Woodson Mountain granodiorite is not found, but a similar and probably related rock, white and lacking potash feldspar, is widespread.

## AGE

The Woodson Mountain granodiorite intrudes the Green Valley and Bonsall tonalites, as in many places it carries inclusions of these rocks and sends dikes into them. In places where it is in contact with these rocks, it is no darker or finer-grained near the contact, but in most places there is no appreciable change. It clearly intrudes the Temescal Wash quartz latite porphyry and the mafic granodiorite east of Temescal Wash, as it carries inclusions of these rocks and sends apophyses into them, but it does not become finer-grained at the contact. It also intrudes the La Sierra tonalite, as it sends many dikes into that tonalite and carries inclusions of it... The Woodson Mountain granodiorite is clearly cut by the dike of granodiorite porphyry in Temescal Wash. It is older than the Roblar leucogranite, the Mt. Hole granodiorite, and the micropegmatite northeast of Corona.

## GENERAL CHARACTER

The Woodson Mountain granodiorite shows lighter-colored outcrops than any other large body of the crystalline rocks. It is rather resistant to erosion and forms many of the hills and mountains of the area. It weathers into distinctive huge "wool sack" boulders of disintegration, many of them over 10 feet high. In areas of low relief, scattered boulders may rise above the soil; but on the steeper slopes the huge boulders are piled one on another with little intervening soil. Some of the mountains, notably Woodson Mountain, are great piles of huge boulders...

Near the contact of the granodiorite, where it intrudes older rocks, it commonly has a more or less banded or even gneissoid structure, and in several places this is very marked for a few feet next to the contact but gradually becomes less prominent, and in the main mass of the rock is entirely lacking. The banding parallels the contact and is everywhere nearly vertical. This gneissoid border zone is more common and wider in some of the smaller bodies and is well shown in the elongated body that is about 2 1/2 miles east of Auld, in the southeastern part of the Elsinore quadrangle. The Woodson Mountain granodiorite everywhere carries scattered inclusions of dark, fine-grained rock. They are mostly irregular in shape and only a few inches across. They are much less abundant than in the Bonsall tonalite, are smaller, and are rarely drawn out into discs. An outcrop 20 feet across will usually show a few such inclusions.

Most of the rock mapped as Woodson Mountain granodiorite is a uniform coarse-grained rock, but some is much finer grained. In part the finer grained rock is younger than the coarser rock; in part it may represent small bodies of the same magma that yielded the coarser rock. The body of granodiorite from the southern part of Elsinore Mountains to San Mateo Creek has numerous small bodies of fine-grained granodiorite that resemble the Roblar leucogranite. These were not separated, due to intimate mixture of the two rocks, the deep weathering, and the impenetrable brush of the area.

## PETROGRAPHY

The typical Woodson Mountain granodiorite is a rather coarse-grained rock, nearly white to pale brownish gray in color in the fresh rock, and flecked with scattered black grains. In the hand specimens, large grains of quartz in places with a bluish color, white plagioclase, and flesh-colored microcline are the chief constituents, and black biotite and hornblende are present in small amounts. The microscope shows that the rock has a moderate variation in grain size from 0.5 to 4 mm. In places near the contacts, the rock is much finer textured. The rocks are so coarse-grained that an ordinary thin section does not give a good sample of the rock. The estimated mineral percentages, as estimated from the better of the thin sections available, are as follows:

|  | Range | Average |
|---|---|---|
| Quartz | 30-40 | 33 |
| Microperthite | 10-30 | 20 |
| Plagioclase | 30-55 | 41 |
| Anorthite content | 20-30 | 25 |
| Biotite | 1-8 | 5 |
| Hornblende | 0-2 | 1 |

The accessories are magnetite, ilmenite, sphene, mostly growing about the ilmenite and associated with biotite, zircon commonly enclosed in the biotite, allanite, and apatite. The plagioclase is zoned, and a diagram of a typical zoned plagioclase is shown in Figure 15. The microperthite tends to intergrow with the quartz. The biotite is in irregular, ragged grains and is in part a late or deuteric mineral. The hornblende is in dark olive-green homogeneous grains and seems to be a primary mineral rather than uralitic. Augite is a rare mineral in the granodiorite. In most places the Woodson Mountain granodiorite is more or less gneissoid near the contacts, and it has a little more mica than usual but is otherwise similar. South of San Pasqual Valley in the southeastern part of the San Luis Rey quadrangle and elsewhere there are up to 100 feet of a fine-grained aplitic rock near the contact, and this grades into the normal rock...

\* \* \* \* \*

The rock carries few dark inclusions except near the contacts. In the hand specimens, the rock shows white plagioclase, flesh-colored microcline, quartz, and a little biotite. It is seriate in texture, and the grains range in size from about 0.5 to 4 mm.

\* \* \* \* \*

## INCLUSIONS AND ASSIMILATION

Dark inclusions are present in nearly all the plutonic rocks except the gabbro. In most of the formations they are only a few inches in maximum dimension and are widely scattered so that they make up only a fraction of a per cent of the mass. In the Bonsall tonalite, the granodiorite west of Lakeview, the Escondido Creek leucogranodiorite, and the Domenigoni Valley granodiorite many of these inclusions have volumes of a cubic foot, and some are much larger than that. These formations everywhere have abundant inclusions a few per cent in most places but 20 per cent or more locally. In the Escondido Creek leucogranodiorite the inclusions are variable in size; they are angular to rounded blocks and are very unevenly distributed in the outcrops. They are chiefly fragments of the Santiago Peak volcanics and other rocks cropping out near the granodiorite. The magma has not strongly reacted on them, and they were probably picked up by the magma after it had reached nearly its final position.

The Lake Wolford leucogranodiorite contains scattered inclusions of metamorphosed sedimentary rocks and amphibolite, some of which are hundreds of feet in length. Such inclusions are found locally in some of the other formations. They are very abundant in the isolated body of Bonsall tonalite about 2 1/2 miles north of Eagan, in the eastern part of the Elsinore quadrangle.

By far the greater part of the inclusions are made up essentially of the same minerals as are in the host, but they contain more hornblende and biotite. Many of them contain remnants of pyroxene and a calcic plagioclase. Hurlbut (1935) has shown that those in the Bonsall tonalite were derived chiefly from the San Marcos gabbro and that they have been so reworked by the magma that they were nearly in equilibrium with it.

The inclusions range from nearly equant, angular to subangular blocks, to flat discs whose maximum dimensions are many times the thickness, to dark streaks in the host. Where the process of reaction and mixing has gone further all evidence of the inclusions is lost. The Green Valley tonalite, which occurs in several large bodies in the San Luis Rey quadrangle, contains few dark inclusions, but everywhere has remnants of pyroxene and calcic plagioclase that must represent remnants of gabbro... The amount of gabbro added may have varied from place to place, and this may be the cause of the moderate variation in the composition of the tonalite.

The Bonsall tonalite, which everywhere has abundant inclusions, underlies about 325 square miles in the three mapped quadrangles and large areas both to the north and south of the mapped areas. An almost identical rock, with very abundant flattened inclusions, is widely distributed in the southern Sierra Nevada. The inclusions must have come from some depth, as they are abundant in all parts of large bodies and show no relation to the neighboring older rocks.

To furnish the inclusions, the wall rock must have been greatly shattered immediately before or during the intrusion of the tonalite, and to account for the rather uniform distribution the magma must have moved for some distance to scatter the inclusions throughout very large masses.

The typical inclusions were in almost perfect equilibrium with the magma, and they would probably persist a very long time in a stagnant magma without mixing with the magma to give a homogeneous rock. Movement of the magma would tend to mix the magma and crystals of the inclusions, and probably accounts for the uniform distribution of relict minerals of the gabbro in the Green Valley tonalite. This tonalite reaction has not gone so far as in the Bonsall tonalite, because in the former rock relicts of augite and calcic plagioclase are commonly scattered through the tonalite, while in the latter rock they are rarely found in the inclusions. To preserve the inclusions from mechanically mixing in the magma of the Bonsall tonalite it must have been necessary for the main movement of the magma and mixing of the inclusions to have taken place before the inclusions were much softened by reaction. The flattening of the inclusions indicates only a small movement of the magma after the inclusions were plastic. The Green Valley tonalite, with remnants of augite and calcic plagioclase, probably underwent considerable movement after the inclusions in it were softened, but before complete reaction had taken place. Superheat in the magma would favor complete reaction on the inclusions while crystallization before complete reaction would tend to armor the feldspar and other minerals by growths of feldspar in equilibrium with the magma and thus protect them from further reaction.

Inclusions of siliceous sediments in the granitic rocks were seen in a few places. They are usually subangular and have not been much reacted on. Widespread light-colored schlieren corresponding to the dark schlieren were nowhere seen. At the surface, the wall rock of the Bonsall tonalite contains a more siliceous sedimentary rock than gabbro, and it seems probable that much of these sedimentary rocks must have originally been incorporated in the intrusive. Inclusions of siliceous rocks (lower in the reaction series of Bowen) would be melted, dissolved, or reworked more rapidly than would those of gabbro, and diffusion and mixing would destroy all evidence of the inclusions.

The evidence is clear that granitic rocks do assimilate appreciable amounts of foreign rock. A similar conclusion was reached for volcanic rocks before eruption by Larsen, and others (1938, p. 429).

## MAGMATIC DIFFERENTIATION

The major differentiation that yielded the different rock types of the mapped units must have taken place in depth, as the contacts of the rock bodies are steep intrusive contacts with the earlier intrusion largely crystalline before the next intrusion was emplaced. Of the rock units, the gabbro, a complex of intrusions, is variable, the tonalites, are moderately variable, and the granodiorites and granites are rather uniform in composition. The different rocks in the gabbro are due largely to separate related injections. Some are due to local differentiation, some to concentration of settled crystals, and the hornblende rocks are due to local concentrations of mineralizers. Some of the different rocks in the tonalites are due to separate injections, some to the injection of an inhomogeneous magma, and some to uneven distribution and assimilation of inclusions.

Nearly all the analyses of rocks from the batholith of Southern California fall near smooth variation curves, a characteristic of most petrographic provinces. In common with those from most petrographic provinces, the mafic-rich rocks are less regular than the intermediate and felsic rocks. Similarly, the amount of the different minerals in the rocks, except for the substitution of biotite and amphibole for pyroxene in some of the mafic rocks, ranges systematically in the rocks from gabbro to granite (Figs. 11, 12). This probably requires that the dominant process that brought about the variations in the rocks was systematic and relatively simple. Since the first intrusion was gabbro, and the later intrusions changed regularly toward granite, the original or primary magma was probably near a gabbro in composition. This is in agreement with the commonly accepted opinion of the present time.

A group of rocks formed by the mixing of a gabbroic and a granitic magma or by the assimilation of a granite by a gabbroic magma would yield a variation diagram with straight lines. For many of the oxides of the California area and for most other petrographic provinces, the variation curves are nearly straight but show some systematic curvature. Hence some process is required other than assimilation or pure mixing of two magmas, especially near the granite ends of the series.

The variation of the rocks in the Southern California batholith from gabbro to granite is best explained as brought about by crystal differentiation, modified by assimilation and probably some other processes. Our knowledge of the course of crystallization of rocks of this kind is fairly satisfactory and is confirmed by the laboratory experiments on simple systems. Quantitatively the course of variation in the rocks proceeds as would be expected from crystal differentiation, as interpreted from the course of crystallization of such rocks. In passing from gabbro toward granite, the mafites decrease in amount and become richer in iron, the plagioclase become poorer in lime, and quartz and orthoclase increase in amount.

* * * * *

## METHOD OF INJECTION OF THE IGNEOUS ROCKS

A great composite batholith, much elongated parallel to the structure of the enclosing bedded rock, could have been emplaced by any one or any combination of the following processes:

the remnants of this intrusion do not suggest this. Moreover, the larger bodies enclosed in schists are crosscutting in large part.

(1) Forcing aside its walls
(2) Forcing the enclosing rock aside in the direction of the structure
(3) Forcing great plugs upwards
(4) Foundering great blocks of the crust
(5) Stoping
(6) Melting or otherwise replacing the country rock

There is little evidence of forceful injection in the Southern California area. Along the western border of the batholith, the great intrusions that project for miles into the older rocks have not appreciably disturbed the structure of the enclosing sediments.

The batholith as a whole is very elongate parallel to the bedding of the rocks, and in the area studied it is near the contact between the Triassic and Paleozoic rocks. This might be interpreted as indicating that the magma had made a place for itself by prying the sediments apart without greatly distorting them. However, the batholith was emplaced as a number of separate intrusions, with considerable time intervals between them. At the present surface some of these intrusions formed single bodies of rock, but many of them formed a number of bodies separated by other rocks. Within the batholith, where one of the granitic rocks is in contact with older sedimentary rocks, the contact is practically parallel to the bedding. This is well illustrated by many of the screens in which the bedding of the rocks is essentially parallel to the regional structure. Along the borders of the batholith, where it intrudes the main body of the sedimentary rocks, the contacts show much less tendency to follow the bedding of the sediments. The main body of the Bonsall tonalite crops out in a wide strip, approximately parallel to the structure and more than 70 miles long. This strip has been intruded by several large bodies of younger rock, and it carries many small and some large inclusions of the older rocks. It may have been emplaced (as a thick dike). However, some bodies of this tonalite are present as irregular masses that cut across the structures in such a way that they could not have been intruded by forcing their walls apart. The great body to the east of the Bonsall tonalite was not studied in detail, but a reconnaissance study shows that it is made up of many separate injections and that the individual bodies are not tabular.

The San Marcos gabbro crops out only as remnants of larger masses. It is confined to the western half of the batholith and may originally have formed a number of lens-like bodies arranged en echelon and have been injected as thick sills or dikes. However, the shape and distribution of the outcrops of the remnants of this intrusion do not suggest this. Moreover, the larger bodies enclosed in schists are crosscutting in large part.

The granodiorites and granites are in large, widely separated bodies that show little tendency to be elongated, and many of them are thick bodies enclosed in older granitic rocks, except where screens are present, or crosscutting the bedded rocks. It is difficult to see how they could have been forcefully injected.

If the injection of the granitic bodies forced the older rocks aside in the direction of their structure, the displacement about some of the great rounded bodies that intrude the sediments in the Santa Ana Mountains and elsewhere must have been measured in miles. Such displacement should have contorted and broken the sediments, but little such distortion or fracturing is present...

Consider a body of magma, such as one of the large individual intrusions of the batholith of Southern California, from the time of its intrusion to the time of solidification. It was probably emplaced rather slowly. At the time of emplacement to approximately its present position, compared with the late stages of solidification it had a higher temperature, lower viscosity, and fewer crystals and hence a lower density. It could, of itself, develop little energy, and it could transmit hydrostatically any forces slowly imposed upon it. When it moved, it must not only have displaced or replaced older rocks for its new position, but older rocks must have moved into the space from which the magma came.

If a magma is injected forcefully it should therefore move in the direction of least resistance. If the bordering rocks are not under tension, the magma, at least when near the surface, should move upward, as the resistance, as the resistance to the load of rock would, in general, be less than the strength of the wall rock. Such upward movement of the magma would result in the shattering of the wall rock and would be favorable for stoping.

If the crust is under tension, the magma might furnish the additional force to bring about movement in a nearly horizontal direction. This would probably give rise to a dikelike intrusion, such as the intrusions described by Cloos (1936) in the Sierra Nevada. It would not form the rounded, nearly equant, partly cross-cutting bodies such as those that make up most of the Southern California batholith.

In the batholith of Southern California, the contacts of the granitic bodies tend to follow the bedding of the older rocks, and the many screens that separate successively injected bodies for the most part are approximately parallel to the regional structure of the older rocks, though a few are in different positions across the regional structure. If the magmas were

dikes and other small intrusive bodies would move into the fractures. In Southern California, where the wall rock of an intrusive body is made up of an older sediment, little evidence of fracturing or distortion of the wall rocks is observed, and very few dikes or other intrusive bodies are present in the sediments. Where the intruded rock is an older granitic rock, there are locally very many small intrusive bodies of aplite, pegmatite, and other rocks in the wall. The San Marcos gabbro appears to be an especially favorable host for such minor intrusions, as shown at Pala and many other places. Moreover, if the individual intrusions were emplaced by forcing the overlying rocks upward, there would have resulted a large number of horsts of irregular shape and a relatively small area with thousands of feet of displacement. In Southern California all evidence of such horsts would have been removed by erosion, but in areas in which batholiths have been only partly uncovered by erosion such horsts should be common and conspicuous.

There is good evidence that some bodies of granitic rock have been formed by a combination of melting and replacing crystalline rock in place. If a liquid formed from such processes, the low-melting rocks such as granite and pegmatite should be the earliest and commonest rocks formed. Most granitic bodies that have been shown to have such an origin are near granite in composition. It is difficult to see how a complex batholith such as that of Southern California could have been formed by such a process. If melting took place in depth, evidence of it would probably be lacking. In the part of the Southern California batholith now exposed, there is good evidence that the granitic rocks are intrusive bodies and that there has been little reaction with the wall rock and no melting or extensive replacement of the wall rock. At a few contacts where one intrusion closely followed another, there is a zone of mixed rock, but this indicates that the early magma had not completely crystallized at the time the later magma was injected. Contacts are sharp, and there is little or no soaking of granitic material into the wall rocks and no evidence that the granitic rock has been contaminated by the wall rock. The older sediments adjacent to the contact have been little changed, and even the contact metamorphism is slight. The granitic rocks rarely have contact zones that differ in composition from the main part of the mass, and in the places where they do differ, the change in composition does not show any systematic relation to the composition of the wall rock.

There is abundant evidence that fragmentation and stoping of relatively small blocks played some part in the emplacement of the magmas. Nearly all the rocks contain scattered inclusions, and the Bonsall tonalite and some other bodies contain abundant inclusions. These inclusions range from little

forcefully injected, the contacts would probably be much more uniformly parallel to the structure than they are. If emplacement was by shattering the roof and stoping, the contact would tend to follow weak surfaces in the rocks. The magma would tend to displace the weaker rocks and leave the massive granitic rocks and the adjoining hardened sediments, thus forming the screens.

Each of the intrusions of the Southern California batholith has its characteristic type and quantity of inclusions. The Escondido Creek leucogranodiorite carries very abundant, angular inclusions, in general less than a few feet across. They have been little resorbed or reacted on by the magma and are composed of the bordering rocks, chiefly the older volcanic rocks. The Lake Wolford leucogranodiorite has scattered relatively large —up to several hundreds of feet across—inclusions of the older schists. The granodiorite west of Lakeview and the Bonsall tonalites carry abundant small, dark inclusions that are commonly disc-shaped, well oriented and almost completely reworked by the magma. The Green Valley tonalite has a rather uniformly distributed foreign calcic cores in the plagioclase and augite cores in the hornblende. The character of these inclusions depends somewhat on the composition and size of the included fragment but in large part on the stage in the cooling of the magma at the time the fragment was picked up and the movement of the magma after the fragment was softened by reaction. Inclusions picked up at the time of the main injection of the magma would be softened and completely reworked by the magma and, if siliceous, they would be completely dissolved. At times when the magma was actively moving they would tend to be dissipated in the magma, partly by solution, partly by mechanical disintegration. At times of quiet they would tend to settle by gravity.

Probably all the inclusions now seen in the granitic rocks were picked up after the magma had become relatively cold, highly viscous, partly crystalline, and nearly frozen in position. Little movement is indicated by the flattened inclusions; much movement would make them streaks and finally would completely disintegrate them. All stages in the streaking out and disintegration of inclusions can be found in the Bonsall tonalite.

There are good reasons for concluding that the various granitic intrusions did not emplace themselves by forcing upward the overlying rocks. Many of the intrusive bodies increase in size downward, although this is not so true in Southern California as it is in many batholiths. To emplace a magma that increases in size downward by pushing upward the overlying rocks would greatly fracture and distort the older rocks about its border, and abundant

granodiorite near Lakeview have similar inclusions and were probably intruded under similar conditions and possibly at about the same time.

Another widespread movement during the tonalite stage shattered the walls and imbedded inclusions throughout the moving magma. The gabbroic inclusions were softened and then, by later movement, disintegrated, and crystals were rather uniformly scattered throughout the magma. Cooling then brought about crystallization before these crystals were completely reworked. Thus was produced the Green Valley tonalite.

After the intrusion of the tonalites no widespread earth movement took place until the upper part of the remaining magma reached the composition of a granodiorite. The Woodson Mountain granodiorite was then injected over much of the length of the mountain range.

The small volume of granite in the batholith may result from a scarcity of residual granite liquid or from the fact that during the granite stage diastrophism was not favorable to movement of the magma toward the surface.

---

altered rock to fragments that have been drawn out and completely worked over by the magma to scattered cores of calcic plagioclase and other minerals that must represent relics of inclusions. The reworking of inclusions no doubt went farther and destroyed all evidence of their former presence. Nearly all of the inclusions that show intense reworking by the magma are of rocks more mafic than the host, and it seems certain that siliceous fragments must also have been originally present and completely incorporated into the magma. Most of the inclusions are small—less than a few cubic feet in volume—but a few are much larger.

*    *    *    *    *

The origin of the rocks of the batholith is believed to have been as follows: A magma of relatively uniform gabbro was formed at depth in the core of the ranges. This gabbro was slowly differentiating, having at any given time in its history an upper part essentially the same throughout the length of the batholith as to chemical composition and kind and amount of suspended crystals. From time to time diastrophism moved the magma toward the surface. This diastrophism may have been local and may have furnished local bodies of magma or it may have been widespread and have emplaced many bodies. The magma resulting from any such movement would be uniform in composition throughout the area. Such features as texture and character of inclusions would be determined by details of the movement such as: whether brief, prolonged, intermittent, with extensive brecciation, *et cetera*.

After the emplacement of the gabbro, little further movement took place until the upper part of the magma had the composition of a tonalite. Then many local and several widespread movements emplaced the several tonalites. The earliest of these movements did not greatly shatter the wall rock, or any shattering that took place was sufficiently early to allow most of the included blocks to be dissipated in the magma.... Several of the tonalites to the east of the mapped area and some of the local bodies of tonalite were emplaced under similar conditions and at about the same time.

A later movement emplaced the Bonsall tonalite, and the later stages of this movement shattered the wall rock and mixed abundant inclusions in the magma. Any siliceous inclusions were probably dissolved, but inclusions of gabbro were reacted on, softened, and flattened by flow shortly before movement of the magma ceased. The Domenigoni granodiorite and the

# Esper S. Larsen, Jr., on the batholith of southern California revisited

Gordon R. Gastil
9435 Alto Drive, La Mesa, California 91941

## INTRODUCTION

In 1948, Esper Larsen, in GSA Memoir 29, described the batholith of southern California. He wrote about the entire batholith from Riverside down the length of Baja California, but specifically addressed only those parts of Orange, Riverside, and San Diego Counties that he and his associates had mapped. His geologic map adds a special dimension to his work.

Although Larsen described the physiography, wall rocks, and metamorphism of the area, the outstanding features of his account are on the nature of the plutonic rocks of this vast region, their magmatic differentiation, and emplacement. Larsen and coworkers systematically mapped, microscopically described, and chemically analyzed a large area of nearly contemporaneous and continuously exposed granitic rocks, ranging in composition from gabbro to granite. It is this aspect of his work that is truly classic in its contribution and that is reprinted here.

Larsen's observations were influenced by his assumption of magmatic differentiation by gravitational settling of crystals and emplacement by stoping of the roof. Because of his "ruling hypotheses," Larsen saw the evolution of gabbro as the first differentiated magma to rise, followed successively by tonalite and granodiorite. He viewed these systematic pulses of intrusion as batholith-wide, and he gave formation names to each pulse (rather like the superunits invoked in Peru by Cobbing and Pitcher [1983]). Recent workers would not view such units as being so widespread and probably would not lump them together with a single name. For example, they would not include all rocks resembling those on Woodson Mountain as "the Woodson Mountain Granodiorite," but rather as "the granodiorite of Woodson Mountain," absent firm evidence that two or more bodies of rock had the same texture, composition, and age.

When Larsen wrote, the age control on the batholith was based on the intrusion of Aptian-Albian strata in Baja California (Woodford and Harriss, 1938). In the 1950s the work of L. T. Silver and coworkers, using isotopic U-Pb dating, began to show that the emplacement was not a single sequence of gabbro to granite differentiation (Banks and Silver, 1968); and soon thereafter that there was a progression of intrusion from west to east across the batholith (Krummenacher et al., 1975; Silver et al., 1979). Thus, tonalites in the east are younger than granodiorites in the west.

Unfortunately, the mapping of Larsen and coworkers did not extend into the substantially different eastern side of the batholith (Todd and Shaw, 1979; Hill, 1984; Walawender et al., 1990). However, the advent of mineral dating, the analyses of stable isotopes, the comparison of rare earth element abundances, and even the paradigm of plate tectonics do not compromise or diminish Larsen's descriptive contribution. Larsen's memoir remains the largest detailed map of a portion of the batholith, and what it says about the individual plutons and their interrelations is still important.

## COMMENTARY

In Memoir 29, Larsen described some 20 different rock units in considerable detail. Because of space limitations, only his comprehensive descriptions of the gabbro (San Marcos), two of the most abundant tonalites (Green Valley and Bonsall), and the Woodson Mountain granodiorite are reprinted here.

Larsen emphasized description of the dark enclaves and schlieren derived from them. Miller (1937) interpreted them to be xenoliths of gabbro and Santiago Peak volcanic rock. This interpretation of the enclaves and schlieren carried implications for the emplacement of plutons, and for the regional tectonics of the batholith.

The conclusions of Larsen are not in themselves difficult to accept, but with the benefit of hindsight, they are surprising for their omissions. Mechanisms of crystal differentiation, stoping, reaction between melt and enclaves, and passive emplacement

Gastil, G. R., 1999, Esper S. Larsen, Jr., on the batholith of southern California revisited, in Moores, E. M., Sloan, D., and Stout, D. L., eds., Classic Cordilleran Concepts: A View from California: Boulder, Colorado, Geological Society of America Special Paper 338.

emphasized by Larsen would also be invoked as appropriate by the current batholithic community. Larsen did not mention, however, other processes now recognized as important, such as simultaneously mobile magma bodies, magma mingling, synplutonic dikes, diapirism, or caldera subsidence.

Dark enclaves are present, especially at the margins, of nearly all plutons in the Peninsular Ranges batholith, including plutons not adjacent to gabbro or older volcanic rocks. For example, Hill (1984) reported such a pluton in the San Jacinto Mountains. Many workers have recently concluded that these dark enclaves and/or schlieren swarms may result either from fracturing and stoping followed by stretching of older gabbro, or in some cases by boudinage of synplutonic dikes.

Dark dikes cross cut, and thus are younger than, more siliceous tonalite and granodiorite. In many places these dikes are also back-veined by the host rock. Thus, mafic magma was not entirely pretonalite, contradicting a hypothesis of exclusively mafic to siliceous progression of magma generation and emplacement.

Larsen's emphasis on stoping implies a passive, nearly vertical movement of roof rock down and melt upward. Such a process would imply the presence of abundant stoped roof rock fragments in all parts of the pluton, which is not the case. Although the mechanisms of crystal differentiation and stoping may apply in the regions mapped by Larsen and his coworkers, they are not applicable to the entire Peninsular Ranges batholith, let alone all batholiths in general. Larsen did have, however, an intuition about the relationship between orogenesis, the regional generation of magma, and upward migration of magma that presaged the subsequent development of ideas about plate tectonics processes and magma generation and emplacement.

## REFERENCES CITED

Banks, P. O., and Silver, L. T., 1969, U-Pb isotopic analyses of zircons from Cretaceous plutons of the Peninsular and Transverse Ranges of southern California, Geological Society of America Special Paper 121, p. 17–18.

Cobbing, E. J., and Pitcher, W. S., 1983, Andean plutonism in Peru, and its relation to volcanism and metallogenesis at a segmented plate edge, *in* Roddick, J. A., ed., Circum-Pacific pluton terranes: Geological Society of America Memoir 159, p. 277–291.

Hill, R. I., 1984, Petrology and petrogenesis of batholithic rocks, San Jacinto Mountains, southern California [Ph.D. thesis]: Pasadena, California Institute of Technology, 700 p.

Krummenacher, D., Gastil, R. G., Brushee, J., and Doupont, J., 1975, K-Ar apparent ages in the Peninsular Ranges batholith, southern California and northwestern Mexico: Geological Society of America Bulletin, v. 86, p. 760–768.

Larsen, E. S., Jr., 1948, Batholith and associated rocks of Corona, Elsinore, and San Luis Rey quadrangles, southern California: Geological Society of America Memoir 29, 182p.

Silver, L. T., Taylor, H. P., Jr., and Chappel, B., 1979, Some petrological, geochemical, and geochronological observations of the Peninsular Ranges batholith near the international border of the U.S.A. and Mexico, *in* Abbott P. L., and Todd, V. R., eds., Mesozoic crystalline rocks: San Diego, California, San Diego State University, Department of Geological Sciences, p. 83–110.

Todd, V. R., and Shaw, S. E., 1979, Metamorphic, intrusive and structural framework of Peninsular Ranges batholith in southern San Diego County, California, *in* Abbot, P. L., and Todd, V. R., eds., Mesozoic crystalline rocks: San Diego, California, San Diego State University, p. 177–232.

Walawender, M. J., Gastil, R. G., Wardlaw, M., Gunn, S. H., Eastman, B. G., McCormick, W. V., and Smith, B. M., 1990, Geochemical, geochronologic, and isotopic constraints on the origin and evolution of the eastern zone of the Peninsular Ranges batholith, *in* Anderson, J. L., ed., The nature and origin of Cordilleran magmatism: Geological Society of America Memoir 174, p. 1–18.

Woodford, A. O., and Harriss, T. F., 1938, Geological reconnaissance across Sierra San Pedro Martir, Baja California: Geological Society of America Bulletin, v. 49, p. 1297–1336.

MANUSCRIPT ACCEPTED BY THE SOCIETY ON NOVEMBER 23, 1998

# Chapter 12
# THE ASSEMBLY OF CALIFORNIA

**Eldridge M. Moores**
*Department of Geology*
*University of California*
*Davis, California 95616*
*E-mail: moores@geology.ucdavis.edu*

**Yildirim Dilek**
*Department of Geology*
*Miami University*
*Oxford, Ohio 45056*
*E-mail: dileky@muohio.edu*

**John Wakabayashi**
*1329 Sheridan Lane*
*Hayward, California 94544*
*E-mail: wako@tdl.com*

# ELDRIDGE M. MOORES

Eldridge Moores on Trinity Ophiolite during first Penrose Ophiolite Conference, summer 1972. Photo by Greg Davis.

Eldridge M. Moores received his bachelors degree in geology from Caltech in 1959, and his doctoral degree from Princeton University in 1963. Moores has been a faculty member in the Geology Department, University of California, Davis, since 1966. His activities have included the positions of Editor of Geology (1982–1987), Science Editor of GSA Today (1991–1995), and Chair of the Ocean Drilling Project Tectonics Panel (1990–1993). Moores served as President of the Geological Society of America in 1996. His awards and honors include the first Geological Association of Canada Medal (1994), Geological Society of America Distinguished Service Award (1988), Fellow of the American Association for the Advancement of Science (1992), Fellow of the California Academy of Science (1996), and doctor of science (*Honoris causa*), College of Wooster, 1996. In addition to his academic activities, he has worked to communicate geologic knowledge to the general public and to call attention to the need to consider geology in public policy and education.

Moores is perhaps best known for recognizing ophiolites as on-land remnants of oceanic crust (Moores and Vine, 1971), one of the major advances of the plate tectonics revolution. The "classic paper" reviewed in this volume, in which the occurrence of ophiolites was used as a tool to interpret orogenic belts, was perhaps even more revolutionary. This paper not only changed the way orogenic belts are viewed, but it radically reinterpreted the western U.S. Cordillera as the product of a series of collisional orogenic episodes. One of Moores' (1991) most significant recent contributions entails probable late Precambrian connections between Antarctica and North America, i.e. the southwest U.S.–east Antarctic ("SWEAT") connection. This hypothesis has far-reaching implications not only for the tectonic evolution of the North American Cordillera, but also for the tectonics of the Appalachians and the Andes, and late Precambrian environmental changes.

Yildirim Dilek and John Wakabayashi

---

In the critical and exciting period, in the late 1960s and early 1970s, when concepts of plate tectonics and its geological corollaries were developing, three seminal papers were written about the U.S. Cordilleran system, one of which was Moores' "Ultramafics and orogeny, with models of the U.S. Cordillera and the Tethys." This paper was the first to enunciate clearly the role of ophiolites as fragments of oceanic lithosphere that mark suture zones between collided continents, as well as of arcs and related ophiolite obduction and suturing to orogenies in both the U.S. Cordillera and the Alpine system as a template for other orogens. A prescient and innovative feature of the paper was the recognition of the successive arrivals of exotic arcs with "suturing" ophiolite obduction and polarity flip to explain the Antler, Sonoma, and Nevadan orogenies, thus foreshadowing modern displaced terrane concepts. An especially neat feature of the paper is the recognition of the Caribbean and Scotia plates as foreign Pacific intrusions into the Atlantic. This is a paper full of original ideas that is still fresh and is still a good read.

John F. Dewey

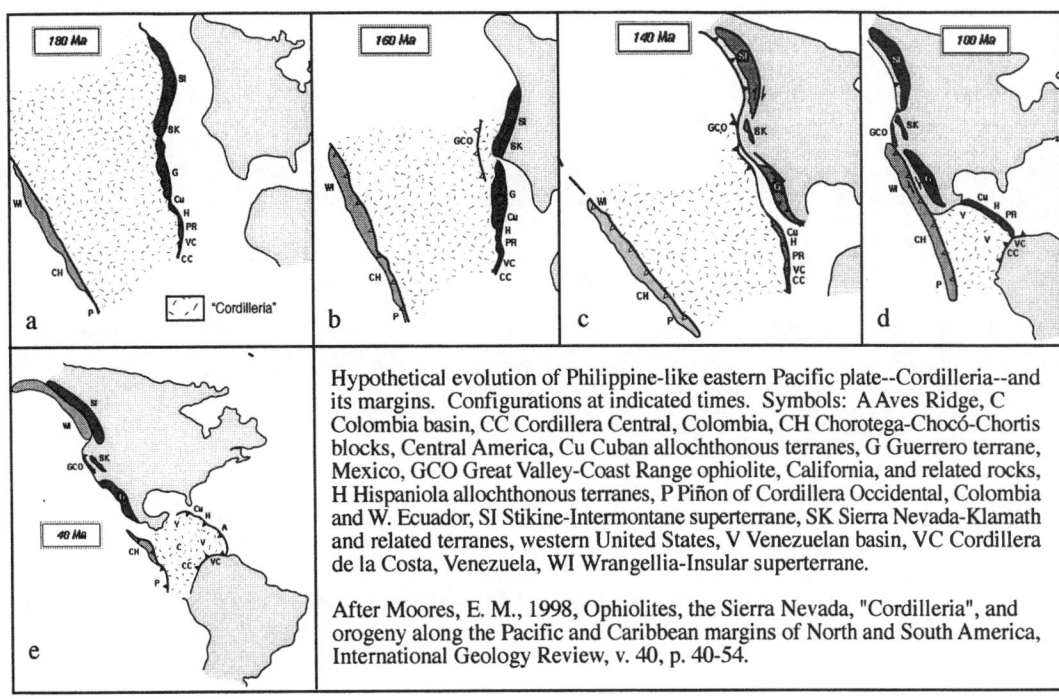

Hypothetical evolution of Philippine-like eastern Pacific plate--Cordilleria--and its margins. Configurations at indicated times. Symbols: A Aves Ridge, C Colombia basin, CC Cordillera Central, Colombia, CH Chorotega-Chocó-Chortis blocks, Central America, Cu Cuban allochthonous terranes, G Guerrero terrane, Mexico, GCO Great Valley-Coast Range ophiolite, California, and related rocks, H Hispaniola allochthonous terranes, P Piñon of Cordillera Occidental, Colombia and W. Ecuador, SI Stikine-Intermontane superterrane, SK Sierra Nevada-Klamath and related terranes, western United States, V Venezuelan basin, VC Cordillera de la Costa, Venezuela, WI Wrangellia-Insular superterrane.

After Moores, E. M., 1998, Ophiolites, the Sierra Nevada, "Cordilleria", and orogeny along the Pacific and Caribbean margins of North and South America, International Geology Review, v. 40, p. 40-54.

# Ultramafics and Orogeny, with Models of the US Cordillera and the Tethys

by
ELDRIDGE MOORES
Department of Geology,
University of California, Davis

An attempt to synthesize regional and structural geology, particularly in the Alpine and Cordilleran systems, during the whole of Phanerozoic time.

THE emplacement of ultramafic rocks in mountain systems provokes a number of interesting and important questions. Hess[1,2] pointed to the existence of belts of these rocks and showed that they were apparently intruded in the initial stages of orogeny. Tethyan occurrences of the ophiolite suite[3,4] may be fragments of oceanic crust and mantle and there is a similar occurrence in Papua[5]. Temple and Zimmerman have recently proposed[6] that emplacement of such oceanic crustal and mantle rocks may be achieved by collision of continental margin with a subduction zone, or lithosphere consumption (Benioff) zone, dipping away from the continent rather than towards it as is more common (see Fig. 1).

This article carries the hypothesis further, applying it to a preliminary analysis of the Alpine and Cordilleran systems. The underlying assumptions in this analysis include: (1) most large ultramafic–mafic sheets represent oceanic crust and mantle[3,4,7-9]; (2) the emplacement of large ultramafic sheets represents the collision of a continent with a subduction zone dipping away from the continental margin (Fig. 1); and (3) an island arc will migrate outward towards its trench[10] so that there will be high heat flow behind the arc as in the Marianas and Tonga–Kermadec arcs[11] and the Japan Arc[12,13].

After such a collision, the buoyancy of the continental material will arrest the process of continental subduction

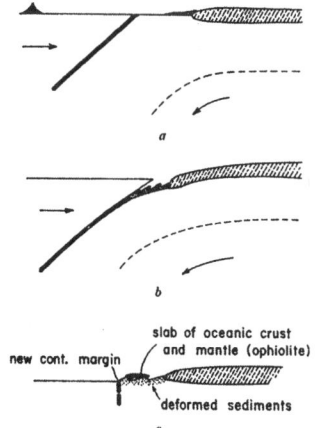

Fig. 1. Schematic diagram of ophiolite emplacement, modified after ref. 6. Diagonal pattern, continent; stippled, marginal sediment; horizontal lines, island arc; heavy lines, subduction zone; dashed line, lithosphere–asthenosphere boundary; arrows, relative motions. a, Situation just before collision; b, at collision; c, after collision. c shows a new continental margin developed oceanward of the deformed sediments and a new slab of oceanic crust and mantle now incorporated on the continental margin.

Reprinted by permission from Nature (228: 837–842) copyright 1970 Macmillan Magazines Ltd.

and the final product will look perhaps as in Fig. 1c. There may be some accretion by the addition of deformed sediments and a slab of oceanic crust and mantle to the continent, but the details of this process will vary from one occurrence to another. One possibility is that the direction of subduction "flips" and the zone continues operation dipping under the continent[6,14]. The important feature of the process, however, will be the asymmetric deformation of the old continental marginal material and basement away from the ocean and towards the foreland. The marginal material will be affected only when it first intersects the subduction zone so that this collision might be called the "intrusion" of ultramafics ". . . during the first great deformation of a mountain belt" (ref. 1, p. 391).

Hess[1] also observed that the age of emplacement of ultramafics, hence the beginnings of orogeny, migrated in time along the strike of a mountain belt. In my model, this migration in time can be seen as a migration of the point of continent–subduction zone collision along the continental margin (Fig. 2). The direction of migration of the collision will depend on the orientation of the subduction zone with respect to the continental margin, as shown in Fig. 2.

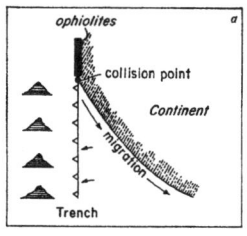 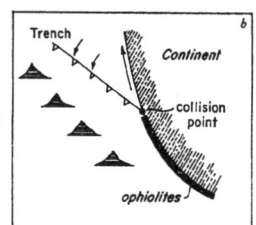

Fig. 2. Model for progression of orogeny along strike in time as migration of collision of continental margin and subduction zone as shown in Fig. 1. Arrows show relative motions. Diagonal ruled, continental margin. Barbs on upper plate of subduction zone. Small arrows indicate underthrusting plate. Horizontal rule, island arc over subduction zone.

By analysing the positions and ages of emplacement of old slices of oceanic mantle and crust exposed in mountain systems, one can establish not only the direction of dip of fossil subduction zones, but also their attitudes and migrations with respect to old continental margins. Petrological and mineralogical studies of such old ocean-mantle thrust sheets and associated metamorphic rocks afford a means of determining the thickness of the overthrust mantle–crust wedge at a given point and possibly the palaeo-geothermal gradients. The distance from this point to the toe of the wedge should provide the dip on the fossil subduction zone.

### Types of Intersection

Two types of collisions between a continental margin and an oceanic subduction zone may take place in which ultramafic material will be emplaced on the continent or the margin of the continent. These collisions differ according to whether the edge of the continent was an Andes-style or Atlantic-style margin (see Fig. 3). In each case, however, the colliding subduction zone must dip away from the continental margin. This situation is relatively unusual for present island arc systems and can be seen only in the Indonesia–Australia area where the Australian plate is at present moving into the Java–New Hebrides Trench System[15]. Such a situation may have existed in New Guinea and New Caledonia in the Tertiary in which ultramafic–mafic mantle and ocean crust wedges were thrust south-west over the margin of the Australian plate[5,16]. A situation resembling Fig. 3b may be represented today by the North Arm of Celebes and Halmahera where two seismic zones dipping away from one another are nearly intersecting[15].

Whether the collision was of a trailing (or Atlantic-style; Fig. 3a) or leading (Andes-style; Fig. 3b) continental edge may be deduced from the pre-collision history of the area in question. Examples of this situation might include: (1) the Bay of Islands Complex, Newfoundland, which sits between the Canadian shield and the principal orogenic belt of Newfoundland[17] (under this model, it would have been emplaced by a subduction zone–continental collision as in Fig. 3a); (2) the Troodos Complex, Cyprus and Vourinos Complex, Greece[3], which under this model would represent the collision of a subduction zone and the African continent or a microcontinent as in Fig. 3a; (3) the Canyon Mountain Complex, Oregon[18], and the Coast Range ultramafic sheet, California[9], which would represent collisions as in Fig. 3b.

### Cordillera System of West–Central United States

The western North American margin is characterized by an orogenic belt, which has as two principal features a "eugeosyncline" and "miogeosyncline" which received deposits during much of the Palaeozoic and early Mesozoic[19-21]. The width of the orogen is anomalously great in the region of the central United States. This "eugeosynclinal–miogeosynclinal" system has been affected three times by major "orogeny": by the Devonian Antler orogeny, the Permian Sonoma orogeny and the Jurassic Nevadan orogeny.

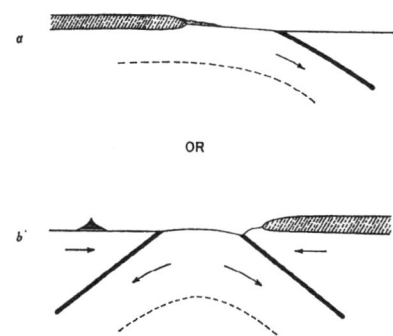

Fig. 3. Schematic diagrams of (a) trailing (Atlantic-type) marginal, and (b) leading (Andes-type) marginal collisions which may give rise to ultramafic emplacement. In b, possibly the island arc zone on the left is more vigorous than the continental marginal zone on the right and overrides it. Possible examples: a, Ordovician Appalachian–Caledonide system, Mesozoic Mediterranean–Tethyan system; b, North American Cordillera. Symbols as in Fig. 1.

All of these "orogenies" have resulted in thrusts and overturned folds directed towards the foreland[19-21]. This alternation of "eugeosynclinal–miogeosynclinal" conditions with orogeny conflicts with the interpretation that the Palaeozoic Cordillera is a simple Andean-type continental margin[22,23], and has prompted Roberts[19-21], among others, to propose a cyclic history driven by sedimentation and mantle phase changes. Much of Roberts's evidence is drawn from Nevada. In California to the west, the Sierra Nevada and Klamath Mountain basement rocks consist of lower Palaeozoic chert, clastic sediments and volcanics of the Shoo Fly and Calaveras sequences[24,25] (and E. M. M. and W. S. Wise, in preparation) succeeded by a Mississippian–Permian volcanic sequence, and a Triassic–Jurassic volcanic–sedimentary sequence. Several large ultramafic masses are also exposed in the Klamaths and Sierra Nevada. The 400 square mile Trinity ultramafic pluton of the Klamath Mountains[26] is pre-Permian and probably Devonian in age[27]. The large Feather River peridotite mass in the northern Sierra is probably pre-

Triassic[24,25] (and E. M. M. and W. S. Wise, in preparation), as is the Canyon Mountain Complex in Oregon[18].

In the California–Oregon Coast Ranges, the Franciscan mélange and associated material probably represent the result of late Mesozoic–early Tertiary underthrusting of the Pacific plate under North America[22]. The Great Valley sequence may be deposited on Jurassic oceanic crust[9,28]. Most late Mesozoic structures of the western Sierra and Coast Ranges are overturned to the west. This westward overturning to the west, coupled with eastward overturning to the east, has prompted Burchfiel and Davis[29] to call attention to the "two-sided" nature of the Cordilleran orogen.

I suggest that all these diverse and apparently conflicting features of the Cordilleran orogen can be fitted into a model in which a North American continent with a Japan-style or Andes-style margin[30] has collided three times in Phanerozoic time with a subduction zone dipping the other way, representing possibly a collision as in Fig. 3b. The Japan or Andes-style continental margin is reflected in the "eugeosynclinal–miogeosynclinal" suite, and the intersections are represented by large ultramafic–mafic masses and the eastward directed orogenies.

The lower Palaeozoic Cordilleran system of west-central United States reflects a Japan Sea-type continental margin (Fig. 4a). A carbonate platform and trough (shelf and miogeosyncline) passed westward into a lower Palaeozoic foredeep clastic basin of Japan Sea-type (the Vinini Formation) and a volcanic arc (the Valmy Formation) in western Nevada[19–21]. The Shoo Fly and Calaveras

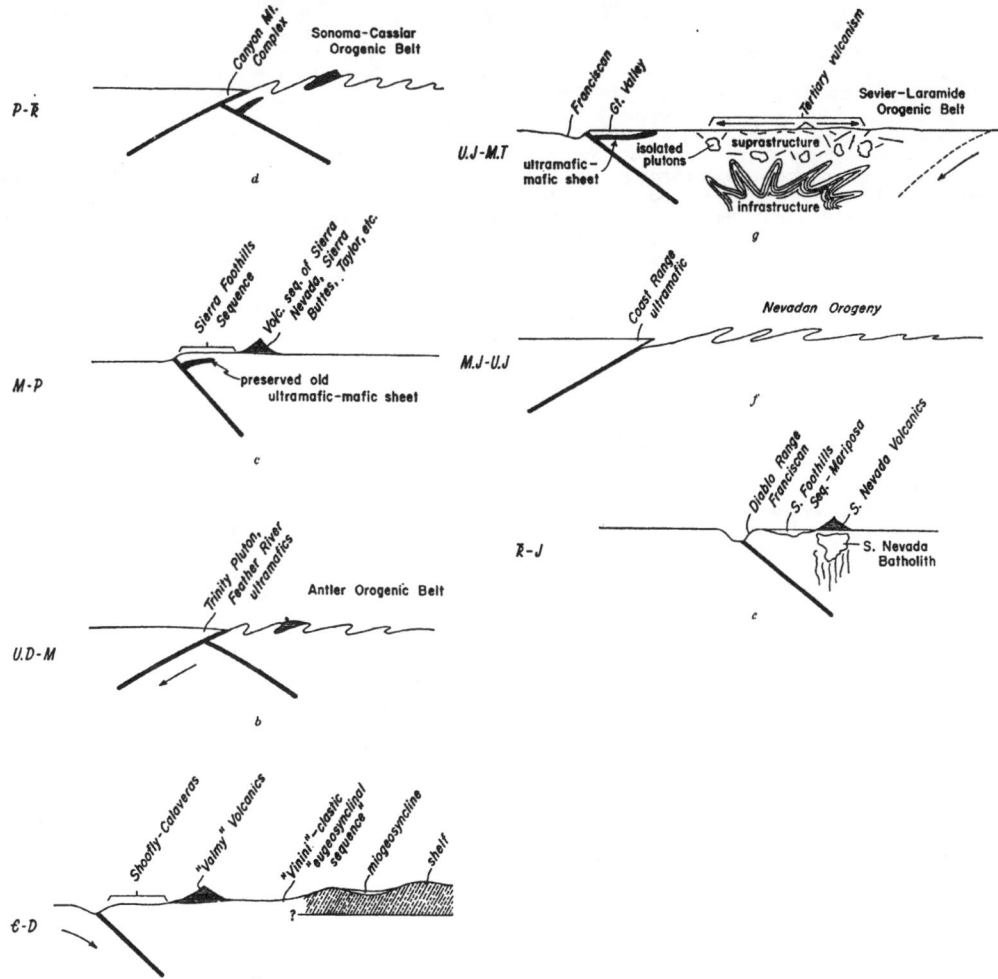

Fig. 4. Model for evolution of western US Cordillera. Heavy line, subduction zone; diagonal ruled area, continent. a, (E–D) Model for development of margin from middle Cambrian to Ordovician and possibly to Devonian. "Miogeosyncline" is thick carbonate deposits in eastern Nevada. "Vinini-Valmy" are "eugeosynclinal" deposits in western Nevada. "Shoofly-Calaveras" are chert, clastics, and associated sediments and volcanics of northern Sierra Nevada[19–21,24–25]. b, (U.D.–M.) Postulated collision of subduction zone in upper Devonian-Mississippian time, emplacing Trinity ultramafic pluton of Klamath Mountains[27,28] and causing Antler orogeny[19–21]. c, (M–P) Andean-type continental margin operative during Mississippian through Permian time. Sierra foothills sequence of chaotically deformed Palaeozoic rocks with lens-like bodies of mafic and ultramafic rocks[48]. d, Collision of subduction zone system which produced Sonoma-Cassiar orogeny and emplaced the Canyon Mountain complex of Oregon[18,32,33]. e, Triassic–lower Jurassic Andean-type continental margin[19–36]. Diablo Range Franciscan is a thick sequence of deformed and metamorphosed greywacke, chert, volcanics, and ultramafics. Sierra foothills sequence is same as in c. f, (M.J.–U.J.) Postulated collision of subduction zone and continental margin resulting in Nevadan orogeny. Coast Range ultramafic is ultramafic–mafic sheet below Great Valley sequence[9,39]. g, (U.J.–M.T.) Inferred development of western Cordillera subsequent to Nevadan orogeny. Subduction zone on left resulted in Franciscan–Great Valley relations[22,39], underthrusting to right produced Sevier-Laramide thrust belt and Outer Rockies[19–21,36,37,45], area between experienced formation of infrastructure of recumbently folded metamorphic rocks and suprastructure of low-angle normal faulting and isolated plutons[38–40] and, in early Tertiary, volcanism[41–43].

sequences of the Sierra Nevada may represent trench or oceanic material. A collision of this continental margin with a subduction zone dipping away from it in late Devonian–early Mississippian time possibly emplaced the Trinity and the Feather River ultramafic bodies and resulted in deformation of the pre-existing continental marginal rocks in the Antler Orogeny (see Fig. 4b). These ultramafic bodies may have been emplaced up over pre-existing volcanic and sedimentary rocks.

After this collision a subduction zone dipping under North America resumed operation, with a shift oceanward of the locus of volcanism (Fig. 4c). This period of activity is reflected in the upper Palaeozoic volcanogenic deposits located in the Klamath Mountains, Sierra Nevada and western Nevada.

Another possible collision of a subduction zone with the continental margin in the Permo–Triassic emplaced an ultramafic sheet now represented by the Canyon Mountain Complex, Oregon[18,31,32], and possibly the Feather River body. The resulting deformation of continental marginal rocks is the Sonoma Orogeny in Nevada (which is probably equivalent to the Cassiar Orogeny in British Columbia; Fig. 4d).

During the Triassic and lower Jurassic, a subduction zone again operated under North America (Fig. 4e). Volcanic and plutonic rocks of this age are found in the Sierra Nevada, the Klamath Mountains and in western Nevada. Trench material for this episode may be represented by the chaotic western Palaeozoic–Mesozoic belts of the Klamath Mountains and the old Franciscan rocks of the eastern Coast Ranges[9,22,23].

In the late middle Jurassic another collision occurred between the North American continental margin and a subduction zone dipping away from it, which may have emplaced the Colebrook Schist in Oregon[33], the glaucophane schist–eclogite terrane of the California Coast Ranges[34] and the Coast Range peridotite sheet[9,25]. This collision deformed the Andean-type margin of North America, and is traditionally called the Nevadan orogeny (Fig. 4f).

From the late Jurassic to Miocene, a subduction zone dipping under the western margin of the continent continued activity[22,23], but in the early middle Cretaceous, possibly as a result of beginning of rapid drift in the Atlantic Ocean, a zone of west-dipping subduction may have developed in eastern Nevada and western Utah (Fig. 4g). This zone resulted in underthrusting of the eastern foreland and gave rise to the Sevier–Laramide overthrust belt[35,36] and the outer Rockies[19–21]. The area in between western Nevada and central Utah thus became a zone between two oppositely dipping subduction zones (Fig. 4g), and, hence, possibly a zone of formation of infrastructure, Jurassic to Miocene plutonism and metamorphism, and low-angle normal faulting of the suprastructure[37–39]. Cessation of the west dipping zone in the Eocene may have caused the east dipping zone to extend under central Colorado as proposed by Lipman and others[40–42], causing volcanism in the San Juan Mountains and Great Basin.

The last effect of the foreland underthrusting was formation of the Wyoming and Colorado Rockies in the late Cretaceous–early Eocene. These dome and wedge uplifts of shield and overlying shelf rocks may represent buckle folds of the continental crust. (Assuming equal viscosities, the dominant wavelength of a fold approaches 3·46 times the thickness of the folding layer[43]. Calculations of the folding layer thickness based on this assumption yield values of 20 km just east of the Laramide thrust front to 46 km for the crust between the Bighorn Mountains and the Black Hills[44]. These values seem plausible thicknesses of continental crust at the time and locations in question.) Late Miocene to Pliocene activity has been in response to a new situation related perhaps to the collision of the East Pacific Rise with North America and the reorientation of spreading in the past 10 m.y.[40–42,45–47]. Isostatic rebound subsequent to termination of subduction possibly gave rise to the Plio–Pleistocene uplift of the Basin and Range and Colorado Plateau provinces.

## Correlation with Other Cordilleran Structures

Clearly, the picture I have presented will vary considerably along strike from the cross-section chosen. In particular, the two postulated late Mesozoic–early Tertiary opposing subduction zones come together in the Idaho Batholith region, and possibly also in the Mojave Desert region of eastern California[48]. In each of these areas one would expect to find a great deal of thermal (batholithic) activity connected with divergent structures. In southeastern California, Burchfiel and Davis[48] describe two major divergent thrust zones separated by a 50 mile wide central terrane of abundantly intruded sediments unaffected by the divergent structure (*Zwischengebirge*). This *Zwischengebirge* may be equivalent to the entire Basin and Range province, and represents an area where the two postulated opposing zones are nearly intersecting.

## The Alpine System

In the Alps proper, the Mesozoic Tethyan sequence usually developed on Hercynian basement and it consists of Permo–Triassic Verrucano red beds and volcanics, Triassic reef complexes, Jurassic–lower Cretaceous deep sea deposits and ophiolites, and Cretaceous carbonates and clastic sediments[49]. This stratigraphic–tectonic sequence can be interpreted as a Triassic opening of a Tethyan seaway in the western Alpine region, foundering of the continental margin areas in late Triassic (formation of large reef complexes), deep sea sedimentation during the Jurassic and lower Cretaceous, followed by closing of the ocean (emplacement of ophiolites) commencing in the upper Jurassic in the west and proceeding eastward in time.

The spatial relations of ophiolites in the Tethyan systems suggests the intersection at various places along strike of both the African and Eurasian continents with subduction zones dipping away from the margins, as shown in Fig. 5.

Fig. 5b, which is very simplified, shows both zones colliding at the same time and in the same position along strike of the deformed belt which are not necessarily true. Furthermore, the ridge may continue to operate during this collision. But whether or not a collision takes place, if subduction continues on zones dipping towards each other, the intervening area will become a zone of high heat flow, volcanism, sedimentation and possibly formation of infrastructure (Fig. 5b). This would result in a *Zwischengebirge* such as the Hungarian Basin[50–52]. Presence of a microcontinent between the two subduction zones, as perhaps the Rhodope massif in Bulgaria and Greece, and the Menderes massif in Turkey, would result in their remobilization and infrastructure formation[53].

If these conditions are developed and the continents continue to approach and finally collide and/or override each other, the result would be the "regurgitation" of the mobilized material between the subduction zones (Fig. 5c), and overthrusting of this crystalline material over the continents, for example, the formation and emplacement of the Penninic and the Pelagonian Nappes.

## Implications for Orogeny

Several important implications for orogeny are corollary.

(1) Batholithic activity, infrastructure formation and development of characteristic early isoclinal recumbent folds are related to operation of a subduction zone[22,23]. Ophiolite emplacement and nappe formation may be related to the attempted subduction of a continent, a continent–continent or a continent–island arc collision.

(2) In the US Cordilleran system, the repeated orogenic episodes and ultramafic emplacement may have consist-

ently resulted from collision of subduction zones with the Pacific margin of North America[10]. Two sources for these zones can be suggested: (a) Island arc systems may be generated off Asia and migrate across the Pacific, ultimately colliding with the American margin[7]. Such a situation may be implied by Karig's[11] documentation of new crust now being formed in the seas behind the Tonga–Kermedec and Marianas arcs. He suggests that possibly all island-arc systems in the western Pacific are migrating away from Asia. (b) A new ridge system might develop in the Pacific or the Pacific basin might begin to close rapidly, thus causing new subduction zones to be formed somewhat marginal to and facing toward both the North American and Asian continents. These new zones would collide with the continental margins of each continent. Remnants of such a system may be represented by the west-facing arcs observed in the south-western Pacific, for example, west of Luzon and Halmahera[15], and by Tertiary emplacement of oceanic crust and mantle slabs in New Guinea[5] and New Caledonia[16].

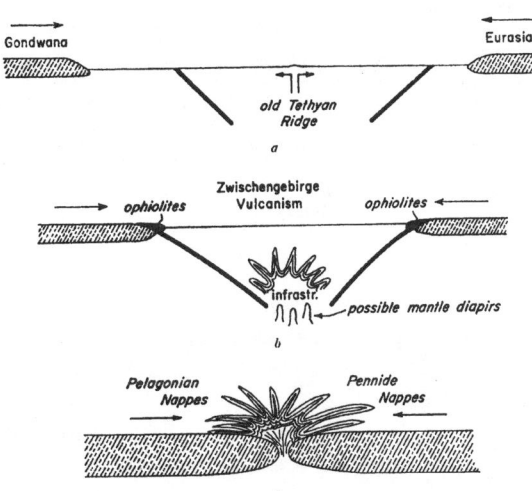

Fig. 5. Model for evolution of some Alpine features. a, Closing of an old Tethyan ocean basin would necessitate formation of subduction zones being placed as shown. Remnants of old Tethyan ridge may be represented by ophiolites, such as in Greece and Cyprus[3,4]. b, Continued subduction after emplacement of ophiolites on zones dipping in a similar manner would give rise to divergent deformed belts, separated by a *Zwischengebirge* in which infrastructure formation and volcanism would proceed, as also seen in Fig. 4g. c, Collision of continental masses would result in squeezing out volcanic material and hot infrastructure as overthrust crystalline nappes over the former Gondwana or Eurasian continent. The Penninic nappes would have come over the latter, the Pelagonian zone nappes over the former.

An important control on which system has operated may be obtained by observation of the old sheets of oceanic crust and mantle. Any evidence of an old island arc should be present in the geologic record as thick andesite accumulations over a peridotite slab. If, however, the "ophiolites" demonstrably contain only oceanic crust and mantle formed by spreading as in Cyprus and possibly Greece[3,4], then the subduction zone which emplaced them onto the continent should have been developed near the continental margin, so that not enough underthrusting would have taken place to give rise to island arc volcanism over the pre-existing oceanic crust (F. J. Vine and E. M. M., in preparation).

(3) Alpine ultramafic rocks in orogenic belts can now be differentiated into several types of occurrences: (a) Preserved overthrust masses of oceanic crust and mantle, for example, Vourinos–Pindos masses, Greece[3,4]; Klamath Mountains mass and Coast Range sheet, California[9,26,28]; Canyon Mountain complex, Oregon[18]; and Bay of Islands complex, Newfoundland[17]. (b) Disrupted parts of overthrust mantle sheets presently incorporated into mélanges, as in Italy[54], Turkey[55], California Coast Ranges[56,57] and Arosa schuppen zone[58]. (c) Small conformable masses of ultramafic and mafic rock in metamorphic belts, as in the Sierra Nevada and Klamath Mountains[24,25,59]; Penninic zone of the Alps[60,61]; Caledonian core region[62]; and southern Appalachians[1,2]. These masses represent either mantle diapirs emplaced in the high temperature region of *Zwischengebirge* infrastructure (Fig. 5b), or metamorphosed mélange sequences. (d) Hot diapiric intrusions with metamorphic aureoles intruded into greenschist terranes, for example, Mt Albert, Quebec[63]; Tinaquillo, Venezuela[64,65]; and Lizard, Cornwall[66]. These masses remain a problem. They may represent diapirs somehow peripheral to the subduction process, but most of these masses are bordered by a fault, and they may represent fragments of a fossil mantle overthrust sheet.

(4) All orogenic belts in the Eurasian–North American region such as the Urals, Caledonide–Taconic, Appalachian–Hercynian, seem to have resulted from the separation and subsequent collision of two continental plates, whereas the Cordilleran deformation as shown here seems to have always been the result of collision of a continent and an island arc system[7]. These relations argue for a fundamental difference between the circum-Pacific region and the Laurasia–Gondwana system during Phanerozoic time. The Pacific may be a permanent feature, which has generated within it ridges and island arcs which migrate and ultimately intersect the surrounding continental margins[67,68]. Some accretion takes place on these margins as a result of subduction zone operation (as in Fig. 4a, c and e), which may be more or less counterbalanced by the shortening resulting from collision of opposing subduction zones (as in Fig. 4b, d and f). At any rate, the evidence from mountain systems strongly implies the action of observed plate tectonic processes—sea-floor spreading and subduction—at least since the Cambrian or Ordovician.

(5) Today, on the west coast of North America, the East Pacific Rise can be seen intersecting with a trench system, which is resulting in right lateral transform fault movement[45-47]. Possibly the North American subduction zone collisions inferred above for the Devonian, Permian and Jurassic periods alternated in some cases with ridge intersections in intervening times. If relative motions between the North American plate and Pacific plate were favourably orientated during these hypothetical collisions, they may have resulted in transform fault movement. Hence the model of sequence of events in the record of the Cordillera should possibly be modified as follows: continental margin subduction system—ridge collision and transform faulting—island arc collision (orogeny)—continental margin subduction system. Thus possibly there should be evidence in the geologic record of three large scale strike-slip displacements, during the Siluro-Devonian, Permian and Jurassic. The 40 miles of pre-Mesozoic displacement on the Tintina trench[69] may be a result of this sort of process. Such recurrent strike-slip movement may also explain partly the Salinian Block emplacement[19-21], possible strike-slip movement on the Texas Lineament[70] and the truncation of Palaeozoic sedimentary trends against the continental margin[19-21] (and personal communication from B. C. Burchfiel).

(6) Where the last postulated arc to migrate toward the Pacific margin of the Americas collided with a continental mass, the result was orogeny, as in North and South America. Where no continent was present, they simply kept going, giving the Scotia and Antillean arcs, as diagrammatically represented in Fig. 6 (ref. 71). If at the same time the distance between North America and South America decreased, the result would be the Venezuelan Coast Ranges and the greater Antilles, with combined crustal shortening and strike-slip movement occurring more or less simultaneously. As this shortening continued,

perhaps the old transform faults (which would have been north of Cuba and in the Venezuelan Coast Ranges) may have been abandoned and the Bartlett Fault would have formed. One implication of this interpretation is that the northern front of this Caribbean plate is the north coasts of Cuba and Hispaniola. Hence the Gulf of Mexico represents ocean between the northern margin of the Caribbean plate and the North American continent. The Gulf therefore is at least as old as the Jurassic. If one accepts Bullard et al.'s[72] fit of the Atlantic Ocean and the idea that the Hercynian–Appalachian System represents a suturing of a proto-Laurasia and proto-Gondwana[73], then the Gulf of Mexico may be a remnant of Palaeozoic ocean.

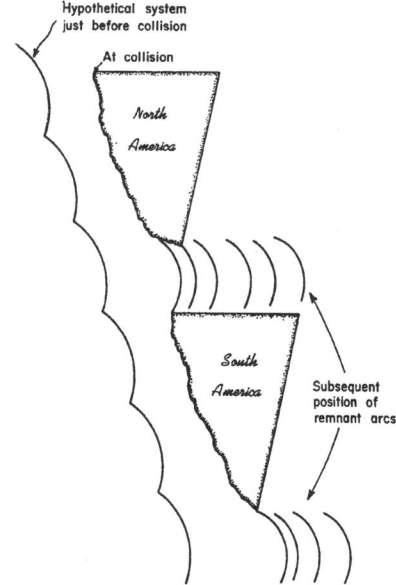

Fig. 6. Model for evolution of Caribbean and Scotia Arcs. Hypothetical island arc system collided with North and South America in Jurassic-Cretaceous time resulting in Nevadan orogeny. Where no continent was present (as between North and South America and south of South America), the remnants of this island arc system simply continued migration, forming the Caribbean and Scotian Seas. Simultaneous convergence of North and South America would result in Greater Antilles and Venezuelan Coast Ranges.

(7) During the early Palaeozoic, the margin of North America seems to have been a Japan Sea-type of continental margin, and for the rest of the Phanerozoic has been Andean-type. Nelson and Temple's model[10] for east flowing mantle mainstream convection holds that Japan Sea-type margins occur on the east sides of continents and Andean-type margins on their west sides. Does the change in margin type in the mid-Palaeozoic for North America signal a 180° change in its orientation relative to an east-flowing convection system? Correlation of such marginal types in the past with palaeomagnetic data should provide an important test of the mainstream model[7].

I acknowledge the late H. H. Hess for stimulating my interest in peridotites, and D. I. Axelrod, B. C. Burchfiel, R. Cowen, W. R. Dickinson, I. D. MacGregor, J. C. Maxwell, T. H. Nelson, D. H. Roeder, J. W. Valentine and F. J. Vine for many helpful discussions. Part of this work was supported by the US National Science Foundation.

Received April 28; revised July 15, 1970.

[1] Hess, H. H., *Spec. Pap. Geol. Soc. Amer.*, 62, 391 (1955).
[2] Hess, H. H., *Proc. Intern. Geol. Cong. Moscow*, 17 (2), 263 (1939).
[3] Moores, E. M., *Spec. Pap. Geol. Soc. Amer.*, 118 (1969).
[4] Moores, E. M., and Vine, F. J., *Phil. Trans. Roy. Soc.*, A (in the press).
[5] Davies, H., *Proc. Twenty-third Intern. Geol. Cong.*, Sect. 1, 209 (1968).
[6] Temple, P. G., and Zimmerman, J., *Abstr. with Prog., Geol. Soc. Amer.*, 221 (1969).
[7] Vine, F. J., and Hess, H. H., in *The Sea*, 4 (in the press).
[8] DeRoever, W. P., *Geol. Rdsch.*, 46, 137 (1956).
[9] Bailey, E. H., Blake, jun., M. C., and Jones, D. L., *Abstr. with Prog., Geol. Soc. Amer.*, 68 (1970).
[10] Nelson, T., and Temple, P. G., *Trans. Amer. Geophys. Un.*, 50, 634 (1969).
[11] Karig, D. E., *J. Geophys. Res.*, 75, 239 (1970).
[12] Lee, W. H. K., and Uyeda, S., *Monog. Amer. Geoph. Un.*, 8, 87 (1965).
[13] Takeuchi, H., Uyeda, S., and Kanamori, H., *Debate About the Earth* (Freeman Cooper, San Francisco, 1967).
[14] McKenzie, D. P., *Geophys. J.*, 18, 1 (1969).
[15] Fitch, T. J., and Molnar, P., *J. Geophys. Res.*, 75, 1431 (1970).
[16] Avias, J., *Tectonophysics*, 4, 531 (1966).
[17] Smith, C. H., *Mem. Geol. Surv. Canad.*, 290 (1958).
[18] Thayer, T. P., *Prof. Pap. US Geol. Surv.*, 475-C, C82 (1963).
[19] Roberts, R. J., *J. Univ. Missouri Rolla*, 1, 101 (1967).
[20] Roberts, R. J., *Abstr. with Prog., Geol. Soc. Amer.*, 286 (1969).
[21] King, P. B., *Evolution of North America* (Princeton Univ. Press, 1959).
[22] Ernst, W. G., *J. Geophys. Res.*, 75, 886 (1970).
[23] Hamilton, W., *Bull. Geol. Soc. Amer.*, 80, 2409 (1969).
[24] McMath, V. E., *Bull. Calif. Div. Mines Geol.*, 190, 173 (1966).
[25] Irwin, W. P., *Bull. Calif. Div. Mines Geol.*, 190, 29 (1966).
[26] Lipman, P. W., *Amer. J. Sci.*, 262, 199 (1964).
[27] Lanphere, M. A., Irwin, W. P., and Hotz, P. E., *Bull. Geol. Soc. Amer.*, 79, 1027 (1968).
[28] Bezore, S. P., *Abstr. with Prog. Geol. Soc. Amer.*, 5 (1969).
[29] Burchfiel, B. C., and Davis, G. A., *Proc. Twenty-third Intern. Geol. Congr.*, 3, 175 (1968).
[30] Mitchell, A. H., and Reading, H. G., *J. Geol.*, 77, 629 (1969).
[31] Thayer, T. P., and Brown, C. E., *Bull. Geol. Soc. Amer.*, 75, 1255 (1964).
[32] Dott, jun., R. H., *Amer. J. Sci.*, 259, 561 (1961).
[33] Coleman, R. G., and Peterman, Z. E., *Abstr. with Prog., Geol. Soc. Amer.*, 12 (1969).
[34] Coleman, R. G., and Lee, D. E., *J. Petrol.*, 4, 260 (1963).
[35] Armstrong, R. L., *Bull. Geol. Soc. Amer.*, 79, 429 (1968).
[36] Misch, P., in *Geology of East Central Nevada*, Salt Lake City, 17 (Intermont. Assoc. Petrol. Geol., 1960).
[37] Armstrong, R. L., and Hansen, E., *Amer. J. Sci.*, 264, 112 (1966).
[38] Moores, E. M., Scott, R. B., and Lumsden, W. W., *Bull. Geol. Soc. Amer.*, 79, 1703 (1968).
[39] Moores, E. M., Scott, R. B., and Lumsden, W. W., *Bull. Geol. Soc. Amer.*, 81, 323 (1970).
[40] Armstrong, R. L., Ekren, E. B., McKee, E. M., and Noble, D. C., *Amer. J. Sci.*, 276, 478 (1969).
[41] Christiansen, R. L., and Lipman, P. W., *Abstr. with Prog., Geol. Soc. Amer.*, 81 (1970).
[42] Lipman, P. W., Prostka, H. S., and Christiansen, R. L., *Abstr. with Prog., Geol. Soc. Amer.*, 112 (1970).
[43] Ramsay, J. G., *Folding and Fracturing of Rocks*, 375 (McGraw-Hill, New York, 1967).
[44] Data taken from King, P. B., *Tectonic Map of North America* (US Geol. Surv., Washington, DC, 1969).
[45] Menard, H. W., and Atwater, T., *Science*, 219, 463 (1968).
[46] Atwater, T., *Bull. Geol. Soc. Amer.* (in the press).
[47] MacKenzie, D. P., and Morgan, W. J., *Nature*, 224, 125 (1969).
[48] Burchfiel, B. C., and Davis, G. A., *Bull. Geol. Soc. Amer.* (in the press).
[49] Trumpy, R., *Bull. Geol. Soc. Amer.*, 71, 843 (1960).
[50] Cermak, V., *J. Geophys. Res.*, 73, 820 (1968).
[51] Korosy, L., *Geol. Sb. L'vov.*, 8, 377 (1968).
[52] Mahel, M., and Buday, T., *Regional Geology of Czechoslovakia* (Schwezerbart'sche Verlagsbuchhandlung, Stuttgart, 1968).
[53] Smith, A. G., and Moores, E. M., in *Data for Orogenic Studies* (Geol. Soc. London, in the press).
[54] Maxwell, J. C., and Azzaroli, A., *Prog. Geol. Soc. Amer.*, 103A (1962).
[55] Bailey, E. B., and McCallien, W. J., *Trans. Roy. Soc. Edin.*, 62 (11), 403 (1953).
[56] Himmelberg, G. R., and Loney, R. A., *Abstr. with Prog., Geol. Soc. Amer.*, 102 (1970).
[57] Bailey, E. H., Irwin, W. P., and Jones, D., *Bull. Calif. Div. Mines Geol.*, 183 (1964).
[58] Peters, T., *Schweiz. Miner. Petrogr. Mitt.*, 43, 531 (1963).
[59] Hietanen, A., *Bull. Geol. Soc. Amer.*, 62, 565 (1961).
[60] O'Hara, M. J., and Mercy, E. L. P., *Earth Planet. Sci. Lett.*, 1, 255 (1966).
[61] Dal Vesco, E., *Schweiz. Miner. Petrogr. Mitt.*, 33, 177 (1953).
[62] Brueckner, H., *Amer. J. Sci.*, 267, 1195 (1969).
[63] Smith, C. H., and MacGregor, I. D., *Bull. Geol. Soc. Amer.*, 71, 1978 (1960).
[64] MacKenzie, D. B., *Bull. Geol. Soc. Amer.*, 71, 303 (1960).
[65] Green, D. H., *Bull. Geol. Soc. Amer.*, 74, 1397 (1963).
[66] Green, D. H., *J. Petrol.*, 5, 134 (1964).
[67] Danner, W. R., *Abstr. with Prog., Geol. Soc. Amer.*, 84 (1970).
[68] Nelson, T. H., and Roeder, D. H., *Trans. Amer. Geophys. Un.* (Abstr.), 51, 421 (1970).
[69] Roddick, J. A., *J. Geol.* 75, 23 (1967).
[70] Muehlberger, W. R., *Trans. NY Acad. Sci.*, 27, 385 (1965).
[71] Wilson, J. T., *Earth Planet Sci. Lett.*, 1, 335 (1966).
[72] Bullard, E. C., Everett, J. E., and Smith, A. G., *Phil. Trans. Roy. Soc. Lond.*, 258A, 41 (1966).
[73] Valentine, J. W., and Moores, E. M., *Nature* (in the press).

# California terranes

**Eldridge M. Moores**
*Department of Geology, University of California, Davis, California 95616*

**Yildirim Dilek**
*Department of Geology, Miami University, Oxford, Ohio 45056*

**John Wakabayashi**
*1329 Sheridan Lane, Hayward, California 94544*

...subjected to the principle of least astonishment, geologic science has always tended to adopt the most static interpretation allowed by the data, and evidence indicating displacements larger than conceivable...has consistently met strong resistance.

Seeber, 1983, p. 1528

A vital lesson of plate tectonics is that there is no validity to any assumption that the simplest and therefore most acceptable interpretation demands a proximal rather than a distant origin.

Coombs, 1997, p. 763

## INTRODUCTION

Two of this volume's coeditors persuaded the third into commenting on his own paper. There are three clear alternatives as a "classic paper" on California terranes. In the first of these, Hamilton (1969) first mentioned the North American Cordillera as a collage of exotic terranes and argued that the continental margin of North America was built out from the shelf edge by material carried to it on a converging oceanic plate. He argued for a process of "ocean-floor sweeping" caused by "underflow" (later called subduction; White et al., 1970) of Pacific mantle beneath the North American margin. Although it was a revolutionary paper, Hamilton (1969) emphasized exclusively east-dipping subduction beneath North America. The single-subduction tectonic model rapidly became a "ruling theory" (Chamberlin, 1890), a position it has held for the past 30 years. In the second alternative paper, Irwin (1972, p. C103) first defined a tectonic terrane as "an association of geologic features...which lend a distinguishing character to a particular tract of rocks and which differ from those of an adjacent terrane."

We, however, base our discussion on the third paper by one of us (Moores, 1970). This paper had its origins at the second Penrose conference, convened by W. R Dickinson and described briefly earlier in this volume (in the biography of Dickinson). One of us (E. M. M.) attended this conference. In 1969, I (E. M. M.) had been working in ophiolites for some six years, first on the Vourinos complex in northern Greece and subsequently on the Troodos complex in Cyprus (Moores and Vine, 1971). At the Penrose conference I reported on the evidence for sea-floor spreading in the Troodos complex, although I had not yet made the connection to all ophiolites. One of the informal breakout sections at the Penrose conference was devoted to discussion of the problem of ophiolite emplacement. One model discussed during this session was emplacement of ophiolites on continents by collision of a continental margin with a subduction zone dipping away from the continent. I had recently read an abstract from the annual meeting of the GSA by Ralph J. Roberts (Roberts, 1969), in which he described the Phanerozoic tectonic history of the U.S. Cordillera as an alternation of eugeosyncline and miogeosyncline depositional conditions along the margin with orogeny involving thrust from the continental margin toward the interior of the continent. As described earlier in this volume, in the final session of the conference, convenor Dickinson summarized his emerging ideas relating ancient geosynclines to modern tectonic environments, a talk I earlier described as "one of the most exciting moments of my scientific life." Sitting there greatly inspired after Dickinson's remarks, I was meditating on the previous evening's discussions of ophiolite emplacement and Roberts' account of Cordilleran history, when in a blinding flash of insight, it came to

Moores, E.M., Dilek, Y., and Wakabashi, J., 1999, California Terranes, *in* Moores, E. M., Sloan, D., and Stout, D. L., eds., Classic Cordilleran Concepts: A View from California: Boulder, Colorado, Geological Society of America Special Paper 338.

me that Roberts' scenario could be explained as an alternation of Japan Sea and Andean continental marginal depositional systems with collisions of oceanward-dipping subduction zones. The ophiolites and island-arc terranes preserved in the Cordillera represented the remains of the colliding blocks.

To say that I was excited is a gross understatement—for weeks, I could hardly contain myself. Suddenly the entire Phanerozoic tectonic history of the U.S. Cordillera seemed crystal-clear. And the importance of ophiolites as tectonic indicators seemed equally clear. *Ophiolites were the key tectonic indicator. Their emplacements represented first-order tectonic events. Because these collisions were the cause of fold-thrust belts directed toward the interior of a continent, the latter, which were and still are much better studied than ophiolites, were derivative features, and thus secondary.* In my mind, this interpretation placed in plate tectonic context the arguments made earlier by Hess (1939, 1955) that "Alpine-type peridotite intrusions" (ophiolites) represented *the* major orogeny-initiating tectonic events. The first words of the title of my (Moores, 1970) paper, "Ultramafics and orogeny" were deliberately paraphrased after that of Hess's (1955) paper: "Serpentines, orogeny, and epeirogeny."

Because Penrose conferences had a no-publication requirement (rescinded in 1997 by act of the GSA Council), a moderate length of time had to pass until ideas generated at these conferences could be submitted for publication. The resulting publication (Moores, 1970) was the first to apply the collisional model of ophiolite and/or island-arc emplacement to an explanation of the multiple orogenic episodes found in California and surrounding regions in the Phanerozoic (Moores, 1970, Fig. 4; see also Moores and Twiss, 1995, Fig. 12.10; Ingersoll, 1997, Fig. 3). (In his evaluation of the paper, one of the *Nature* reviewers, F. J. Vine, commented that it was the first paper to invoke the existence of plate tectonics in pre-Mesozoic time.)

This was the first paper expressly to suggest the existence of a role in Cordilleran tectonics for one or more subduction zones dipping away from the continent. In this model, the "two-sided" Cordilleran orogen (see Şengör, this volume) developed by collision of these seaward-dipping subduction zones with the North American continental margin on the downgoing plate and a subsequent polarity flip. The ophiolites or island arcs were not explicitly called "terranes" as such, but it was clear that they originated elsewhere than their present position. Although the analysis was essentially two-dimensional, possible diachroneity of collision (orogeny) along the continental margin was explicitly discussed, and margin-parallel transform faulting was mentioned as an important alternative possibility.

## DEVELOPMENT OF IDEAS

The "conceptual revolution" visualized by Moores (1970) was slow to gather steam. Early subsequent developments included a synthesis of the relationship between plate tectonics and orogenic belts (Dewey and Bird, 1970), an authoritative and influential synthesis of Phanerozoic history from regional information (Burchfiel and Davis, 1972, 1975) and a synthesis of the "non-collisional" tectonic development of western North America (Coney, 1972). Coleman's (1971) model of the emplacement of ophiolites by "obduction" was quickly and generally accepted. Most workers assumed that obduction was antithetic to subduction beneath a continental margin, in concert with the single subduction zone "ruling theory." By contrast, under the collisional model of ophiolite emplacement, obduction amounts effectively to aborted subduction of a continental margin (e.g., Moores, 1982, 1998).

Terrane tectonic analyses burgeoned after Irwin's (1972) paper (e.g., Coney et al., 1980; Schermer et al., 1984, and references therein). Maps that formerly displayed rock units and fault slices were reissued as terrane maps. A Circumpacific terrane map (Howell et al., 1985) contains a total of 333 separately defined units (~ 30 in California and contiguous states).

The terrane concept has added immeasurably to our understanding of orogenic belts and their evolution (e.g., Howell, 1989), and it has made it clear that a multidisciplinary approach is necessary in order to understand orogenic belts. Terrane analysis has focused attention on the existence of far-traveled crustal pieces that have ended up in various locations, as well as on the existence in orogenic belts of large-scale transcurrent, or strike-slip, displacements. The approach has catalyzed much research in all orogenic regions, particularly in the Circumpacific region. Paleomagnetic investigations of individual terranes (e.g., Alvarez et al., 1980; Debiche et al., 1987; Butler et al., 1997; Irving et al., 1996; and many others) have enriched our understanding of the trajectory of individual terranes. New tectonic terms, such as "superterrane," "docking," and "stitching" (e.g., Monger et al., 1982), have appeared. Silver and Smith (1983) drew a clear analogy between Mesozoic western North America and the present-day western Pacific. Careful structural analysis along the soles of ophiolites has elucidated the kinematic history of these complexes (e.g., Cannat and Boudier, 1985; Harper et al., 1990).

Problems exist, however, with the terrane concept as it has been applied in western North America. Five of these are as follows.

1. A few terranologists have somewhat piously proclaimed that the approach is objective (e.g., Jones, 1990), a misconception roundly and rightly criticized by Şengör and Dewey (1990), but one often held about geologic mapping (Moores, 1985).

2. Terrane analysis as commonly practiced is essentially a map-view analysis of an orogen; the shape of terranes in three dimensions has seldom been considered. This fact has led to sweeping, somewhat grandiose pronouncements about the significance of terranes in the growth of continents, assuming that each terrane is a block of crustal thickness. Careful geophysical surveys (Clowes et al., 1993) and geologic work (Hansen and Dusel-Bacon, 1998) have shown in southern Canada and Alaska, respectively, that terranes there are not complete crustal blocks but thin sheets, and that the volumetric addition to the continental crust in many cases is minor. In other words, these terranes are thrust sheets or nappes, rather than continental blocks, and are thus amenable to comparison with the nappes in the Alpine-Himalayan terrane that have been known for more than a century

(Şengör and Dewey, 1990, and references therein; Hansen and Dusel-Bacon, 1998; Moores and Day, 1984).

3. Many tectonic models for the western United States have considered only east-dipping subduction zones, and the role of terranes in structural development of an orogen has seldom been considered explicitly. There have, however, been important exceptions. Tempelman-Kluit (1979) presented a very clear description of an ophiolite-related arc-continent collision interpretation for the northern Omineca belt in Canada and Alaska that closely resembles the models of Moores (1970). Arc-continent collision models have been invoked for the mid-Mesozoic Nevadan and related orogenies (e.g., Schweickert and Cowan, 1975; Ingersoll and Schweickert, 1986; Dilek et al., 1988), and to explain Paleozoic tectonics of the Sierra Nevada and surrounding region (Schweickert and Snyder, 1981). Ehlig (1981) associated a Paleocene, west-dipping subduction with the development of the Pelona-Orocopia schists. Cannat and Boudier (1985) suggested north- and west-dipping subduction for ophiolitic zones of Paleozoic and Mesozoic age, respectively, in the Klamath Mountains. Henderson et al. (1984) modeled the Laramide orogeny as a collision of a large oceanic plateau on the Farallon plate with North America. Dilek and Moores (1992) suggested a model of oceanic arcs separated by a sinistral transform fault for terranes in the Klamaths and Sierra Nevada prior to their collision with North America. Cloos (1993) analyzed the relationship between the size of accreted blocks and structure, although his analysis considered only a single subduction zone, as did Henderson et al. (1984). Wakabayashi (1992, Fig. 3) considered the impact of accretion of much smaller-scale Jurassic and Cretaceous Franciscan Complex terranes on the structural evolution of the Franciscan Complex and coeval tectonic events in the North American Cordillera (including a possible collision in western Nevada). Maxson and Tikhoff (1996) suggested a "hit-and-run" collisional model for the Laramide orogeny caused by arrival and subsequent northward shift of the Wrangell-Insular superterrane.

4. With a few exceptions (e.g., Tarduno et al., 1985; Debiche et al., 1987; Wakabayashi, 1992), terrane analyses in western North America have mostly considered only margin-parallel movement of terranes for latitudinal transport (see summary of accounts in Cowan et al., 1997). This restricted view of latitudinal transport of terranes may have resulted from a model of strain partitioning of oblique subduction, wherein all of the tangential component of the relative plate motion is partitioned into strike-slip faulting on the continent (e.g., Jayko and Blake, 1993). McCaffrey (1996) showed that the fraction of the tangential component of oblique subduction partitioned into strike-slip faulting ranges from 0% to 100%, depending on the subduction zone, so the 100% partitioning model applied in the Cordillera represents the *maximum* possible amount of strike-slip faulting. Paleomagnetic data constrain latitude, but not longitude; few data from terranes fix their longitudinal positions during transport relative to the western margin of North America.

5. Most terrane analyses have considered only dextral translation of terranes relative to North America. Exceptions, however, include Avé Lallemant (1995), Oldow et al. (1993), Wakabayashi (1992), and Moores (1998). Plate-motion models indicate both sinistral oblique and dextral oblique plate convergence between the North American and Farallon plates during the late Mesozoic (Engebretson et al., 1985), although these plates may not have been in contact with each other until the Late Cretaceous (Moores, 1998). Hansen (1996) has invoked what she terms "yo-yo tectonics" for motions of Wrangellia and Stikinia, involving Permian–Early Jurassic paleolatitudes concordant with North America, and Late Jurassic–Cretaceous southward and mid-Cretaceous and younger northward motions, respectively, relative to North America. Motions of entirely oceanic plates in Early Jurassic and earlier time are constrained only by paleomagnetic evidence from ophiolites around the Pacific Rim, because of the lack of preserved ocean floor of that age.

## TERRANES AND CALIFORNIA TECTONIC DEVELOPMENT: CURRENT STATUS

We summarize here tectonic events related to terranes in California, proceeding forward in time. The reader should bear in mind that at present no consensus exists, even among ourselves, as to the nature of some of the most fundamental geologic contacts in California, or the number and polarity of subduction zones represented in the Sierra Nevada and surrounding regions (Fig. 1) (cf. Moores, 1970, 1998; Ingersoll, 1997; Moores and Twiss, 1995, Fig. 12.10; Ingersoll and Schweickert, 1986; Schweickert and Snyder, 1981; Burchfiel and Davis, 1972, 1975; Saleeby, 1982; Sharp, 1988). When appropriate, we use the more traditional nomenclature (e.g., nappe, block, thrust fault, allochthon, ophiolite) in order to enhance the precision of the discussion.

The principal terrane-related tectonic events for California and surroundings include the following.

1. Rifting of the western North American margin occurred during the latest Proterozoic–earliest Cambrian, possibly from Antarctica-Australia (e.g., Schweickert and Snyder, 1981; Moores, 1991). The Neoproterozoic-Paleozoic Trinity ophiolite (also called the Trinity terrane) in the Klamath Mountains may, but need not be a remnant of the oceanic crust formed during this rifting event. This complex includes two or three separate blocks with different histories sutured together, presumably in an ocean in Neoproterozoic-Ordovician time (N. Lindsley-Griffin, 1998, written commun.). Ophiolite formation was followed by Silurian-Devonian arc development (Wallin et al., 1995), and Devonian thrusting of the ophiolite complex over the "Central Metamorphic Belt." The present east-over-west direction of thrusting becomes north-northeast over south-southwest in present coordinates if 100° clockwise rotation during Late Jurassic time of the eastern Klamaths is applicable (e.g., Cannat and Boudier, 1985; Fagin and Gose, 1983).

2. The Roberts Mountain allochthon of deep-water sedimentary and volcanic rocks was thrust over miogeoclinal rocks from central Nevada to the east-central Sierra Nevada roof pendants (Burchfiel et al., 1992; Saucier, 1997). Foredeep strata, the Diamond Peak–Chainman Formations, deposited in the miogeocline, attest to significant topography to the west (present coordinates)

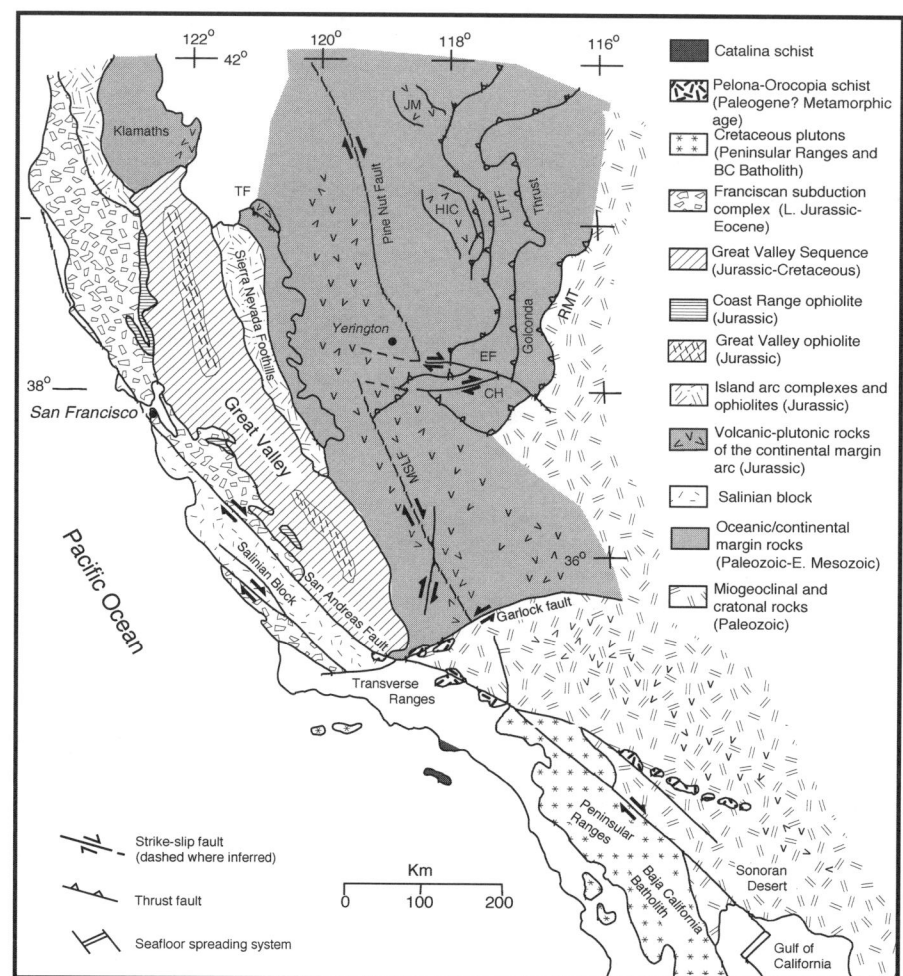

Figure 1. Simplified tectonic map of part of the western U.S. Cordillera (modified from Dilek and Moores, 1993). Key to lettering: CH—Coaldale fault, EF—Excelsior fault, HIC—Humboldt igneous complex, JM—Jackson Mountains, LFTF—Luning-Fencemaker thrust fault, MSLF—Mojave–Snow Lake fault, RMT—Roberts Mountains thrust, TF—Taylorsville fault.

after thrusting. No Paleozoic ophiolitic sequence has been found in Nevada, although possibly correlative deposits are present in the eastern belt of the Sierra Nevada, as described in the following (Varga and Moores, 1981; Schweickert and Snyder, 1981).

3. The "Northern Sierra terrane" or Sierra Nevada Eastern Belt developed. Thrusts and melange in the lower Paleozoic Shoo Fly Complex include volcanic and metasedimentary complexes, and serpentinite (Saleeby et al., 1989; Hansen et al., 1993). Overlying are three island-arc complexes, of Devonian-Mississippian, Permian-Triassic, and Jurassic age (Day et al., 1985). Unconformities are present within the Shoo Fly Complex, between Jurassic and older rocks, and possibly at the base of the Triassic sequences. The Shoo Fly melange and serpentinite could be related to the Roberts Mountain allochthon, *but there is no a priori requirement that the Sierra rocks were adjacent to North America prior to Middle Jurassic time.* Harwood and Murchey (1990) argued for proximity of the Sierra Nevada and the Havallah sequence of the Golconda allochthon of western Nevada (event 6 here) based on stratigraphic and paleontologic similarity. The stratigraphic evidence is not compelling; widely scattered arcs of the western Pacific show similar stratigraphies (e.g., Kennett et al., 1977; Moores and Twiss, 1995, p. 195). Arguments of proximity based solely on paleontologic similarity are simply incorrect, as any observation of present global biostratigraphic patterns demonstrates (e.g., Valentine and Moores, 1974). For example, the Indo-Pacific faunal province extends 15,000 km from east Africa to the Tuamotus. This often-made correlation between faunal similarity and proximity is another "ruling theory" in the sense of Chamberlin (1890).

4. Arc development in the eastern Klamaths occurred during the mid-late Paleozoic. These arc sequences have often been correlated with those of the northern Sierra Nevada, but chemical differences argue for a distinct origin (Roure and Lapierre, 1989).

5. The formation of the Feather River–Devil's Gate ophiolite complex, northern Sierra Nevada, occurred during the mid-late Paleozoic (Edelman et al., 1989a; Day et al., 1985; Saleeby et al., 1989). This enigmatic body may represent the remnant of ocean

lithosphere preserved by out-of-sequence thrusting during collision of two opposing-polarity colliding arcs in Jurassic time (Moores, 1998), or it may represent a remnant of a long history of mostly east-dipping subduction and suturing.

6. The Golconda allochthon of late Paleozoic deep-water sedimentary and volcanic rocks was emplaced. Rocks of this allochthon previously were deposited oceanward of the Roberts Mountain allochthon and were thrust over it in a Permian-Triassic Sonoma orogeny representing an island arc–continental margin collision, followed possibly by arc reversal (Speed, 1978; Schweickert and Snyder, 1981; Ingersoll, 1997). No foredeep deposit has been identified corresponding to that of the Antler event, however. Possibly correlative structures of comparable age are present in the White-Inyo Mountains, and in the eastern Sierra Nevada roof pendants (e.g., Lahren and Schweickert, 1989).

7. Formation of the late Paleozoic–early Mesozoic chaotic deposits (either melanges or olistostromes) containing Carboniferous-Permian limestone blocks of Tethyan affinity, early Mesozoic chert-argillite sequences, and Late Triassic basalt (Irwin, 1981; Edelman et al., 1989a). These deposits crop out in the Western Paleozoic-Triassic belt of the Klamaths and the central belt of the Sierra Nevada.

8. An early Mesozoic continental arc formed in west-central Nevada, followed to the west by a basinal sequence and an allochthonous ophiolite-arc assemblage, the Humboldt complex (Dilek and Moores, 1995).

9. Early Jurassic volcanic arcs are present in the eastern Klamath Mountains and Sierra Nevada. These may have been, but are not required to be, continuous with each other. They may or may not have been contiguous with Nevada.

10. The basement rocks of the western Klamath Mountains volcanic arcs formed; the rocks appear to be a multiple series of oceanic-affinity terranes (Hacker et al., 1993). They may have, but need not have, formed near the North American continental margin.

11. Formation of "old ophiolitic" basement of the Smartville, Slate Creek, and other western-central belt ophiolites of the Sierra Nevada occurred, probably in an oceanic environment at about 200 Ma (Dilek et al., 1990; Edelman et al., 1989b).

12. The mid-Mesozoic (ca. 170 Ma) arrival of early Mesozoic and older arc-ophiolite assemblages in northern Sierra (old Smartville, Slate Creek), and possibly the western Klamath Mountains (e.g., Chetco, Preston Peak), was part of an east-facing Stikine-Intermontain superterrane (e.g., Moores, 1998).

13. During the late Mesozoic (ca. 160 Ma) there was development of the west-facing arc and backarc basin in the Klamath Mountains (Josephine ophiolite) and spreading in the northwest Sierra Nevada within the collided block and adjacent oceanic region (Smartville dikes, Coast Range, Great Valley [?] ophiolites; Dilek and Moores, 1992).

14. The late Mesozoic (150–160 Ma) emplacement of the Great Valley and possibly the Coast Range ophiolites (Godfrey and Klemperer, 1998), and possibly the Josephine ophiolite (Harper et al., 1996), occurred over the western edge of the continent along a west-dipping subduction zone. There was deformation of the northern Sierra Nevada, emplacement of the Humboldt complex in western Nevada (e.g., Wakabayashi, 1992, Fig. 6), possible arc-arc collision followed by westward thrusting in the Klamath Mountains (Harper et al., 1996), and arrival of western volcanic terrane in the Peninsular Ranges of Baja California and southern California (Dilek and Moores, 1993). The inception of Franciscan Complex subduction and possibly Coast Range ophiolite emplacement was about 165–160 Ma. The general pattern possibly was reminiscent of the present-day Philippines (e.g., Moores, 1998; Silver and Smith, 1983).

15. Sinistral faulting in the Sierra Nevada and adjacent Nevada and major 800 km displacement along Mojave-Sonora megashear occurred during the late Mesozoic (150–140? Ma).

16. During the late Mesozoic (150–80 Ma), development of Franciscan subduction complex, including nappe formation, arrival of exotic terranes of various dimensions, and metamorphism of blueschists (e.g., Wakabayashi, 1992, and many others) occurred. Initiation of subduction associated with development of the Catalina schist occurred about 110 Ma (cf. Mattinson, 1986; Grove and Bebout, 1995). There was dextral displacement on the Mojave–Snow Lake and related faults (Lahren and Schweickert, 1989; Maxson and Tikoff, 1996).

17. The Wrangell-Insular superterrane arrived in "Baja BC" at the latitude of California (Irving et al., 1996) in the late Cretaceous (ca. 80–70 Ma). If this superterrane collided with the Franciscan subduction zone (Maxson and Tikoff, 1996), subsequent translation northward may have taken place either inboard (east) of the Franciscan Coastal Belt (which arrived later) along the present Coastal Belt–Central Belt boundary, or entirely west of the Franciscan subduction zone.

18. The latest Cretaceous–Tertiary (70–25 Ma) emplacement of the Franciscan Coastal Belt was near its present location. Initiation of subduction associated with the development of the Pelona-Orocopia schists occurred at about 65 Ma; the polarity of subduction is controversial (Ehlig, 1981; Jacobsen, 1990). Juxtaposition of the Salinian block with the Franciscan occurred between 80 and 50 Ma along a contact distance of 150 km or more (Wakabayashi and Moores, 1988; Wakabayashi and Hengesh, 1995).

19. Collision of a spreading ridge–transform system with the Franciscan subduction zone occurred at 25 Ma, and was followed by progressive development (to present) of the San Andreas fault system (e.g., Atwater, 1970, and many others).

## DISCUSSION

Although we reiterate that there is no consensus on the tectonic development of California, we can draw the following conclusions.

1. West of the Cordilleran miogeocline, the multiplicity of terranes implies a complex mosaic of intraoceanic and continent-oceanic interactions reminiscent of the modern western Pacific (Silver and Smith, 1983; Moores, 1998). The Paleozoic and early Mesozoic tectonic histories of the Sierra Nevada and Klamath Mountains involve multiple tectonic interactions of distal continent-derived debris, oceanic island arcs, ophiolites, and associated

features. Despite the Antler and Sonoma orogenic events, there is no a priori evidence that *requires* that the terranes of the Klamath Mountains and Sierra Nevada developed close to the North American continent or were next to it prior to Middle Jurassic time.

2. Much of the Mesozoic and associated Paleozoic terranes may represent part of a major oceanic-island arc complex, similar to the Philippines, comprising the Stikine-Intermontain-Guerrero terranes; the complex collided with the continental margin about 170 Ma and was involved in subsequent simultaneous dextral and sinistral strike-slip faulting. This complex strike-slip faulting situation recalls that of the Himalayan-Tibetan tectonic region, to which a rigid-plastic slip-line tectonic model has been applied (Moores, 1998; Tapponnier and Molnar, 1976). This scenario implies that the Sierra Nevada and Klamath Mountains were decoupled from Nevada during much of the Mesozoic, and that the early to mid-Mesozoic Nevada thus did not necessarily constitute a backarc environment to the continental margin arc (Dilek and Moores, 1993), a view expressed by Hamilton (1969) and entrenched deeply in the "ruling theory" Cordilleran-type models.

3. The recently documented presence of a 600-km-long ocean crust-mantle slab beneath the Great Valley (Godfrey et al., 1997; Godfrey and Klemperer, 1998) is a major new result that should be taken into account in all future tectonic scenarios for California.

4. In the three decades since Moores (1970), there have been many important advances in the methodology of analyzing orogenic belts. Nevertheless, the significance of ophiolites in large-scale crustal tectonics and terrane amalgamation has not yet been universally grasped. *To reiterate, ophiolites emplaced by collision represent major first-order tectonic structures; associated decollement fold-thrust belts are secondary.* The major paradigm shift, dreamed of by Moores (1970) following Hess (1939, 1955), that ophiolite emplacements are the tectonic events of prime importance, has yet to happen. Why? Possibly because most regional structural geologists and/or tectonicists are trained mainly in the traditionally dominant fields of structure and stratigraphy. Ophiolite workers have tended to be petrologists and geochemists, less concerned with details of regional tectonics. The true tectonic significance of ophiolites represents an interdisciplinary issue, and thus falls into the "interdisciplinary oceans of ignorance" lamented by Ziman (1996), and fostered by traditional discipline-based educational curricula. Further progress in tectonics of orogenic belts will be made primarily by exploration of this other type of "ocean."

**Note on computer animation of the southwest Pacific–Southeast Asia region.** The actual complex patterns of terrane interaction in continental orogenic belts such as the North American Cordillera may never be known because of the inherent gaps in the record, particularly the oceanic history of various pieces. The modern region most reminiscent of the Paleozoic-Mesozoic Cordillera is the southwest Pacific (Silver and Smith, 1983; Hamilton, 1979; Moores, 1998). There the tectonic interactions that will eventually lead to continental orogenic terranes are still ongoing. To accompany his tectonic reconstruction of this region (Hall, 1996), Robert Hall has developed a computer animation of the complex, even counterintuitive motions of the southwest Pacific–Southeast Asian region. This animation is reproduced on the compact disk included with this volume. It should serve as a conceptual model of the motions that acted to produce the North American Cordillera.

## ACKNOWLEDGMENTS

We are deeply indebted to many of our coworkers for many stimulating discussions and ideas over the years. It is not possible to name them all. Chief among them are, however, W. S. Alvarez, B. C. Burchfiel, R. Coleman, D. Cowan, G. A. Davis, W. R. Dickinson, W. G. Ernst, R. A. Schweickert, R. J. Twiss, and J. R. Unruh. We thank V. L. Hansen, N. Lindsley-Griffin, S. Roeske, R. A. Schweickert, D. Sloan, and R. J. Twiss for helpful reviews of an earlier version of the manuscript.

## REFERENCES CITED

Alvarez, W. S., Kent, D. V., Premoli Silva, I., Schweickert, R. A., and Larson, R. L., 1980, Franciscan-complex limestone deposited at 17° S. latitude: Geological Society of America Bulletin, v. 91, part I, p. 476–484.

Atwater, T., 1970, Implications of plate tectonics for the Cenozoic tectonic evolution of western North America: Geological Society of America Bulletin, v. 81, p. 3513–3536.

Avé Lallemant, H. G., 1995, Pre-Cretaceous tectonic evolution of the Blue Mountains province, northeastern Oregon: U.S. Geological Survey Professional Paper 1438, p. 271–304.

Burchfiel, B. C., and Davis, G. A., 1972, Structural framework and evolution of the southern part of the Cordilleran orogen, western United States: American Journal of Science, v. 272, p. 97–118.

Burchfiel, B. C., and Davis, G. A., 1975, Nature and controls of Cordilleran orogenesis, western United States: Extensions of an earlier synthesis: American Journal of Science, v. 272, p. 97–118.

Burchfiel, B. C., Cowan, D. S., and Davis, G. A., 1992, Tectonic overview of the Cordilleran orogen in the western United States, *in* Burchfiel, B. C., Lipman, P. W., and Zoback, M. L., eds., The Cordilleran orogen, Conterminous U.S.: Boulder, Colorado, Geological Society of America, Geology of North America, v. G-3, p. 407–479.

Butler, R. F., Gehrels, G. E., and Bazard, D. R., 1997, Paleomagnetism of Paleozoic strata of the Alexander terrane, southeastern Alaska: Geological Society of America Bulletin, v. 109, p. 1372–1388.

Cannat, M., and Boudier, F., 1985, Structural study of intra-oceanic thrusting in the Klamath Mountains, northern California: Implications on accretion geometry: Tectonics, v. 4, p. 435–452.

Chamberlin, T. C., 1890, The method of multiple working hypotheses: Science, v. 15, p. 92–96.

Cloos, M., 1993, Lithospheric buoyancy and collisional orogenesis—subduction of oceanic plateaus, continental margins, island arcs, spreading ridges, and seamounts: Geological Society of America Bulletin, v. 105, p. 715–737.

Clowes, R. M., Zelt, C. A., Amor, J. R., and Ellis, R. M., 1993, Lithospheric structure in the southern Canadian Cordillera from a network of seismic refraction lines: Canadian Journal of Science, v. 32, p. 1485-1513.

Coleman, R. G., 1971, Plate tectonic emplacement of upper mantle peridotites along continental edges: Journal of Geophysical Research, v. 76, p. 1212–1222.

Coney, P. J., 1972, Cordilleran tectonics and North American plate motion: American Journal of Science, v. 272, p. 603–628.

Coney, P. J., Jones, D. L., and Monger, J. W. H., 1980, Cordilleran suspect terranes: Nature, v. 288, p. 329–333.

Coombs, D. S., 1997, A note on the terrane concept, based on an introduction to the Terrane '97 conference, Christchurch, New Zealand, February, 1997: American Journal of Science, v. 297, p. 762–764.

Cowan, D. S., Brandon, M. T., and Garver, J. I., 1997, Geologic tests of hypotheses for large coastwise displacements—A critique illustrated by the Baja British Columbia controversy: American Journal of Science, v. 297, p. 117–173.

Day, H. W., Moores, E. M., and Tuminas, A. C., 1985, Structure and tectonics of the northern Sierra Nevada: Geological Society of America Bulletin, v. 96, p. 436–450.

Debiche, M. G., Cox, A., and Engebretson, D., 1987, The motion of allochthonous terranes across the north Pacific Basin: Geological Society of America Special Paper 207, 49 p.

Dewey, J. F., and Bird, J. M., 1970, Mountain belts and the new global tectonics: Journal of Geophysical Research, v. 75, p. 2625–2647.

Dilek, Y., and Moores, E. M., 1992, Island arc evolution and fracture zone tectonics in the Mesozoic Sierra Nevada, California, and implications for transform offset of the Sierran/Klamath convergent margins, in Bartholomew, M., Hyndman, D.W., Mogk, D.W., and Mason, R., eds., Basement tectonics 8, Characterization and comparison of ancient and Mesozoic continental Margins—Proceedings of the 8th International Conference on Basement Tectonics, Butte, Montana, 1988: The Netherlands, Kluwer Academic Publishers, p. 179–196.

Dilek, Y., and Moores, E. M., 1993, The across-strike anatomy of the Cordilleran orogen at 40°N latitude and implications for the Mesozoic paleogeography of the western United States, in Dunne, G., and McDougall, K., eds., Mesozoic paleogeography of the western United States-II: Pacific Section, SEPM, Book 71, p. 333–346.

Dilek, Y., and Moores, E. M., 1995, Geology of the Humboldt igneous complex, Nevada, and tectonic implications for the Jurassic magmatism in the Cordilleran orogen, in Miller, D. M., and Busby, C., eds., Jurassic magmatism and tectonics of the North American Cordillera: R. L. Armstrong Memorial Volume: Geological Society of America Special Paper 299, p. 229–248.

Dilek, Y., Moores, E. M., and Erskine, M. C., 1988, Ophiolitic thrust nappes in western Nevada: Implications for the Cordilleran orogen: Geological Society of London Journal, v. 145, p. 969–975.

Dilek, Y., Thy, P., Moores, E. M., and Grundvig, S., 1990, Late Paleozoic–early Mesozoic oceanic basement of a Jurassic arc terrane in the northwestern Sierra Nevada, California, in Harwood, D., and Miller, M.M., eds., Paleozoic and early Mesozoic paleogeographic relations; Sierra Nevada, Klamath Mountains, and related terranes: Geological Society of America Special Paper 255, p. 351–370.

Edelman, S. H., Day, H. W., Moores, E. M., Zigan, S. M., Murphy, T. P., and Hacker, B. R., 1989a, Structure across a Mesozoic ocean-continent suture zone in the northern Sierra Nevada, California: Geological Society of America Special Paper 224, 56 p.

Edelman, S. H., Day, H. W., and Bickford, M. E., 1989b, Implications of U-Pb zircon ages for the tectonic settings of the Smartville and Slate Creek Complexes, northern Sierra Nevada: Geology, v. 17, p. 1032–1035.

Ehlig, P., 1981, Origin and tectonic history of the basement terrane of the San Gabriel Mountains, central Transverse Ranges, in Ernst, W. G., ed., The geotectonic development of California: Englewood Cliffs, New Jersey, Prentice-Hall, p. 253–283.

Engebretson, D. C., Cox, A., and Gordon, R. G., 1985, Relative motions between oceanic and continental plates in the Pacific basin: Geological Society of America Special Paper 206, 59 p.

Fagin, S. W., and Gose, W. A., 1983, Paleomagnetic data from the Redding section of the eastern Klamath belt, northern California: Geology, v. 11, no. 9, p. 505–509.

Godfrey, N. J., and Klemperer, S. L., 1998, Ophiolitic basement to a forearc basin and implications for continental growth: The Coast Range/Great Valley ophiolite, California: Tectonics, v. 17, no. 4, p. 558-570.

Godfrey, N. J., Beaudoin, B. C., and Klemperer, S. L., 1997, Ophiolitic basement to the Great Valley forearc basin, California, from seismic and gravity data: Implications for crustal growth at the North American continental margin: Geological Society of America Bulletin, v. 109, p. 1536–1562.

Grove, M., and Bebout, G. E., 1995, Cretaceous tectonic evolution of coastal southern California: Insights from the Catalina schist: Tectonics, v. 14, p. 1290–1308.

Hacker, B. R., Donato, M. M., Barnes, C. G., McWilliams, M. O., and Ernst, W. G., 1993, Timescales of orogeny—Jurassic construction of the Klamath Mountains: Tectonics, v. 14, p. 677–703.

Hall, R., 1996, Reconstructing Cenozoic SE Asia, in Hall, R., and Blundell, D., eds., Tectonic evolution of Southeast Asia: Geological Society of London Special Publication 106, p. 153–184.

Hamilton, W., 1969, Mesozoic California and underflow of the Pacific mantle: Geological Society of America Bulletin, v. 80, p. 2409–2430.

Hamilton, W., 1979, Tectonics of the Indonesian region: U.S. Geological Survey Professional Paper 1078, 345 p.

Hansen, R. E., Girty, G. H., Harwood, D. S., and Schweickert, R. A., 1993, Devonian and Jurassic volcano-plutonic association in the northern Sierra terrane: Implications for arc evolution and Mesozoic deformation—Part I, in Lahren, M. M., Trexler, J. H., Jr., and Spinosa, C., eds, Crustal evolution of the Great Basin and Nevada (Geological Society of America Cordilleran–Rocky Mountain field-trip guidebook): Reno, University of Nevada Department of Geological Sciences, p. 97–128.

Hansen, V. L., 1996, DNAG reaches Alaska: Journal of Structural Geology, v. 18, p. 715–716.

Hansen, V. L., and Dusel-Bacon, C., 1998, Structural and kinematic evolution of the Yukon-Tanana upland tectonites, east-central Alaska: A record of late Paleozoic to Mesozoic crustal assembly: Geological Society of America Bulletin, v. 110, p. 211–230.

Harper, G. D., Grady, K., and Wakabayashi, J., 1990, A structural study of a metamorphic sole beneath the Josephine ophiolite, western Klamath terrane, California-Oregon in Harwood, D. S., and Miller, M. M., eds., Paleozoic and early Mesozoic paleogeographic relations, Sierra Nevada, Klamath Mountains, and related terranes: Geological Society of America Special Paper 255, p. 379–396.

Harper, G. D., Grady, K., and Coulton, A. J., 1996, Origin of the amphibolite "sole" of the Josephine ophiolite: Emplacement of a cold ophiolite over a hot arc: Tectonics, v. 15, p. 296–313.

Harwood, D. S., and Murchey, B. L., 1990, Biostratigraphic, tectonic, and paleogeographic ties between upper Paleozoic volcanic and basinal rocks in the northern Sierra terrane, California, and the Havallah sequence, Nevada in Harwood, D. S., and Miller, M. M., eds., Paleozoic and early Mesozoic paleogeographic relations, Sierra Nevada, Klamath Mountains, and related terranes: Geological Society of America Special Paper 255, p. 157–173.

Henderson, L. J., Gordon, R. G., and Engebretson, D. C., 1984, Mesozoic aseismic ridges on the Farallon plate and southward migration of the shallow subduction during the Laramide orogeny: Tectonics, v. 3, p. 121–132.

Hess, H. H., 1939, Island arcs, gravity anomalies, and serpentinite intrusions, a contribution to the ophiolite problem: International Geological Congress, Moscow, 1937, report 17, v. 2, p. 263–283

Hess, H. H., 1955, Serpentines, orogeny, and epeirogeny in Poldervaart, A., ed., Crust of the Earth: Geological Society of America Special Paper 62, p. 391–408.

Howell, D. G., 1989, Tectonics of suspect terranes: New York, Chapman and Hall, 232 p.

Howell, D. G., Schermer, E. R., Jones, D. L., Ben-Avraham, Z., and Scheibner, E., 1985, Preliminary tectonostratigraphic terrane map of the circum-Pacific region, in Howell, D. G., ed., Tectonostratigraphic terranes of the circum-Pacific region: Houston, Texas, Circum-Pacific Council for Energy and Mineral Resources, Earth Science Series 1, 581 p.

Ingersoll, R. V., 1997, Phanerozoic tectonic evolution of central California and environs: International Geology Review, v. 39, p. 957–972.

Ingersoll, R. V., and Schweickert, R. A., 1986, A plate-tectonic model for Late Jurassic ophiolite genesis, Nevadan orogeny, and forearc initiation, northern California: Tectonics, v. 5, p. 901–912.

Irving, E., Wynne, P. J., Thorkelson, D. J., and Schiarizza, P., 1996, Large (1000 to 4000 km) northward movements of tectonic domains in the northern Cordillera, 83 to 45 Ma: Journal of Geophysical Research, v. 101, p. 17,901–17,916.

Irwin, W. P., 1972, Terranes of the western Paleozoic and Triassic belt in the southern Klamath Mountains, California: U.S. Geological Survey Professional Paper 800-C, p. C103–C111.

Irwin, W. P., 1981, Tectonic accretion of the Klamath Mountains, *in* Ernst, W. G., ed., Geotectonic development of California (Rubey Volume I): Englewood Cliffs, New Jersey, Prentice-Hall, p. 29–49.

Jacobsen, C. E., 1990, The $^{40}Ar/^{39}Ar$ geochronology of the Pelona schist and related rocks, southern California: Journal of Geophysical Research, v. 95, p. 509–528.

Jayko, A. S., and Blake, M. C., Jr., 1993, Northward displacements of forearc slivers in the Coast Ranges of California and southwest Oregon during the late Mesozoic and early Cenozoic, *in* Dunne, G., and McDougall, K., eds., Mesozoic paleogeography of the western United States—II: Pacific Section, SEPM, Book 71, p. 19–36.

Jones, D. L., 1990, Synopsis of late Paleozoic and Mesozoic terrane accretion within the Cordillera of western North America, *in* Dewey, J. F., Gass, I. G., Curry, G. B., Harris, N. B. W., and Şengör, A. M. C., eds., Allochthonous terranes: Cambridge, Cambridge University Press, p. 23–29.

Kennett, J. P., McBirney, A. R., and Thunell, R. C., 1977, Episodes of Cenozoic volcanism in the circum-Pacific region: Journal of Volcanology and Geothermal Research, v. 2, p. 145–163.

Lahren, M. M., and Schweickert, R. A., 1989, Proterozoic and Lower Cambrian miogeoclinal rocks of Snow Lake pendant, Yosemite-Emigrant Wilderness, Sierra Nevada, California: Evidence for major Early Cretaceous dextral translation: Geology, v. 17, p. 156–160.

Mattinson, J. M., 1986, Geochronology of high-pressure–low-temperature Franciscan metabasites: A new approach using the U-Pb system, *in* Evans, B.W., and Brown, E. H., eds., Blueschists and eclogites: Geological Society of America Memoir 164, p. 95–106.

Maxson, J., and Tikoff, B., 1996, Hit-and-run collision model for the Laramide orogeny, western United States: Geology, v. 24, p. 963–972.

McCaffrey, R., 1996, Estimates of modern arc-parallel strain rates in fore arcs: Geology, v. 24, p. 27–30.

Monger, J. W. H., Price, R. A., and Tempelman-Kluit, D. J., 1982, Tectonic accretion and the origin of the two major metamorphic and plutonic welts in the Canadian Cordillera: Geology, v. 10, p. 70–75.

Moores, E. M., 1970, Ultramafics and orogeny, with models of the US Cordillera and the Tethys: Nature, v. 228, p. 837–842.

Moores, E. M., 1982, Origin and significance of ophiolites: Reviews of Geophysics and Space Physics ,v. 20, p. 735–760.

Moores, E. M., 1985, Editorial: On geologic inquiry: Geology, v. 13, p. 3.

Moores, E. M., 1991, Southwest U.S.–East Antarctic (SWEAT) connection: A hypothesis: Geology, v. 19, p. 425–428.

Moores, E. M., 1998, Ophiolites, the Sierra Nevada, "Cordilleria," and orogeny along the Pacific and Caribbean margins of North and South America: International Geology Review, v. 40, p. 40–54.

Moores, E. M., and Day, H. W., 1984, Overthrust model for the Sierra Nevada: Geology, v. 12, p. 416–419.

Moores, E. M., and Twiss, R. J., 1995, Tectonics: New York, W.H. Freeman, 415 p.

Moores, E. M., and Vine, F. J., 1971, The Troodos massif, Cyprus and other ophiolites as oceanic crust; evaluation and implications, *in* Bullard, E., Cann, J. R., and Matthews, D. H., eds., A discussion on the petrology of igneous and metamorphic rocks from the ocean floor: Royal Society of London Philosophical Transactions, ser. A, v. 268, p. 443–466.

Oldow, J. S., Satterfield, J. I., and Silberling, N. J., 1993, Jurassic to Cretaceous transpressional deformation in the Mesozoic marine province of the northwestern Great Basin, *in* Lahren, M. M., Trexler, J. H., Jr., and Spinosa, C., eds., Crustal evolution of the Great Basin and Nevada (Geological Society of America Cordilleran–Rocky Mountain field-trip guidebook): Reno, University of Nevada Department of Geological Sciences, p. 129-166.

Roberts, R. J., 1969, The Cordilleran continental margin—Continental collisions vs. geotectonic cycles: Geological Society of America Abstracts with Programs for 1969, part 7, p. 286–288.

Roure, O., and Lapierre, H., 1989, Comparison between two Paleozoic island-arc terranes in northern California (eastern Klamath and northern Sierra Nevada): Geodynamic constraints: Tectonophysics, v. 169, p. 341–349.

Saleeby, J. B., 1983, Accretionary tectonics of the North American Cordillera, Annual Review of Earth and Planetary Science, v. 11, p. 45-73.

Saleeby, J. B., Shaw, H. F., Niemeyer, S., Moores, E. M., and Edelman, S., 1989, U/Pb, Sm/Nd and Rb/Sr geochronological and isotopic study of northern Sierra Nevada ophiolitic assemblages, California: Contributions to Mineralogy and Petrology, v. 102, p. 205–220.

Saucier, A. E., 1997, The Antler thrust system in northern Nevada, *in* Perry A. J., and Abbott, E. W., eds., The Roberts Mountains thrust, Elko and Eureka Counties, Nevada: Reno, Nevada Petroleum Society 1997 Field Trip Guidebook, , p. 1–16.

Schermer, E., Howell, D. G., and Jones, D. L., 1984, The origin of allochthonous terranes: Perspectives on the growth and shaping of continents: Annual Review of Earth and Planetary Sciences, v. 12, p. 107–132.

Schweickert, R. A., and Cowan, D. S., 1975, Early Mesozoic tectonic evolution of the western Sierra Nevada, California: Geological Society of America Bulletin, v. 86, p. 1329–1336.

Schweickert, R. A., and Snyder, W. S., 1981, Paleozoic plate tectonics of the Sierra Nevada and adjacent regions, *in* Ernst, W. G., ed., Geotectonic development of California (Rubey Volume I): Englewood Cliffs, New Jersey, Prentice-Hall, p. 182–202.

Seeber, L., 1983, Large scale thin-skin tectonics: Reviews of Geophysics and Space Physics, v. 21, p. 1528–1538.

Şengör, A. M. C., and Dewey, J. F., 1990, Terranology: Vice or virtue?, *in* Dewey, J. F., Gass, I. G., Curry, G. B., Harris, N.B.W., and Şengör, A. M. C., eds., Allochthonous terranes: Cambridge, Cambridge University Press, p. 23-29.

Sharp, W. D., 1988, Pre-Cretaceous crustal evolution in the Sierra Nevada region, California: *in* Ernst, W. G., ed., Metamorphism and crustal evolution, western United States (Rubey Volume VII): Englewood Cliffs, New Jersey, Prentice-Hall, p. 824–864.

Silver, E. A., and Smith, R. B., 1983, Comparison of terrane accretion in modern Southeast Asia and the Mesozoic North American Cordillera: Geology, v. 11, p. 198–202.

Speed, R. C., 1978, Collided Paleozoic platelet in the western United States: Journal of Geology, v. 87, p. 279–292.

Tapponnier, P., and Molnar, P., 1976, Slip-line field theory and large-scale continental tectonics: Nature, v. 264, p. 319–324.

Tarduno, J. A., McWilliams, M. O., Debiche, M. G., Sliter, W. V., and Blake, M. C., Jr., 1985, Franciscan Complex Calera limestones: Accreted remnants of Farallon oceanic plateaus: Nature, v. 317, p. 345–347.

Tempelman-Kluit, D. J., 1979, Transported cataclasite, ophiolite, and granodiorite in Yukon: Evidence of arc-continent collision: Geological Survey of Canada Paper 79–14. 27 p.

Valentine, J. W., and Moores, E. M., 1974, Plate tectonics and the history of life in the oceans: Scientific American, v. 230, p. 80–89.

Varga, R. J., and Moores, E. M., 1981, Age, origin and significance of an unconformity that predates island-arc volcanism in the northern Sierra Nevada: Geology, v. 9, p. 512–518.

Wakabayashi, J., 1992, Nappes, tectonics of oblique plate convergence, and metamorphic evolution related to 140 million years of continuous subduction, Franciscan complex, California: Journal of Geology, v. 100, p. 19–40.

Wakabayashi, J., and Hengesh, J. V., 1995, Distribution of late Cenozoic displacement on the San Andreas fault system, northern California, *in* Sangines, E. M., Andersen, D. W., and Buising, A. W., eds., Recent geologic studies in the San Francisco Bay area: Pacific Section, SEPM (Society for Sedimentary Geology) Book 76, p. 19–30.

Wakabayashi, J., and Moores, E. M., 1988, Evidence for the collision of the Salinian block with the Franciscan subduction zone, California: Journal of Geology, v. 96, p. 245–253.

Wallin, E. T., Coleman, D. S., Lindsley-Griffin, N., and Potter, A. W., 1995, Silurian plutonism in the Trinity terrane (Neoproterozoic and Ordovician), Klamath Mountains, California, United States: Tectonics v. 14, p. 1007–1013.

White, D. A., Roeder, D. H., Nelson, T. H., and Crowell, J. C., 1970, Subduction: Geological Society of America Bulletin, v. 81, p. 3431–3432.

Ziman, K., 1996, Is science losing its objectivity?: Nature, v. 382, p. 751–754.

MANUSCRIPT ACCEPTED BY THE SOCIETY NOVEMBER 23, 1998.

Printed in U.S.A.

Section IV

# WATER AND OIL—THE VITAL FLUIDS OF CALIFORNIA

# Water and Oil: Introduction

## INTRODUCTION

Water and oil have been, and are, the lifeblood of California. Water has figured prominently in different ways in California history since the discovery of gold. The discovery and production of oil in the early twentieth century added greatly to California's prosperity and fostered the "automobile cult" for which California is so famous.

California has a Mediterranean climate; it is in mid-latitudes, extending from the dry horse latitudes in the south to the Boreal wet regions in the north. Precipitation is bimodal, falling mainly in the winter, and summers are dry . Rainfall is generally abundant in the northern mountains but decreases progressively toward the south. The mountains of California, principally the Sierra Nevada, and to a lesser extent the Klamath Mountains to the north and the Transverse Ranges to the south, produce a rain-shadow effect. Winds coming off the North Pacific rise and drop rain or snow on the western slopes; the eastern slopes are dry.

It was in one of the modern streams draining the western slope of the Sierra Nevada that gold was first discovered. Water—too much or too little—has been of prime concern ever since. Sacramento was established near the confluence of the American and Sacramento Rivers with little regard for the flood hazard. It got its comeuppance in the winter of 1861–1862, when the entire Great Valley was flooded from north to south (Farquahar, 1966), and the entire downtown section had to be raised 10 feet (an action still visible).

Water abundance was important in California development in two ways. First, the development of hydraulic mining, wherein entire streams were diverted to attack the Auriferous Gravels, producing catastrophic downstream environmental consequences leading to the ban on such practices (see Introduction to Section II). Second, the settlement of the Great Valley and the Imperial Valley, leading to California's development as the United States' most productive agricultural state would never have happened without abundant water for irrigation.

Water scarcity has also been and continues to be of crucial importance. As Powell (1879) first noted, the entire western part of North America, including California, lacks precipitation adequate for agriculture without irrigation (see also Stegner, 1954). Development of large population centers in southern California would not have been possible without importation of water, first from Owens Valley, east of the Sierra Nevada, then from the Colorado River, and finally from northern California. These huge engineering structures transfer water from one watershed basin to another on a scale never before attempted (e.g., Reisner, 1993).

Nevertheless, periodic droughts are a continuing problem as available water supplies are increasingly stretched to the limit. During the twentieth century, California droughts have included the periods 1928–1934, 1975–1977, and 1987–1992 (Stine, 1994). The latter periods caused considerable disturbance of agricultural and urban water-use habits.

These short-term dislocations, however, pale by comparison with the longer term record gleaned from indirect means. Fritts (1984) amassed from tree-ring evidence a 400 year estimate of California precipitation (Fig. 1). The outstanding features of this figure are: (1) that the average precipitation for the twentieth century exceeds that of the entire period since 1600; and (2) from 1760 until 1830, not once was the estimated precipitation what we consider normal levels. Even more ominously, from study of relict tree stumps from lakes, marshes, and streams, Stine (1994) documented periods of extreme drought in the Sierra Nevada lasting for more than 200 years before about A.D. 1100, and for more than 140 years prior to about A.D. 1350. One shudders to contemplate what effects any future drought of such duration would have on the California economy.

Water is, however, only one of California's two vital fluids. Oil has been essential to California's twentieth century devel-

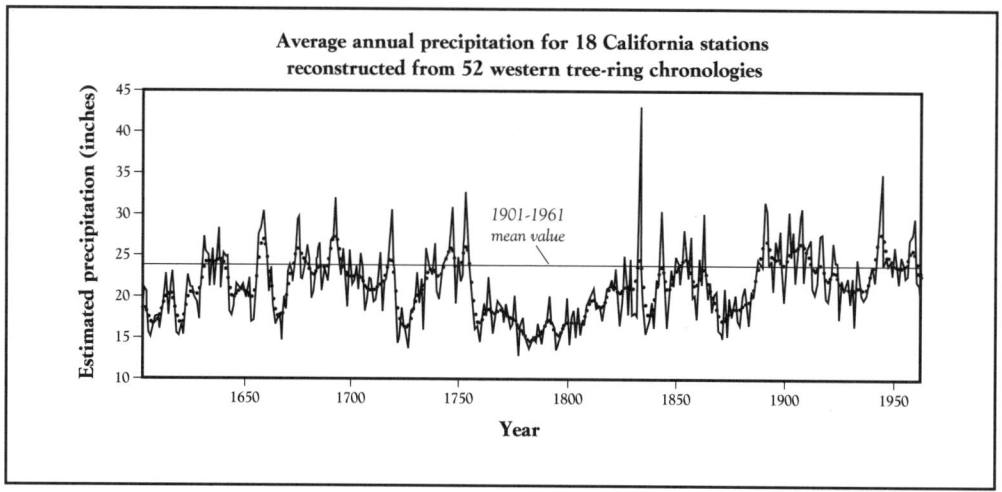

Figure 1. Average annual precipitation for 18 California stations reconstructed from 52 western tree-ring chronologies. Dots represent eight-year weighted averages. The horizontal line corresponds to the 1906–1961 mean value (after Fritts, 1984, Fig. 1.18, p. 46).

opment as the world's leading motor-vehicle culture. Ironically, it didn't start out that way. After the impetus of the Gold Rush and Comstock Lode, California's development was fueled by construction of railroads; the gift of large tracts of land to the railroad companies was incentive for their construction, and the patterns of development of the Great Valley and southern California resulted as the railroad companies (principally the Southern Pacific and Santa Fe companies) strived to profit from these large land holdings. In the early twentieth century, California transportation was dominated as elsewhere by railroads, the major railroads being augmented by well-developed urban and interurban surface railway systems. It is said that in 1910, the Los Angeles basin had the world's best electrified interurban mass-transit system, the Pacific Electric. Other systems characterized cities in the San Francisco Bay area and in the valley. These systems were bought up in the 1930s to 1950s by National City Lines, a consortium of Standard Oil of California (now Chevron), General Motors, and Firestone Tire (now a subsidiary of the Japanese tire maker Bridgestone). The scrapping of the private urban railway systems and the large-scale construction of public highways and the California freeway system led to the present almost absolute dependence of the California economy on cars and trucks.

In the early twentieth century, rapid development of oil deposits in the Los Angeles, Ventura, and southern San Joaquin basins made California an oil-exporting state, a position it gradually lost as the deposits declined, population burgeoned, and dependence on the automobile increased.

Studies of water and oil together have led to several basic conceptual advances in our understanding of the geologic role of these vital fluids; the studies span a range of considerations of the geologic role of surface water, the origin of oil and its stratigraphic implications, and the long-term fluctuations of California's water supply and the factors that have an impact on it.

These concepts are laid out in five chapters. We begin with a chapter based upon part of G. K. Gilbert's study of California hydraulic mining, *The Transportation of débris by running water*, which forms the basis of the contribution by E. A. Keller. Gilbert's study grew out of his assignment to study the effects of California hydraulic mining, which itself grew out of pressure from the mining industry on Congress to rescind the court order banning hydraulic mining from the California mountains. Gilbert spent about two decades spanning the turn of the century on this study, documenting the total scale and effects of hydraulic mining. Keller's essay focuses on a small part of that effort—Gilbert's pioneering experiments on the transport of debris by running water. The remainder of Gilbert's paper documents the mind-boggling effects of hydraulic mining, during which some $10^9$ m$^3$ of debris were washed off the mountains into the Great Valley and Sacramento–San Joaquin Delta.

Two chapters related to oil follow. Kleinpell's paper on Miocene stratigraphy of California forms the focus for W. B. N. Berry's essay. Bramlette's 1946 paper on the Miocene Monterey Formation, the principal source rock of California oil, forms the basis of the essay by R. J. Behl. We end this section with a chapter on the California Water Project, the most recent and most ambitious of all the interbasin water transfer engineering projects in California. An excerpt from the California Water Plan proposal, published in 1957, is the "classic" focus of this chapter. The essay by R. T. Bean and E. M. Weber, two players in development of this plan, now retired, presents the "company view" of the plan and the present state of water in the various California watershed regions. The final essay, by C. J. Hauge, a leading California hydrogeologist, presents an up-to-date and authoritative assessment of California's current water situation. In particular, Hauge discusses the complex issues of water storage, ground-water extraction and recharge, land subsidence, and the ominous issue of the fragility of the Sacramento–San Joaquin Delta and recent attempts to come to grips with its challenges.

## REFERENCES CITED

Farquahar, F., ed., 1966, Up and down California in 1860–1864, The journal of William H. Brewer, Professor of Agriculture in the Sheffield Scientific School from 1864 to 1903: Berkeley, University of California Press, 583 p.

Fritts, H. C., 1984, Discussion, *in* Engelbert, E. A., and Scheuring, A. F., eds., Water scarcity impacts on western agriculture: Berkeley and Los Angeles, University of California Press, p. 44–48.

Powell, J. W., 1879, Report on the lands of the arid region of the United States (second edition): Washington, D. C., Government Printing Office, 195 p.

Reisner, M., 1993, Cadillac desert: The American West and its disappearing water (revised and updated): New York, Penguin Books, 582 p.

Stegner, W. E., 1954, Beyond the hundredth meridian: John Wesley Powell and the second opening of the West: Boston, Houghton, Mifflin, 438 p.

Stine, S., 1994, Extreme and persistent drought in California and Patagonia during medieval time: Nature, v. 369, p. 546–549.

MANUSCRIPT ACCEPTED BY THE SOCIETY NOVEMBER 23, 1998.

# Chapter 13
# GILBERT'S HYDRAULIC EXPERIMENTS

**E. A. Keller**
*Department of Geological Sciences*
*University of California, Santa Barbara*
*Santa Barbara, California 93106*
*E-mail: keller@magic.ucsb.edu*

# GROVE KARL GILBERT

Grove Karl Gilbert was born on May 6, 1843, in Rochester, New York. Gilbert graduated from the University of Rochester at age 19 and began his geologic career with an apprenticeship with the Geological Survey of Ohio in 1869. In 1871, Gilbert joined the U.S. Army's Geographic Survey and later (1875) joined the Powell Survey, exploring the west. He was employed by the U.S. Geological Survey in 1879 and completed many important studies leading to publications that presented fundamental concepts of geology to American science. Gilbert helped found the Geological Society of America and is the only person to be elected twice as president (1892 and 1909). He was elected to the National Academy of Sciences in 1883. Gilbert's contributions to the U.S. Geological Survey were deemed so important that a symposium on his career was held as part of celebrations for the centennial of the U.S. Geological Survey. Gilbert died on May 27, 1918, at the age of 75 while in the midst of planning future research projects (Pyne, 1980).

I chose Professional Paper 86, *The transportation of débris by running water*, by Grove Karl Gilbert as a "classic paper" in geology and geomorphology for several reasons. First, it is a creative paper, the vitality of which demands continuous rediscovery; and second, the paper provides us with a framework and terminology for exploring fluvial geomorphology that is still in use today while demonstrating the value of experimentation and quantitative analysis of complex geomorphic systems.

E. A. Keller

DEPARTMENT OF THE INTERIOR
UNITED STATES GEOLOGICAL SURVEY
GEORGE OTIS SMITH, DIRECTOR

PROFESSIONAL PAPER 86

THE

# TRANSPORTATION OF DÉBRIS BY RUNNING WATER

BY

GROVE KARL GILBERT

BASED ON EXPERIMENTS MADE WITH THE ASSISTANCE OF

EDWARD CHARLES MURPHY

WASHINGTON
GOVERNMENT PRINTING OFFICE
1914

APPARATUS ON CAMPUS OF UNIVERSITY OF CALIFORNIA.

# ABSTRACT.

*Scope.*—The finer débris transported by a stream is borne in suspension. The coarser is swept along the channel bed. The suspended load is readily sampled and estimated, and much is known as to its quantity. The bed load is inaccessible and we are without definite information as to its amount. The primary purpose of the investigation was to learn the laws which control the movement of bed load, and especially to determine how the quantity of load is related to the stream's slope and discharge and to the degree of comminution of the débris.

*Method.*—To this end a laboratory was equipped at Berkeley, Cal., and experiments were performed in which each of the three conditions mentioned was separately varied and the resulting variations of load were observed and measured. Sand and gravel were sorted by sieves into grades of uniform size. Determinate discharges were used. In each experiment a specific load was fed to a stream of specific width and discharge, and measurement was made of the slope to which the stream automatically adjusted its bed so as to enable the current to transport the load.

*The slope factor.*—For each combination of discharge, width, and grade of débris there is a slope, called competent slope, which limits transportation. With lower slopes there is no load, or the stream has no capacity[1] for load. With higher slopes capacity exists; and increase of slope gives increase of capacity. The value of capacity is approximately proportional to a power of the excess of slope above competent slope. If $S$ equal the stream's slope and $\sigma$ equal competent slope, then the stream's capacity varies as $(S-\sigma)^n$. This is not a deductive, but an empiric law. The exponent $n$ has not a fixed value, but an indefinite series of values depending on conditions. Its range of values in the experience of the laboratory is from 0.93 to 2.37, the values being greater as the discharges are smaller or the débris is coarser.

*The discharge factor.*—For each combination of width, slope, and grade of débris there is a competent discharge, $\kappa$. Calling the stream's discharge $Q$, the stream's capacity varies as $(Q-\kappa)^o$. The observed range of values for $o$ is from 0.81 to 1.24, the values being greater as the slopes are smaller or the débris is coarser. Under like conditions $o$ is less than $n$; or, in other words, capacity is less sensitive to changes of discharge than to changes of slope.

*The fineness factor.*—For each combination of width, slope, and discharge there is a limiting fineness of débris below which no transportation takes place. Calling fineness (or degree of comminution) $F$ and competent fineness $\phi$, the stream's capacity varies with $(F-\phi)^p$. The observed range of values for $p$ is from 0.50 to 0.62, the values being greater as slopes and discharges are smaller. Capacity is less sensitive to changes in fineness of débris than to changes in discharge or slope.

*The form factor.*—Most of the experiments were with straight channels. A few with crooked channels yielded nearly the same estimates of capacity. The ratio of depth to width is a more important factor. For any combination of slope, discharge, and fineness it is possible to reduce capacity to zero by making the stream very wide and shallow or very narrow and deep. Between these extremes is a particular ratio of depth to width, $\rho$, corresponding to a maximum capacity. The values of $\rho$ range, under laboratory conditions, from 0.5 to 0.04, being greater as slope, discharge, and fineness are less.

*Velocity.*—The velocity which determines capacity for bed load is that near the stream's bed, but attempts to measure bed velocity were not successful. Mean velocity was measured instead. To make a definite comparison between capacity and mean velocity it is neces-

---

[1] *Capacity* is defined for the purposes of this paper as the maximum load of a given kind of débris which a given stream can transport. See page 35.

sary to postulate constancy in some accessory condition. If slope be the constant, in which case velocity changes with discharge, capacity varies on the average with the 3.2 power of velocity. If discharge be the constant, in which case velocity changes with slope, capacity varies on the average with the 4.0 power of velocity. If depth be the constant, in which case velocity changes with simultaneous changes of slope and discharge, capacity varies on the average with the 3.7 power of velocity. The power expressing the sensitiveness of capacity to changes of mean velocity has in each case a wide range of values, being greater as slope, discharge, and fineness are less.

*Mixtures.*—In general, débris composed of particles of a single size is moved less freely than débris containing particles of many sizes. If fine material be added to coarse, not only is the total load increased but a greater quantity of the coarse material is carried.

*Modes of transportation; movement of particles.*—Some particles of the bed load slide; many roll; the multitude make short skips or leaps, the process being called saltation. Saltation grades into suspension. When particles of many sizes are moved together the larger ones are rolled.

*Modes of transportation; collective movement.*—When the conditions are such that the bed load is small, the bed is molded into hills, called dunes, which travel downstream. Their mode of advance is like that of eolian dunes, the current eroding their upstream faces and depositing the eroded material on the downstream faces. With any progressive change of conditions tending to increase the load, the dunes eventually disappear and the débris surface becomes smooth. The smooth phase is in turn succeeded by a second rhythmic phase, in which a system of hills travel upstream. These are called antidunes, and their movement is accomplished by erosion on the downstream face and deposition on the upstream face. Both rhythms of débris movement are initiated by rhythms of water movement.

*Application of formulas.*—While the principles discovered in the laboratory are necessarily involved in the work of rivers, the laboratory formulas are not immediately available for the discussion of river problems. Being both empiric and complex, they will not bear extensive extrapolation. Under some circumstances they may be used to compare the work of one stream with that of another stream of the same type, but they do not permit an estimate of a river's capacity to be based on the determined capacities of laboratory streams. The investigation made an advance in the direction of its primary goal, but the goal was not reached.

*Load versus energy.*—The energy of a stream is measured by the product of its discharge (mass per unit time), its slope, and the acceleration of gravity. In a stream without load the energy is expended in flow resistances, which are greater as velocity and viscosity are greater. Load, including that carried in suspension and that dragged along the bed, affects the energy in three ways. (1) It adds its mass to the mass of the water and increases the stock of energy pro rata. (2) Its transportation involves mechanical work, and that work is at the expense of the stream's energy. (3) Its presence restricts the mobility of the water, in effect increasing its viscosity, and thus consumes energy. For the finest elements of load the third factor is more important than the second; for coarser elements the second is the more important. For each element the second and third together exceed the first, so that the net result is a tax on the stream's energy. Each element of load, by drawing on the supply of energy, reduces velocity and thus reduces capacity for all parts of the load. This principle affords a condition by which total capacity is limited. Subject to this condition a stream's load at any time is determined by the supply of débris and the fineness of the available kinds.

*Flume transportation.*—In the experiments described above—experiments illustrating stream transportation—the load traversed a plastic bed composed of its own material. Other experiments were arranged in which the load traversed a rigid bed, the bottom of a flume. Capacities are notably larger for flume transportation than for stream transportation, and their laws of variation are different. Rolling is an important mode of progression. For rolled particles the capacity increases with coarseness, for leaping particles with fineness. Capacity increases with slope and usually with discharge also, but the rates of increase are less than in stream transportation. Capacity is reduced by roughness of bed.

*Vertical velocity curve.*—The vertical distribution of velocities in a current is controlled by conditions. The level of maximum velocity may have any position in the upper three-fourths of the current. In loaded streams its position is higher as the load is greater. In unloaded streams its position is higher as the slope is steeper, as the discharge is greater, and as the bed is rougher.

*Pitot tube.*—The constant of the Pitot velocity gage—the ratio between the head realized and the theoretic velocity head—is not the same in all parts of a conduit, being less near the water surface and greater near the bottom or side of the conduit.

A. A SIERRA CANYON BED IN ITS NATURAL CONDITION.
The water channel occupies the entire width.

B. A SIERRA CANYON CLOGGED BY MINING DÉBRIS.
The surface of the débris constitutes a broad plain that is covered by water only when the stream is in flood. The low-water channels traverse this plain. The photograph was made in 1908. At that time most of the canyon deposits had been removed. (See Pl. VII, A.) This one is the head of a piedmont deposit. (See pp. 28, 52.)

CHANGE OF CONDITION CAUSED BY MINING DÉBRIS.

from G. K. Gilbert, 1917, Hydraulic-Mining Débris in the Sierra Nevada, U.S. Geological Survey Professional Paper 105.

# CHAPTER XIII.—APPLICATION TO NATURAL STREAMS.

## INTRODUCTION.

The flow of a river is a complex phenomenon. The transportation of débris by it involves intricate reactions. The quantity of débris transported depends on a variety of conditions, and these conditions interact one on another. Direct observation of what takes place at the base of the current is so difficult that the body of information thus obtained is small. In the work of the Berkeley laboratory the attempt was made to study the influence of each condition separately, and to that end all the conditions were subjected to control. This involved the substitution of the artificial for the natural; and while the principles discovered are such as must enter into the work of natural streams, their combinations there are different from the combinations of the laboratory. It is the province of the present chapter to consider the differences between the laboratory streams and natural streams, and in view of those differences the applicability of the laboratory results to problems connected with natural streams.

## FEATURES DISTINGUISHING NATURAL STREAMS.

### KINDS OF STREAMS.

Classification necessarily involves a purpose, or point of view, and there are in general as many scientific, or natural, or otherwise commendable classifications as there are functions to be subserved. The classification of streams here given has no other purpose than to afford a terminology convenient to the subject of débris transportation.

When the débris supplied to a stream is less than its capacity the stream erodes its bed, and if the condition is other than temporary the current reaches bedrock. The dragging of the load over the rock wears, or abrades, or corrades it. When the supply of débris equals or exceeds the capacity of the stream bedrock is not reached by the current, but the stream bed is constituted wholly of débris. Some streams with beds of débris have channel walls of rock, which rigidly limit their width and otherwise restrain their development. Most streams with beds of débris have one or both banks of previously deposited débris or alluvium, and these streams are able to shift their courses by eroding their banks. The several conditions thus outlined will be indicated by speaking of streams as *corrading*, or *rock-walled*, or *alluvial*. In strictness, these terms apply to local phases of stream habit rather than to entire streams. Most rivers and many creeks are corrading streams in parts of their courses and alluvial in other parts.

Whenever and wherever a stream's capacity is overtaxed by the supply of débris brought from points above a deposit is made, building up the bed. If the supply is less than the capacity, and if the bed is of débris, erosion results. Through these processes streams adjust their profiles to their supplies of débris. The process of adjustment is called gradation; a stream which builds up its bed is said to aggrade and one which reduces it is said to degrade.

An alluvial stream is usually an aggrading stream also; and when that is the case it is bordered by an alluvial plain, called a flood plain, over which the water spreads in time of flood.

If the general slope descended by an alluvial stream is relatively steep, its course is relatively direct and the bends to right and left are of small angular amount. If the general slope is relatively gentle, the stream winds in an intricate manner; part of its course may be in directions opposite to the general course, and some of its curves may swing through 180° or more. This distinction is embodied in the terms *direct alluvial* stream and *meandering* stream. The particular magnitude of general slope by which the two classes are separated is greater for small streams than for large. Because fineness is one of the conditions determining the general slope of an alluvial plain, and because the gentler slope go with the finer alluvium, it is true in the main that meandering streams are associated with fine alluvium.

### FEATURES CONNECTED WITH CURVATURE OF CHANNEL.

As nearly all the laboratory experiments were performed with straight channels, and as all natural channels are more or less curved, the features resulting from curvature constitute differences of which account must be taken in applying laboratory results. Some of these differences have been mentioned in connection with the short series of experiments with curved and bent channels, but a fuller account is desirable.

In a straight channel the current is swifter near the middle than near the sides and is swifter above mid-depth than below. On arriving at a bend the whole stream resists change of course, but the resistance is more effective for the swifter parts of the stream than for the slower. The upper central part is deflected least and projects itself against the outer bank. In so doing it displaces the slow-flowing water previously near that bank, and that water descends obliquely. The descending water displaces in turn the slow-flowing lower water, which is crowded toward the inner bank, while the water previously near that bank moves toward the middle as an upper layer. One general result is a twisting movement, the upper parts of the current tending toward the outer bank and the lower toward the inner.[1] Another result is that the swiftest current is no longer medial, but is near the outer or concave bank. Connected with these two is a gradation of velocities across the bottom, the greater velocities being near the outer bank. The bed velocities near the outer bank are not only much greater than those near the inner bank, but they are greater than any bed velocities in a relatively straight part of the stream. They have therefore greater capacity for traction, and by increasing the tractional load they erode until an equilibrium is attained. On the other hand, the currents which, crossing the bed obliquely, approach the inner bank are slackening currents, and they deposit what they can no longer carry.

It results that the cross section on a curve is asymmetric, the greatest depth being near the outer bank. As the winding stream changes the direction of its curvature from one side to the other, the twisting system of current filaments is reversed, and with it the system of depths, but the process of change includes a phase with more equable distribution of velocities, and this phase produces a shoal separating the two deeps. The shoal does not cross the channel in a direction at right angles to its sides but is somewhat oblique in position, tending to run from the inner bank of one curve to the inner bank of the other. In meandering streams it is usually narrow and is appropriately called a bar. In direct alluvial streams, where bends are apt to be separated by long, nearly straight reaches, it is usually broad and may for a distance occupy the entire width of the channel. In navigated rivers the locality of the bar is usually called a crossing, being the place where the thalweg, the line of strongest current, and the route of travel cross from side to side; and the name is often applied also to the bar itself.

The twisting current attacks the outer bank, being swifter at contact with that bank than in any other part of the wetted perimeter. If the bank consists of alluvium there is erosion, the amount being determined in part by resistances arising from roots, or adhesion of alluvial particles, or incipient cementation; and the eroded material, so far as it joins the bed load, helps to satisfy the bed current and limit downward erosion. In alluvial streams the erosion from concave banks offsets the deposition under convex banks, so that the channel may gradually shift its position without change of sectional area.

The sectional area may be either greater or less at a curve than on a reach, but the differences are normally[1] of small amount. Therefore the mean velocity does not vary greatly. The current in a curved channel, as compared to that in a straight channel, is characterized by diversity. Its bed velocities are both higher and lower, and the same is true of velocities along the banks. This diversity is favorable to traction, because capacity for traction varies with a high power of velocity; but the advantage to traction is partly offset by the fact that increase of velocity affects a smaller portion of the wetted perimeter than

---

[1] The system of movements here described has been observed by many students of rivers. They were demonstrated by the aid of a model channel by J. Thomson, in connection with an explanation which differs somewhat from that of the present text. See Roy. Soc. London Proc., pp. 5–8, 1876, and 356–357, 1877; also Inst. Mech. Eng. Proc., pp. 456–460, 1879.

[1] That is to say, they are of small amount when the system of depths is adjusted to the discharge, as explained on a later page.

is affected by reduction of velocity. There is also diversity in the directions followed by elements of current, and this diversity includes not only the twisting movement but various minor eddies and swirls. Diversified movements, by including upward movements, promote suspension, and in conjunction with diversified velocities they modify the partition of the load between traction and suspension. On the whole, suspension claims more in a diversified current, but it is also true that the line of separation between suspension and traction shifts to and fro in such a current. Much débris which is suspended in the swift water under the concave bank joins the bed load in passing the shoal between deeps, and the suspended load is still more restricted in passing the shoal of the convex bank. Deposition on the latter shoal includes both tractional and suspensional materials.

### FEATURES CONNECTED WITH DIVERSITY OF DISCHARGE.

All streams vary in volume from season to season and from year to year. In a stream fed by springs the changes may be slight. At the opposite extreme are creeks and even rivers which exist only during storms. In most large streams the discharge at flood stage is many times greater than at low stage. Usually flood stages continue only for brief periods and in the aggregate occupy but a small fraction of the year.

It is broadly true that streams give shape to their own channels, and among alluvial streams there are few exceptions. It is broadly true also that the shapes of channels, including cross sections and plans, are the same for large streams as for small. But the large stream requires and develops a larger channel—broader, deeper, and winding in larger curves. Through variation of discharge the same stream is alternately large and small, so that its needs are different at different stages. At each stage it tends to fit its channel to the needs of the particular discharge. The formative forces residing in the current are so much stronger with large discharge than with small that the greater features of channel are adjusted to large discharge, and this despite the fact that floods are of brief duration. The feebler forces of smaller discharges modify the flood-made forms but do not succeed in completing their work of adjustment before it is interrupted by another flood. The deeps of high stage are pools at low stage and have currents too feeble for traction. As the reduced stream passes from pool to pool it crosses the shoal formed at high stage with quickened current. The velocities are still diversified, but the greater and smaller velocities have exchanged places. The slope of water surface is more diversified than at high stage, being lower at the pools and higher between them. Traction is restricted to the shoals, and the loads are small. The load at each shoal is obtained from the shoal itself and is deposited in the next pool, and in this way shallow channels are developed from pool to pool.

In contrasting the features of high and low stages, it has been convenient to use the terms as if high stage and low stage were specific and definite phases of stream activity, thereby ignoring the actual diversity in fluctuations of discharge. Floods are of all magnitudes, and each flood presents not only a maximum discharge but a continuous series of changing discharges. At each instant the stream contains a system of currents of which the details depend not only on the discharge but on the shapes of channel created by the work of previous discharges. So long as the discharge continues, its currents are eroding and depositing in such way as to remodel the channel for its own needs, and so long as the work of remodeling continues the loads and capacities at different cross sections are different.

With the changes in the values and distribution of velocities go changes in those values of competent fineness which on one side limit traction and on the other separate traction from suspension. With maximum discharge all the coarser grades of débris within the domain of the stream are in transit along the path of highest activity, and that path includes the deeps and the intervening shoals. With lessening discharge the coarsest material stops, but it stops chiefly in the deeps, because the change in bed velocity is there greatest. At the same time the coarsest of the suspended load escapes from the body of the stream and joins the bed load. By this double change the mean fineness of the tractional load is increased, and so also is the mean fineness of the suspended load. With continued reduction of discharge the tractional load in the deeps becomes gradually finer and at last ceases to move, while the graduated deposit caused by its arrest receives a final contribution from the suspended load. The tractional load on the shoals changes less in mean fineness and may cease to change altogether when the supply from the deep is cut off. It is then derived wholly from the subjacent bed and is greatly reduced in quantity. Soon the derivation becomes selective, the finer part being carried on while the coarser remains, with the result that the shallow channels on the bars come to be paved with particles which the enfeebled currents can not move.

If the section of the alluvium underlying a shoal be afterwards exposed, it is seen to be in the main heterogeneous but veneered at the top by a layer of its coarser particles. The typical section of a deposit in a deep shows the coarsest débris below and the finest at top, with a gradual change.

With the return of large discharge the modeling work of smaller discharges is rapidly obliterated, and the débris deposited in the pools rejoins the tractional and suspended loads.

### SECTIONS OF CHANNEL.

Rock-walled channels result from the aggradation of corraded channels. Often they are recurrent temporary conditions of corraded channels. Their widths have been developed in connection with the work of corrasion and are less than the widths of alluvial streams. In the fact that their sides are immobile they resemble the laboratory channels, and their types of cross section are illustrated by the experiments with crooked channels. The channels of all alluvial streams are strongly asymmetric at the bends, and in the meandering streams the bends constitute the greater part of the course. Departure from symmetry is less pronounced in the reaches of direct alluvial streams, but even there a close approximation to symmetry is exceptional.

Alluvial streams tend to broaden their channels by eroding one or both banks. The influence of vegetation opposes this tendency. Often the erosion of the bank exposes roots, and some trees extend rootlets into the water. At low stages the bared parts of the flood channel are occupied by young plants. In these ways vegetation creates obstacles which retard the current at its contact with the bank and thus oppose erosion. If the current is strong erosion is merely retarded, not prevented; if the current is weak deposition may be induced. As a meandering stream encroaches on its concave bank, the convex bank encroaches on the stream, and channel width is maintained. A large stream is less affected than a small stream by the opposition of vegetation and maintains a channel of relatively small form ratio.

Some streams aggrade so rapidly that vegetation does not secure a foothold. By erosion of its banks such a stream broadens its channel and reduces its depth until the slackened current clogs itself by deposition of its load. The built-up bed becomes higher than the adjacent alluvial plain, and the stream takes a new course. Before the assumption of the new course the banks are overtopped by shallow distributaries which deposit their loads on the banks, thus building them up, until the stream is made to flow on a sort of elevated conduit; and when the main body of water at last leaves this pathway, it is apt to start its new course with a steepened slope and scour for itself a relatively narrow channel.

The building up of the bank by deposition from overflow is more pronounced in the presence of vegetation. The ridge thus created is called a *natural levee*. Its crest separates the channel from the flood plain and delimits at flood stage two provinces in which the conditions affecting transportation are strongly contrasted. In both provinces the general slope of the water surface is the same, but the broad sheet covering the plain has so little depth that its currents are sluggish. Between the banks are the normal channel depths and currents, and transportation is active, alike by traction and suspension. Beyond them transportation is effected almost wholly by suspension, and the coarser particles of the suspended load are deposited. As the flood subsides the lateral sheets of water are returned to the main channel by a draining process which involves the making and maintenance of small channels within the plain.

When the channel of a river is fully adjusted to the discharge the same load is transported through each section. All sections are then equally adapted to transportation, though in different ways. The most symmetric has a wide space at the bottom devoted to traction.

## APPLICATION TO NATURAL STREAMS.

The least symmetric has a relatively narrow tractional space, but traction is there relatively active. The partition of the load between traction and suspension is not the same for the two sections, the tractional load having the greater range in the symmetric section and the suspended load in the asymmetric.

There is reason to believe that the sectional area is about the same in different parts of an adjusted channel. At low stages, when form is least adjusted to discharge, the sectional area is much larger for the asymmetric sections. At higher stages the contrast is less, and the greater area may be associated with either type of section. It is also true, if attention be restricted to the channel proper and the expansions over flood plains be excluded, that the variations in width from point to point of an adjusted channel are not of large amount. If it were strictly true that both sectional areas and widths are equal in different parts of an adjusted channel, it would follow (1) that mean depths are equal, and (2) that form ratios are equal, provided form ratio be defined as the ratio of mean depth to width. Such a generalization, while crude and doubtless subject to important qualification, nevertheless warrants the selection of mean depth rather than maximum depth as the quantity to be used in applying the conception of form ratio to rivers.

Assuming the generalization as an approximation to the actual fact and connecting with it the fact that all sections of an adjusted river are equally efficient for transportation, we are able to make a general application of the laboratory results on optimum form ratio to rivers. The ratio of mean depth to width in alluvial rivers, as a class, is very much smaller than in the laboratory examples by means of which the optimum ratio was discussed in Chapter IV. It is so much smaller that the range of form ratio for alluvial rivers overlaps but slightly the range observed in the laboratory. This disparity indicates, though without demonstrating, that the form ratios of the rivers are less than the optimum, and that their tractional capacities would be greater if they were narrower and deeper. As the optimum ratio is the one which enables a stream to transport its load with the least expenditure of head, it is probable that the slopes of most alluvial rivers can be lessened by artificially reducing their widths.

# The transportation of débris by running water, by Grove Karl Gilbert

**E. A. Keller**

*Department of Geological Sciences, University of California, Santa Barbara, California 93106; e-mail: keller@magic.ucsb.edu*

## INTRODUCTION

In the preface to U.S. Geological Survey (USGS) Professional Paper 86, Gilbert (1914) stated that 35 years prior to his experiments at the University of California, Berkeley, he completed a study of streams and their role in shaping the landscape. The work that Gilbert alluded to was probably his research on the geology of the Henry Mountains (Gilbert, 1877). He stated that his work included a qualitative and partly deductive investigation of factors that control the transport of debris by running water and included a desire for quantitative data. Gilbert realized that quantitative data he could use would most likely be obtained by flume experimentation, wherein conditions could be controlled and the effects of variables, including channel slope, discharge, and bed-material characteristics, could be determined.

Gilbert, with E. C. Murphy, carried out flume experiments from 1907 to 1909 at the University of California, Berkeley Campus. During part of that time Gilbert was ill, and some planned experiments were done by Murphy. The main experimental device consisted of a flume (wood trough) ~ 9.6 m long by 0.6 m wide and 0.5–0.3 m in height.

Gilbert's paper on the transport of debris by running water (USGS Professional Paper 86, 1914) is a classic paper because of the creative concepts presented. The paper was also an early attempt to quantify fluvial geomorphology and to understand some of the basic relationships that control the movement of bed load, and in particular, relationships among the capacity, slope, velocity, discharge, and characteristics of the bed load (what Gilbert referred to as degree of commutation, by which he meant the stream processes that physically reduce the size of bed material the stream is transporting). More important, Professional Paper 86 made a number of fundamental contributions to the understanding of stream processes and transport of bed load, including the introduction of several primary factors related to stream capacity; the introduction of the concept of total stream power (defined by Gilbert as energy of a stream); and the application of results of flume experiments to the understanding of natural streams (in particular channel form and process). I believe that the understanding of natural stream processes was Gilbert's primary goal. Gilbert was also motivated by the controversy over the impacts of hydraulic mining for gold on the Sacramento River system (especially the Yuba River) from headwaters in the foothills of the Sierra Nevada downstream to the San Francisco Bay (Pyne, 1980). Gilbert recognized that understanding the "debris wave" of sediment produced by the dredging and sluicing of gold-bearing gravels would be assisted by a better quantitative understanding of river processes.

The purpose of this paper is to provide an introduction to Professional Paper 86 and Gilbert's major contributions in that paper. Because the paper is 263 pages long, our discussions here focus on two particular parts: the abstract, which included some of the basic equations and relationships between stream capacity and slope, discharge, bed-material size, channel form, and velocity; and the first few pages (with summary and conclusions) of chapter XIII—Application to natural streams, which provided significant insight into relationships between hydrologic variability and channel form and process. Thus, the abstract, the first few pages of chapter XIII, and the summary and conclusions of that chapter are reprinted as they appear in Professional Paper 86. The following is a discussion of the contributions Gilbert made and their relevance to our understanding of stream processes since Professional Paper 86 was published in 1914.

## COMMENTARY

Professional Paper 86 made several major contributions that have influenced how we study rivers. The first contribution was in the general area of bed-load transport and specifically

---

Keller, E. A., 1999, The transportation of débris by running water by Grove Karl Gilbert, *in* Moores, E. M., Sloan, D., and Stout, D. L., eds., Classic Cordilleran Concepts: A View from California: Boulder, Colorado, Geological Society of America Special Paper 338.

involved the concepts of stream competency and capacity, thresholds, and the roles of discharge, velocity, channel form, and sediment size in the transport process. Second, Gilbert introduced the concept of total stream power as a significant factor to understanding stream processes. Third, Gilbert discussed in very creative ways the application of the results of his flume studies at Berkeley to natural channels. Of particular importance is the role of hydrologic variability in channel form and process. Finally, Gilbert's flume experiments demonstrated the value of the experimental approach to river studies that continues today (e.g., Friedkin, 1945; Simons and Richardson, 1966; Schumm, 1977; Wohl and Ikeda, 1997).

### Bed-load transport

Much of the discussion by Gilbert concerning bed-load transport involved the concept of stream competency and stream capacity, the latter being defined by Gilbert as the maximum load a stream can carry. The primary purpose of the flume experiments was to explain what controls the movement of bed load and to apply this to understand and predict bed-load transport in natural streams. The primary objective, Gilbert readily admitted, was not achieved, but what was learned has helped future researchers better understand sediment transport in rivers. His ideas for Professional Paper 86 were crucial in the development of the concept of the graded river (Mackin, 1948), and to better understanding of the behavior of alluvial rivers through the interaction of variables important in sediment transport (Maddock, 1969).

Gilbert's approach in the Berkeley flume experiments was to measure quantitatively several factors: slope factor, discharge factor, and form factor, and to relate these to stream capacity. He also evaluated the role of velocity in determining capacity, finding a strong relationship between transport rate and mean velocity. Gilbert's (1914) primary results from flume experiments (that mean velocity is directly related to sediment discharge) were confirmed nearly a half century later by Colby (1961), as reported in Leopold et al. (1964).

A creative aspect of Gilbert's slope and discharge factors was the recognition that there is a slope or discharge necessary to transport bed load of a particular size. Gilbert reasoned that there must be a threshold value of slope or discharge which, if exceeded, would result in sediment transport. He also suggested that using empirical relationships based upon his flume studies could quantitatively predict stream capacity. These relationships are power functions of the difference between a stream's slope and the slope necessary to transport bed load. These relationships are stated in the abstract of Professional Paper 86. The important point is that many of the bed-load functions used today are based on the same basic principle: that transport goes as the difference between a driving factor and its threshold value, taken to some power. Examples include the Du Bois–type bed-load formula based upon critical shear stress; the Schoklitsch-type equation based upon critical discharge; and the Bagnold-type relationship based upon critical stream power necessary to transport the bed load (Knighton, 1984).

With his discussion of bed-load transport, Gilbert (1914, see abstract) also introduced what he calls the form factor, defined as the ratio of depth to width of a stream channel. Gilbert's form factor is the inverse of the width to depth ratio that has evolved to be one of the primary form factors for studying rivers. For example, the width to depth ratio is one of the primary discriminating factors in classifying channel types (Schumm, 1981), as well as sediment characteristics of channel banks (Schumm, 1960).

### Stream power

The concept of time rate of stream-energy expenditure or total stream power was also introduced by Gilbert in Professional Paper 86. He defined this energy expenditure as the product of discharge (mass per unit time), slope, and acceleration of gravity. His discussion of total stream power and its relationship to the load a stream carries was perhaps one of the first such discussions in the American literature. The concept of stream power has been used to develop relationships to predict sediment discharge (Bagnold, 1966; Yang, 1976) and a stream's tendency to aggrade or degrade (Bull, 1979). Both Bull's and Bagnold's use of stream power emphasized the existence of a threshold stream power necessary to transport the sediment load. Yang (1976) developed the hypothesis that streams adjust slope, velocity depth, and roughness so that sediment may be transported with a minimum unit stream power (power per unit width of channel). Thus, these ideas follow directly the ideas set out in Professional Paper 86, in which Gilbert defined stream capacity in terms of stream power, and the existence of threshold values of variables important in causing sediment transport.

### Channel form and process

Gilbert's creative insights and observations in chapter XIII of Professional Paper 86, while being more speculative, have opened up a variety of paths followed by research since its publication in 1914. Of particular significance are his discussions of features related to the variability of discharge in rivers. He was apparently influenced by the work of McMath (1881, cited in Gilbert, 1914) who, in Gilbert's words, argued that with rising river stage the river scours from the deeps (pools) to deposit bed material on shoals (riffles or crossings), and with falling stage scours from the shoals and deposits in the deeps. Gilbert's thoughts concerning processes in pools and riffles are eloquently stated in chapter XIII, p. 221, of Professional Paper 86. Gilbert's observations were largely unrecognized until the mid to late twentieth century, when several studies (Lane and Borland, 1954; Leliavsky, 1955; Wolman, 1955; Leopold and Wolman, 1957; Leopold et al., 1964; Richards, 1976; Dietrich et al., 1979; and Montgomery and Buffington, 1997) further explored the phenomena of the role of variability of discharge in channel form and process. Perhaps the first data set to confirm that velocity in pools at high, channel-forming flows exceeds that of riffles was that of Keller (1971). The term "hypothesis of velocity reversal" was suggested by Keller (1971) as an explanation of Gilbert's (1914) observations. Since its initiation, the hypothesis of velocity reversal has been controversial. The basic concept that Gilbert (1914) suggested was that processes at relatively high flows are very different from those at low flows and that the pattern of scour and fill in shallows (riffles or crossings) and deeps (pools) may be explained by these differences. Lane and Borland (1954) argued that because the channel geometry of the pool is asymmetric (triangular) and the riffle or crossing is more symmetric (rectangular), during low flow the cross-sectional area of the pool is greater than that of the riffle, and the mean velocity is therefore less than the riffle. However, at high flow the cross-sectional area of the riffle exceeds that of the pool, and the mean velocity of the pool is greater than that of the riffle. Lane and Borland (1954) went on to suggest that this process was responsible for the scouring of pools and deposition on riffles during high flow as well as the deposition or filling in pools and scouring of finer bed material on riffles during low flow. Based on observations of rivers in the mid-western United States, Dittbrenner (1954) restated the concepts suggested by Lane and Borland. The pattern of scouring of pools and deposition on riffles at high flow reported by Lane and Borland was confirmed by studies of Leopold and Wolman (1960), with additional information that the mean shear stress was greater on the riffle at low flow but was greater in the pool at high flow. During the past 25 years the hypothesis of velocity reversal has been supported by several studies that have measured reversals in flow strength (Andrews, 1979, mean velocity; Lisle, 1979, mean shear stress; O'Connor et al., 1986, mean stream power). The velocity reversal hypothesis was further tested by Keller and Florsheim (1993), who modeled the flow through several pool-riffle sequences. That study supported the velocity reversal hypothesis, but also suggested that all pool-riffle sequences do not undergo a velocity reversal at the same stage. More detailed explanation and sophisticated models of the hydraulics associated with riffle-pool sequences are necessary to delineate more robustly the processes first recognized by Gilbert (1914).

Gilbert, in Professional Paper 86, also acknowledged the importance of bank vegetation in the role of channel form and process. He speculated that bank vegetation can inhibit erosion by increasing the shear strength of bank materials through the presence of plant roots. Since the publication of Professional Paper 86, the role of vegetation in understanding channel form and process has evolved into the field of forest geomorphology. This work has extended from the understanding of the effects of plant roots on stream-bank stability to the occurrence of large, woody debris in streams that greatly affects channel processes of erosion and deposition (e.g., Heede, 1972; Keller and Swanson, 1979; Gregory et al., 1985; Keller and MacDonald, 1995).

## SUMMARY OF CONTRIBUTIONS FROM PROFESSIONAL PAPER 86

The major contributions from Professional Paper 86 are not so much the equations Gilbert developed from the flume experiments as the ideas he generated. Gilbert moved fluvial geomorphology from qualitative description to quantitative analysis with experiments, and by doing so, placed emphasis on river processes. His work was followed by a number of important contributions, resulting in better understanding of river processes from a quantitative perspective. This includes the work of Strahler (1957), who focused on the analysis of entire watersheds, and that of Leopold et al. (1964), whose synthesis of fluvial processes in geomorphology is an intellectual landmark.

Gilbert (in Professional Paper 86) defined and quantitatively discussed concepts of vertical velocity gradients, competency, and the development of armoring of the channel bed. Gilbert recognized differences between the transport of relatively fine-grained and coarse-grained particles and their importance to channel form and process. This distinction is integral to the work of Mackin (1948) and is the subject of a recent paper that suggests a two-fraction (sand and gravel) model of sediment transport in gravel-bed rivers (Wilcock, 1998).

Gilbert's work introduced the concept of a threshold that must be exceeded if bed-load transport is to occur. The concept of thresholds in fluvial geomorphology was later emphasized and expanded by the work of Stan Schumm (1973, 1977). Gilbert's introduction of the concept of stream power assisted later workers who developed the concept to explain bed-load transport (Bagnold, 1966; Yang, 1976; Leopold and Emmett, 1976). Bull (1979) further developed the concept of threshold with respect to stream power to explain the tendency of streams to aggrade or degrade.

Gilbert emphasized the importance of hydrologic variability or how river processes change with discharge. His ideas concerning riffles and pools pointed the way toward explaining the origin and maintenance of these primary features of meandering alluvial rivers. This insight has led to a better understanding of meandering (Parker, 1976; Dietrich et al., 1979); development of the flood plain (Wolman and Leopold, 1957); significance of events of varying magnitude (Wolman and Miller, 1960); and scour and fill patterns of rivers in a variety of studies (Friedkin, 1945; Leopold and Wolman, 1957; Lane and Borland, 1954; Keller, 1971; Andrews, 1979; Lisle, 1979; O'Connor et al., 1986; Carling, 1991; Clifford and Richards, 1992).

In conclusion, Grove Karl Gilbert contributed to our understanding of fluvial geomorphology through Professional Paper 86 in several important ways (Robert N. Anderson, 1998, personal written commun.):

• Gilbert suggested power law relationships between important driving variables and the response of the fluvial system, allowing (if the power is greater than 1) for nonlinear behavior emphasizing the role of large discharges.

• He raised the issue of thresholds by stating his equations in terms of excess over thresholds, providing a framework in which to evaluate many geomorphic systems.

- Gilbert suggested that in order to understand river processes we need to acknowledge the importance of variable discharge, including the annual hydrograph as well as the variability of the hydrograph from year to year and event to event, in evaluating the response of the river system.
- Gilbert illustrated the relevance of quantitative studies that focus on process in the understanding of how river systems function.
- He moved the field of geomorphology into the realm of experimental science by demonstrating the relevance of experiments to understanding geomorphic systems.

## ACKNOWLEDGMENTS

Suggestions for improvement of this paper by Robert Anderson, Tom Lisle, and Doris Sloan are appreciated.

## REFERENCES CITED

Andrews, E. D., 1979, Scour and fill in a stream channel, East Fork River, western Wyoming: U.S. Geological Survey Professional Paper 1117, 49 p.

Bagnold, R. A., 1966, An approach to the sediment transport problem from general physics: U.S. Geological Survey Professional Paper 422-I, 37 p.

Bull, W. B., 1979, Threshold of critical power in streams: Geological Society of America Bulletin, v. 90, p. 453–464.

Carling, P. A., 1991, An appraisal of the velocity-reversal hypothesis for stable pool-riffle sequences in the River Severn, England: Earth Surface Processes and Landforms, v. 16, p. 19–31.

Clifford, N. J., and Richards, K. S., 1992, The reversal hypothesis and the maintenance of riffle-pool sequences: A review and field appraisal, in Petts, G. E., ed., Fluvial Dynamics of Lowland River Channel and Floodplain Systems: British Geomorphological Research Group Symposium Series, pages.

Colby, B. R., 1961, Effect of depth of flow on discharge of bed material: U.S. Geol. Survey Water-Supply Paper 1498-D, p. 1–10.

Dietrich, W. E., Smith, J. D., and Dunne, T., 1979, Flow and sediment transport in a sand bedded meander: Journal of Geology, v. 87, p. 305–315.

Dittbrenner, E. E., 1954, Discussion: American Society of Civil Engineers Transactions, v. 119, p. 1080–1087.

Friedkin, J. F., 1945, A laboratory study of the meandering of alluvial rivers: U.S. Waterways Engineering Experimental Station, 40 p.

Gilbert, G. K., 1877, Report on the geology of the Henry Mountains: Washington, D. C., General Printing Office, U.S. Geological and Geographical Survey, Rocky Mountain region, 170 p.

Gilbert, G. K., 1914, The transportation of débris by running water: U.S. Geological Survey Professional Paper 86, 221 p.

Gregory, K. J., Gurnell, A. M., and Hill, C. T., 1985, The permanence of debris dams related to river channel processes: Hydrological Science Journal, v. 30, p. 371–381.

Heede, B. H., 1972, Influences of a forest on the hydraulic geometry of two mountain streams: Water Research Bulletin, v. 8, p. 523–530.

Keller, E. A., 1971, Areal sorting of bedload material: The hypothesis of velocity reversal: Geological Society of America Bulletin, v. 82, p. 753–756.

Keller, E. A., and Florsheim, J. L., 1993, Velocity-reversal hypothesis: A model approach: Earth Surface Processes and Landforms, v. 18, p. 733–740.

Keller, E. A., and MacDonald, A., 1995, River channel change: The role of large woody debris, in Gurnell, A., and Petts, G., eds., Changing river channels, Chichester, John Wiley & Sons, p. 217–235.

Keller, E. A., and Swanson, F. J., 1979, Effects of large organic material on channel form and fluvial processes: Earth Surface Processes, v. 4, p. 361–380.

Knighton, D., 1984, Fluvial forms and processes: London, Edward Arnold, 218 p.

Lane, E. W., and Borland, W. M., 1954, River-bed scour during floods: American Society of Civil Engineers Transactions, v. 119, p. 1069–1079.

Leliavsky, S., 1955, An introduction to fluvial hydraulics: London, Constable, 257 p.

Leopold, L. B., and Emmett, W. W., 1976, Bedload measurements, East Fork River, Wyoming: National Academy of Sciences Proceedings, v. 73, p. 1000–1004.

Leopold, L. B., and Wolman, M. G., 1957, River channel patterns; braided meandering and straight: U.S. Geological Survey Professional Paper 282-B, p. 39–85.

Leopold, L. B., and Wolman, M. G., 1960, River meanders: Geological Society of America Bulletin, v. 71, p. 769–794.

Leopold, L. B., Wolman, M. G., and Miller, J. P., 1964, Fluvial processes in geomorphology: San Francisco, W. H. Freeman, 522 p.

Lisle, T. E., 1979, A sorting mechanism for a riffle-pool sequence: Geological Society of America Bulletin, v. 90, p. 1142–1157.

Mackin, J. H., 1948, Concept of the graded river: Geological Society of America Bulletin, v. 59, p. 463–512.

Maddock, T., 1969, The behavior of straight open channels with movable beds: U.S. Geological Survey Professional Paper 622A, 70†p.

Montgomery, D. R., and Buffington, J. M., 1997, Channel-reach morphology in mountain drainage basins: Geological Society of America Bulletin, v. 109, p. 596–611.

O'Connor, J. E., Webb, R. H., and Baker, V. R., 1986, Paleohydrology of pool-and-riffle pattern development: Boulder Creek, Utah: Geological Society of America Bulletin, v. 97, p. 410–420.

Parker, G., 1976, On the cause and characteristic scales of meandering and braiding in rivers: Journal of Fluid Mechanics, v. 76, p. 457–480.

Pyne, S. J., 1980, Grove Karl Gilbert: A great engine of research: Austin, University of Texas Press, p. 1–14.

Richards, K. S., 1976, The morphology of riffle-pool sequence: Earth Surface Processes, v. 1, p. 71–88.

Schumm, S. A., 1960, The shape of alluvial channels in relation to sediment type: U.S. Geological Survey Professional Paper 352-B, 30 p.

Schumm, S. A., 1973, Geomorphic thresholds and complex response of drainage systems, in Morisawa, M., ed., Fluvial geomorphology: Binghamton NY, New York State University Publications in Geomorphology, p. 299–309.

Schumm, S. A., 1977, The fluvial system: New York, Wiley-Interscience 338 p.

Schumm, S. A., 1981, Evolution and response of the fluvial system, sedimentologic implications in Ethridge, F. G., and Flores, R. M., eds., Recent and ancient nonmarine depositional environments: Models for exploration: The Society of Economic Paleontologists and Mineralogists Special Publication 31, p. 19–29.

Simons, D. B., and Richardson, E. V., 1966, Resistance to flow in alluvial channels: U.S. Geological Survey Professional Paper 422J, 61 p.

Strahler, A. N., 1957, Quantitative analysis of watershed geomorphology: Eos Transactions, (American Geophysical Union), v. 38, p. 913–920.

Wilcock, P. R., 1998, Two-fraction model of initial sediment motion in gravel-bed rivers: Science, v. 280, p. 410–412.

Wohl, E., and Ikeda, H., 1997, Experimental simulation of river incision into a cohesive substrate at varying gradients: Geology, v. 25, p. 295–298.

Wolman, M. G., 1955, The natural channel of Brandywine Creek, Pennsylvania: U.S. Geological Survey Professional Paper 271, 56 p.

Wolman, M. G., and Leopold, L. B., 1957, River flood plains: Some observations on their formation: U.S. Geological Survey Professional Paper 282-C, p. 87–109.

Wolman, M. G., and Miller, J. P., 1960, Magnitude and frequency of forces in geomorphic processes: Journal of Geology, v. 68, p. 54–74.

Yang, C. T., 1976, Minimum unit stream power and fluvial hydraulics: American Society of Civil Engineers, Journal of the Hydraulics Division, v. 102, HY7, p. 919–934.

Manuscript Accepted by the Society November 23, 1998

Printed in U.S.A

Geological Society of America
Special Paper 338
1999

Chapter 14

# APPLIED STRATIGRAPHY IN THE COAST RANGES

**William B. N. Berry**
*Department of Geology and Geophysics*
*University of California*
*Berkeley, California 94720-4767*
*E-mail: bberry@uclink4.berkeley.edu*

## ROBERT MINSSEN KLEINPELL

Robert M. Kleinpell was born in Chicago on September 13, 1905. His family moved to the Los Angeles area when he was six. Inasmuch as he lived in that area, he began his university education at the University of California, Los Angeles. After a year there, however, he transferred to Occidental because that college had outstanding programs in history and archeology. He completed a dual major in those subjects, and, as certain of his contemporaries remember, enjoyed playing football, entertaining at parties, and playing the piano. He entered the geology graduate program at Stanford University in the fall of 1927, where he was guided by Hubert G. Schenck and J. P. Smith. During his career as a graduate student, Kleinpell acquired practical geological experience working with the U.S. Geological Survey and Richfield Oil Company. These experiences as a graduate student led him to later urge students and academic colleagues to achieve a similar close association between practical experience and academic research.

Kleinpell's doctoral dissertation was both a pioneering effort in documenting the usefulness of analyzing the stratigraphic distribution of fossil foraminfers and a mature synthesis of an enormous amount of stratigraphic and paleontologic data. A part of the work was presented at the 1934 Cordilleran Section meeting of the Geological Society of America.

The published version of his dissertation (*Miocene Stratigraphy of California*) became a classic in the study of likely oil-bearing strata in complex geologic terrain. The stratigraphic paleontological methodology documented in it was used by numbers of petroleum geologists and engineers to discover several millions of barrels of oil, many of them in California's oil fields.

After completing his doctorate in 1933, Kleinpell married fellow Stanford student Dariel ("Jerry") Shively and opened a consulting practice in Bakersfield. His clients included many of the most successful exploration programs in California's oil industry. He became Visiting Professor of Geology at California Institute of Technology in 1939, but left that position within a year to join the National Development Company of the Philippines. While in the Philippines, he was taken prisoner by the Japanese Army and held in concentration camps from 1941 until 1945. Upon returning home after the war, he accepted a position as Professor of Paleontology at the University of California, Berkeley, and remained there until 1971. He spent his last academic and his retirement years in Santa Barbara, where he gave guest lectures and courses at the University of California Santa Barbara. He died in Santa Barbara March 13, 1986.

The methodology Kleinpell documented in his dissertation constituted a monumental leap forward in the development of the discipline of stratigraphic paleontology. Throughout the rest of his life, Kleinpell was a leader in developing stratigraphic paleontology as a field of scientific inquiry of great value both in resource exploration (metals as well as petroleum) and in academic research. He contended that the discipline involved the careful fusion of relevant aspects of evolutionary biology, ecology, and systematic paleontology with stratigraphic, structural, and sedimentary geology. Although he published relatively little personally, his contributions to his discipline were achieved through untold numbers of professional conversations with colleagues and students. He added many passages to nearly all of the works his students and certain of his colleagues wrote and published. Although he could have added his name to those of his students and colleagues in nearly two-score publications, Kleinpell elected not to do so and to let his associates and students develop their own reputations, even though such development involved work he had enhanced substantially.

The excerpt taken from Kleinpell's *Miocene Stratigraphy of California* describes the essentials of his methodology. That work was selected for citation because it constitutes one of the cornerstones of stratigraphic paleontology, the discipline that Kleinpell expanded and used so successfully to enhance not only geology, but also the growth of California's economy. Publications with so broad a spectrum of impacts are rare in any field of endeavor.

### *Note concerning German text*

Kleinpell cited a number of pages of German text in his book. He believed that original meaning could be lost if he provided a translation. Furthermore, in his day the citation of foreign language text was relatively common in many scientific publications. Most doctoral students in paleontology were required to pass foreign-language reading examinations as part of the doctoral degree requirements until about the mid-1980s. Scientists who no longer read languages other than English have lost a certain ability to understand developments in their field of inquiry. The German passages included here in the excerpt from Kleinpell's work provide the reader with much of the basic information that Kleinpell meant to convey. Some of the German text that Kleinpell included has been omitted. The reader is referred to Kleinpell's book to read the entire material he cited.

William B. N. Berry

# MIOCENE STRATIGRAPHY
## OF
# CALIFORNIA

*By*

ROBERT M. KLEINPELL

PUBLISHED BY
THE AMERICAN ASSOCIATION OF
PETROLEUM GEOLOGISTS
TULSA, OKLAHOMA, U.S.A.

LONDON: THOMAS MURBY & CO., 1, FLEET LANE, E.C. 4
1938

lower and upper portion of a unit than between the separate units themselves. Within the limits of these major units, nevertheless, there are several distinctive joint occurrences of species; these serve to define units which may be recognized over wide areas as distinct subdivisions of the major units. Even much smaller subdivisions may be differentiated locally though not upon the basis of a unique association of species.

### Biostratigraphic Nomenclature

Although it might have been desirable to refer to the afore-mentioned chronologic and biostratigraphic units in terms of the stratigraphic nomenclature already in use in California, an examination of the type localities of a number of the established Formations and "zones" has revealed that most of the Formations are such in the Brongniartian or lithologic-genetic, rather than chronologic, sense. This Brongniartian usage is the usage adopted by the International Geological Congress in Bologna in 1881 for the term, "formation." The limits of the California Formations do not, in many instances, correspond with the horizons which on paleontologic grounds appear to be chronologically the most significant. Furthermore, excellent cartographic units though they be, the Formations are hopelessly inadequate as stratigraphic units with accurate chronologic connotation, and their use in this sense has led to much confusion.[69] Only a few of many examples need be cited. The "post-Monterey disturbance," based on the supposed presence of a marked unconformity between the Monterey and the overlying Santa Margarita Formations, was considered one of the major revolutions in the geologic history of the Coast Ranges until 1925, when the unconformity on which it was based was shown to be nonexistent in most of the sections from which it had been reported.[70] Since this time, the Monterey and the Santa Margarita Formations have been found to be lithologically distinct but contemporaneous facies of a chronologic-stratigraphic unit which occupies a considerable portion of the Middle Tertiary column in California; thus, it now becomes extremely difficult to place a finger upon this old "major revolution." Another example of unnecessary confusion is the controversy over the relative age of the Vaqueros and Temblor Formations; this controversy has ebbed and flowed for a generation, and the top of the "Vaqueros Formation" is still being moved up and down within the

---

[66] George L. Richards, unpublished manuscript. *See* Table II and accompanying discussion.

[67] Bradford C. Adams and George L. Richards, personal communication.

[68] The term biostratigraphy applies to that phase of the science of paleontology which relates fossils to the containing strata, in distinction from "paleobiology" which is the more essentially systematic phase of the science. Diener has reviewed these two phases of paleontology in the preface of *Grundzüge der Biostratigraphie*, Franz Deuticke, publisher, Leipzig and Vienna (1925).

[69] The facies, rather than chronologic, significance of several California Miocene Formations has been emphasized by Louderback, "The Monterey Series in California," *Univ. California Pub., Dept. Geol.*, Vol. 7, No. 10 (1913), pp. 177–241.

[70] R. D. Reed, "The Post-Monterey Disturbance in the Salinas Valley, California," *Jour. Geol.*, Vol. 33 (August–September, 1925).

Cyrus F. "Chief" Tolman near an outcrop of the Claremont Formation, Claremont Rd., Berkeley, Apr. 13, 1938.

Photo from U.S. District Court of Northern California; courtesy of Geohistory Archives, School of Earth Sciences, Stanford University.

### Chronologic-Biostratigraphic[68] Classification of Marine Middle Tertiary ("Miocene") of California

With a complete stratigraphic sequence of foraminiferal faunas established, as summarized above, one may evaluate the faunal changes which are found successively from bottom to top through this sequence, and thus may arrive at a natural chronologic-biostratigraphic classification of the California Miocene. The entire sequence may be grouped readily into six distinct major units; the upper and lower portions of these units are sufficiently distinct to be differentiated and recognized throughout the province, but it is more difficult to draw a sharp line between the

## CORRELATION AND AGE

type section of the Temblor Formation, as more and more of the facies upon which the recognition of the Vaqueros Formation depends is encountered along the strike of the type section of the essentially deeper-water-facies Temblor Formation.[71] The disastrous cumulative effect of similar misconceptions, furthermore, has recently been well illustrated by the paleogeographic sketch and accompanying discussion appearing in a recent book by A. Morley Davies,[72] showing the regressive Briones-Cierbo as compared with the "transgressive Santa Margarita." Confusion is apparent immediately on reflecting that the type Santa Margarita Formation includes not only beds of post-Cierbo age in the upper part, as inferred by the usage under discussion, but that it also includes beds of pre-Briones age in the lower part. Not only does this complicate the matter of what actually constitutes a "transgressive Santa Margarita," however. It is difficult, of course, to ascertain exactly what the author had in mind in using the term "Santa Margarita," but it is to be inferred that he had in mind some post-Cierbo horizon, which in comparison to the horizons of the Cierbo itself certainly are transgressive. That the Briones, too, is regressive in comparison to the post-Cierbo "Santa Margarita," however, depends entirely upon the assumption that, since the Briones Formation (essentially a sandstone) is comparatively restricted in its geographic distribution, it therefore represents a marine regression. On the contrary, when it is borne in mind that the shale-facies equivalents of the Briones, that is, those shales of the Hercules shale Member of the Briones, which are correlatives of the Monterey Formation are among the most widely distributed and most strongly overlapping, that is, transgressing, marine sedimentary units in the California Tertiary, this assumption is found to be unsupported by fact. Thus, in spite of the regressive aspects of the Cierbo in whole or in part, the two paleogeographic maps presented by Davies are in exactly reverse order in so far as the relative regressive and transgressive aspects of the "Briones-Cierbo" and the post-Cierbo "Santa Margarita" are concerned, for the Briones and its age-equivalents represent a much greater area of marine deposition and transgression than do the age-equivalents of the post-Cierbo "Santa Margarita."

Not only are the established Formations of spurious chronologic value; in addition, only a few of the "zones" in use are Zones in the sense of Oppel, who first used the term as one of definite biostratigraphic as well as chronologic significance. It is preferable therefore to adopt, on the following grounds, the term Stage for the time-stratigraphic units of largest magnitude, the term Zone for the more clear-cut subdivisions of these Stages, and the term Zonule for locally recognizable smaller biostratigraphic units.

The concept of Stages and Zones as stratal units independent of lithologic and faunal facies, and thus independent of Formations and of the presence of locally conspicuous fossil species and fossil ecologic associations, dates from the time of d'Orbigny.[73] He, however, appears to have used the terms almost interchangeably. It remained for Oppel, a student of Quenstedt, to apply the concept to tangible and valid stratal units.[74] Oppel set up 33 Jurassic Zones, grouping them into eight "etagen oder zonengruppen," each Stage containing from two to seven Zones. The International Geological Congress, meeting in Berlin in 1883, adopted the two terms as applicable to stratal units of chronologic significance (and of lesser magnitude than a Series) in the same sense and relationship as Oppel employed them in his first zonal table for the Jurassic.[75] Employed as they were in a classification of the Jurassic System, the terminology of Oppel and that of d'Orbigny were applied to stratal units differentiated on the basis of ammonites. From this it follows that a classification essentially consistent with that of Oppel may be established for the remainder of the Mesozoic and for the late Paleozoic as well. In 1893 Munier-Chalmas and de Lapparent used the terms Stage and Zone in a stratigraphic classification which included the Tertiary as well as the older Systems. The usage of the term Stage, as explained by Munier-Chalmas and de Lapparent, follows.

Pour la définition des étages, nous l'avons basée, partout où cela était possible, sur la considérations des Cephalopodes, Goniatitides, et Ammonitides. Ce criterium est devenu d'une application facile pour la partie supérieure des terrains paléozoiques ainsi que pour tout l'ensemble du groupe secondaire.

---

[71] *See* Hubert G. Schenck, "What Is the Vaqueros Formation of California and Is It Oligocene?" *Bull. Amer. Assoc. Petrol. Geol.*, Vol. 19, No. 4 (1935), p. 523.

[72] A. Morley Davies, *Tertiary Faunas. Vol. II The Sequence of Tertiary Faunas* (London, 1934), p. 188, Fig. 25 ("after Bruce Clark, generalized").

[73] Alcide d'Orbigny, *Paléontologie française; terrains jurassiques*, Vol. I (1842–9). p. 604. D'Orbigny adopted a Stage nomenclature "…in order to put an end to this jumbled nomenclature, based on the local lithology, which varies so greatly according to the place, and on the fossils which happen to be predominant at one locality, but may be wanting elsewhere," according to the translation given by Arkell, *The Jurassic System in Great Britain* (Oxford, 1933). p. 9.

[74] Albert Oppel, *Die Juraformation Englands, Frankreichs und des Südwestlichen Deutschlands* (Stuttgart, 1856–1858).

[75] For a review of the history of stratal terminology the reader is referred to the preface and first chapter of W. J. Arkell, *The Jurassic System in Great Britain* (Oxford Press, 1933), and to the eighth chapter of C. Diener, *Grundzüge der Biostratigraphie* (Leipzig and Vienna, 1925).

## CORRELATION AND AGE

Quant au groupe tertiaire, ses divisions ont été fondées sur les grands changements de faunes marines, mises en concordance avec les mouvements orogéniques correspondants, ainsi que sur l'évolution des Mammifères.[76]

Thus the term Stage is clearly applicable in a chronologic and biostratigraphic classification of a portion of the Tertiary System such as the one made in the present paper, which is based on the great changes apparent in the marine foraminiferal faunas of a province as large as California. The supplementary criteria of orogeny and mammalian evolution mentioned by Munier-Chalmas and de Lapparent are as yet difficult to apply with any degree of accuracy in formulating the present classification. There is definite evidence that strong and widespread diastrophic, if not truly orogenic, movements have taken place in California at horizons corresponding to the greatest faunal changes.[77] This evidence will be summarized and evaluated later in a discussion of the individual Stages. Evidence also exists indicating changes in vertebrate faunas corresponding to those noted in the foraminiferal faunas. From recent correspondence with Remington Kellogg of the United States National Museum; the writer learns that three markedly distinct phases of mammalian evolution are represented in pelagic mammals found with foraminiferal faunas of the Saucesian, Relizian, and Luisian Stages in California.

The use of established formational names even as roots for Stage names has also been avoided, primarily because of the confusion of cartography with chronology, which has already been discussed. Furthermore, such names as Vaqueros, Temblor,[78] Monterey,[79] and Rincon, as well as the names of a number of other Formations the type sections of which have not been studied in detail, apply to units of greater magnitude than Stages; that they were originally coined for cartographic purposes largely

---

[76] Munier-Chalmas and de Lapparent in *Bull. Soc. Géol. France*, 3 sér., t. 21 (1893), p. 439.

[77] Four of the orogenic disturbances alluded to have recently been described by R. D. Reed, "Miocene Orogenies in the California Coast Ranges" (abstract), *Geol. Soc. America*, Cordilleran Section, list of papers with abstracts (1935), pp. 6, 7. See also a more recent paper by R. D. Reed and J. S. Hollister, "Structural Evolution of Southern California," *Bull. Amer. Assoc. Petrol. Geol.*, Vol. 20, No. 12 (1936), pp. 1533-1721, particularly pp. 1589-92.

[78] For a review of the Vaqueros and Temblor Formations see L. W. Wiedey, "Notes on the Vaqueros and Temblor Formations of California with Descriptions of New Species," *Trans. San Diego Soc. Nat. Hist.*, Vol. 5, No. 10 (1928), pp. 95-182, in addition to papers on the Vaqueros Formation by Loel and Corey and by Schenck, already mentioned.

[79] For a review of the Monterey Formation see George D. Louderback, "The Monterey Series in California," *Univ. California Pub., Bull. Dept. Geol.*, Vol. 7, No. 10 (1913), pp. 177-241.

---

on the basis of a roughly homogeneous lithology and similar origin and not upon the basis of an accurately delimited stratigraphic relation to a geological time scale has already been mentioned. Other names such as Gould, Claremont, Oursan, Tice, Hambre, Rodeo, Briones, and San Pablo have been applied to units of smaller magnitude than a Stage; with three possible exceptions they have been based *entirely* on a certain lithologic homogeneity. These unit names are best preserved as denoting lithologic-genetic homogeneity, applying in the case of the last mentioned group, to Members of Formations. New names have therefore been coined for the six California Miocene Stages that have been recognized.

The term Zone, which has been applied here to subdivisions of Stages, has been used in West Coast stratigraphy as indiscriminately as the term Stage has been used sparingly. That a Zone may well be considered as a subdivision of a Stage is clearly demonstrated by the use of Oppel's Zones as subdivisions of his Stages. Munier-Chalmas and de Lapparent employ this usage in their classification of the geological column, and the International Geological Congress has adopted and in 1901 codified the same usage. It may be well therefore to review at this point the meaning of the term Zone as it was used by Oppel. Unfortunately, nowhere in his book did Oppel specifically define his usage of the term. His examples, however, speak for themselves and the following quotations from his discussion give a clear picture of the stratigraphic conditions with which he was faced, and the reasons for the application thereto of the nomenclature he employed. In fact, the passages here quoted[80] could, in principle, have been written as well of the Miocene foraminiferal sequence in California as of the Jurassic cephalopod sequence in northwestern Europe to which they actually referred.

Vorrede, p. 3:

... Es wurden immer bloss ganze Schichtengruppen mit einander parallelisirt, nicht aber gezeigt, dass in jeglicher Horizont, der an dem einen Orte durch eine Anzahl für ihn constanter Species markirt wird, auch in der entferntesten Gegend mit derselben Sicherheit wieder zu finden sei. Diese Aufgabe ist zwar eine schwierige, aber nur durch ihre Erfüllung kann eine genaue Vergleichung ganzer Systeme gesichert werden. Es wird dabei nöthig gemacht, mit Hintansetzung der mineralogischen Beschaffenheit der Schichten, die verticale Verbreitung jeder einzelnen Species an den verschiedensten Orten zu erforschen, hernach diejenigen Zonen hervorzuheben, welche durch stätes und alleiniges Auftreten gewisser Arten sich von den angrenzenderf als bestimmte Horizonte absondern. Man erhält dadurch ein ideales Profil, dessen Glieder gleichen Alters

---

[80] From Albert Oppel, *Die Juraformation Englands, Frankreichs und des Südwestlichen Deutschlands* (Stuttgart, 1856-1858).

## CORRELATION AND AGE

in den verschiedenen Gegenden immer wieder durch dieselben Arten charakterisirt werden. Eine solche Theilung habe ich versucht und sie bei den meisten Etagen ausführbar gefunden; bei andern halte ich sie noch für unvollendet. Die Schwierigkeit dabei hängt hauptsächlich an der ungenügenden Zahl gut beschriebener Arten. Je schärfer die Species getrennt sind, desto genauer können auch die Schichten eingetheilt werden. . . .

p. 4:       ... Wenn schon die einzelnen Horizonte sich oft genauer unter einander begrenzen als eine ganze Etage gegen die andere, so habe ich doch die Gruppirung der Juraformation in Etagen auch noch beibehalten, da hiedurch besonders die Zussammenstellung der weniger bekannten Fossile erleichtert wird. Am Schluss der Betrachtung jeder Etage führe ich diejenigen Species an, auf welche sich die Eintheilung und Vergleichung der Schichten vorzugsweise stützt. Ich gebe in jedem Anhange zugleich die Synonymik der einzelnen Arten so weit es die beabsichtigte Kürze der Arbeit zuliess. . . .

Erster Abschnitt, p. 13:

... wenn ich die Zone de Am. raricostatus ... als das oberste Gliede unteren Lias aufstelle. Darüber beginnt der erste Paxillose (Bel. elongatus), sowie noch andere Arten, welche die unterste Zone des mittleren Lias ... charakterisiren. Obschon die paläontologischen Unterschiede zwischen den Grenzschichten zweier Etagen selten ausgesprochener sind, als die von zwei benachbarten Zonen derselben Etage, so lässt sich doch hier die Trennung in den meisten Fällen mit Leichtigkeit ausführen. . . .

p. 15:      ... Jeder der einzelnen Zonen sind immer diejenigen Arten beigeschrieben, welche sie besonders charakterisiren und noch in keiner anderen Schichte gefunden wurden. Dass nicht sämmtliche Species des unteren Lias hier genau eingetheilt werden konnten, versteht sich von selbst, es haben ja sogar ganze Etagen einzelne Arten mit einander gemein, wie viel mehr sollten nicht in zwei angrenzenden Zonen solche Uebergänge vorkommen. Ich habe zwar die Verbreitung der einzelnen Species ... so genau als möglich angegeben, doch führe ich hier einige besonders an, welche je für mehrere Zonen von Wichtigkeit sind: Nautilus striatus, Spirifer Walcotti, Terebratula Rehmanni sind im ganzen unteren Lias mit Ausnahme des Bonebeds zu Hause. Gryphaea arcuata ist zwar am häufigsten in den Schichten des Am. Bucklandi kommt aber gleichfalls bezeichnend mit Am. angulatus vor. Bel. acutus findet sich in und an der Basis der vier obersten Zonen des untern Lias, Gryphaea obliqua beginnt etwas höher und geht noch in die unteren Schichten des mittleren Lias hinauf. . . .

Dreizehnter Abschnitt, p. 813:

Schlussbetrachtungen über die Eintheilung der jurassischen Ablagerungen und deren Vergleiche nach verschiedenen Ländern.

Paragraph 118:

Benennung der Hauptabtheilungen, Etagen, und Zonen. Zonen. . . .

## MIOCENE STRATIGRAPHY OF CALIFORNIA

Herein he chooses the use of species names for Zones in preference to geographic names, which are used for Stages.

p. 814:     ... Vorerst hätten wir somit 33 jurassische Zonen zu unterscheiden. Manche derselben zeigten auf dem ganzen hier betrachteten Terrain eine merkwürdige Uebereinstimmung, andere sind dagegen z. Thl. durch den Wechsel der vorwaltenden Facies an manchen Localitäten sehr schwierig wiederzuerkennen, einige endlich (wie insbesondere die Zone der Terebratula digona und die Zone der Diceras arietina) konnten nur in gewissen Districten aufgenommen werden, indem sie in anderen Gegenden diejenigen Charactere verleugnen, welche uns dort zu ihrer Unterscheidung dienten, so dass ich sogar bei einer derselben (Zone der T. digona) noch nicht gewiss bin, ob wir sie nicht später wiederum aus der Reihe der übrigen Zonen zu streichen und sie nur als locale Unterabtheilung zu betrachten haben werden. . . .

Etagen. Die von d'Orbigny eingeführten Etagen sollten ursprünglich nur Stufen oder Zonen darstellen, erst später hat es sich ergeben, dass sich die Mehrzahl seiner Etagen wiederum ebenso bestimmt in weitere Zonen abtrennen lasse. D'Orbigny hat zur Bezeichnung seiner jurassischen Etagen beinahe ausschliesslich Localitätsnamen gewählt. Nur bei den "E. liasien und corallien" machte er eine Ausnahme.

Oppel therefore renames it Pliensbach grupper instead.

p. 824:     ... so finden wir die stätige Entwicklung zwar dann und wann durch grössere und plötzliche Veränderungen unterbrochen, allein dieselben beschränken sich auf enger begrenzte Districte, besitzen gewöhnlich einen localen Character und lassen sich meist auch durch locale Einflüsse erklären. Dagegen verlieren diese plötzlich eingetreten Erscheinungen schon auf dem hier betrachteten Terrain ihre Allgemeinheit. Mit letzterer geht aber auch die Schärfe der Abschnitte verloren, auf deren Unterscheidung sich eine Eintheilung nach getrennten Perioden doch gründen müsste. Wir haben beim Studium der Grenzschichten zwischen Lias und mittlerem Jura . . . gesehen, wie nahe sich hier diese beiden Hauptabtheilungen treten und wie verhältnissmässig klein die Zahl der massgebenden Charactere ist, auf welche sich unsere Unterscheidung gründet. Dieselben Uebergänge finden wir auch bei den übrigen Etagen und Etagengruppen. Dennoch wird aber eine übereinstimmende Art der Abtrennung der Hauptabtheilungen schon durch ihre Anwendung auf geognostische Karten u. s. w. nöthig, so dass wir es uns zur Aufgabe zu machen haben, die erstmalige Grundlage für die weitere Gliederung der Juraformation in den verschiedenen Ländern in consequenter Weise durchzuführen.

Was ferner die von mir beibehaltene Etageneintheilung betrifft, so lege ich auch ihr nur denjenigen Werth bei, welchen sie in Beziehung auf leichtere Handhabung und schnellere Verständigung über die Ablagerungen verdient. Je grösser das Terrain ist, welches wir untersuchen, desto gleichmässiger wird die Aufeinanderfolge der unter analogen Verhältnissen entstandenen Organismen sich zeigen und desto mehr werden die etwa seither noch zu Grund gelegten

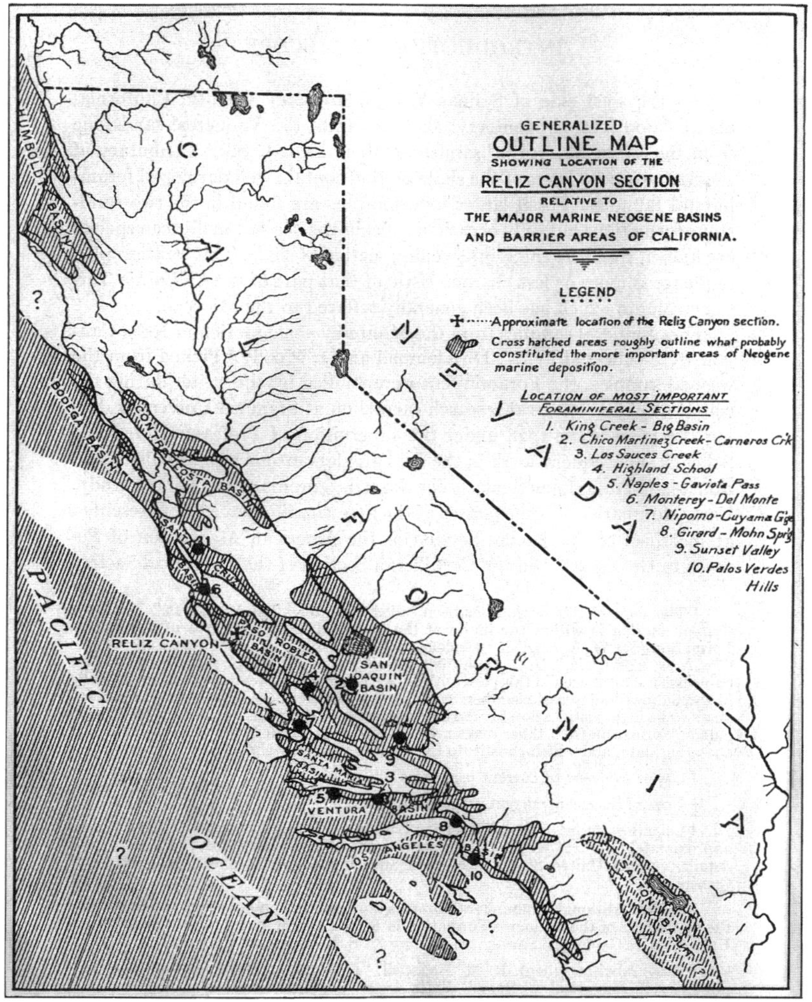

Fig. 2.—Outline sketch of California, showing locations of some important Miocene foraminiferal sections with relation to approximate distribution of marine Neogene.

[This map shows the classic areas where Kleinpell measured stratigraphic sections and collected the samples on which his zones and stages were based. — W.B.N. Berry]

## CORRELATION AND AGE

Unterschiede zwischen zwei Etagen verschwinden, oder wenigstens durch anderweitige Einflüsse erklärt werden können, im Vergleiche mit den Verschiedenheiten, welche oft mitten in der Etage jeder einzelne Horizont gegen den angrenzenden zeigt. Selbst unsere Zonen und Horizonte werden später eine natürlichere Form erhalten, manche Schlüsse, welche sich durch locale Beobachtungen zu ergeben schienen, werden durch neue locale Untersuchungen umgestossen oder vermehrt werden und wir werden später, statt uns an diese Zonen zu binden, die ganze Entwicklung, sowohl der lithologischen Niederschläge, als der früheren Bewohner unserer Erdoberfläche, zu verfolgen und mit den Verhältnissen des Raumes und der Zeit in Verbindung zu bringen haben. Das Resultat der Arbeit ist somit kein abgeschlossenes, es sollte keine vollendete Eintheilung erzielt werden, deren Glieder von nun an, gleichsam als neues System, der Nummer nach zusammengestellt, ein unveränderliches Ganzes bilden, im Gegentheile wir wollen damit beginnen, die Schranken der Systeme, durch welche die jurassischen Bildungen oft auf die unnatürlichste Weise von einander abgetrennt wurden, nur als mechanischen Stützpunkt zu betrachten, dagegen eine immer weiter ins Detail gehende, zugleich aber alle Erscheinungen berücksichtigende Forschung als Zielpunkt unserer Bestrebungen wählen.

In summarizing a discussion of stratal nomenclature Arkell has quoted a definition of the term Zone given by Professor J. E. Marr in *Principles of Stratigraphical Geology* (1898), p. 68.

Zones are belts of strata, each of which is characterized by an assemblage of organic remains, of which one abundant and characteristic form is chosen as index.

To this should be added the vital point that these "belts of strata" have a dimension of time in addition to the three spacial dimensions; for example, a Zone is also the rock equivalent of a Secule. The fact that Oppel's Zones were often named after species which ranged beyond the stratigraphic limits of the Zone in question, as is to be inferred from Marr's definition, is further emphasized by Arkell when he states (p. 21):

. . . a Zone is not a strictly biological unit, but a *bed* or group of *beds* characterised by an *assemblage* of organisms, one of which is chosen as index species, but need not necessarily be either confined to its 'zone or found throughout every part of the 'Zone.

This statement brings into strong relief an item in the definition of the term Zone which remains the basis for considerable confusion and has perhaps been too little discussed, namely, the meaning of the term *assemblage*, upon which it seems to be agreed that the recognition of a Zone depends. The terms *assemblage* and *association* are perhaps unfortunately used in this case, as they immediately bring to mind the *assemblage* or *association* of organisms used by the ecologist in classifying environmental phenomena,—terms used thus in a technical or semi-technical sense and

## MIOCENE STRATIGRAPHY OF CALIFORNIA

applying to *faunal facies* units. It should be borne in mind, on the contrary, that the terms *Zone* and *Stage* were originally coined and used in an attempt to arrive at a *chronological* classification of rocks *independent of both lithologic and faunal facies*. Therefore, the particular type of assemblage or association upon which depends the recognition of a *Zone* could better be described as *the joint occurrence in the same strata of two or more species having different geologic ranges*. Such a *joint occurrence* may involve a very large or a very small percentage of the total number of species or specimens which constitute the particular assemblage which (in whole or in part) is the local expression of the Zone. The need for differentiating *Zones* on the basis of such *joint occurrences* is due to the fact that most species of invertebrates, mollusks, as well as smaller Foraminifera,[81] are of longer range in geologic time than the duration of any one Zone, and the few species which are found to be restricted to one Zone are generally either restricted in their geographic distribution or restricted to one particular facies, and thus in either case are unsatisfactory for purposes of chronologic correlation. Diener[82] also, in discussing Oppel's Zones, has not distinguished between the possible interpretations of his zonal key term, his "gesamtfaunen." Nevertheless, the following conclusions of Diener will in other important respects further serve to clearly define the term Zone in its original meaning.

...Jede Schichte oder Schichtgruppe, die ihm (Oppel) durch eine ihr eigentümliche Fauna charakterisiert erschien und deren räumliche Ausdehnung über eine weitere Erstreckung sich nachweisen liess, nannte er (Oppel) Zone.... Die Zone im Sinne Oppels ist also ein Terminus für einen räumlichen stratigraphischen Begriff....

...doch darf nicht jede believige Schichte oder Bank in einem Lokalprofil, in der ein bestimmtes Fossil häufig vorkommt, ohne Rücksicht auf ihren sonstigen Schichtverband herausgegriffen und als Zone nach dem Namen jenes Fossils bezeichnet werden. Oppels Zonen haben vielmehr die faunistische Untersuchung eines grösseren Schichtkomplexes zur Voraussetzung. Sie stehen als Unterabteilungen des letzteren insofern in einem bestimmten Zusammenhang, als jede eine eigene Fauna enthält, deren wichtigste Arten sich durch konstante Merkmale von jenen ihrer Nachbarzonen trennen lassen. ... Die Zonen sind nicht nach einzelnen, herausgegriffenen Arten, sondern nach den Gesamtfaunen, beziehungsweise nach den vertikal am wenigsten, dagegen horizontal möglichst weit verbreiteten Elementen in den letzteren zu beurteilen. ... Oppels Jurazonen sind Faunenzonen. Jede derselben hat eine ihr eigentümliche Fauna. Wenn auch viele Arten durch mehrere Zonen hindurchgehen, kennzeichnet

[81] See A. Morley Davies, *Tertiary Faunas*: Vol. II. *The Sequence of Faunas* (Thomas Murby and Company, London, 1934), Chapter 2.

[82] C. Diener, *Grundzüge der Biostratigraphie* (Franz Deuticke, Leipzig and Vienna, 1925), Chapter 8.

## CORRELATION AND AGE

doch jede Fauna durch ihr besonderes Gepräge eine Phase in der Entwicklungsgeschichte der organischen Welt. Die Zonenfauna wird nicht bezeichnet durch eine bestimmte einzelne Art, sondern durch eine bestimmte Vergesellschaftung von Arten. Die Namengebung nach einem einzelnen Fossil darf daher nur eine Abkürzung bedeuten.

From this discussion it is clear that many of the Miocene "zones" referred to in West Coast stratigraphy do not represent Zones in the sense that Oppel used the term. The "*Turritella inezana* Zone," in some respects comparable to a Zone as used by Oppel, unless used as a Biozone, is in magnitude comparable to a Stage, as is the "*Vaðulineria californica* Zone" *sensu stricto* and also perhaps the *Chione esmerensis* Zone of the Lower Pliocene and the *Turritella variata* Zone of the Upper Eocene or Lower Oligocene. The "*Turritella ocoyana* Zone," "*Arca montereyana* Zone," and "*Vaðulineria californica* Zone" *sensu lato* are units of even larger magnitude and actually constitute Biozones[83] for the species whose name they bear, rather than true Zones. A still different usage of the term "zone" is that recently employed by Loel and Corey in their discussion of the Vaqueros Formation,[84] this usage is also distinct from that of Oppel, in as much as the "zones" of Loel and Corey represent *beds* ("schichten ... in einem Lokalprofil" of Diener), or perhaps Zonules in some cases, rather the true Zones. The various echinoid Zones recognized in the Upper Miocene of California, however, seem to represent biostratigraphic units comparable in all important respects with the faunal Zones of Oppel. Two of the three foraminiferal "zones" set up by Cushman and Laiming (26) in the Lower Miocene also represent faunal Zones in this sense. Where they appear to be warranted additional foraminiferal Zones are set up in the present paper. They are noted in the summary of Stages, which follows.

In 1928, Fenton[85] coined and defined the term Zonule as follows:

A zonule is the strata or stratum which contains a faunule, its thickness and area being limited by the vertical and horizontal range of that faunule.

In the same paper, Fenton considered a faunule as

...an assemblage of fossil animals associated in one or a few contiguous strata and dominated by the representatives of one, community, commonly either an association, or a layer society.

[83] See W. J. Arkell, *op. cit.*, pp. 21–22.

[84] Wayne Loel and W. H. Corey, "The Vaqueros Formation. Lower Miocene of California. I. Paleontology," *Univ. California Pub., Bull. Dept. Geol. Sci.*, Vol. 22, No. 3 (1932), pp. 31–410.

[85] C. L. and A. F. Fenton, "Ecologic Interpretation of Some Biostratigraphic Terms," *Amer. Midland Nat.*, Vol. 11, No. 1 (January, 1928), pp. 20–22.

Fenton also notes that "faumules of essentially equivalent constitution" may be found at distinctly different horizons, that is, that they may be "recurrent."

The "*Haplophragmoides trullissata* Zone" of Cushman and Laiming (26) apparently represents a Zonule rather than a true Zone. The association of the particular species constituting the assemblage characteristic of this "zone" in Los Sauces Creek is not unique in geologic time as in a Zone; these species also occur jointly at horizons stratigraphically both higher and lower, although their association in Los Sauces Creek, as well as throughout the larger part of the western Ventura Basin, is characteristic of a particular horizon; such a condition may be considered of essentially local ecologic significance, and to this type of biostratigraphic unit, distinct though not distinguishable on the basis of its contained species alone,—Fenton's term of Zonule has been applied in the present paper. A number of Zonules are recognized in Reliz Canyon, and one is noted in Los Sauces Creek; work in other Miocene areas has not been sufficiently detailed to permit their description here. A great number of Zonules have been recognized in the Pliocene and Pleistocene of southern California; those from the Dominguez oil field have recently been summarized in great detail by Wissler, who refers to them variously as "marker zones . . . major foram zones . . . subzones . . . horizons . . . marker horizons," *et cetera*.[86]

Before leaving the subject of biostratigraphic nomenclature a final word should be said in explanation of the use of new Stage names instead of using Stage names already in use in Europe and elsewhere. This procedure results from the need for an objective chronologic terminology suitably applicable to data which may be adequate for a part of the earth's surface but at the same time inadequate for the entire world. The synchronization of stratal units as small as Stages in areas as geographically remote as Europe and California may well be considered extremely hazardous in the light of present knowledge. Similar synchronization between either of these areas and the intermediate Caribbean area is also hardly to be considered reliable within the limits of error of a Stage or of a Zone. Arkell, in discussing d'Orbigny's Stages, noted that

few of the geologists working in other countries at first adopted d'Orbigny's stages, for with their more detailed local knowledge they were unable to recognize any such ten divisions which might have been "delineated by nature with bold strokes across the whole earth";

[86] Stanley G. Wissler, "Foraminiferal Zones of the Dominguez Oil Field, Los Angeles County, California," (abstract), *Amer. Assoc. Petrol. Geol.*, 22nd annual meeting (March 17–19, 1937), list of papers with abstracts, p. 75.

and the difficulty of relatively fine-spun interregional correlation in the Jurassic alluded to by Arkell has an analogue in similar correlation within the Tertiary. The consequent impossibility of applying the same detailed chronologic terminology to the various areas does not, however, imply that the use of a local chronology is similarly undesirable. In this regard it is noteworthy that Oppel's Stages and Zones embodied a recognition and application of d'Orbigny's more essentially theoretical concepts through a restricted geographic area roughly comparable in magnitude to a Recent molluscan province such as recognized by Woodward,—a restricted area throughout which these Stages and Zones could be recognized as tangible time-stratigraphic entities. The area of Tertiary marine sediments embracing California, Oregon, and Washington is a similarly comparable area or province. With chronologies established for each such area, the way is open for the use of the time-dimension in the geologic problems of each area, and at the same time it also remains open for the ultimate synchronization of the various provincial chronologies whenever there become available the world-wide data which, because of the huge geographic gaps in the sedimentary record, are necessary as control for such a cosmopolitan chronology. And in the meantime it is thereby possible to avoid the confusion and misconceptions inherent in the failure to distinguish time-stratigraphic (System, Series, Stage, Zone) from rock-stratigraphic (Formation, Member, *et cetera*) units everywhere.

---

*Time*

If I were asked as a geologist what is the single greatest contribution of the science of geology to modern civilized thought, the answer would be the realization of the immense length of time. So vast is the span of time recorded in the history of the earth that is generally distinguished from the more modest kinds of time by being called "geologic time".

Adolph Knopf, GSA President, 1944

# Stratigraphic paleontology: From oil patch to academia

**William B. N. Berry**
*Department of Geology and Geophysics, University of California, Berkeley, California 94720-4767;
e-mail: bberry@uclink4.berkeley.edu*

To R. M. ("Doc") Kleinpell, paleontology involved more than finding, describing, illustrating, and discussing evolutionary development of fossils. To him, fossils occurred in a rock matrix and stratigraphic successional context, and the paleontologist should use that context in interpretations of fossils. Kleinpell's *Miocene Stratigraphy of California* (1938) exemplified his notion of paleontologic inquiry, for it not only describes fossil foraminifers, but it also discusses the contexts in which they were found. The work incorporated use of that information to propose small-scale divisions of geologic time and to discuss limits to their use. Most paleontologists lack the breadth of understanding of stratigraphic and sedimentary geology fully to comprehend Kleinpell's ideas for appropriate paleontologic investigations. Accordingly, his publications received little notice from the majority of paleontologists.

Kleinpell's ideas developed at the same time that California's oil production was in its early phase. The initial development of oil recovery from Tertiary strata in California presented petroleum geologists with the challenge of how to correlate the stratigraphic sequences in the subsurface that were seen only in cores taken from wells. Solution of that challenge was needed, if the state's economy was to be enhanced by development of an oil industry.

The forces of industry-driven economic expansion became a primary stimulus to the quest for and use of scientific principles in resource recovery, as it had in Europe during the 1800s (Berry, 1968, 1987). In English geology, for example, William Smith developed and used principles of stratal superposition and faunal succession to enhance recovery of coal and the positioning of canals through which that resource could be transported in a manner that would result in the greatest profits (Berry, 1987). Similarly, as oil became acknowledged to be a valued resource in developing California's economy, elucidation of those principles inherent in techniques to improve oil recovery became important (Kleinpell, 1972). The surface geology of many parts of California had been mapped by the late 1920s, but oil production required an understanding of the stratigraphic record that lay underground and, therefore, out of sight.

Surface mapping of California Tertiary strata led to a basic understanding of the succession of lithostratigraphic units, or formations. Some faunas, primarily clams and snails and some echinoids (sand dollars), had been collected and described from several formations (Kleinpell, 1972). Lithologic aspects of the rocks, as well as their contained fossils, indicated that most of them were sandstones, siltstones, and some shales that had been deposited in relatively nearshore and strandline, inner continental shelf environments. Oil, however, was being found in relatively finer grained strata deposited in more offshore, outer continental shelf and continental slope sediments (Kleinpell, 1962–1972, oral commun.)

Mappable units, the formations, could be traced laterally along strike, but significant changes in lithologic aspects (facies changes) were noted (Kleinpell, 1938, 1972). These facies changes created uncertainties for correlation in areas in which rock units could not be traced laterally along strike for significant distances. In his historical review of the context for division of the West Coast Tertiary strata by petroleum geologists, Kleinpell (1972, p. 94) described this facies problem in his uniquely colorful and rambling style, as follows.

> For example, although the term Repetto in the sense of its original time-stratigraphic usage was eventually rescued by Natland, the term Poncho Rico, originally a rock-stratigraphic term, had an even more checkered and confusing career. When in Prohibition days oil geologists, shuttling back and forth between Taylor Hotel and Paderewski's old favorite spa, the Paso Robles Inn with its elegant decor and its high-ceilinged rooms and elaborate cut-glass chandeliers, tried for mapping purposes to match the mid-Neogene marine sequence on the

Berry, W. B. N., 1999, Stratigraphic paleontology: From oil to academia, *in* Moores, E. M., Sloan, D., and Stout, D. L., eds., Classic Cordilleran Concepts: A View from California: Boulder, Colorado, Geological Society of America Special Paper 338.

more "cratonic" east side of the Salinas Valley with the more readily mappable formations in the essentially geosynclinal sequence on the west side, use of the term Pancho Rico as a solution to their east side areal mapping terminological problems was agreed upon in a smoke-filled room. There, on the east side, extensions of the Gabilan Mesa were apt to reveal great expanses of the non-marine Paso Robles Formation, with the marine sequences best exposed on the dissected and areally narrow canyon walls. This was fine for handling in section but not in plan, and especially so since lithologies comparable to McLure Shale, Reef Ridge Shale, Santa Margarita Sandstone and Jacalitos Sandstone were to be found there, often highly lenticular, and in their local sequences not necessarily in the same homotaxial relationship from place to place nor as in the formational type areas. What to do?

In addition to the problem of facies, another factor that hindered subsurface analyses was created by the procedure by which the California Tertiary stratal succession had been divided. Although Charles Lyell's names for divisions of the Tertiary were used (Lyell, 1833), based primarily upon analysis of molluscan faunas, divisions of the Lyellian units were simply formations or groups of formations (Kleinpell, 1972). The formations recognized in surface mapping could not be identified in the subsurface with assurance. Furthermore, the molluscan faunas that enabled use of Lyell's names were found only rarely in the subsurface. For petroleum recovery to develop into a growth industry, new insights were required. The problems encountered in working with the unseen subsurface strata and with the issues created by the significant differences in depositional environments of the oil-bearing strata from those of strata mapped on the surface had to be resolved. Accordingly, the search for oil required solution of these problems: (1) how to correlate mudrocks and siltstones that did not bear mollusks with those that did; (2) how to correlate stratal sequences encountered by drilling a small-diameter hole into the subsurface; and (3) how to divide the Tertiary succession into relatively short duration time units using fossils, as was a European practice. The answers to these questions were debated at petroleum geologists meetings during the 1920s and early 1930s without resolution (Kleinpell, 1972, and 1964–1970, oral commun.) until the insightful work of R. M. Kleinpell, who documented the basic biostratigraphic data and used certain basic principles to enhance oil recovery from West Coast Tertiary strata. The basic principles used and promoted by Kleinpell permitted California's oil industry to grow into a major force in the state's economy. Kleinpell turned what was initially an academic, essentially theoretical, exercise into a logical and reasonable solution to the oil industry's great challenge of the late 1920s and early 1930s.

During the latter part of the 1920s, Kleinpell was employed as a geologist in petroleum exploration. He examined thousands of feet of well core, much of it stored in the basement of old buildings in San Francisco (Kleinpell, 1972). The reason behind these studies was to determine if anything of value could be identified in the cores. If not, then the cores were to be thrown away prior to relocation of the business to Los Angeles (Kleinpell, 1972). Kleinpell recorded lithologic aspects of the stratal succession in the cores, as well as the occurrences of certain fossils, primarily the calcium carbonate shells of benthic foraminifers. Subsequently, these studies were augmented by similar examinations of many other well cores obtained by a number of companies (Kleinpell, 1972, and 1962–1970, oral commun.). During his investigations, Kleinpell became acquainted with Ralph Reed and his book, *Geology of California* (Reed, 1933). Reed suggested to Kleinpell that a German paleontologist who worked with Jurassic ammonites in western Europe might have established the basic scientific guidelines required to enhance oil recovery in California. That German paleontologist was Albert Oppel and his mentor was the stratigraphic geologist August Quenstedt. Quenstedt and Oppel worked with ammonite-bearing Jurassic stratal successions in western Europe in the 1830s and 1840s. They wrote penetrating discussions of the need for precise examination of stratigraphic sequences and of placing every fossil found in its stratigraphic context (Oppel, 1856–1858; Quenstedt, 1856–1858).

Kleinpell entered graduate study at Stanford University under the guidance of Hubert Schenck, a specialist in the study of California mid-Tertiary rocks and faunas. Within that academic setting, Kleinpell decided to, as he put it, "look into" Reed's suggestion. He read the works of Quenstedt and Oppel, as well as those of the French paleontologist and stratigrapher, Alcide d'Orbigny. D'Orbigny was Oppel's predecessor in analysis in the use of fossil faunas found in documented stratigraphic superpositional relationships for short-duration temporal division of the stratigraphic record. Albert Oppel's proposals for short-duration temporal divisions were founded upon collecting large numbers of fossils of one major group (in Oppel's work, these were ammonites) from precisely documented stratigraphic positions or levels, in many carefully measured stratigraphic sections that represented relatively continuous deposition in many different environments over a broad geographic area. Oppel demonstrated that the pattern of co-occurrence of fossil taxa in certain stratal layers recognized through analysis of the overlapping stratigraphic ranges of many taxa could be used to identify discrete stratal (and, by implication, time) intervals (Oppel, 1856–1858). Oppel called his shortest-duration interval a zone. Zones, Oppel indicated, could be clustered into groups of zones. Oppel's groups of zones ("Zonengruppen" in Oppel's, 1856–1858 discussion) are similar to the units that d'Orbigny (1842, 1849–1852) termed stages. Oppel's zone and stage intervals were based upon the evolutionary development of fossil taxa present in the stratal succession under study. Because they were, time's arrow could be identified and so, too, could unique time intervals characterized by the successive co-occurrences of two or more taxa. Commonly, most of the taxa in a zone's fauna had different stratigraphic ranges. Over time, use of Oppel's methodology resulted in selection of the lowest stratigraphic occurrence of a widely found taxon in the zone's fauna as the base of the zone. Each zone was terminated by the occurrence of that taxon used to denote the base of the superjacent zone.

Upon reading Oppel, Kleinpell realized that he already had a considerable understanding of the basic ingredients of Oppel's methodology. He had carried out relatively extensive studies of the California Tertiary succession encountered in many wells, and he had engaged in a number of surface mapping projects in the search for oil (Kleinpell, 1972, and 1962–1972, oral commun.). In conducting his investigations, he had become well acquainted with the occurrences of fossil foraminifers in the many stratigraphic sequences he examined. He used the work of Joseph Cushman (1926), the foremost student of fossil benthic foraminifers of the day, to enhance his knowledge of these organisms. Kleinpell recognized, as the excerpts from Oppel cited in his book indicate, that Oppel's methodology provided the insights needed to make refined time divisions of the mollusk-poor but benthic foraminifer–rich California Tertiary strata. These strata from which oil was being produced had accumulated in continental shelf and slope environments. Kleinpell applied Oppel's methodology in an analysis of the stratigraphic ranges of fossil foraminifers collected in the stratigraphic sections he measured in outcrop as well as from strata he had seen in the many cores from wells drilled into the subsurface. Kleinpell realized that Oppel's methodology would be most useful if as many samples as possible from as many stratal sequences as possible could be included in the basic database before it was analyzed. Kleinpell turned to many friends and acquaintances engaged in measuring stratigraphic sections, in examining well cores, and in collecting fossil foraminifers for additional data (Kleinpell, 1972, and 1962-1972, oral commun.). When it was realized that the foraminiferal calcium carbonate shells could be released from rocks through disaggregation of the matrix containing them, large numbers of them were collected by students of many different oil-producing and potential oil-producing horizons (the "bug" men and women of the petroleum business; see Stinemeyer, 1972, for photographs of some of them).

Collecting was enhanced by taking samples from well cuttings and from well side-wall cores (Kleinpell, 1972). The shells are small enough to be recovered in considerable numbers from small-sized samples obtained in the process of well drilling. As he began his analysis of the stratigraphic ranges of fossil benthic foraminifers obtained from California mid-Tertiary strata, Kleinpell had at hand significantly more data than Oppel had in his studies of Jurassic ammonite-bearing strata. Fossil foraminifers were very abundant in most samples taken, and they were obtained from many carefully measured stratal successions encountered in the subsurface. These foraminifer occurrence data augmented those obtained from measured surface sections. Accordingly, an extensive data set was available. Following procedures used by Oppel, Kleinpell analyzed the stratigraphic ranges of scores of foraminiferan taxa obtained from nearly 200 stratigraphic sections in a study of California mid-Tertiary (essentially Miocene) strata. He identified unique associations of taxa that typified discrete stratal intervals, as Oppel had done.

Oppel's zonal analysis was not the only such study well known at the time Kleinpell was conducting his of California Tertiary strata. Gertrude Elles and Ethel M. R. Wood, under the guidance of Charles Lapworth, had erected a succession of Ordovician and Silurian zones based upon graptolites obtained from many British stratal sequences. They used Oppel's methods of collecting and analysis (Elles and Wood, 1914). They pointed out that each of their zones was based upon an association of species that is unique in the stratigraphic record. Also, Dutch petroleum geologists recognized zones and stages as divisions of Tertiary strata in the former Dutch East Indies in the late 1920s (Kleinpell, 1960-1972, oral commun.).

Kleinpell's initial results were presented in an abstract at the 1933 annual meeting of the Geological Society of America (Kleinpell, 1933). He documented his characterization of six stages within the California Miocene based on overlapping stratigraphic ranges of fossil benthic foraminifers in his Stanford doctoral dissertation that was completed in 1933. Each stage was divided into two zones, following the fundamental procedures used by Oppel. With help from Ralph Reed, his dissertation was published in 1938 by the American Association of Petroleum Geologists as *Miocene Stratigraphy of California*. The title seemingly promised only a recitation of California Miocene stratigraphic relationships. However, it actually delivered far more. It documented the validity of basic scientific principles by which strata of any age may be divided and precise time correlations among them established. It also presented an impressive array of systematic paleontological descriptions of benthic foraminifers. The principles of time division using the patterns provided by analysis of the overlapping stratigraphic ranges of many fossil foraminiferan species were carefully described and demonstrated. In addition, Kleinpell set forth the limitations of his zones and stages. Because he considered species in his analyses and because most of these species seemed to be restricted to a certain geographic area (in Kleinpell's day, this area was essentially California), he suggested that zones and stages were restricted to a biogeographic province. Ecologically, his zones and stages could be applicable only to those environments on the sea floor in which the specific organisms under investigation lived. In the case of most of the species that characterize Kleinpell's zones and stages, these environments were primarily outer shelf and slope settings. Kleinpell's zones and stages were used by him and many other geologists involved in California petroleum exploration during the 1930s and later (see Stinemeyer, 1972). Leaving his methodology in the hands of associates, Kleinpell went to the Philippines in the early 1940s to explore the possibilities of expanding oil production there.

While there, Kleinpell's career in the petroleum business was interrupted by Japanese expansion into the Philippines during World War II. Kleinpell was taken prisoner and interred in a prisoner of war camp by the Japanese for about four years. During that time, he developed a considerable fondness for and skill in teaching. Camp inmates taught each other their areas of expertise as a means of passing time productively. After the war ended,

Kleinpell returned to California to resume his petroleum consulting practice, using the basic methodology he had established prior to the war. His interest in teaching developed in the camp became a passion. He commenced instruction in the Department of Paleontology at the University of California Berkeley in 1946. "Doc," as he became known fondly to generations of students and professional colleagues, drew upon both his wartime experiences in instruction and his vast knowledge of fossil foraminifers and California Tertiary stratigraphy to enchant his many students with the intricacies of reasoning logically, as well as with an understanding of benthic foraminiferan biostratigraphy. His students and professional associates carried on studies of potential petroleum-bearing strata using the basic principles of zonation that their mentor established in his *Miocene Stratigraphy of California*. Among them were V. Standish Mallory, who documented stages and zones for the West Coast Paleocene and Eocene (Mallory, 1959), and C. R. Haller, who outlined Pliocene zones and stages in northern California (*in* Kleinpell, 1980).

Kleinpell recognized that zones and stages could be documented using fossils other than foraminifers in the search for oil in Tertiary strata. Accordingly, he sent Frank Sullivan to work with M. N. Bramlette on fossil coccolithophores. Walter W. Wornardt studied with G. Dallas Hanna at the California Academy of Sciences and became very familiar with diatoms. Based on that familiarity and on an extensive knowledge of the California Tertiary stratigraphic succession, Wornardt (1967) developed a Tertiary diatom biostratigraphic chronology. A consultant, A. D. Warren, was urged to document correlations between zones based on planktic foraminifers and the units based on benthic foraminifers. Kleinpell and his students Ann Tipton and Donald Weaver addressed questions involved in equating the foraminiferan-based stages and zones with California molluscan faunal successions, especially those of the Oligocene (see Kleinpell and Weaver, 1963). Tipton et al. (1973) documented a number of foraminiferan phyletic lineages useful in precise zonation. Kleinpell influenced his one-time graduate school associate, Hollis Hedberg, to include Oppel zones in the codes of stratigraphic terminology he wrote. He encouraged Z. M. Arnold to study the life habits of modern benthic foraminifers to enhance understanding of the significances of environmental influences on shell morphology. His insights were profoundly valuable to me in writing the history of the geological time scale's division, *Growth of a Prehistoric Time Scale* (Berry, 1968, 1987). Kleinpell exercised considerable influence on those involved in California's petroleum business. His influence was unnoticed by all but his associates because so much of it was promoted orally through lengthy discussions. That influence was enhanced through conferences with colleagues at professional meetings and through lengthy discussions in "Doc's" office. "Doc's" practice at professional meetings was to ensconce himself at a prominent table in the most noted bar in the area and, in essence, hold court. He would be visited by numbers of people, all seeking advice on how to solve some problem encountered in the search for oil.

Kleinpell's influence on and stimulus to petroleum exploration in California and to California Tertiary stratigraphy were recognized in a convention of the Pacific Section of the Society of Economic Paleontologists and Mineralogists held in Bakersfield, California, in March 1972 (Stinemeyer, 1972). The papers presented at that meeting resulted in the recognition that Kleinpell's many students and colleagues had generated a substantial volume of biostratigraphic data that modified and enhanced Kleinpell's original work with the California Miocene. Accordingly, Kingsley Davis, President of the Pacific Coast Section, encouraged the American Association of Petroleum Geologists to ask Kleinpell to write an update of his original *Miocene Stratigraphy of California*. Reluctantly, and with the assistance of associates and former students, Kleinpell did so in the work The *Miocene Stratigraphy of California Revisited*. It was published in 1980 and included C. R. Haller's "Pliocene Biostratigraphy of California" (Haller, *in* Kleinpell, 1980). That 1980 publication demonstrated the depth and breadth of Kleinpell's influence on Tertiary paleontology and biostratigraphy and in using biostratigraphic principles to advantage in resource recovery.

Kleinpell's contributions to the California oil business include not only the solution to the challenge that plagued it before his work, but also his influence on essentially countless students of the California "oil patch." What was not realized during his career was that he not only had enhanced California's and the nation's economy through application of scientific principles, but also that he had demonstrated that Oppel's methodology was in fact sound scientific principle applicable to any part of the fossiliferous stratigraphic record.

Though never recognized as more than a practitioner of applied science, Kleinpell was more than that. He was an accomplished scholar who possessed great breadth of knowledge achieved through reading extensively on a broad range of subjects. Not only did he possess a photographic memory, but he was also able to place every topic he discussed into context. His intellectual strength lay in understanding the geological context for both his own observations and those made by others. He espoused going beyond descriptions of fossils to an analysis of them using the sedimentologic and stratigraphic contexts in which they occur. He used that understanding to demonstrate the validity of certain principles inherent in the recognition of refined, short-duration intervals of the geologic time scale based upon analysis of patterns in fossil occurrence. He documented not only the general applicability of the refined temporal divisions he recognized, but also the limitations imposed upon them by organismal ecological and biogeographical relationships. The pages reproduced here from his book for historic interest present the basic ingredients inherent in Kleinpell's remarkable career in development of certain scientific principles and their application to enhancement of that part of California's and the nation's economy founded upon petroleum recovery.

# REFERENCES CITED

Berry, W. B. N., 1968, Growth of a prehistoric time scale: San Francisco, W. H. Freeman and Company, 158 p.

Berry, W. B. N., 1987, Growth of a prehistoric time scale (revised edition): Palo Alto, California, Blackwell Scientific Publications, 202 p.

Cushman, J. A., 1926, Foraminifera of the typical Monterey of California: Cushman Laboratory of Foraminiferal Research Contribution, v. 2, p. 53–69.

d' Orbigny, A. D., 1842, Paleontologie Francaise, Terrraines Jurassiques, Part 1, Cephalopodes: Paris, Masson, 623 p.

d' Orbigny, A. D., 1849–1852, Cours elementaire de paleontologie et de geologie stratigraphiques: Paris, Masson, 2 volumes and atlas.

Elles, G. L., and Wood, E. M. R., 1914, Part 10, in A monograph of British graptolites: London, Palaeontographical Association, p. 487–526.

Kleinpell, R. M., 1933, Miocene foraminifera from Reliz Canyon, Monterey County, California: Geological Society of America Bulletin, v. 44, p. 165.

Kleinpell, R. M., 1938, Miocene stratigraphy of California: Tulsa, Oklahoma, American Association of Petroleum Geologists, 450 p.

Kleinpell, R. M., 1972, Some of the historical context in which a micropaleontological stage classification of the Pacific Coast middle Tertiary developed, in Stinemeyer, E. H., ed., The proceedings of the Pacific Coast Miocene biostratigraphic symposium: Pacific Section, Society of Economic Paleontologists and Mineralogists, p. 89–110.

Kleinpell, R. M., 1980, The Miocene stratigraphy of California revisited: American Association of Petroleum Geologists Studies in Geology 11, 338 p.

Kleinpell, R. M., and Weaver, D. W., 1963, Oligocene biostratigraphy of the Santa Barbara embayment, California. I. Foraminiferal faunas from the Gaviota and Alegria Formations by R. M. Kleinpell and D. W. Weaver. II. Mollusca from the *Turritella variata* Zone by D. W. Weaver and R. M. Kleinpell: University of California Publications in Geological Sciences, v. 43, 250 p.

Lyell, C., 1833, Principles of Geology, London, J. Murray, v. 3, 398 p.

Mallory, V. S., 1959, Lower Tertiary biostratigraphy of the California Coast Ranges: Tulsa, Oklahoma, American Association of Petroleum Geologists, 416 p.

Oppel, A., 1856–1858, Die Juraformation Englands, Frankreichs und des sudwestlichen Deutschlands: Wurttemberg Naturwissenschaften Verein Jahreshefte, v. xii–xiv, p. 1–438, 1856; p. 439–694, 1857; p. 695–857, 1858, Stuttgart.

Quenstedt, F. A., 1856–1858, Der Jura: Tubingen, H. Laupp.

Reed, R. D., 1933, Geology of California: Tulsa, Oklahoma, American Association of Petroleum Geologists, 355 p.

Stinemeyer, E. H., 1972, The proceedings of the Pacific Coast Miocene Biostratigraphic symposium presented at the Forty Seventh Annual Pacific Section S. E. P. M. Convention, March 9-10, 1972: Bakersfield, California, Pacific Section, Society of Economic Paleontologists and Mineralogists, 364 p.

Tipton, A., Kleinpell, R. M., and Weaver, D. W., 1973, Oligocene biostratigraphy, San Joaquin Valley, California: University of California Publications in Geological Sciences, v. 105, 81 p.

Wornardt, W. W., 1967, Miocene and Pliocene marine diatoms from California: California Academy of Sciences Occasional Paper, 108 p.

Manuscript Accepted by the Society November 23, 1998.

Geological Society of America
Special Paper 338
1999

# Chapter 15
# THE MONTEREY FORMATION: THE SOURCE OF OIL

**Richard J. Behl**
*Department of Geological Sciences
California State University, Long Beach
Long Beach, California 90840
E-mail: behl@csulb.edu*

## MILTON NUNN BRAMLETTE

Milton Nunn Bramlette (1896–1977) was one of the extraordinary sedimentary geologists of the twentieth century. His expertise and key contributions spanned sedimentary petrology, stratigraphy, and micropaleontology. After graduating with high honors from the University of Wisconsin (Madison) in 1921, he joined the U.S. Geological Survey (USGS) at a time when new geologists were encouraged to discover their own research interests. There, he delved into the alteration of volcanic deposits, and the origin and character of clays, zeolites, and phosphatic sediments, making important advances in each of these fields. In the late 1920s, Bramlette completed graduate work toward his doctoral degree at Yale (completed in 1936) and explored for oil in Latin America for the Gulf Oil Company before returning to the USGS in 1930. In close association with W. P. Woodring and others, he completed a series of fundamental studies of California geology, including that of the Palos Verdes Hills, Santa Maria Valley, and Kettleman Hills oil field, in addition to starting his classic work on the Monterey Formation. Bramlette's move to the University of California Los Angeles from 1940 to 1951 was interrupted by a two-year return to the USGS during World War II to investigate the geology and origin of strategically important bauxite ore deposits. In 1951, Bramlette joined the faculty of the Scripps Institution of Oceanography, where he began pioneering studies in the modern distribution and biostratigraphy of calcareous nannoplankton that continued beyond his official retirement in 1962. In recognition of his contributions, Bramlette was awarded the Distinguished Service Medal of the Department of the Interior, and elected to the National Academy of Sciences, which awarded him the Thompson Medal in 1964. Milton Nunn Bramlette died on March 31, 1977.

M. N. Bramlette's (1946) classic paper on the Monterey Formation was selected for this volume because its significance extended far beyond the North American Cordillera by advancing new concepts of silica diagenesis and of the influence of depositional environment on the character of hemipelagic sediments. Bramlette's work is internationally cited as evidence for the diagenetic origin of chert and other hard siliceous rocks. He linked the characteristic organic-matter enrichment and silica enrichment of the Monterey Formation to modern observations of marine upwelling, planktonic productivity, and oxygen deficiency of the depositional environment. Bramlette documented the details and possible origins of the Monterey's distinctive rhythmic bedding and related them to climatic and oceanographic change. He correlated discontinuous and separately named deposits to determine the regional and even pan-Pacific extent and significance of the Monterey Formation and its stratigraphic equivalents, and thereby suggested the relationship between widespread paleoceanographic conditions and the development of prolific petroleum source rocks along continental margins.

Richard J. Behl

UNITED STATES DEPARTMENT OF THE INTERIOR
J. A. Krug, Secretary

GEOLOGICAL SURVEY
W. E. Wrather, Director

Professional Paper 212

# THE MONTEREY FORMATION OF CALIFORNIA AND THE ORIGIN OF ITS SILICEOUS ROCKS

BY

M. N. BRAMLETTE

UNITED STATES
GOVERNMENT PRINTING OFFICE
WASHINGTON : 1946

For sale by the Superintendent of Documents, U. S. Government Printing Office, Washington 25, D. C.
Price 60 cents

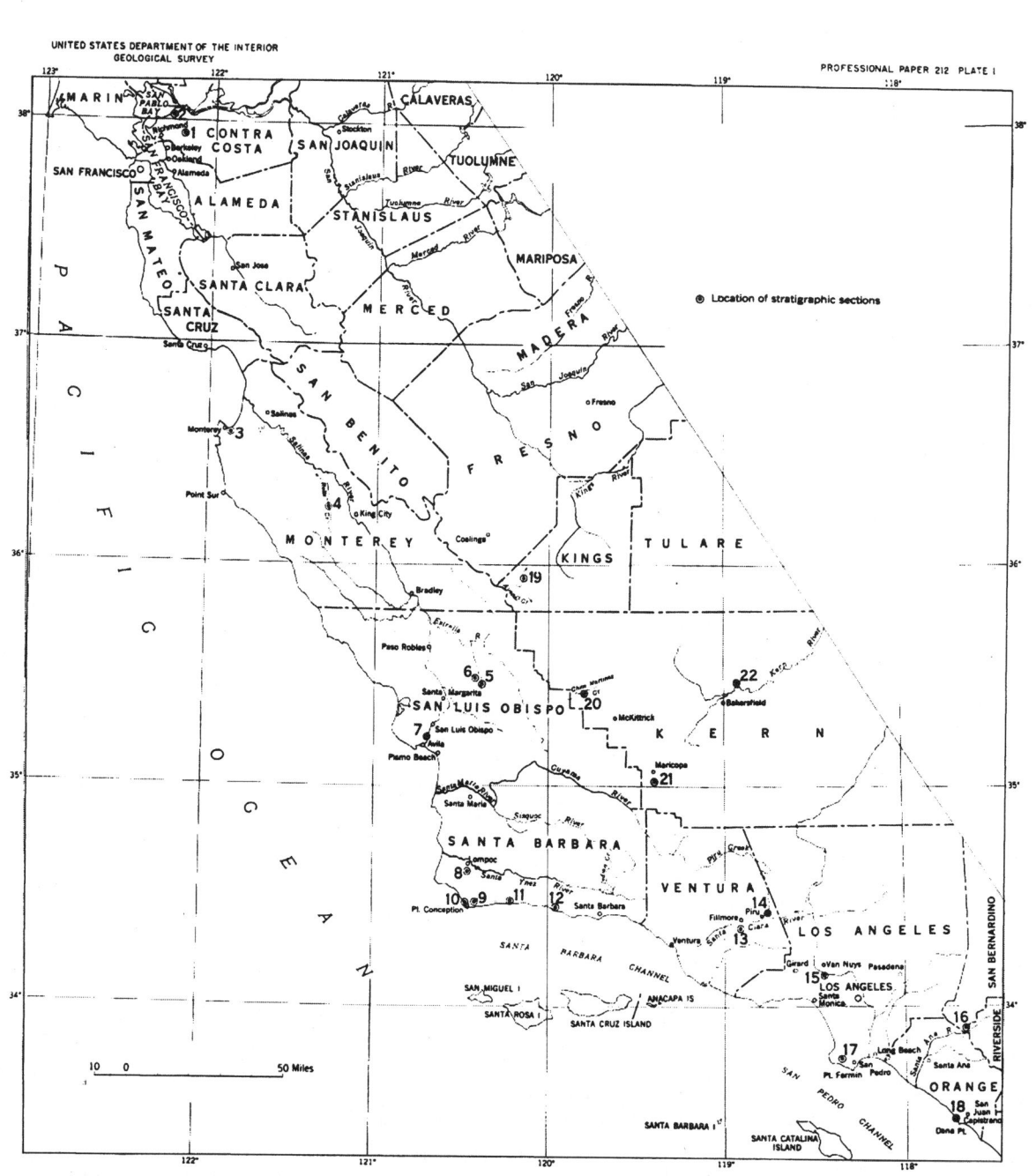

INDEX MAP SHOWING LOCATION OF STRATIGRAPHIC SECTIONS OF THE MONTEREY FORMATION IN CALIFORNIA

# THE MONTEREY FORMATION OF CALIFORNIA AND THE ORIGIN OF ITS SILICEOUS ROCKS

By M. N. Bramlette

## ABSTRACT

The Monterey formation is so thick and extensive, and has so much economic as well as scientific importance, that it constitutes a major element in the geology of California. The Miocene strata that consist predominantly of highly siliceous rocks have received a number of names in different areas, but intensive stratigraphic work in recent years, particularly that done by the oil companies, indicates that most of these locally named stratigraphic units are essentially equivalent. It appears advisable, therefore, to return to an early and convenient usage by extending the term Monterey formation to include many of these locally named units. This report contains stratigraphic sections representing most of the areas where local names have been applied, with their suggested correlation.

The siliceous rocks which characterize the formation belong to several widely varied but intergrading types. Diatomite and less pure diatomaceous rocks are conspicuous in the upper part of the formation in many areas. Harder siliceous rocks classed as porcelanite, porcelaneous shale, cherty shale, and chert constitute a large part of the formation. Preserved siliceous organisms are rare in the porcelaneous and cherty rocks, but various lines of evidence indicate that these silica-cemented rocks were formed in major part through an alteration of originally diatomaceous, rocks similar to those now present in the upper part of the formation. The alteration seems to have been a process of solution of the relatively unstable silica of the delicate opaline shells and reprecipitation of the silica as a cement to form the porcelaneous rocks. Beds or lentils of dense chert were formed by the same process in parts of the strata that were originally purer diatomaceous rocks.

The time at which the alteration occurred and the fundamental causes of the alteration are problems not yet fully solved. These sediments appear to have been undergoing alteration at varying rates ever since they were deposited on the sea floor. Much of it occurred at an early stage in the compaction of the deposits and may be termed diagenetic; most of it, however, was subsequent to most of the compaction and seems to have been effected, in part at least, through load and deformation during late Miocene and early Pliocene time.

Sandstone and the finer clastic sediments of the formation are more briefly considered than some less abundant rocks such as the carbonate beds and concretions, because the carbonate concretions include preserved diatoms and show evidence of a diagenetic process of formation that bears on the origin of the associated siliceous rocks. The widespread occurrence of tuffaceous material also bears on the origin of siliceous rocks. Several earlier writers have considered the possible significance of the close association of volcanic rocks, particularly of tuffs, with highly siliceous deposits in many regions. Additional evidence is presented in this report that the alteration of tuffs may result in silicification of adjacent beds, and that an abundant supply of silica from tuffs commonly results in an unusually large development of the siliceous organisms.

Large parts of the formation are bituminous and are generally recognized as important source beds of many California oil fields. The character of the bedding and its mode of formation suggest a significant relationship with the conditions for formation of petroliferous strata, in that the thin rhythmic bedding or lamination was evidently formed at depths below that affected by appreciable wave or current action, and such conditions are favorable for the accumulation of the organic matter from which petroleum is formed. The bedding and fossil content differ from those of formations deposited in shallower or more agitated waters and indicate two distinct genetic types of rock. The rhythmically bedded type generally indicates greater crustal instability and geosynclinal deposition.

## INTRODUCTION
### NATURE OF THE INVESTIGATION

The wide extent of the Monterey formation in California, its distinctive lithologic character, and its importance to the petroleum industry as one of the major oil-producing formations of California, make a comprehensive study of the formation desirable. A study of both local details of character and regional features was accordingly undertaken in 1931 as one of the projects made possible through certain research funds then available to the United States Geological Survey. Subsequent plans made it impossible to carry out this program in full, but some of the data and conclusions obtained before the interruption of the work are presented in this report.

\* \* \* \* \*

### GENERAL CHARACTER AND DISTRIBUTION OF THE FORMATION

The Monterey formation of California includes the Miocene strata characterized by an unusually high proportion of silica. The formation is widely distributed in and near the Coast Ranges from a latitude north of San Francisco to one south of Los Angeles, and in many areas it is several thousand feet thick. The siliceous rocks are more than a mile thick over some areas many square miles in extent, and about half a mile thick in much greater areas; and their total volume thus amounts to thousands of cubic miles. It shows remarkably rapid variations in thickness and lithologic character that permit few generalizations on the formation as a whole. But, despite all this variation, its siliceous character makes it one of the most distinctive and easily recognized of the formations in the thick Tertiary system of California.

The siliceous rocks locally include thick diatomaceous members, more widespread and in general thicker members of the hard but not very dense silica-cemented rocks termed porcelanite and porcelaneous shale, and large amounts of the harder and denser silica rocks classed as chert and cherty shale. Although the formation is characterized by these highly siliceous rocks, it includes, in many areas, large amounts of interbedded rocks of other types, particularly of normal clastic shale, mudstone, and sandstone. The more siliceous rocks also grade laterally into strata that are made up dominantly of normal clastic rocks, and where these clastic rocks predominate the name Monterey formation does not seem appropriate.

A diatomaceous member is not everywhere present in the formation, but where present it forms the upper part of the Monterey siliceous rocks. This upper diatomaceous member, however, is not everywhere of the same age. Porcelaneous and cherty rocks of some areas are equivalent in age to diatomaceous rocks of other areas, even though the diatomaceous rocks consistently form the upper part of the siliceous rocks of all areas where present. The diatomaceous deposits reach a maximum thickness of about 1,000 feet in an area south of Lompoc, Santa Barbara County, and they are several hundred feet thick in other places, including the type area near the town of Monterey, the Palos Verdes Hills of Los Angeles County, the southwestern San Joaquin Valley, and parts of the Salinas Valley.

Porcelanite and porcelaneous shale are more widespread, and in most areas, including the type area near Monterey, they are the dominant siliceous rocks of the formation. In many areas these rocks are several thousand feet thick, and at two widely separated localities, on Chico Martinez Creek and in Reliz Canyon, they are more than 5,000 feet thick.

Chert and cherty shale, though less abundant than the porcelaneous rocks, are of almost equally wide distribution and occur in most areas of the formation. They commonly form thin beds, alternating with beds of less siliceous rocks. The cherty rocks are particularly abundant in parts of the Berkeley Hills, as in the Claremont Canyon exposures (pl. 6, C), from which the local name Claremont shale was adopted.

Although the siliceous rocks and the associated normal clastic rocks constitute most of the formation, interbedded deposits of pyroclastic materials are numerous and widespread. The purer beds of pyroclastic materials occur at many horizons, but they are generally thin and form only a small percentage of the total thickness. In a few areas, however, certain of these beds are scores of feet thick, and in part of San Luis Obispo County one of the beds is locally several hundred feet thick. The pyroclastic beds consist of unaltered vitric tuff or volcanic ash and of tuffs in various stages of alteration to beds of bentonite. Lava flows and associated intrusive bodies, mostly basaltic in composition, occur in the formation in a number of areas. Most of these bodies are sills, which apparently were intruded under little cover, and it is not always easy to distinguish the sills from the flows.

Carbonate rocks form a minor part of the formation but are widespread. They occur as dolomitic and calcareous beds and concretions. Many of the calcareous beds consist largely of Foraminifera, and much of the formation is more or less calcareous because of the disseminated remains of foraminiferal shells.

Parts of the formation are highly bituminous, and much of the dark color of the unweathered siliceous rocks is due to organic matter, which is recognizable in thin sections and fragments under the microscope. The rocks that have been lightened in color by weathering contain less recognizable organic matter, most of the organic matter having apparently been removed by oxidation and leaching. Oil accumulations are known in the formation in a number of widely separated areas, and in some other areas the formation is considered as the probable source of the oil in adjacent formations.

A widespread and conspicuous feature of the formation is a thin rhythmic bedding, which appears to have been an important factor in the markedly incompetent behavior of the rocks under deformation. The close folding and very complicated minor structural features commonly present in many areas (pl. 4, A, B) are probably in part a result of this bedding, which also makes these features especially conspicuous.

## STRATIGRAPHY AND CORRELATIONS
### STRATIGRAPHIC RELATIONS

The marked and abrupt lateral variations in lithology and the scarcity of macro-fossils have made correlations difficult in much of the Miocene of California, with the result that many local formation names have been used. The complex stratigraphic relations were well summarized in 1913 by Louderback.[1] Intensive study, in recent years of the abundant micro-fossils, especially of the Foraminifera, has greatly clarified the correlations, however, and the results have largely confirmed the views presented so long ago by Louderback.

In general the lower Miocene consists largely of clastic sediments, with sandstone dominant toward the base, and with a rather consistent decrease in grain size to mudstone in the upper part of the lower Miocene (Saucesian stage of Kleinpell). Calcareous shale and mudstone are commonly dominant in the lower part of the middle Miocene and siliceous rocks become much more common and widespread in the upper part of the middle Miocene. In many areas, particularly in the coast ranges and in southern California, the Monterey siliceous rocks are thus of late middle Miocene and upper Miocene age. The underlying mudstones and calcareous shales would preferably receive other formational names, as the Rincon mudstone and the Sandholdt formation. Though their contacts are gradational, these stratigraphic units are distinct and mappable over large areas.

Detailed paleontologic and stratigraphic studies tend to increase the number of recognized stratigraphic units in the vertical sequence, but they tend at the same time to eliminate some unit names that have been applied locally. The data now available seem to justify the extension of the early and well-known name Monterey formation to include a number of locally named formations that are lithologically similar to it but that, because of their occurrence in separate areas, are not demonstrably continuous with it and until recently were

---

[1] Louderback, G. D., The Monterey series in California: Univ. California Pub., Dept. Geol. Sci. Bull., vol. 7, no. 10, pp. 177-241, 1913.

# THE MONTEREY FORMATION OF CALIFORNIA AND THE ORIGIN OF ITS SILICEOUS ROCKS

not definitely known to be approximately equivalent to it. Woodring,[2] in 1940, summarized the confusing status of the stratigraphic names in the California Miocene and advocated the wider extension of the name Monterey formation.

\* \* \* \* \*

### PALEOGEOGRAPHIC INTERPRETATIONS

... Maps by Reed showing the distribution and thickness of the lower, middle, and upper Miocene deposits in the southern part of California were published,[31] in 1936, and these maps, though necessarily generalized, embody enough surface and subsurface data to indicate some pertinent facts regarding the Miocene seas. The sea was less extensive in lower Miocene time than in the middle and upper Miocene, but throughout the Miocene it included deep basins wherein great thicknesses of sediment accumulated. These basins were more or less separated by land areas, or by shallower sea areas where much less sediment was accumulated. The basins show little relation to the present coast line, and apparently were connected in part with the open ocean by relatively shallow and narrow seaways. The deeper basins and shallower-water divides off the present coast in the region of the Channel Islands,[32] constitute such conditions of bottom topography as are postulated for the Monterey seas...

The mere shape and distribution of the areas of maximum sedimentary accumulation, and their changes and shiftings through Miocene time, suggest that they were tectonic basins of subsidence. The thickness of the deposit in a given place appears to reflect the amount of subsidence that permitted accumulation rather than the supply of sedimentary material locally available, and this seems even more probable for that large part of the sediment which is not of clastic origin. Here as in many other regions, it seems more probable that the thick deposition was largely a result of the accumulating load of sediment. The shifts of the areas of maximum deposition indicate that, in general, the basins of subsidence were not originally depressed to their total depth and gradually filled with all subsequent Miocene deposition...

Seas that had deeper basins, formed by local subsidence, within which all the sediment available from land or areas of shallower water could accumulate but which received relatively small supplies of terrigenous material, would be particularly favorable for the accumulation of thick organic deposits. The relatively small proportion of clastic material in these deposits may have resulted, however, from an unusually rapid supply of the diatomaceous sediment as well as from the relatively small supply of terrigenous sediment. The origin of the diatomaceous deposits is considered on p. 37 and leads to the conclusion that the organic accumulation was rapid as compared with the rate commonly assumed, and that it probably was not very much slower than the rate of accumulation of many clastic deposits. Among the factors favoring an unusually large supply of diatoms for this rapid accumulation, perhaps the most important was the drifting of this micro-plankton by currents from the open ocean into catchment areas of deeper water and into cul-de-sacs along the coast. J. P. Smith suggested that an oceanic circulation similar to that now active along the west coast of North America would have favored a drifting of diatoms by currents from the open ocean to the northwest and accumulation in cul-de-sacs in the Monterey sea.[39]

## LITHOLOGY OF THE FORMATION
### SILICEOUS ROCKS

The various types of siliceous rocks, all distinguished by a much higher content of silica than normal clastic shale or mudstone, constitute a large and characteristic part of the formation. The terminology used by Cayeux[40] for the many types of siliceous rocks described in his comprehensive volume on the siliceous rocks of France is not followed in this paper, because of the rather artificial distinction between many of the types, few of which are recognized by common American usage. As there is complete gradation between the various types of siliceous rocks and from them to the ordinary sedimentary rocks that have a normal content of silica, any classification within this gradational series must be rather arbitrary. An obvious distinction can be made only between the diatomaceous rocks, which are comparatively soft and contain abundant well preserved remains of diatoms, and the porcelaneous and cherty rocks, which are hard and cemented by silica, and in which well preserved diatoms are rare.

Chemical analyses of some of the various types of siliceous rocks are given in the following table. These indicate the high proportion of silica compared to alumina. An exception is the sample from Graciosa Ridge (column No. 3), which is more properly classed as a diatomaceous mudstone—a massive rock, much less than half of which generally consists of diatoms. This rock from Graciosa Ridge is now assigned to a formation unconformably overlying the Monterey, but similar diatomaceous mudstone is also common in the Monterey formation. The calcium carbonate which is present in some of the samples and abundant in a few is eliminated by recalculation, because it is clearly an admixed constituent and the object of making and tabulating the analyses was to compare the significant ratio of silica to alumina.

---

[2] Woodring, W.P., Stewart, Ralph, and Richards, R.W., Geology of the Kettleman Hills oil field, Calif.: U. S. Geol. Survey Prof. Paper 195, pp. 117-118, 1940.
[31] Reed, R. D., and Hollister, J. S., Structural evolution of southern California: Am. Assoc. Petroleum Geologists Bull., vol. 20, no. 12, figs. 17, 19, and 20, 1936.
[32] Trask, P.D., Sedimentation in the Channel Islands region, California: Econ. Geology, vol. 26, no. 1, pp. 24-43, 1931.

[39] Louderback, G. D., The Monterey series in California: Univ. California Pub., Dept. Geol. Sci. Bull., vol. 7, no. 10, p. 235, 1913.
[40] Cayeux, L., Les Roches Sedimentaires de France—Roches Siliceuses, Imprimerie Nationale, Paris, 1929.

## LITHOLOGY OF THE FORMATION

*Chemical analyses of siliceous rocks of the Monterey formation of California*

| | 1 | 2 | 3 | 4 | 5 | 6 | 7 | 8 | 9 | 10 | 11 |
|---|---|---|---|---|---|---|---|---|---|---|---|
| $SiO_2$ | 71.80 | 73.71 | 72.50 | 87.20 | 84.45 | 78.70 | 82.55 | 88.90 | 86.89 | 86.92 | 92.37 |
| $Al_2O_3$ | 5.02 | 7.25 | 11.71 | 1.86 | 4.14 | 5.83 | 4.82 | 2.28 | 2.32 | 4.27 | 2.46 |
| $Fe_2O_3$ | 2.45 | 2.63 | } 2.35 | 1.06 | 1.48 | 1.92 | .90 | .87 | } 1.28 | | |
| FeO | .35 | .44 | | .33 | .51 | .62 | .13 | .20 | | | |
| MgO | 1.69 | 1.47 | .83 | 1.14 | .52 | .71 | | | Trace | Trace | |
| CaO | [1]1.45 | 1.72 | .32 | [1]1.05 | [1]1.25 | [1]2.59 | 1.30 | 1.05 | 1.43 | 1.60 | 1.70 |
| $Na_2O$ | 1.81 | 1.19 | } 1.88 | .53 | .46 | .56 | } 1.09 | .49 | 3.58 | 2.48 | |
| $K_2O$ | 3.55 | 1.00 | | 2.75 | .64 | 1.06 | | | | | |
| $H_2O-$ | 4.64 | 2.88 | } 9.54 | 2.78 | 3.25 | 3.64 | 5.53 | 3.10 | } 4.89 | 5.13 | 2.74 |
| $H_2O+$ | 5.50 | 6.94 | | 2.23 | 3.11 | 4.26 | 2.90 | 2.80 | | | |
| $TiO_2$ | | .50 | | | .35 | .35 | | | | | |
| $CO_2$ | [1] | Trace? | | [1] | [1] | [1] | | | | | |
| $P_2O_5$ | .17 | .24 | | .22 | .28 | .34 | | | | | |
| $SO_3$ | | .16 | | | .18 | Trace? | | | | | |
| Cl | | | | | .16 | | .81 | .27 | | | |
| Organic carbon | Small | .00 | | Small | .12 | .23 | | | | | |
| | 99.77 | 100.13 | 99.13 | 100.31 | 100.74 | 100.81 | 100.03 | 100.06 | 100.39 | 100.40 | 99.27 |
| $CaCO_3$ (calculated[1]) | 9.20 | | | 5.16 | 13.51 | 29.1 | | | | | |

[1] $CO_2$ and a corresponding proportion of CaO to combine as $CaCO_3$ are calculated out of the composition.

1. Composite sample of 10 diatomaceous rocks of the Monterey formation. R. C. Wells, analyst.
2. Diatomaceous shale from road near Hollywood Country Club: U. S. Geol. Survey Prof. Paper 165-C, p. 108, 1931.
3. "Diatomaceous shale", Graciosa Ridge, 3 miles southeast of Orcutt, Santa Barbara County, Calif.: U. S. Geol. Survey Bull. 322, p. 45, 1907.
4. Composite sample of 10 cherty shales of the Monterey formation. R. C. Wells, analyst.
5. Cherty shale, Mulholland Highway, Santa Monica Mountains: U. S. Geol. Survey Prof. Paper 165-C, p. 108, 1931.
6. Cherty shale ("hard platy shale"), Mulholland Highway, Santa Monica Mountains: U. S. Geol. Survey Prof. Paper 165-C, p. 108, 1931.
7. "Fairly soft cherty shale", Palos Verdes Hills, Los Angeles County, Calif.: Unpublished dissertation at California Institute of Technology, by Hampton Smith. G. Eisenhauer of Los Angeles, analyst.
8. "Very hard cherty shale", Palos Verdes Hills, Los Angeles County, Calif.: Unpublished dissertation at California Institute of Technology, by Hampton Smith. G. Eisenhauer of Los Angeles, analyst.
9. "White shale", Monterey, Monterey County, Calif.: Univ. California, Dept. Geology Bull., vol. 1, p. 25, 1893.
10. "White porcelain shale", region of Point Sal, Santa Barbara County, California: Univ. California, Dept. Geology Bull., vol. 2, no. 1, p. 12, 1896.
11. "Opaque flint", Point Sal, Santa Barbara County, Calif.: Univ. California, Dept. Geology Bull., vol. 2, no. 1, p. 12, 1896.

### DIATOMACEOUS ROCKS

Diatomaceous rocks are common in the upper part of the Monterey formation in many areas, and the formation includes the thickest diatomaceous deposits known. These show all gradations from nearly pure deposits of diatoms, through diatomaceous shales, diatomaceous mudstones, and siltstones, to rocks without appreciable quantities of these siliceous organisms. The name diatomite is used for the purer diatomaceous rocks in commercial as well as geological usage, though the term implies no very definite degree of purity and is often used loosely by geologists for any of the soft "punky" rock in which diatoms are conspicuously present. Diatom shells are the predominant siliceous remains in these deposits, but the remains of radiolaria and silico-flagellates are also common. Sponge spicules are usually present, but are scarce except in some of the more silty deposits; where silt is abundant the spicules are commonly also abundant. A fine lamination is common, particularly in the purer diatomaceous deposits (pl. 5, A), though the thickness and distinctness of the laminae are variable.

The porosity of the purer diatomite is very high because of the minute pore spaces within and between the diatoms. Air-dried samples of this material have a specific gravity of only about 0.5, whereas the less pure diatomaceous shales and mudstones commonly have a specific gravity between 0.8 and 1.0. The silica content of a composite sample of ten diatomaceous rocks from different areas is 65.20 percent, the alumina content only 4.56 percent, and the $CaCO_3$ content 9.20 percent. The $CaCO_3$ is largely in Foraminifera. Some deposits, however, contain many beds of diatomite that are more nearly pure silica. Microscopic examination shows that some of this material contains only about 10 percent of impurity, which consists largely of minute clay particles and includes no calcium carbonate.

Most of the diatomaceous deposits are at present of no commercial value, the unusually pure deposits in a few areas being so large that the only important economic factor controlling production is the size of the market. Only three quarries have been operating during recent years in diatomite deposits of the Monterey formation, though several others have operated at times on a small scale. The properties and uses of diatomite have been described in many publications; one of the most comprehensive, which includes summaries of the known deposits, is that by Eardley-Wilmot[41]...

### PORCELANEOUS ROCKS

Porcelanite is the name used by Taliaferro[44] and others and adopted here for designating the silica-cemented rocks that are less hard, dense, and vitreous than chert. Such rock has minute pore spaces, which usually give it a dull or matte lustre resembling that of unglazed porcelain. Porcelanite is commonly of light color in surface exposures, but its range of color is great, and some of the variation is obviously due to bleaching and leaching effects of surface alteration. No color, therefore, is implied in the term. The rock is not conspicuously laminated or fissile, though it is commonly rather thin bedded; siliceous beds from less than one inch to several inches in thickness alternate with thin partings of less siliceous rock.

Porcelaneous rocks that are finely laminated or fissile are classed as porcelaneous shale, although all gradations to a non-fissile porcelanite are found, so that the distinction must commonly be arbitrary. The finely laminated porcelaneous rocks are not necessarily fissile and perhaps should not be classed as porcelaneous shale, but no distinction seems practicable, because the fissility of the laminated rocks is accentuated by surface leaching and weathering. By an increase in the proportion of clay and silt, the porcelanite grades into porcelaneous mudstone, which resemble crude pottery more than porcelain, and this in turn grades into normal, unsilicified mudstone. The porcelaneous rocks constitute the major part of the siliceous rocks of the formation in the type area near Monterey and in most other areas. Their intergradation makes it difficult in many areas to distinguish the several varieties of porcelaneous rocks, however, the distinctive terms can usefully be applied to particular strata in certain areas...

The porcelaneous rocks consist essentially of a mixture of clay or silty clay with a large but variable proportion of opaline silica. The chemical composition is accordingly variable, but the ratio of silica to alumina, as indicated in the analyses on page 13 is much higher than in ordinary clay shale or mudstone. Calcium carbonate, occurring in part as Foraminifera, varies greatly in amount; it may be considered an admixed accessory, best eliminated by recalculation in comparing analyses of the siliceous rocks. The relatively light weight of the porcelaneous rocks indicates their high porosity, though their pore spaces are very minute. The specific gravity of some representative samples range from 0.9 to 1.4, the lighter ones thus being no heavier than some of the less pure diatomaceous rocks. With decreasing pore space, there is complete gradation to the denser and more vitreous types classed as chert and cherty shale. Impressions or molds of the largest discoid diatoms, such as *Cosinodiscus*, are often observable with the hand lens in the porcelaneous rocks, and commonly are abundant on bedding surfaces of the more fissile porcelaneous shales. Examination of many of these samples in the laboratory shows that the original diatoms are not preserved in these rocks, but the molds are abundant. Microscopic examination, with high magnification, makes it obvious that much of the fine porosity of these porcelaneous rocks is due to the molds of various small diatoms and other siliceous organisms. The finer pore spaces are obscure, however, in thin sections, because of the impregnation of the fine pore spaces with Canada balsam and the intimate mixture of fine clay particles and opaline matrix; only by reheating the Canada balsam and introducing some air bubbles by lifting an edge of the cover glass are the shapes of the finer pore spaces, thus filled with air, made easily recognizable. In this way molds of even the very delicate forms, such as the silico-flagellates, may be seen.

The porcelaneous shale that shows the most abundant impressions of diatoms is usually of a markedly fissile type, relatively soft, porous, and of light weight, and is therefore classed by some field geologists as diatomaceous shale. But the distinction from true diatomaceous shale containing well-preserved diatoms is ordinarily not difficult to make in the field with a hand lens, and the distinction is obviously significant. The so-called "poker chip shale" of the drillers is a platy porcelaneous shale which splits from the cylindrical core samples into thin discs. These core samples are generally dark in color, largely because of organic matter, and so are the porcelaneous rocks of surface samples that have been exposed for so short a time as to have undergone little surface alteration or weathering (p. 36).

Calcareous Foraminifera range from abundant to entirely absent in the porcelaneous rocks, as in other types of siliceous rocks of the Monterey formation. In some of the porcelaneous rocks the Foraminifera originally present have been leached and occur only as molds. This leaching seems to have been effected largely by ground waters related to the present topographic surface, and occurs more commonly in the porcelaneous than in the cherty rocks. Fish scales are abundant in the porcelaneous rocks, as in the other types of siliceous rocks. Further details of similarity and contrast with the other siliceous rocks, which bear on the origin of these rocks, will be discussed later.

---

[41] Eardley-Wilmot, V. L., Diatomite, its occurrence, preparation, and uses: Canada Dept. Mines, Mines Branch, Pub. no. 691, 1928.
[44] Taliaferro, N. L., Contraction phenomena in cherts: Geol. Soc. America Bull., vol. 45, p. 196, 1934.

### CHERT AND CHERTY SHALE

The name chert is applied to the relatively pure silica rocks of the Monterey formation that are dense and vitreous, regardless of whether they consist mainly of opal or mainly of chalcedony, and regardless of their color, which varies greatly. Some of the darker chert resembles rock that has been termed flint, but there seems to be no satisfactory basis for the definition of flint, and the more widely used term chert seems preferable for all such rocks in the Monterey formation. The chert is generally thin-bedded and has partings of clay shale, carbonate (usually dolomite) or more rarely sandstone. The chert beds, in general ranging from less than one inch to several inches in thickness, usually show a fine lamination (pl. 5, D) but little or no tendency to part along the laminae. The term cherty shale is used for somewhat less pure chert rocks in which some of the fine laminae contain enough non-siliceous material to make the rock split into plates. The cherty shale is almost as dense and vitreous as the chert, and it is not commonly practicable to distinguish the two, because of the large influence that the degree of weathering seems to have on the extent of parting along laminae. Weathering also tends to make the cherty shale resemble porcelaneous shale in some surface outcrops. The thin-bedded chert and cherty shale are similar to rocks termed Kieselschiefer in Germany and phthanite in France.

The partings of shale or other rock are usually more prominent in the thin-bedded chert and cherty shale than in the porcelaneous rocks and diatomaceous rocks; the partings in them are not uncommonly as thick as the chert beds and occasionally are thicker. The beds of chert and cherty shale are usually persistent and regular as far as any continuous exposure extends (pl. 6, A), which is a hundred feet or more in some outcrops, but in places they are lenticular or vary greatly in thickness, as in Claremont Canyon (pl. 6, C), and in the sea cliffs northwest of Pismo (pl. 6, B).

The chert and cherty shale usually contain a larger proportion of chalcedonic silica and cryptocrystalline quartz than the porcelaneous rocks, but in general they are dominantly composed of opaline silica. Some of the purer cherts are more than 90 percent silica, as indicated in the chemical analyses on page 13. The interiors of Foraminifera and of the larger diatoms, and other original pore spaces, have been filled with chalcedony and cryptocrystalline quartz in much of the cherty rock, whose density stands in contrast with the high porosity of the porcelaneous rocks; and dark-colored chalcedonic silica commonly fills small cross-cutting fissures and partings along bedding planes in the cherty rocks. In some specimens of the laminated chert and cherty shale, an occasional lamina about 0.1 millimeter thick of nearly amorphous but minutely granular material, which seems to have been originally an especially clayey lamina, contains delicate diatoms preserved in their original opaline state.

In the lower part of some of the thickest Monterey sections, as in those of Chico Martinez Creek and Bixby Canyon, the silica matrix of the cherty shale and chert is largely chalcedony and cryptocrystalline quartz rather than opal. Some of the calcareous Foraminifera in these rocks have been replaced by silica, and their outlines are made visible only by slight differences in index of refraction between the chalcedony and the cryptocrystalline quartz…

### THE SILICA MINERALS OF THE SILICEOUS ROCKS

The most abundant form of silica in the Monterey siliceous rocks is opal, though chalcedony is common and cryptocrystalline quartz is usually associated with it. No entirely reliable means of distinguishing, in the feld, between the opaline and chalcedonic silica was found, but the predominance of opal is evident from microscopic study of a large number of thin sections and of a yet larger number of thin slivers chipped from samples. A slight difference in lustre between the opal and chalcedony is usually observable, and the mode of occurrence is in part suggestive; chalcedony more commonly has filled openings and fractures, though veinlets of opal occur also. Color apparently has no significance, for both chalcedony and opal range from nearly black to nearly white.

Only three distinct types, grouped as opal, chalcedony, and quartz, are distinguished, with no attempt to differentiate minerals within the chalcedonic type. Some chalcedony (quartzine) elongated perpendicular to the C axis was recognized, and perhaps other forms of silica that are distinguished optically, especially by French petrographers, are present. These varieties are, however, far less abundant in these rocks than normal chalcedony, and are closely associated with it insofar as they occur; just which of these varieties are present, and how abundant each one is, therefore, seems unimportant. The greater stability of the less hydrous forms of silica probably accounts for the relative scarcity of opal as compared with chalcedony and quartz in preTertiary siliceous formations, particularly in those of Paleozoic age; and this tendency of opal to alter to the more stable forms of silica seems to be well shown in the Monterey formation.

The term opal is used to designate all of the apparently amorphous and isotropic silica, though it is now recognized that opal commonly includes small quantities of cristobalite, which can be determined only by X-ray diffraction patterns…

Chalcedony and cryptocrystalline quartz, though distinctly less abundant than opal in most of the Monterey rocks, are the dominant silica minerals in the cherty shale of the lower part of the formation in some areas where the formation is unusually thick, as on Chico Martinez Creek and in Bixby Canyon. In the chert and cherty shale of Claremont Canyon, the rocks seem to be exceptionally altered and indurated, as might be expected from their occurrence in almost vertical strata next to a

# THE MONTEREY FORMATION OF CALIFORNIA AND THE ORIGIN OF ITS SILICEOUS ROCKS

fault of large throw,[53] and they there contain more than the usual proportion of chalcedony and quartz...

### CLASTIC SHALE, MUDSTONE, AND SANDSTONE

Although the characteristic siliceous rocks constitute most of the Monterey formation, large quantities of normal clastic rocks are interbedded with them in a number of areas, and where the siliceous rocks grade laterally into strata consisting dominantly of these clastic rocks the name Monterey is not appropriate. Besides the decidedly siliceous rocks and the normal clastic rocks, there are intermediate or gradational rocks, such as the porcelaneous mudstone, which may form such large stratigraphic units as the McLure shale member of the Monterey formation. These intermediate rocks are commonly very fine-grained, but in a few places the siliceous rocks grade into sandstone. In the lower part of the Pismo formation, for example, the sandstones are commonly opal-cemented and grade into beds of sandy chert (p. 6). As the normal clastic rocks of the Monterey formation are not, in general, significantly different from the usual sedimentary rocks of these types, little discussion of them is necessary here, though certain significant features of their bedding are considered in some detail in connection with rhythmic bedding...

### CARBONATE BEDS AND CONCRETIONS

Limestone is remarkably scarce in the thick Tertiary deposits of California, but impure calcareous and dolomitic rocks occur as thin beds, or more commonly as concretions, in a large part of the Monterey formation. In an area including parts of Santa Barbara and San Luis Obispo Counties, a limestone member constitutes the lower part of the formation and is rather thick and conspicuous in some localities,[59] though it is thin or absent in intermediate areas. This limestone occurs near the contact of the lower and middle Miocene foraminiferal zones, and thin beds of less pure limestone and dolomite are common at various horizons in the middle Miocene part of the formation in many areas. Most of the calcium carbonate in the formation, however, is disseminated, and much of it occurs as Foraminifera shells, which form the greater part of some thin beds. In fresh exposures, as in the sea cliffs along the coast of the Palos Verdes Hills area and in the measured section near Naples, impure carbonate beds are unusually conspicuous and form hard reefs a foot or more in thickness, alternating with thicker zones that contain much less carbonate. In most of the hard beds the shells of Foraminifera are relatively rare and the carbonate occurs largely as rhombic crystals. Much of the carbonate is more or less magnesian, and part of it is dolomite. The varying proportions of calcite and dolomite in concretions and beds make it necessary to use the broad term carbonate rock, especially since it is often difficult, because of the obscuring effect of silicification, to distinguish the carbonates in the field by applying hydrochloric acid. Additional data on the calcium-magnesium ratios in these rocks, with relation to their areal and stratigraphic distribution, might well prove significant. Carbonate with considerable iron seems to be relatively scarce, though some of the cherty rocks near intrusive contacts of basalt contain crystals of ankerite, obviously formed as a result of contact alteration...

Although these concretions are secondary, it seems clear that they were formed during the early stages of compaction and lithification and may thus be termed diagenetic products. The fact that the concretions occur at definite stratigraphic horizons indicates that they are not late secondary developments, related to present topographic and ground-water conditions. The individual beds are thicker in the concretions than in adjacent strata, and thus give evidence that the concretions were formed before the compaction of the beds was far advanced. When the enclosing rock is fine-grained, the beds are usually about three to four times as thick in the concretions as in the adjacent rock, but where the enclosing rock is sandy there is less difference in thickness...

### PYROCLASTIC MATERIAL IN THE MONTEREY FORMATION

Pyroclastic material occurs in much of the Monterey formation. It consists of unaltered vitric ash or tuff beds, partially altered tuffaceous beds, and the more thoroughly altered tuffaceous material known as bentonite, composed largely of the clay mineral montmorillonite. Some pyroclastic material is also disseminated in various kinds of sedimentary rocks. The character and stratigraphic relations of these pyroclastic materials not only bear on the problem of the origin of the siliceous rocks in the formation, but offer some additional information on the ways in which tuffaceous deposits may become altered. When more is known regarding their areal and vertical distribution, they should help in the making of more accurate stratigraphic correlations.

\* \* \* \* \*

## RHYTHMIC BEDDING

A conspicuous feature of the Monterey formation in many areas is a rhythmic bedding that shows remarkably well the regular recurrence or alternation in the character of the sedimentary deposition. The Pliocene of California also shows rhythmic bedding in some areas, as in the thick deposits of the Ventura basin, and such bedding is common in the thick Cretaceous sediments in the Coast Range[85] The rhythmic bedding in the Monterey formation seems worthy of being discussed in some

---

[53] Lawson, A. C., U. S. Geol. Survey Geol. Atlas, San Francisco folio (no. 193), 1914.
[59] Arnold, Ralph, and Anderson, Robert, Geology and oil resources of the Santa Maria oil district, Santa Barbara County, Calif.: U. S. Geol. Survey Bull. 322, p. 34, 1907.

[85] Reed, R. D., Geology of California, p. 105, Am. Assoc. Petroleum Geologists, 1933.

Plate 14. A. Larger alternation superimposed on the thinner rhythmic bedding on the Topanga Canyon Road south of Girard, Los Angeles County. Note sandstone dike that is offset along bedding plane slips. B. Details of bedding at same locality. Note change from rhythmites that are dominantly siliceous rock to those of a zone in which they are dominantly clastic sediment. Scale is almost 6 1/2 inches long.

detail. Even though the discussion may lead to no final conclusions regarding the processes or geological agencies of periodic character that produce this type of bedding, the formation illustrates particularly well many features that should throw some light on the origin of such bedding that is likewise characteristic of many other formations.

Several distinct types of rhythmic bedding are indicated by the superposition of those of relatively great thickness upon others of much less thickness (pl. 14, *A*). The thinnest laminae are to be measured in fractions of a millimeter. In general, however, beds about one inch thick are most conspicuous and best show the characteristic features of the rhythmic bedding, and they are therefore described first. This type of rhythmic bedding is well developed along the north side of the Santa Monica Mountains, and is well exposed there on the many road cuts in the small canyons south of Ventura Boulevard...

These couplets of a clastic layer and an organic layer, each a unit of rhythmic bedding, indicate a repeated process, initiated rather abruptly by the accumulation of clastic material, which gradually decreased in grain size and was succeeded by the accumulation of more dominantly organic sediment, apparently deposited in clearer water.

These beds resemble varves formed in glacial lakes, but though the term varve may be applied to marine beds,[86] as its definition implies only a seasonal sequence of sedimentary accumulation forming annual deposits, a broader term, without any time connotation, is needed for similar beds whose mode of formation is uncertain. The word "rhythmite," used by Sander[87] for the individual units of rhythmic beds, will be adopted here as a brief term to designate the couplet of distinct sedimentary types of rock, or the graded sequence of sediments, that form a unit bed or lamina in rhythmically bedded deposits. No definite limit on the thickness of the bed or lamina called a rhythmite seems desirable, but it seems preferable not to let the term include those more complex units, consisting of several distinct beds or groups of beds, which have been termed cyclothems.[88] A varve would be a rhythmite of a special class, known or believed to consist of the sediment deposited in a year.

The rhythmites illustrated in plate 13 show well the character of the recurrent sequence of sedimentary accumulation. The lower part of each rhythmite consists of clastic material, whose base is sharply defined, though there is seldom any evidence that the underlying deposit has been appreciably scoured. The grain size of the clastic material decreases upward, and the clastic material is overlain by sediment that is dominantly organic. The presence of a thin layer of mud just below the base of the organic layer indicates that the organic material was deposited in clearer water than the mud, rather than that the rate of deposition of organic sediment was increased. Some significant variations in the character of the rhythmites should be described, however, before considering possible causes of such a recurrent process...

Superimposed on the rhythmic sequence of beds one or two inches thick is a much thicker cyclic bedding which, though conspicuously developed in relatively few places, has a significant bearing on the origin of the several types of rhythmic bedding. It is well seen in road-cut exposures in a canyon just east of Beverly Glen Boulevard, on the north side of the Santa Monica Mountains. Several hundred feet of strata here show a regular alternation of zones five to eight feet thick. Zones including about forty to fifty rhythmites consisting dominantly of fine clastic material alternate with zones including a similar number of rhythmites that consist dominantly of organic sediment...

Very fine laminae, usually a small fraction of a millimeter in thickness, are common in both the diatomaceous and the harder siliceous rocks of the formation, and these fine laminae also represent a type of rhythmite, as they show a definite alternation of layers that contain abundant organic matter with those that contain less. Lamination that apparently is only a fissility due to orientation of the sedimentary particles parallel with the bedding—a kind of lamination that tends to form in many fine-grained sediments during their accumulation and compaction[90]—is also common in the Monterey formation, but it is not to be confused with the fine laminae illustrated in plate 5, *A* and *D*, which represent a rhythmic alternation of different kinds of sediment. In thin sections (pl. 15, *B*) and thin chips of the fresher siliceous rocks, it may be seen that the laminae consist of thin layers relatively rich in brown organic matter alternating with layers, usually somewhat thicker, that contain relatively little organic matter. The couplets forming this type of rhythmite are variable in thickness, but their thickness is generally a small fraction of a millimeter and probably averages between 0.1 and 0.2 millimeter. They appear identical with the rhythmic laminae interpreted as marine varves by Bradley.[91] Indeed, one specimen from Los Angeles County examined and so classed by him was from the Modelo formation, now called Monterey.

The fine rhythmic lamination is common in the diatomaceous deposits, but is more apparent in the cherty rocks of the formation, particularly those that also show the larger type of rhythmic bedding. These thin rhythmites are of a distinctly lower order of magnitude from those that average between one and two inches—many of these finer rhythmic laminae occur within a single one of the larger rhythmites—and it is thus evident that they were formed by a different process. The thin laminae, the more conspicuous one- to two-inch rhythmites, and the alternations several feet thick represent three superposed rhythms.

---

[86] Bradley, W. H., Nonglacial marine varves: Am. Jour. Sci., 5th ser., vol. 22, p. 318, 1931.
[87] Sander, Bruno, Beitrage zur Kenntnis der Anlagerungsgefuge: Mineralogische und petrographische Mitteilungen, vol. 48, pp. 27-139, Leipzig, 1936.
[88] Wanless, H.R., and Weller, J. M., Correlation and extent of Pennsylvania cyclothems: Geol. Soc. America Bull., vol. 43, no. 4, p. 1003, 1932.

[90] Lewis, J. V., Fissility of shale and its relation to petroleum: Geol. Soc. America Bull., vol. 35, no. 3, pp. 557-590, 1924.
[91] Bradley, W. H., op. cit., p. 324.

# THE MONTEREY FORMATION OF CALIFORNIA AND THE ORIGIN OF ITS SILICEOUS ROCKS

The remarkably clear-cut sequence of sedimentation shown in some of the rhythmic bedding in the Monterey formation, the variations in its character, and the superposition of different rhythms, all seem to offer an unusually good opportunity for determining the mode of formation of such bedding. Similar bedding is common in many other formations, but in them it seldom shows such well-defined and varied characteristics. A diagenetic or later rearrangement of the originally deposited sediments seems to have played a part in the formation of some types of rhythmic bedding, but the bedding here described shows textural gradations that obviously cannot be thus accounted for and that must be due to the nature of the depositional process.

It is in the rhythmites an inch or two in thickness that the characteristic features of such beds are especially obvious; these rhythmites, therefore, seem to offer the best opportunity to ascertain what processes were involved in their formation. They show clearly a repeated sedimentary sequence that involves a periodic influx of clastic material to the area of deposition; this influx gradually decreases as the grain size of the material decreases, and is usually succeeded by an increase in the proportion of organic remains. A new cycle is then begun by the next influx of clastic material. The general lack of disturbance of the underlying layers, illustrated in plate 13, *B, C*, indicates that each layer was formed by an actual influx of additional sediment to the area rather than a mere agitation and resettling of bottom sediment. It seems evident that these beds were deposited below depths of much wave or current action, for each bed commonly shows remarkable uniformity of both character and thickness in exposures that may be traced continuously for distances of a hundred feet or more. Sudden pinching or cutting out of beds, ripple marks, and signs of scour by waves or currents are uncommon. Plate 15, *C*, illustrates an exceptional example of scour. Only in the rhythmic beds that are dominantly sandy is such scour and ripplemarking fairly common (see pl. 12, *C*), and here the wave or current action was not sufficient to prevent the accumulation and preservation of thin beds with graded texture.

Processes that might produce rhythmic bedding have been discussed by a number of writers. These have been reviewed and analyzed by Rubey,[92] and they will therefore be considered here only briefly, with regard to their possible application to the origin of such bedding in the Monterey. Among the more definitely periodic processes, those involved in the annual change of seasons or of several longer climatic cycles are recognized. Other processes that seem capable of producing the individual graded beds include exceptionally heavy storms, shiftings of ocean currents, unusually heavy rain storms on the adjacent land areas, and slumping of bottom sediment on steep submarine slopes. In these processes, however, no long-continued periodicity has been recognized, nor would it be expected. Topographic changes or earth movements are obviously inapplicable as a cause of the repeated sequence of such thin beds as those considered here.

Since the beds generally show little or no sign of scour, storm waves could not have produced this bedding through a mere stirring up of the bottom and resettling of sediments, and the upper, chiefly organic, layer of one bed would not supply enough sand through such wave disturbance to form the succeeding bed. But where the sea floor was of irregular topography, as it is off the present coast of California in the Channel Islands region, and conditions of sedimentation were essentially as described in that area by Trask,[93] unusual storm waves might act on some of the ridges down to depths not ordinarily affected, carry off relatively coarse sediment, and drop it from suspension in adjacent waters of greater depth. Such a process could perhaps produce the type of graded bed described, and other evidence (p. 11) suggests that such physiographic conditions on the sea floor were not improbable in the Monterey sea. However, the regular repetition of these beds of nearly uniform character and thickness would require a periodicity in the intervals between larger storms that does not exist at present. A more probable effect of storm wave action would be to form sandy layers, alternating with layers of finer or more organic sediment, at rather widely varying intervals.

Rhythmic bedding might likewise be produced by water flowing over an uneven sea floor similar to that postulated for the action of storm waves. Under these conditions, the ridges might supply relatively coarse sediment that the submarine currents would carry on and drop in nearby basins of deeper water. Off the coast of South America there is some evidence of a periodic shifting of current,[94] but it seems improbable that currents could shift in particular areas for a long period of geologic time with such regularity as is indicated in the rhythmites of the Monterey formation.

Periods of unusually heavy rainfall on adjacent lands, other than those determined by annual or other climatic cycles, would also presumably come at too irregular intervals to produce this regular and uniform bedding. Even greater irregularity would be expected in submarine slides or slumping of sediments on the steeper slopes of a basin.

The annual change of seasons is an established cause of rhythmic bedding in glacial lake deposits. Varves are forming, because of seasonal changes, in some modern lakes where the relationship to glacial action seems a minor if not negligible factor. Nipkow[95] seems to have demonstrated very clearly that the laminated sediments in some of the Swiss lakes are varves formed by seasonal changes. Stagnant parts of some of the Norwegian fiords are also accumulating laminated sediment that Strøm[96]

---

[92] Rubey, W. W., Lithologic studies of fine-grained Upper Cretaceous sedimentary rocks of the Black Hills region: U. S. Geol. Survey Prof. Paper 165, p. 40, 1930.

[93] Trask, P. D., Sedimentation in the Channel Islands region, California: Econ. Geology, vol. 26, no. 1. pp. 24–43, 1931.
[94] Murphy, R. C., Oceanic and climatic phenomena along the west coast of South America during 1925: Geogr. Rev., vol. 16, pp. 26-54 1926.
[95] Nipkow, F., Uber das verhalten der Skelette planktisher Kieselalgen in geschichteten Tiefenschlamm des Zurich- und Baldeggersees: Revue d'Hydrologie Annales vol. 4, nos. 1-2, pp. 71-120, 1927.
[96] Strøm, K. M., Land-locked waters: Skrifter utgitt av Det Norske vidensk.-akad. Oslo, Mat.-Naturv. Klasse, no. 7, pp. 67-69, 1936.

has interpreted as varves. In a core from the little-compacted sediment of Drammensfjord, these varves average about 2.6 mm. in thickness in the upper 130 mm. of the core, about 0.8 mm. in the lower part, and about 1 mm. for the whole core. Similar varves might form in more or less stagnant marine basins under certain conditions. In a region of marked seasonal rainfall, the rivers would carry out far greater loads of clastic material during the rainy season, when they were swollen, than during the dry season. Comparatively little of the sediment would be carried in suspension into deep sea water, but some of the finer portion might be thus transported to relatively deep basins near shore. The gradual decrease in the volume of the rivers and their load of sediment in the dry season would be accompanied by a relative increase of organic material. The uniformity of the rhythmic bedding therefore seems most plausibly explained as the result of such a seasonal weather cycle, or of longer climatic cycles.

These long cycles are of course less well established than the seasonal ones; several of them, however, have been suggested as explaining certain types of rhythmic bedding, among them being the Brückner cycle of about 33 years, the 21,000 year cycle of the precession of the equinoxes, and others. Rhythmic bedding in a Jurassic formation of the Alps was attributed by Winkler[97] to some one of these longer climatic cycles, because he believed the annual cycle too brief for the deposition of beds about 7 centimeters thick. The precessional period of about 21,000 years was suggested by Gilbert[98] for a cyclic sequence with alternations several feet thick in the Cretaceous of Colorado. The annual climatic cycle, however, is the only one that has yet been definitely shown to be of sufficient intensity to produce a clear record in the rocks deposited in certain periods and areas. The thickness of the rhythmic beds, within a range of a small fraction of a millimeter to several centimeters, seems to be an inadequate criterion for distinguishing an annual from some longer climatic cycle; varying conditions might cause annual deposits, even in marine sediments, to differ that much in thickness. Pleistocene glacial varves at least 14 centimeters in thickness have been described,[99] but they could hardly be formed in marine waters except possibly in very localized accumulations.

Even so much as an inch of marine deposition in a year would seem possible only under exceptional and localized conditions—where, for example, a small basin of accumulation is supplied through by-passing of sediment from large areas of shallower water. Rough calculations from the measured load of sediment in the Colorado river indicate that this river system carries about 2,180 tons of sediment per year for each square mile of drainage area, or about 43,600 cubic feet, figured at 20 cubic feet per ton. Assuming that all this sediment is deposited over an area equal to that drained by the river, it would form a layer .019 inch thick each year. An annual deposit one inch thick would therefore require a localized area of accumulation about 1/60 as large as the drainage area. The wide distribution of the Monterey formation seems inconsistent with such extremely localized accumulation, and, conversely, the assumption of a very extensive area of supply for the sediments seems inconsistent with the hypothesis of a disintegrated drainage system during much of Miocene time (p. 12). Though diatomaceous deposits from a large part of the Monterey, they apparently accumulated at a rate not far different from that of the clastic sediments (p. 38).

But if the rhythmic beds that are one or two inches thick are assumed to be annual deposits, the accumulation of the entire 3,000 feet of Monterey strata on the north flank of the Santa Monica Mountains, which represent upper Miocene time, would have required only about 24,000 years. This figure does not appear to be even of the right order of magnitude; and it cannot be very greatly increased, for in no more than a small part of these strata is the rhythmic bedding so obscure that it might represent a much slower rate of deposition, nor does the section show evidence of long periods of non-deposition. It thus appears improbable that these most conspicuous rhythmites are annual deposits, however strongly their regularity and sharp definition suggest a correlation with the annual weather cycle with which we are familiar.

We therefore seem led to the conclusion that the annual cycles are represented by the thinnest rhythmic beds or laminae, whose average thickness lies between 0.1 and 0.2 mm. If these are assumed to be annual deposits, the 3,000 feet of upper Miocene strata would represent about 7,000,000 years if all the strata were laminated. Actually, not more than about one-third shows this lamination, because of the interbedded sand, which presumably accumulated more rapidly, so that about 2,000,000 or 3,000,000 years would probably be a closer estimate. This is of the right order of magnitude, at least, for the latest estimates indicate that the whole of Miocene time amounted to about 15,000,000 or 20,000,000 years.[1] ...

If the thinnest laminae represent annual deposits it is necessary to account for the thicker and more conspicuous rhythmites by some cyclic process whose period was of the order of magnitude of 100 years, and that was more conspicuous in its effect on the accumulating sediments than the annual weather cycle. A year or two with very extraordinary rains and floods occurs occasionally along the west coast of South America. The rainy periods appear to be related to a shift of the marine currents off the coast,[3] and there is a slight suggestion that they are periodic; the last three, at least, have occurred at intervals of about 35 years. Some such climatic change, if it occurred at fairly regular intervals near 100 years, might account for the one- to two-inch rhythmites in which the lower part consists of graded clastic sediment and the upper part consists of finely-

---

[97] Winkler, A., Zum Schichtungsproblem; Ein Beitrag aus den Sudalpen: Neues Jahrb., Beilage-Band 53, Abt. B, p. 309, 1926.
[98] Gilbert, G. K., Sedimentary measurement of Cretaceous time: Jour. Geology, vol. 3, pp. 121-127, 1895.
[99] Sauramo, Matti, Studies on the Quaternary varve sediments in Southern Finland: Comm. geol. Finlande Bull. 60, p. 85, 1923.

[1] Urry, W. D., Ages by the helium method: Geol. Soc. America Bull., vol. 47, no. 8, p. 1229, 1936.
[3] Murphy, R. C., op. cit., pp. 26-54.

# THE MONTEREY FORMATION OF CALIFORNIA AND THE ORIGIN OF ITS SILICEOUS ROCKS

laminated organic sediment. Incidentally, the steep submarine slope along the coast of Ecuador and Peru, with deep water relatively near land, would seem a particularly favorable environment for the formation of such beds. The yet larger cyclic alternation evident in some parts of the Monterey seems to be the result of some more gradual climatic oscillation, with a period of perhaps 2,000 or 3,000 years, for it is represented by about that many of the thinnest rhythmic beds...

### EVIDENCE ON CONDITIONS IN DEPOSITIONAL BASINS

Although the precise nature of the periodic processes that caused the rhythmic bedding in the Monterey formation is problematic, the mere presence of such bedding marks a distinctive type of sedimentary accumulation, which contrasts in many important features with the more common type that has either massive or irregular bedding. The formation and preservation of the rhythmic bedding implies deposition below wave base and below any marked current action, though not necessarily in very deep water, for the amount of water agitation and movement is often more closely related to local topographic irregularities of the bottom than to absolute depth. A great thickness of rhythmically bedded sediment must have accumulated in a basin that either was originally very deep or that underwent long-continued subsidence. The rate of subsidence need not have kept in close balance with the rate of deposition, but must merely have been so rapid that the basin was generally not filled with sediment up to the level of wave and current action...

The deposits that show regular rhythmic bedding were obviously not subjected to much scour and by-passing of sediment; they represent a maximum of accumulation, limited only by the rate of supply of sediment to the basin of deposition. Under these conditions of deposition the mollusks and other benthonic organisms would have a somewhat unfavorable habitat, for water below wave or current action would be poorly aerated and there might be some smothering of the less actively moving bottom dwellers by uncompacted ooze, which would be abundant on a sea floor where there was no by-passing of the finest sediment. Such conditions may account for the paucity of macro-fossils in the Monterey and other rhythmically bedded formations. The San Pablo group and the Santa Margarita sandstone of central California represent a different facies of the upper Miocene deposits, wherein the sandstones do not show graded bedding but are characteristically massive or irregularly bedded, and these strata contain more macro-fossils.

Larger fossils are thus apt to be relatively rare under the conditions of deposition below wave base or much current action, but micro-fossils are often abundant, because the lack of water agitation favors the accumulation of the small pelagic organisms, which are not swept on to greater depths. The remains of the abundant plankton organisms that had no hard parts and that are therefore not apparent as fossils would accumulate with the diatoms and other micro-fossils. The lack of aeration would largely prevent oxidation and destruction of this organic matter. Conditions resulting in the formation and preservation of rhythmic bedding or lamination would thus also be favorable for the accumulation of deposits that, being unusually rich in organic matter, may constitute source beds of petroleum (p. 37).

Rhythmic bedding or lamination is common in thick formations in regions of long-continued crustal instability such as the coastal region of California, and in regions of geosynclinal deposition such as the former seaway of Tethys in Eurasia. Regions of less crustal unrest, in which there have been few or no deep and rapidly subsiding basins of sedimentary deposition, are illustrated by the Atlantic Coastal Plain. There the Tertiary sediments are massive or irregularly bedded, having evidently accumulated in shallow seas on the continental shelf, where by-passing and discontinuous deposition were usual. The contrast between the two types of formation with respect to character of bedding has been emphasized by Bailey;[6] and the contrast in their fossil content and other features likewise indicates that they were deposited under very different conditions, which reflect the tectonic history.

## SOURCE BEDS OF PETROLEUM

Many of the earlier papers on the Monterey formation describe it as consisting in large part of bituminous shale, and most of the oil fields that were discovered early were in Monterey strata or closely associated with them. These relationships apparently led to an early view that the Monterey was the only important source of petroleum in California, and its importance was thus overemphasized. The many questions that remain unsettled about such factors as the extent of lateral and vertical migration of petroleum make it difficult to evaluate the importance of particular strata as a possible source of oil, but the discovery of many important oil fields in thick Pliocene strata, particularly in the Los Angeles and Ventura Basins, has made it seem probable that these Pliocene rocks are the source of much or most of the oil within them. Oil accumulations in pre-Miocene formations of the California Tertiary are also becoming increasingly well known. These discoveries, together with a natural reaction against the early view, may have tended for a time to depreciate unduly the importance of the Monterey as a source of oil. The fact seems to remain that a great deal of oil originated in this formation, and it certainly contains many of the greatest oil pools in California...

Strata that may be termed bituminous shale because of their high content of organic matter are common in parts of the Monterey formation in several areas, and other parts of the formation are commonly much higher in organic matter than might be supposed from their appearance when weathered. The dark color usual in the

---

[6] Bailey, E. B., New light on sedimentation and tectonics: Geol. Mag., vol. 67, p. 77, 1930.

unweathered porcelaneous and cherty shales of the formation is due primarily to a high content of organic matter, but weathering leaches or oxidizes much of the organic matter, so that the rocks become light-colored. Thin sections of the dark, unweathered siliceous rocks show abundant small areas of dark-brown organic matter, either disseminated or largely concentrated in definite laminae, but organic matter is less conspicuous or largely absent in weathered rock. Where the dense opaline rocks have been weathered only in a thin outer zone (pl. 17, C), individual laminae also show this local removal of organic matter by weathering. When the dark and fresher rocks are leached with sodium or potassium hydroxide solution, they are lightened as they are by weathering (pl. 5, D), and the caustic solution is darkened with the organic matter that has been extracted. This result, together with study of thin sections, indicates that the dark color is due in major part to organic matter. Black iron sulphide is noted in thin sections, but is usually much less abundant than recognizable organic matter, though the two materials are apt to be associated and thus combine to cause the dark color. The bleaching effected by caustic solutions and by weathering indicates that organic matter is generally more abundant than iron sulphide in these rocks, and the buff stains—due to iron oxide derived from the iron sulphide—that are common on the weathered surfaces of many dark strata are not usually conspicuous in the light-weathering porcelaneous rocks.

Perhaps some of the porcelaneous and cherty rocks of the formation were originally of light color, because of their containing little admixed organic matter at the time they were deposited, but the surface leaching and bleaching of originally darker rocks is evidently responsible for much of the nearly white to mauve-colored rock that is so common in surface exposures. For example, the Monterey rocks as mapped at the surface in the Palos Verdes Hills are generally light colored and do not appear to be particularly high in organic matter, but where these same strata were encountered in the tunnel recently driven through the hills for the Los Angeles sewage-disposal project they were all dark gray to almost black, and it is found by study of thin sections and chips of these fresh rocks that the color is largely due to organic matter. Heavy tar or asphalt is common in the fractured siliceous rocks in the tunnel, though it is not conspicuous at the surface. The beds of cherty and porcelaneous rock in the lower parts of sea-cliff exposures that are being actively eroded by the waves are likewise relatively dark, as they contain more brown organic matter than the equivalent but more thoroughly weathered beds that extend up the cliffs to old terrace surfaces, and the dark rock is found, by study of rock chips and thin sections, to contain much organic matter. Localities on the coast at which these relations were observed include San Juan Capistrano Point, Palos Verdes Hills, Naples, and Point Concepcion; other such localities where sections were not measured include South Point west of Pismo, Point Buchon, a locality near Santa Cruz, and a locality near Point Reyes. The same relation is noted in some narrow deep canyons, where erosion has been so rapid as to expose beds in the canyon bottoms that have not been conspicuously weathered. Drill cores from many areas show that the cherty and porcelaneous Monterey rocks are consistently darker far below the surface than they are in the surface exposures directly above.

The diatomaceous shales, also, are usually darker in unweathered than in weathered exposures, as may be seen in the cliffs at Malaga Cove in the vicinity of Palos Verdes Hills, and at Naples, but this difference is due in part to saturation with water; samples of this rock are a much lighter gray when dried than when moist, though not as nearly white as the more weathered diatomaceous deposits. Although he believed that diatoms may be in part the source of Monterey oil, Tolman noted that the diatomaceous members in general contain less organic matter than much of the harder siliceous rock. This may be due in part to the high porosity and permeability of the diatomaceous deposits, which facilitates leaching of the original organic material by surface or ground-water solutions. Many beds and lentils of opaline chert are dark from their organic content, though immediately adjacent and equivalent beds in the porous diatomite are nearly white and obviously contain very little organic matter (pl. 7, A). It seems probable, however, that little of the diatomaceous shale was ever as rich in organic matter as the porcelaneous and cherty rocks commonly called bituminous shale, which occur lower in the Monterey formation in certain areas. These bituminous, siliceous shales are believed to have been originally diatomaceous (p. 50), but the importance of the diatoms as a source of the bituminous matter remains an open question, for abundant microplankton organisms without hard parts would also be expected to accumulate under the same conditions as the diatoms.

The strata that are most conspicuously bituminous and that show the most organic matter in thin sections are in general the distinctly laminated porcelaneous and cherty shales. The sands that are impregnated with free oil or its inspissated products are obviously to be excepted; they are reservoirs in which oil has accumulated, rather than source beds. Fine lamination or other thin rhythmic bedding may, as previously pointed out, be a common characteristic of source beds of petroleum, since the conditions that allowed these features to be formed and preserved would be favorable for the accumulation of unusual amounts of organic matter (p. 35); the lack of appreciable wave or current action indicated by the preserved lamination implies a minimum renewal of oxygen in the bottom waters and, consequently, little destruction of finely-divided organic matter. Furthermore, the organic matter would also be supplied to the bottom sediment more rapidly here than in more agitated waters, where it would tend to be swept away...

The most conspicuously bituminous strata are commonly associated with phosphatic material. Hoots[12] has emphasized the close association of petroliferous beds and bituminous matter with the phosphatic shale of

---

[12] Hoots, H. W., Blount, A. L, and Jones, P. H., Marine oil shale, source of oil in Playa del Rey field, Calif: Am. Assoc. Petroleum Geologists Bull., vol. 19, no. 2, pp. 172-205, 1935.

the Monterey formation near the base of the upper Miocene in the Playa del Rey oil field. Other places in which this association has been observed are the Palos Verdes Hills, the Grimes Canyon area, the Naples section, and the Bixby Canyon section. It has been suggested that large accumulations of phosphate are favored in areas of rapid destruction of marine life where surface currents of different temperatures meet, as along the Agulha Banks off the South African coast.[13] Such a situation would seem equally favorable for increased accumulation of organic matter, especially if the currents met where the depth of the water and the character of the bottom topography were such as to prevent such active movement of bottom currents as would result in rapid oxidation.

## ORIGIN OF THE SILICEOUS ROCKS
### ORIGIN OF THE DIATOMACEOUS DEPOSITS

Some data bearing on the mode of formation of the diatomite and less pure diatomaceous rocks that form extensive deposits in the upper part of the Monterey have already been presented in describing these materials, and this problem was also considered more broadly in discussing paleogeographic conditions and certain lithologic features such as rhythmic bedding. The problem will now be considered more explicitly.

Large diatomaceous or other organic deposits can have accumulated only under one of the following conditions: (1) growth of the organisms in such unusually great abundance as to completely dominate normal deposition of clastic sediments; (2) deposition of clastic sediment so nearly lacking that organic material dominates even though it accumulates very slowly; or (3) a combination of 1 and 2, large development of the organisms being combined with the deposition of relatively little clastic material. As the following discussion will show, the diatomaceous deposits of the Monterey formation appear to be due to the combination of conditions last mentioned rather than to either of the first two conditions alone.

The larger diatomaceous deposits of modern seas—those, for example, of the north and south Pacific Ocean—occur in deep water and seem to represent relatively slow accumulations in areas nearly free from terrigenous material. Hence it has been inferred that ancient diatomaceous deposits accumulated under similar conditions, this view being implied in the following statement on the Monterey formation by Fairbanks:[14] "The time required for the deposition of 4,000 feet of such material, which so far as we know accumulates at an exceedingly slow rate, must have been enormous." Very little more is known at present about the absolute rate of accumulation of such sediments than when Fairbanks made this statement. Reasons have been given, however (pp. 11-12), for believing that these diatomaceous deposits, though laid down where little clastic material was being deposited, were not formed in abyssal waters far from land; and some details of occurrence suggest that the accumulation of the organic material was relatively rapid.

The discussion of rhythmic bedding indicated that, whatever the agency producing such bedding, the alternation is so regular that some process of nearly uniform period is involved. The rhythmic beds, like the superimposed larger cycles (p. 31), are not markedly thinner in material consisting largely of diatoms than in associated material that is largely composed of silt and fine sand. Only in the coarser sandstone members of the formation are the rhythmites, or graded beds of sandstone, generally and conspicuously thicker than in the diatomites. No systematic relation is apparent between the thickness of the formation at a given place and the proportion of clastic sediment in it at that place. Some of the thickest sections of the Monterey deposits, such as those at Point Concepcion, Chico Martinez Creek, and Reliz Canyon, consist largely of the non-clastic sediments. Other thick sections representing about the same period of time consist largely of clastic sediment, as in the Puente Hills and Modelo Canyon areas. These relations indicate that the clastic sediments did not ordinarily accumulate many times faster than the nonclastic siliceous sediments; in general they apparently did not accumulate even twice as fast. The thousands of feet of rhythmically bedded Miocene deposits in many basin areas indicate relatively little by-passing of sediment and comparatively rapid accumulation in those areas, of the entire formation.

As already pointed out, the rapid accumulation of the thick diatomaceous deposits may have been due in part to a concentration of the slowly settling diatoms by current drifting. Coastal currents might thus carry these diatoms from a large area of the open sea and deposit them in embayments or in the less agitated bottom waters of the deeper basins along the coast. The importance of current drift in some diatom deposits is indicated by Philippi,[15] who shows that in a part of the South Pacific affected by northward-flowing currents, the extensive bottom accumulations of diatoms lie north of the areas in which these organisms are growing most abundantly in the surface waters. Branner[16] suggested that the large diatomaceous deposits of the Monterey formation might represent material that was drifted by colder currents flowing southward from Alaska and that was caught in cul-de-sacs, such as probably existed in the area now occupied by the southern part of the San Joaquin Valley. This view seems to be supported by the statement of Dr. Mann that the diatoms in the Lompoc deposit are characteristic of northern waters (p. 10)...

Most oceanographers, however, including Bigelow, seem to agree that under normal conditions in modern seas the development of phytoplankton as a whole is limited in the main by the available phosphate and nitrogen in sea water, but that for the diatoms in particular the supply of silica may more commonly limit development. No agreement has been reached even in the

---

[13] Murray, John, Challenger Report, Deep sea deposits, p. 396, 1891.
[14] Fairbanks, H. W., U. S. Geol. Survey Geol. Atlas, San Luis folio (no. 101), p. 10, 1904.

[15] Philippi, E., Die Grundproben der Deutschen Sudpolar Expedition: Deutsche Sudpolar Exp., Bd. 2, Hft. 6, p. 614-15, 1912.
[16] Branner, J. C., Influence of wind on the accumulation of oil bearing rocks (abstract) Geol. Soc. America Bull., vol. 24, p. 95, 1913.

studies of modern seas as to which of these three nutrients is generally scarcest relatively to the needs of the diatoms and thus plays the critical part in limiting their development. Before considering the significance of the oceanographic data available, therefore, some geologic evidence on the question, suggested by the mode of occurrence and the associations of diatomaceous deposits, may be summarized.

The frequent association of siliceous organisms and volcanic ash was noted by Ehrenberg[22] as early as 1844. In 1867, J. D. Whitney[23] emphasized this common association in the many diatomaceous deposits of California and adjacent states, and suggested that the unusual supply of silica available in the ash might have favored the growth of diatoms. A similar view was advanced by de Lapparent[24] in 1923. More recently, in 1933, Taliaferro[25] has reviewed the many examples of the association of diatomaceous and volcanic rocks, which include most of the larger diatomaceous deposits known, and he also concluded that this association was due to the large supply of silica made easily available for the development of the siliceous organisms...

One of the earliest and quantitatively most important effects of the alteration of vitric pyroclastics, is a loss of silica. Great quantities of silica would thus be dissolved in sea water from volcanic ash, supplying one element highly favorable to the growth of diatoms. This condition seems a probable cause of the common association of diatoms with volcanic ash, and some support for this view is found in the results of recent oceanographic studies.

The most abundant diatom planktons of modern oceans occur most widely in high latitudes, where their development seems to be favored, in general, by the low temperature and salinity of the sea water. More temperate waters, however, show an equally great abundance of diatoms in certain places where streams from the land or the upwelling of deeper ocean waters rapidly replenish the supply of nutrients. An abundance of nutrients is thus shown to be of paramount importance...

The problem of which nutrient material may have been critical for the long-continued development of the diatoms that have formed thick deposits seems less complex than the question of which nutrients may temporarily or locally limit production in the modern seas. The siliceous deposits make it evident that much of the silica was being permanently removed from the sea water, so that in a long period of time, replenishment of silica was more likely to have been critical than replenishment of any other nutrient. Such equally necessary nutrients as the less stable phosphate and nitrogenous compounds more largely move in a cycle of organic and inorganic forms, these constituents being returned to the near-surface waters, where they could be used again by the phytoplankton, through oceanic circulations, such as upwelling and turbulence...

### ORIGIN OF THE PORCELANEOUS AND CHERTY ROCKS

Porcelanite, porcelaneous and cherty shales, and cherts constitute in the aggregate a much larger part of the Monterey formation than the diatomaceous rocks. Their mode of formation is more obscure than that of the rocks consisting largely of recognizable siliceous organisms but is of more general interest, because similar cherty rocks, which might have had a similar origin, occur in many other regions in formations ranging in age from pre-Cambrian to Tertiary. Chemically, the distinctive feature of these rocks is a ratio of silica to alumina much higher than is usual in fine-grained clastic rocks, most of which are largely composed of clay and other silicate minerals. The unusual proportion of silica is due to its original presence in abnormal amount or to its later introduction, since neither theoretical considerations nor observed facts indicate that it is due to removal of any part of the chemically more inert alumina.

### REVIEW OF ALTERNATE THEORIES OF ORIGIN

A great many papers have been written regarding the source of the silica in various siliceous formations, and the conclusions from this extensive literature on the subject indicate that siliceous rocks probably have been formed in various ways. It would be impracticable to review all these papers here, but a summary, in outline form, of the various theories that have been proposed—most of them in various interpretations of the Monterey siliceous rocks—is presented below in order to consider their applicability to the Monterey formation.

*Tentative genetic grouping of the various theories regarding the source of silica in siliceous formations*

I. Inorganic source of silica
   A. Deposition of sediment unusually high in clastic silica
   B. Inorganic precipitation from siliceous waters
      1. Syngenetic
         (a) Siliceous emanations from volcanic rocks
         (b) Silica in solutions and as colloids introduced by streams
      2. Epigenetic
         (a) Secondary introduction of silica by ground or surface waters
   C. Chemical alteration and redistribution of silica of tuffaceous sediments
      1. Syngenetic
         (a) Halmyrolysis or "submarine weathering"
      2. Epigenetic

---

[22] Ehrenberg, C. G., On the remains of infusorial animalcules in volcanic rocks: Geol. Soc. London Quart. Jour., pp. 73-91, Aug. 1846
[23] Whitney, J. D., On the fresh water infusorial deposits of the Pacific coast and their connections with the volcanic rocks: California Acad. Nat. Sci. Proc., vol. 3, pp. 319-324, 1867.
[24] de Lapparent, J., Lecons de Petrographie, p. 322, Masson et Cie, Paris, 1923.
[25] Taliaferro, N. L., the relation of volcanism to diatomaceous and associated siliceous sediments: Univ. California Publ. Dept. Geol. Sci. Bull., vol. 23, no. 1, 1933.

(a) During compaction and lithification or later

II. Organic source of silica
  A. Organic precipitation and accumulation of siliceous organisms
  B. Chemical alteration and redistribution of silica of organisms
    1. Syngenetic
      (a) Halmyrolysis or "submarine weathering"
    2. Epigenetic
      (a) Diagenetic alteration during compaction and lithification
      (b) Metamorphic alteration during deformation and igneous intrusion
      (c) Alteration by ground or surface waters

\* \* \* \* \*

#### SILICA DERIVED FROM SOLUTION OF SILICEOUS ORGANISMS

Many deposits of siliceous rock in which the siliceous organisms preserved are not sufficient to class them as organic deposits are believed by some investigators to have derived their silica from the siliceous organisms originally present in the rocks. Some of these same deposits have been interpreted by other investigators as of inorganic origin, the preserved siliceous fossils in them being regarded as more or less incidental. Evidence for either interpretation seems inconclusive in many cases. The many papers on an organic source of the silica of other deposits will therefore not be reviewed, but some of the facts and conclusions given in these studies will be considered in so far as they bear on the eviddnce of origin derived from the examination of the Monterey siliceous rocks.

Fairbanks[57] was apparently the first to interpret the cherts and other hard siliceous rocks of the Monterey formation as altered diatomaceous sediments, and his interpretation was accepted by Arnold and Anderson,[58] though Fairbanks presented no evidence for such an origin. More recently, in 1927, Tolman[59] has suggested the same origin for the porcelaneous rocks or "cemented opal shales" of the Santa Maria region. Microscopic examination of some of these rocks led him to conclude that they contained abundant diatom debris. The writer's examination of porcelaneous and cherty rocks of the Monterey from many areas has led him to conclude (p. 15), on the contrary, that recognizable diatoms or diatom debris are not in general abundantly preserved in the indurated or silica-cemented rocks, though molds of diatoms are commonly very abundant.

Some of the upper part of the Monterey formation is obviously of organic origin, being dominantly composed of diatoms and other siliceous organisms, but the greater part of the formation contains relatively few recognizable siliceous organisms. This part may be considered of organic origin, however, in the sense that it represents altered deposits of siliceous organisms, even though it has reached its present state through an inorganic process of solution and reprecipitation of silica within the rocks.

### EVIDENCE OF FORMATION FROM DIATOMACEOUS ROCKS

#### EVIDENCE FROM THE CHERT OCCURRING IN THE DIATOMACEOUS DEPOSITS

Nodular concretions and lenticular beds of opaline chert are rather common in some of the diatomaceous deposits of the Monterey formation. Their mode of occurrence is illustrated in plates 6, *D*, and 7, *A*. They occur most commonly in the lower part of the diatomaceous deposits and in a transition zone where the diatomaceous rocks are interbedded with the underlying porcelaneous and cherty rocks.

Polished surfaces showing the relations of these lentils of dense opaline chert to the adjacent diatomite are illustrated in plate 16, *C*. In thin sections, differences in the texture and constitution of the fine laminae may be seen to pass from the diatomite into the chert without appreciable distortion of the bedding planes, a relation that clearly shows the equivalence of the two sorts of rock. Well-preserved large diatoms are abundant in the white porous diatomite, but they are not recognizable in the chert, even when the thin sections are examined under the microscope with dark-field illumination. The obvious continuity of the laminae, without appreciable changes in thickness or in their clastic constituents, strongly suggests that the cherty lentils were formed by impregnation of the diatomite with additional opal. Such continuity of laminae probably would not exist if the chert represented an inorganically precipitated mass of silica gel that was surrounded and buried by the diatomaceous sediment. Equally improbable is the view that silica was inorganically precipitated as thin laminae in local small areas on a sea floor that elsewhere accumulated similar laminae of diatomite, especially as the small area of the different deposits must first have increased and then decreased to have formed the lenticular masses of chert (pl. 16, *C*). The lack of recognizable diatom outlines in the opaline chert might conceivably be due to similarity in refringence between the delicate organic forms and the opaline matrix: the mean index of refraction of the opal in the matrix is about $1.45\pm.003$, which is not very different from that of the organic opal—about $1.440\pm.003$. It seems more probable, however, that the diatoms originally present may have completely lost their identity in an intimate intergrowth with the secondary opal. Etching with a caustic solution brings out the fine lamination of the chert, but it does not reveal any such distinction between opaline matrix

---

[57] Fairbanks, H. W., U. S. Geol. Survey Geol. Atlas, San Luis folio (no. 101), p. 4, 1904.
[58] Arnold, Ralph, and Anderson, Robert, Geology and oil resources of the Santa Maria oil district: U. S. Geol. Survey Bull. 322, pp. 45-47,. 1907.
[59] Tolman, C. F., Biogenesis of hydrocarbons by diatoms: Econ. Geology, vol. 22, no. 5, pp. 454-74, 1927.

and original opaline diatoms as might be expected if the tests of these organisms had been preserved...

### DIATOMACEOUS STRATA AND EQUIVALENT STRATA OF PORCELANEOUS AND CHERTY ROCKS

A change from diatomaceous to cherty rocks on a small scale, along observable stratigraphic planes, is illustrated by the lenticular concretions described above. A similar change on a much larger scale can be shown to have occurred in certain areas, and additional detailed stratigraphic work seems likely to prove that such lateral change is commoner than is now generally recognized. In field mapping, however, the lack of easily identifiable stratigraphic markers usually makes it necessary to assume that the contact of diatomaceous and harder silica rocks represents a definite stratigraphic horizon. Lateral variation in the stratigraphic position of this contact in the Palos Verdes Hills has been mentioned in describing the alteration of the Miraleste tuff bed (in the Altamira shale member of the Monterey) (p. 27). Several hundred feet of diatomaceous shale in the northeastern part of these hills is represented on the south side of the hills, less than a mile distant, by strata consisting of cherty and porcelaneous shales without recognizable diatoms. A similar relation seems to exist, though it is less clearly established, in a nearby area at the town of San Pedro, where diatomaceous strata in the western part of the town are apparently represented by porcelaneous rocks to the southwest.[62]

A more obvious and well established example is found in the middle Miocene strata of the Highland Monocline area. Here about 400 to 500 feet of hard porcelaneous rocks exposed on Indian Creek (pl. 2) are traceable into soft diatomaceous beds with only a few lenticular and concretionary masses of cherty rock, exposed in a small canyon about four miles to the west of Indian Creek. This latter place was taken by Kleinpell as the type for his Luisian stage.[63] These two sections are particularly well correlated by abundant foraminiferal faunas and by distinctive lithologic sequences recognizable in both areas.

A lateral variation is also made evident by comparing two sections along the Santa Barbara County coast. In the Naples section (pl. 2), the middle Miocene strata are soft diatomaceous and phosphatic shales interbedded and alternating with zones of cherty and porcelaneous shales. The equivalent strata in the section at the mouth of Gaviota Creek, about 16 miles to the west, consist of cherty, porcelaneous, and phosphatic shales without preserved diatoms. The correlation of these two sections is well established both by the faunal and the lithologic sequence...

### VERTICAL CHANGES IN STRATIGRAPHIC SECTIONS

At every place where diatomaceous deposits are present in the Monterey, they occur in what constitutes the upper part of the formation at that particular place, although, as is shown in plates 2 and 3, the diatomaceous deposits at one place may differ in age from those at another. In no area do the diatomaceous rocks form the lower part of the formation and the porcelaneous and cherty rocks form the upper part. The few areas in which diatomaccous deposits occur in the lower part of the formation offer no exceptions to this rule, for in these areas the overlying strata are not dominantly composed of the porcelaneous and cherty rocks. The same is true of some areas illustrated by the sections at Naples and Lompoc (pl. 2), where the diatomaceous upper part of the section and the cherty and porcelaneous shales of the lower part are separated by an unusually thick transitional zone, in which diatomaceous and cherty rocks are interbedded but in comparatively thin layers. In most areas no such thick intermediate zone of the interbedded rocks is present, the transition from diatomaceous strata to underlying harder rocks being so abrupt that there is little difficulty in mapping the contact locally, although the basal part of the diatomaceous member commonly includes a zone several feet or a few tens of feet thick that contains lentils and beds of chert, which increase downward in abundance.

Another type of transitional change downward from the diatomaceous deposits is found in some areas—for example, along Chico Martinez Creek (pl. 3) and in the nearby North Beldridge oil field. The nearly continuous sequence of cored strata from the Bear State No. 23 well gave a particularly good set of samples for examination. In these sections the upper diatomaceous strata are underlain by a rather light and porous porcelaneous shale in which impressions of diatoms are common, especially molds of the large discoid types, which are visible under a hand lens. Although the diatoms are represented only by molds, the heavier siliceous tests of radiolaria and sponge spicules are at least partly preserved. With increasing depth in the section, the diatom molds are increasingly difficult to recognize and the sponge spicules increasingly attacked by solution, their axial canals being commonly so much enlarged that their walls are very thin, and finally the spicules are entirely dissolved, so that only molds remain. The porcelaneous and cherty rocks in the lower part of the thick Chico Martinez Creek section consist largely of chalcedonic rather than opaline silica, and in these rocks no siliceous organisms, nor even molds of them, are recognizable, though they are present in the associated carbonate concretions.

### DIATOMS IN CARBONATE CONCRETIONS

The carbonate concretions were formed after the deposition of the enclosing beds (pp. 20-21), though before the beds had been much compacted and lithified. The distinctive constituents and variations in beds extend laterally into the concretions, so that those in diatomaceous deposits include abundant diatoms.

The carbonate concretions which occur abundantly in the porcelaneous and cherty shales at many places, likewise generally contain diatoms, although the

---

[62] Woodring, W. P., Bramlette, M. N., and Kew, W. S. W., Geology and paleontology of the Palos Verdes Hills, Calif.: U. S. Geol. Survey Prof. Paper 207, p. 31, 1946.
[63] Kleinpell, R. M., Miocene stratigraphy of California, p. 122, fig. 11, Am. Assoc. Petroleum Geologists, 1938.

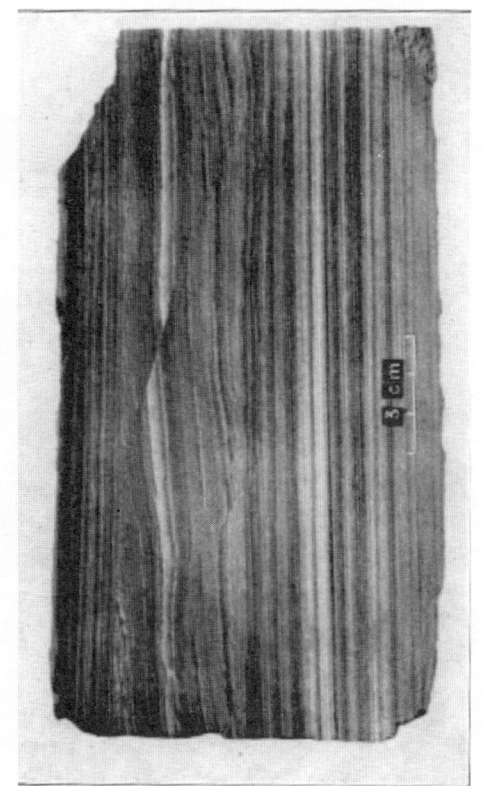

Plate 5. A. Small step faults confined within certain beds in laminated diatomite. B. Low angle faults in laminated diatomite.

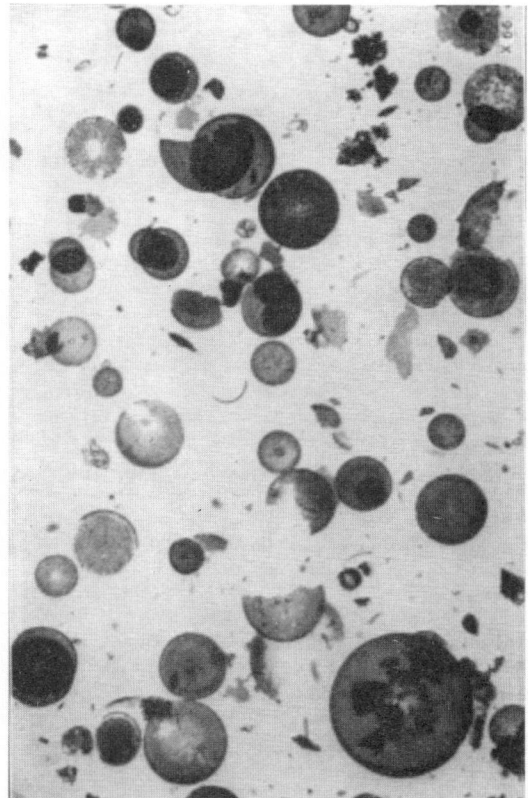

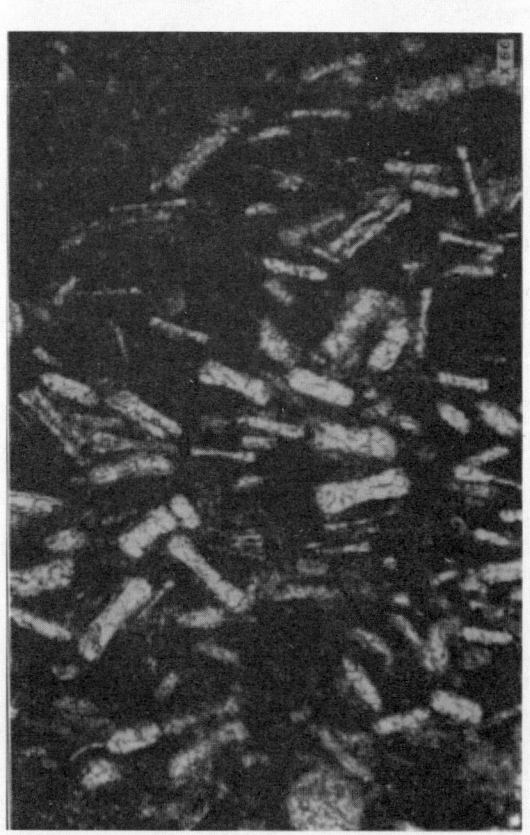

Plate 11. A. Diatoms in thin section from the calcareous concretion shown in Plate 10, B, with clear and more coarsely crystalline calcite filling the interior of diatoms. B. Opaline diatoms obtained from the digestion with acid of part of the same specimen shown in 11 A.

# THE MONTEREY FORMATION OF CALIFORNIA AND THE ORIGIN OF ITS SILICEOUS ROCKS

adjacent beds contain only molds of diatoms or are entirely devoid of their recognizable remains. Other characteristics of the bedding, however, may be clearly traced from the enclosing beds into the concretions. Plate 10, *B*, shows a carbonate concretion in cherty shale. Thin sections from this concretion and from equivalent beds in the cherty rock show that there are no recognizable diatoms in the cherty rock but many in the calcareous concretion. Diatoms in a thin section are illustrated in plate 11, *A*, and some of those obtained by digestion with acid are shown in plate 11, *B*. Diatoms are generally preserved in the calcareous concretions within the porcelaneous and cherty strata, presumably because of the relative impermeability of these concretionary masses to the solutions and their resistance to the pressures that have affected the enclosing beds. The carbonate concretions in the Monterey formation thus yield good diatom floras for micro-paleontologists to study, although great masses of the cherty and porcelaneous rocks that enclose the concretions are otherwise largely barren of identifiable diatoms. Rarely, as in a concretion from the Claremont shale of the type locality in the Berkeley Hills, the diatoms in concretions have been replaced or altered to chalcedony, though they have retained their form. In this occurrence the strata are vertical, and are next to a fault of large displacement. However, the strata appear to be more altered than usual and the chert beds themselves are largely composed of chalcedony and quartz.

### TEXTURAL AND STRUCTURAL SIMILARITIES BETWEEN DIATOMACEOUS AND CHERTY ROCKS

The diatomaceous rocks are remarkably similar in many details to the porcelaneous and cherty rocks. Perhaps no one of these resemblances is very striking in itself, but the consistency with which a number of the distinctive features of the diatomaceous rocks are duplicated in the harder silica-cemented rock types is highly significant.

The very fine lamination of cherty and porcelaneous shales, so common in many areas and well illustrated in the polished and etched specimen of plate 5, *D*, is apparently identical with the fine lamination common in the diatomite and diatomaceous shale (pl. 5). Some of the diatomaceous deposits are not finely laminated but consist of massive diatomaceous mudstone, and similarly massive porcelaneous mudstone is common in certain areas and certain parts of the formation. In the few areas where diatomaceous rocks are known to be exactly equivalent to hard siliceous rocks occurring near-by, the bedding has the same character in one rock that it has in the other.

In many areas a well marked rhythmic bedding an inch or two in thickness is found both in diatomaceous strata and in porcelaneous and cherty strata. The gradation within each bed is similar in both kinds of rock except that the upper part of each rhythmite in the diatomaceous sediment is rich in diatoms, whereas in the cherty rocks it is hard silica-cemented material in which no diatoms have been preserved (pl. 15, *A, C*).

Step faults on a very small scale, generally confined to particular beds and apparently due to slip movements along the bedding, are rather common in the diatomaceous shale of many areas (pl. 5, *A*). Similar faults are also found in both the porcelaneous and cherty rocks (pl. 18, *C, D*), but there they are not likely to be obvious except in weathered exposures or in specimens that have been etched with caustic solution.

Foraminifera, abundant fish scales, and some other organic remains are equally abundant, and show the same mode of occurrence, in the diatomaceous and cherty strata. In both types of rock the Foraminifera may be well preserved or may be leached and be represented only by molds. Foraminifera, or molds of them, appear to be in general less common in the diatomaceous strata than in the harder strata, though notable exceptions are found in some areas; but the distribution of these organisms appears to be correlated with stratigraphic position rather than lithologic character. Foraminifera are less abundant, and unaltered diatomite more abundant, in the upper part of the Monterey than in the lower part. Where diatomite occurs in the middle Miocene—that is, in the lower part of the Monterey—it contains abundant Foraminifera; and where, as in Reliz Canyon, porcelaneous and cherty rocks extend to the uppermost part of the Monterey, they contain few Foraminifera.

\* \* \* \* \*

### PORCELANEOUS AND CHERTY ROCKS FORMED CHIEFLY BY ALTERATION OF DIATOMACEOUS ROCKS

Evidence suggesting that the harder siliceous rocks were formed by alteration of diatomaceous rocks was observed in many localities and different tpes of occurence. In some beds of the cherty rock the added silica was evidently derived from adjacent altered tuffs, but such beds appear to be few and relatively thin. Lateral transition of a thick body of diatomaceous strata to cherty and porcelaneous strata can be demonstrated in only a few places, but thousands of instances of the same change on a smaller scale can be seen in the lenticular masses of chert in diatomaceous deposits. Carbonate concretions containing well-preserved diatoms are common throughout great thicknesses of cherty and porcelaneous rocks that otherwise contain few or no recognizable diatoms. Equally common and widely observed are the striking similarities of the diatomaceous and the cherty rocks in details of lamination, large rhythmic bedding, and other features. The conclusion thus seems justified that most of the cherty and porcelaneous rocks were formed by rearrangement of the silica of originally diatomaceous rocks...

### TIME OF ALTERATION

Evidence that most of the porcelaneous and cherty rocks of the Monterey were formed from diatomaceous rocks is more adequate than the evidence on the precise nature of the process of alteration and the time at which it occurred. The time of alteration cannot be limited entirely to a particular period during or following the deposition of the sediments; there is some evidence that the process has gone on more or less continuously since deposition, or since a time soon thereafter. It seems probable, however, that most of the alteration occurred in one or more definite stages and was largely completed in Pliocene time. Pebbles and boulders of typical porcelaneous and cherty rocks of the Monterey occur abundantly in Pliocene and Pleistocene conglomerates at many places and form the dominant constituent of some of these conglomerates, such as those in the extensive Paso Robles formation. The alteration obviously took place in large part before the accumulation of these conglomerates. Pebbles of the cherty rocks are also found, though less abundantly, in the upper Miocene sandstones which in some areas overlie Monterey formation—in the Santa Margarita sandstone of the southern Salinas valley, for example, and in the Pismo formation of the San Luis quadrangle. Plate 18, *A*, illustrates an occurrence of cherty shale pebbles at the contact of the Pismo formation with the Monterey formation. Many of these pebbles are angular and show little or no rounding by attrition.

Intraformational conglomerates composed largely of chert fragments have been seen at a few places in the Monterey formation. Such a conglomerate is very conspicuous in the coastal cliffs just west of the mouth of Gaviota canyon (pl. 18, *B*). Here the chert fragments are obviously derived from the Monterey rocks, as shown by lithologic similarity and included Foraminifera. Many of the angular fragments have sharp edges and must have been deposited as brittle cherty rock and not as soft diatomaceous rock that was later altered to chert. At approximately the same horizon in the Naples section, diatomaceous beds with some interbedded chert contain a few scattered chert pebbles of Monterey formation. One of these pebbles, shown in plate 17, *D*, contains borings like those made by present-day boring mollusks in pebbles of chert and other rocks along this coast. After the holes were bored the pebble was buried in the diatomaceous sediment, which was then compacted around it and squeezed into the cavities. This chert pebble is underlain by the finely laminated bed shown in the illustration, which after compaction was altered to chert. The soft diatomaceous sediment which rested on the upper part had been partly eroded, so that the specimen was exposed in the wave-cut cliff, but enough of the upper part of the chert pebble was still embedded in the diatomaceous shale to show that the diatomite likewise was compacted around the pebble and filled some of the bored cavities. These relations, as well as the occurrence of the specimen above present high-water levels, make it evident that the boring was not done by modern organisms living on the present coast. A fragment of chert of the Monterey that had similar borings was found embedded in the Malaga mudstone member of the Monterey (upper Miocene) of the Palos Verdes Hills.

At least a part of the alteration of diatomite to chert was thus early enough to furnish pebbles that were reworked into succeeding Miocene deposits. It should be emphasized, however, that these intraformational pebbles consist of dense cherty rock; pebbles of the more common porcelaneous rocks are absent, though they are dominant in many Pliocene conglomerates in areas where these overlie Monterey rocks. So far as the writer has observed, the strata classed as Santa Margarita in the Salinas Valley that contain abundant pebbles of porcelaneous shale are of Pliocene age; pebbles of such rock are rare or generally absent in the part of the Santa Margarita that contains upper Miocene fossils. The pebbles of dense chert in the intraformational conglomerate at Gaviota are strikingly different from the immediately underlying beds of porcelaneous shale. This distribution of the pebbles suggests that only the beds of chert were available as hard rock to form the intraformational conglomerates, and that the porcelaneous rocks were formed in a later period of alteration (pp. 53-54).

The beautifully preserved lamination of the porcelaneous and cherty rocks indicates that they were altered after their accumulation as laminated diatomaceous deposits, for it seems improbable that an opal gel formed on the sea floor would possess a fine lamination identical with that of the diatomite. Identical lamination would be expected, on the other hand, if additional opal simply impregnated the laminated diatomite...

### PHYSICO-CHEMICAL CONDITIONS OF ALTERATION

That diatoms are to some extent dissolved by sea water is indicated by the work of Cooper,[65] and also by that of Brockmann,[66] Hustedt,[67] and others, all of whom have noted that many common pelagic diatoms, most of which have unusually delicate shells, are not commonly found among the diatoms of the bottom sediment. Even in the relatively shallow water of the North Sea the more delicate of the pelagic forms are completely dissolved during their slow descent to the bottom. The very much thicker shells of some other pelagic forms, and of most shallow-water benthonic forms, accumulate in the bottom sediment without being noticeably corroded, though removal of a nearly uniform thin surface film from these diatoms would perhaps be unrecognizable. Appreciable solution of opaline organic remains by sea water is also indicated by Schulze's[68] observation that the axial canals in spicules from benthonic sponges are often seen to be enlarged by solution. This solution by bottom waters would perhaps be favored, as Hustedt[69] has suggested, by relatively sparse distribution of opaline shells in a highly calcareous bottom sediment. Solution by bottom waters is evidently not sufficient, however, to affect appreciably the large diatoms, at least where they accumulated rapidly or in great abundance, since many large deposits of them remain...

---

[65] Cooper, L. H. N., Chemical constituents of biological importance in the English Channel, November 1930 to January 1932: Jour. Marine Biol. Assoc. of United Kingdom, new ser., vol. 18, no. 2, p. 697 1933.
[66] Brockmann, Chr., Diatomeen und Schlick im Jade-Gebiet: Senckenbergischen Naturf. Gesellschaft, Abh. 430, p. 18, 1935.
[67] Hustedt, F., Vorlaufige Ergebnisse vergleichender Untersuchungen der Diatomeen holsteinischer Seen: Vorhandlungen der Intern. Vereinigung fur theoretische und angewandte Limnologie, Kiel, p. 103, 1922.
[68] Schulze, F. E., Report on the Hexactinellida, Challenger Rept., Zoology, vol. 21, pp. 26-27, 1887.
[69] Hustedt, F., op. cit., p. 103.

# THE MONTEREY FORMATION OF CALIFORNIA AND THE ORIGIN OF ITS SILICEOUS ROCKS

Alterations by included sea water within the sediment may occur early, and may therefore be essentially contemporaneous with processes occurring at the surface of the bottom sediment, but the results produced in the two environments may be quite different. The physico-chemical condition would certainly differ. The relatively small quantity of much less mobile water within the sediment would inevitably become more nearly saturated with some constitutents than the more mobile overlying sea water. Under these conditions, simultaneous solution and reprecipitation of the opaline silica in the sediment are not incompatible processes. Particularly vulnerable to solution is the finely divided opal in the small diatoms that have a very delicate mesh structure, so that their surface area is large relative to their mass. What Harker[72] says of crystals would apply to diatoms:

Crystals of the same kind but of different sizes, in presence of their saturated solution and within the range of effective diffusion, constitutue a sensitive system, in that a slight cause may suffice to bring about corrosion of some crystals with correlative addition of material to others. Such a cause is found in surface tension. Since the pressure due to surface tension is proportional to the curvature of the surface, a small crystal is under greater stress than a larger one. Increased stress, as we shall have occasion to point out later, causes increased solubility. Material is, therefore, dissolved from the smaller crystals and deposited upon larger ones in their neighborhood, until the smaller have disappeared.

Under such chemical conditions, alteration within some of the diatomaceous sediments seems to have begun with their deposition, and some opaline chert was formed at this early stage in some of the Monterey deposits. There are, however, large deposits of diatomite showing little or no such alteration, and that fact indicates that conditions were not always favorable for this diagenetic alteration. What conditions favored early alteration is not yet apparent, but some suggestions may be offered. Among the various factors that might further a diagenetic alteration of the diatomaceous deposits, the relative proportion of associated carbonate may prove to be the most significant. In general, the unaltered diatomaceous deposits are less rich in carbonates than the porcelaneous and cherty rock parts of the formation, though there are many exceptions to this rule. The proportion of organic matter is likewise generally greater in the silica-cemented rocks than in the diatomaceous strata, though the correlation may originally have been less close than it is now, because the more porous diatomaceous beds may have been more largely leached of their original organic matter (p. 36) than the denser siliceous rocks...

The vitric pyroclastic beds in the diatomaceous deposits are little altered, whereas those in the porcelaneous and cherty strata are largely converted to bentonite. This contrast suggests that the loss of silica and alkalies from the tuffs during their alteration is related to the alteration of the diatomaceous deposits. The conditions that caused an alteration of the pyroclastics may therefore have been of primary importance, though the same factors may possibly have influenced both these alterations about equally and yet independently. The amount of silica lost from the pyroclastics in the formation appears to be quite inadequate to account for most of the silica cement of the rocks of the formation as a whole. The alkalis which were also liberated from the tuffs early in the process of alteration, though much less in quantity than the free silica, may be a more important factor, because of their influence on the solubility of the opaline organisms within the sediments.

### ALTERATION DUE TO LOAD AND DYNAMIC METAMORPHISM

The continuance of bedding laminae without warping from diatomite to chert (pl. 16, C) indicates that part of the silicification occurred later than most of the compaction of the sediments. The small step faults in cherty and porcelaneous rocks (pl. 18, C, D), obviously formed before the silica cementation of these rocks, show that part of the silicification was subsequent to some deformation of the beds. These small faults at certain horizons may however have been formed soon after the beds were deposited, for there is reason to believe that some deformation of the Miocene deposits took place during their accumulation, though much of this seems to have been only local and minor deformation, perhaps largely related to slumping of sediment on the sea floor (pl. 19, B)...

The fact that the diatomaceous deposits occur only in the upper part of the Monterey formation, and never underlie the dominantly porcelaneous and cherty part, suggests that depth of burial and age may be important factors in much of the alteration. In many areas the formation is several thousand feet thick, and where the thickness of the post-Miocene deposits does not greatly exceed the thickness of the Monterey, the relative load on the upper and the lower parts of the Monterey would be very different. During deformation—and there are few areas in which the Monterey is not considerably deformed—this difference in overlying load might be an important factor in the low-grade metamorphism that would probably suffice to change the diatomaceous sediment to more stable silica rocks...

The limited data known to the writer indicate that diatomite or highly diatomaceous and "punky" rocks are not encountered in wells at depths much greater than 3,000 or 3,500 feet. But even though thickness of overburden probably influences the vertical distribution of the hard and soft rocks, other important factors have doubtless been involved, such as relative intensity of deformation, permeability of the beds to included solutions, and composition of the solutions. The suggested relationship of alteration to load and intensity of deformation seems nevertheless to justify further consideration as data become available, for the significance of such a possible relationship in many of the structural and stratigraphic problems regarding these rocks is obvious...

Diatoms are not known to have existed earlier than mid-Mesozoic time, but other siliceous micro-fossils are

---

[72] Harker, Alfred, Metamorphism, p. 20, Methuen & Co., London, 1932.

known in strata as old as pre-Cambrian. It seems significant that in these older formations the siliceous fossils are preserved only in a dense, cherty matrix, usually of chalcedonic silica or quartz. Softer, "punky" rocks that contain siliceous remains in an uncemented matrix from which the siliceous shells may be separated are not known in strata older than Cretaceous, and nearly all of them are Tertiary or Quaternary.

The general relation, in the Monterey, of alteration to depth suggests that moderate load and dynamic metamorphism tend to produce the porcelaneous shale, in which the original diatoms are seldom recognizable except as molds, and that greater metamorphism tends to produce the denser, cherty rocks containing a greater proportion of chalcedonic silica. Some alteration, however, has been shown to occur in the early stages of compaction and lithification of the rocks, through what may be considered a diagenetic process. No sharp separation seems possible between the processes or between their products, but some tentative conclusions regarding them are suggested.

### SUGGESTED PROCESS OF ALTERATION

The alteration of the siliceous rocks is believed to have been effected largely through redistribution of silica, the relatively unstable finely divided opal in the porous diatomaceous deposits having been dissolved, then re-precipitated near by to form denser and more stable silica rocks. Where the composition or other properties of the water were favorable, this process began soon after the deposition of the sediment. Because of abundant included solutions and high permeability at this early stage, parts of the diatomaceous sediment were thoroughly impregnated with the opal from adjacent sediment and converted into dense opaline chert and cherty shale. The diatoms appear in general to have been dissolved most readily in those thin beds or laminae in which they were relatively scarce, and the silica derived from them was reprecipitated in adjacent beds in which these organisms were relatively abundant (pl. 16, B). Ellipsoidal masses of dense opaline chert were formed in some of the more massive deposits of diatomite, whose relatively uniform composition and lack of bedding variations tended to favor this normal form of concretionary growth.

Strata not much affected by the diagenetic or early alteration—including the diatomaceous deposits remaining as such up to the present time—were subjected to varying degrees of metamorphism during subsequent periods of deformation. Where load and intensity of deformation were sufficient to further the tendency of the finely divided opal of the diatom shells to assume a more stable form, a redistribution of silica without extensive movement is believed to have formed the porcelaneous types of rock. During this low-grade metamorphism the rocks would be less permeable, and would contain a much smaller quantity of solutions, than in the early diagenetic period of alteration. Less transfer of silica to other beds or centers of precipitation might be expected to occur, therefore, at this stage than occurred in the early period of alteration that resulted in the formation of dense cherty rocks. The porcelaneous rocks show relatively little differentiation into distinct beds of greater and less silica content. They seem to have been formed by solution of the opaline shells and redeposition of the silica as a cementing matrix without appreciable transfer and redistribution, and they are minutely porous, because they contain abundant molds of dissolved siliceous shells. Some light is thrown on this process by the following statement of Harker.[75]

Even recrystallization of a single mineral, where no chemical reaction is implied, most usually be brought about by solution. The presence of some solvent medium pervading the rocks is, therefore, to be presumed as an essential part of the mechanism of metamorphism of any kind. It is no less important to observe, however, that the solvent must be present in general only in very exiguous quantity. The kind of solution to which we appeal is a local and temporary solution.

Harker says further:[76]

In general the conditions prohibit flowing movement of the solvent medium, and any redistribution of material must be effected, not by molar, but by molecular flux, that is by diffusion.

Field observations appear to support the view that there were two rather distinct kinds and periods of alteration. The siliceous rock fragments in intraformational conglomerates within the Monterey formation are of dense chert such as seems to have formed in the diagenetic alteration, whereas the siliceous rock pebbles of the Monterey in later conglomerates (Pliocene and Pleistocene) are chiefly of porcelaneous rocks. The gradational change that is apparent in the porcelaneous rocks is evidently related to depth. In areas where the formation is exceptionally thick, the porcelaneous rocks become increasingly dense and hard with depth, and the molds of siliceous tests become correspondingly more obscure.

However, this distinction between two periods of alteration and between the resulting types of siliceous rocks is not evident in all occurrences. The later metamorphism has been superimposed on any earlier diagenetic alteration, and in some areas the two were probably not far separated in time, and were therefore less distinct in character and results. Where metamorphism has been greatest, the lower parts of the Monterey contain a relatively large proportion of cherty rock and are distinctly differentiated into beds of greater and less silica content. Here, also, the proportion of chalcedony and cryptocrystalline quartz gradually increases downward, and that of opal decreases. Some of the dense cherty rocks thus appear to have been formed from the porcelaneous rocks by increased metamorphism, and these are not always distinguishable from the chert formed by diagenetic alteration. The relative amounts of opal and crystalline quartz may be suggestive evidence, however, even for specimens with-

---

[75] Harker, Alfred, Metamorphism, p. 14, Methuen & Co., London, 1932.
[76] Idem, p. 18.

# THE MONTEREY FORMATION OF CALIFORNIA AND THE ORIGIN OF ITS SILICEOUS ROCKS

out relation to their field occurrence, as the early-formed chert consists largely of opal, which in dense beds or ellipsoidal masses would be relatively little altered to crystalline quartz.

Other factors have doubtless influenced the degree of alteration and the extent to which the reprecipitated silica has been segregated into distinct beds or masses. These factors include differences in the original proportion of admixed clay, which would have affected the permeability, and of silica. Some of the rhythmites that have been little altered differ very markedly in silica content...

**COMPARISON WITH OTHER BEDDED CHERT DEPOSITS**

The conclusion that most of the porcelaneous and cherty rocks of the Monterey formation were formed through an alteration that consisted largely of a rearrangment of the silica of originally diatomaceous deposits is based almost entirely on direct evidence derived from examination of these rocks. No principles of physical chemistry, or theories derived from them, that might be of general application to other siliceous deposits have been recognized. It seems worth considering, however, whether certain older siliceous rocks, resembling those of the Monterey in many respects, may not be similarly interpreted, inasmuch as the evidence for such an origin becomes increasingly obscure with age and alteration. Direct evidence on the original nature of some of these older and more highly altered formations would almost necessarily be scanty, even as the evidence becomes more obscure in the lower part of the Monterey siliceous rocks, especially wherever a great thickness of overlying strata and great structural disturbances have obliterated most of the transitional stages of their alteration.

Davis[78] has pointed out the many similarities between the cherts of the Franciscan formation and those of the Monterey formation, and has suggested that the study of the Monterey might offer the best clues to an interpretation of the origin of the older formation. He presented some objections to various hypotheses of origin, including evidence that the thin-bedded cherts were not formed by an epigenetic alteration of deposits of siliceous organisms. These objections, however, were largely confined to alteration effected at the present surface or near it, and do not exclude the possibility of an early alteration within the strata. Intermediate stages of alteration would not be expected in these older rocks, and therefore no method of testing this hypothesis is apparent. Judging from alteration evident in the Monterey formation, however, it seems possible, and even probable, that the Radiolaria preserved in the cherts of the Franciscan formation represent only the heavier-shelled forms of the Radiolaria or other siliceous organisms originally present. The difficulty, if not impossibility, of determining the original character of some of the more ancient and altered deposits of bedded chert is thus emphasized.

---

[78] Davis, E. F., The Radiolarian cherts of the Franciscan group: Univ. California Pub. Dept. Geol. Sci., Bull., vol. 11, no. 3, pp. 235-432, 1918

# Since Bramlette (1946): The Miocene Monterey Formation of California revisited

**Richard J. Behl**
*Department of Geological Sciences, California State University, Long Beach, California 90840-3902; e-mail: behl@csulb.edu.*

## INTRODUCTION

For more than a century the Miocene Monterey Formation has fascinated geologists with its uniquely siliceous composition, complex diagenesis, and importance as both source and reservoir of oil in California. The Monterey's extensive and excellent outcrops, exposed at different stages of alteration, have served as laboratories for countless studies of silica, clay, carbonate, phosphate, organic matter, and petroleum. Bramlette's U.S. Geological Survey Professional Paper 212 (1946) served as the foundation for all of these studies and provided the detailed sedimentology, stratigraphy, and petrology to give them context and meaning. For the most part, Bramlette had it right, and an explosion of new research since the 1970s advanced and refined understanding without disproving many of Bramlette's fundamental observations and assertions. One hypothesis that did eventually fall was that abundant siliceous volcanism was the essential source of the silica incorporated in the frustules of diatoms and in the sediment of the Monterey Formation; we have since learned that within zones of intense upwelling, diatoms or radiolarians can extract enough silica from normal seawater to produce highly siliceous sediments when undiluted by other sedimentary components (Calvert, 1966, 1968).

Research since Bramlette's has broadly focused on diagenesis (especially that of silica, carbonate, and organic matter), petroleum generation and reservoirs, dating and stratigraphic correlation, and the oceanographic context of deposition of the Monterey Formation. Much of this work benefited from technological advances in X-ray diffraction, stable isotopic analysis, electron microscopy, and the results of the Deep Sea Drilling Project (DSDP) and Ocean Drilling Program (ODP). A burst of research, initially proprietary, began about 1970 as oil companies sought to explain and exploit major offshore discoveries in the Monterey Formation following the first sale of Federal Outer Continental Shelf leases in the Santa Barbara Channel in 1966. A tremendous amount of this work was published in a series of American Association of Petroleum Geologists (AAPG) and Society of Economic Paleontologists and Mineralogists (SEPM; now the Society for Sedimentary Geology) special publications and symposium volumes in the 1980s and 1990s (Isaacs, 1981a; Garrison and Douglas, 1981; Williams and Graham, 1982; Isaacs and Garrison, 1983; Garrison et al., 1984; Surdam, 1984; Casey and Barron, 1986; Dunham and Blake, 1987; Schwalbach and Bohacs, 1992; Hornafius, 1994a; Eichhubl, 1998) that coincided with the upturn in industry interest in the petroleum potential of the offshore Monterey (Isaacs, 1984; Crain et al., 1985). An additional major volume focusing on the organic geochemistry of the Monterey Formation (Isaacs and Ruellkötter, 1999) should be published by the time this review is published.

## GEOLOGIC SETTING

The Miocene Monterey Formation was deposited along the North American plate boundary during the transition of the California margin from a convergent to transform setting (Blake et al., 1978; Barron, 1986a). Resulting tectonic subsidence and landward transgression of the shoreline during the late Oligocene to middle Miocene led to the development of middle bathyal depocenters in which the Monterey sediments accumulated (Figs. 1 and 2) (Ingle, 1980, 1981a). Presedimentary and synsedimentary tectonic deformation (chiefly extension, shearing, and rotation) during the Miocene has been overprinted by Pliocene-Pleistocene shortening, making palinspastic reconstruction of the location and extent of the Neogene sedimentary basins extremely challenging (Ingersoll and Ernst, 1987; Crouch and Suppe, 1993; Fritsche, 1998; Isaacs, 1999). In

---

Behl, R. J., 1999, Since Bramlette (1946): The Miocene Monterey Formation of California revisited, *in* Moores, E. M., Sloan, D., and Stout, D. L., eds., Classic Cordilleran Concepts: A View from California: Boulder, Colorado, Geological Society of America Special Paper 338.

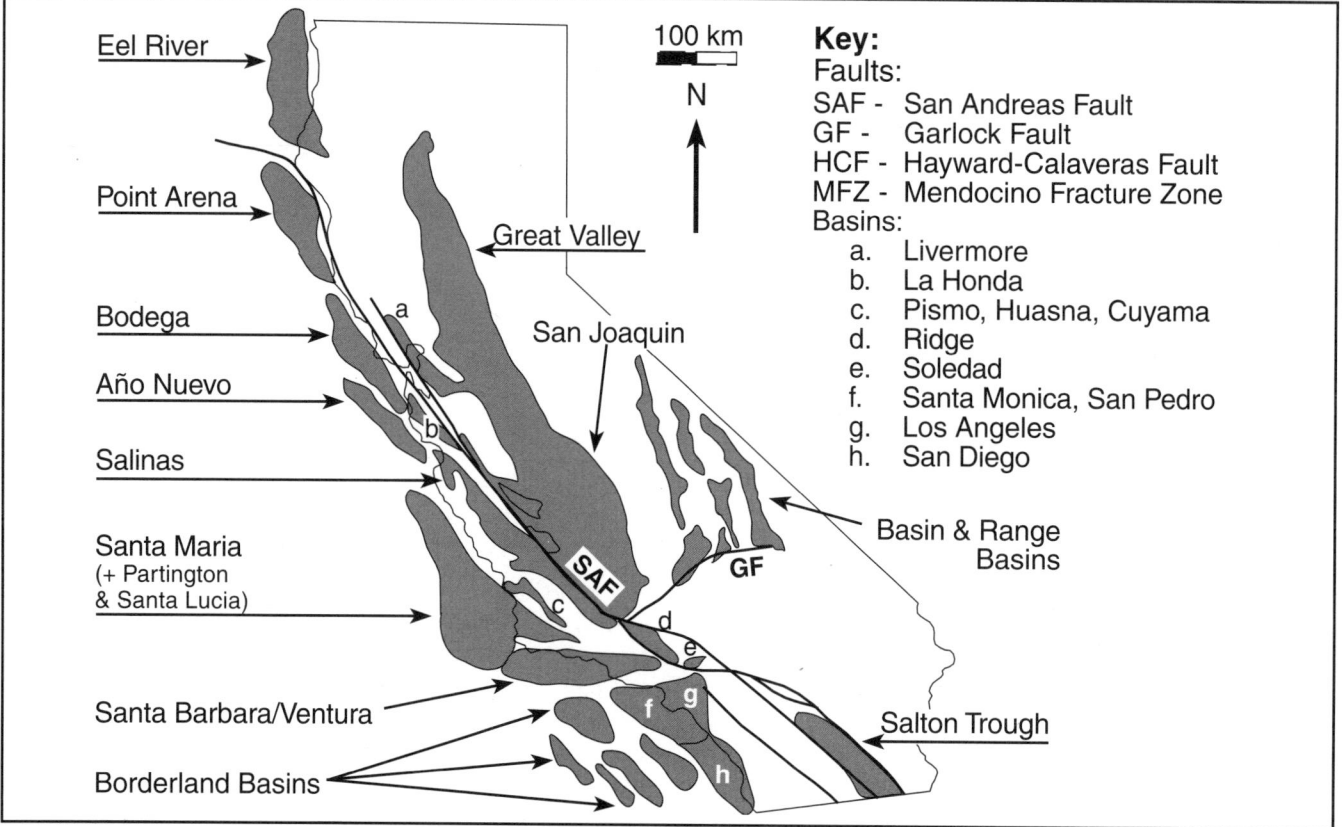

Figure 1. Present location of Neogene depocenters or sedimentary basins (after Biddle, 1991; Dunkel and Piper, 1997).

many cases, the geometry and bathymetry of individual depocenters evolved from the Miocene through the Pleistocene, with earlier deposited sediments, including the Monterey Formation, now forming the folded and faulted flanks of the Pliocene and Quaternary depocenters (Teng and Gorsline, 1989; Blake, 1991).

## GEOGRAPHIC EXTENT

The Monterey Formation is part of a discontinuous belt of fine-grained, notably siliceous (diatomaceous) sediments that accumulated around the north Pacific Rim chiefly during the Miocene (ca. 16–4 Ma) (Ingle, 1973, 1980, 1981b). On land, well-studied Monterey strata form extensive outcrops and subcrops in the Coast Ranges and western parts of California (Bramlette, 1946; Pisciotto and Garrison, 1981), extending as an irregular blanket some 1700 km north and south along the continental margin. Offshore equivalents of Monterey siliceous sediments have been cored by deep-sea drilling as far as 300 km seaward from the modern coast and in water as deep as 4200 m (ODP Sites 1010, 1016, 1021) (Lyle et al., 1997). The formation is typically 300–500 m thick on land, but is locally much thinner and thicker (Bramlette, 1946; Isaacs and Petersen, 1987).

## AGE OF THE MONTEREY FORMATION

Like most lithostratigraphic units, the age of the middle to late Miocene Monterey Formation varies with location, as sedimentation characteristic of the formation commenced and terminated at different times in separate depocenters. If a typical duration could be specified, it would be from about 16 Ma to 6 Ma (Barron, 1986b). Initiation of Monterey deposition started as early as 17.8 Ma (Saucesian stage, Naples Beach) (DePaolo and Finger, 1991) or as late as 15 Ma (e.g., Relizian, Palos Verdes Hills, Berkeley Hills, Monterey; Obradovich and Naeser, 1981). The youngest Monterey strata at any one location range from about 13 Ma (Luisian, Berkeley Hills) to <5 Ma (Delmontian, Pliocene, Palos Verdes Hills; Woodring et al., 1946; Obradovich and Naeser, 1981). In the Cuyama basin, the Saltos Shale and Whiterock Bluff Shale, often assigned as members of the Monterey Formation (Hill et al., 1958), were apparently deposited entirely before initiation of Monterey sedimentation in the type area (Obradovich and Naeser, 1981).

### Biostratigraphy

Microfossils provide the primary basis for biostratigraphy within the fine-grained Monterey Formation, with benthic

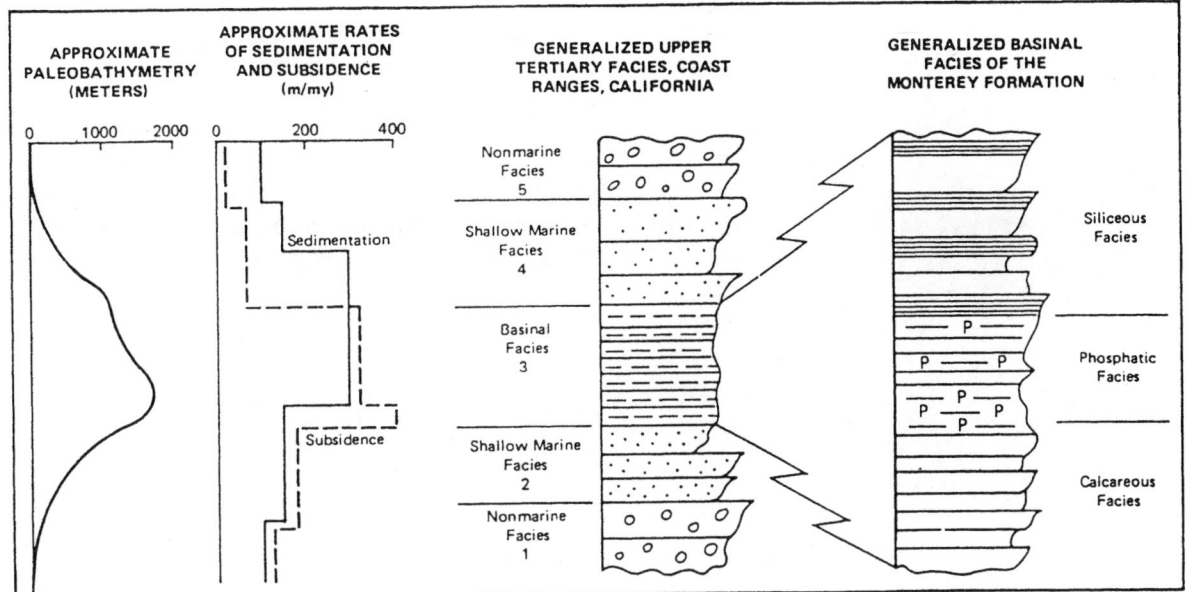

Figure 2. Generalized upper Tertiary sedimentary facies of the California Coast Ranges, showing the position and facies of the Monterey Formation (Pisciotto and Garrison, 1981).

foraminifers remaining the most commonly used taxa for correlation. Because of downsection silica phase transformations and upsection loss or dissolution of carbonate, none of the major biostratigraphic groups are generally useful through the entire formation.

Monterey strata span the late Saucesian, Luisian, Mohnian, and locally, the early Delmontian benthic foraminiferal stages of California (Kleinpell, 1938, 1980). Since development of these Neogene stages, however, it has become evident that benthic assemblages were influenced by local paleobathymetry, character of the impinging water mass, and benthic sedimentology, making them time transgressive and often provincial in nature (Crouch and Bukry, 1979; Ingle, 1980; Obradovich and Naeser, 1981). Although still quite useful within individual fields or basins because of their abundance (Finger, 1995; Blake, 1991), benthic foraminifers had to be integrated with planktonic foraminifers (Keller and Barron, 1981), diatoms (Barron, 1986b; Barron and Isaacs, 1999), nannofossils (Poore et al., 1981), magnetostratigraphy (Omarzai, 1992), radiometric geochronology (Obradovich and Naeser, 1981), and chemostratigraphy (DePaolo and Finger, 1991; Flower and Kennett, 1993, 1994).

## LITHOLOGY AND COMPOSITION

The Monterey Formation is distinguished by its overall highly biogenic composition, in which the average contributions of silica (chiefly the tests and frustules of diatoms and radiolarians), carbonate (coccoliths and foraminifers), organic matter (mostly type II kerogen, marine algae) and their diagenetic equivalents greatly exceed those of other Neogene fine-grained sedimentary units (Isaacs, 1985; Isaacs and Petersen, 1987). Although the highly diatomaceous and organic-rich deposit has been interpreted to record unusually great planktonic productivity along the eastern Pacific margin (Barron, 1986a; Ingle, 1980, 1981b; Pisciotto and Garrison, 1981), mass accumulation rates show that the purest biogenic intervals reflect decreased terrigenous input, and consequently less dilution of the biogenic component (Isaacs, 1985, 1999). Overall, the Monterey Formation records sediment starvation in conjunction with surface productivity associated with upwelling along the California Current system. These sedimentary conditions increased the relative proportions of silica, organic matter, phosphate, or carbonate with respect to fine-grained detritus— mainly illite-smectite mixed-layer clay minerals, feldspars, and quartz (Isaacs, 1980; Pollastro, 1990; Compton, 1991). Periods of extremely slow pelagic sedimentation, undiluted by much fine-grained detritus and during which most of the primary biogenic hard components ($SiO_2$ or $CaCO_3$) dissolved or winnowed away, resulted in the extreme organic richness characteristic of some condensed intervals (e.g., the carbonaceous marl–phosphatic shale facies of the Santa Barbara coastal area) (Isaacs, 1985).

Bramlette (1946) described the typical Monterey lithologies—diatomite, diatomaceous and siliceous mudrocks, porcelanite, chert, calcareous and phosphatic mudrocks, dolostone, and limestone—in remarkable completeness and detail. He also recognized the significance of graded, clastic to biogenic "rhythmites" before the importance of fine-grained turbidites in deep water was generally understood or accepted. Where clastic siltstone and sandstone are common, they are usually assigned to another formation or to a distinct member of the Monterey (e.g., Point Sal Formation, Santa Maria basin, or the Stevens Sands, southern San Joaquin basin) (Williams and Graham,

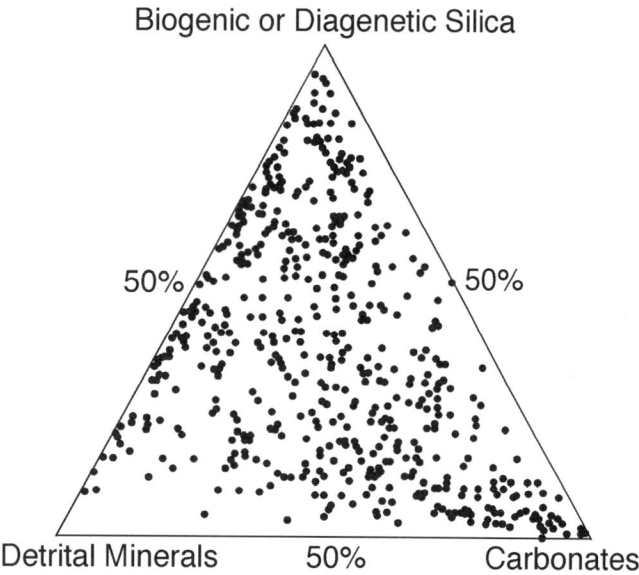

Figure 3. Diagram showing the wide range of sedimentary compositions in the Monterey Formation of the Santa Maria and Santa Barbara basins (Isaacs, 1985).

1982). Conglomerates are even more rare (Garrison and Ramirez, 1989). At scales from less than 1 mm to hundreds of meters, the lithologies of the Monterey are characterized by great compositional variability, making any individual sample usually unrepresentative of its own stratigraphic interval (Fig. 3) (Isaacs, 1985). Compositional variation is expressed by rhythmic alternation of clastic-biogenic, massive-laminated, or diagenetically distinct lithologic cycles (Pisciotto and Garrison, 1981; Govean and Garrison, 1981; Isaacs, 1985). Even with such lithologic variation, large-scale trends in average composition, both vertically and laterally, are relatively consistent within individual regions, giving rise to a number of local stratigraphic subdivisions into informal members (e.g., Canfield, 1939; Woodring et al., 1943; Foss and Blaisdell, 1968; Isaacs, 1981b, 1983; Pisciotto and Garrison, 1981; MacKinnon, 1989a). Although there is a broad similarity to some of the compositional trends—for example, the widespread occurrence of middle Luisian to Relizian calcareous mudstones and late Luisian to early Delmontian diatomaceous sediments—member-scale facies are clearly time transgressive when compared between regions (Blake, 1981; White, 1989; Hornafius, 1991, 1994b; Schwalbach and Bohacs, 1995).

## DEPOSITIONAL ENVIRONMENTS

The Monterey Formation was chiefly deposited in lower middle bathyal (1500–2300 m) to upper middle bathyal (500–1500 m) environments (Fig. 2) (Ingle, 1973, 1980; Isaacs, 1999), which shallowed upward in most sequences. Preservation of organic matter, abundance of fine varve-like laminations, and presence of dysaerobic benthic foraminifers indicate that the Monterey was commonly deposited in or associated with an oxygen-deficient environment. Consequently, likely depositional environments for the Monterey include basin plains, slopes, banktops, and shelf edges where they intersect or are influenced by the mid-water oxygen minimum zone (Calvert, 1966; Garrison et al., 1979; Lagoe, 1981; Pisciotto and Garrison, 1981). Possible modern analogues for these settings occur beneath upwelling zones associated with the Southern California Borderland, the Gulf of California, and the Peru and Pakistan margins (Calvert, 1966, 1968; Donegan and Schrader, 1981; Soutar et al., 1981). Although the silled basins of the California Continental Borderland have been most frequently cited as present-day examples, there is little direct evidence for the existence of such steep-sided, silled basins during deposition of the Monterey (Isaacs, 1999).

Thin, millimeter-scale laminations are only intermittently present in the Monterey Formation. They are rare in the predominantly massive and thin- to thick-bedded lower portion of the Monterey, and become increasing prevalent upsection (Mohnian stage), while remaining rhythmically or irregularly interbedded with massive (bioturbated or redeposited) strata (Pisciotto and Garrison, 1981; Govean and Garrison, 1981; Isaacs, 1985; Ozalas et al., 1994). Such alternation suggests continuously fluctuating levels of paleo-oxygenation during deposition (Behl and Kennett, 1996). The overall upward increase in lamination indicates either that bottom water was progressively (if inconsistently) depleted of oxygen through time or that the Monterey Formation depositional environment shoaled into the heart of the mid-water oxygen minimum zone with progradation of the Miocene California margin (Isaacs et al., 1996).

## DIAGENESIS

The highly reactive biogenous components of the Monterey Formation (i.e., opaline silica, calcite, phosphate, and organic matter) have undergone a complex paragenetic sequence of alteration with time, burial, and tectonic deformation. Although it is simpler to examine mineralogic and chemical systems in isolation, every stage of dissolution, precipitation, or alteration influenced simultaneous and subsequent reactions by altering pore-water chemistry, water-rock ratios, and permeability (Kastner et al., 1984; Eichhubl and Behl, 1998). Diagenetic modification by chemical migration can enhance or suppress the physical and compositional contrasts that already existed in the originally heterogeneous Monterey sediments, making it a wonderfully complicated unit to work with (Pisciotto and Garrison, 1981; Govean and Garrison, 1981; Grivetti, 1982; Murray and Jones, 1992; Behl, 1992).

Figure 4. Scanning electron micrograph of nascent opal-CT lepispheres growing within a partially dissolved opal-A test of a diatom. Field of view = 15 μm.

## Silica

Although Bramlette (1946) clearly documented the alteration of soft diatomaceous sediments to hard porcelanite and chert, we now know considerably more about the nature, controls, and distribution of silica diagenesis. The sequence of mineralogic alteration involves two steps of complete dissolution and reprecipitation. The first is from biogenic opal-A (hydrous silica that is crystallographically amorphous to X-ray diffraction) to diagenetic opal-CT (hydrous silica composed of interlayered cristobalite and tridymite) (Fig. 4). The second is from opal-CT to diagenetic quartz (generally cryptocrystalline, microcrystalline, or chalcedonic quartz). Transformation is controlled by temperature and burial depth (Murata and Nakata, 1974; Murata and Larson, 1975; Isaacs, 1981c; Pisciotto, 1981a), bulk composition (Isaacs, 1982; Behl and Garrison, 1994), and rock properties, such as porosity and permeability (Behl, 1998; Eichhubl and Behl, 1998). Within sediments of common compositions for the Monterey Formation (i.e., diatomaceous or siliceous mudstones and porcelanites), silica phase conversion takes place within two relatively narrow temperature ranges and burial depths (~40–50 °C and ~0.5–2 km for opal-A to opal-CT and ~65–80 °C and ~1.5–3 km for opal-CT to quartz; Fig. 5) (Pisciotto, 1981a; Keller and Isaacs, 1985). Within an individual stratigraphic sequence, however, the silica phase transformation may not be abrupt, but can occur across a broad transition zone, to 300 m thick, of interbedded lithologies containing different silica phases (Fig. 6) (Isaacs, 1982).

The stratigraphic co-occurrence of silica phases with different thermal stabilities and solubilities is explained by compositionally controlled variation in the kinetics of the phase transformations, in which the opal-A to opal-CT transition is retarded and the opal-CT to quartz transition is accelerated in more detrital- or clay-mineral–rich sediments (Kastner et al., 1977; Isaacs, 1981c, 1982; Williams et al., 1985). The purest siliceous sediments undergo diagenesis even earlier than predicted (Bohrmann et al., 1994), with hard, brittle opal-CT and quartz cherts forming at temperatures as low as 2–33 °C and 36–76 °C, respectively (Fig. 5) (Behl, 1992; Behl and Garrison, 1994). On a larger scale, boundaries between silica phase zones are locally discordant to stratigraphy, reflecting lateral variation in sediment accumulation and burial depth, geothermal gradient, or tectonic deformation (Figs. 6 and 7) (Bramlette, 1946; Murata and Larson, 1975; Pisciotto, 1981a). Within each diagenetic zone, silica becomes increasingly well ordered with depth, temperature, or time, even though there may not be any lithologic change. Opal-A becomes less soluble as higher surface area diatoms dissolve and smaller submicroscopic mineralogic domains give way to larger ones (Williams et al., 1985). Ordering of opal-CT is revealed by decreased spacing of the $d_{101}$ lattice planes (Murata and Nakata, 1974; Murata and Larson, 1975; Cady et al., 1996) and increased crystallite size with growth (Grivetti, 1982; Williams et al., 1985; Behl and Meike, 1990). Progressive growth of crystallite domains in diagenetic quartz is shown by the height and sharpness of X-ray diffraction peaks in the quartz crystallinity index (Murata and Norman, 1976).

Complete dissolution and reprecipitation at the two silica phase transitions produce dramatic changes in the physical properties of the sediment as the rigid, but porous framework collapses, or as internal pore spaces are filled with

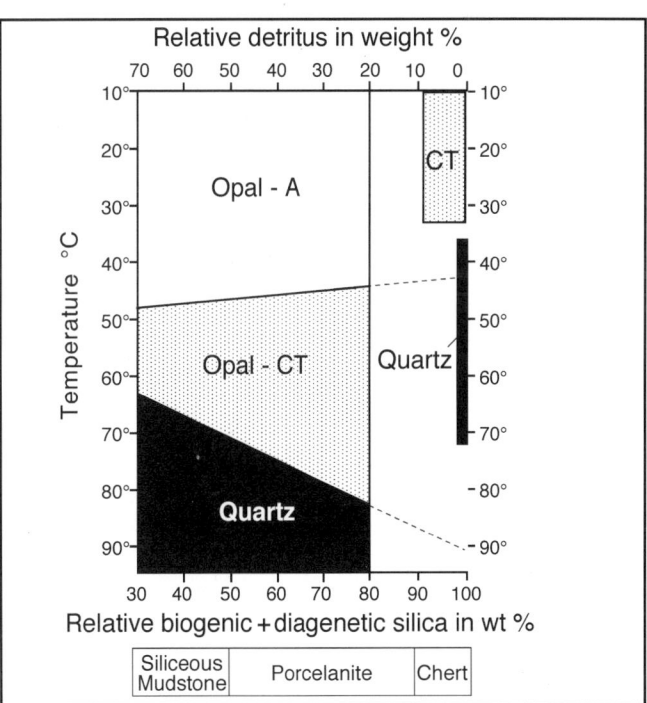

Figure 5. Diagram showing the relative timing and temperatures of silica phase changes (Keller and Isaacs, 1985), modified to include data on the purest diatomites and cherts (Behl, 1992; Behl and Garrison, 1994).

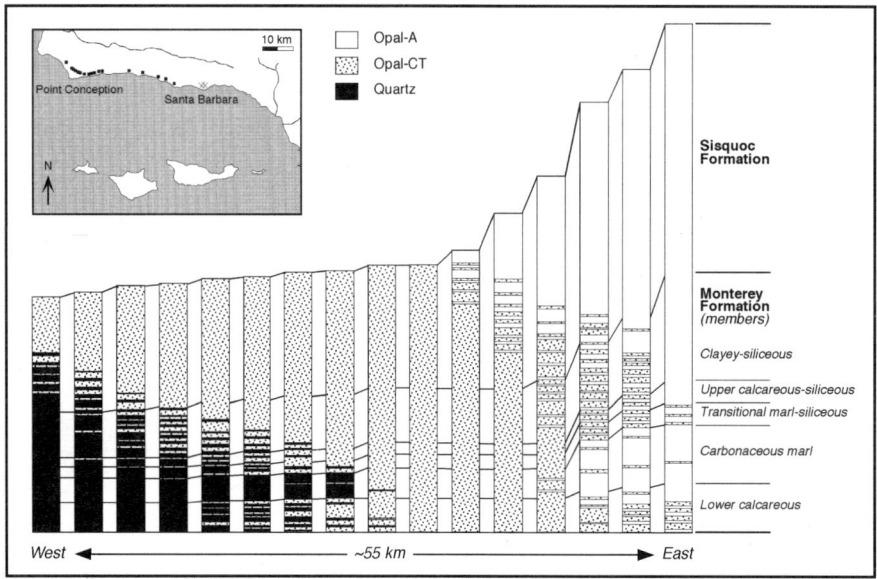

Figure 6. Schematic view of Santa Barbara coastal area, showing silica phase zones cutting across lithostratigraphic boundaries, the interbedded nature of the transition zones, and typical compaction with increased physical and chemical diagenesis (modified from Isaacs, 1981).

silica cement (Fig. 4). These abrupt changes in bulk density can be imaged locally by seismic methods as extensive crosscutting reflectors in the subsurface (Fig. 7) (Mayerson and Crouch, 1994) and are associated with the expulsion, migration, and trapping of hydrocarbons as well as the potential for forming fractured petroleum reservoirs (McGuire et al, 1983; MacKinnon, 1989b; Mayerson et al., 1995).

## *Carbonate*

Carbonate diagenesis in the Monterey Formation has been studied by a wide variety of geochemical, isotopic, and sedimentological means to determine its paragenesis with organic matter and silica (Murata et al., 1967, 1972; Friedman and Murata, 1979; Pisciotto, 1981b; Garrison et al., 1984; Burns and Baker, 1987; Malone et al., 1996; Eichhubl and Boles, 1998). Although primary carbonate components are mainly calcitic coccoliths and foraminifers, the dominant secondary carbonate phase in the Monterey is calcium-rich dolomite, whether occurring as finely disseminated rhombs, cross-cutting veins, or as tightly cemented concretions and beds (Pisciotto, 1981b). Dolomite forms in anoxic or dysoxic conditions related to the diagenesis of organic matter, within or below the zone of sulfate reduction (Pisciotto and Mahoney, 1981). Low sedimentation rates during early burial diagenesis tend to increase the concentration of dolomite by providing better conditions for continued precipitation in the zone of sulfate reduction (Pisciotto and Mahoney, 1981; Burns and Baker, 1987).

## *Phosphate*

Diagenetic sedimentary phosphate (cryptocrystalline carbonate fluorapatite) forms chiefly with the shallow degradation of organic matter, probably via a number of physical, chemical, and biological mechanisms (Garrison et al., 1990; Föllmi et al., 1991). Most carbonate fluorapatite precipitation occurs within a few tens of centimeters of the sediment surface and during slow sedimentation or depositional hiatuses (Garrison et al., 1994). The most prominent phosphatic facies in the bathyal Monterey Formation are laminated, organic-rich phosphatic marlstones that developed as the condensed residue of slowly deposited, calcareous-siliceous muds and oozes (Garrison et al., 1987) during sediment starvation (Isaacs, 1985, 1999). In this facies, carbonate fluorapatite occurs as small nodules, lenses, laminations, and peloids that formed in place with little or no subsequent reworking. Shelfal and banktop phosphoritic sands occur interbedded with hemipelagic sediments, and consist mostly of phosphatic peloids (Garrison et al., 1987, 1994). Conglomerates and hardgrounds composed of dense, dark phosphatic pebbles, nodules, and concretions are less common in the Monterey Formation, but record repeated episodes of phosphatization, exhumation, winnowing, and reworking by currents, slumping, and sea-level change (Föllmi et al., 1991; Garrison et al., 1994).

## SOURCE OF PETROLEUM

The Miocene Monterey Formation is widely considered to be the primary source rock for hydrocarbons in California (Woodring and Bramlette, 1950; Crawford, 1971; Taylor, 1976; Lillis and Lagoe, 1983; Isaacs and Petersen, 1987). Total organic carbon (TOC) in the Monterey can be as high as 23% (34% organic matter by weight), but averages between 2% and 5%, with large sample to sample variation, depending on lithology and depositional setting (Isaacs and Petersen, 1987).

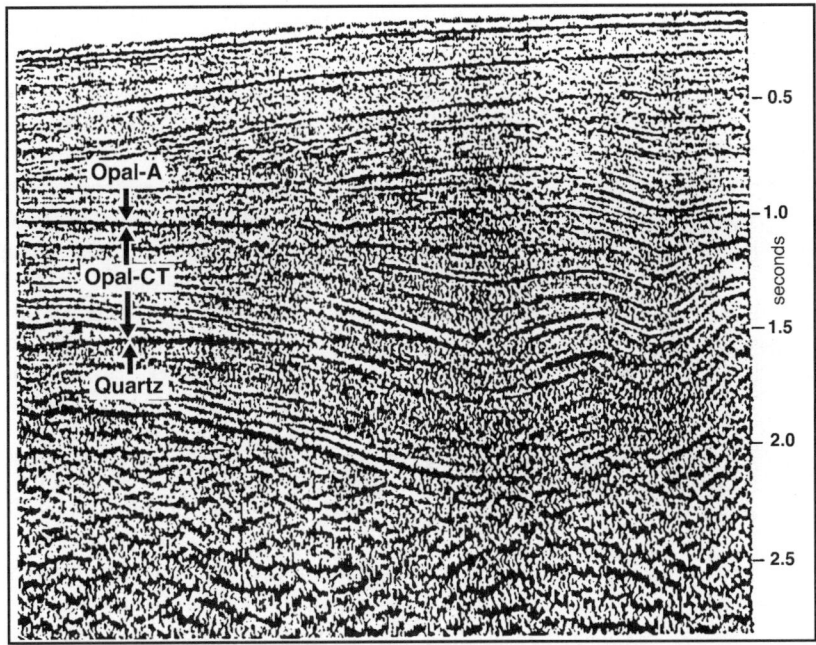

Figure 7. Seismic-reflection profile showing the near-horizontal opal-A to opal-CT and opal-CT to quartz silica phase transitions that cut across stratigraphy. After Crouch, Bachman, and associates, 1991.

Organic matter is overwhelmingly amorphous marine algal debris, but locally includes significant portions of terrestrial origin (Isaacs and Magoon, 1984; Graham and Williams, 1985). Biomarkers in both Monterey oil and rocks also indicate that the organic matter is largely marine (King and Claypool, 1983; Curiale et al., 1985). Kerogens in the highly biogenic Monterey sediments are mostly sulfur-rich, oil-prone type II-S (Surdam and Stanley, 1981; Kruge, 1983; Orr, 1986; Isaacs, 1988; Ruellkötter and Isaacs, 1996).

Much of the oil sourced in the Monterey was generated in rocks considered to be immature or marginally mature by conventional methods of assessment, for example, vitrinite reflectance ($R_o$ <0.4), thermal alteration index (TAI <2.3), Rock-eval pyrolysis ($T_{max}$, variable and problematic), sapropel fluorescence, hydrogen/carbon ratios, and silica diagenetic grade (Taylor, 1976; McCulloh, 1979; Kablanow and Surdam, 1984; Global Geochemistry Corporation, 1985; Petersen and Hickey, 1987; Ruellkötter and Isaacs, 1996), although some of these indicators may not be reliable indicators of maturity in Monterey-type rocks (Walker et al., 1983). Initiation of catagenesis as early as 60–80 °C is likely related to the high sulfur content (up to 9% by weight) of the kerogen and the weakness of its carbon-sulfur bonds (Hunt, 1979; Orr, 1984, 1986; Isaacs and Petersen, 1987). The generally low API gravity (<20 API°) of Californian oil is also related to early generation, low maximum temperatures, and bacterial degradation, both as organic matter and as hydrocarbons (Petersen and Hickey, 1987; Ruellkötter and Isaacs, 1996). The co-occurrence of both in situ kerogen and migrated hydrocarbons within the rock matrix presents difficulties in assessing the true maturity of source rocks in the Monterey Formation as well as the relative contributions of oil from adjacent or distant (deeper) sources within the formation (Dunham et al., 1991).

Although much of the Monterey Formation has sufficiently high TOC and H/C ratios to be classified as good oil-prone source rock, a proportionally large amount of the oil may come from organic-rich carbonaceous marl (phosphatic shale) strata (Orr, 1984; Dunham et al., 1991; Isaacs and Ruellkötter, 1999) at whatever stratigraphic level and location it is best developed.

## PETROLEUM RESERVOIRS

The Monterey Formation is unusual in that it is both source and reservoir of oil (Crawford, 1971; Isaacs and Petersen, 1987). Typically, fine-grained organic-rich rocks lack the effective porosity and permeability to provide commercial petroleum reservoirs. Consequently, petroleum reservoirs generally consist of either adjacent or interfingered sandstone beds, members, or formations, or they consist of naturally fractured, brittle diagenetic siliceous and dolomitic rocks. Oil is also locally produced from highly porous, opal-A diatomite in western parts of the San Joaquin basin through natural and artificially induced fractures.

The high diagenetic potential of the Monterey's fine-grained components (chiefly of silica, carbonate, and organic matter), diagenetic embrittlement (of chert, porcelanite, and dolomite) with burial, and location in a tectonically active set-

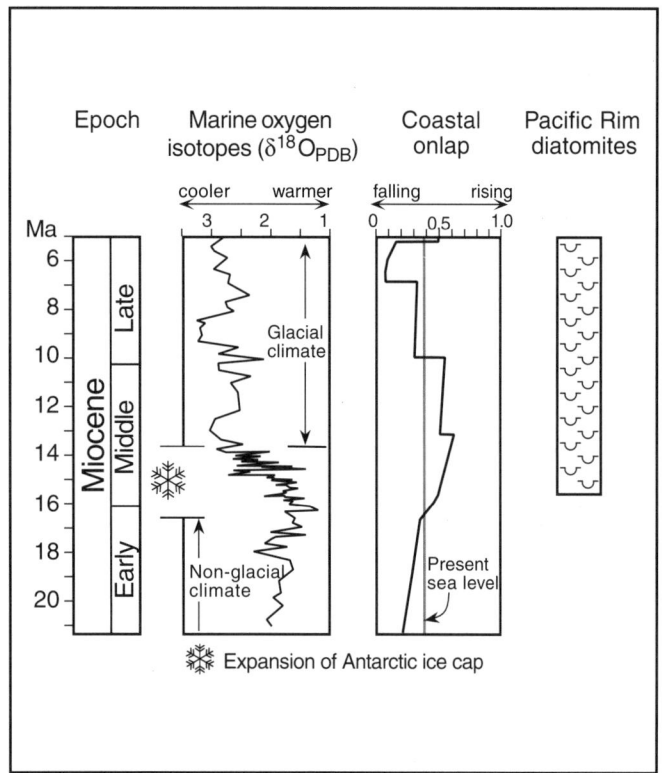

Figure 8. Global Miocene climatic and eustatic events and occurrence of Miocene diatomites in bathyal sequences around the north Pacific margin (modified from Ingle, 1981b). PDB—Peedee belemnite.

ting combined to create many highly fractured or brecciated oil reservoirs in the subsurface (Regan and Hughes, 1949; McGuire et al., 1983; Dunham and Blake, 1987; MacKinnon, 1989b; Eichhubl and Behl, 1998). Depending on original depositional constraints, different lithologies may make up the important fractured reservoirs in individual fields. Whereas fractured siliceous shale and porcelanite provide important production in the San Joaquin basin, chert and dolomite breccias form the most important reservoirs in the onshore and offshore Santa Maria basin (Redwine, 1981; Roehl, 1981; McGuire et al., 1983; Crain et al., 1985; Dunham et al., 1991). In all cases, fractures are critical for fluid flow in the otherwise extremely low permeability (<1 md) Monterey lithologies. The distribution and density of fractures vary with rock type, diagenetic grade, bed thickness, location on tectonic structures, and the regional stress field (Snyder et al., 1983; Belfield et al., 1983; Snyder, 1987; MacKinnon, 1989b; Narr, 1991; Gross, 1993; Gross et al., 1997; Finkbeiner et al., 1997) and are also related to large-scale faulting (Eichhubl, 1997).

In addition to microscopic and macroscopic fracture porosity and permeability, most highly siliceous rocks have substantial (10%–35%) matrix porosity (Isaacs, 1981d), which can form the major part of reservoir storage, but also contributes to a complex production behavior.

While much of the oil generated in the Monterey is produced from associated or overlying clastic reservoirs, fractured reservoirs are locally very important. For example, Monterey fractured reservoirs account for ~75% of the oil produced in the Santa Maria area (Crawford, 1971). In the most recent assessment of hydrocarbon resources of the Pacific Outer Continental Shelf region, fractured Monterey Formation or equivalent strata are estimated to contain more than one-half of the undiscovered conventionally recoverable oil (5.96 billion barrels) and more than one-third of the undiscovered conventionally recoverable gas (6.32 trillion cubic feet) for a total of 7.08 billion barrels of oil equivalent (Dunkel and Piper, 1997).

## PALEOCEANOGRAPHIC AND PALEOENVIRONMENTAL SIGNIFICANCE

Deposition of the Monterey and its equivalents coincided with or followed major changes in Miocene ocean circulation, global climate, and tectonics (Ingle, 1980, 1981b; Pisciotto and Garrison, 1981; Vincent and Berger, 1985; Barron, 1986a). The diatomaceous and organic-rich Monterey sediments were deposited following a major switch in marine thermohaline circulation into approximately the modern configuration where deep water that forms in the North Atlantic and circum-Antarctic regions principally upwells in the Pacific and Indian Oceans (Kennett, 1977; Keller and Barron, 1983; Woodruff and Savin, 1989). Monterey deposition also encompassed the important middle Miocene cooling step in which the Southern Hemisphere cryosphere expanded into western Antarctica (Fig. 8) (Kennett, 1977; Miller et al., 1987). Regional intensification of upwelling and increased affinity with higher latitude assemblages in the late Miocene is indicated by most planktonic taxa (Ingle, 1973, 1981b; Weaver et al., 1981; Barron, 1986a). The co-occurrence of all these events in the middle to late Miocene has led many to attribute or relate the character of Monterey deposits to this important reorganization of the Earth's cryosphere-hydrosphere-atmosphere system, both as cause and as effect (Ingle, 1981b; Pisciotto and Garrison, 1981; Vincent and Berger, 1985; Barron, 1986a). In particular, middle Miocene accumulation of organic matter in marine sediments was great enough to perturb the carbon balance of the global ocean and atmosphere and produce a prominent positive excursion in carbon isotopes that has been recognized in deep-sea and Monterey sequences (Vincent and Berger, 1985; Compton et al., 1990; Flower and Kennett, 1993;,1994; Raymo, 1994). Although the accumulation of organic carbon in the Monterey Formation alone was probably insufficient to account for this shift, the Monterey was clearly deposited within the context of an important transition in Cenozoic cooling associated with cryospheric expansion, thermohaline circulation reorganization (Fig. 8), and possibly accelerated flux of nutrients to the ocean related to Himalayan uplift (Richter et al., 1992). The widespread lower calcareous mudstone facies of the Monterey was largely deposited during an interval of early to middle Miocene gradual warming. The phosphatic and organic-rich facies correlate with a middle Miocene sea-level rise and highstand that occurred prior to expansion of the Antarctic ice sheet (Pisciotto and Garrison, 1981), thus are in effect, condensed, "transgressive shales" (Isaacs, 1999).

Recently, major member-scale stratigraphic shifts in bulk composition in the Monterey have been reinterpreted to reflect shoaling and shoring of the Monterey depositional environment as part of a prograding margin, modified by eustatic sea-level changes, rather than regional or global changes in paleoceanography and climate (Isaacs et al., 1996; Isaacs, 1999). In this model, the time-transgressive nature of major compositional lithofacies reflects proximity to loci of coastal or banktop upwelling, sources of terrigenous detritus, as well as periods of sediment starvation (Isaacs et al., 1996; Isaacs, 1999). For example, the generally most siliceous middle to upper members of the Monterey (late Miocene, Mohnian stage) are interpreted to reflect deposition within the direct influence of shallow (~500 m or less) coastal (~20 km from shore) upwelling or bathymetrically induced upwelling, such as that adjacent to shallowly submerged banks. This interpretation is difficult to reconcile, however, with the presence of highly diatomaceous middle to late Miocene deposits in offshore locations from Baja California to the Oregon border that were deposited and remain at middle to lower bathyal depths and are >100 km away from the modern prograded shoreline (Ingle, 1973, 1980; Barron, 1986a, 1986b; Blake, 1981; Lyle et al., 1997). The wide spatial distribution of the important and unusual Monterey-type deposits likely reflects the unique co-occurrence of paleoceanographic, paleoclimatic, and tectonic events during the Miocene epoch.

## ACKNOWLEDGMENTS

This review could only cite a few of the many studies of the Monterey Formation. I am greatly indebted to all of the "friends of the Monterey" who by their efforts have given us so much over the past century. Any omission is due to space limitations and my incomplete efforts—I hope not to offend anyone whose work I left out. The manuscript was greatly improved by the reviews of Tom Wright and Bob Garrison. I also thank the editors, Dottie Stout, Eldridge Moores, and Doris Sloan, for encouraging me to complete this manuscript. Acknowledgment is made to the Donors of The Petroleum Research Fund, administered by the American Chemical Society, for support of this research.

## REFERENCES CITED

Barron, J. A., 1986a, Paleoceanographic and tectonic controls on deposition of the Monterey Formation and related siliceous rocks in California: Palaeogeography, Palaeoclimatology, Palaeoecology, v. 53, p. 27–45.

Barron, J. A., 1986b, Updated diatom biostratigraphy for the Monterey Formation of California, in Casey, R. E., and Barron, J. A., eds., Siliceous microfossil and microplankton of the Monterey Formation and modern analogs: Los Angeles, Pacific Section, Society of Economic Paleontologists and Mineralogists, p. 105–120.

Barron, J. A., and Isaacs, C. M., 1999, Updated chronostratigraphic framework for the California Miocene, in Isaacs, C. M., and Ruellkötter, J., eds., The Monterey Formation: From rocks to molecules: New York, Columbia University Press (in press).

Behl, R. J., 1992, Chertification in the Monterey Formation of California and Deep-Sea sediments of the West Pacific [Ph.D. thesis]: Santa Cruz, University of California, 287 p.

Behl, R. J., 1998, Relationships between silica diagenesis, deformation, and fluid flow in Monterey Formation cherts, Santa Maria Basin, California, in Eichhubl, P., ed., Diagenesis, deformation, and fluid flow in the Miocene Monterey Formation: Pacific Section, SEPM (Society for Sedimentary Geology) Special Publication 83, p. 77–83.

Behl, R. J., and Garrison, R. E., 1994, The origin of chert in the Monterey Formation of California (USA), in Iijima, A., Abed, A., and Garrison, R., eds., Siliceous, phosphatic and glauconitic sediments of the Tertiary and Mesozoic: Utrecht, International Geological Congress Proceedings, Part C: p. 101–132.

Behl, R. J., and Kennett, J. P., 1996, Brief interstadial events in the Santa Barbara basin, NE Pacific, during the past 60 kyr: Nature, v. 379, p. 243–246.

Behl, R. J., and Meike, A., 1990, Cryptocrystalline relationships of silica phases in chert: Particulate Science and Technology, v. 8, p. 111–122.

Belfield, W. C., Helwig, J., La Pointe, P., and Dahleen, W. K., 1983, South Ellwood Oil Field, Santa Barbara Channel, California, a Monterey Formation Fractured Reservoir, in Isaacs, C. M., and Garrison, R. E., eds., Petroleum generation and occurrence in the Miocene Monterey Formation, California: Los Angeles, Pacific Section, Society of Economic Paleontologists and Mineralogists, p. 213–221.

Biddle, K. T., 1991, The Los Angeles Basin: An overview, in Biddle, K. T., ed., Active margin basins: American Association of Petroleum Geologists Memoir 52, p. 1–24.

Blake, G. H., 1981, Biostratigraphic relationship of Neogene benthic Foraminifera from the southern California Outer Continental Borderland to the Monterey Formation, in Garrison, R. E., and Douglas, R. G., eds., The Monterey Formation and related siliceous rocks of California: Los Angeles, Pacific Section, Society of Economic Paleontologists and Mineralogists, p. 1–14.

Blake, G. H., 1991, Review of the Neogene biostratigraphy and stratigraphy of the Los Angeles Basin and implications for basin evolution, in Biddle, K. T., ed., Active margin basins: American Association of Petroleum Geologists Memoir 52, p. 135–184.

Blake, M. C., Campbell, R. H., Dibblee, T. W., Jr., Howell, D. G., Nilsen, T. H., Normark, W. R., Vedder, J. C., and Silver, E. A., 1978, Neogene basin formation in relation to plate-tectonic evolution of the San Andreas Fault System, California: American Association of Petroleum Geologists Bulletin, v. 62, p. 344–372.

Bohrmann, G., Abelmann, A., Gersonde, R., Hubberten, H., and Kuhn, G., 1994, Pure siliceous ooze, a diagenetic environment for early chert formation: Geology, v. 22, p. 207–210.

Bramlette, M. N., 1946, The Monterey Formation of California and the origin of its siliceous rocks, U.S. Geological Survey Professional Paper 212, 57 p.

Burns, S. J., and Baker, P. A., 1987, A geochemical study of dolomite in the Monterey Formation, California: Journal of Sedimentary Petrology, v. 57, p. 128–139.

Cady, S. L., Wenk, H. R., and Downing, K. H., 1996, HRTEM of microcrystalline opal in chert and porcelanite from the Monterey Formation, California: American Mineralogist, v. 81, p. 1380–1395.

Calvert, S. E., 1966, Accumulation of diatomaceous silica in the sediments of the Gulf of California: Geological Society of America Bulletin, v. 77, p. 569–596.

Calvert, S. E., 1968, Silica balance in the ocean and diagenesis: Nature, v. 219, p. 919–920.

Canfield, C. R., 1939, Subsurface stratigraphy of Santa Maria Valley oil field and adjacent parts of Santa Maria Valley, California: American Association of Petroleum Geologists Bulletin, v. 23, p. 45–81.

Casey, R. E., and Barron, J. A., 1986, Siliceous microfossil and microplankton of the Monterey Formation and modern analogs: Los Angeles, Pacific Section, Society of Economic Paleontologists and Mineralogists, 147 p.

Compton, J. S., 1991, Origin and diagenesis of clay minerals in the Monterey Formation, Santa Maria Basin area, California: Clays and Clay Minerals, v. 39, p. 449–466.

Compton, J. S., Snyder, S. W., and Hodell, D. A., 1990, Phosphogenesis and weathering of shelf sediments from the southeastern United States: Implications for Miocene $\delta^{13}C$ excursions and global cooling: Geology, v. 18, p. 1227–1230.

Crain, W. E., Mero, W. E., a nd Patterson, D., 1985, Geology of the Point Arguello discovery: American Association of Petroleum Geologists Bulletin, v. 69, p. 537–545.

Crawford, F. D., 1971, Petroleum potential of Santa Maria province, California, in Cram, I. H., ed., Future petroleum provinces of the United States—Their geology and potential: Tulsa, American Association of Petroleum Geologists, p. 316–328.

Crouch, J. A., and Bukry, D., 1979, Comparison of Miocene provincial foraminiferal stages to coccoliths in the California continental borderland: Geology, v. 7, p. 211–215.

Crouch, J. K., and Suppe, J., 1993, Late Cenozoic tectonic evolution of the Los Angeles basin and inner California borderland: A model for core complex-like crustal extension: Geological Society of America Bulletin, v. 105, p. 1415–1434.

Crouch, Bachman, and associates, 1991, Structure and stratigraphy of the Monterey Formation and adjacent rocks, central California: A field seminar, Part I: Descriptive text and guidebook, in Lewis, L., Hubbard, P., Heath, E., and Pace, A., eds., Southern Coast Ranges, Annual field trip guidebook 15, Santa Ana, South Coast Geological Society, p. 189–217.

Curiale, J. A., Cameron, D., and Davis, D. V., 1985, Biological marker distribution and significance in oils and rocks of the Monterey Formation, California: Geochimica Cosmochimica Acta, v. 49, p. 271–288.

DePaolo, D. J., and Finger, K. L., 1991, High-resolution strontium-isotope stratigraphy and biostratigraphy of the Miocene Monterey Formation, central California: Geological Society of America Bulletin, v. 103, p. 112–124.

Donegan, D., and Schrader, H., 1981, Modern analogues of the Miocene diatomaceous Monterey Shale of California: Evidence from sedimentologic and micropaleontologic study, in Garrison, R. E., and Douglas, R. G., eds., The Monterey Formation and related siliceous rocks of California: Los Angeles, Pacific Section, Society of Economic Paleontologists and Mineralogists, p. 149–157.

Dunham, J. B., and Blake, G. H., 1987, Guide to coastal outcrops of the Monterey Formation of western Santa Barbara county, California, in Dunham, J. B., ed., Guide to coastal outcrops of the Monterey Formation of western Santa Barbara County, California: Los Angeles, Pacific Section, Society of Economic Paleontologists and Mineralogists, Special Publication v. 53, 36 p.

Dunham, J. B., Bromley, B. W., and Rosato, V. J., 1991, Geologic controls on hydrocarbon occurrence within the Santa Maria basin of western California, in Gluskoter, H. J., Rice, D. D., and Taylor, R. B., eds., Economic geology, U.S.: Boulder, Colorado, Geological Society of America, Geology of North America, p. 431–446.

Dunkel, C. A., and Piper, K. A., 1997, 1995 National assessment of United States oil and gas resources: Assessment of the Pacific Outer Continental Shelf Region: U.S. Department of the Interior Minerals Management Service MMS 97-0019, 207 p.

Eichhubl, P., 1997, Scale, rates, and timing of fracture-related fluid flow in the Miocene Monterey Formation, coastal California [Ph.D. thesis]: Santa Barbara, University of California, 298 p.

Eichhubl, P., ed., 1998, Diagenesis, deformation, and fluid flow in the Miocene Monterey Formation: Pacific Section, SEPM (Society for Sedimentary Geology) Special Publication 83, 98 p.

Eichhubl, P., and Behl, R. J., 1998, Diagenesis, deformation, and fluid flow in the Miocene Monterey Formation, in Eichhubl, P., ed., Diagenesis, deformation, and fluid flow in the Miocene Monterey Formation: Pacific Section, SEPM (Society for Sedimentary Geology) Special Publication 83, p. 5–13.

Eichhubl, P., and Boles, J. R., 1998, Vein formation in relation to burial diagenesis in the Miocene Monterey Formation, Arroyo Burro Beach, Santa Barbara, California, in Eichhubl, P., ed., Diagenesis, deformation, and fluid flow in the Miocene Monterey Formation: Pacific Section, SEPM (Society for Sedimentary Geology) Special Publication 83, p. 15–36.

Finger, K. L., 1995, Recognition of middle Miocene foraminifers in highly indurated rocks of the Monterey Formation, coastal Santa Maria province, central California: U.S. Geological Survey Bulletin, v. 1995L, p. L1–L30.

Finkbeiner, T., Barton, C. A., and Zoback, M. D., 1997, Relationships among in-situ stress, fractures and faults, and fluid flow: Monterey Formation, Santa Maria Basin, California: American Association of Petroleum Geologists Bulletin, v. 81, p. 1975–1999.

Flower, B. P., and Kennett, J. P., 1993, Relations between Monterey Formation deposition and middle Miocene global cooling: Naples Beach section, California: Geology, v. 21, p. 877–880.

Flower, B. P., and Kennett, J. P., 1994, Oxygen and carbon isotopic stratigraphy of the Monterey Formation at Naples Beach, California, in Hornafius, J. S., ed., Field guide to the Monterey Formation between Santa Barbara and Gaviota, California: Los Angeles, Pacific Section, American Association of Petroleum Geologists, p. 59–66.

Föllmi, K. B., Garrison, R. E., and Grimm, K. A., 1991, Stratification in phosphatic sediments: Illustrations from the Neogene of California, in Einsele, G., Ricken, W., and Seilacher, A., eds., Cycles and events in stratigraphy: Berlin, Springer-Verlag, p. 492–507.

Foss, C. D., and Blaisdell, R., 1968, Stratigraphy of the west side southern San Joaquin Valley, in Karp, S. E., ed., Guidebook, geology and oilfields west side southern San Joaquin Valley: Bakersfield, Pacific Section, American Association of Petroleum Geologists, p. 33–43.

Friedman, I., and Murata, K. J., 1979, Origin of dolomite in Miocene Monterey Shale and related formations in the Temblor Range, California: Geochimica Cosmochimica Acta, v. 43, p. 1357–1365.

Fritsche, A. E., 1998, Miocene palinspastic restoration of southwestern California: American Association of Petroleum Geologists Bulletin, v. 82, p. 847–848.

Garrison, R. E., and Douglas, R. G., 1981, The Monterey Formation and related siliceous rocks of California: Los Angeles, Pacific Section, Society of Economic Paleontologists and Mineralogists, 327 p.

Garrison, R. E., and Ramirez, P. C., 1989, Conglomerates and breccias in the Monterey Formation and related units as reflections of basin margin history, in Colburn, I. P., Abbott, P. L., and Minch, J., eds., Conglomerates in basin analysis: A symposium dedicated to A.O. Woodford: Pacific Section, Society of Economic Paleontologists and Mineralogists, p. 189–206.

Garrison, R. E., Stanley, R. G., and Horan, L. J., 1979, Middle Miocene sedimentation on the southeastern edge of the Lockwood high, Monterey County, California, in Graham, S. A., ed., Tertiary and Quaternary geology of the Salinas Valley and Santa Lucia Range, Monterey County, California: Pacific Coast paleogeography field guide: Los Angeles, Pacific Section, Society of Economic Paleontologists and Mineralogists, p. 51–65.

Garrison, R. E., Kastner, M., and Zenger, D. H., 1984, Dolomites of the Monterey Formation and other organic-rich units: Los Angeles, Pacific Section, Society of Economic Paleontologists and Mineralogists, 215 p.

Garrison, R. E., Kastner, M., and Kolodny, Y., 1987, Phosphorites and phosphatic rocks in the Monterey Formation and related Miocene units, coastal California, in Ingersoll, R. V., and Ernst, W. G., eds., Cenozoic basin development of coastal California (Rubey Volume VI): Englewood Cliffs, New Jersey, Prentice-Hall, p. 348–381.

Garrison, R. E., Kastner, M., and Reimers, C. E., 1990, Miocene phosphogenesis in California, in Burnett, W. C., and Riggs, S. R., eds., Phosphate deposits of the world; Neogene to modern phosphorites: New York, Cambridge University Press, p. 285–299.

Garrison, R. E., Hoppie, B. W., and Grimm, K. A., 1994, Phosphates and dolomites in coastal upwelling sediments of the Peru margin and the Monterey Formation (Naples Beach section), California, in Hornafius, J. S., ed., Field guide to the Monterey Formation between Santa Barbara and Gaviota, California: Bakersfield, Pacific Section, American Association of Petroleum Geologists, p. 67–84.

Global Geochemistry Corporation, 1985, The Geochemical and Paleoenvironmental history of the Monterey Formation—Sediments and hydrocarbons: Canoga Park, California, Global Geochemistry Corporation, 459 p.

Govean, F. M., and Garrison, R. E., 1981, Significance of laminated and massive diatomites in the upper part of the Monterey Formation, California, *in* Garrison, R. E., and Douglas, R. G., eds., The Monterey Formation and related siliceous rocks of California: Los Angeles, Pacific Section, Society of Economic Paleontologists and Mineralogists, p. 181–198.

Graham, S. A., and Williams, L. A., 1985, Tectonic, depositional, and diagenetic history of Monterey Formation (Miocene), central San Joaquin basin, California: American Association of Petroleum Geologists Bulletin, v. 69, p. 385–411.

Grivetti, M. C., 1982, Aspects of stratigraphy, diagenesis, and deformation in the Monterey Formation near Santa Maria–Lompoc, California [Master's thesis]: Santa Barbara, University California, 155 p.

Gross, M. R., 1993, The origin and spacing of cross joints: examples from the Monterey Formation, Santa Barbara coastline, California: Journal of Structural Geology, v. 15, p. 737–751.

Gross, M. R., Gutierrez-Alonso, G., Bai, T., Wacker, M. A., Collinsworth, K. B., and Behl, R. J., 1997, Influence of mechanical stratigraphy and kinematics on fault scaling relations: Journal of Structural Geology, v. 19, p. 171–183.

Hill, M. L., Carlson, S. A., and Dibblee, T. W., Jr., 1958, Stratigraphy of the Cuyama Valley and Caliente Range area, California: American Association of Petroleum Geologists Bulletin, v. 42, p. 2973–3000.

Hornafius, J. S., 1991, Facies analysis of the Monterey Formation in the northern Santa Barbara Channel: American Association of Petroleum Geologists Bulletin, v. 75, p. 894–909.

Hornafius, J. S., 1994a, Field guide to the Monterey Formation between Santa Barbara and Gaviota, California: Bakersfield, Pacific Section, American Association of Petroleum Geologists, v. GB-72, 123 p.

Hornafius, J. S., 1994b, Overview of the stratigraphy of the Monterey Formation along the coastline between Santa Barbara and Gaviota, California, *in* Hornafius, J. S., ed., Field guide to the Monterey Formation between Santa Barbara and Gaviota, California: Bakersfield, Pacific Section, American Association of Petroleum Geologists, p. 1–15.

Hunt, J. M., 1979, Petroleum geochemistry and geology: San Francisco, W.H. Freeman, 617 p.

Ingersoll, R. V., and Ernst, W. G., 1987, Cenozoic basin development of coastal California (Rubey Volume VI): Englewood Cliffs, New Jersey, Prentice-Hall, 496 p.

Ingle, J. C., Jr., 1973, Summary comments on Neogene biostratigraphy, physical stratigraphy, and paleo-oceanography in the marginal northeastern Pacific Ocean, *in* Initial reports of the Deep Sea Drilling Project, Washington, D.C., U.S. Government Printing Office, p. 163–195.

Ingle, J. C., Jr., 1980, Cenozoic paleobathymetry and depositional history of selected sequences within the southern California continental borderland: Cushman Foundation for foraminiferal Research Special Publication 19, p. 163–195.

Ingle, J. C., Jr., 1981a, Cenozoic depositional history of the northern continental borderland of southern California and the origin of associated Miocene diatomites, *in* Isaacs, C. M., ed., Guide to the Monterey Formation in the California coastal area, Ventura to San Luis Obispo: Los Angeles, Pacific Section, American Association of Petroleum Geologists Special Publication 52, p. 1–8.

Ingle, J. C., Jr., 1981b, Origin of Neogene diatomites around the north Pacific rim, *in* Garrison, R. E., and Douglas, R. G., eds., The Monterey Formation and related siliceous rocks of California: Los Angeles, Pacific Section, Society of Economic Paleontologists and Mineralogists, p. 159–179.

Isaacs, C. M., 1980, Diagenesis in the Monterey Formation examined laterally along the coast near Santa Barbara, California [Ph.D. thesis]: Stanford, California, Stanford University, 329 p.

Isaacs, C. M., 1981a, Guide to the Monterey Formation in the California coastal area, Ventura to San Luis Obispo: Los Angeles, Pacific Section, American Association Petroleum Geologists, 91 p.

Isaacs, C. M., 1981b, Lithostratigraphy of the Monterey Formation, Goleta to Point Conception, Santa Barbara coast, California, *in* Isaacs, C. M., ed., Guide to the Monterey Formation in the California coastal area, Ventura to San Luis Obispo: Los Angeles, Pacific Section, American Association Petroleum Geologists, p. 9–23.

Isaacs, C. M., 1981c, Outline of diagenesis in the Monterey Formation examined laterally along the Santa Barbara coast, California, *in* Isaacs, C. M., ed., Guide to the Monterey Formation in the California coastal area, Ventura to San Luis Obispo: Los Angeles, Pacific Section, American Association Petroleum Geologists, p. 25–38.

Isaacs, C. M., 1981d, Porosity reduction during diagenesis of the Monterey Formation, Santa Barbara coastal area, California, *in* Garrison, R. E., and Douglas, R. G., eds., The Monterey Formation and related siliceous rocks of California: Los Angeles, Pacific Section, Society of Economic Paleontologists and Mineralogists, p. 257–271.

Isaacs, C. M., 1982, Influence of rock composition on kinetics of silica phase changes in the Monterey Formation, Santa Barbara area, California: Geology, v. 10, p. 304–308.

Isaacs, C. M., 1983, Compositional variation and sequence in the Miocene Monterey Formation, Santa Barbara coastal area, California, *in* Larue, D. K., and Steel, R. J., eds., Cenozoic marine sedimentation, Pacific margin: Pacific Section, Society of Economic Paleontologists and Mineralogists Special Publication 28, p. 117–132.

Isaacs, C. M., 1984, The Monterey—Key to offshore California boom: Oil & Gas Journal, p. 75–81.

Isaacs, C. M., 1985, Abundance versus rates of accumulation in fine-grained strata of the Miocene Santa Barbara basin, California: Geo-Marine Letters, v. 5, p. 25–30.

Isaacs, C. M., 1988, Marine petroleum source rocks and reservoir rocks of the Miocene Monterey Formation, California, U.S.A., *in* Wagner, H. C., Wagner, L. C., Wang, F. F. H., and Wong, F. L., eds., Petroleum resources of China and related subjects: Houston, Texas, Circum-Pacific Council for Energy and Mineral Resources Earth Science Series, v. 10, p. 825–848.

Isaacs, C. M., 1999, Depositional framework of the Monterey Formation, California, *in* Isaacs, C. M., and Ruellkötter, J., eds., The Monterey Formation: From rocks to molecules: New York, Columbia University Press (in press).

Isaacs, C. M., and Garrison, R. E., 1983, Petroleum Generation and Occurrence in the Miocene Monterey Formation, California: Los Angeles, Pacific Section, Society of Economic Paleontologists and Mineralogists, 228 p.

Isaacs, C. M., and Magoon, L. B., 1984, Thermal indicators of organic matter in the Sisquoc and Monterey Formations, Santa Maria basin, California: Society of Economic Paleontologists and Mineralogists Annual Midyear Meeting Abstracts, 40 p.

Isaacs, C. M., and Petersen, N. F., 1987, Petroleum in the Miocene Monterey Formation, California, *in* Hein, J. R., ed., Siliceous sedimentary rock-hosted ores and petroleum: Evolution of ore fields: New York, Van Nostrand Reinhold, p. 83–116.

Isaacs, C. M., and Ruellkötter, J., eds., 1999, The Monterey Formation: From rocks to molecules: New York, Columbia University Press (in press).

Isaacs, C. M., Baumgartner, T. R., Tennyson, M. E., Piper, D. Z., and Ingle, J. C., Jr., 1996, A prograding margin model for the Monterey Formation, California: American Association of Petroleum Geologists and Pacific Section SEPM (Society for Sedimentary Geology) Annual Meeting Abstracts, v. 5, p. 69.

Kablanow, R. I., II, and Surdam, R. C., 1984, Diagenesis and hydrocarbon generation in the Monterey Formation, Huasna basin, California, *in* Surdam, R. C., ed., A guidebook to the stratigraphic, tectonic, thermal, and diagenetic histories of the Monterey Formation, Pismo and Huasna basin, California: Tulsa, Society of Economic Paleontologists and Mineralogists, p. 53–68.

Kastner, M., Keene, J. B., and Gieskes, J. M., 1977, Diagenesis of siliceous oozes—I. Chemical controls on the rate of opal-A to opal-CT transformatio—An experimental study: Geochimica et Cosmochimica Acta, v. 41, p. 1041–1059.

Kastner, M., Mertz, K., Hollander, D., and Garrison, R., 1984, The association of dolomite-phosphorite-chert: Causes and possible diagenetic sequences, *in*

Garrison, R. E., Kastner, M., and Zenger, D. H., eds., Dolomites of the Monterey Formation and other organic-rich units: Los Angeles, Pacific Section, Society of Economic Paleontologists and Mineralogists, p. 75–86.

Keller, G., and Barron, J. A., 1981, Integrated planktic foraminiferal and diatom biochronology for the northeast Pacific and the Monterey Formation, in Garrison, R. E., and Douglas, R. G., eds., The Monterey Formation and related siliceous rocks of California: Los Angeles, Pacific Section, Society of Economic Paleontologists and Mineralogists, p. 43–54.

Keller, G., and Barron, J. A., 1983, Paleoceanographic implications of Miocene deep-sea hiatuses: Geological Society of America Bulletin, v. 94, p. 590–613.

Keller, M. A., and Isaacs, C. M., 1985, An evaluation of temperature scales for silica diagenesis in diatomaceous sequence including a new approach based on the Miocene Monterey Formation, California: Geo-Marine Letters, v. 5, p. 31–35.

Kennett, J. P., 1977, Cenozoic evolution of Antarctic glaciation, the circum-Antarctic Ocean, and the impact on global paleoceanography: Journal of Geophysical Research, v. 82, p. 3843–3860.

King, J. D., and Claypool, G. E., 1983, Biological marker compounds and implications for generation and migration of petroleum in rocks of the Point Conception deep-stratigraphic test well, OCS-Cal 78-164 No. 1, offshore California, in Isaacs, C. M., and Garrison, R. E., eds., Petroleum generation and occurrence in the Miocene Monterey Formation, California: Los Angeles, Pacific Section, Society of Economic Paleontologists and Mineralogists, p. 191–200.

Kleinpell, R. M., 1938, Miocene stratigraphy of California: Tulsa, American Association of Petroleum Geologists, 450 p.

Kleinpell, R. M., 1980, Miocene stratigraphy of California revisited: American Association of Petroleum Geologists Studies in Geology 11, 349 p.

Kruge, M. A., 1983, Diagenesis of Miocene biogenic sediments in Lost Hills oil field, San Joaquin basin, California, in Isaacs, C. M., and Garrison, R. E., eds., Petroleum generation and occurrence in the Miocene Monterey Formation, California: Los Angeles, Pacific Section, Society of Economic Paleontologists and Mineralogists, p. 39–51.

Lagoe, M. B., 1981, Subsurface facies analysis of the Saltos Shale member, Monterey Formation (Miocene) and associated rocks, Cuyama Valley, California, in Garrison, R. E., and Douglas, R. G., eds., The Monterey Formation and related siliceous rocks of California: Los Angeles, Pacific Section, Society of Economic Paleontologists and Mineralogists, p. 199–211.

Lillis, P. G., and Lagoe, M. B., 1983, Regional patterns of oil gravity in the Monterey Formation, Santa Maria basin, California; implications for petroleum exploration and tectonic history, in Isaacs, C. M., and Garrison, R. E., eds., Petroleum generation and occurrence in the Miocene Monterey Formation, California: Los Angeles, Pacific Section, Society of Economic Paleontologists and Mineralogists, p. 225.

Lyle, M., Koizumi, I., Richter, C., et al., eds., 1997, Proceedings of the Ocean Drilling Program, Initial reports 167: College Station, Texas, Ocean Drilling Program, 1378 p.

MacKinnon, T. C., 1989a, Origin of the Miocene Monterey Formation in California, in MacKinnon, T. C., ed., Oil in the California Monterey Formation: Fieldtrip guidebook T311: Washington, D.C., American Geophysical Union, p. 1–10.

MacKinnon, T. C., 1989b, Petroleum geology of the Monterey Formation in the Santa Maria and Santa Barbara coastal and offshore areas, in Mackinnon, T. C., ed., Oil in the California Monterey Formation: Fieldtrip guidebook T311: Washington, D.C., American Geophysical Union, p. 11–27.

Malone, M. J., Baker, P. A., and Burns, S. J., 1996, Hydrothermal dolomitization and recrystallization of dolomite breccias from the Miocene Monterey Formation, Tepusquet area, California: Journal of Sedimentary Research, v. 66A, p. 976–990.

Mayerson, D., and Crouch, J., 1994, The opal-CT/quartz diagenetic boundary within the Monterey Formation of the California offshore Santa Maria Basin; an untapped exploration target: American Association of Petroleum Geologists Bulletin, v. 78, p. 669–670.

Mayerson, D. A., Dunkel, C. A., Piper, K. A., and Cousminer, H. L., 1995, Identification and correlation of the opal-CT/quartz phase transition in offshore central California [abs]: American Association of Petroleum Geologists Bulletin, v. 79, p. 592.

McCulloh, T. H., 1979, Implications for petroleum appraisal, in Cook, H. E., ed., Geologic Studies of the Point Conception deep stratigraphic test well OCS-CAL 78-164 No. 1, Outer Continental Shelf, southern California, United States: U.S. Geological Survey Open-File Report 79-128, p. 26–42.

McGuire, M. D., Bowersox, J. R., and Earnest, L. J., 1983, Diagenetically enhanced entrapment of hydrocarbons—Southern Lost Hills fractured shale pool, Kern County, California, in Isaacs, C. M., and Garrison, R. E., eds., Petroleum generation and occurrence in the Miocene Monterey Formation, California: Los Angeles, Pacific Section, Society of Economic Paleontologists and Mineralogists, p. 171–183.

Miller, K. G., Fairbanks, R. G., and Mountain, G. S., 1987, Tertiary oxygen isotope synthesis, sea level history, and continental margin erosion: Paleoceanography, v. 2, p. 1–19.

Murata, K. J., and Larson, R. R., 1975, Diagenesis of Miocene siliceous shales, Temblor Range, California: U.S. Geological Survey Journal of Research, v. 3, p. 553–566.

Murata, K. J., and Nakata, J. K., 1974, Cristobalitic stage in the diagenesis of diatomaceous shale: Science, v. 184, p. 567–568.

Murata, K. J., and Norman, M. B., 1976, An index of crystallinity for quartz: American Journal of Science, v. 276, p. 1120–1130.

Murata, K. J., Friedman, I. I., and Madsen, B. M., 1967, Carbon-13-rich diagenetic carbonates in Miocene formations of California and Oregon: Science, v. 156, p. 1484–1485.

Murata, K. J., Friedman, I., and Cremer, M., 1972, Geochemistry of diagenetic dolomites in Miocene marine formations of California and Oregon: Washington, D.C., U.S. Government Printing Office, 12 p.

Murray, R. W., and Jones, D. L., 1992, Diagenetic formation of bedded chert: Evidence from chemistry of the chert-shale couplet: Geology, v. 20, p. 271–274.

Narr, W., 1991, Fracture density in the deep subsurface: Techniques with application to the Point Arguello oil field: American Association of Petroleum Geologists Bulletin, v. 75, p. 1300–1323.

Obradovich, J. D., and Naeser, C. W., 1981, Geochronology bearing on the age of the Monterey Formation and siliceous rocks in California, in Garrison, R. E., and Douglas, R. G., eds., The Monterey Formation and related siliceous rocks of California: Los Angeles, Pacific Section, Society of Economic Paleontologists and Mineralogists, p. 87–95.

Omarzai, S. K., 1992, Monterey Formation of California at Shell Beach (Pismo basin): Its lithofacies, paleomagnetism, age, and origin, in Schwalbach, J. R., and Bohacs, K. M., eds., Sequence stratigraphy in fine-grained rocks: Examples from the Monterey Formation: Santa Fe Springs, Pacific Section, SEPM (Society for Sedimentary Geology), p. 47–65.

Orr, W. L., 1984, Sulfur and sulfur isotope ratios in Monterey oils of the Santa Maria basin and Santa Barbara channel area: Society of Economic Paleontologists and Mineralogists, Annual Midyear Meeting Abstracts, v. 1, p. 62.

Orr, W. L., 1986, Kerogen/asphaltene/sulfur relationships in sulfur-rich Monterey oils: Organic Geochemistry, v. 10, p. 499–516.

Ozalas, K., Savrda, C. E., and Fullerton, R. R., Jr., 1994, Bioturbation oxygenation-event beds in siliceous facies: Monterey Formation (Miocene), California: Palaeogeography, Palaeoclimatology, Palaeoecology, v. 112, p. 63–83.

Petersen, N. F., and Hickey, P. J., 1987, California Plio-Miocene oils: Evidence of early generation, in Meyer, R. F., ed., Exploration for heavy crude oil and natural bitumen: American Association of Petroleum Geologists Studies in Geology 25, p. 351–359.

Pisciotto, K. A., 1981a, Diagenetic trends in the siliceous facies of the Monterey Shale in the Santa Maria region, California: Sedimentology, v. 28, p. 547–571.

Pisciotto, K. A., 1981b, Review of secondary carbonates in the Monterey Formation, California, *in* Garrison, R. E., and Douglas, R. G., eds., The Monterey Formation and related siliceous rocks of California: Los Angeles, Pacific Section, Society of Economic Paleontologists and Mineralogists, p. 273–283.

Pisciotto, K. A., and Garrison, R. E., 1981, Lithofacies and Depositional Environments of the Monterey Formation, California, *in* Garrison, R. E., and Douglas, R. G., eds., The Monterey Formation and related siliceous rocks of California: Los Angeles, Pacific Section, Society of Economic Paleontologists and Mineralogists, p. 97–122.

Pisciotto, K. A., and Mahoney, J. J., 1981, Isotopic survey of diagenetic carbonates, DSDP Leg 63, *in* Orlofsky, S., Yeats, R. S., Haq, B. U., Barron, J. A., Bukry, D., Crouch, J., Denham, C., Douglas, A. G., Grechin, V. I., Leinen, M., Niem, A. R., Verma, S. P., Pisciotto, K. A., Poore, R. Z., Shibata, T., and Wolfart, R., eds., Initial reports of the Deep Sea Drilling Project: Washington, D.C., U.S. Government Printing Office, p. 595–609.

Pollastro, R. M., 1990, Geothermometry from smectite and silica diagenesis in the diatomaceous Monterey and Sisquoc Formations, Santa Maria basin, California: American Association of Petroleum Geologists Bulletin, v. 74, p. 742.

Poore, R. Z., McDougall, K., Barron, J. A., Brabb, E. E., and Kling, S. A., 1981, Microfossil biostratigraphy and biochronology of the type Relizian and Luisian Stages of California, *in* Garrison, R. E., and Douglas, R. G., eds., The Monterey Formation and related siliceous rocks of California: Los Angeles, Pacific Section, Society of Economic Paleontologists and Mineralogists, p. 15–41.

Raymo, M. E., 1994, The Himalayas, organic carbon burial, and climate in the Miocene: Paleoceanography, v. 9, p. 399–404.

Redwine, L., 1981, Hypothesis combining dilation, natural hydraulic fracturing, and dolomitization to explain petroleum reservoirs in Monterey Shale, Santa Maria area, California, *in* Garrison, R. E., and Douglas, R. G., eds., The Monterey Formation and related siliceous rocks of California: Los Angeles, Pacific Section, Society of Economic Paleontologists and Mineralogists, p. 221–248.

Regan, L. J., and Hughes, A. W., 1949, Fractured reservoirs of Santa Maria District, California: American Association of Petroleum Geologists Bulletin, v. 33, p. 32–51.

Richter, F. M., Rowley, D. B., and DePaolo, D. J., 1992, Sr isotope evolution of seawater: The role of tectonics: Earth and Planetary Science Letters, v. 109, p. 11–23.

Roehl, P. O., 1981, Dilation brecciation—proposed mechanism of fracturing, petroleum expulsion, and dolomitization in Monterey Formation, California, *in* Garrison, R. E., and Douglas, R. G., eds., The Monterey Formation and related siliceous rocks of California: Los Angeles, Pacific Section, Society of Economic Paleontologists and Mineralogists, p. 285–315.

Ruellkötter, J., and Isaacs, C. M., 1996, Monterey source rock facies and petroleum formation; a synthesis of results of the cooperative Monterey organic geochemistry study: American Association of Petroleum Geologists and Pacific Section SEPM (Society for Sedimentary Geology) Annual Meeting Abstracts, v. 5, p. 123.

Schwalbach, J. R., and Bohacs, K. M., 1992, Sequence stratigraphy in fine-grained rocks: Examples from the Monterey Formation: Santa Fe Springs, Pacific Section, Society of Economic Paleontologists and Mineralogists (SEPM), Special Publication 70, p. 80.

Schwalbach, J. R., and Bohacs, K. M., 1995, Stratigraphic sections and gamma-ray spectrometry from five outcrops of the Monterey Formation in southwestern California; Naples Beach, Point Pedernales, Lion's Head, Shell Beach, and Point Buchon: U.S. Geological Survey Bulletin 1995, p. Q1–Q39.

Snyder, W. S., 1987, Structure of the Monterey Formation: Stratigraphic, diagenetic, and tectonic influences on style and timing, *in* Ingersoll, R. V., and Ernst, W. G., eds., Cenozoic basin development of coastal California (Rubey Volume VI): Englewood Cliffs, New Jersey, Prentice-Hall, p. 321–347.

Snyder, W. S., Brueckner, H. K., and Schweickert, R. A., 1983, Deformational styles in the Monterey Formation and other siliceous sedimentary rocks, *in* Isaacs, C. M., and Garrison, R. E., eds., Petroleum generation and occurrence in the Miocene Monterey Formation, California: Los Angeles, Pacific Section, Society of Economic Paleontologists and Mineralogists, p. 151–170.

Soutar, A., Johnson, S. R., and Baumgartner, T. R., 1981, In search of modern analogs to the Monterey Formation, *in* Garrison, R. E., and Douglas, R. G., eds., The Monterey Formation and related siliceous rocks of California: Los Angeles, Pacific Section, Society of Economic Paleontologists and Mineralogists, p. 123–147.

Surdam, R. C., 1984, Stratigraphic, tectonic, thermal, and diagenetic histories of the Monterey Formation, Pismo and Huasna Basin, California: Tulsa, Oklahoma, Society of Economic Paleontologists and Mineralogists Guidebook 2, 94 p.

Surdam, R. C., and Stanley, K. O., 1981, Diagenesis and migration of hydrocarbons in the Monterey Formation, Pismo syncline, California, *in* Garrison, R. E., and Douglas, R. G., eds., The Monterey Formation and related siliceous rocks of California: Los Angeles, Pacific Section, Society of Economic Paleontologists and Mineralogists, p. 317–327.

Taylor, J. C., 1976, Geologic appraisal of the petroleum potential of offshore southern California: The borderland compared to onshore coastal basins: U.S. Geological Survey Circular 730, 43 p.

Teng, L. S., and Gorsline, D. S., 1989, Late Cenozoic sedimentation in California continental borderland basins as revealed by seismic facies analysis: Geological Society of America Bulletin, v. 101, p. 27–41.

Vincent, E., and Berger, W. H., 1985, Carbon dioxide and global cooling in the Miocene: The Monterey hypothesis, *in* Sundquist, E. T., and Broecker, W. S., eds., The carbon cycle and atmospheric $CO_2$: Natural variations Archean to present: American Geophysical Union Geophysical Monographs 32, p. 455–468.

Walker, A. L., McCulloh, T. H., Petersen, N. F., and Stewart, R. J., 1983, Anomalously low reflectance of vitrinite, in comparison with other petroleum source rock maturation indices, from the Miocene Modelo Formation in the Los Angeles Basin, California, *in* Isaacs, C. M., and Garrison, R. E., eds., Petroleum generation and occurrence in the Miocene Monterey Formation, California: Los Angeles, Pacific Section, Society of Economic Paleontologists and Mineralogists, p. 185–190.

Weaver, F. M., Casey, R. E., and Perez, A. M., 1981, Stratigraphic and paleoceanographic significance of early Pliocene to middle Miocene radiolarian assemblages from northern to Baja California, *in* Garrison, R. E., and Douglas, R. G., eds., The Monterey Formation and related siliceous rocks of California: Los Angeles, Pacific Section, Society of Economic Paleontologists and Mineralogists, p. 71–86.

White, L. D., 1989, Chronostratigraphic and paleoceanographic aspects of selected chert intervals in the Miocene Monterey Formation, California [Ph.D. thesis]: Santa Cruz, University of California, 236 p.

Williams, L. A., and Graham, S. A., 1982, Monterey Formation and associated coarse clastic rocks, central San Joaquin basin, California: Los Angeles, Pacific Section, Society of Economic Paleontologists and Mineralogists, 95 p.

Williams, L. A., Parks, G. A., and Crerar, D. A., 1985, Silica diagenesis, I. Solubility controls: Journal of Sedimentary Petrology, v. 55, p. 301–311.

Woodring, W. P., and Bramlette, M. N., 1950, Geology and paleoecology of the Santa Maria District, California: U.S. Geological Survey Professional Paper 222, 185 p

Woodring, W. P., Bramlette, M. N., and Kew, W. S., 1946, Geology and paleontology of Palos Verdes Hills, California: USGS Professional Paper 207, 145 p.

Woodring, W. P., Bramlette, M. N., and Lohman, K. E., 1943, Stratigraphy and paleontology of Santa Maria District, California: American Association of Petroleum Geologists Bulletin, v. 27, p. 1335–1361.

Woodruff, F., and Savin, S. M., 1989, Miocene deepwater oceanography: Paleoceanography, v. 4, p. 87–140.

MANUSCRIPT ACCEPTED BY THE SOCIETY NOVEMBER 23, 1998.

Geological Society of America
Special Paper 338
1999

# Chapter 16

# WATER MANAGEMENT: SLAKING CALIFORNIA'S THIRST

**Robert T. Bean**
*650 W. Harrison Avenue*
*Claremont, California 91711*

**Ernest M. Weber**
*2726 Timberlake Drive*
*La Crescenta, California 91214*

A portion of the team that prepared "The California Water Plan." Sitting front and center, flanked by four members of the secretarial staff, is William L. Berry, Division Chief, who directed preparation of the plan.

## WATER PLAN TEAM

The "classic paper" chosen, covering both ground water and surface water throughout California, is "The California Water Plan," Bulletin 3 of the California Department of Water Resources, published in 1957. This bulletin was prepared by a team of 121 professionals, most of whom were civil engineers. Eleven geologists are listed.

Ground water is the principal province of the hydrogeologist, and therefore this paper emphasizes the treatment of ground water in "The California Water Plan." However, since all naturally potable water stems from precipitation, surface water in the form of runoff is also considered.

Robert T. Bean and Ernest B. Weber

STATE OF CALIFORNIA
DEPARTMENT OF WATER RESOURCES
DIVISION OF RESOURCES PLANNING

# Bulletin No. 3

# The CALIFORNIA WATER PLAN

GOODWIN J. KNIGHT
*Governor*

May, 1957

HARVEY O. BANKS
*Director of Water Resources*

# CHAPTER II
# WATER PROBLEMS OF CALIFORNIA

## SURFACE WATER

\* \* \* \* \*

### Runoff

Runoff is defined as that portion of precipitation which drains from the land through surface channels. . . Most of the runoff in California originates on mountain and foothill lands, and debouches from these watersheds onto adjoining valley floors. . .

Because runoff in California is derived from precipitation, it generally reflects similar monthly and seasonal variations, particularly in those portions of the State where precipitation occurs as rainfall. . . Thus, with the exception of snow-fed streams, runoff in California is generally sporadic in nature, with short, intense floods followed by long periods of little or no flow.

A substantial portion of California's precipitation occurs in the form of snow in the Sierra Nevada and parts of the Cascade and Siskiyou Ranges, which contributes a modifying effect on runoff. This water accumulates during the winter in extensive snow fields at high elevations and is released, as runoff, months later during the late spring and early summer snowmelt period. This flow is far more uniform than runoff resulting directly from rainfall and its value is greatly enhanced by its more or less predictable nature and the fact that it is sustained well into the growing season when precipitation is negligible. . .

As previously stated, the bulk of the total seasonal natural runoff in California occurs in the northern portion of the State, with more than 40 per cent from the North Coastal Area and about 32 per cent stemming from the Sacramento Valley. In contrast, only 2 per cent of the total runoff occurs in the South Coastal and Colorado Desert Areas. . .

### Ground Water

The extensive ground water basins of California provide natural regulation for runoff from tributary drainage areas and for precipitation directly on overlying lands. Some 250 ground water basins having valley floor areas of about 5 square miles or larger have been identified in California. A large part of the surface runoff from tributary mountain and foothill watersheds that would otherwise waste to the ocean is retained in these basins and conserved for later utilization. In effect, these ground water reservoirs provide a means for natural regulation of stream flow in much the same manner as is accomplished by surface reservoirs.

Sufficient data on the ground water basins of California are available to permit an estimate of gross storage capacity within certain depth limits for 211 valley floor areas. The areas for which such storage capacities were estimated comprise 96 per cent of the total valley floor area of all basins of the State. The depth limits vary from basin to basin, but the average weighted interval is approximately 185 feet, or generally between the depths of about 15 and 200 feet. The gross storage capacity within this depth interval is about 450,000,000 acre-feet. The Central Valley alone contains over 130,000,000 acre-feet of this total in approximately the same depth interval.

Only a portion of the gross storage capacity is usable storage, largely because of the presence of saline water or other waters of deleterious mineral quality. These waters either limit the depth to which ground water levels may be lowered or, in many areas, preclude the use of ground water. Enough information is presently at hand to estimate the usable storage capacity for only 80 ground water basins, comprising 43 per cent of the total valley floor area of the State. In the Central Valley, usable capacity in the depth interval from 15 to 200 feet aggregates about 100,000,000 acre-feet.

More than half the water presently consumptively used in California comes from underground sources. Many of these ground water basins have been intensively developed. In the San Joaquin Valley and parts of southern California particularly, the ready availability of ground water has been primarily responsible for supporting rapid expansion of agriculture and industry far beyond the firm capabilities of water resource developments. This has been accomplished by utilizing the vast reserves of water stored in these underground reservoirs, in many cases at rates greatly exceeding their replenishment. Presently available data concerning ground water are far less comprehensive than for surface water resources. Much more study will be necessary to

## WATER PROBLEMS OF CALIFORNIA

evaluate reasonably accurately the capability of ground water resources of the State.

\* \* \* \* \*

In addition to. . . surface water facilities, ground water is developed in Petaluma, Sonoma and Napa Valleys, and the Fairfield area of Solano County. It is estimated that these ground water basins have the aggregate yield equivalent to the present (1950) draft therefrom, or about 18,000 acre-feet per season. The potential for additional development of ground water in these basins is limited. In localized areas in each of the basins excessive pumping has lowered ground water elevations below sea level, so that sea-water intrusion has become an active threat.

\* \* \* \* \*

As a result of heavy long sustained drafts on ground water resources in the Southeast Bay Group, and the continuing trend toward increasing municipal, industrial and irrigation demands, the ground water basins of Livermore Valley, Santa Clara Valley, and southern Alameda County are seriously overdrawn at the present time. Ground water pumping levels in the vicinity of San Francisco Bay are substantially below sea level in the latter two areas, with the resultant threat of destruction of the ground water resources by intrusion of sea water from beneath the bay. In fact, sea water has already intruded into the upper aquifer in southern Alameda County, rendering the water unsuitable for use, and has entered the lower aquifer, largely through abandoned or defective wells. Overdrafts on ground water in these areas presently (1955) aggregate an estimated 41,000 acre-feet per season.

\* \* \* \* \*

Water supplies in the Central Coastal Area are presently obtained principally from ground water sources. although these water supplies are physically meeting present requirements, most of the ground water basins are highly developed, and in some cases are overdrawn. Such conditions have developed in the Pajaro and southern Santa Clara Valleys, the Hollister area, and the Salinas, Santa Maria, and Cuyama River Valleys. As a result of these overdraft conditions, degradation of quality of ground water occurred in the Monterey coastal area of the Salinas and Pajaro Valleys, and perennial lowering of ground water levels is manifested in the San Benito, Pajaro and lower Salinas Valleys. the overdraft on the ground water resources in the Central Coastal Area is presently (1955) estimated to aggregate some 65,000 acre-feet per season.

\* \* \* \* \*

With the exception of small amounts of direct surface diversion, the presently utilized water supplies in the Ventura Group are obtained by pumping from several major ground water basins which underlie most of the developed area. It is estimated that the total usable ground water storage capacity in these basins is over 1,000,000 acre-feet, of which about 400,000 acre-feet has been utilized to date. Off-stream spreading works have been operated by the Santa Clara Water Conservation District and its successor, the United Water Conservation District, since about 1927.

\* \* \* \* \*

Water supplies required by . . . urban and agricultural development were first obtained by direct surface diversions and some small surface storage developments. Continuation of the development resulted in intensive utilization of the large ground water basins underlying the valley and coastal plain lands. Of the total underground storage capacity, about 7,000,000 acre-feet of capacity is considered usable, on the basis of those factors of basin configuration, economic pumping lift, and others, as described in the appendix on ground water. The utilization of about 4,500,000 acre-feet of this storage historically has resulted in the development of a safe yield from the local water supplies of 780,000 acre-feet per season.

In the light of the great importance of ground water basins to the economy of the Los Angeles-Santa Ana Group, many steps have been undertaken to assure the fullest practicable utilization of these basins. Percolation in natural stream channels is augmented by spreading operations . . . At the present time there are 70 artificial recharge projects in the Los Angeles-Santa Ana Group, with a capacity sufficient to spread a continuous flow of about 17,000 second-feet. An additional 55 artificial recharge projects, with capacity of about 4,000 second-feet, are proposed for construction by various local agencies in the group. . . The Los Angeles County Flood Control District is constructing spreading works throughout pervious areas of the county to enhance natural percolation. . . The district is also presently engaged in injecting Colorado River water into confined aquifers in the Manhattan Beach area of the West Coast Basin to create a pressure ridge along a portion of the coast line, in an effort to repel sea water. Additional Colorado River water is spread in the forebay areas of the Los Angeles and Orange Counties coastal plains.

\* \* \* \* \*

Water supplies in the San Diego Group are obtained from numerous small ground water basins, from 12 major and several lesser surface storage

## WATER PROBLEMS OF CALIFORNIA

developments, and from importations through the two-barreled San Diego Aqueduct. The ground water basins in the area have relatively small capacity and limited recharge.

\* \* \* \* \*

... The most serious and widespread problem is the inadequacy of the available water supply... to sustain and protect the vast agricultural wealth of the basin during periods of drought... The present annual water deficiency, estimated to be about 2,300,000 acre-feet in 1957, is so great that if a severe drought period, such as those experienced in the past, should now occur, the necessary water conservation and conveyance works could not be constructed fast enough to prevent widespread havoc and economic disaster...

At the present time there are large overdrafts in all valley hydrographic units in the Tulare Lake Basin and in several of the valley units of the San Joaquin River Basin. These deficiencies are particularly serious in the western and southern portions of the valley and in certain areas of the eastern portion...

With the continued recession of the ground water levels, amounting to as much as 30 feet per year in some instances, water supplies in some areas have become almost exhausted, while in others pumping lifts have become so excessive as to be nearly economically prohibitive.

... Other serious problems of water quality are developing on the west side of the San Joaquin Valley. In much of this area the usable aquifer for pumping is found between overlying unusable perched water and underlying connate brines, and improperly constructed wells permit a commingling of the waters in the three zones to the detriment of the usable water.

\* \* \* \* \*

It is probable that the future development of ground water resources in the Honey Lake area may provide an appreciable portion of the ultimate water requirements of the area. Some development of this source has taken place in the past, and ground water is now being utilized for domestic, industrial, and agricultural purposes.

\* \* \* \* \*

... The City of Los Angeles purchased some 300,000 acres of lands in Owens-Mono Basin to obtain water rights for its project ... which now delivers 320,000 acre-feet per annum, approximately its full capacity.

No plans have been prepared for further local development in the Mono-Owens Group as the City of Los Angeles claims rights to the use of most of the waters of these basins.

\* \* \* \* \*

Water quality problems are inextricably connected with the development of the native water resources of the Mojave Group and the provision of additional imported supplies. Poor-quality ground water is presently found in many of the individual ground water basins...

Future development of available ground water storage capacity, involving the utilization of large quantities of imported water supplies, would require adequate control over the maintenance of salt balance. This is a serious and aggravated problem under conditions of internal drainage such as are found in the Mojave Group ... Salt balance in the usable ground water reservoirs must be maintained by providing facilities to export, or transfer, ... as great a quantity of salts as is added in the processes of use and re-use.

\* \* \* \* \*

The continued use of ground water is vital to the existing urban and agricultural development in portions of the Colorado Desert Area, although quantities of ground water available are small ... The primary sources of ground water in the area are seepage from the Colorado River into basins bordering the river, precipitation, and percolation of runoff from tributary drainage areas. Ground water use for agricultural purposes is centered principally in the Coachella, Borrego, and Lucerne Valleys. The safe annual yield of these developed ground water basins, however, is only about 78,000 acre-feet...

Ground water quality varies greatly both in composition and concentration throughout the Colorado Desert Area, and often within the individual ground water basin. In general, ground water quality is suitable for all uses except in the Imperial Valley, Chuckawalla Valley, and the ground water basins bordering the Salton Sea on the east and west. However, localized areas of poor-quality water are encountered throughout the area.

\* \* \* \* \*

...Estimates of the storage capacity existing in the alluvium of the Central Valley, made by the United States Geological Survey and the Department of Water Resources, indicate that some 133,000,000 acre-feet of gross storage capacity is available within 200 feet of the land surface. Taking into consideration areas of questionable water quality and areas where rates of recharge and extraction might present problems, it is indicated that the usable storage capacity might amount to about 98,000,000 acre-feet.

# WATER PROBLEMS OF CALIFORNIA

* * * * *

## CONCLUSIONS

It is concluded that:

1. The future growth of California will depend in large measure upon the early acceptance and implementation of a coordinated, state-wide, multipurpose program of water control, conservation, protection, and utilization.

2. The California Water Plan constitutes such a program, and should be accepted and implemented now as the master plan to guide and coordinate the planning, construction, and operation by all agencies of works required for the control, protection, conservation, and distribution of the water resources of California for all people and beneficial uses in all areas of the State.

3. Critical and increasing needs for supplemental water supplies, for flood control, and for preservation and protection of water resources now exist in many areas of California.

4. The waters originating in California, together with the rights of California in and to the waters of the Colorado River, are adequate in quantity and quality to satisfy all water requirements of the State after it has reached full development, if the waters are properly controlled, conserved, protected, and distributed.

5. The control of floods to provide protection to the growing population and expanding economy of the State must be attained and at all times maintained at a degree commensurate with the need therefor.

6. The quality of waters which are available to meet the full ultimate requirements of all parts of California must be protected and maintained at requisite high levels to make this achievement possible.

7. Minimum standards of well construction and proper procedures for the abandonment of wells should be enforced in order to protect adequately the quality of the State's ground waters.

8. Water development works to satisfy present and future needs of local areas are an essential part of any comprehensive plan for solution of the water problems of the State.

9. Solution of California's water problems must assure adequate provision for municipal, industrial, and agricultural water supplies, quality of water control and protection, flood control and protection, drainage, navigation, hydroelectric power generation, and protection and enhancement of recreation and fish and wildlife resources, and other related water use activities.

10. The authorized Feather River Project, the first unit of The California Water Plan, should be financed and constructed at so vigorous a rate as will assure delivery of water to the San Joaquin Valley not later than 1963 and to southern California not later than 1970.

11. The California Water Development Program should be financed and prosecuted on a continuing basis adequate to provide plans for meeting the growth in demand for water resource development in California.

12. Immediate action should be taken by the legislature and the people of the State of California to provide the constitutional amendment, and by the Legislature to provide the enabling legislation necessary for early and orderly implementation of The California Water Plan.

13. The Legislature should provide for the financing, on an adequate and continuing basis, of the State's share of costs of construction, operation, and maintenance of projects under The California Water Plan, as such projects are authorized by the Legislature.

## RECOMMENDATIONS

It is recommended that:

1. The California Water Plan be accepted by the Legislature as the general and coordinated master plan for the progressive and comprehensive future development of the water resources of California by all agencies, subject to: (a) more detailed investigation and study of component features of the Plan to determine their need, engineering feasibility, economic justification, financial feasibility, and recommended priority of construction; and (b) continuing review, modification, and improvement of the Plan in the light of changing conditions, advances in technology, additional data, and future experience.

3. Adequate funds be provided by the Legislature, on an assured and continuing basis, for the support of the California Water Development Program, comprising: (a) continuation of the compilation and publication of basic water resource data necessary for implementation of The California Water Plan; (b) more detailed investigation and study of component features of the Plan, to determine their need, engineering feasibility, economic justification, financial feasibility, and recommended priority of construction; and (c) continuing review, modification, and improvement of The California Water Plan in the light of changing conditions, advances in technology, additional data, and future experience.

4. Research programs to supply needed basic and experimental data concerning hydrology, hydraulics, water quality, and other pertinent matters be given authorization and adequate financial support.

5. The efforts of all agencies and entities engaged in the planning, financing, construction, and operation of water development projects be

coordinated within the framework of The California Water Plan to the end that maximum ultimate objectives may be achieved.

6. The quality of the water resources of the State be protected against unreasonable deterioration from all sources of impairment. In the administration of the statues governing the disposal of sewage and industrial wastes to waters of the State, consideration should be given not only to the present uses of the waters concerned but also to the future developments and uses envisioned in The California Water Plan. In planning for future urban and industrial developments, consideration should be given to the necessity of adequate waste disposal without endangering the future utility of the State's waters.

7. Proper watershed management practices and methods be formulated and followed to protect and enhance the State's water resources.

8. Positive assurances, to the maximum practicable extent, be provided, by constitutional amendment and legislative enactments, that water required to meet all future beneficial uses in all areas of the State will be available in adequate quantity and quality, when and where needed, and on a dependable basis.

9. A long-range water development fund and enabling policies to assure the financing and construction of needed water development works in California on a continuing, progressive basis be established by the Legislature at the earliest practicable date.

10. The financing and construction of the authorized Feather River Project, the initial unit of The California Water Plan, be expedited in order that urgently needed flood protection will be provided at an early date, and in order that supplemental water supplies will be available to areas of serious water deficiency in the San Francisco Bay Area and in the San Joaquin Valley not later than 1963, and in southern California not later than 1970. The financing and construction of other presently needed water development works should likewise be undertaken immediately.

11. Study be initiated now for the additional legislation that will be necessary for progressive implementation of The California Water Plan, and that such legislation be enacted when and as required. This includes policy recognition of the interests of recreation, fish, and wildlife as important and necessary factors in water development, and the maintenance of live stream flow in the interests of fish, wildlife, and recreation as a beneficial use of water. It further includes: provisions authorizing and implementing administration of ground water development and utilization; the planned operation of ground water basins as storage reservoirs, when necessary in the public interest; the enforcement of minimum standards of well construction and of adequate procedures for abandonment of wells; and legislation to simplify and strengthen the current procedures for the determination of water rights.

California Aqueduct crossing the Tehachapi Mtns.    Photo by Doris Sloan

# The California Water Plan

**Robert T. Bean**
*650 West Harrison Avenue, Claremont, California, 91711*
**Ernest M. Weber**
*2726 Timberlake Drive, La Crescenta, California 91214*

## INTRODUCTION

The basic background for both "The California Water Plan," (California Department of Water Resources [CDWR], 1957), the "classic paper" treated here, and the State Water Project, the primary implementation of that plan, is the same: the maldistribution of the state's natural water supplies. Of the runoff from precipitation in the state, 72% occurs in the North Coast and Sacramento Valley drainage basins, and only 2% occurs in the combined South Coast and Colorado Desert regions (CDWR, 1957; see Figs. 1 and 2).

The California Water Plan (CWP) "describes a comprehensive master plan for the control, protection, conservation, distribution, and utilization of the waters of California, to meet present and future needs for all beneficial purposes in all areas of the State to the maximum feasible extent" (CDWR, 1957, p. xxv). Thus, the attempt was to be made to balance agricultural, domestic, municipal, and industrial needs, as well as needs for fish and wildlife, as fairly as possible to bring the greatest good to the greatest number of people. Existing projects were to be incorporated insofar as possible, and existing agreements were to be honored.

Many major developments with respect to water have taken place in California since publication of the plan in 1957. In all of these, geologic conditions and their evaluation have been fundamental in carrying out the development.

Certainly the most spectacular development has been the construction of the vast State Water Project. Prior to publication of the CWP (1957), the California Legislature, recognizing that the federal Central Valley Project was inadequate to meet the growing water needs in the southern part of the state, authorized the State Water Project. In 1960, the voters of the state authorized implementation of the project by approving the necessary bond issue. The State Water Project now brings water from the Feather and Sacramento river watersheds to the San Francisco Bay area and southern California via an aqueduct system 561 miles in length.

Ground-water management has been developed very extensively in California since the publication of the plan in 1957. Management has taken place at a local level with little or no guidance from state or federal officials. Where problems in a ground-water basin, such as drastically falling water levels or a serious water-quality problem, were recognized and local action was taken, the normal result eventually was a management program. The management of numerous basins in southern California has been developed to a high degree.

Conjunctive use, where surface water is artificially recharged to ground water in years of normal to above-normal runoff and then is withdrawn by wells for use in drier years, is practiced in several areas, notably in the deep sedimentary basins of the South Coast region, and in the southern San Joaquin Valley. The most effective recharge facilities, or "spreading grounds," are located in alluvial fans having high vertical hydraulic conductivity at the base of surrounding mountains.

In Los Angeles and Orange Counties, artificial recharge by deep well injection (CDWR, 1957) has been greatly expanded in protecting key coastal aquifers from seawater intrusion. Artificial recharge of treated waste water is also taking place, principally by infiltration in surface artificial recharge facilities; however, waste water that has undergone tertiary treatment is also being injected at depth to assist in protecting aquifers in Orange County from seawater intrusion.

Active element analog-digital models have been developed for many ground-water basins. These models quantify ground-water movement and storage, and in some cases water quality as well.

Water rights have been clarified in several areas by landmark decisions. In the San Fernando Reference, the California

---

Bean, R. T., and Weber E. M., 1999, The California Water Plan, *in* Moores, E. M., Sloan, D., and Stout, D. L., eds., Classic Cordilleran Concepts: A View from California: Boulder, Colorado, Geological Society of America Special Paper 338.

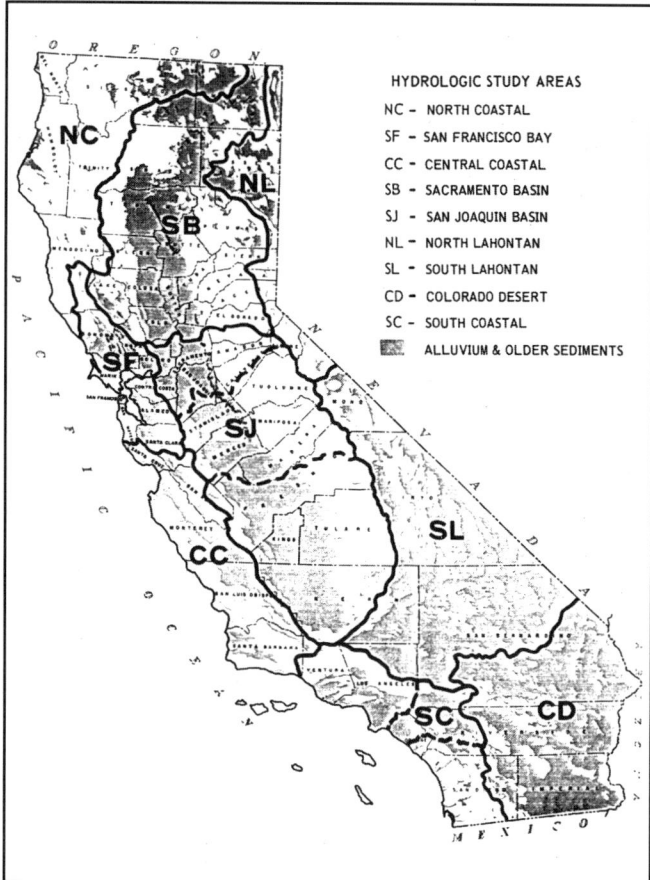

Figure 1. California ground-water basins (California Department of Water Resources, 1975).

State Supreme Court in 1975 established the right of the city of Los Angeles to the ground water and surface water of San Fernando Valley. The right of a public agency to withdraw water which it has put into ground-water storage by artificial recharge was established by the same decision. Import of waters from the Colorado River by the Metropolitan Water District of Southern California has been limited by the U.S. Supreme Court, and the right of Los Angeles to obtain water from the Owens-Inyo basin has been reduced by the State Supreme Court.

The two major problems with which users and managers of ground water in California must cope are problems of quantity and quality. The state's ground-water basins are estimated to be overdrawn about 1.46 million acre-feet (maf) in an average year (CDWR, 1998), but as much as 6 maf during droughts (Kennedy, 1990). However, many of the state's major ground-water basins are not overdrawn at all, but are operating at perennial yield, or safe yield (Table 1). Problems of poor-quality ground water have become paramount during the past 20 years, particularly in the southern part of the state.

Most of the information in the present paper has been obtained from CDWR (1994). More recent information is mostly not available within the time constraints of the present paper. However, where data could be obtained from a draft of CDWR (1998), the more recent information has been included here.

## STATE WATER PROJECT

Geologic factors were fundamental in the design and construction of the vast State Water Project. This project was begun in earnest shortly after publication of the CWP (1957), and is now essentially complete, delivering water from the more humid north to the more arid south. Oroville Dam, the key structure of the project, is located on the Feather River in rocks identified before and during construction as massive amphibolite.* The dam foundation was thoroughly explored and geologic conditions evaluated before and during construction. Similar geologic exploration took place for the other 13 dams in the project, and for the underground power plant, 6 tunnels, 21 pumping plants, 11 power recovery stations, and 561 mi of aqueduct. In all of these, geologists worked with their engineer colleagues in location, design, and construction of the facility. Location of suitable construction materials was a responsibility of the geologists. Geologic problems that were dealt with included faults, landslides, expansive clays, and land subsidence (see James et al., 1997).

## CENTRAL VALLEY PROJECT

Construction of the Shasta, Friant, and Folsom dams of the Central Valley Project preceded publication of the CWP (1957) and construction of the State Water Project. Additional dams and distribution systems make the Central Valley Project today the largest storage and delivery system in the state (CDWR, 1998). Most Central Valley Project facilities are in the northern and central parts of the state, including the east side of the San Joaquin Valley.

## LAND SUBSIDENCE

Land subsidence of two kinds was a threat to the State Water Project. Shallow subsidence, or hydrocompaction, took place in certain low-density soils that subsided locally on application of water. Deep subsidence was due to reduction of head in confined aquifers as a result of heavy pumping. Each type had to be dealt with for proper construction and operation of the California Aqueduct along the west side of the San Joaquin Valley.

The portions of the aqueduct route subject to hydrocompaction were first delineated by test ponds and then presubsided with ponded water prior to construction. The threat of deep subsidence due to overpumping was solved as soon as water from

---

*Editors' Note: This so-called amphibolite clearly includes identifiable dikes and screens of gabbro that are a part of the Smartville Complex of the northwest Sierra foothills.

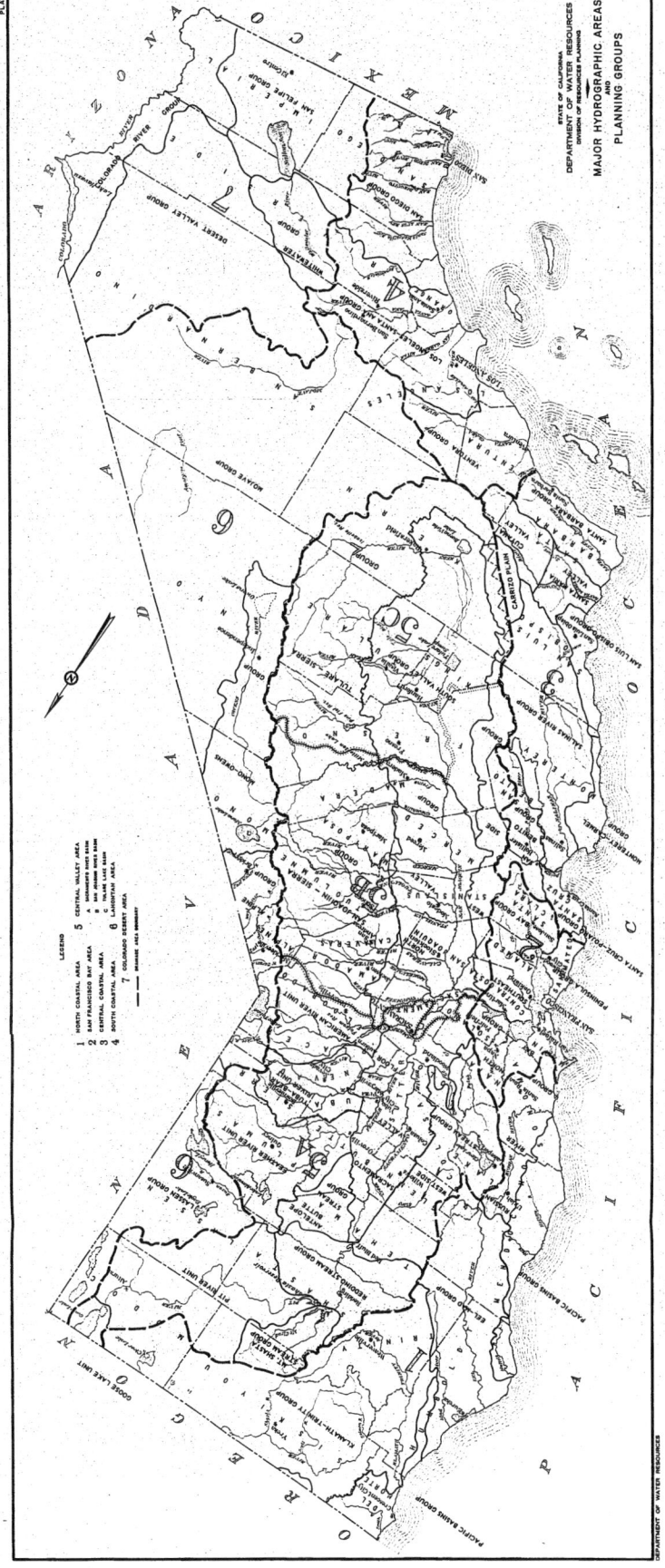

Figure 2. Major hydrographic areas and planning groups. Plate 3, CDWR, 1957.

## TABLE 1
## GROUND-WATER DEVELOPMENT IN CALIFORNIA
Quantities are normalized estimates as of 1990 level of development
(Only major basins listed)

| Counties / Basins | Major Water-Bearing Materials | Perennial Yield (acre-feet/year) | Overdraft (acre-feet/year) | Usable Storage (acre-feet) | Management Activities / Problems |
|---|---|---|---|---|---|
| **South Coast Region (4 on Fig. 2)** | | | | | |
| Ventura, San Bernardino, Los Angeles, Orange, Riverside | | | | | |
| Ventura-Santa Clara | Younger and older alluvium, older sediments | 121,000 | 22,000 | Unknown | Partial management by local agency / Sea water intrusion |
| San Fernando Valley | Younger and older alluvium, older sediments | 104,000 | 0 | 600,000 | Adjudicated. Management by local agency — Superfund site |
| Central Basin | Younger and older alluvium, Pleistocene and Pliocene sediments | 217,380 | 0 | 1,300,000 | Adjudicated. Management by local agency |
| West Coast Basin | Younger and older alluvium, Pleistocene and Pliocene sediments | 64,470 | 0 | 756,000 | Adjudicated. Management by local agency |
| Raymond Basin | Younger and older alluvium | 30,620 | 0 | 400,000 | Adjudicated. Management by local agency |
| San Gabriel Basin | Younger and older alluvium, Pleistocene and Pliocene sediments | 254,000 | 0 | 8,600,000 | Adjudicated. Management by local agency — Superfund site |
| San Bernardino Basin | Younger and older alluvium | 232,100 | 0 | 1,500,000 | Adjudicated. Management by local agencies |
| Chino Basin | Younger and older alluvium | 145,000 | 0 | 1,200,000 | Adjudicated. Management by local agency |
| Riverside Basin | Younger and older alluvium | 50,110 | 0 | Unknown | Adjudicated. Management by local agency |
| Orange County Coastal Plain | Younger and older alluvium, Pleistocene and Pliocene sediments | 262,000 | 0 | 800,000 | Adjudicated. Management by local agency |
| **Central Coast Region (3)** | | | | | |
| Monterey, San Benito, San Luis Obispo, Santa Barbara | | | | | |
| Pajaro Basin | Alluvium and older sediments | 53,000 | 11,000 | 600,000 | Management by local agency / Sea water intrusion |
| Salinas Valley | Alluvium and older sediments | 500,000 | 50,000 | 5,500,000 | Management by local agency / Sea water intrusion |
| Santa Clara-Hollister | Alluvium and older sediments | 75,000 | 0 | 1,800,000 | Monitoring program |
| Santa Maria | Alluvium and older sediments | 66,000 | 63,000 | 1,000,000 | Managment plan under development |
| Cuyama Valley | Alluvium and older sediments | 16.000 | 12,000 | 400,000 | None |

## TABLE 1 (continued)
## GROUND WATER DEVELOPMENT IN CALIFORNIA
Quantities are normalized estimates as of 1990 level of development
(Only major basins listed)

| Counties<br>Basins | Major Water-Bearing Materials | Perennial Yield (acre-feet/year) | Overdraft (acre-feet/year) | Usable Storage (acre-feet) | Management Activities-Problems |
|---|---|---|---|---|---|
| **San Francisco Bay Region (2)** | | | | | |
| Marin, Sonoma, Napa, Solano, Santa Clara, Alameda | | | | | |
|   Petaluma Valley | Alluvium & older seds. | Unknown | 0 | Unknown | Water quality problems |
|   Napa-Sonoma Valley | Alluvium & older seds. | Unknown | 0 | Unknown | Water quality problems |
|   Santa Clara Valley | Alluvium & older seds. | Unknown | 0 | Unknown | Management by local agency Superfund site |
|   Livermore Valley | Alluvium & older seds. | Unknown | 0 | 200,000 | Management by local agency |
| **San Joaquin and Tulare Lake Regions (5B and 5C)** | | | | | |
| Counties in San Joaquin Valley Proper | | | | | |
|   Total 15 basins | Alluvium and older sediments | 7,859,000 | 854,000 | 52,592,000 | Management by local agencies in many basins |
| **Sacramento Valley Region (5A)** | | | | | |
| Counties in Sacramento Valley Proper | | | | | |
|   Total 10 basins | Alluvium and older sediments, Volcanics | 2,418,000 | 33,000 | 46,000,000 | Local planning in most basins |
| **North and South Lahontan Regions (6 N and 6 S)** | | | | | |
| Lassen | | | | | |
|   Honey Lake | Alluvium and volcanics | Unknown | Unknown | 4,000,000 | Management by local agencies |
| Mono, Inyo, San Bernardino, Los Angeles | | | | | |
|   Owens Valley | Younger and older - alluvium, glacial deposits | 110,000 | 0 | Unknown | Cooperative management, local agencies and Los Angeles Dept. of Water & Power |
|   Antelope Valley | Alluvium and older seds. | 58,000 | 0 (?) | 20,000,000 | Limited management by local agencies |
|   Mojave River Valley | Alluvium and older seds. | 72,000 | 57,000 | 4,370,000 | Recent adjudication |
| **Colorado Desert Region (7)** | | | | | |
| San Bernardino, Riverside, San Diego, Imperial | | | | | |
|   Chuckwalla Valley | Alluvium and older seds. | 4,000 | 23,000 | Unknown | None |
|   Coachella Valley | Alluvium and older seds. | 33,000 | 52,000 | 3,600,000 | Management by local agencies |

Sources (modified): CDWR (1975) and (1994)  (1 acre-foot = 1,233.5 cubic meters)

the aqueduct became available. Farmers reduced, and in some cases eliminated, their pumping, and the head of the deep aquifers recovered and land subsidence slowed and stopped. Unfortunately, there may have been some reactivation of deep subsidence in recent years due to further decline in head.

## GROUND-WATER MODELS

The hydrogeologist has come to rely on models of ground-water basins to gain an understanding of how and why the basin behaves as it does, and how the system may react in the future in response to proposed management plans. Early development of basin-wide ground-water models in the late 1950s was done by Skibitzke of the U.S. Geological Survey (USGS). These were resistor-capacitance networks able to evaluate two-dimensional flow in a ground-water basin.

In the early 1960s the California Department of Water Resources built on the work of the USGS. R. G. Thomas and E. M. Weber (CDWR, 1961) developed a model of the Los Angeles Coastal Plain using a general purpose analog computer (electronic differential analyzer). The Coastal Plain model was used in analyzing the various alternative plans for conjunctive use of local and imported water.

By the mid 1960s relatively fast digital computers with large memory banks plus sophisticated programming languages (FORTRAN) became attractive. Tyson and Weber (1964) developed a finite difference digital model of the Los Angeles Coastal Plain. Based on this early digital model, the Department of Water Resources developed models for the major ground-water basins of the South Coastal area and utilized these models in cooperation with local entities to develop water-resource management plans.

Today, models are common tools of the hydrogeologist and hydrologist. However, the ground-water basin model is not the end point, but serves as a tool for further evaluation of basin operational studies. Most of the major and many of the smaller basins have mathematical models; a significant number evaluate water quality in addition to the original components of quantity. Numerous specialty models also cover specific situations related to basin management and contaminant problems.

## SOUTH COAST REGION

The ground-water basins of the South Coast region, from Ventura to San Diego Counties, have been key factors in the robust economic growth of California and its rise to become the most populous state. The water supplies essential for growth in this region of low and seasonal rainfall have come basically from ground water. All the major basins today store not only local runoff, but water imported from the Colorado River and/or the State Water Project. The vast usable storage capacity of the major basins, now estimated to total more than 15 maf (Table 1), makes it possible to pump ground water as needed, and in wet to normal years, to recharge the ground-water body during the rainy season (December through March.) The major artificial recharge facilities, called "spreading grounds," are mostly located in alluvial fans at the base of surrounding mountains.

Ground water in the Central basin, West Coast basin, San Fernando Valley, Raymond basin, and San Gabriel basin is governed by court decree or stipulated judgement. Management of these basins is accomplished by local agencies, management boards, or a watermaster. The basins along the Santa Ana River are managed individually by local agencies in compliance with a stipulated judgement. In all managed basins, pumpage is limited in accordance with the court decisions, water levels, anticipated amounts of local recharge, and necessary imported water.

Basin management has made it possible to balance input and outgo for the basins listed here. All these basins are being operated within perennial yield (safe yield), and thus have no overdraft in an average year. Table 1 lists the perennial yield from ground water of each of these basins.

The Orange County Water District is at the forefront in using advanced management techniques (Bean and Brown, 1992). A pumping trough and pressure ridge barrier has been established to repel seawater from the coastal plain aquifers along the lower Santa Ana River. The aquifer system underlying the plain was described after a detailed investigation by CDWR (1966). Utilizing this information, the Orange County Water District installed a line of 23 multipoint injection wells approximately parallel to the coast and inland from a line of 7 extraction wells. Recharge can be injected into any specific aquifer in the ridge needing replenishment. Furthermore, the seaward trough makes it possible to operate the protected ground-water basin at levels actually below sea level. Additional management techniques of the Orange County District include a pump tax and in lieu delivery of water—surface water delivered in lieu of being recharged and then pumped.

Some of the water injected at the Orange County barrier is treated waste water. After receiving treatment from a local plant, waste water is given advanced tertiary treatment, including reverse osmosis. Treated water is blended with fresh water and injected. An average total injection of 20 million gallons per day, or 22,000 acre-feet per year, maintains the pressure ridge.

Treated waste water is also artificially recharged in the Montebello forebay of the Central basin of the Coastal Plain. A very detailed and extensive health-effects study found that there were no undesirable effects on populations using well water which contains a component of treated waste water derived through surface recharge facilities. As much as 20% of the water entering the Central basin through the forebay is reclaimed water.

Although many studies have proved that pathogenic organisms die out in the oxidizing environment of the vadose zone above the water table, nitrates unfortunately are found below the water table where recharge is occurring by water percolating through organic sources. Nitrates above the maximum contaminant level of 45 mg/L are a problem in parts of many South Coast basins. Wells have had to be shut down, or their waters blended.

Since 1979, contamination of ground water by volatile organic compounds has increasingly become a problem in a number of basins, including the San Gabriel and San Fernando Valleys. Wastes from a great variety of industries have been permitted to infiltrate ground water for many tens of years. Contamination of these great basins has become so severe that the federal Environmental Protection Agency has put them on the National Priority List, designating them as Superfund sites. Cleanup, if funding can be made available, would probably take 30 to 50 years.

The pioneer fresh-water injection barrier of the 1950s (CDWR, 1957) protected the West Coast basin from seawater intrusion. This barrier has been expanded, and additional barriers have been constructed at Dominguez gap, Alamitos gap, and at Santa Ana gap in Orange County. The complex series of Pleistocene and Holocene aquifers in the West Coast and Central basins, originally described by the USGS and CDWR (1961), present a challenge for effective design and operation of the fresh-water barriers. A gentle hydraulic gradient is maintained from the crest of the ridge to the sea, to minimize fresh-water loss, while a steeper landward gradient moves more water into the fresh water basin.

The major aquifer serving this thoroughly urban area is the Silverado aquifer of the Pleistocene San Pedro Formation. The quality of water in the Silverado is protected from poor-quality water in a shallow aquifer by an overlying aquiclude. Hydrocarbons, resulting from activity by oil refineries, have contaminated large areas of the shallow aquifer. To date, satisfactory water quality in the Silverado has been maintained (L. C. Nagler, 1998, personal commun.).

*Ventura County.* The principal ground-water basin in Ventura County underlies the Oxnard Plain. This basin extends inland from the coast south of Ventura, and has a long history of seawater intrusion. The 1000 mg/L isochlor line was almost 3 mi from the coast in 1966, and has continued to advance inland.

After many years of disagreement, ground-water management was activated in 1982. Pumpage is now limited in all subbasins, and the attempt is being made to reduce overdraft, estimated at 22,000 acre-feet/year in 1990, and to bring the basin to safe yield by the year 2010.

*San Diego County.* San Diego County does not have the large ground-water basins with deep alluvial fill such as those present in the northern part of the South Coast region. The fill along rivers is mostly narrow and subject to seawater intrusion. Most inland basins have little alluvial fill, and ground-water pumpage is largely limited to domestic wells drawing from granitic residuum and/or crystalline basement complex.

As precipitation in the county is generally low, water import is essential. Colorado River water has been obtained for some years through the Metropolitan Water District of Southern California, and negotiations are currently underway between the San Diego County Water Authority and Imperial Water District to purchase additional Colorado River water for which the Imperial Irrigation District holds the water right. Waste-water recycling is currently being pursued to a limited extent, and desalination is in the planning stage.

## CENTRAL COAST REGION

Water supplies in the Central Coast region were obtained principally from ground water at the time of publication of the CWP (1957), and the situation is the same today. Under average conditions as of 1990, ground-water supplied 82% of the water used, local surface water supplied 12%, imports supplied 5%, and reclaimed water supplied 1%.

Most of the Central Coast basins extend inland from the coast, and are thus subject to seawater intrusion when ground-water levels fall below sea level. Salinas Valley, the largest of the basins and a very important agricultural area, has had a progressive seawater-intrusion problem since before publication of the CWP (1957). The Monterey County Water Agency is currently investigating management alternatives to relieve the overdraft in Salinas Valley, estimated as 50,000 acre-feet/year as of 1990. In addition to this figure, Table 1 shows that three additional major basins are also in overdraft, and the ground water is thus being mined. The region, with its increasing population and economy, faces a very uncertain future unless additional water supplies can be produced. Desalination, although very expensive, may eventually be required.

## SAN FRANCISCO BAY REGION

The need for additional water supplies to permit continuing population growth was recognized in the San Francisco Bay region long before publication of the CWP (1957). The two early major import projects were the Hetch-Hetchy project bringing water to San Francisco, and the Mokelumne project serving the East Bay area.

As of average conditions in 1990, imports supplied about 66% of the urban water needs of the San Francisco Bay region (CDWR, 1994). The imports include Central Valley Project water, additional federal project water, water from the State Water Project, and water from a few smaller projects. Only 2% of the total supply is ground water, and about 1% is reclaimed water.

*North Bay group.* The larger ground-water basins of the North Bay group are Napa-Sonoma Valley, Petaluma Valley, and Suisun-Fairfield Valley. Although the safe yield of these basins has not been evaluated, in part because of the threat of seawater intrusion, in 1990 ground water supplied about 24,000 acre-feet (in an average year). The gross storage capacity of these basins is about 1.7 maf (CDWR, 1994), but the usable proportion has not been estimated. Salt-water intrusion has been a problem in the lower parts of the basins bordering the bay. Ground-water quality is otherwise generally good, except for localized areas of high nitrate, boron, iron, hardness, and total dissolved solids.

*South Bay group.* The Santa Clara Valley bordering southern San Francisco Bay and the inland Livermore Valley are the two principal ground-water basins of the South Bay group.

Their total storage capacity, plus that of the Pittsburg plain south of Suisun Bay, is about 6.5 maf (CDWR, 1994).

Salt-water intrusion and land subsidence due to reduction of head in confined aquifers were severe problems in Santa Clara Valley before publication of the CWP (1957). However, both problems have since been ameliorated by progressive management techniques. The Santa Clara Valley Water District augments natural recharge in stream beds and percolation ponds along major watercourses as part of an extensive conjunctive use program. Land subsidence has been arrested and is believed to be under control. The Alameda County Water District (1998) is conducting an aquifer reclamation program that includes an extraction trough and an injection barrier. In the "Silicon Valley" area west of the bay, organic solvents used in electronics manufacturing have contaminated the ground water. A cleanup operation is under way, including steps to prevent further degradation of the basin.

## NORTH COAST REGION

In the North Coast region, which has the highest rainfall in the state, 99% of the annual water supply is from surface sources and only 1% is from ground water (1990 level, average conditions). In most of the region, rivers are protected from development by the California Wild and Scenic Rivers Act of 1972.

CDWR (1994) lists 18 ground-water basins. However, there was no significant coverage of ground-water in any of these basins in CDWR (1957), and limitations of space do not permit description of these basins in the present paper.

## SAN JOAQUIN VALLEY

The dire prediction made in the CWP (1957, p. 126) that a drought period would bring "widespread havoc and economic disaster" to the agricultural interests of the San Joaquin Valley has fortunately proved to be untrue, so far. Ground water has saved the economy. Growers have augmented their diminished surface supplies by increasing pumpage from old and newly drilled wells during droughts, and the crops have been saved.

The price that is paid is reduction of the amount of ground-water in storage. According to CDWR (1994, v. 1, Fig. 4-4), ground water in storage in the San Joaquin Valley decreased 13 maf between 1988 and 1994, which includes six drought years.

This loss of ground-water supplies would have been even greater except for the availability of State Project water from the California Aqueduct, which extends along the west side of the valley, near the edge of the alluvium. Deliveries of water from the aqueduct have significantly reduced the draft on ground-water.

Water management at an advanced level is practiced by a number of water districts, particularly in the dry Tulare Lake region of southern San Joaquin Valley. Conjunctive use (called "water banking" by some agencies) is practiced, with the recharge water including runoff (principally from the Kern River), water imported from the State Water Project and the Central Valley Project, and treated waste water. Water transfers between districts are common practice.

The average annual recharge in the 25 year period ending in 1995 at 35 artificial recharge sites in Kern County as reported by the Kern County Water Agency (1998) was about 421,000 acre-feet. This compares with average pumpage of just under 1.4 maf for the same period.

Ground-water quality problems tend to increase with time, particularly in the southern part of the San Joaquin Valley. In the confined aquifer, two areas where concentrations of salts are >1000 mg/L occur along the west side and in the extreme southwest part of the valley sediments. In the overlying unconfined aquifer, these areas of high salt concentration are more extensive, and additional patches occur elsewhere in the valley.

Even with the progressive management techniques (Bachman et al., 1997) being utilized in many parts of the San Joaquin Valley, the numbers in Table 1, taken from CDWR (1994), are disquieting with regard to the ultimate future of the valley. A continued annual overdraft of >850,000 acre-feet will result in lowering water levels, with increased pumping costs, continuing degradation of water quality, and eventual depletion of the ground-water supply.

## SACRAMENTO VALLEY REGION

Surface-water runoff in the Sacramento River drainage basin averages 22.4 maf annually. This amounts to nearly one-third of California's total natural runoff. Export of a significant portion of this runoff furnishes water for much of the state's urban and agricultural needs.

The filled trough of Sacramento Valley is estimated to have a usable ground-water storage capacity of about 40 maf (CDWR, 1994) stored in assorted clastic sediments and some volcanic material. CDWR (1994) lists no overdraft in Sacramento Valley except in Sacramento County, where the amount is 33,000 acre-feet annually. Export of ground water from Sacramento Valley is a possibility, and if this should take place, falling water levels can be expected, as well as possible overdraft and land subsidence.

The characteristics of 29 ground-water basins, including several that are subdivisions of Sacramento Valley proper, are listed in CDWR (1994). However, none of these basins outside of the valley are described in CDWR (1957) except in connection with export of water. This fact, plus limitations of space, makes it necessary to omit treatment of basins outside the valley from this paper.

## NORTH LAHONTAN REGION

The eastern strip of California north of the Mono Lake watershed and extending to the Oregon border is the North Lahontan region. Annual precipitation varies from about 4 in in Surprise Valley on the north to more than 70 in at the crest of the Sierra Nevada on the south. Of the region's water supply, 74% comes from surface sources, 23% is from ground water, 2% is reclaimed water, and 1% is imported.

The plan was optimistic about ground-water supplies from the Honey Lake area providing an "appreciable proportion" of the ultimate water requirements of that area. CDWR (1998) reported that water supply and demand from all sources were equal as of 1995 for the North Lahontan region except for drought years, but that a shortage is anticipated for the future. An additional complication is the fact that the Honey Lake basin is partially in Nevada.

## SOUTH LAHONTAN REGION

Precipitation in the South Lahontan region, which contains the highest and lowest elevations in the conterminous United States, ranges from <2 in annually in Death Valley to >50 in (mostly snow) at the crest of the Sierra Nevada. Water-supply sources for average conditions as of 1990 were as follows: ground water, 52%; surface water, 33%; imports, 13%; and reclaimed water, 2%.

Many years of litigation have characterized the export of water by the city of Los Angeles from Owens Valley and the Mono basin. In 1991 a long-term water management agreement was reached involving ground-water pumping and surface-water diversions. This agreement was challenged; however, a court Memo of Understanding in 1997 has resulted in progress in implementing the agreement (CDWR, 1998). In 1997, the Great Basin Unified Air Pollution Control District issued an order to modify dry Owens Lake bed, in part by flooding, because of severe dust storms from the lake bed. The Los Angeles Department of Water and Power has recently agreed, subject to approval by the City Council, to progressively implement this agreement in revised form until air-pollution standards are met.

The water level of Mono Lake was severely reduced by diversion of tributary streams as part of the Los Angeles import project. After much litigation, the State Supreme Court in 1994 issued a decision setting flow requirements for these streams so as to maintain the lake level.

In the Mojave River basin, ground-water overdraft and conditions of poor quality water have been problems since before publication of the CWP (1957). The Mojave River furnishes about 80% of ground-water recharge in the basin. A court-ordered stipulated judgement was reached in 1996, in which the Mojave Water Authority was named Watermaster to alleviate overdraft by conservation procedures and purchase of supplemental water (CDWR, 1998).

## COLORADO DESERT REGION

This region, called the Colorado River region in recent CDWR reports, receives 95% of its water supply from the Colorado River, 4% from ground-water, and 1% from imports (including State Project water).

The large Coachella Valley, extending northwest from the Salton Sea, is urbanized in its upper portion and agricultural in the lower. There is an annual ground-water overdraft of 52.000 acre-feet in the valley. A management plan reducing this overdraft is in preparation by the Coachella Valley Water District. Management procedures being considered include basin adjudication, water recycling, conservation, and recharge with water from the Colorado River or the State Water Project.

The remainder of the ground-water basins in this region, including Chuckwalla Valley, have very limited recharge, except for basins along the Colorado River. In the latter basins, pumpage is limited because recharge from the river is considered river water for allocation purposes.

## WHAT OF THE FUTURE?

Although local, state, and federal agencies have made tremendous strides in management and solution of California's water problems since publication of the CWP (1957), continuing concerns about the future remain. The most obvious of these is the possibility of water shortage, particularly as the population of the state increases to an estimated 47.5 million by 2020 (CDWR, 1998). Any shortage would be greatly accentuated in the event of an extended drought of 10 years or more, which clearly has taken place in the past according to paleoclimatologic evidence.

Vast ground-water supplies are still available for drought emergencies in many basins of the state (Table 1). However, quality problems in some of these, such as San Gabriel Valley, would require water treatment before domestic use. All management options including conjunctive use, water conservation, waste-water reclamation, and water transfers must be utilized to the maximum extent possible.

The ultimate resource, particularly for coastal areas where conveyance costs would be low, is the sea. Seawater conversion is highly energy consumptive, and costs $900 to $2000 per acre-foot (CDWR, 1994), which is several times more costly than water from most other sources, if available. The city of Santa Barbara, for example, has constructed a seawater conversion plant after rejecting an opportunity to be served by the State Water Project.

Perhaps the Achilles heel of the State Water Project at this time is the Sacramento–San Joaquin Delta. Subsidence is a threat to the levees protecting the islands of the delta, and levee breach during flood has already caused complete island loss. An earthquake would compound the tragedy. A bypass from the Sacramento River around the delta is needed to protect delivery of water to the California Aqueduct and the southern part of the state. Voters rejected the peripheral canal when first proposed in 1982.

However, a joint state-federal planning organization called CALFED, established in 1995 to develop a bay-delta solution, has recently issued a draft report (McClurg, 1998). After innumerable hours of negotiation among environmental, agricultural, and urban interests, CALFED recommends a solution that includes a bypass canal. It is certainly to be hoped that this solution will be accepted by the voters of California.

## ACKNOWLEDGMENTS

The authors wish to acknowledge with thanks the valuable information received from a number of water agencies, including knowledgeable individuals and recent agency reports. All agencies contacted were very cooperative in furnishing information.

## REFERENCES CITED

Alameda County Water District, 1998, Survey report on groundwater conditions, February, 1998: Fremont, California, Alameda County Water District, 17 p.

Bachman, S., Hauge, C., Neese, K., and Saracino, A., 1997, California groundwater management: Sacramento, Groundwater Resources Association of California, 145 p.

Bean, R. T., and Brown, G. A., 1992, Groundwater and its management in southern California, in Pipkin, B., and Proctor, R., eds., Engineering geology practice in southern California: Long Beach, California, Association of Engineering Geologists, p. 385–409.

California Department of Water Resources, 1957, The California Water Plan: Sacramento, California Department of Water Resources, Bulletin 3, 246 p.

California Department of Water Resources, 1966, Ground water basin protection projects, Santa Ana Gap salinity barrier, Orange County, California Department of Water Resources, Bulletin 147-1.

California Department of Water Resources, 1975, California's ground-water: Sacramento, California Department of Water Resources, Bulletin 118, 135 p.

California Department of Water Resources, 1994, California Water Plan update: Sacramento, California Department of Water Resources, Bulletin 160-93, unpaginated.

California Department of Water Resources, 1998, California Water Plan update, public review draft: Sacramento, California Department of Water Resources, Bulletin 160-98, unpaginated.

James, L. Richter, R., and Bean, R., 1997, History of engineering geology in the California Department of Water Resources: Environment and Engineering Geoscience, v. 3, no. 1, p. 89-110.

Kennedy, D. N., 1990, Are California's water resources adequate to meet the state's projected growth? in Coping with water scarcity: The role of ground water, 17th Biennial Conference on Ground Water: Davis, University of California Water Resources Center, p. 19-23.

Kern County Water Agency, 1998, Water supply report 1995: Bakersfield, Kern County Water Agency, 100 p.

McClurg, S., 1998, Delta debate: Western Water, March/April, p. 4–17.

Tyson, H. M., and Weber, E. M., 1964, Groundwater management for the nation's future—Computer simulation of groundwater basins: American Society of Civil Engineers Journal of Hydraulics Division, v. 90, no. HY4, p. 59–73.

MANUSCRIPT ACCEPEPTED BY THE SOCIETY NOVEMBER 23, 1998.

# Water in California—1998: A brief update

**Carl J. Hauge**
*California Department of Water Resources, Sacramento, California 95814; e-mail: chauge@water.ca.gov*

## INTRODUCTION

The amount of water available in California is dependent on the climate, which is defined as arid to semiarid Mediterranean type. The southern two-thirds of the state receives little rainfall, and all the precipitation usually occurs between November and March, with the exception of a limited number of summer thundershowers. A significant amount of precipitation falls in the mountain ranges and in the northern third of the state as snow, which, during the spring melt, flows downstream either to the ocean or into one of the "dry lakes" in the Basin and Range geomorphic province. Rainfall in the northern part of the state is significantly greater than in the southern part. Natural variations in California climate include extended periods of below average precipitation and some years when the average is significantly exceeded.

Historically, the snowmelt runoff provided a lot of water in the spring, but there was not much flow in many streams late in the summer when crops were maturing. Early in California history residents recognized the need to store water in the spring for use later in the summer during the irrigation season, just as they recognized the importance of ground water for the expansion of irrigated agriculture. By the late 1800s many people had also begun to think about importing surface water from other watersheds. The history of water development in California is well documented (Blomquist, 1992; Gottlieb and FitzSimmons, 1991; Harding, 1960; Hundley, 1975; Kahrl, 1982; and others).

## THE CALIFORNIA WATER PLAN

In 1957 the California Department of Water Resources published Bulletin 3, The California Water Plan, which outlined a plan for the development of all of California's water resources. The California Water Plan included water collection, storage, and delivery systems that would provide a supply of water in addition to the water supplied by the federal government's Central Valley Project. In 1957, many of the Central Valley Project's facilities had already been constructed and other facilities were being added.

Bulletin 3 pointed out that more than 70% of the precipitation is produced north of the latitude of Sacramento, whereas 80% of the use and projected future need for water is south of that latitude. Large differences in the amount of precipitation and runoff from year to year, the occurrence of that runoff primarily in the winter and spring, and the frequency of multiyear droughts made surface-water reservoirs for flow-regulating purposes look very attractive. While such surface water reservoirs would provide a year-round water supply, the frequent flooding that occurred after heavy precipitation and runoff made the same dams look very attractive for flood-control purposes. Each use conflicts with the other use. Full reservoirs provide a good water supply, but if they are full when surface runoff peaks, they are not useful as flood-control reservoirs. Since most dams and reservoirs are multiuse projects, maintenance of adequate flood control space and an adequate water supply for dry-season use relies on correct interpretation of the hydrologic record.

At the time of publication, the California Water Plan was considered to be a master plan to guide the activities of all agencies in planning, constructing, and operating facilities to control, develop, protect, conserve, distribute, and use California's water resources for the benefit of all areas of the state and for all beneficial uses. "Surplus" water was to be transferred from areas with an abundant water supply to areas of need. The plan envisioned that 11.6 million acre-feet (maf) of water could be made available from the north coast and 10.3 maf could be made available from the Sacramento Valley. One acre-foot is equal to 325,851 gallons. Local surface water projects would be built that would store 77 maf of water in addition to the 1957 existing surface-

---

Hauge, C. J., 1999, Water in California—1998: A brief update, *in* Moores, E. M., Sloan, D., and Stout, D. L., eds., Classic Cordilleran Concepts: A View from California: Boulder, Colorado, Geological Society of America Special Paper 338.

water storage capacity of 20 maf. Operation of surface-water facilities in conjunction with ground-water basins to increase water supply was a significant part of the plan. Total cost for the entire plan was estimated at $12 billion in 1955 dollars. Using the consumer price index to estimate 1998 dollars, the cost would be ~ $75 billion.

The Feather River Project, the initial unit of the California Water Plan, was implemented, and water deliveries began in the late 1960s and early 1970s. That unit includes Oroville dam and the California aqueduct, and in water-sufficient years delivers about 2.2 maf of water to parts of Sacramento Valley, San Francisco Bay area, San Joaquin Valley, and southern California. The project was built to carry 4.4 maf of water per year, but so far that total has not been available.

What happened to the California Water Plan envisioned in 1957 and what is the picture in 1998? The assumption embodied in the California Water Plan was that the pace of water resources development established in the years before and after World War II would continue indefinitely. But that pace slackened for a number of reasons.

First, construction costs escalated rapidly. Second, sources of funding to pay for such projects became harder to find. Third, dams had already been built on the best sites. Fourth, residents in areas from which water was to be exported became more concerned and more involved in the process. Fifth, and probably the most important factor, was the realization that water-development projects were impacting environmental resources much more detrimentally than had been anticipated when the earlier projects had been built.

The demand for water continued to increase, but the source of additional water to satisfy that demand continued to go unidentified. Various species were listed as threatened or endangered under the federal or state statutes, providing additional threats to existing sources of water. The 1987–1992 drought made it clear that contractual obligations for water deliveries could not be met after such a prolonged dry spell.

## THE NORTH-SOUTH LINK

San Francisco Bay and the delta formed by the confluence of the Sacramento and San Joaquin Rivers have been recognized as the key to any long-term transfer of water from northern to southern California. Despite this recognition, federal, state, and local agencies have been unable to agree on a plan that would maintain delta water quality, levee integrity, and maintain or improve aquatic, riparian, and terrestrial habitat.

The Central Valley Project Improvement Act was passed by Congress in 1992 and was intended to modify procedures for renewing water agency contracts with the U.S. Bureau of Reclamation that had originally been signed in the early 1900s. The act was controversial with growers, who tried unsuccessfully to rewrite it the following year.

As a means of resolving the controversial questions surrounding water-supply issues relating to the bay and delta, the state and federal governments signed an agreement in 1994 creating the CALFED Bay-Delta Program. While the mission of the CALFED Bay-Delta Program (1998) is to restore the ecological health of the Bay-Delta system, it is clear that any solution will have consequences that will reach far upstream on each river, as well as into southern California.

CALFED principles require that any solution must be affordable, equitable, implementable, durable, must reduce conflicts, and must not redirect significant impacts. The CALFED primary objectives are as follows.
• Water quality—To provide good water quality for all beneficial uses
• Ecosystem quality—To improve ecological functions in the bay-delta to support sustainable populations of diverse and valuable plant and animal species
• Water supply—To reduce the mismatch between bay-delta water supplies and current and projected beneficial uses dependent on the bay-delta system
• Vulnerability of delta functions—To reduce the risk to land use and associated economic activities, water supply, infrastructure, and the ecosystem from catastrophic breaching of delta levees.

These principles and objectives have resulted in six common program elements:
• Ecosystem restoration
• Water quality to lower toxicants in the system
• Water-use efficiency
• Levee-system integrity
• Water transfers
• Coordinated watershed management

Although the problem identification is limited to the bay-delta, solutions to those problems will benefit almost every water user in the state.

In 1998, California finds itself with a rapidly growing population and increasing demands for water. CDWR Bulletin 160-98, the update of the California water plan, estimates that California will need an additional 3 maf per year by 2020. With no large surface-water storage projects in the planning stages, interest has turned to increased efficiencies in ground-water basin use. While several potential sites for surface water storage have been identified, strong interest has been focused on greater use of ground-water basins in the north part of the state. Specifically, that focus is on conjunctive management of surface water and ground water, water banking, water marketing, water transfers, water conservation, and water recycling.

## DEFINITIONS

**Conjunctive use** is the operation of a ground-water basin in combination with a surface-water storage and conveyance system to maximize water supply. The three common forms of conjunctive use are listed in the following.

**Incidental conjunctive use** occurs when an area relies on surface water when it is available, and on ground water when surface water is not available. This is the basic level of conjunctive use. Management techniques may be used to define the timing and location of surface-water deliveries and ground-water pumping to maximize water-supply reliability.

**In-lieu recharge** brings additional surface water into an area using ground water or both surface water and ground water. The additional surface water is used to irrigate in lieu of ground water, thereby allowing ground-water levels to recover. The replenished ground-water supply can then be retrieved during dry years, easing the burden on surface-water supplies.

**Direct recharge** involves conjunctive use programs incorporating artificial recharge methods require a source of surface water that is not needed for immediate use. The surface water is placed directly into the ground by various means, including spreading ponds and injection wells. The water stored in the aquifer is then available for use in dry years. Water managers in many basins in California have operated conjunctive use programs of various kinds for years, beginning in the 1800s.

**Ground-water overdraft** is the intentional or inadvertent withdrawal of water from an aquifer in excess of the amount of water that recharges the basin over a period of years during which water-supply conditions approximate average, which, if continued over time, could eventually cause the underground supply to be exhausted, cause subsidence, cause the water table to drop below economically feasible pumping lifts, or cause a detrimental change in water quality. Ground-water overdraft is a synonym for ground-water mining.

**Water banking** is a water management program whereby water is allocated for current use or stored in surface-water reservoirs or in aquifers for later use. Water banking is a means of handling surplus water resources.

**Water marketing** is the selling or leasing of water rights in an open market.

**Long Term** is a long-term contract that shall be for any period in excess of one year (California Water Code Section 1735).

**Water transfer** is the conveyance of ground water or surface water from one area to another that involves crossing a political or hydrologic boundary. It also means a voluntary change in a point of diversion, place of use, or purpose of use that may involve a change in water rights.

## GROUND-WATER MANAGEMENT IN CALIFORNIA

There are six types of ground-water management that can be implemented in California. Each type of management requires recognition of the political, institutional, legal, and technical constraints and opportunities that are available in each basin. In most basins more than one form of management has usually been implemented. For example, overlying landowners, local agencies, a specially legislated district, and an AB 3030 ground-water management plan may all affect ground-water management in the same basin. This overlap of different management authority provides room for disagreement and opportunities for building consensus. A brief explanation of each of the six types of management follows.

**1. Individual management.** Overlying property rights allow anyone in California to build a well and extract their correlative share of ground water, which is not defined until the basin is adjudicated. The availability and use of ground water has increased local prosperity in some areas, and in some cases, has provided enough money to construct a water project that can convey surface water into the local area. Even though the management of ground water may not have been closely coordinated between landowners, this has been called a form of management.

**2. Local agency management.** Local agencies are identified in the California Water Code with specific statutory provisions to manage surface water. Some of these 23 types of identified agencies also have statutory authority to develop ground-water management plans and programs. Many of these agencies have done so.

**3. Adjudicated basin management.** Adjudicated basins are those basins in which one or more water users have asked the court to adjudicate the basin (e.g., Alhambra vs. Pasadena) and decide the ground-water rights of all the overlying property owners and appropriators. The court also decides: (1) who the extractors are; (2) how much ground water those well owners can extract; and (3) who the watermaster will be to ensure that the basin is managed in accordance with the court's decree. The watermaster must report periodically to the court. There are 16 adjudicated ground-water basins in California. In one basin the decision came 24 years after the case was filed.

In the early adjudications, the concern was primarily about water quantity. Water quality was a secondary consideration, if it was considered at all. More recently concern about water quality has prompted those involved to include quality in any decisions.

**4. Water management agencies.** Special legislation has been enacted to form ground-water management districts, or water management agencies in some parts of California. This legislation allows such districts to enact ordinances to limit or regulate extraction. There are nine of these water management agencies in California, but only a few have been effective in ground water management. Three other agencies have amended appropriate sections of the water code and have obtained similar authority.

**5. AB 3030 ground-water management.** Sections 10750 through 10756 of the California Water Code (AB 3030) provide a systematic procedure for an existing local agency to develop a ground-water management plan. This section of the code provides such an agency with the powers of a water replenishment district to raise revenue to pay for facilities to manage the basin (extraction, recharge, conveyance, quality). As of 1998, 149 agencies had adopted an AB 3030 ground-water management plan.

**6. Local ordinances.** Ordinances are being enacted by local governments to develop ground-water management plans. In 1995 the California Supreme Court declined to review a lower court decision (Baldwin vs. Tehama County) that holds that state law does not occupy the field of ground-water management and does not prevent cities and counties from adopting ordinances to manage ground water. Butte, Glenn, Imperial, Inyo, Kern, Sacramento, San Benito, San Diego, San Joaquin, Shasta, Tehama, and Yolo Counties have adopted ordinances relating to ground-water management. The nature and extent of the police power of cities and counties to regulate ground water is currently uncertain.

The ordinances generally require that a proponent of a project that would export ground water from the county must obtain a permit from the Board of Supervisors. To obtain an export permit, the proponent must show the board that the export would not cause subsidence, ground-water quality degradation, or deplete the ground water in storage.

## CONCLUSION

The pace of development of water resources in California has slowed dramatically in the last 25 years, but demand for water continues to increase. Solution of delta issues, which the CALFED Bay-Delta Program is specifically designed to accomplish, would facilitate development of additional water supplies. Developing an adequate water supply will remain a major issue in the future as the burgeoning population of California continues to strain the available water resources.

## REFERENCES CITED

Blomquist, W., 1992, Dividing the waters: Governing ground water in Southern California: San Francisco, California, Institute for Contemporary Studies Press, 415 p.

CALFED bay-delta Program, 1998, Programmatic EIS/EIR, executive summary: Sacramento, California, CALFED Bay-Delta Program, 30 p.

California Department of Water Resources, 1957, The California Water Plan: Sacramento, California Department of Water Resources Bulletin 3, 246 p.

California Department of Water Resources, 1998, Draft of the California Water Plan update: Sacramento, California Department of Water Resources, Bulletin 160-98 (unpaginated).

Gottlieb, R., and FitzSimmons, M., 1991, Thirst for growth: Water agencies as hidden government in California: Tucson, The University of Arizona Press, 286 p.

Harding, S. T., 1960, Water in California: Palo Alto, California, N-P Publications, 231 p.

Hundley, N., 1975, Water and the West: Berkeley, University of California Press, 395 p.

Kahrl, W., 1982, Water and power: Berkeley, University of California Press, 583 p.

MANUSCRIPT ACCEPTED BY THE SOCIETY NOVEMBER 23, 1998.

# Section V
# TECTONICS IN THE DESERT

# Tectonics in the California Desert: Introduction

## INTRODUCTION

Deserts characterize the entire eastern side of California. Throughout most of California's length, from the Oregon border to the southern Sierra Nevada, the deserts are east of the mountains, and they represent a "cold" desert formed by the orographic (rain shadow) effect. South of the Sierra Nevada, the orographic effect combines with the increasing overall dryness of the northern "Horse Latitudes" to give rise to the "hot" deserts of southeastern California.

In a real way, the deserts of the West set California off from the rest of the country, giving it a unique biota and landscape distinct from any other part of the United States. Deserts were a major hurdle to the westward migrations of the mid-nineteenth century. The first overland party, the Bartleson-Bidwell expedition of 1841, made the trek without untoward difficulty (Stewart, 1962). Others did not fare so well. The Humboldt River route, which became the principal California Trail, the line of the first railroad, and of one of the major highways (I-80), presented many difficulties—its small size, the overtaxing of fuel, feed, and water, and finally its petering out in the deserts of western Nevada. J. C. Frémont's 1843–1844 expedition first documented this fact. The story goes that one of his assignments was to map the course of the previously recognized Humboldt River from its headwaters near the Nevada-Utah state line to the Pacific. However, the Humboldt ends in the Humboldt Sink, near present-day Lovelock, Nevada. When Frémont reached that point, he was lost and in trouble. After spending some time searching vainly for a passage of the Sierra Nevada, one of his guides, Kit Carson, who had been to California before, climbed a peak near a pass to try to find a way. He looked west, and exclaimed, "There's Mt. Diablo. I know where I am." (Frémont named the pass after Carson.)

The rigors of the waterless passage from the Humboldt Sink to the Truckee River led to several attempts to find an alternative, better route. One of the least successful of these, the Lassen "Cutoff," named for the Danish-American early settler who advocated it, extended northwest from the lower Humboldt River to Fandango Pass, on the Warner Range, in the extreme northeast corner of California, and thence to the northern Sacramento Valley, thereby adding several hundred desert miles to the route. The Humboldt route remained the main California Trail. Records of the total migration along it during the 1849–1852 Gold Rush peak suggest an average of one wagon every 15 minutes past several constrictions or "choke points" along the route (W. Fisher, 1980, oral commun.).

Geologically, all the California deserts are part of the Basin and Range province, that region of north-northwest–trending ranges and valleys that separates the mountains of California from the Colorado Plateau and Rocky Mountains to the east. This region has been referred to as "an army of caterpillars marching north out of Mexico." Perhaps more appropriately, the Basin and Range might be called "the stretch marks of the continent," because in this region the continent is actively extending.

This section provides a geologic context for this diverse, historic, and fascinating region. It contains three chapters. In the first, the paper by R. L. Armstrong forms the focus of the essay on widespread low-angle "denudation" faults by Brian Wernicke and Jon Spencer. The existence of these features was highly controversial when first proposed. Many early workers still bear scars from the fierce reaction the idea first engendered (e.g., one of us, E. M. Moores; see Moores et al., 1970). As described by Wernicke and Burchfiel, Armstrong's paper comprehensively laid out the abundant regional evidence for low-angle normal faulting. These features have proven to be of worldwide significance, despite the difficulty of reconciliation of them with present-day seismicity and faulting theory (e.g., Parsons and Thompson, 1993a, 1993b; Yin, 1993; Yin and Dunn, 1992).

In the second chapter in this section, Walker and Glazner base their essay entitled "Tectonic development of the southern California deserts" on the "classic paper" entitled "Garlock fault:

An intracontinental transform structure, southern California" by B. C. Burchfiel and G. A. Davis. The Garlock fault is arguably the second most famous fault in California, after the San Andreas. Walker and Glazner's essay gives a good summary of the state of knowledge of this complex region.

The final chapter in this section is an essay by L. A. Wright and B. W. Troxel, entitled "Levi Noble's Death Valley, a 58-year perspective." It focuses on Noble's 1941 paper on part of this region. Death Valley is of special interest because it is the lowest of the basins of the Basin and Range province. The valley also is a romantic part of the California legend. It was the scene of an early mishap in the great westward migration (the ill-fated Manley-Brier party of December 1849–January 1850 that allegedly named it), it was a site of early economic exploitation of evaporite deposits, and it is one of the most popular national parks.

Together these three chapters give a good flavor of the nature of the terrain along California's eastern margin, and the geologic processes that produced it.

## REFERENCES CITED

Moores, E. M., Scott, R. B., and Lumsden, W., 1970, Structure and tectonics of the White Pine Range–Grant Range region, east central Nevada, a reply to H. Drewes: Geological Society of America Bulletin, v. 81, p. 323–330.

Stewart, G., 1962, The California trail; an epic with many heroes: New York, McGraw-Hill, 339 p.

Parsons, T., and Thompson, G. A., 1993a, Does magmatism influence low-angle faulting?: Geology, v.21, p. 247–250.

Parsons, T., and Thompson, G. A. 1993b, Does magmatism influence low-angle faulting?: Reply: Geology, v. 21, p 957–958.

Yin, A., 1993, Does magmatism influence low-angle normal faulting?: Comment: Geology, v. 21, p. 956.

Yin, A., and Dunn, J. F., 1992, Structural and stratigraphic development of the Whipple-Chemehuevi detachment fault system, southeastern California: Implications for the geometrical evolution of domal and basinal low-angle normal faults: Geological Society of America Bulletin, v. 104, p. 659–674.

Manuscript Accepted by the Society on November 23, 1998.

Geological Society of America
Special Paper 338
1999

# Chapter 17
# BASIN AND RANGE EXTENSION

**Brian Wernicke**
*Division of Geological and Planetary Sciences*
*California Institute of Technology*
*Pasadena, California 91108*
*E-mail: brian@gps.caltech.edu*

**Jon Spencer**
*Arizona Geological Survey*
*416 West Congress Street, #100*
*Tucson, Arizona 85701*
*E-mail: jspencer@geo.arizona.edu*

## RICHARD L. ARMSTRONG

Richard Lee Armstrong was born in 1937 in Seattle, Washington, and left in 1955 to attend Yale University, where he spent the next 18 years, first as a student (bachelor of science degree, 1959; doctoral degree, 1964), and later as an assistant and then associate professor of geology. In 1973 he left Yale for the University of British Columbia, where he was an associate and then full professor of geology until his untimely death from cancer in 1991.

Dick Armstrong's interests in geology were broad and diverse. According to his former thesis advisor Karl Turekian, his passion "was to understand the earth" (Parrish, 1992). He and his students produced an enormous amount of isotopic geochronologic data that clarified the timing of numerous events in Cordilleran geologic history (see summaries and citations in Armstrong and Ward, 1991, 1993). Not content merely to date geologic events, Dick Armstrong took on some of the major problems of Cordilleran geology and produced syntheses of large scope and penetrating insight (e.g., Armstrong, 1968a, 1972, 1975). He was best known internationally for proposing and defending the concept of Early Archean crust-mantle differentiation followed by crustal recycling, as opposed to gradual extraction of crustal materials from the mantle over the course of Earth history (Armstrong, 1968b, 1981, 1991).

## RATIONALE

By 1972, 100 years of geologic research in the Great Basin of the western United States had identified high-angle normal faulting as the dominant mechanism of upper crustal extension. Enigmatic low-angle faults and penetrative footwall deformation were commonly attributed to Mesozoic, thrust-related orogenesis, and in some cases were attributed to denudation at the uphill end of giant gravity slides that had produced the Cordilleran fold and thrust belt at the downhill end. Richard Lee Armstrong recognized that many low-angle faults and associated deformation must be middle Cenozoic and fundamentally related to extension. His 1972 paper in the GSA *Bulletin*, the topic of this retrospective, contributed enormously to the conceptual disengagement of thrust and extension tectonics in the eastern Great Basin. With additional field studies in the mid 1970s, the geologic community embraced low-angle extensional tectonics as a fundamental component of Cenozoic Basin and Range tectonics, and in the late 1970s the concept of extensional metamorphic core complexes was born.

Jon Spencer

RICHARD LEE ARMSTRONG  *Department of Geology and Geophysics, Yale University, New Haven, Connecticut 06520*

# Low-Angle (Denudation) Faults, Hinterland of the Sevier Orogenic Belt, Eastern Nevada and Western Utah

## ABSTRACT

Low-angle faults that place younger strata on older are the distinctive structural feature of the hinterland of the Sevier orogenic belt in Nevada. Although shown on many maps as Mesozoic thrust faults, these low-angle faults may in many places be extension, denudation, and gravitational gliding features of Tertiary age. The complexity of later Tertiary deformation caused by crustal extension in the hinterland has not been sufficiently emphasized.

Several contrasting interpretations of the relation between hinterland and Sevier belt structures exist. Whitebread, Hose, Roberts, and Crittenden advocate gravitational gliding models that differ in detail, but in general they correlate extension in the hinterland with thrusting toward the foreland during the Cretaceous. Misch, Nelson, Fritz, Miller, and Woodward assign a Mesozoic age to the low-angle faults which they relate to a regional décollement. The frontal breakout of this décollement is proposed by them to be west of, and older than, the thrust faults of the Sevier belt or perhaps equivalent in age to them. Other geologists (Armstrong, Burchfiel, Davis, and Fleck) argue for a compressional origin for the Sevier belt that involves considerable crustal shortening. To them the low-angle hinterland faults are unrelated to Sevier belt thrusting. Geometric and chronologic problems are created by the first two interpretations, making the third interpretation most attractive.

## INTRODUCTION, GENERAL SETTING

The Basin and Range province is characterized by tilted fault blocks bounded by normal faults of post-Eocene age (Nolan, 1943; Mackin, 1960; Hamilton and Meyers, 1966) that trend generally north-south. The blocks expose deformed Paleozoic strata that display a variety of structural styles but most distinctive in the eastern Great Basin are low-angle faults that juxtapose younger rocks on older (Misch, 1960). These faults, which will often be referred to in this paper as denudation faults (because strata originally overlying the footwall have been removed tectonically), have been interpreted in various ways. Their origin and significance is today a topic of dispute and disagreement. I believe that the models linking denudation faults with thrust faults of Cretaceous age that lie to the east in the Sevier orogenic belt are unlikely and that the denudation faults are predominantly of Tertiary age and related to Basin and Range faulting. This involves reinterpretation by me of certain structures, but does not mean that I disagree with the distribution of rock units as shown on any maps. The central focus will be on structures of east-central Nevada where denudation faults are abundant and many areas are well mapped. The most useful large-scale geologic maps for the eastern Great Basin are the western half of the state geologic map of Utah (1:250,000) (Stokes, 1963; Hintze, 1963) and the county maps of Nevada (1:250,000 and approximately, 1:125,000-preliminary form) (Longwell and others, 1965; Tschanz and Pampeyan, 1970; Hose and Blake, 1970; Kleinhampl and Ziony, 1967).

Some of the topics discussed in this paper were touched upon in my papers on the Sevier orogenic belt (Armstrong, 1968a, 1968b). Those papers should be consulted for premonitions, elaboration, or clarification of ideas, and more thorough documentation of facts cited.

The eastern Great Basin (Fig. 1) is underlain by Paleozoic strata of the Cordilleran miogeosyncline. The zone of transition between miogeosyncline and platform, the Wasatch zone, is the locus of the Sevier orogenic belt. In this zone, approximately 100 mi wide, the strata of the miogeosyncline were thrust over platform strata during the Cretaceous. Here, Tertiary normal faults bound fault blocks that contain older-over-younger listric thrust faults and related folds which conform to the rules of structural style established in the Wyoming, Montana, and Canadian Rocky Mountains (Rubey and Hubbert, 1959; Armstrong and Oriel, 1965; Mudge, 1970; Price and Mountjoy, 1970; Dahlstrom, 1970). At least 40 mi of tectonic contraction is expressed by these Mesozoic compressional structures of the Sevier belt.

The hinterland of the Sevier belt (Fig. 2) has a complex tectonic history and is the site of many extant problems, including, but not restricted to, the denudation faults discussed in detail in this paper. Between the Sevier orogenic belt and the frazzled eastern edges of the Antler orogenic belt in central Nevada (Roberts and others, 1958) is a zone 100 to 150 mi wide of heterogeneous structural style. The structure of individual mountain ranges in the hinterland of the Sevier orogenic belt may be relatively simple homoclines or broad folds, broken only by steep normal faults. Many ranges display low-angle faults placing younger rocks on older and older on younger, but the former predominate. A few ranges (Snake, Southern Deep Creek, Grant, Ruby, and Raft River ranges and the Wood Hills are spectacular examples) expose metamorphosed and tightly folded lower Paleozoic and Eocambrian strata with metamorphic fabrics produced during the kinematic penetrative deformation during the Mesozoic (Misch, 1960; Misch and Hazzard, 1962; Nelson, 1966; Bick, 1966; Cebull, 1970; Howard, 1971; Thorman, 1970; Compton, 1969). Older Precambrian basement rocks recrystallized and remobilized during metamorphism of geosynclinal strata, are exposed only in the Raft River-Albion Ranges in northwestern Utah-southern Idaho and in California (Armstrong and Hills, 1967; Lanphere and others, 1964). Elsewhere in the hinterland, the presence of exposed older Precambrian rocks is disputable, although such ancient rocks probably underlie most of the region at depth. The contact between metamorphosed and unmetamorphosed strata in Nevada and Utah is usually a fault that developed during or after the waning stages of regional metamorphism (Misch, 1960; Misch and Hazzard, 1962; Nelson, 1969; Misch, 1971). Commonly it juxtaposes younger rocks over older.

## PRE-TERTIARY HINTERLAND STRUCTURES

Evidence sufficient to date any orogenic structures as pre-Tertiary is rare in the hinterland. A conformable sequence of Eocambrian through Lower Jurassic strata indicates that virtually all tectonic complexity in the region developed at some time during the last 200 m.y., but it is only at those few places where Late Jurassic–Early Cretaceous or mid-Cretaceous plutons crosscut deformed rocks that a significant refinement in dating is possible. In Elko County, plutons at least 150 m.y. old crosscut complex imbricate thrust faults in the H-D range (Riva, 1970), deformed and metamorphosed strata in the Ruby Mountains (Kistler and Willden, 1969) and deformed miogeosynclinal strata at Whitehorse Mountain (Adair and Stringham, 1957). A slightly younger pluton cuts deformed strata in Dolly Varden area (Snow, 1964). In White Pine County, metamorphism and deformation in the Snake Range precedes 120 m.y. and may predate 165 m.y. (Armstrong and Hansen, 1966; Lee and others, 1970). At Ely, faulting of upper Paleozoic strata, including younger-on-older style faults, predates plutons that have been accurately dated as 109 m.y. (Bauer and others, 1966; McDowell and Kulp, 1967).

Mesozoic thrusting, much of it older than the Sevier belt in Nevada and Utah is also recognized in the Inyo and Clark Mountain regions of California (Stevens, 1969; Olson, 1970; Burchfiel and others, 1970; Burchfiel and Davis, 1971). Thrust faults in the Nevada test site (Barnes and Poole, 1968; Hinrichs, 1968) and farther north in western Lincoln County and in northern Nye County (Tschanz and Pampeyan, 1970; Dodge, 1970) may be equally as old.

The amount of displacement on various Mesozoic thrust faults in the hinterland can only be estimated in rare cases where contrasting facies of units such as Mississippian clastic rocks, derived from the Antler orogenic belt, are juxtaposed (Brew, 1971; Thorman, 1970).

One Mesozoic structure that is widespread, and perhaps universal in the Grant-White Pine, Egan and Shell Creek Ranges is localization of bedding-plane faults along Mississippian shale. Rocks below this fault zone are usually relatively unfaulted. The Pennsylvanian and younger strata in the hanging wall are severely broken by high-angle faults that do not extend below the detachment zone, as shown particularly well in the southern Egan Range. Structures of this zone are locally imbricate (for example, the Becky Peak thrust fault of the northern Schell Creek Range described by Dechert, 1967); in some places, such as the White Pine Range (Drewes and others, 1970) Tertiary volcanic rocks postdate movement along such faults within the Mississippian strata, but in some areas, the Tertiary rocks are as broken as the underlying upper Paleozoic strata. A reasonable interpretation is that weak zones, such as the Mississippian shale, were sites of considerable movement at one or more times during the Mesozoic. These same zones of weakness were reactivated during the Tertiary. At any given place, the movement at different times may have been in different directions (generally eastward movement of upper plates being probable during the Mesozoic; down-dip movement likely during the Tertiary); structural confusion is an obvious consequence.

Cretaceous sedimentary rocks at a few localities in the hinterland unconformably overlie older rocks but do not provide evidence for dating specific structural features. The pre-Tertiary unconformity (as shown by the paleogeologic maps of Armstrong, 1968b; Fig. 3) truncates many broad folds, some of considerable length and magnitude and a few faults; most of these faults have stratigraphic displacements of less than one geologic system. Mapping of the Schell Creek and Snake Range areas by Dechert (1967) and Nelson (1966) shows Tertiary volcanic rocks overlying Cambrian rocks. This is in contrast to the Mississippian or younger strata that usually underlie the Tertiary in the hinterland. These relations could be disregarded, or argued to be due to unrecognized structures of Tertiary age, but they may indicate pre-Tertiary periods of denudation in the region, a possibility I would prefer to leave open, particularly in the light

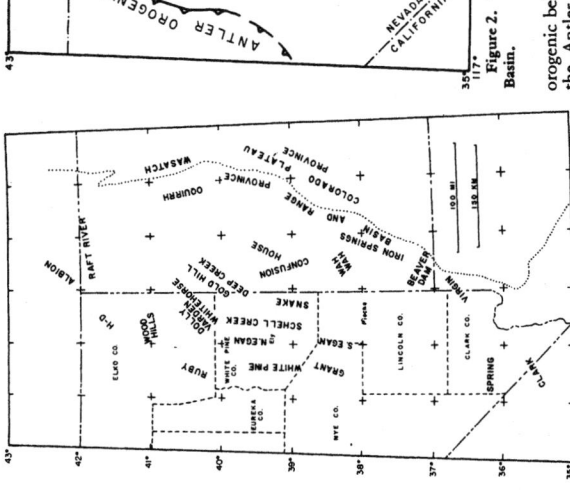

Figure 1. Geographic names of selected topographic and geographic features in the eastern Great Basin of Nevada and Utah.

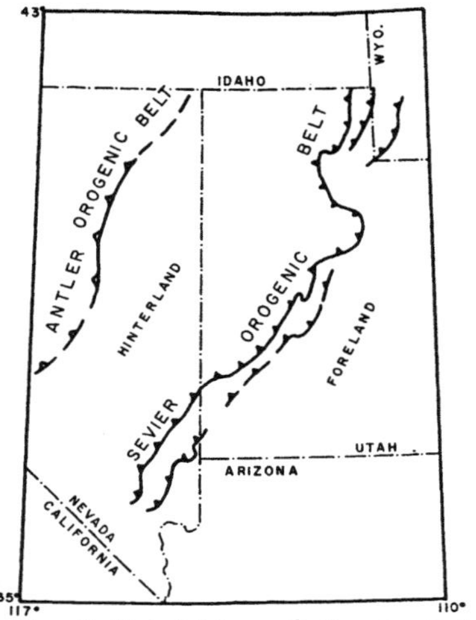

Figure 2. Tectonic index map for the eastern Great Basin.

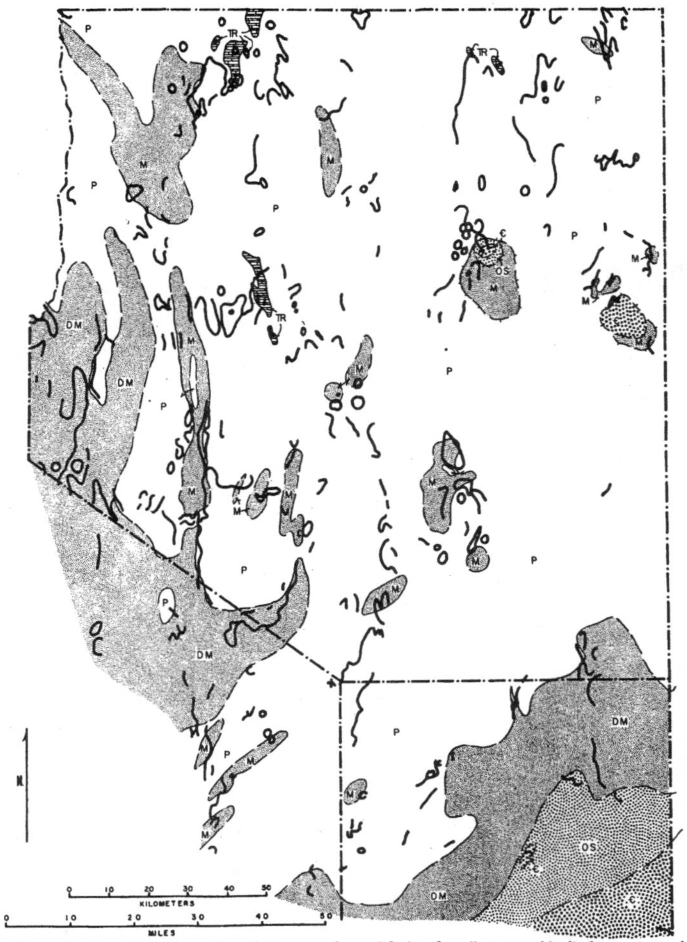

of the uncertain age of many denudation-type structures and the complex tectonic history of the region.

Although we can be certain that some metamorphism of geosynclinal strata occurred in the Ruby Mountain and Snake Range areas before the Cretaceous, it is possible that elsewhere, presumably at deeper structural levels or in hot spots, that deep-seated flow of rocks occurred much more recently. Indeed some flow and the consequent development of flaser gneisses seem to have taken place within the Tertiary in the Raft River–Albion Range areas (R. R. Compton, R. E. Zartman, and Armstrong, work in progress). Any crustal shortening model for the Sevier orogenic belt necessarily requires flow or faulting at depth in the hinterland—deformation that must be synchronous with the eastward thrusting of miogeosynclinal strata in the Wasatch zone.

## GLIDE BLOCKS IN THE SEVIER OROGENIC BELT

Gravity slides of Late Cretaceous(?) and Tertiary age have been recognized at several localities within the Sevier belt. Typically these displaced blocks lie with low-angle contacts on older strata but some lie on rocks as young as Tertiary. The rocks are usually intensely brecciated near tectonic contacts; the entire upper plate may be shattered, but some plates are not. Underlying rocks are deformed only in a thin zone close to the tectonic contact. An example of the scale and style of these glide blocks is shown in Figure 4, taken from Seager's (1970) study of the Virgin Mountains. Farther south, glide blocks in the Spring Mountains have been described by Secor (1963a, 1963b) and in the Clark Mountains by Burchfiel and Davis (1971). In the Pioche district north of the Virgin Mountains, several glide structures override Tertiary sedimentary and volcanic rocks (Armstrong, 1964; Gemmill, 1968); to the northeast, similar structures are reported in the Beaver Dam Mountains (Cook, 1960; Jones, 1963) and Iron Springs district (Mackin, 1960) in Utah.

An example of a fault that was cited as a thrust fault (Baker, 1964), but which I think is a large glide structure is the Big Baldy fault of the southern Wasatch Mountains. As illustrated on Figure 5, a simple interpretation

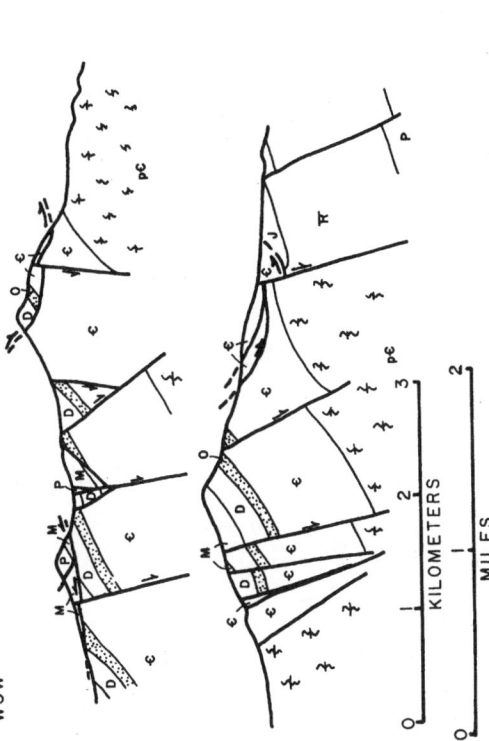

Figure 4. Cross sections of the northern Virgin Mountains (Seager, 1970) showing gravity glide structures. Geologic systems indicated by letter symbols are: pЄ, Precambrian; Є, Cambrian; O, Ordovician; D, Devonian; M, Mississippian; P, Pennsylvanian and Permian; Ŗ, Triassic; J, Jurassic.

Figure 3. Pre-mid-Tertiary paleogeologic map of White Pine and portions of Nye and Lincoln Counties, eastern Nevada. Heavy lines indicate distribution of significant contacts as shown on the county maps (Hose and Blake, 1970; Kleinhampl and Ziony, 1967; Tschanz and Pampeyan, 1970). Contacts for the paleogeology are shown by a light line. Geologic systems indicated by letter symbols are: Є, Cambrian; OS, Ordovician and Silurian; D, Devonian; M, Mississippian; P, Pennsylvanian and Permian; TR, Triassic. The paleogeology shows that by the end of the Mesozoic, the area was characterized by broad gentle folds, and faults of small stratigraphic displacement. At the surface, upper Paleozoic strata were extensive. Only two exceptional areas where lower Paleozoic carbonate rocks have been reported to underlie Oligocene volcanic strata in White Pine County are evident. The surface of this region was not broken by any major through-going thrust faults of comparable magnitude to those in the Sevier belt, which lies only a few miles to the east and southeast, nor was it broken by structures expressing tens of miles of extension as the present-day geology of the region does.

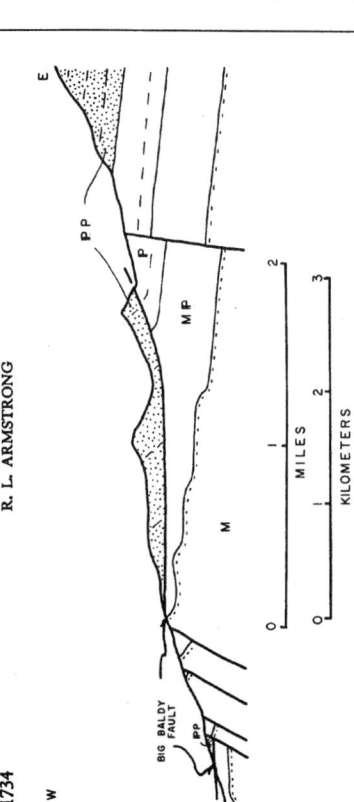

Figure 5. Cross section of the Big Baldy fault of the southern Wasatch Mountains (Baker, 1964). This appears to be a normal fault that bounds a simple glide block rather than a thrust fault. Geologic systems indicated by letter symbols are: M, Mississippian; ℙ, Pennsylvanian; P, Permian.

of the Big Baldy block is that it moved down and to the west and was rotated as it slid along a curved fault surface. Faulting along the Wasatch front would have provided the original free surface that permitted failure; later faulting along the front displaced the glide surface slightly. An alternate proposal, that it is a thrust fault, requires a complex ad hoc explanation in which a thrust fault rises stratigraphically, descends, and rises again, resulting in a stratigraphic displacement much less than total displacement and the observed younger-on-older relation. Such gymnastics are unnecessary, and certainly not needed to explain the structures known to occur west of the Wasatch fault in this area. I have chosen the Big Baldy fault for reinterpretation because I think it illustrates the potential for conflicting interpretations that is widespread in the structurally more complex hinterland.

The examples cited demonstrate that gliding, often provably related to relief developed as a consequence of later Tertiary faulting, is a phenomenon of common occurrence within the miogeosynclinal strata of the eastern Great Basin. It is reasonable to expect similar structures within the hinterland and surprising that such an explanation has not been considered for all the structures of similar style and scale known to occur there.

## DENUDATION STRUCTURES OF THE HINTERLAND

Robert Scott used the term "denudation fault" for the widespread tectonic contacts that place younger on older rocks (Moores and others, 1968; Drewes and others, 1970), and I think this is a useful term with minimal genetic connotations attached. It is concordant with the thinking of most workers concerned with the hinterland (tectonic removal of hanging-wall strata) regardless of how they think these faults are related to the big picture. Peter Misch and his students prefer to use the term décollement (shearing off) for many of the same features and believe in a compressional origin for most of these faults, although Misch (1960) admits that Tertiary gravitational gliding is important locally, as around Sacramento Pass in the Snake Range. Where younger rocks are observed lying on older, the simplest geometric model for the development of such a structure involves extension—as along high- or low-angle normal faults (Dahlstrom, 1970). This is illustrated in Figure 6. An important concept, popular in the Canadian Cordillera and I think useful in the Great Basin, is the listric normal fault—one that alternately cuts across and parallels stratification. Such faults, as illustrated on Figure 6, result in juxtaposition of younger-on-older rocks as occur on many of the low-angle faults of the Sevier hinterland. If strata are upright and not tightly folded, the only possible geometric explanations for compressional faulting of older rocks on younger are: (1) low-angle faults that decapitate folds (this mechanism produces at most only 50 percent younger-on-older contacts, usually less); (2) imbrication of already imbricate rocks by a second generation of

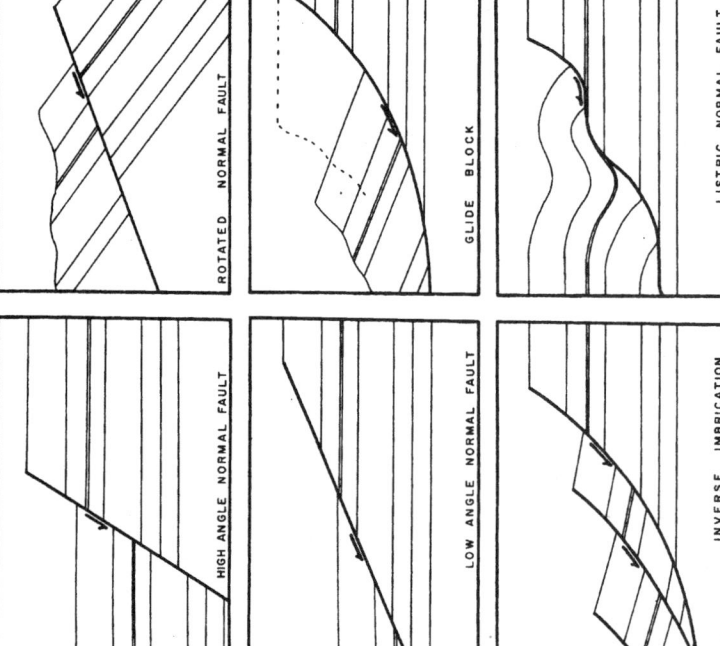

Figure 6. Ideal geometries of extension and denudation faults that place younger rocks on older. Many younger-on-older faults of the hinterland of the Sevier orogenic belt can be interpreted as conforming to these simple models or to more complex permutations and combinations.

thrusts steeper, relative to bedding, than the first generation (this likewise produces many more older-on-younger contacts than vice versa); and (3) by erosion thrusting or thrusting along an unconformity, which can proceed with completely flexible geometry, depending on the landscape or unconformity being overridden. None of these geometric mechanisms seems to be applicable to the hinterland structures. Geometric logic is a necessary prerequisite to any acceptable model for structural evolution.

A recent great increase in the acceptance of gravitational gliding and extensional mechanisms to explain denudation faults of the hinterland is evident. As long ago as 1945, Longwell described low-angle faults in Clark County, Nevada, but the idea did not become popular until the late 1960s. Hunt and Mabey (1966) interpreted the structure of the Panamint Range as shown on Figure 7 as a deeply exposed normal fault, an inverse-imbricate denudation structure. Wright and Troxel (1969) interpreted the Amargosa chaos as drag below normal faults that flatten at depth, inverse imbricate listric normal faults in the jargon of this paper. Fleck (1970a) suggested that the chaos may be megaslides (large-scale versions of

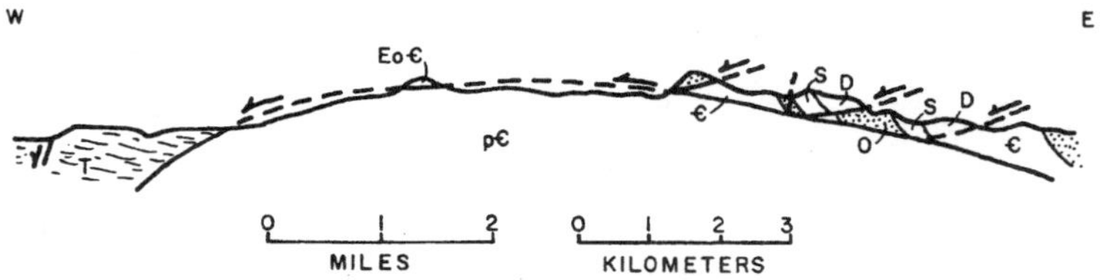

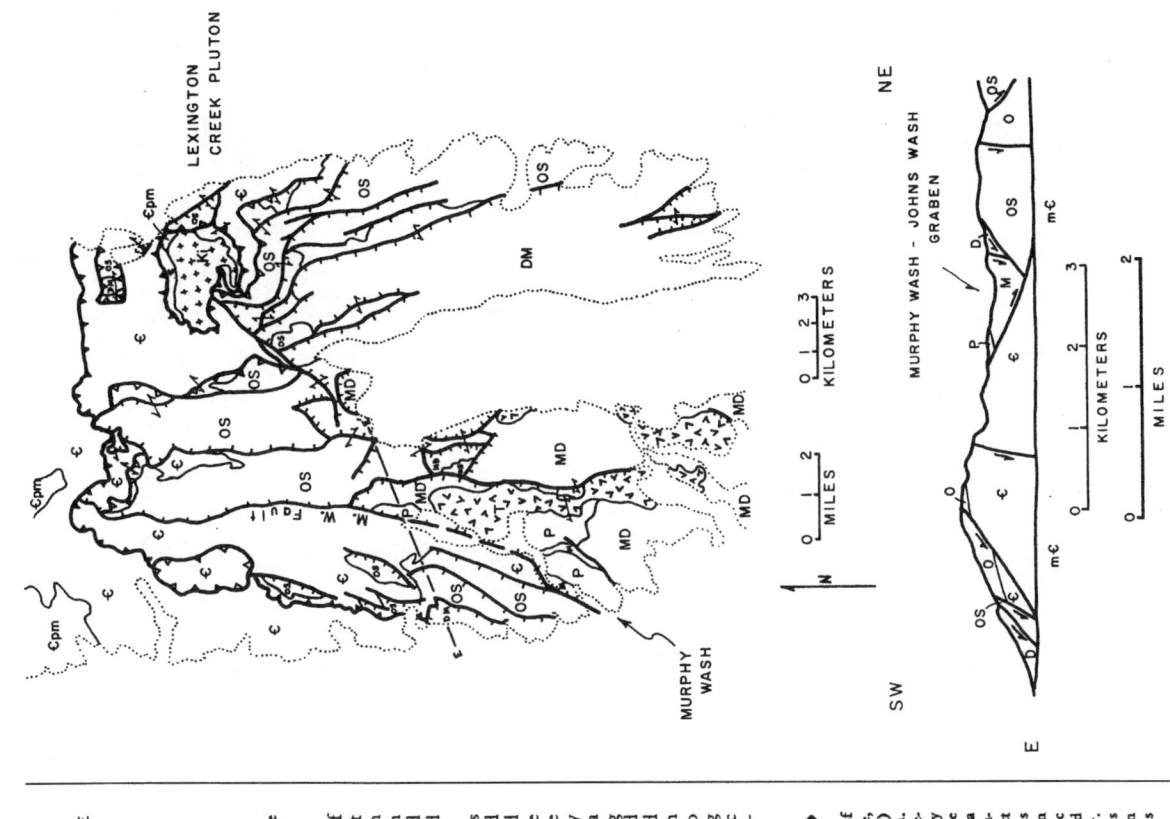

Figure 7. Cross section of the Tucki Mountain fenster (Hunt and Mabey, 1966), an example of an inverse-imbricate denudation structure of Tertiary age in the Death Valley region.

Figure 8. Geologic map of the southernmost part of the Snake Range (Whitebread, 1969; Hose and Blake, 1970) and a cross section (E-E' of Whitebread, 1969) that includes the Murphy Wash-Johns Wash Graben. The numerous faults above the Snake Range décollement merge or truncate against it implying that they are synchronous or older. The fact that some of these faults displace Tertiary volcanic strata suggests a Tertiary age for the décollement. Geologic units indicated by letter symbols are: €pm, Cambrian Prospect Mountain quartzite; €, Cambrian carbonate rocks and shale; OS, Ordovician and Silurian; DM, Devonian and Mississippian; P, Pennsylvanian; Ki, Mesozoic intrusive rocks; T, mid-Tertiary volcanic rocks. Dotted lines are contacts of bedrock with Quaternary cover. Faults are annotated: hachures for high-angle faults (ticks on upper, downdropped plate) and teeth on upper plate of low-angle faults. These same conventions apply to all the maps in this paper.

the Tin Mountain landslide described by Burchfiel, 1966). There is general agreement that these structures are Tertiary, as they postdate and involve Tertiary rocks. Once popular and widely quoted compressional-thrust explanations for these same features are falling out of favor. Hunt and Mabey (1966) cite the reason for their interpretation: extreme attenuation of units, maintenance of stratigraphic order within the fault mosaic, abundance of small normal faults which show a tendency to merge with the master fault, and absence of imbrication and folding. These aspects are precisely those of many of the low-angle fault complexes of east-central Nevada.

Extension and gravitational gliding explanations for low-angle faults in east-central Nevada were suspected by Young (1960) and strongly advocated by Drewes (1964, 1965, 1967), Moores (1968), Moores and others (1968), Drewes and others (1970), Whitebread (1968), Hose and Danés (1968), and Hose and Blake (1969), but the age and significance of the extension is not agreed upon. In order to discuss the interpretations of various authors and to develop my own ideas, it is useful to discuss specific examples going from east to west across east-central Nevada.

## Snake Range

First, and perhaps most important of all, is the Snake Range. This area inspired Misch's (1960) idea of a décollement thrust fault separating complexly sliced upper plate rocks (Cambrian and younger) from a relatively intact autochthon of Cambrian and older strata that had been metamorphosed during the Mesozoic (prior to emplacement of massive granites in the vicinity of Wheeler Peak). Misch (1960) and Nelson (1966, 1969) describe the décollement in great detail and on indirect evidence (unconformities below Cretaceous strata that lie more than 100 mi away from

the Snake Range) conclude that it is of Mesozoic, pre-Laramide age. I believe that evidence for dating the faults may be found in the southern Snake Range area shown on Figure 8, recently mapped by Whitebread (1969) and Hose and Blake (1970) and studied by Lee and others (1970).

As can be seen on Figure 8A, normal faults that displace Oligocene volcanic rocks (dated 29.7 m.y. by Armstrong, 1970) can be traced northward where they are observed to merge with the basal décollement. This suggests to me that the décollement itself may be of Tertiary age, at least in part. Further evidence of a young age for the décollement is the dating studies of Lee and others (1971) that showed total argon loss from cataclastic rocks, produced during movement on the décollement, less than 20 m.y. ago. These authors interpreted this to mean that the most recent movement along the décollement, which cuts the Mesozoic granites, is of Tertiary age. I agree wholeheartedly.

## Deep Creek Mountains and Gold Hill

North of the Snake Range, at the southwestern end of the Deep Creek Mountains, Nelson (1966, 1969) mapped a structurally complex area of upper Paleozoic rocks. The inverse imbrication and brecciation associated with deformation suggest a shallow extensional genesis for his structures. A critical factor in unraveling this area will be a detailed study of Tertiary volcanic stratigraphy and structure, but such studies are still incomplete (work in progress by M. C. Blake and R. K. Hose). If the volcanic structure is complex, it will be necessary to attribute at least some of the low-angle faults mapped by Nelson to the Tertiary. A principle that must be observed in the study of such areas where both Tertiary volcanic and older rocks are present was emphasized by Mackin (1960): *Tertiary strata, including layered volcanic rocks are the guides to the amount of later Tertiary deformation that has taken place.* This younger deformation must be subtracted before a reasonable interpretation of Mesozoic structure is possible. Casualness on this point has contributed much confusion to the literature concerning the eastern Great Basin.

Just north of the Deep Creek Mountains, at Gold Hill, Nolan (1935) first recognized the complex tectonic history of the hinterland. At a time when the geology of the surrounding region was largely unknown, he recognized several periods of compression and extension that he could demonstrate were pre-Eocene in age. In addition, he described complex younger-on-older fault relations and high-angle block faulting of post-Eocene age. Such a chronology fits well with the regional history as known even today, the only qualification being that the post-Eocene younger-on-older faults are more likely extension-denudation-gliding structures than the result of regional compression during later Tertiary time.

## Schell Creek Range

The next range to the west is the Schell Creek Range. Dechert (1967), Young (1960), and Drewes (1967) are the principal contributors to knowledge of the geology there. Drewes recognized the complexity of the tectonic history of the area and carefully qualified the interpretations he gave his structures. He was one of the first workers in the region to show large Tertiary glide blocks containing Tertiary strata and recognize that Mesozoic faults may have been sites of Tertiary movement. I go only a bit further than he in interpreting faults in his area as Tertiary. In the Cave Mountain area (Fig. 9), a denudation fault has been mapped. The fault begins at a normal stratigraphic contact between massive upper Cambrian and Ordovician limestone (€Ol) and underlying shale (€l) (known to be prone to slumping and landsliding) and, as followed eastward, shows increasing stratigraphic discordance so that eventually the limestone (€Ol) overlies middle Cambrian limestone (€pc). The strata designated €Ol are upright and dip steeply into the fault. The geometric relations are precisely correct for an extensional-gravitational, perhaps mega-landslide, origin for this structure. It is most easy to imagine this happening during the Tertiary, during or after uplift of the range. A mirror-image symmetrical structure involving the same stratigraphic units occurs southwest of Wheeler Peak in the southern Snake Range on the opposite side of Spring Valley. This "twin" of the Cave Mountain structure, visible on Figure 8 (Whitebread, 1969), is part of the southern Snake Range décollement for which a Tertiary origin by extension has already been discussed.

Farther south in the area mapped by Drewes in the vicinity of Connors Pass, I would extend the hypothesis of a Tertiary age to the "Schell Creek Range thrust." Figure 10 shows Drewes'

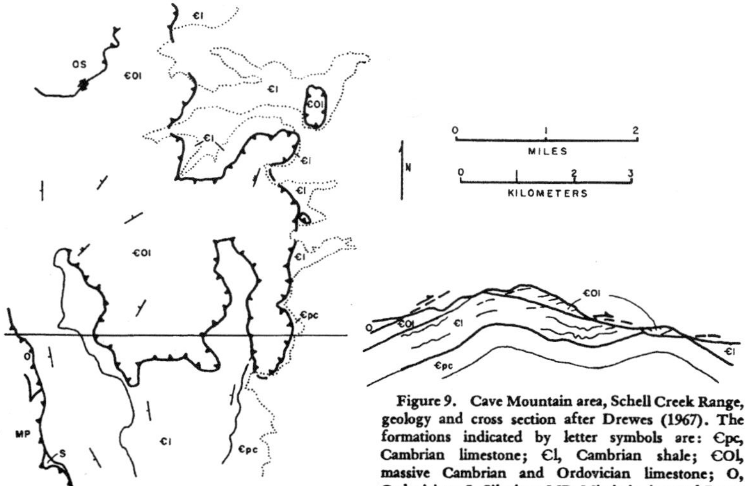

Figure 9. Cave Mountain area, Schell Creek Range, geology and cross section after Drewes (1967). The formations indicated by letter symbols are: €pc, Cambrian limestone; €l, Cambrian shale; €Ol, massive Cambrian and Ordovician limestone; O, Ordovician; S, Silurian; MP, Mississippian and Pennsylvanian. The large allochthon of €Ol may be a Tertiary glide block, a mirror image of similar structures in the southern Snake Range.

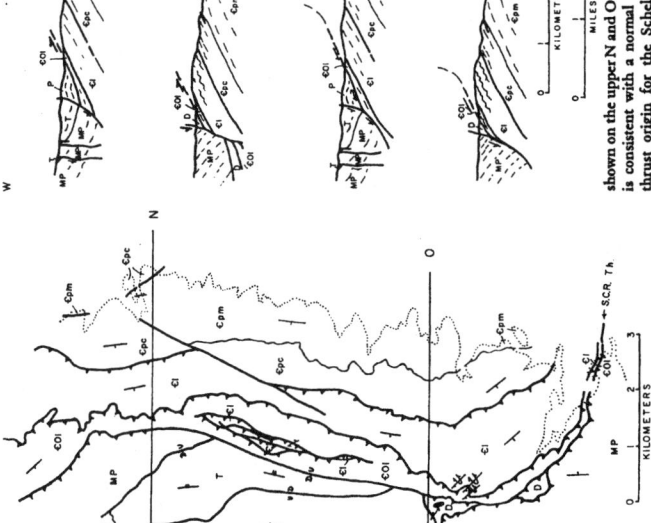

Figure 10. Geologic map and sections, southern Schell Creek Range in the vicinity of Connors Pass (Drewes, 1967). The lower sections, N' and O', are an alternate interpretation to the published structure, shown on the upper N and O sections. The map pattern is consistent with a normal displacement, rather than thrust origin for the Schell Creek Range "thrust" (S.C.R. Th. on map). Geologic formations indicated by letter symbols are: €pm, Cambrian Prospect Mountain quartzite and overlying Pioche shale; €pc, Cambrian limestone; €l, Cambrian shale; €Ol, Cambrian and Ordovician limestone; D, Devonian; MP, Mississippian through Permian; T, Tertiary.

map and sections and my interpretation. The map pattern is actually more consistent with a Tertiary age for the Schell Creek "thrust" because a large area of Tertiary rocks is down faulted against the €Ol limestone between sections N and O. The published map suggests that the Schell Creek Range thrust fault is cut by the younger normal fault that forms the southeast boundary of the east dipping Tertiary strata. The problem, as I see it, is that this Tertiary normal fault has the same trend, direction of stratigraphic displacement, and magnitude of displacement as the fault called the Schell Creek Range thrust fault in areas to the north and south. The minimum displacement of the normal fault is one-half mile, yet it dies out on the map within a few hundred feet under a small patch of alluvium just south of its projected intersection with cross section N (Fig. 10). This is a geometric impossibility — a fault displacement cannot change along strike by more than the distance along strike (the maximum would be along a fault with scissors movement — displacement away from the hinge point would be equal to the distance from the hinge times the rotation on the hinge in radians — this would seldom approach a 1:1 relation). Continuing on the other side of the same patch of alluvium with the same trend and sense of stratigraphic displacement is the Schell Creek Range thrust fault. To me, the simplest interpretation is that the "thrust" and the Tertiary normal fault are one and the same feature—structures due to extension and resulting in denudation of the half-dome east of Connors Pass. Most of the structure in this area can be viewed as Tertiary denudation rather than the effects of Mesozoic compression or extension. Indeed, several striking drag folds observed near Connors Pass (Fig. 10) indicate down-dip movement, the fault zone between the limestone (€Ol) and shale (€l) itself being deformed by these folds. Drewes (1967) commented on this evidence in his report but felt that, in general, fold attitudes gave ambiguous results for direction of displacement along faults in the area. He felt the evidence slightly favored eastward movement, but he made the uncertainties involved in his determination explicit.

### Egan Range

An area where the structural significance of Tertiary volcanic rocks (Mackin, 1960) has been ignored is the northern Egan Range. Figure 11 shows the geology of part of the area mapped by Fritz (1960, 1968). The published map shows many faults, and on cross section C-C′ they appear as nearly horizontal "thrusts" to which a Mesozoic age is assigned in the descriptive text (Fritz, 1960). The steeply dipping Tertiary volcanic strata intersected by cross section C-C′ *are not* related to any structure capable of producing their steep dip in the Tertiary! In fact, they dip westward toward a steeply east-dipping contact mapped as a normal stratigraphic overlap. Insight into the true nature of the structure is gained by rotating the volcanic rocks back to horizontal (as they must have been approximately 35 m.y. ago). This has been achieved on Figure 12 by simply rotating Fritz's section C-C′. The Mesozoic "thrusts" immediately become normal faults (consonant with the direction and magnitude of the stratigraphic displacements along them) that postdate, and thus displace the volcanic rocks. The western contact of the volcanic rocks must, of course, be itself a fault.

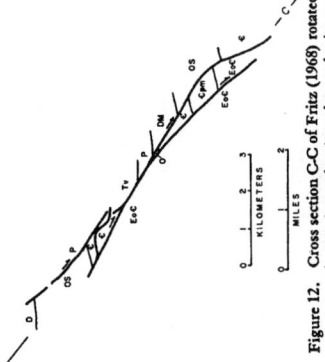

Figure 12. Cross section C-C′ of Fritz (1968) rotated so as to bring the Tertiary volcanic rocks to a horizontal position. The true nature of many of the "thrust" faults shown on Fritz's map is revealed — they are normal faults rotated to a near horizontal attitude during later Tertiary time.

The only normal stratigraphic contact is with the underlying Pennsylvanian strata — with only a small angular discordance being evident. The structure here is rather similar to the Panamint Range structure of Hunt and Mabey (1966). In the Egan Range, we have normal faults rotated to a near horizontal present-day attitude by tilting of the range — tilting proven by the attitude of the volcanic strata.

The patch of volcanic rocks near the north end of the area shown on the map likewise seems to me to violate the principle discussed by Mackin (1960); a major fault between the volcanic rocks and the Eocambrian strata to the north is required. Such an interpretation is in agreement with the work of Adair (1961, and 1963, personal commun.) but denied by R. K. Hose (1971, personal commun.). The contact in dispute is quite steeply dipping but not well enough exposed to be uniquely identified as depositional or tectonic. I would connect the volcanic-Eocambrian contact with a major later Tertiary fault that forms the contact of the Cherry Creek pluton with Paleozoic rocks south of it (Adair, 1961). This Tertiary normal fault would include, in part, a younger-on-older "thrust" that Misch identified as part of his Mesozoic regional décollement. There need be no such structure here at all.

This reinterpretation of parts of Fritz's map does not mean to imply that his entire structural complex is a Tertiary feature. There are many faults, including imbricate thrust faults that cannot be reasonably interpreted as anything but pre-Tertiary structures. The point to make is that the Mesozoic thrust interpretation has been carried too far, and the Tertiary structures have been considerably underrated.

Another example of proliferation of Mesozoic thrust faults in violation of Mackin's principle occurs farther south, in the southern Egan Range, in an area mapped by Brokaw and Shawe (1965; Fig. 13).

On the published map and section, a younger-on-older thrust fault is shown along the west side of the eastward-tilted Egan Range fault block. Slightly farther west, a normal fault of negligible stratigraphic displacement is shown. The Tertiary strata lying on the east side of the range, dipping eastward, are the same units as those lying nearly horizontally west of the range and west of the two faults mentioned above. These two areas were part of an originally extensive horizontal and continuous blanket of Tertiary strata. The displacement *proven* by present attitude of the Tertiary strata amounts to thousands of feet of warping or faulting. There is no structure on the map, as published, to account for this displacement. However, the orientation of the "thrust" and amount of stratigraphic throw along it are exactly as needed to explain the Tertiary deformation. My solution to this dilemma is that the Mesozoic "thrust" is a low-angle normal fault of later Tertiary age—the low angle being in part a consequence of eastward tilting of the southern Egan Range.

### Grant–White Pine Ranges

Moores and others (1968) emphasized the importance of gravitationally driven Tertiary gliding and denudation structures in the northern Grant and southern White Pine Ranges. They could not fail to be impressed by large masses of Paleozoic strata enclosed within, and overlying, Tertiary volcanic and sedimentary rocks in their area. Yet in the same ranges, at deeper levels, structures of pre-Tertiary age are present. Cebull (1970) has described the spectacular effects of regional metamorphism, large-scale recumbent folding, and imbrication that occur around Troy, south of the area mapped by Moores and others (1968). Some of the structural complications described by Cebull may be of Tertiary age; denudation faults related to range uplift would be so similar in appearance as to be indistinguishable from some of Cebull's thrusts, but a Mesozoic age seems reasonable for most.

To the north of the area mapped by Moores and others (1968), in the White Pine District, the published map of Humphrey (1960) shows Mesozoic thrust faults as well as several faults which can be interpreted as Tertiary denudation features, one in particular being the Lampson "thrust" as shown on Figure 14. The striking feature of this "thrust" is that it dies out completely near the north edge of the figure where it is joined by a wrench fault, the Eberhardt fault. Geometric logic suggests that the increasing stratigraphic displacement southward is an expression of increasing normal displacement along a fault hinged at its northern end. But this is the reverse of the published interpretation! The Tsv strata that abut against the Lampson fault scarp provide further suggestion, following Mackin's principle, that displacement along the fault is normal rather than reverse. To me, the map pattern suggests that the upper plate on which the Tertiary strata lie has moved downward and westward, at the same time rotating clockwise about 10°. The slice of Silurian[1] rock thrust on nearly overturned Devonian strata near the south side of the area shown on Figure 14 remains to be explained. This structure is presumably the reason a thrust origin for the Lampson fault was initially inferred. These structural complications at depth suggest to me that the normal fault has merged with, and followed a décollement in the Mwp shale before Tertiary faulting began. This suggestion allows a reconciliation of the evidence for normal and reverse movements along the same fault. The map pattern around Mount Hamilton, also on Humphrey's (1960) map, likewise contains suggestions of possible reinterpretation of Mesozoic thrust faults as composite structures involving reverse *and* normal displacements of various times along the same faults. Unfortunately, these structures are not datable and thus could be Mesozoic or Tertiary. It is impressive, however, that two illustrations (his Figs. 12 and 13) in Humphrey's (1960) report show drag suggesting down-dip movement

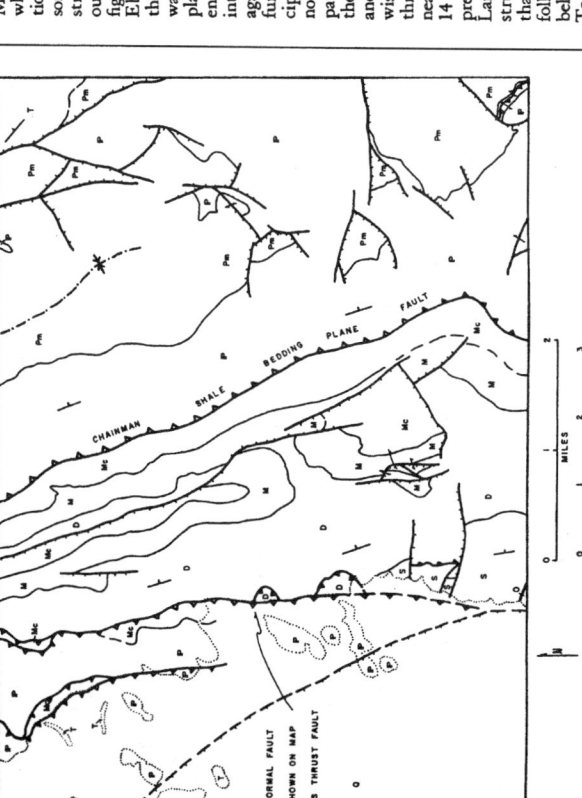

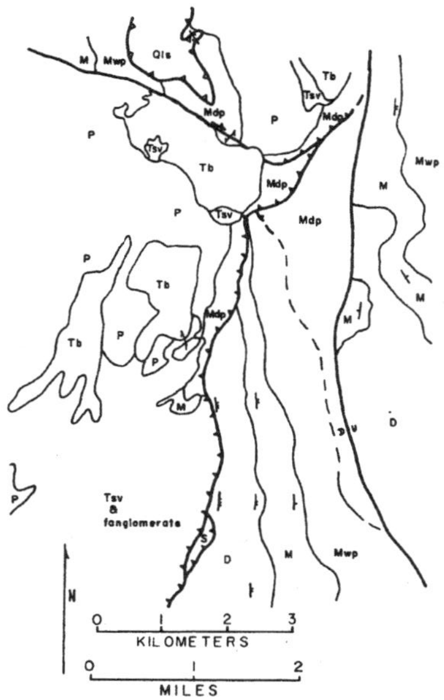

Figure 14. Geologic map of the Lampson "thrust" White Pine district (Humphrey, 1960). Stratigraphic displacement on the "thrust" increases southward, suggesting normal, rather than thrust movement, but some features observed toward the south (imbrication, near-overturning of Devonian strata) suggest thrusting. This may be reconciled by proposing that a Tertiary normal fault intersects and joins at moderate depth a Mesozoic listric thrust. This would be an example of reversal of movement along a given fault surface during two stages of deformation, a normal expectation in a region with complex tectonic history.

---

[1] R. K. Hose (Hose and Blake, 1970, and 1971, personal commun.) reports that this is probably Devonian Simonson Dolomite, and so the structure may not be as complex as indicated by Humphrey (1960).

Figure 13. Geologic map of the northern part of the Ely 3 SW. quadrangle (Brokaw and Shawe, 1965), Egan Range, just southwest of Ely, Nevada. The thrust fault shown along the west side of the range must be the Tertiary normal fault required to explain the uplift and eastward tilting of the range. Geologic systems indicated by letter symbols are: O, Ordovician; S, Silurian; D, Devonian; M, Mississippian, mostly limestone; Mc, Mississippian shale; P, Pennsylvanian; Pm, Permian; T, mid-Tertiary sedimentary and volcanic rocks.

along a fault mapped as a thrust fault (Monte Cristo fault with up-dip movement, but younger-on-older stratigraphic displacements). Similar complexity is probably applicable to many other hinterland structures. Re-use of Mesozoic faults during Tertiary deformation was an important conclusion of Drewes (1964, 1967), and has been well documented in the Canadian Rockies (Dahlstrom, 1970; Bally and others, 1966).

### Ruby Mountains

In the Ruby Mountains, Willden and others (1967) and Willden and Kistler (1969) discovered a klippe of brecciated Devonian rocks overlying an Oligocene granitic pluton, itself slightly brecciated near the tectonic contact. They invoked a period of compressional thrusting during the Tertiary and related it to K-Ar dates on crystalline rocks north of the klippe that reach a minimum of 21 m.y. along the west side of the range, the side where the klippe lies. The evidence could as well be interpreted, as in the Snake Range, to indicate tectonic denudation of the range during Tertiary time. The westward decrease in K-Ar dates may even be a clue to the direction of movement of the cover and the rate it was pulled (or slid) off the Ruby Range into the valley on the west side. The klippe of Devonian rocks would be merely a remnant of the block that moved off, resulting in rapid cooling of rocks that had up to that time remained too hot to retain argon. To interject a period of regional compression during the later half of Tertiary time unnecessarily complicates the tectonic history; it can only be accomplished by a rather involved manipulation of known examples of datable displacements of Tertiary strata that are convincing evidence of extension which occurred throughout the region after the effusion of Tertiary volcanic rocks began, approximately 40 m.y. ago.

### GLIDING VERSUS PUSHING IN THE SEVIER OROGENIC BELT—HOW DOES THE HINTERLAND FIT THE BIG PICTURE?

Several authors (Whitebread, 1968; Roberts, 1968; Roberts and others, 1965; Hose and Danés, 1968; Hose and Blake, 1969) have suggested that the denudation (attenuation of Hose and Danés) structures of the hinterland are genetically linked with thrusting in the Sevier orogenic belt. They invoke a gravitational gliding mechanism whereby denudation (attenuation) in the hinterland is synchronous with thrusting along the Wasatch zone; extension in the hinterland must match compression in the Wasatch zone if crustal shortening has not occurred. Gravitational gliding mechanisms have also been invoked as explanations for other sectors of the Cordilleran fold and thrust belt (Scholten, 1968; Crosby, 1968, 1969; Mudge, 1970). It should be emphasized that the scale of the glide structures advocated to explain the Sevier belt and its lateral continuation is at least an order of magnitude larger than the scale of glide blocks related to local uplifts during Tertiary [and Mesozoic(?)] time. Large-scale gliding involves movement of sheets of sedimentary strata that are at most 5 mi thick and more than 100 mi in width over distances of 40 to 100 mi. Opposition to large-scale gliding has been expressed, however, by other geologists who advocate crustal shortening as a genetic explanation for the fold and thrust belt (Armstrong, 1968B; Burchfiel and Davis, 1968; Fleck, 1970b). For the Canadian Cordillera, Price (1969) and Price and Mountjoy (1970, 1971) propose an attractive model of crustal thickening in the hinterland and consequent lateral flow of supra-crustal material to produce the fold and thrust belt. This model involves crustal shortening of a magnitude similar to, if not identical with, the compression expressed in the fold and thrust belt (more than 100 mi in southern Canada), and although it is referred to as a gravitational mechanism (gravity being a very important part of the driving force), it significantly differs from the mechanism proposed by the gliders. The thrust faults are projected westward, at depth, into deep zones of intracrustal flow. Plates move uphill (but down lithostatic gradients), so that an eastward slope of the movement planes is not required at any time. Such a mechanism does not require that the rear of the moving plates be a zone of extension at the earth's surface so that a region of extension and denudation in the hinterland is not necessary—if lacking it is not a detriment for the model. If present, such a zone of extension can have an explanation independent of the genesis of the fold and thrust belt; there is no necessity that events in zones of extension and compression be synchronous and displacements in the two belts of comparable magnitude.

Armstrong (1968b), Burchfiel and Davis (1968), and Fleck (1970b), do not consider the denudation structures of the hinterland of the Sevier belt to be directly related to the genesis of the fold and thrust belt. A different model created by Misch (1960; Misch and Hazzard, 1962) and supported by his students (*including* Fritz, 1960; Woodward, 1964; Miller, 1966; Nelson, 1966, 1969) proposes a link between hinterland and Wasatch-zone structures that has to be discussed separately as it does not represent a combination of any of the "glide" or "push" models already outlined. Misch envisages the younger-over-older faults of the hinterland as décollements between a relatively rigid "basement" (deep seated geosynclinal strata that have been subject to Mesozoic synkinematic regional metamorphism prior to thrusting) and a severly tectonized cover that was being pushed[2] eastward relative to the "basement" at some time late in the Mesozoic but *prior* to the "Laramide," deformation in the Sevier orogenic belt (Misch, 1960, 1971). Another zone of thrusting, lying geographically between the Sevier belt and the hinterland, but now buried by Tertiary valley fill, is proposed by Misch. This hidden thrust zone is the one Misch relates to the hinterland structures. He does not recognize hinterland structures of "Laramide" age that would correspond with Sevier orogenic belt structures in age and be linked with them in genesis. Moreover, Misch believes the denudation faults to be of compressional origin, not structures resulting from extension.

Misch believed the décollement faults he described were Mesozoic, because he felt that the deformation style was in contrast with the style of faults that displace Tertiary volcanic rocks, but this might be simply a change in style with depth along the same fault. Some of the faults upon which our disagreement focuses were buried at depths of at least 5 mi, if the paleogeologic map and sections are at all correct, so that we both agree that the slicing, mylonitization, and retrogression observed in lower Paleozoic rocks was not a surficial phenomenon, but rather one that took place at lithostatic pressures of several kilobars. The intense brecciation noted by Nelson (1966, 1969) where low-angle faults displace upper Paleozoic rocks is exactly the style of deformation associated with Tertiary glide blocks in the Mojave Desert region (Burchfiel, 1966; Burchfiel and Davis, 1971).

Obviously a diversity of opinion exists as to genesis and significance of hinterland and Wasatch-zone structures. My own preference has been, admittedly, for pushing rather than gliding as an origin for the Cretaceous Sevier orogenic belt and, as outlined in this paper, for denudation and gliding, much of it Tertiary, on a local scale to explain the hinterland structures. I think it worthwhile to outline all the reasons that I think make the large-scale gravitational gliding model untenable, and why I differ with Misch on interpretation although we have no disagreement on geologic observations or tectonic sequence in the hinterland. Some objections are geometric, others chronologic. Arguments about mechanism are, for the most part unsatisfactory and inconclusive, as they require knowledge we do not have on the large-scale behavior of rocks. I prefer to analyze what has happened, geometrically and chronologically, rather than to try to answer abstract questions of how it happened (mechanism and driving force). I am willing to grant anyone a process regardless that it may conflict with intuition or an evolving body of theory, if it can be proved to have happened.

Gliding models require a net downhill slope between area of uplift, source of moving plate, and area of imbrication of the plate. Mudge (1970) imagines an uplift of 60,000 ft in western Montana to propel his moving plate, yet there is neither stratigraphic evidence within the area of uplift for such extremely elevated mountains (needless to speculate on the paleoclimatic consequences of such mountains that exceed the highest point in the Himalaya Mountains by a factor of two), nor is there evidence on either side that requires such an uplift. In fact, this supposed source area is known to have been the site of basins accumulating sediments at moderate elevations (a few thousand feet at most) by the later part of Eocene time (Price, 1962, 1971). In the eastern Great Basin, it is difficult to find evidence of great uplift during later Mesozoic time. What we do know is that the entire region was near sea level up to the Jurassic. A few localities within the region contain fluviatile and lacustrine sediments of approximately middle Cretaceous age (this should be near the time of maximum uplift) which have yielded

---

[2] Misch (1960, p. 41) describes the process as "westward underthrusting," of a non-yielding foreland. He does not consider attenuation in the hinterland to be significant (Roberts and others, 1965).

paleobotanical remains indicative of relatively low elevations (Easton, 1954), not alpine highlands. Over much of the hinterland, the earliest datable Tertiary deposits, Eocene lake-beds containing faunas that indicate relatively modest elevations (Winfrey, 1960), overlie upper Paleozoic strata (*compare* paleogeologic map of Armstrong, 1968 and Fig. 3). The fact that these upper Paleozoic strata are preserved over much of the hinterland strongly suggests that the area never lay more than a few thousand feet above sea level. Only an average of a few thousand feet of strata have been eroded from the geosynclinal pile of the hinterland, and this could have taken place at low elevations during the 100 m.y. of available geologic time. Had there been high mountains in the hinterland, one would expect some evidence of them. As it is, they are needed to support the gliding theory—but are not independently documentable. The available facts support a history during which the hinterland is only moderately uplifted. The Sevier fold and thrust belt that lay to the east was the principal source of clastic material that filled the Rocky Mountain geosyncline.

The estimates of cumulative thrust and fold displacements in the Sevier orogenic belt range from a precisely documented figure of more than 175 mi in southern Canada to conservative estimates of 40 to 60 mi in Utah and southern Nevada. Extension of comparable magnitude is simply not present in the hinterland. This is a bold assertion, but a defensible one. The width of the hinterland belt does not usually exceed 150 mi. In order to preserve a geometrically *necessary* equality between extension and compression would require 30 to 100 percent extension of the hinterland during the Mesozoic. Nothing of this magnitude has been shown by existing maps. The pre-Tertiary paleogeology (Fig. 3) of the region is relatively simple, consistent with only a few extension faults having a cumulative displacement of no more than a few miles. An aggregate of a few miles of pre-Tertiary extension is the most that the hinterland geology allows—insufficient by an order of magnitude to be an inverse expression of the Sevier orogenic belt. The extension needed is of greater magnitude, in the hinterland zone itself, than that extension known to be responsible for the Tertiary structure of the Great Basin! There is simply not such a basin-and-range structure evident in the pre-Tertiary geology, yet without it the glide model is untenable.

Comparison of Mudge's (1970) cross sections and map of his "source area" reveal a similar geometric inconsistency that makes his glide model unsatisfactory (Price, 1971). There is no extension structure in his "source area" that approaches the magnitude of the compression that he, himself, documents in thrusts and folds along the eastern margin of his supposed glide block. Individual stratigraphic units in the Belt Series can be followed across his hinterland. The maximum gaps shown on his map would allow (but don't even require) about 10 mi of extension. He needs more than 30 mi of extension to claim that his model satisfies the most elementary geometric rules of conservation of volume and area (Dahlstrom, 1969).

Further geometric difficulties of the glide hypothesis are the seemingly discontinuous nature of suitable denudation structures. The glide model or Misch's pushing model require continuity of a movement surface over the entire hinterland, yet a suitable fault or fault zone is difficult to follow from range to range and seemingly lacking in many places. For example, contrast the northern Egan Range (denudation-décollement structures reported by Woodward, 1964, and Fritz, 1960, 1968) with the southern Egan Range (continuous lower Paleozoic sections reported by Kellogg, 1960, 1964; and Playford, 1961) or the Snake Range (denudation-décollement structures reported by Misch, 1960; Nelson, 1966; and Hose and Blake, 1970) with the northern Deep Creek Mountains (continuous lower Paleozoic section reported by Bick, 1966; and Nolan, 1935). Either the denudation-décollement structures are discontinuous (negating the glide or décollement thrust hypothesis) or they jump up and down stratigraphically to levels not observed in presently exposed rocks (an *ad hoc* excuse to retain the tenability of the glide-décollement hypotheses). This is very much in contrast to the zone of compression where major faults parallel individual stratigraphic horizons for hundreds of miles.

An even more devastating argument against the models that correlate hinterland denudation-décollement structures with the fold and thrust belt is their provincialism. Most participants in this debate would agree that the same genetic model must apply throughout the length of the fold and thrust belt (no one has proposed that all three models are correct but apply in different areas that lie on strike with one another). The denudation-décollement structures that are the focal point of this paper exist only in the Basin and Range province. Comparable or correlative features are not recognized elsewhere in the North American Cordillera. The inspiration and support for a glide-décollement model is lacking entirely from some sectors of the Cordillera. My own inclination is to seek a Basin and Range-related explanation for these structures confined to the Basin and Range province. A plausible genetic connection between extension and compression structures cannot be taken seriously when it proves inapplicable elsewhere along the same structural belt.

Further geometric difficulties posed by the glide-décollement models can be shown by trying to draw a cross section from the southern Snake Range through the Wah Wah Range to the Colorado Plateau platform (Fig. 15) (between sections C and D of Armstrong, 1968). The "autochthon" (according to the gliders) of the Snake Range lies only about 60 mi from the Wah Wah thrust trace where Eocambrian and lower Cambrian quartzites overlie imbricate Paleozoic carbonate strata. Before Tertiary regional extension, the two areas were probably even closer. If the shortening in the Wah Wah area is on the order of 50 mi as is suggested for areas along strike (the stratigraphic displacements in the Wah Wah area exceed 5 mi—and this is for faults essentially parallel to bedding planes where now observed so that actual displacements much be much larger) then that entire upper plate east of the Snake Range must be allochthonous. This would *require* a source area *west* of the Snake Range (and not on top of it). This, in turn, creates insurmountable stratigraphic problems. The nature of the Cambrian quartzite and overlying carbonate strata of the Wah Wah allochthon requires that they come from the east of the Snake Range. "autochthonous" sections. Wah Wah Cambrian is transitional between Snake Range and platform sections in paleogeographic setting and sedimentary facies (Robison and Bentley, 1958; Robison, 1960; Palmer, 1960). Wah Wah Lower Cambrian is not the least bit affected by regional metamorphism whereas the Lower Cambrian of, and west of, the Snake Range is distinctly metamorphosed (chlorite grade) and we have already agreed that the metamorphism is regional, and prior to the Cretaceous in the Snake Range and vicinity (Misch, 1960; Misch and Hazzard, 1962; Armstrong and Hansen, 1966; Lee and others, 1970). The unsuitability of the glide model is shown by these purely geometric arguments. The fact that the same Cambrian formations occur in both upper and lower plates in the Snake Range is further evidence that displacements involved are small. The Eocambrian rocks of the Snake Range that underlie the "décollement" are fully as allochthonous as the Wah Wah upper plate. The "décollement" structures are younger and an order of magnitude smaller in displacement than the Sevier Belt thrust faults. Alternate interpretations that have been proposed to date are in conflict with the facts shown on geologic maps!

Finally, I reject the glide-décollement models on chronological grounds. I have attempted to show that few, if any, of the hinterland denudation structures are synchronous with deformation in the Sevier orogenic belt. Some can be shown to be definitely older; many, including type examples of denudation-décollement structures are *probably* Tertiary in age and most *may* be, although this proposition is debatable because the age of many faults simply cannot be established.

## CONCLUSIONS

Roberts and others (1965), Roberts (1968), Hose and Danes (1968), Whitebread (1968), and Hose and Blake (1969) conceive of a broadly arched-up hinterland within which denudation and attenuation occurred as Paleozoic strata glided down an eastward gradient toward the Sevier belt, a zone of compression tectonics during the Cretaceous. I have attempted to show that the age and magnitude of displacement of denudation faults in the hinterland are unsuitable for tectonic correlation with thrusting in the Sevier belt, that the two processes may be unrelated. My objections to Misch's model are that the predominantly young age and lack of continuity of the décollement structures make it unnecessary, and indeed impossible, to relate them to a deformed belt that has never been observed, or to the Sevier belt structures. We both agree that the décollement structures are of deep-seated origin involving retrograde metamorphism and cataclastic deformation of rocks that had been affected by regional metamorphism at some

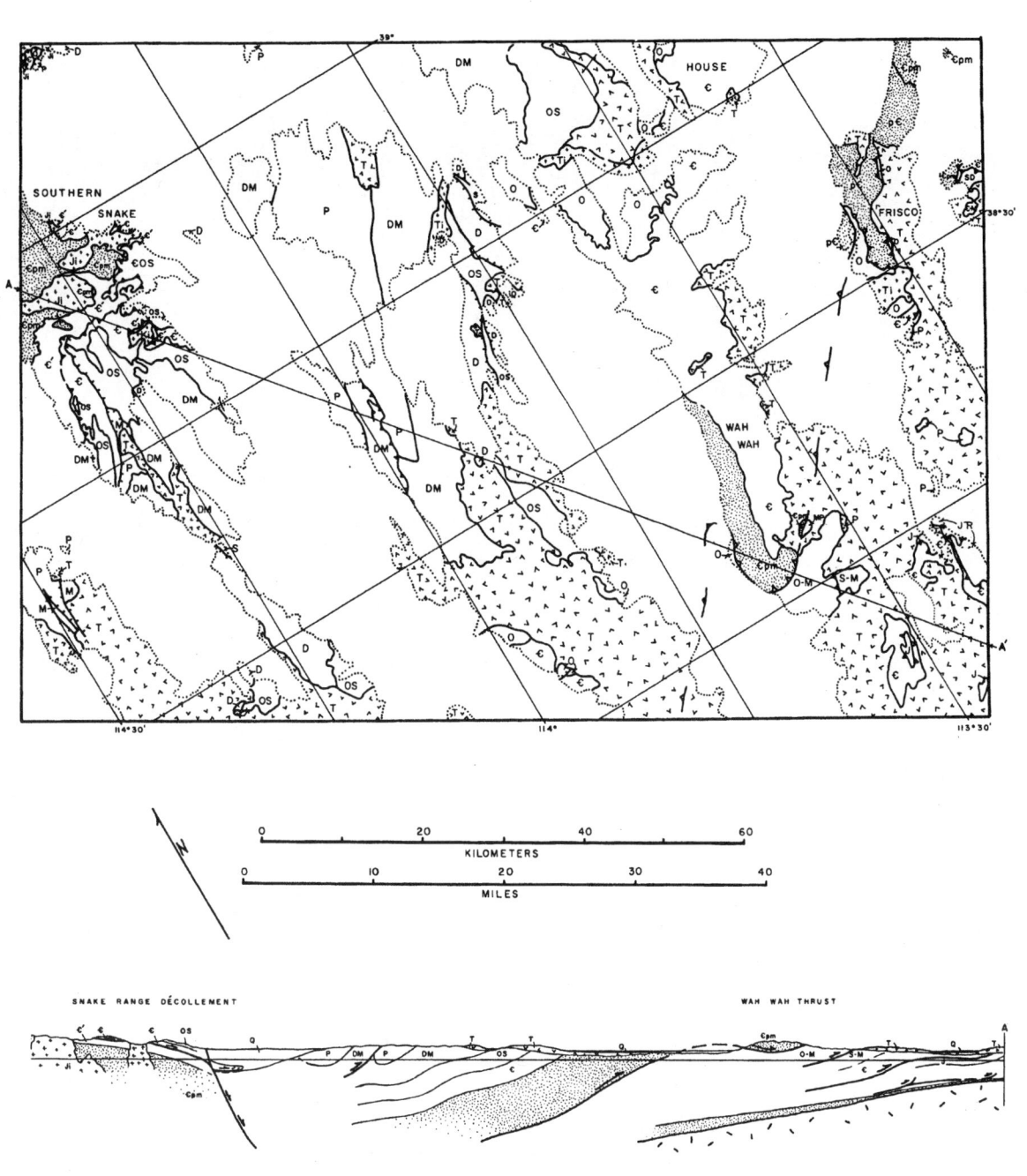

Figure 15. Geologic map and section for the region including the Snake Range, site of denudation-décollement structures of the hinterland, and the Wah Wah Range, a part of the Sevier orogenic belt zone of imbricate thrusts. This area of closest approach of the two types of structure clearly demonstrates the difficulties of genetically relating the two types of structures. The Snake Range is not a logical or adequate source for the Wah Wah allochthon and a connection between the thrust and décollement faults does not seem likely. The interpretation offered is that the Snake Range is as allochthonous as the Wah Wah upper plate, the décollement structures being much smaller and related to Tertiary extension and denudation in the Snake Range area. Letter symbols for geologic units are: pЄ and Єpm, Eocambrian and lower Cambrian basal clastic sequence; Є′, Cambrian of Snake Range "autochthon"; Є, Cambrian; O, Ordovician; S, Silurian; D, Devonian; M, Mississippian; P, Pennsylvanian and Permian; Ji and Ki, Mesozoic intrusive rocks; Ti, Tertiary intrusive rocks; T, mid-Tertiary volcanic strata.

earlier time (at least as early as the Jurassic in some areas), but we disagree in that I concur with the advocates of gliding in attributing them to extension, while he argues for a compressional genesis.

Figure 16 contains three cross sections of the crust of the western United States illustrating a possible tectonic history for the past 200 m.y. The sections have been constructed so as to preserve length and area, allowing for some removal of strata by erosion, but preserving the amount of crystalline crust. Until the middle of the Jurassic, the Cordilleran geosyncline remained close to sea level, as illustrated on section C. The crustal thickness chosen is that typical of areas of low elevation and shallow seas today (James and Steinhart, 1966). By this time, considerable tectonic complexity existed in areas to the west of east-central Nevada as a consequence of Paleozoic and Triassic orogenies.

During the remainder of the Mesozoic and until about 50 m.y. ago, the region of the miogeosyncline was subject to one or more protracted periods of orogeny. A metamorphic infrastructure formed as high-temperature isotherms rose and engulfed deep geosynclinal strata. Regional compression led to thickening of the mobile, deforming crust, principally in the zones of high-grade metamorphism and, as a consequence of thickening, resultant isostatic uplift, and regional compression, the suprastructure pushed up over the edges of the stable craton, creating a foredeep immediately east of a fold and thrust belt (Sevier orogeny; section B). Clastic sediments largely derived from the fold and thrust belt accumulated in the foredeep. At the end of the Cretaceous the platform itself broke up into basins and uplifted blocks (Laramide orogeny) (Sales, 1968).

About 40 m.y. ago, volcanism spread into the Great Basin from the north, and distension of the crust began. During the last 20 m.y. distension has widened the area east of the Antler belt to near its early Jurassic width (section A). An anomalously low-density upper mantle keeps the region today at higher elevation than would seem normal for its observed crustal thickness (Prodehl, 1970). At the surface, the distension is expressed by tilted fault blocks. Internally these blocks are cut by faults of various ages. Faults formed at depth during earlier stages of extension and denudation are cut by higher angle faults formed during later stages. The most recent large normal faults bound the present ranges.

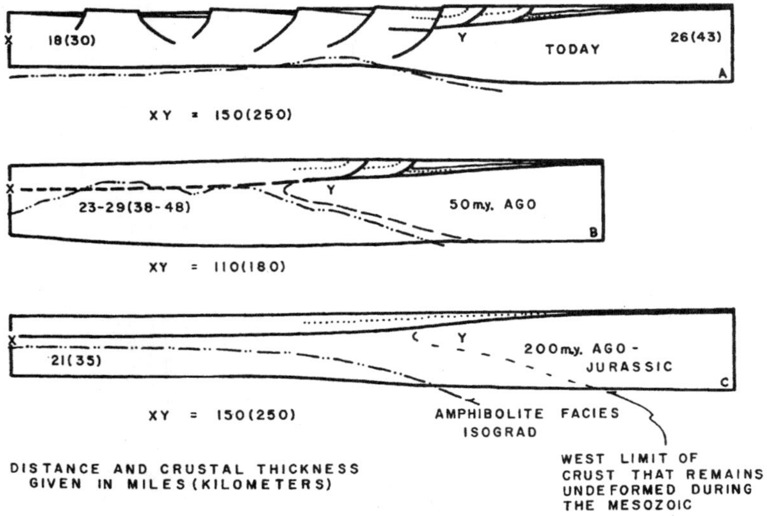

The hinterland is thus a polyorogenic region of heterogeneous structural style. Unconformities and cross-cutting granitic bodies of pre-Tertiary age prove that regional metamorphism of deep parts of the miogeosyncline and folding and faulting of shallower strata occurred during the Mesozoic over the entire region, but there is no evidence for extensive stretching in the hinterland prior to 40 m.y. ago. Regional extension during later Tertiary time is largely responsible for the observed high- and low-angle normal faulting, vertical and rotational movement of blocks, arching of uplifted blocks, and tectonic denudation on a variety of scales—up to several miles of displacement. Low-angle faults of this period tend to follow incompetent strata and pre-existing low-angle faults, thus complicating and mimicking the Mesozoic structures. Although it is difficult to date many low-angle faults, it is likely that many, once considered to be Mesozoic, are actually Tertiary. Several examples of low-angle faults, mapped as thrust faults, that may be interpreted as Tertiary denudation structures, have been described. The fact that many low-angle faults in the hinterland cannot be logically considered as anything but Mesozoic should not detract from the conclusion that extension of the hinterland is predominantly a late Tertiary phenomenon. Thus the extension cannot be genetically linked with thrusting in the Sevier orogenic belt.

## ACKNOWLEDGMENTS

My Great Basin research has been supported by National Science Foundation Grants GA4192, GP 5383, and GA 1694. I have benefited from discussion with many geologists but especially Max D. Crittenden, Ralph J. Roberts, Harald Drewes, Richard K. Hose, B. C. Burchfiel, Keith A. Howard, Robert B. Scott, and John Rodgers. Richard Hose provided a copy of the White Pine County map, an essential help in this discussion, and both he and Ralph Roberts provided helpful comments on the manuscript. Nevertheless, they remain skeptical of some of my arguments.

## REFERENCES CITED

Adair, D. H., 1961, Geology of the Cherry Creek District, Nevada [M.S. thesis]: Salt Lake City, Univ. Utah, 125 p.

Adair, D. H., and Stringham, B., 1957, Whitehorse quartz monzonite, eastern Nevada [abs.]: Geol. Soc. America Bull., v. 68, p. 1857.

Armstrong, F. C., and Oriel, S.S, 1965, Tectonic development of Idaho-Wyoming thrust belt: Am. Assoc. Petroleum Geologists Bull., v. 49, p. 1847–1866.

Armstrong, R. L., 1964, Geochronology and geology of the eastern Great Basin in Nevada and Utah [Ph.D. thesis]: New Haven, Yale Univ., 202 p.

—— 1968a, The Cordilleran miogeosyncline in Nevada and Utah: Utah Geol. and Mineralog. Survey Bull., v. 78, p. 58.

—— 1968b, The Sevier orogenic belt in Nevada and Utah: Geol. Soc. America Bull., v. 79, p. 429–458.

—— 1970, Geochronology of Tertiary igneous rocks, eastern Basin and Range province, western Utah, eastern Nevada, and vicinity, U.S.A.: Geochim. et Cosmochim. Acta, v. 34, p. 203–232.

Armstrong, R. L., and Hansen, E., 1966, Cordilleran infrastructure in the eastern Great Basin: Am. Jour. Sci., v. 264, p. 112–127.

Armstrong, R. L., and Hills, F. A., 1967, Rubidium-strontium and potassium argon geochronologic studies of mantled gneiss domes, Albion Range, southern Idaho, U.S.A.: Earth and Planetary Sci. Letters, v. 3, p. 114–124.

Baker, A. A., 1964, Geology of the Orem quadrangle, Utah: U.S. Geol. Survey Geol. Quad. Map GQ-241.

Bally, A. W., Gordy, P. L., and Stewart, G. A., 1966, Structure, seismic data, and orogenic evolution of southern Canadian Rocky Mountains: Bull. Canadian Petroleum Geology, v. 14, p. 337–381.

Barnes, Harley, and Poole, F. G., 1968, Regional thrust-fault system in Nevada Test Site and vicinity: Geol. Soc. America Mem. 110, p. 233–238.

Bauer, H.L., Jr., Breitrick, R. A., Cooper, J. J., and Anderson, J. A., 1966, Porphyry copper deposits of the Robinson mining district, Nevada, in Geology of the porphyry copper deposits, southwestern North America: Tucson, Univ. Arizona Press, p. 232–244.

Bick, K. F., 1966, Geology of the Deep Creek Mountains, Tooele and Juab Counties, Utah: Utah Geol. and Mineralog. Survey Bull., v. 77, p. 120.

Brew, D. A., 1971, Mississippian stratigraphy of the Diamond Peak area, Eureka County, Nevada: U.S. Geol. Survey Prof. Paper 661, p. 84.

Brokaw, A. L., and Shawe, D. R., 1965, Geologic map and sections of the Ely 3 SW quadrangle, White Pine County, Nevada: U.S. Geol. Survey Misc. Geol. Inv. Map, I-449.

Burchfiel, B. C., 1966, Tin Mountain landslide,

Figure 16. Crustal cross sections of the region between Eureka, Nevada, and the Colorado Plateau, showing changes in crustal thickness and width during Mesozoic and Tertiary deformation of the region. The profiles conserve area, except for some removal of supracrustal rocks by erosion between sections C and B

southeastern California, and origin of megabreccia: Geol. Soc. America Bull., v. 77, p. 95-100.

Burchfiel, B. C., and Davis, G. A., 1968, Two-sided nature of the Cordilleran orogen and its tectonic implications: Internat. Geol. Cong., 23rd, Prague 1968, Proc. sec. 3, p. 175-184.
—— 1971, Clark Mountain thrust complex in the Cordillera of southeastern California: Geologic summary and field trip guide: Geol. Soc. America Guidebook (Cordilleran Sec.), 1971 mtg., p. 1-28.

Burchfiel, B. C., Pelton, P. J., and Sutter, J., 1970, An early Mesozoic deformation belt in south-central Nevada-southeastern California: Geol. Soc. America Bull., v. 81, p. 211-215.

Ceboll, S. E., 1970, Bedrock geology and orogenic succession in southern Grant Range, Nye County, Nevada: Am. Assoc. Petroleum Geologists Bull., v. 54, p. 1828-1842.

Compton, R. R., 1969, Thrusting in northwest Utah: Geol. Soc. America Abs. with Programs for 1969, Pt. 5 (Rocky Mountain Sec.), p. 15.

Cook, E. F., 1960, Breccia blocks (Mississippian) of the Welcome Springs area, southwest Utah: Geol. Soc. America Bull., v. 71, p. 1709-1712.

Crosby, Gary W., 1968, Vertical movements and isostasy in western Wyoming overthrust belt: Am. Assoc. Petroleum Geologists Bull., v. 52, p. 2000-2015.
—— 1969, Radial movements in the western Wyoming salient of the Cordilleran overthrust belt: Geol. Soc. America Bull., v. 80, p. 1061-1078.

Dahlstrom, C.D.A., 1969, Balanced cross sections: Canadian Jour. Earth Sci., v. 6, p. 743-757.
—— 1970, Structural geology in the eastern margin of the Canadian Rocky Mountains: Bull. Canadian Petroleum Geology, v. 18, p. 332-406.

Dechert, C. P., 1967, Bedrock geology of the northern Schell Creek Range, White Pine County, Nevada [Ph.D. thesis]: Seattle, Univ. Washington, 266 p.

Dodge, Harry W., Jr., 1970, Klippen of Devonian eastern carbonates on upper Paleozoic clastics in central Nevada: Geol. Soc. America, Abs. with Programs (Cordilleran Sec.), v. 2, no. 2, p. 87-88.

Drewes, H., 1964, Diverse recurrent movement along segments of a major thrust fault in the Schell Creek Range near Ely, Nevada: U.S. Geol. Survey Prof. Paper, 501-B, p. 20-24.
—— 1965, Thrust faults and glide faults in the Schell Creek Range near Ely, Nevada: Geol. Soc. America, Abs. for 1964, Spec. Paper 82, p. 249.
—— 1967, Geology of the Connors Pass quadrangle, Schell Creek Range, east central Nevada: U.S. Geol. Survey Prof. Paper 557, p. 93.

Drewes, H., Moores, E. M., Scott, R. B., and Lumsden, W. W., 1970, Tertiary tectonics of the White Pine-Grant Range region, east-central Nevada, and some regional implications: Discussion and reply: Geol. Soc. America Bull., v. 81, p. 319-330.

Easton, W. H., 1954, Geology of the Illipah quadrangle: Am. Assoc. Petroleum Geologists Pacific Sec. Newsletter, v. 8.

Fleck, R. J., 1970a, Age and tectonic significance of volcanic rocks, Death Valley area, California: Geol. Soc. America Bull., v. 81, p. 2807-2816.
—— 1970b, Tectonic style, magnitude, and age of deformation in the Sevier orogenic belt in southern Nevada and eastern California: Geol. Soc. America Bull., v. 81, p. 1705-1720.

Fritz, W. H., 1960, Structure and stratigraphy of the northern Egan Range, White Pine County, Nevada [Ph.D. thesis]: Seattle, Univ. Washington, p. 178.
—— 1968, Geologic map and sections of the southern Cherry Creek and northern Egan Ranges, White Pine County, Nevada: Nevada Bur. Mines Map 35.

Gemmill, Paul, 1968, The geology of the ore deposits of the Pioche district, Nevada, *in* Ore deposits of the United States, 1933-1967: New York, Am. Inst. Mining Metallurgy, and Petroleum Engineering, v. 2, p. 1128-1147.

Hamilton, Warren, and Meyers, W. B., 1966, Cenozoic tectonics of the western United States: Rev. Geophysics, v. 4, p. 509-549.

Hinrichs, E. N., 1968, Geologic structure of Yucca Flat area, Nevada: Geol. Soc. America Mem. 100, p. 239-246.

Hintze, L. F., compiler, 1963, Geologic map of southwestern Utah: Utah Geol. and Mineralog. Survey.

Hose, R. K., and Blake, M. C., Jr., 1969, Structural development of the eastern Great Basin during the Mesozoic: Geol. Soc. America Abs. with Programs for 1969, Pt. 5 (Rocky Mountain Sec.), p. 34.
—— 1970, Geologic map of White Pine County, Nevada: U.S. Geol. Survey open-file map.

Hose, R. K., and Danés, Z. F., 1968, Late Mesozoic structural evolution of the eastern Great Basin: Geol. Soc. America, Abs. for 1967, Spec. Paper 115, p. 102.

Howard, Keith A., 1966, Structure of the metamorphic rocks of the northern Ruby Mountains, Nevada [Ph.D. thesis]: New Haven, Yale Univ., p. 170.
—— 1971, Paleozoic metasediments in the northern Ruby Mountains, Nevada: Geol. Soc. America Bull., v. 82, p. 259-264.

Humphrey, F. L., 1960, Geology of the White Pine mining district, White Pine County, Nevada: Nevada Bur. Mines Bull. 57, p. 119.

Hunt, C. B., and Mabey, D. R., 1966, Stratigraphy and structure of Death Valley, California: U.S. Geol. Survey Prof. Paper 494-A, p. 162.

James, David E., and Steinhart, John S., 1966, Structure beneath continents—a critical review of explosion studies 1960-1965, *in* The earth beneath the continents: Am. Geophys. Union Geophys. Mon., p. 293-333.

Jones, R. W., 1963, Gravity structures in the Beaver Dam Mountains, southwestern Utah: Intermountain Assoc. Petroleum Geol. Guidebook, 12th Ann. Field Conf., p. 90-95, 1963.

Kellogg, H. E., 1960, Geology of the southern Egan Range, Nevada: Intermountain Assoc. Petroleum Geol. Guidebook, 11th Ann. Field Conf., p. 189-197.
—— 1964, Cenozoic stratigraphy and structure of the southern Egan Range, Nevada: Geol. Soc. America Bull., v. 75, p. 949-968.

Kistler, R. W., and Willden, Ronald, 1969, Age of thrusting in the Ruby Mountains, Nevada: Geol. Soc. America, Abs. with Programs for 1969 (Rocky Mountain Sec.), pt. 5, p. 40-41.

Kleinhampl, F. J., and Ziony, J. I., 1967, Preliminary geologic map of northern Nye County, Nevada: U.S. Geol. Survey open-file map.

Lanphere, M. A., Wasserburg, G.J.F., Albee, A. L., and Tilton, G. R., 1964, Redistribution of strontium and rubidium isotopes during metamorphism, World Beater complex, Panamint Range, California, *in* Craig, H., Miller, S. L., and Wasserburg, G.J.F., eds., Isotopic and cosmic chemistry: Amsterdam, North-Holland Pub. Co., p. 269-320.

Lee, D. E., Marvin, R. F., Stern T. W., and Peterman, Z. E., 1970, Modification of potassium-argon ages by Tertiary thrusting in the Snake Range, White Pine County, Nevada: U.S. Geol. Survey Prof. Paper 700-D, p. 92-102.

Longwell, C. R., 1945, Low angle normal faults in the Basin and Range province: Am. Geophys. Union Trans., v. 26, p. 107-118.

Longwell, C. R., Pampeyan, E. H., Bowyer, Ben, and Roberts, R. J., 1965, Geology and mineral deposits of Clark County, Nevada: Nevada Bur. Mines Bull. 62, p. 218.

Mackin, J. H., 1960, Structural significance of Tertiary volcanic rocks in southwestern Utah: Am. Jour. Sci., v. 258, p. 81-131.

McDowell, F. W., and Kulp, J. L., 1967, Age of intrusion and ore deposition in the Robinson mining district of Nevada: Econ. Geology, v. 62, p. 905-909.

Miller, G. M., 1966, Structure and stratigraphy of southern part of Wah Wah Mountains, southwest Utah: Am. Assoc. Petroleum Geologists Bull., v. 50, p. 858-900.

Misch, Peter, 1960, Regional structural reconnaissance in central-northeast Nevada and some adjacent areas: observations and interpretations: Intermtn. Assoc. Petroleum Geologists 11th Ann. Field Conf. Guidebook, p. 17-42.
—— 1971, Geotectonic implications of Mesozoic décollement thrusting in parts of eastern Great Basin: Geol. Soc. America, Abs. with Programs (Cordilleran Sec.), v. 3, p. 164-165.

Misch, P., and Hazzard, J. C., 1962, Stratigraphy and metamorphism of late Precambrian rocks in central northeastern Nevada and adjacent Utah: Am. Assoc. Petroleum Geologists Bull., v. 46, p. 289-343.

Moores, E. M., 1968, Mio-Pliocene sediments, gravity slides and their tectonic significance, east-central Nevada: Jour. Geology, v. 76, p. 88-98.

Moores, E. M., Scott, R. B., and Lumsden, W. W., 1968, Tertiary tectonics of the White Pine-Grant Range region, east-central Nevada, and some regional implications: Geol. Soc. America Bull., v. 79, p. 1703-1726.

Mudge, M. R., 1970, Origin of the disturbed belt in northwestern Montana: Geol. Soc. America Bull., v. 81, p. 377-392.

Nelson, R. B., 1966, Structural development of northernmost Snake Range, Kern Mountains, and Deep Creek Range, Nevada and Utah: Am. Assoc. Petroleum Geologists Bull., v. 50, p. 921-951.
—— 1969, Relation and history of structures in a sedimentary succession with deeper metamorphic structures, eastern Great Basin: Am. Assoc. Petroleum Geologists Bull., v. 53, p. 307-339.

Nolan, T. B., 1935, The Gold Hill mining district, Utah: U.S. Geol. Survey Prof. Paper 177, 172 p.
—— 1943, The Basin and Range province in Utah, Nevada, and California: U.S. Geol. Survey Prof. Paper 197-D, p. 172-184.

Olson, R. C., 1970, A major thrust fault in the northwestern Inyo Mountains, Inyo County, California: Geol. Soc. America, Abs. with Programs (Cordilleran Sec.), v. 2, p. 128.

Palmer, A. R., 1960, Some aspects of the early Upper Cambrian stratigraphy of White Pine County, Nevada, and vicinity: Intermtn. Assoc. Petroleum Geologists 11th Ann. Field Conf. Guidebook, p. 53-58.

Playford, P. E., 1961, Geology of the Egan Range near Lund, Nevada [Ph.D. thesis]: Stanford, Stanford Univ., 249 p.

Price, R. A., 1962, Fernie map-area, east half, Alberta and British Columbia: Canada Geol. Survey Paper 61-24, 65 p.
—— 1969, The southern Canadian Rockies and the role of gravity in low-angle thrusting, foreland folding, and the evolution of migrating foredeeps: Geol. Soc. America, Abs. with Pro-

grams for 1969, pt. 7 (Ann. Mtg.), p. 284–286.
―― 1971, Gravitational sliding and the foreland thrust and fold belt of the North American Cordillera: Geol. Soc. America Bull., v. 82, p. 1133–1138.

Price, R. A., and Mountjoy, E. W., 1970, Geologic structure of the Canadian Rocky Mountains between Bow and Athabasca Rivers—a progress report: Geol. Assoc. Canada Spec. Paper 6, p. 7–25.

―― 1971, The Cordilleran foreland thrust and fold belt in the southern Canadian Rockies: Geol. Soc. America, Abs. with Programs (Rocky Mountain Sec.), v. 3, p. 404–405.

Prodehl, Claus, 1970, Seismic refraction study of crustal structure in the western United States: Geol. Soc. America Bull., v. 81, p. 2629–2646.

Riva, John, 1970, Thrusted Paleozoic rocks in the northern and central H-D Range, northeast Nevada: Geol. Soc. America Bull., v. 81, p. 2689–2716.

Roberts, R. J., Hotz, P. E., Gilluly, J., and Ferguson, H. G., 1958, Paleozoic rocks of north-central Nevada: Am. Assoc. Petroleum Geologists Bull., v. 42, p. 2813–2857.

Roberts, R. J., Crittenden, M. D., Jr., Tooker, E. W., Morris, H. T., Hose, R. K., and Cheney, T. M., 1965, Pennsylvanian and Permian basins in northwestern Utah, northeastern Nevada and south-central Idaho: Am. Assoc. Petroleum Geologists Bull., v. 49, p. 1926–1956.

Roberts, Ralph J., 1968, Tectonic framework of the Great Basin, in A coast to coast tectonic study of the United States: Univ. Missouri, Rolla, Jour., p. 101–119.

Robison, R. A., 1960, Lower and Middle Cambrian stratigraphy of the eastern Great Basin: Intermtn. Assoc. Petroleum Geologists 11th Ann. Field Conf. Guidebook, p. 43–52.

Robison, R. A., and Bentley, C., 1958, Upper Cambrian stratigraphy in central and western Utah: Geol. Soc. America Bull., v. 69, p. 1702–1703.

Rubey W. W., and Hubbert, M. K., 1959, Overthrust belt in geosynclinal area of western Wyoming in light of fluid-pressure hypothesis: Geol. Soc. America Bull., v. 70, p. 167–206.

Sales, John K., 1968, Crustal mechanics of Cordilleran foreland deformation: a regional and scale-model approach: Am. Assoc. Petroleum Geologists Bull., v. 52, p. 2016–2044.

Scholten, R., 1968, Model for evolution of Rocky Mountains east of Idaho batholith: Tectonophysics, v. 6, p. 109–126.

Seager, W. R., 1970, Low-angle gravity glide structures in the northern Virgin Mountains, Nevada and Arizona: Geol. Soc. America Bull., v. 81, p. 1517–1538.

Secor, D. T., Jr., 1963a, Geology of the central Spring Mountains, Nevada [Ph.D. thesis]: Stanford, Stanford Univ., 197 p.

―― 1963b, Structure of the central Spring Mountains, Nevada: Geol. Soc. America Spec. Paper 73, p. 63–64.

Snow, G. G., 1964, Mineralogy and geology of the Dolly Varden Mountains, Elko County, Nevada [Ph.D. thesis]: Salt Lake City, Univ. Utah, 187 p.

Stevens, C. H., 1969, Middle to Late Triassic deformation in the Inyo, White, and northern Argus Mountains, California: Geol. Soc. America, Abs. with Programs for 1969, pt. 5 (Rocky Mountain Sec.), p. 78.

Stokes, W. L., compiler, 1963, Geologic map of northwestern Utah: Utah Geol. and Mineralog. Survey.

Thorman, C. H., 1970, Metamorphosed and nonmetamorphosed Paleozoic rocks in the Wood Hills and Pequop Mountains, northeast Nevada: Geol. Soc. America Bull., v. 81, p. 2417–2448.

Tschanz, C. M., and Pampeyan, E. H., 1970, Geology and mineral deposits of Lincoln County, Nevada: Nevada Bur. Mines Bull., v. 73, 188 p.

Whitebread, D. H., 1968, Snake Range décollement and related structures in the southern Snake Range, eastern Nevada: Geol. Soc. America, Abstracts for 1966, Spec. Paper 101, p. 345–346.

―― 1969, Geologic map of the Wheeler Peak and Garrison quadrangles, Nevada and Utah: U.S. Geol. Survey Misc. Geol. Inv. Map I-578.

Willden, Ronald, and Kistler, R. W., 1969, Geologic map of the Jiggs quadrangle, Elko County, Nevada: U.S. Geol. Survey Geol. Quad. Map 859.

Willden, Ronald, Thomas, H. H., and Stern, T. W., 1967, Oligocene or younger thrust faulting in the Ruby Mountains, northeastern Nevada: Geol. Soc. America Bull., v. 78, p. 1345–1358.

Winfrey, W. M., 1960, Stratigraphy, correlation and oil potential of the Sheep Pass Formation, east central Nevada: Intermtn. Assoc. Petroleum Geologists 11th Ann. Field Conf. Guidebook, p. 126–133.

Woodward, L. A., 1964, Structural geology of central northern Egan Range, Nevada: Am. Assoc. Petroleum Geologists Bull., v. 48, p. 22–39.

Wright, L. A., and Troxel, B. W., 1969, Chaos structure and Basin and Range normal faults: evidence for a genetic relationship: Geol. Soc. America Abs. with Programs for 1969, pt. 7 (Ann. Mtg.), p. 242.

Young, J. C., 1960, Structure and stratigraphy in north-central Schell Creek Range: Intermtn. Assoc. Petroleum Geologists 11th Ann. Field Conf. Guidebook, p. 158–172.

MANUSCRIPT RECEIVED BY THE SOCIETY JUNE 17, 1971
REVISED MANUSCRIPT RECEIVED DECEMBER 7, 1971

# Retrospective on "Low-angle (denudation) faults, hinterland of the Sevier orogenic belt, eastern Nevada and western Utah" by Richard Lee Armstrong

**Brian Wernicke**
*Division of Geological and Planetary Sciences, California Institute of Technology, Pasadena, California 91108;
e-mail: brian@gps.caltech.edu*

**Jon Spencer**
*Arizona Geological Survey, 416 W. Congress St., #100, Tucson, Arizona 85701;
e-mail: jspencer@geo.arizona.edu*

The 1960s and early 1970s brought sweeping new syntheses of the Cordilleran orogen as having developed primarily in response to the motions of oceanic plates to the west. The main conclusions of these syntheses were that the Cordilleran miogeocline resulted from Neoproterozoic continental rifting; magmatic systems are a signature of subducting slabs beneath the continent; and variably oblique plate convergence was responsible for coastwise shearing and progressive accretion of large, sometimes far-traveled crustal fragments onto the continental margin through most of Phanerozoic time. The novelty of these concepts is clear from the pre-1960s literature, which contains only rare allusions to ideas that might be considered similar. Afterward, the nature of the debate illustrated their profound acceptance. Are the magmas arc or backarc? Was there just one Neoproterozoic rifting event? How far had the westernmost accreted blocks traveled? That none of these comparatively second-order issues are resolved after some three decades of research is humbling testimony to the progress represented by the early syntheses.

It is all the more surprising that out of an era focused on refining the role of plate tectonics in Cordilleran evolution, there emerged what may be the most "classical" of all concepts associated with the Cordilleran orogen, or at least, a tectonic element first discovered in the Cordillera and subsequently identified in every major orogen on the globe, namely, the geological signature of large-magnitude continental extension. Under rubrics such as "denudation faults," "low-angle normal faults," "detachment faults" and "Cordilleran metamorphic core complexes," these features were initially regarded by most geologists as oddities. However, by the early 1990s analogous structures had been reported as fundamental tectonic elements of nearly every major orogenic system in the world (Alpine, Caledonian, Himalayan, Hercynian, Grenvillian, and many others) and of many passive margins and intracratonic rift systems (Burg and Chen, 1984; Lister et al., 1984, 1986; Wernicke, 1985; Seranne and Seguret, 1987; Selverstone, 1988; Zheng et al., 1988; Doblas and Oyarzun, 1989; Jolivet et al., 1990; Mezger et al., 1991; Hill et al., 1992; Mpodozis and Allmendinger, 1993). There was, and continues to be, much discussion regarding the implications of plate reconstructions on extensional tectonics in the Cordillera and elsewhere. But it is not the plate tectonic setting of extension that is the main "export" from the Cordillera to other orogenic systems, it is the structural expression of large-magnitude extension. Its discovery was not inspired by plate tectonics, but was instead the product of intense debate over geological field relations and the timing and nature of regional metamorphism.

That debate began in earnest in the late 1960s and early 1970s, at the same time plate tectonics was sweeping the globe. The debate was well articulated in a paper published in the Geological Society of America *Bulletin* in 1972 entitled "Low-angle (denudation) faults, hinterland of the Sevier orogenic belt, eastern Nevada and western Utah," by Richard Lee Armstrong. Somewhat out of character, the paper contained no new data, but was mainly a series of simple line drawings of geologic maps from the "hinterland" of the east-vergent Mesozoic foreland fold and thrust belt in Utah. The structure of the hinterland was dominated by areally extensive low-angle faults developed within the thick, conformable Cordilleran miogeocline. In contrast to the thrust belt, these low-angle faults characteristically omit stratigraphic section, placing younger strata on older. In addition, the footwalls of the faults in a number of instances were metamorphosed to greenschist or amphibolite facies, suggesting that higher structural levels were emplaced on lower structural levels. The paper

---

Wernicke, B. and Spencer, J., 1999, Retrospective on "Low-angle (denudation) faults, hinterland of the Sevier orogenic belt, eastern Nevada and western Utah" by Richard Lee Armstrong, *in* Moores, E. M., Sloan, D., and Stout, D. L., eds., Classic Cordilleran Concepts: A View from California: Boulder, Colorado, Geological Society of America Special Paper 338.

challenged the prevailing view, championed by Peter Misch of the University of Washington and Richard Hose of the U.S. Geological Survey, that the faulting was genetically related to the thrust belt to the east.

The Misch and Hose conceptions of the hinterland differed significantly. Misch and his students, who discovered the faults and contributed the bulk of their detailed mapping, viewed them as a direct expression of Mesozoic crustal shortening, more or less coeval with thrusting to the east (e.g., Misch, 1960). Stressing that the faults consistently omit section and are therefore normal faults, Hose and colleagues argued that they were the headward area of a huge gravity slide complex, the "toe" area of which was the thrust belt to the east (Hose and Danes, 1973).

The objective of Armstrong's paper was to demonstrate that the structures were related to upper crustal thinning (in agreement with Hose and colleagues), but were mainly Tertiary normal faults or local gravity slide blocks formed in response to Basin and Range faulting. He first plotted all known sub-Tertiary unconformities in the region on a map and showed that their depositional substrate was almost entirely Mississippian or younger. If widespread thrusting or attenuation had occurred in the Mesozoic, leading to what are now widespread exposures of Eocambrian through Devonian strata thousands of meters below the Mississippian, then unconformities between Tertiary strata and pre-Mississippian strata should be widespread. Using simple line drawings of geologic maps, he enumerated specific instances where geometrical and kinematic arguments indicated that the structures were almost certainly Tertiary. For example, a large normal fault in the southern Snake Range (Murphy Wash fault) was mapped as offsetting Tertiary strata on its south end and being truncated by the Snake Range decollement to the north, one of the most areally extensive low-angle faults in the hinterland.

One of the most instructive examples described was the northern Egan Range. There, entire sections of the miogeocline are tilted moderately to steeply westward and broken by low-angle faults with thousands of meters of stratigraphic throw. The pre-Tertiary strata are in contact with two outcrop areas of Tertiary volcanic strata, also moderately to steeply tilted to the west. The volcanic strata were mapped as unconformable on all surrounding rocks, but the map relations show both homoclines striking at a high angle into portions of the contact mapped as depositional. Armstrong (1972) showed in cross-section view (his Fig. 12) that subtracting the tilting of the Tertiary strata indicated that the faults were both normal and postvolcanic, not prevolcanic thrusts, as had been previously supposed.

The paper concluded with a general discussion of the role of hinterland structure in the evolution of the Cordilleran orogen, including a full-throated critical evaluation of the gravity sliding concept for the thrust belt—from which the concept never recovered.

Although a watershed paper on the basis of its ideas alone, the tone of the paper is almost singular in relating a contagious passion for the subject to the reader. Wernicke vividly remembers, as a second-year graduate student at MIT fresh from a summer mapping younger-on-older low-angle faults in Nevada as thrusts, cheering out loud as he read through it for the first time. Here, amid the mountain of bone-dry literature one must absorb to get up to speed in any field, was a lively, bare-knuckled intellectual attack, more in the classic American style of a Mencken or Twain essay than in the stupefying genre we as a community of scientific writers, referees, and editors work so hard to perfect.

In the review process for most major journals, and certainly the staid and conservative GSA *Bulletin*, authors are usually required to address referee and editor admonitions about "toning down" passages where the ideas of others are questioned, especially if they are widely held or "consensus" views. Avoid the first person. Keep references in parentheses. Talk about ideas, not people. Don't preach or patronize the reader! Avoid sarcasm or statements that, even unintentionally, might imply contempt or ridicule. Present interpretations and conclusions only at the end of the manuscript, after an unbiased rendering of facts has allowed the interpretation to "sell itself." In this paper, however, Armstrong would have none of it. The editor of the *Bulletin* at the time, Bennie W. Troxel, himself an early advocate of a Tertiary extensional origin for younger-on-older faults, wouldn't either.

Any decorum that might have been afforded through polite anonymity is crushed outright in the abstract: "Whitebread, Hose, Roberts and Crittenden advocate gravitational gliding models...," "Misch, Nelson, Fritz, Miller and Woodward..." relate the hinterland structures to Mesozoic compression, and "Armstrong, Burchfiel, Davis and Fleck..." argue they are unrelated to the thrust belt. The suspects were all in the lineup, and the curtain was drawn open.

The first paragraph of the text frankly states, "I believe that models linking denudation faults with thrust faults of Cretaceous age...are unlikely, and that the denudation faults are predominantly of Tertiary age and related to Basin and Range faulting" (Armstrong, 1972, p. 1729). In regard to interpreting a younger-on-older fault in the Wasatch Range as a thrust fault, he noted that it would require "a complex ad hoc explanation...," concluding that "Such gymnastics are unnecessary, and certainly not needed to explain the structures known to occur west of the Wasatch fault in this area" (Armstrong, 1972, p. 1734). After summarizing several mechanisms where thrusts might result in local younger-on-older structures, he concludes "None of these geometric mechanisms seems applicable to the hinterland structures. Geometric logic is a necessary prerequisite to any acceptable model for structural evolution" (Armstrong, 1972, p. 1735). In the following paragraph, the reader is informed that "Once popular and widely quoted compressional-thrust explanations for these ....features are falling out of favor" (p. 1736). At this point we're only 8 pages into a 26-page paper, and the jury has been led right on down to the gallows.

No gloves are donned for the descriptive body of the paper (Armstrong, 1972, p. 1734–1744). The basis for Misch's interpretation of a Cretaceous age for the Snake Range decollement is described as "unconformities below Cretaceous strata that lie more than 100 mi. away from the Snake Range" (p. 1736), whereas the author believes "evidence for dating the faults may be found in the southern Snake Range." After describing the Murphy Wash fault, he cites the conclusion of Lee et al. (1970) that the latest movement on the decollement was Tertiary, based on 20 Ma argon ages from cataclasites in the footwall, and simply concludes, "I agree wholeheartedly" (Armstrong, 1972, p. 1736).

As to map relations in the Deep Creek Mountains and near Gold Hill, it is stressed, following Mackin (and Hutton), that Tertiary strata are the guides to Tertiary deformation, and that "Casualness on this point has contributed much confusion to the literature concerning the eastern Great Basin" (Armstrong, 1972, p. 1738). After pointing out that a fault relationship on Harold Drewes's map of the southern Schell Creek Range is "a geometric impossibility—a fault displacement cannot change along strike by more than the distance along strike," he concludes, "To me, the simplest interpretation is that the 'thrust' and the Tertiary normal fault [that putatively cuts it] are one and the same feature..." (Armstrong, 1972, p. 1740, 1741).

The discussion of the northern Egan Range (Armstrong, 1972, p. 1741) begins, "An area where the structural significance of Tertiary volcanic rocks (Mackin, 1960) has been ignored is the northern Egan Range. Figure 11 shows the geology of part of the area mapped by Fritz (1960, 1968)," and, in discussing a contact mapped as depositional between Tertiary and Eocambrian that he reinterprets as a normal fault, he states "Such an interpretation is in agreement with the work of Adair (1961, and 1963, personal commun.) but denied by R. K. Hose (1971, personal commun.)" (p. 1741). Allowing that some structures in Fritz's map area could be Mesozoic, the Mesozoic thrust interpretation had nonetheless "been carried too far, and the Tertiary structures have been considerably underrated." The next sentence and paragraph begins, "Another example of proliferation of Mesozoic thrusts faults in violation of Mackin's principle occurs farther south, in the southern Egan Range, in an area mapped by Brokaw and Shawe (1965; Fig. 13)" (Armstrong, 1972, p. 1742); Brokaw and Shawe were also tried and convicted on grounds of geometric impossibility.

After reinterpreting several other "thrusts" (put in quotation marks no fewer than eight times from pages 1740 to 1743) and praising the conclusion by Moores et al. (1968) that many of the low-angle faults in the Grant and White Pine ranges were Tertiary denudation structures, he began the general discussion of the role of hinterland structure in the development of the Cordillera. Pages 1744–1747 in Armstrong (1972) constitute what is perhaps the most withering assault ever launched on a mainstream concept for Cordilleran evolution, not merely on account of rhetorical style, but in the clarity and persuasiveness of the arguments.

The discussion is prefaced with perhaps the best-known remark in the paper, which is something of a mantra among modern field geologists studying the problem of low-angle normal faults and similar phenomena. In the 1960s and 1970s, the "gravity slide" model of thin-skinned thrusting was advocated, and the compressional, "push-from-the-rear" model discredited, with the argument that it is mechanically impossible to push a thin sheet of rock from behind—only a body force acting on the entire thrust mass could allow it to remain coherent while overcoming frictional resistance along its base. Indeed, even today there is vigorous opposition to the existence of active low-angle normal faults on mechanistic grounds. Armstrong's position? "Arguments about mechanism are, for the most part, unsatisfactory and inconclusive, as they require knowledge we do not have on the large-scale behavior of rocks. I prefer to analyze what happened, geometrically and chronologically, rather than try to answer abstract questions of how it happened (mechanism and driving force). I am willing to grant anyone a process regardless that it may conflict with intuition or an evolving body of theory, if it can be proved to have happened" (Armstrong, 1972, p. 1745).

He went on to slay the gravity sliding dragon with abandon and flair. There is no possibility of a net downhill slope for the slide mass. Extension in the hinterland is insufficient to account for the minimum amount of shortening in the Sevier belt, without which "the glide model is untenable" (Armstrong, 1972, p. 1746). The thrust belt is continuous from Canada to southern California with major faults paralleling individual stratigraphic horizons for hundreds of kilometers. No feature of comparable stratigraphic and structural continuity is present behind the thrust belt, unless one supposes the level of gliding "jumps up and down stratigraphically to levels not observed in the presently exposed rocks (an ad hoc excuse to retain the tenability of the glide-decollement hypothesis)" (Armstrong, 1972, p. 1746). "An even more devastating argument..." is the provincialism of the Basin and Range denudation structures relative to the entire thrust belt. Gravity sliding "cannot be taken seriously" if it applies to only one segment of the thrust belt. "Further geometric difficulties..." are apparent in a regional cross section from the thrust belt to the Snake Range. "Alternate interpretations that have been proposed to date are in conflict with facts shown on geologic maps! Finally, I reject the glide-decollement models on chronological grounds" (Armstrong, 1972, p. 1747).

The concluding section begins by once again identifying the thrusters and gliders by name and dragging them around the block one last time prior to burial. Armstrong goes on to describe a generalized, large-scale cross section showing the thermal and structural state of the east half of the Cordillera (his Fig. 16) that few would take issue with today. It shows thinned crust under the miogeocline, ~100 km of compressional shortening and heating during the Mesozoic, and Tertiary extension of equal magnitude to the shortening to produce the thin crust of the modern Basin and Range. Contemporary ideas regarding the genesis of metamorphic core complexes and detachment systems have evolved far beyond this initial synthesis of denudation structures in the Sevier hinterland. But as in the case of the plate tectonic syntheses of the late 1960s and early 1970s, one can only be humbled in comparing the progress represented by defining the context of the hinterland structures in the Cordillera with that of subsequent refinements.

The significance and magnitude of Cenozoic extension in the Basin and Range province were generally underestimated at the time of Armstrong's 1972 paper, although Hamilton and Myers (1966) had argued for province-wide extension of as much as 100%. Documentation in the 1970s of highly distended arrays of steeply tilted fault blocks consisting largely of mid-Tertiary volcanic rocks (Anderson, 1971; Proffett, 1977) and additional areas of Tertiary low-angle normal faults that had uncovered mylonitic and metamorphic rocks (Davis, 1975; Compton et al., 1977) added support to the idea of large-magnitude Cenozoic extension and raised more questions about the age and significance of penetrative deformation. The association of low-angle normal faults and footwall mylonitic and metamorphic rocks, an association termed "metamorphic core complexes," was the focus of vigorous debate at a 1977 GSA Penrose conference (Coney, 1980a; Crittenden et al., 1980). The dragon of thrust tectonics in extensional guise that Armstrong (1972) sought to slay lived on in metamorphic core complexes as some conference participants insisted that Cenozoic low-angle normal faults were reactivated thrust faults on top of thrust-related mylonites (Drewes, 1977). Questions rather than answers took center stage because (1) some footwall-block deformation and metamorphism really were related to older thrust tectonics (Coney, 1980b); (2) shear-sense indicators in mylonitic rocks (Simpson and Schmid, 1983; Lister and Snoke, 1984) were not yet understood or had not been applied to metamorphic core complexes; and (3) isotope thermochronologic and geochronologic studies had not yet produced an overwhelming amount of high-quality data implicating mid-Tertiary mylonitization.

Metamorphic core complexes are characterized by a set of features that had not been identified in thrust belts, including well-lineated mylonitic fabrics overprinted by distributed brecciation and chloritic alteration, in turn overprinted by localized fault-related fracturing and crushing. Similar features were described in many thorough articles in GSA Memoir 153 (Crittenden et al., 1980), which provided a firm basis for the remarkable Cordilleran revelations that metamorphic core complexes were produced by progressive plastic to brittle deformation during exhumation and cooling of the footwall blocks of large low-angle normal faults (Davis et al., 1986), and that the normal faults accommodated crustal extension and resembled thrust faults in gross geometry (Wernicke, 1981). These insights, now applied around the world, were gained from a large amount of geologic field work in some of the most structurally complicated and lithologically diverse rocks in the Cordillera. Isotope thermochronologic and microstructural studies were instrumental in supporting this new model of extensional tectonism. Furthermore, the concept that thrust sheets are too thin and weak to push from behind and therefore must have been emplaced by gravity sliding was shown to be wrong, or at least unnecessary (Dahlen, 1984), which eliminated one of the original rationales for seeking synthrusting extensional structures and denuded areas in the hinterland of thrust belts. By the late 1980s, the dragon of thrust tectonics in extensional guise that Armstrong wounded so badly had been largely laid to rest.

Low-angle normal faults and metamorphic core complexes were only locally disentangled from thrust faults and areas of crustal thickening, as regionally they are clearly associated. Crustal thickening and associated surface-elevation increase and Moho depression increases crustal gravitational potential energy, and this energy is released by extension and crustal thinning (Molnar and Lyon-Caen, 1988). Greater crustal thickening should therefore be capable of driving greater extension. Metamorphic core complexes have been interpreted as products of unusually large magnitude extension that occurred where earlier crustal thickening had been unusually great (Coney and Harms, 1984). It has generally not been possible, however, to demonstrate that individual metamorphic core complexes correspond to local zones of greater crustal thickening except in the Harcuvar and Whipple complexes in the lower Colorado River trough (Spencer and Reynolds, 1990), and the geologic factors responsible for the spotty distribution of metamorphic core complexes remain enigmatic.

As is typically the case with major scientific insights, the fallout is enough to keep many scientists busy for decades. The denuded footwalls of major low-angle normal faults, now known as "detachment faults," are commonly arched, and some are fluted and grooved, with wavelengths of kilometers to tens of kilometers (Rehrig and Reynolds, 1980; Frost, 1981). Arching along axes perpendicular to extension direction was attributed to isostatic uplift following tectonic denudation (Hyndman, 1980; Howard et al., 1982; Spencer, 1984). Soon after this insight it became apparent that detachment-fault footwalls must have had so little flexural strength during exhumation that the flaccid footwall block rose up from the mid-crust and filled in the space vacated by the laterally traveling hanging-wall block, much like water fills in the space behind a large, slow-moving ship (Buck, 1988; Wernicke and Axen, 1988; Spencer and Reynolds, 1991). Recognition of the fluid-like behavior of the mid-crust during detachment faulting and core complex uplift was a major new insight into extensional tectonic processes (e.g., Block and Royden, 1990; Wernicke, 1990, 1992).

The corrugated footwalls of submarine low-angle normal faults were recognized in the mid-1990s on the inside corners of ridge-transform intersections, a tectonic setting where magmatism is subdued and intermittent along slow-spreading ridges (Tucholke and Lin, 1994; Cann et al., 1997; Tucholke et al., 1998). Metamorphism, mylonitization, and hydrothermal alteration that affected some rocks dredged from below these corrugated surfaces appear to have occurred during progressive plastic to brittle deformation associated with tectonic exhumation and cooling, the same basic processes that produced Cordilleran metamorphic core complexes (Cannat et al., 1992; Jaroslow et al., 1996; Tucholke et al., 1998). The arched and corrugated surfaces probably consist largely of serpentinized peridotite extruded directly from beneath the adjacent mid-ocean ridge as part of the plate spreading process, and form new oceanic crust with little or no basalt and gabbro (e.g., Lagabrielle and Cannat, 1990).

Finally, Armstrong's disregard for assertions of physical impossibility is still tenable 25 year later when, with vastly more computational firepower and additional decades of analysis, geophysicists still can't agree on viable mechanical conditions for detachment-fault initiation (Wills and Buck, 1997) and movement (cf. Xiao et al., 1991, and Scott and Lister, 1992). Seismologists have also had difficulty accommodating low-angle normal faulting because so few first-motion determinations support low-angle slip (Jackson and White, 1989), although a few low-angle events have been detected (Abers, 1991), and their rarity may be related to the greater effectiveness of low-angle normal faults in accommodating crustal extension (Wernicke, 1995). In 1972 Dick Armstrong provided monumental insight into a line of inquiry in Cordilleran geology that, 25 years later, continues to yield insights into tectonic processes and lithospheric behavior around the world.

## REFERENCES CITED

Abers, G. A., 1991, Possible seismogenic shallow-dipping normal faults in the Woodlark-D'Entrecasteaux extensional province, Papua New Guinea: Geology, v. 19, p. 1205–1208.

Anderson, R. E., 1971, Thin skin distension in Tertiary rocks of southeastern Nevada: Geological Society of America Bulletin, v. 82, p. 43–58.

Armstrong, R. L., 1968a, Sevier orogenic belt in Nevada and Utah: Geological Society of America Bulletin, v. 79, p. 429–458.

Armstrong, R. L., 1968b, A model for Pb and Sr isotopic evolution in a dynamic earth: Reviews of Geophysics, v. 6, p. 175–199.

Armstrong, R. L., 1972, Low-angle (denudation) faults, hinterland of the Sevier orogenic belt, eastern Nevada and western Utah: Geological Society of America Bulletin, v. 83, p. 1729–1754.

Armstrong, R. L., 1975, Precambrian (1500 m.y. old) rocks of central Idaho—The Salmon River arch and its role in Cordilleran sedimentation and tectonics: American Journal of Science, v. 275A, p. 437–467.

Armstrong, R. L., 1981, Radiogenic isotopes: The case for crustal recycling on a near-steady-state no-continental-growth Earth: Royal Society of London Philosophical Transactions, v. 301, p. 443–472.

Armstrong, R. L., 1991, The persistent myth of crustal growth: Australian Journal of Earth Sciences, v. 38, p. 613–630.

Armstrong, R. L., and Ward, P., 1991, Evolving geographic patterns of Cenozoic magmatism in the North American Cordillera: The temporal and spatial association of magmatism and metamorphic core complexes: Journal of Geophysical Research, v. 96, p. 13,201–13,224.

Armstrong, R. L., and Ward, P., 1993, Late Triassic to earliest Eocene magmatism in the North American Cordillera: Implications for the Western Interior Basin, in Caldwell, W. G. E., and Kauffman, E. G., eds., Evolution of the Western Interior Basin: Geological Association of Canada Special Paper 39, p. 49–72.

Block, L., and Royden, L. H., 1990, Core complex geometries and regional scale flow in the lower lithosphere: Tectonics, v. 9, p. 557–567.

Buck, W. R., 1988, Flexural rotation of normal faults: Tectonics, v. 7, p. 959–973.

Burg, J. P., and Chen, G. M., 1984, Tectonics and structural zonation of southern Tibet, China: Nature, v. 311, p. 219–223.

Cann, J.R., Blackman, D. K., Smith, D. K., McAllister, E., Janssen, B., Mello, S., Avgerinos, E., Pascoe, A. R., and Escartin, J., 1997, Corrugated slip surfaces formed at ridge-transform intersections on the Mid-Atlantic Ridge: Nature, v. 385, p. 329–332.

Cannat, M., Bideau, D., and Bougault, H., 1992, Serpentinized peridotites and gabbros in the Mid-Atlantic Ridge axial valley at 15°37'N and 16°52'N: Earth and Planetary Science Letters, v. 109, p. 87–106.

Compton, R. R., Todd, V. R., Zartman, R. E., and Naeser, C. W., 1977, Oligocene and Miocene metamorphism, folding, and low-angle faulting in northwestern Utah: Geological Society of America Bulletin, v. 88, p. 1237–1250.

Coney, P. J., 1980a, Introduction, in Crittenden, M. D., Jr., Coney, P. J., and Davis, G. H., eds., Cordilleran metamorphic core complexes: Geological Society of America Memoir 153, p. 3–6.

Coney, P. J., 1980b, Cordilleran metamorphic core complexes: An overview, in Crittenden, M. D., Jr., Coney, P. J., and Davis, G. H., eds., Cordilleran metamorphic core complexes: Geological Society of America Memoir 153, p. 7–31.

Coney, P. J., and Harms, T. A., 1984, Cordilleran metamorphic core complexes: Cenozoic extensional relicts of Mesozoic compression: Geology, v. 12, p. 550–554.

Crittenden, M. D., Jr., Coney, P. J., and Davis, G. H., eds., 1980, Cordilleran metamorphic core complexes: Geological Society of America Memoir 153, 490 p.

Dahlen, F. A., 1984, Noncohesive critical Coulomb wedges: An exact solution: Journal of Geophysical Research, v. 89, p. 10,125–10,133.

Davis, G. A., Lister, G. S., and Reynolds, S. J., 1986, Structural evolution of the Whipple and South Mountains shear zones, southwestern United States: Geology, v. 14, p. 7–10.

Davis, G. H., 1975, Gravity-induced folding off a gneiss dome complex, Rincon Mountains, Arizona: Geological Society of America Bulletin, v. 86, p. 979–990.

Doblas, M., and Oyarzun, R., 1989, Neogene extensional collapse in the western Mediterranean (Betic-Rif Alpine orogenic belt): Implications for the genesis of the Gibraltar Arc and magmatic activity: Geology, v. 17, p. 430–433.

Drewes, H., 1977, Geologic map and sections of the Rincon Valley quadrangle, Pima County, Arizona: U.S. Geological Survey Miscellaneous Investigations Series Map I-997, scale 1:48,000.

Frost, E. G., 1981, Structural style of detachment faulting in the Whipple Mountains, California and Buckskin Mountains, Arizona: Arizona Geological Society Digest, v. 13, p. 25–29.

Hamilton, W., and Myers, W. B., 1966, Cenozoic tectonics of the western United States: Reviews of Geophysics, v. 4, p. 509–549.

Hill, E. J., Baldwin, S. L., and Lister, G. S., 1992, Unroofing of active metamorphic core complexes in the D'Entrecasteaux Islands, Papua New Guinea: Geology, v. 20, p. 907–910.

Hose, R. K., and Danes, Z. F., 1973, Development of the late Mesozoic to early Cenozoic structures of the eastern Great Basin, in DeJong, K. A., and Scholten, R., eds., Gravity and tectonics: New York, John Wiley and Sons, p. 429–442.

Howard, K. A., Stone, P., Pernokas, M. A., and Marvin, R. F., 1982, Geologic and geochronologic reconnaissance of the Turtle Mountains area, California: West border of the Whipple Mountains detachment terrane, in Frost, E. G., and Martin, D. L., eds., Mesozoic-Cenozoic tectonic evolution of the Colorado River region, California, Arizona, and Nevada: San Diego, Cordilleran Publishers, p. 377–392.

Hyndman, D. W., 1980, Bitterroot dome–Sapphire tectonic block, an example of a plutonic-core gneiss-dome complex with its detached suprastructure, in Crittenden, M. D., Jr., Coney, P. J., and Davis, G. H., eds., Cordilleran metamorphic core complexes: Geological Society of America Memoir 153, p. 427–443.

Jackson, J. A., and White, N. J., 1989, Normal faulting in the upper continental crust: Observations from regions of active extension: Journal of Structural Geology, v. 11, p. 15–36.

Jaroslow, G. E., Hirth, G., and Dick, H. J. B., 1996, Abyssal peridotite mylonites: Implications for grain-size sensitive flow and strain localization in the oceanic lithosphere: Tectonophysics, v. 256, p. 17–37.

Jolivet, L., Dubois, R., Fournier, M., Goffe, B., Michard, A., and Jourdan, C., 1990, Ductile extension in alpine Corsica: Geology, v. 18, p. 1007–1010.

Lagabrielle, Y., and Cannat, M., 1990, Alpine Jurassic ophiolites resemble the modern central Atlantic basement: Geology, v. 18, p. 319–322.

Lee, D. E., Marvin, R. F., Stern, T. W., and Peterman, Z. E., 1970, Modification of potassium-argon ages by Tertiary thrusting in the Snake Range, White

Pine County, Nevada: U.S. Geological Survey Professional Paper 700-D, p. 92–102.

Lister, G. S., and Snoke, A. W., 1984, S-C mylonites: Journal of Structural Geology, v. 6, p. 617–638.

Lister, G. S., Banga, G., and Feenstra, A., 1984, Metamorphic core complexes of Cordilleran type in the Cyclades, Aegean Sea, Greece: Geology, v. 12, p. 221–225.

Lister, G. S., Etheridge, M. A., and Symonds, P. A., 1986, Detachment faulting and the evolution of passive continental margins: Geology, v. 14, p. 246–250.

Mezger, K., van der Pluijm, B. A., Essene, E. J., and Halliday, A. N., 1991, Syn-orogenic collapse: A perspective from the middle crust, the Proterozoic Grenville orogen: Science, v. 254, p. 695–698.

Misch, P., 1960, Regional structural reconnaissance in central-northeast Nevada and some adjacent areas: Observations and interpretations, in Boettcher, J. W., and Sloan, W. W., Jr., eds., Guidebook to the geology of east-central Nevada: Intermountain Association of Petroleum Geologists, 11th Annual Field Conference Guidebook, p. 17–42.

Molnar, P., and Lyon-Caen, H., 1988, Some simple physical aspects of the support, structure, and evolution of mountain belts, in Clark, S. P., Jr., Burchfiel, B. C., and Suppe, J., eds., Processes in continental lithosphere deformation: Geological Society of America Special Paper 218, p. 179–207.

Moores, E. M., Scott, R. B., and Lumsden, W. W., 1968, Tertiary tectonics of the White Pine–Grant Range region, east-central Nevada, and some regional implications: Geological Society of America Bulletin, v. 79, p. 1703–1726.

Mpodozis, C., and Allmendinger, R. W., 1993, Extensional tectonics, Cretaceous northern Andes, northern Chile (27°S): Geological Society of America Bulletin, v. 105, p. 1462–1477.

Parrish, R. R., 1992, Memorial to Richard Lee Armstrong: Boulder, Colorado, Geological Society of America, p. 63–67.

Proffett, J. M., Jr., 1977, Cenozoic geology of the Yerington district, Nevada, and implications for the nature of Basin and Range faulting: Geological Society of America Bulletin, v. 88, p. 247–266.

Rehrig, W. A., and Reynolds, S. J., 1980, Geologic and geochronologic reconnaissance of a northwest-trending zone of metamorphic core complexes in southern and western Arizona, in Crittenden, M. D., Jr., Coney, P. J., and Davis, G. H., eds., Cordilleran metamorphic core complexes: Geological Society of America Memoir 153, p. 131–157.

Scott, R. J., and Lister, G. S., 1992, Detachment faults: Evidence for a low-angle origin: Geology, v. 20, p. 833–836.

Selverstone, J., 1988, Evidence for east-west crustal extension in the eastern Alps: Implications for the unroofing history of the Tauren window: Tectonics, v. 7, p. 87–105.

Seranne, M., and Seguret, M., 1987, The Devonian basins of western Norway: Tectonics and kinematics of an extending crust, in Coward, M. P., Dewey, J. F., and Hancock, P. L., eds., Continental extensional tectonics: Geological Society of London Special Publication 28, p. 537–548.

Simpson, C., and Schmid, S. M., 1983, An evaluation of criteria to deduce the sense of movement in sheared rocks: Geological Society of America Bulletin, v. 94, p. 1281–1288.

Spencer, J. E., 1984, Role of tectonic denudation in warping and uplift of low-angle normal faults: Geology, v. 12, p. 95–98.

Spencer, J. E., and Reynolds, S. J., 1990, Relationship between Mesozoic and Cenozoic tectonic features in west-central Arizona and adjacent southeastern California: Journal of Geophysical Research, v. 95, p. 539–555.

Spencer, J. E., and Reynolds, S. J., 1991, Tectonics of mid-Tertiary extension along a transect through west-central Arizona: Tectonics, v. 10, p. 1204–1221.

Tucholke, B. E., and Lin, J., 1994, A geological model for the structure of ridge segments in slow spreading ocean crust: Journal of Geophysical Research, v. 99, p. 11,937–11,958.

Tucholke, B. E., Lin, J., and Kleinrock, M. C., 1998, Megamullions and mullion structure defining oceanic metamorphic core complexes on the Mid-Atlantic Ridge: Journal of Geophysical Research, v. 103, p. 9857–9866.

Wernicke, B., 1981, Low-angle normal faults in the Basin and Range Province: Nappe tectonics in an extending orogen: Nature, v. 291, p. 645–648.

Wernicke, B., 1985, Uniform-sense normal simple shear of the continental lithosphere: Canadian Journal of Earth Sciences, v. 22, p. 108–125.

Wernicke, B., 1990, The fluid crustal layer and its implications for continental dynamics, in Salisbury, M., and Fountain, D. M., eds., Exposed cross sections of the continental crust: Dordrecht, Holland, Kluwer Academic Publishers, p. 509–544.

Wernicke, B., 1992, Cenozoic extensional tectonics of the U.S. Cordillera, in Burchfiel, B. C., Lipman, P. W., and Zoback, M. L., eds., The Cordilleran orogen: Coterminous U.S.: Boulder, Colorado, Geological Society of America, Geology of North America, v. G-3, p. 553–581.

Wernicke, B., 1995, Low-angle normal faults and seismicity: A review: Journal of Geophysical Research, v. 100, p. 20,159–20,174.

Wernicke, B., and Axen, G. J., 1988, On the role of isostasy in the evolution of normal fault systems: Geology, v. 16, p. 848–851.

Wills, S., and Buck, R. W., 1997, Stress-field rotation and rooted detachment faults: A Coulomb failure analysis: Journal of Geophysical Research, v. 102, p. 20,503–20,514.

Xiao, H.-B., Dahlen, F. A., and Suppe, J., 1991, Mechanics of extensional wedges: Journal of Geophysical Research, v. 96, p. 10,301–10,318.

Zheng, Y., Wang, Y., Liu, R., and Shao, J., 1988, Sliding-thrusting tectonics caused by thermal uplift in the Yunmeng Mountains, Beijing, China: Journal of Structural Geology, v. 10, p. 135–144.

MANUSCRIPT ACCEPTED BY THE SOCIETY NOVEMBER 23, 1998.

Geological Society of America
Special Paper 338
1999

# Chapter 18
# MOJAVE—BASIN AND RANGE BOUNDARY

**J. Douglas Walker**
*Department of Geology*
*University of Kansas*
*Lawrence, Kansas 66045*
*E-mail: jdwalker@kuhub.cc.ukans.edu*

**Allen F. Glazner**
*Department of Geology*
*University of North Carolina*
*Chapel Hill, North Carolina 27599*
*E-mail: afg@unc.edu*

## GREGORY A. DAVIS

The geologic career of Greg Davis has touched almost every area and aspect of Cordilleran geology. Davis received his undergraduate degree at Stanford University (where he attended field camp with Burchfiel) and his doctorate from the University of California, Berkeley. His dissertation work was a pioneer study of the south-central Klamath Mountains. From that basis Davis made one of the original and seminal post-plate tectonic correlations of Klamath and Sierra Nevada rocks. Davis has spent his academic career at the University of Southern California. He is an energetic, enthusiastic field geologist, who has written dozens of well-documented, thought-provoking studies of Cordilleran geology, from detailed studies of areas in California, British Columbia, the Cordillera in general, and most recently in China.

## B. CLARK BURCHFIEL

Clark Burchfiel has had a long and distinguished career in the geosciences and in the Cordilleran section of the GSA. Although his research focuses on the tectonic development of many areas of the world (most recently Asia and Europe), his earliest studies were of the U.S. Cordillera. Burchfiel received his bachelor and master of science degrees from Stanford University and his doctoral degree from Yale University. His dissertation work was on an extended area of southern Nevada and addressed problems of the Las Vegas Valley shear zone. He was at Rice University for many years and has been at the Massachusetts Institute of Technology since 1977, where he currently is Schlumberger Professor of Geology. An accomplished linguist fluent in many languages, Burchfiel has received many honors, including the Structural Geology and Tectonic Division's Career Contribution Award, and election as Fellow of the National Academy of Sciences.

Clark Burchfiel and Greg Davis, August 1966, after hike across Grand Canyon.

## SIGNIFICANCE OF PAPER

Perhaps the most significant structure in the California deserts as a whole is the Garlock fault. The Garlock fault forms the principal physiographic and geologic boundary between the Basin and Range province and the Mojave Desert. In addition, it has been clear for some time that the fault has significant strike-slip displacement (64 km or more; Smith, 1962). However, how the fault fits into the overall geologic development and ongoing activity of the deserts was unclear for many decades.

One of the first papers to address Garlock complexity and present a model to reconcile Basin and Range with Mojave Desert Cenozoic tectonics was "Garlock fault: An intracontinental transform structure, southern California" by Gregory A. Davis and B. Clark Burchfiel, published in 1973. The main focus of this work was the nature of the Garlock fault as a unifying tectonic feature in the California deserts. Davis and Burchfiel interpreted the Garlock fault as a major intracontinental strike-slip fault that separates the actively extending Basin and Range province from the unextending Mojave Desert region. This led them to the interpretation that the Garlock fault accommodated the strain gradient between extensional deformation in the Basin and Range and nonextensional strike-slip faulting in the Mojave Desert, analogous to the tectonics of an oceanic transform fault.

J. Douglas Walker

GREGORY A. DAVIS *Department of Geological Sciences, University of Southern California, Los Angeles, California 90007*
B. C. BURCHFIEL *Department of Geology, Rice University, Houston, Texas 77001*

# Garlock Fault: An Intracontinental Transform Structure, Southern California

## ABSTRACT

The northeast- to east-striking Garlock fault of southern California is a major strike-slip fault with a left-lateral displacement of at least 48 to 64 km. It is also an important physiographic boundary since it separates along its length the Tehachapi-Sierra Nevada and Basin and Range provinces of pronounced topography to the north from the Mojave Desert block of more subdued topography to the south. Previous authors have considered the 260-km-long fault to be terminated at its western and eastern ends by the northwest-striking San Andreas and Death Valley fault zones, respectively.

We interpret the Garlock fault as an intracontinental transform structure which separates a northern crustal block distended by late Cenozoic basin and range faulting from a southern, Mojave block much less affected by dilational tectonics. Earlier ideas that the Garlock fault terminates eastward at the Death Valley fault zone appear to us to be in error, although right-lateral offsetting of the Garlock along that zone by about 8 km is necessary. Displacement along the Garlock fault must increase westward from its eastern terminus, a point of zero offset now buried beneath alluvial deposits in Kingston Wash to the east of the Death Valley fault zone. Much of the displacement on the Garlock fault due to east-west components of basin and range faulting appears to have been derived from block faulting in the area between Death Valley and the Nopah Range. Westward displacement of the crustal block north of the Garlock by extensional tectonics within it totals 48 to 60 km in the Spangler Hills–Slate Range area and probably continues to increase westward at least as far as the eastern frontal fault of the Sierra Nevada. Westward shifting of the northern block of the Garlock has probably contributed to the westward bending or deflection of the San Andreas fault where the two faults meet.

Many earlier workers have considered that the left-lateral Garlock fault is conjugate to the right-lateral San Andreas fault in a regional strain pattern of north-south shortening and east-west extension, the latter expressed in part as an eastward displacement of the Mojave block away from the junction of the San Andreas and Garlock faults. In contrast, we regard the origin of the Garlock fault as being directly related to the extensional origin of the Basin and Range province in areas north of the Garlock. Recent models for development of that province related to intracontinental spreading east of an east-dipping subduction zone along the Cenozoic margin of western North America may best account for the differential east-west extension which has occurred in the crustal blocks to the north and south of the Garlock fault.

Other possible examples of intracontinental transform faults in the southwestern Cordillera with geometries similar to that of the Garlock fault include the left-lateral Santa Cruz–Sierra Madre fault zone along the southern margin of the western Transverse Ranges, and the right-lateral Las Vegas shear zone and Agua Blanca fault of Baja California.

## INTRODUCTION

The Garlock fault zone of southern California strikes northeastward to eastward and has been recognized as a major structural element of this region for nearly 50 years. It is equally important as a physiographic boundary, for it separates along its length the Tehachapi–Sierra Nevada and Basin and Range provinces

Geological Society of America Bulletin, v. 84, p. 1407–1422, 5 figs., April 1973

1407

Digital shaded-relief map of the western United States. From Thelin and Pike, 1991, Landforms of the Conterminous United States–a digital shaded-relief portrayal. U.S. Geological Survey Map I-2206; reprinted in GSA Today, vol. 1, no. 11, Nov. 1991, p. 252-3.

of pronounced topography to the north from the Mojave Desert block of more subdued topography to the south. All previous authors have considered the 260-km-long left-lateral fault to be terminated at its western and eastern ends by the northwest-striking San Andreas and Death Valley fault zones, respectively (Fig. 1). This fault geometry, although somewhat unusual, posed no serious problems until Smith (1962) offered evidence for a 64-km left-lateral displacement along the Garlock fault based on an offset late Mesozoic dike swarm. The absence of an equivalent total offset of the Garlock's bounding fault zones—the San Andreas and Death Valley zones—posed a serious geometric problem, especially in light of Smith's estimate of displacement along the Garlock as being equal to one-quarter of its entire length.

Although the San Andreas fault zone swings westward near its junction with the Garlock fault, it is the Garlock–Death Valley fault zone junction that presents the most serious geometric problems. Here, offset features suggesting up to 64 km of left-lateral displacement along the Garlock fault are found on the south side of the Garlock within 16 km of its junction with the right-lateral Death Valley fault zone, thus producing radically different breadths of terrane on the two sides of the fault between these offset features and the Death Valley fault zone. Smith recognized the geometric problems posed by the inferred geometry of the Garlock and its bounding faults, but he concluded that "the evidence that the two dike swarms document 40 mi of lateral displacement is stronger than the evidence that the tectonic pattern at the ends of the fault precludes it" (Smith, 1962, p. 103). We agree, and offer in this paper an explanation for the geometry and origin of the Garlock fault which is compatible with regional geologic relations in the southern Death Valley–eastern Mojave Desert region.

## EARLY STUDIES

The Garlock fault was discovered and named by Hess (1910) during studies in the Randsburg area of the north-central Mojave Desert. Hulin (1925) mapped the fault in his study of the Randsburg quadrangle; he recognized displacement of Quaternary and older rock units along it, and concluded that total displacements of left-lateral nature amounted to approximately 8 km. Hulin (1925) and Noble (1926) both regarded the Garlock fault as hav-

ing major regional significance, although the important paper by Hill and Dibblee (1953) first gave the Garlock its present level of recognition. Hill and Dibblee suggested that the left-lateral Garlock and Big Pine faults had once been continuous before being disrupted along the San Andreas, and that displacements along these two faults have produced the westward bending of the San Andreas near its intersection with them (Fig. 1). They interpreted the Garlock and Big Pine faults as left-lateral shears conjugate in a regional strain pattern to right-lateral faults of the San Andreas system. Others, for example, Kupfer (1968), have also considered the Garlock and San Andreas faults as conjugate shears and have attempted regional strain and stress analyses for southern California on this basis.

The idea that the Garlock and San Andreas faults are regionally related has led some geologists (for example, Hill and Dibblee, 1953; Bucher, 1955; Hewett, 1955, Pl. 1; D. L. Anderson, 1971) to conclude that the wedge-shaped Mojave Desert structural block, bounded on the west by the northwest-striking, right-lateral San Andreas fault and by the northeast-striking, left-lateral Garlock fault (Fig. 1), must be undergoing eastward displacement away from the intersection of the two faults. This conclusion raises serious structural problems at the previously inferred eastern end of the Garlock fault where the presumably "active" Mojave block is bounded by the Garlock fault to the north and the Death Valley fault zone to the east. These problems are discussed in detail below.

## DISPLACEMENTS ACROSS THE GARLOCK FAULT

Smith's study (1962) of the northwest-trending dike swarms in Mesozoic granitic terranes north and south of the Garlock fault zone led him to conclude that approximately 64 km of offset has occurred along the fault since the presumed late Mesozoic emplacement of the dikes (Fig. 1, AA'). This displacement, although equivalent to one-quarter of the length of the known fault trace, has been strongly supported by subsequent field studies. Smith and others (1968) described a previously unrecognized thrust fault in the southern Slate Range to the north of the Garlock fault (Fig. 1, B). Here, cataclasized Mesozoic granitic rocks and gneisses of Precambrian age overlie Mesozoic metavolcanic and granitic rocks along

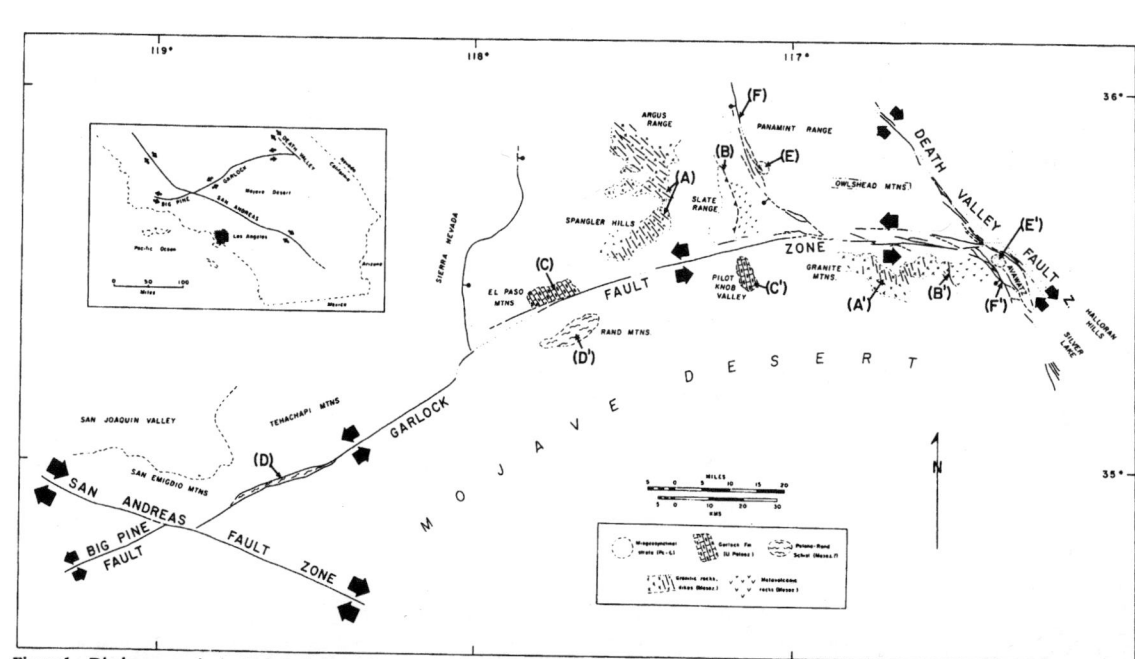

Figure 1. Displacement criteria, Garlock fault, southern California. Displaced features, for example AA', BB', are explained and referenced in text. Base map from Geologic Map of California, scale, 1/250,000 :Los Angeles sheet (Jennings and Strand, 1969), Bakersfield sheet (Smith, 1964), and Trona sheet (Jennings and others, 1962). Inset: Location map of Garlock, Big Pine, and San Andreas faults in southern California.

the west-dipping (30° to 40°) Layton Well thrust. In reconnaissance studies south of the Garlock fault in the eastern Granite Mountains (Fort Irwin Military Reservation) and only 13 km west of its junction with the Death Valley fault zone, we have found a major thrust fault which we regard as being part of the same thrust (Fig. 1, B'). Here, as in the Slate Range, a section of sheared and phyllonitized crystalline rocks several thousands of feet thick overlies Mesozoic metavolcanic rocks along a west-dipping (40°) thrust contact.[1] This thrust lies some 56 to 64 km east of the Layton Well thrust in the Slate Range, but this separation distance may not represent true strike-slip offset. Because of the moderate dip of the thrust, dip-slip components of displacement along the Garlock might significantly affect the horizontal separation distance observed between the two thrust segments.

Smith and Ketner (1970) present evidence for a 48- to 64-km left-lateral offset of the lithologically distinctive, and at least in part Permian, Garlock Formation from north of the fault in the El Paso Mountains (Fig. 1, C) to the south of the fault in Pilot Knob Valley (Fig. 1, C'). The correlative strata in both areas strike north to northwest and because they dip steeply eastward they are well oriented for definition of strike-slip separation along this portion of the Garlock fault.

Other possible but less convincing large offsets of stratigraphic units or sequences across the Garlock fault have been inferred. Dibblee (1967a, p. 115) suggested that low-grade metamorphic rocks of the Rand Schist near Randsburg (Fig. 1, D') and of the "Pelona Schist" within the Garlock fault zone to the west (Fig. 1, D) may once have been adjacent; they are now 72 km apart. Finally, Jahns and others believe that sequences of late Precambrian and early Paleozoic miogeosynclinal strata in the Avawatz Mountains south of the Garlock fault (Fig. 1, E') and in the southern Panamint Range to the north of the fault (Fig. 1, E) may be offset some 65 km from each other (Jahns and others, 1971; 1968, oral commun.).

In summary, a left-lateral offset of about 56 km of pre-Cenozoic structures and rocks across the central part of the Garlock fault zone appears well established. Offset elements on the south side of the fault can be traced to within 13 km of the presumed termination of the Garlock fault at its junction with the northwest-striking Death Valley fault zone. Near this junction the Avawatz and easternmost Granite Mountains correlate geologically with displaced terranes in the westernmost Panamint Range and the Slate Range. But where to the south of the Garlock fault, if the Garlock terminates at the Death Valley fault zone, is the offset equivalent of the 32- to 40-km-wide Owlshead Mountains plutonic terrane (shaded area, Fig. 1) which lies between the southern Panamint Range and the Death Valley fault zone? This geologic and geometric enigma is the heart of the Garlock problem.

## RELATIONS BETWEEN THE GARLOCK AND DEATH VALLEY FAULT ZONES

Smith (1962) reviewed various attempts to resolve the geometric problems at the junction of the Garlock and Death Valley fault zones. Hewett once suggested (1955) that at its eastern end the Garlock fault turns southward and becomes a thrust fault dipping westward under the Avawatz Mountains. The Mojave block to the south of the Garlock was thus viewed by Hewett as being a thrust plate uplifted with respect to northern provinces and displaced eastward intermittently through Tertiary and into Quaternary time. The presence of a Garlock "thrust" fault beneath the Avawatz Mountains has since been discounted (Jahns and Wright, 1960).

Three general possibilities remain for resolving the problems at the Garlock–Death Valley fault zone junction:

1. The Garlock fault terminates at the right-lateral Death Valley fault zone and the 48- to 64-km displacement along the Garlock is accomodated in the junction area.

2. The Garlock fault once continued east of the junction, but its eastward extension has been offset in a right-lateral sense along the younger, northwest-striking Death Valley fault zone and displaced many miles to the south.

3. The Garlock fault crosses the Death Valley fault zone in the junction area and continues eastward past it, despite conclusions to the contrary by other geologists.

Hypotheses pertinent to the first possibility have been raised by several authors. Grose (1959) and Jahns and Wright (1960) suggested that great vertical movements coupled with strike-slip displacements in the junction area may have accommodated the merger of the Garlock and Death Valley fault zones. These suggestions were made, however, prior to Smith's recognition of tens of kilometers of displacement along the Garlock and are no longer compatible with local geology in light of such large displacements. The junction area of the Garlock and Death Valley fault zones is characterized by many fault splays and an apparent southward turning of some east-west faults in the Garlock zone, for example, the Avawatz Mountain block which bounds the Avawatz Mountain block on its western side (Fig. 1, F'). Jahns and others (1971) suggest that the Arrastre Spring fault "may well be the major easterly expression of the Garlock zone." We believe that local structural relations do not support this contention. The Garlock fault would have to curve through an angle of 50° to become the Arrastre Spring fault. Large left-lateral displacement along a curvilinear Garlock fault would necessitate rotations of 50° of structural elements. Smith (1962) suggested that the dike swarm south of the Garlock fault in the Granite Mountains has been rotated in a clockwise direction by 20° to 25° from its offset extension in the Spangler Hills north of the Garlock. However, several kilometers to the east of the dike swarms the northwest strike of the Layton Well thrust in the Slate Range and its offset extension in the eastern Granite Mountains show no discernible relative rotation, even though the latter locality is within 8 km of the junction of the Garlock and Arrastre Spring faults (Fig. 1, B' and F').[2]

The nature of the Garlock fault as a boundary between diverse structural and physiographic provinces has prompted past attempts to explain some or all of its displacement by calling upon differential extensional behavior in crustal blocks to the north and south of it. Eaton (1932) was probably the first to develop this line of reasoning. He concluded that the Sierra Nevada had shifted westward in post-Miocene time, with a maximum displacement of up to 96 km at its southern end. Shifting of the Sierran block was thought to have been accompanied by extensional faulting through-

out the Great Basin province to the east (Eaton, 1932). Eaton also believed that the westward shifting (sliding) of the Sierra Nevada crustal block was responsible for the westward bending of a once straighter San Andreas fault and for the compressive, post-Miocene deformation in the southern Coast Ranges to the west of the San Andreas (1932, Fig. 4).

More recently, Hamilton and Myers (1966) attempted to explain some of the displacement along the Garlock fault by noting the presence of extensional (basin and range) structure north of the fault and west of Death Valley and its absence in the Mojave structural block to the south:

The Garlock and Death Valley faults merge and mutually deflect each other (Jahns and Wright, 1960), but the Garlock does not continue east beyond the Death Valley zone, nor does it appear farther southeast in a position offset along the Death Valley fault. South of the Garlock is the Mojave Desert block of generally old, low mountain blocks, largely buried by basin deposits in the west. North of the Garlock fault is a region of high and exceedingly active fault-block ranges, the Sierra Nevada, Argus, Panamint, and Black Mountains. The Garlock thus has the position and relative offset required by the explanation that the block faulting to the north is due to crustal extension, because the fault marks the south end of the currently stretching mass, although the total amount of lateral displacement is too large to be accounted for by this mechanism alone. The aggregate topographic height of the major range-front scarps north of the Garlock fault is 7 or 8 km; if the bedrock relief is twice the topographic relief, at most 15 km of extension could have accompanied the

---

[1] The thrust lies to the northeast of Drinkwater Lake approximately along 116° 30' W. long. It is not the thrust fault depicted on the Trona Sheet of the Geologic Map of California (Jennings and others, 1962) to the north of Drinkwater Lake. Rocks designated as Mesozoic metavolcanic rocks immediately to the east of this mapped thrust are sheared and phyllonitized granitic rocks in the basal portion of the thrust plate we have described briefly above.

[2] The Arrastre Spring fault zone might be an offset equivalent of the Panamint Valley fault zone (Fig. 1, G) which separates the Slate and Panamint Ranges and appears to cross the Quail Mountains just north of the Garlock fault. In both areas, the faults separate a Mesozoic metavolcanic and plutonic terrane on the west from a late Precambrian–Paleozoic miogeosynclinal terrane intruded by Mesozoic plutonic rocks on the east. The eastward bend of the southern part of the Panamint Valley fault zone complements the westward bend of the northern part of the Arrastre Spring zone, and both bends could be attributed to left-lateral drag along the Garlock fault. If a correlation exists, dip-slip displacements along the Panamint–Arrastre Spring fault zone must have occurred both prior to inception of the Garlock fault *and* subsequently, since the present Panamint Valley and Arrastre Spring faults involve Pliocene and Quaternary sediments, and at least one of the two faults, the Panamint Valley, is still active.

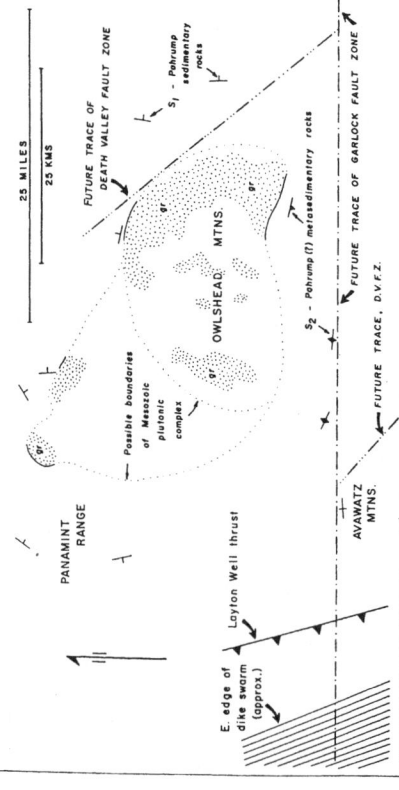

Figure 2. Reconstruction of possible pre-Garlock relations between Panamint, Owlshead, and Avawatz terranes assuming original continuity of the Mesozoic dike swarm and the Layton Well thrust across the Garlock fault and 56 km of subsequent left-lateral displacement along it. The reconstruction does not attempt to undo the distensional effects of Cenozoic normal faulting in the terrane north of the Garlock (see text).

formation of the present ranges. Strike-slip displacement due to this process would increase westward.

Although Hamilton and Myers thus concluded that block faulting to the north of the Garlock could account for 15 km or one-third of the total displacement along it, they were at a loss to account for the remainder. B. W. Troxel and L. A. Wright (1969, oral commun.) modified the Hamilton-Myers concept by inferring a different geometry for the range-front faults north of the Garlock. Troxel and Wright assume that the range-front faults become progressively flatter at depth before merging into a horizontal and deep-seated basal detachment surface beneath the block bounded on the south by the Garlock and on the east by Death Valley. This inferred block geometry could, they believe, allow a crustal distension north of the Garlock equivalent to the total 48- or 64-km displacement along it. They, too, suggest that the Garlock terminates at its intersection with the Death Valley fault zone on the northeastern side of the Avawatz Mountains. The principal drawback to their alternative, in our opinion, is that there are not enough normal faults in the terrane between the Slate Range and Death Valley to account for 48 to 64 km of crustal extension in a terrane now only 64 to 80 km wide.

To support this contention we have prepared a reconstruction of pre-Garlock relations between the Avawatz, Panamint, and Owlshead terranes by reversing 56 km of left-lateral slip along the Garlock and thus aligning the two portions of Smith's offset dike swarm and the Layton Well thrust and its eastern Granite Mountains equivalent (Fig. 2). Despite the alignment of these features it can be seen that the traces of the Death Valley fault zone east of the Avawatz and Owlshead Mountains are still separated by about 56 km. To bring these traces into alignment by undoing the extensional effects of basin and range faulting in the terrane north of the Garlock and west of the Death Valley fault zone requires that the breadth of the northern terrane be reduced by 300 percent. Normal faults do exist within this terrane (Fig. 3), but they are far too few and too limited in dip-slip displacement to have produced such an enormous dilation of the Slate Range to Death Valley region.

We therefore conclude that most of the Owlshead Mountains must originally have been situated east of the longitude of the Avawatz Mountains (as in Fig. 2). The westward shifting by extensional tectonics of the Owlshead terrane to its present position northwest of the Avawatz Mountains requires that the Garlock fault also extends east of the eastern edge of the present trace of the Death Valley fault zone (Fig. 2). The problem of finding the offset counterpart of the Owlshead Mountains south of the Garlock fault is complicated by the nature of the Owlshead area. The Owlshead Mountains are underlain largely by Mesozoic granitic rocks (Gastil and others, 1967) that comprise an apparently concordant plutonic complex at least 32 km wide in an east-west direction (Fig. 2). This complex lies entirely north of the Garlock fault and accordingly does not have a direct offset counterpart to the south, although metamorphic country rocks south of the plutonic complex have been cut by the Garlock. Slices of this country rock terrane, represented by marbles and other metamorphic rocks, may be present *within* the Garlock fault zone 13 to 19 km east of the southeastern Owlshead Mountains (Noble and Wright, 1954), although offset counterparts south of the zone and still farther east have not been found. The likely present location for most of the country rock terrane which once lay south of the Owlshead Mountains plutonic complex is east of the present Avawatz Mountains and Death Valley fault zone in the area now covered by Quaternary alluvial deposits of Silurian lake basin.

We thus regard the first general possibility for explaining geometric relations at the eastern end of the Garlock fault, that is, that the Garlock fault terminates at the Death Valley fault zone and never extended farther east, as geologically untenable. This then brings us to the second possibility—that the Garlock fault once extended east of the Death Valley fault zone, but that its eastern extension has been offset many miles by right-lateral strike-slip faulting along the younger Death Valley fault zone. Recent right-lateral displacements along this zone have been documented by Hill and Troxel (1966), but the amount of total displacement has been the subject of controversy. Hamilton and Myers (1966) hypothesized 48 km of right-lateral displacement on the Death Valley fault zone, although Wright and Troxel (1967) have presented geologic evidence favoring less than 8 km of offset along it. Our con- clusions from geologic reconnaissance along projections of the Death Valley fault zone south of its intersection with the Garlock are in accord with the estimates of Wright and Troxel. There seems to be no basis, for example, for postulating large strike-slip displacements between areas of outcrop along the southward projection of the Death Valley fault zone in the Silver Lake area (Fig. 1, Avawatz Mountains, the hills west of Silver Lake, and the Halloran Hills to the east).

Even if tens of kilometers of displacement have occurred along the Death Valley fault zone, no problems pertinent to the Garlock fault are resolved since we can find no offset Garlock fault tens of kilometers south of the junction. Mapping by Hewett (1956), ourselves (unpub. data), and others in the area southeast of the junction of the Garlock and Death Valley fault zones reveals a continuous pre-Cenozoic geologic terrane extending from the Silurian Hills south to Kelso Valley, north of the Providence Mountains (Fig. 3). A left-lateral fault probably does lie beneath Kelso and Ivanpah Valleys, east of the south-projected trace of the Death Valley fault zone, but this fault is too far to the south (96 km), its trend too northerly (N. 55° E.), and its likely displacement (11 to 13 km) too small to be compatible with its being an offset eastern segment of the Garlock.

We, therefore, regard theories postulating a *major* right-lateral offset of the Garlock fault along the Death Valley fault zone as lacking both evidence for large lateral displacements on the latter, and an offset equivalent of the Garlock fault in areas to the south and east.

### EASTERN EXTENSION OF THE GARLOCK FAULT

As an alternative to the geologically unsatisfying hypotheses that the Garlock fault either ends at the Death Valley fault zone or that it has been displaced many miles along it, we propose that the Garlock fault must cross the Death Valley fault zone in the junction area and continue eastward past it. Minor right-lateral displacement of the Garlock fault by faults in the Death Valley fault zone is required by the geometry of bedrock exposures in the junction area, but certainly no more than 8 km. An offset of the Garlock fault by this amount would place its eastern continuation beneath the Quaternary deposits of Kingston Wash

(Fig. 3). Continuity of the pre-Cenozoic geologic terrane in areas east of Kingston Wash, including the Nopah and Kingston Ranges, Mesquite Mountains, Shadow Mountain, Clark Mountain and Mescal Ranges, and other areas to the south (Burchfiel and Davis, 1971), require the termination of the Garlock fault *beneath* the wash. Cambrian strata in the Nopah fault block continue southeastward into the uppermost (Winters Pass) thrust plate of the Clark Mountain thrust complex. Thus, the Nopah fault block is part of the continuous geologic terrane east of Kingston Wash. We conclude that the Garlock fault either terminates beneath the alluvial deposits of Kingston Wash along the south-projected trace of the Nopah Range frontal fault or that it extends several miles farther east to where other, less important, high-angle faults can be inferred to exist beneath Quaternary alluvium (Fig. 3).

The location of the Garlock fault beneath the western part of Kingston Wash is restricted by the areas of exposed bedrock present in this area. It must pass north of the Silurian Hills and south of the Salt Spring Hills.[3] Both areas are characterized by the distinctive southern facies of the Kingston Peak Formation of the Pahrump Group (Troxel, 1967, p. 35) which contrasts with other local occurrences of this formation. This similarity and the anomalous geosynclinal sections of the two areas—both sections are relatively thin and lack the late Precambrian Noonday Dolomite (Wright and Troxel, 1966)—suggest that the two areas may be offset from each other along the buried Garlock fault.

No trace of the Garlock fault can be seen on aerial photographs of the western part of Kingston Wash or to the south of the Salt Spring Hills, in contrast to its prominence on photographs in the Leach Lake playa area 32 km to the west. It is likely that the Garlock fault is inactive in the former areas, although the possibility exists that the absence of surficial faulting can be attributed to the high rates of Quaternary sedimentation in these areas and to a general parallelism in Kingston Wash of stream channels and the inferred position of the fault.

## GARLOCK FAULT AS A TRANSFORM STRUCTURE

Our proposed extension of the Garlock fault beneath Kingston Wash coincides with an eastward continuation across the Death Valley fault zone of the boundary between basin and range structure to the north and a Mojave-style area to the south where major range-front faults are lacking. In areas to the north of our inferred Garlock trace and *east* of Death Valley are numerous large, east-tilted fault blocks, including the southern Black Mountains, Saratoga-Ibex Hills, the Saddle Peak, Salt Spring, and Dublin Hills, and the Resting Spring and Nopah Ranges (Fig. 3).

In postulating an eastward extension of the Garlock fault and by designating an inferred terminus for it, we are in a position to reanalyze the geometry of terranes north and south of the eastern third of the Garlock fault. We have drawn a northwest-trending reference line parallel to regional structural trends and through the continuous geologic terrane east of Kingston Wash (Fig. 3). Measurements westward from this line to northwest-trending features which were also formerly continuous, but have since been offset by the Garlock, enable us to compare the present breadths of once-equivalent terranes now north and south of the fault, for example, the western limit of the vertical dike swarms mapped by Smith (1962) north and south of the Garlock fault. To the north of the Garlock fault, the east-west breadth of terrane between the western edge of the dike swarm and our eastern reference line of continuous terrane is approximately 160 km. The breadth of what we interpret to be the equivalent terrane south of the Garlock fault (since it has the same western and eastern boundaries) is approximately 96 km, or 64 km less. We believe that the increased breadth of the northern terrane and the left-lateral strike-slip displacement along the Garlock fault can be explained by adopting and expanding upon the basic ideas of Hamilton and Myers (1966) and Troxel and Wright (B. W. Troxel, 1969, oral commun.) that the northern terrane has been distended by basin and range faulting, whereas the southern (Mojave) terrane has not. By relating the Garlock fault to basin and range structure to the east of Death Valley (as well as to the west, in contrast to earlier

---

[3] Strata in the northern Salt Spring Hills strike nearly north, whereas strata in the southern Salt Spring Hills strike about N. 80° W. The progressive southward change in strike between the two areas may indicate drag effects along the left-lateral Garlock fault; it is not compatible with right-lateral displacements within the nearby Death Valley fault zone.

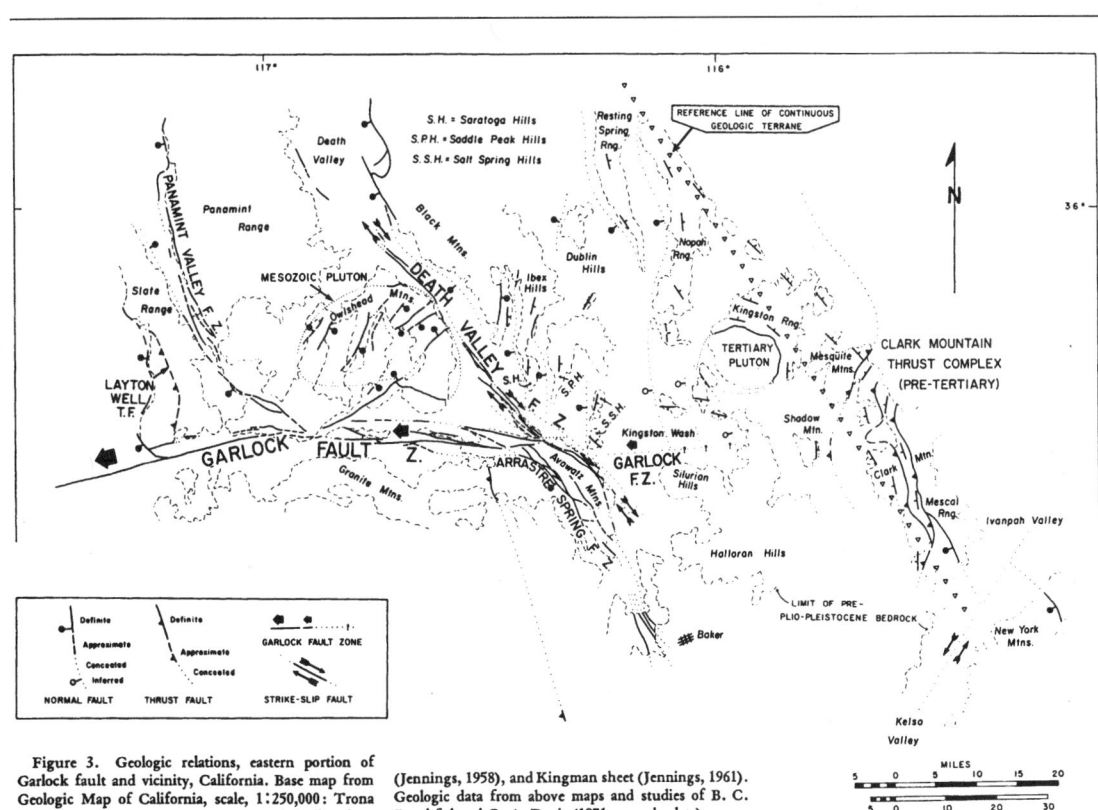

Figure 3. Geologic relations, eastern portion of Garlock fault and vicinity, California. Base map from Geologic Map of California, scale, 1:250,000: Trona sheet (Jennings and others, 1962), Death Valley sheet (Jennings, 1958), and Kingman sheet (Jennings, 1961). Geologic data from above maps and studies of B. C. Burchfiel and G. A. Davis (1971; unpub. data).

hypotheses, many more normal faults are available to produce the extra breadth of terrane north of the Garlock. Much of the displacement on the Garlock fault, in fact, appears to have been derived from block faulting in the area between Death Valley and the Nopah Range.

The stratigraphic throw on some of the range-front faults to the east of Death Valley is considerable. For example, the late Precambrian to Paleozoic miogeosynclinal section in the east-tilted Nopah Range fault block has a thickness in excess of 6,500 m (Hazzard, 1937). It is largely repeated by normal faulting in the adjacent Resting Spring Range to the west. Assuming, therefore, a stratigraphic throw of six and one-half km and a westward dip of 55° on the Nopah frontal fault, an east-west extension (the horizontal component of down-dip displacement) due to this fault alone is 3.7 to 4.5 km, depending upon the inferred eastward dip of the stratigraphic section prior to faulting (40° to 0°). Other fault configurations, for example, a flatter dip or flattening of this and other frontal faults at depth (Moore, 1960; Wright and Troxel, 1969; R. E. Anderson, 1971), can produce even greater amounts of extension.

According to our analysis, the terrane south of the Garlock between its eastern terminus and the western edge of the Granite Mountains dike swarm is 72 km wide, whereas the width of the equivalent terrane to the north of the Garlock is 138 km or nearly 100 percent wider. There is a precedent in other recent studies of basin and range structure for postulating such a large percentage of distension. Proffett (1971) has concluded that east-west extension due to normal faulting in the Yerington district of Nevada exceeds 100 percent. Even greater distension is proposed by R. E. Anderson (1971, unit NII; Fig. 2, AA') within normal faults in southern Nevada. Hamilton (1969, p. 2421) suggested that extension within the entire Basin and Range province might be several times the total amount of dip-slip displacement; his Figure 3 indicates total Cenozoic extensions of up to 50 percent. Nevertheless, our postulated 48- to 64-km extension within the terrane north of the Garlock fault between the Nopah Range and the Spangler Hills (Fig. 1) is comparable in magnitude to that inferred for the *entire* Great Basin province by other authors, for example, Thompson (1959, 48 km), and Stewart (1971, 72 km). Unfortunately we lack the necessary structural and stratigraphic data to make a fault-by-fault summation of east-west extension across the terrane in question, but we believe that our geometric analysis of relations at the eastern end of the Garlock fault and our estimate of total extension within the northern terrane are both sound and geometrically inescapable. The area between the Nopah Range and the Spangler Hills-Argus Range may, therefore, be one of extreme distension in comparison with the Great Basin as a whole.

The Garlock fault thus appears to be an intracontinental transform structure since it is a boundary or narrow zone of accommodation between one crustal block which has grown wider because of late Cenozoic normal faulting and another block, the Mojave, which has not (or which has not been dilated as much). The eastern end of the Garlock fault is thought to lie east of the Death Valley fault zone and to be a fixed point of zero strike-slip displacement. Extensional strain within the terrane north of the Garlock and west of this fixed point has been accommodated regionally by a westward deflection of the San Andreas fault and the crustal block to the west of it. These relations are shown diagrammatically in Figure 4. Like certain other types of transform faults (Wilson, 1965) the length of the Garlock is not constant, but should increase with time as distension continues in areas to the north of it, unless, as Ross (1970) has suggested, portions of the western end of the northern block (the San Emigdio Mountains) are sliced off along the San Andreas fault and transported northward along it.

Evidence for an increase in strike-slip displacement westward along the Garlock fault as required by the transform hypothesis proposed here is, on the basis of available data, meager and inconclusive.[4] There is, however, no evidence against this postulate. All geologic criteria which have been used for displacement estimates (Fig. 1, A-E) lie considerably west of the inferred eastern terminus of the Garlock, and all by their geologic nature are somewhat inconclusive when individually considered. A 48- to 64-km Cenozoic offset of the steeply dipping Garlock Formation (CC') and the vertical Mesozoic dike swarm (AA') does appear well established, but apparently similar displacement values of features farther east are

---

[4] Unfortunately, access to the Garlock fault zone between the Slate Range and the Avawatz Mountains and to the geologic terranes adjacent to it is severely restricted because this critical 80-km-long area lies within the closed boundaries of military reservations.

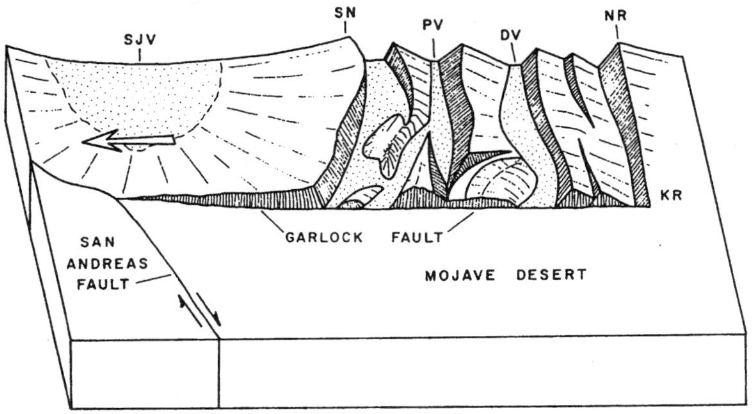

Figure 4. Northward diagrammatic view of Garlock fault, southern California, as a boundary between a northern, distended crustal block (Basin and Range province) and a southern, nondistended crustal block (Mojave Desert). Topographic relations are highly generalized and are shown only north of the Garlock fault. Geographic localities: SJV = San Joaquin Valley, SN = Sierra Nevada, PV = Panamint Valley, DV = Death Valley, NR = Nopah Range, KR = Kingston Range.

in question. As mentioned previously, the strike-slip offset of the Layton Well thrust from its southern continuation in the eastern Granite Mountains may be less than the 56 km measurable from Figure 1 (BB') because of the low dip of the thrusts and the possibility of dip-slip displacement along the Garlock. The 56- to 64-km sinistral separation of late Precambrian-Cambrian miogeosynclinal strata in the Avawatz Mountains and Panamint Range may be due in part to forceful shouldering aside of country rocks during emplacement of the extensive and apparently concordant Owlshead Mountains plutonic complex (see Fig. 2); under any circumstances, displacement estimates based on the present distribution of originally widespread stratigraphic units must be viewed with extreme caution.

## GEOPHYSICAL EVIDENCE

Our interpretation of the Garlock fault as a major structural boundary between crustal blocks with radically different late Cenozoic tectonic histories is supported by available, but limited, geophysical data. Although many geologists assign the Mojave Desert to the Basin and Range (Great Basin) province, a recent reinterpretation of seismic refraction profiles across the southwestern United States (Prodehl, 1970) indicates that the northern and central Mojave block lacks the anomalous low-velocity upper mantle so characteristic of the Basin and Range province (see, for example, Scholz and others, 1971). Prodehl interprets the Basin and Range province north of the Garlock fault as having a crust 31 to 36 km thick, a thickness defined by the depth to the strongest velocity gradient beneath this region, and as having an anomalous upper mantle with $P_n$ velocities less than 8.0 kmps. In contrast, the crust beneath the central and northern Mojave Desert is interpreted as being 29 to 31 km thick (Scholz and others, 1971, Fig. 11) and as being underlain by mantle with velocities equal to or greater than 8.0 kmps (Scholz and others, 1971, Table 2); these "normal" upper mantle velocities are found elsewhere in the western United States only beneath the central Rocky Mountains and the California Coast Ranges.

Our transform model for the Garlock fault accords well with conclusions regarding the Garlock based on regional seismicity. Allen and others (1965, p. 577) concluded from a study of the relations between recent seismicity and structure in "southern California that "the entire Garlock fault zone . . . seems to have served more as a boundary between seismic provinces than as a locus of seismic activity." They regard the western and northern Mojave Desert block between the San Andreas and Garlock faults as seismically and, therefore, tectonically, stable with respect to adjacent blocks, although, as they recognized, the time span of their investigation (1934 to 1963) may be inadequate to support such a conclusion.

Our tentative conclusion, expressed earlier, that that part of the Garlock fault to the east of Death Valley is inactive, whereas the Garlock to the west shows physiographic indications of Holocene (if not historic) activity, requires that Death Valley should also be a boundary between an area of active basin and range faulting to the west, and an area of inactivity between Death Valley and the Nopah Range to the east. This relation appears to be confirmed by the freshness of fault scarps cutting alluvial fans along the western fronts of the Panamint and Slate Ranges and the Black Mountains (Fig. 3), coupled with the absence of such scarps along mountain fronts farther east. A westward shift with time in late Cenozoic basin and range faulting has been documented by geochronologic studies of associated Great Basin volcanism (Armstrong, 1970; Fleck, 1970a; McKee, 1971; Scholz and others, 1971). Most workers agree that basin and range structure in this portion of the province began to develop in the Miocene, perhaps as long ago as 16 to 17 m.y. (Stewart, 1971; McKee, 1971).

The low level of current seismic activity along the Garlock fault to the west of Death Valley (Allen and others, 1965) may reflect (1) a recurrence interval for basin and range faulting in areas to the north of the Garlock which is longer than the short period of historic time within southern California, or (2) a changeover in the tectonic regimen of this portion of California from one of east-west extension to one of northwest-aligned right-lateral strain. The Death Valley area itself shows, for example, a present preponderance of right-lateral strike-slip motion along active or recently active faults (Hill and Troxel, 1966; Burchfiel and Stewart, 1967).

## FAULTS ANALOGOUS TO THE GARLOCK FAULT

It is likely that the Garlock fault is not a unique transform structure within the southwestern United States, but perhaps only one of many strike-slip faults which separate crustal blocks with differential extensional behavior. East-west transcurrent (strike-slip) faults which bound terranes of north-striking, closely spaced, low-angle normal faults have recently been described by R. E. Anderson (1971) in the Basin and Range province of southern Nevada. His studies provide, on a much smaller scale of development than the Garlock fault, interesting analogs to the relations between basin and range structure and the Garlock fault postulated here.

A possible example of a Garlock-type intracontinental transform fault elsewhere in southern California is the Santa Cruz–Malibu Coast–Santa Monica–Raymond–Sierra Madre fault zone which separates the offshore Continental Borderland and the onshore Peninsular Ranges from the east-trending Transverse Ranges to the north (Fig. 5). This fault zone bears a number of geometric similarities to the Garlock. Northwest-trending basin and range structure is highly developed in Borderland areas south of it and is absent in the Transverse Ranges to the north (Clements and Emery, 1947; Allen and others, 1965, Pl. 1; Moore, 1969, Pl. 13, Fig. 19). Left-lateral displacement along the portion of this zone which extends from the Pacific Ocean to the San Andreas fault is estimated by Yeats (1968) and Yerkes and Campbell (1971) as totalling 88 km between the Santa Monica Mountains to the north and the Los Angeles Basin and Santa Ana Mountains to the south.[5] If initial strike-slip displacement on this fault zone was related to extensional tectonics within the Continental Borderland, then the left-lateral nature of faulting suggests that fault displacement would increase eastward from an offshore point of zero displacement. This hypothesis is supported by the apparent lack of left-lateral offset of the continental margin west of Santa Cruz (see Moore, 1969, Pl. 13; compare the 5,000-ft and 10,000-ft isobaths; Fig. 5), and by the pronounced eastward deflection of strands of the San Andreas fault (North and South Branches, Mill Creek, Mission Creek, and Banning faults) near the intersection of the Santa Cruz–Sierra Madre fault zone with the San Andreas (Dibblee, 1968). It is interesting that the two major deflections in trend of the San Andreas fault in California occur in close proximity to

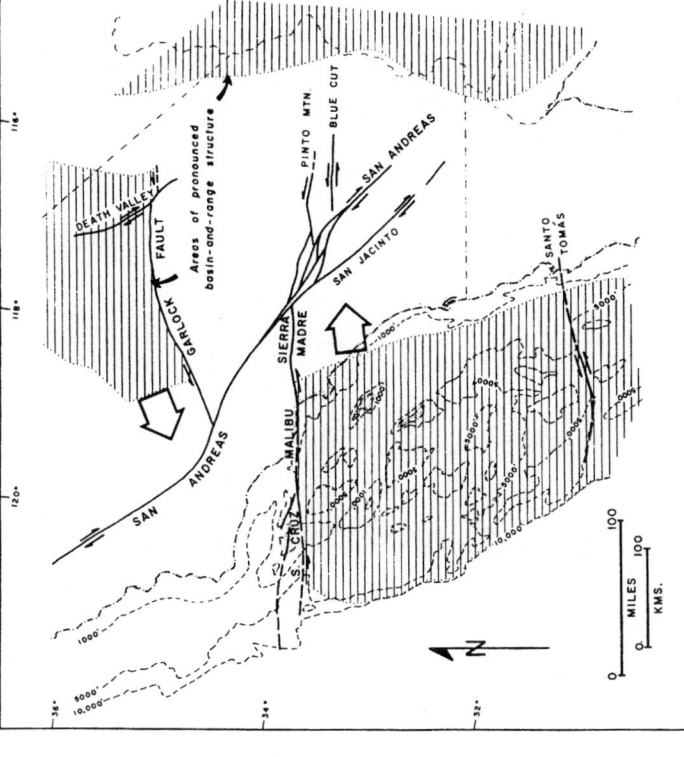

Figure 5. Areas of pronounced basin and range structure in the southern California region and the positions of possible left-lateral transform faults of Garlock-type. The offshore area south of the Santa Cruz–Sierra Madre fault zone comprises the California Borderland province. Bathymetry in this offshore area is from Moore (1969, Pl. 13).

junctions of that fault with the east-striking Garlock and Santa Cruz–Sierra Madre fault zones—both of which bound extensional geologic provinces of basin and range type (Fig. 5).[6]

[5] The San Jacinto fault does not exhibit the deflection shown by adjacent portions of the San Andreas near the eastern end of the Santa Cruz–Sierra Madre fault zone (Fig. 5). It can be argued that the San Jacinto fault, by far the most active member of the San Andreas fault system in southern California (Allen and others, 1965), is younger than the San Andreas of this region and that it represents a tectonic bypassing of the bent and possibly locked portion of the San Andreas fault in the southern and eastern Transverse Range area.

[6] Present fault displacements along the eastern half of the Santa Cruz–Sierra Madre fault zone, where it defines the southern margin of the uplifted San Gabriel Mountains crystalline block, are of reverse type (north side up).

Still other possible examples of intracontinental transform faults of Garlock type within the southern California area are the east-west Pinto Mountain, Blue Cut, and Santo Tomas faults (Fig. 5), all of left-lateral type and with displacements of 9 to 16 km (Dibblee, 1967b), 5 to 6 km (Hope, 1970), and 14 km (Krause, 1965), respectively. Each has geologic and geometric characteristics suggestive of an origin at least in part related to differential amounts of extension by normal faulting in terranes to the north and south of the faults.

The recent literature also provides several probable examples of right-lateral, intracontinental transform faults in the southwestern

Cordillera. Burchfiel and Stewart (1966) proposed that extensional "pull-apart" of the central segment of Death Valley is related to right-lateral strike-slip movement along the bounding Furnace Creek and Death Valley fault zones. Fleck (1970b) has suggested that the right-lateral Las Vegas shear zone is a transform fault separating two centers of late Cenozoic volcanism (and extension)—one to the north of the shear zone in the Nevada Test Site and the other to the east and south of the shear zone along the Colorado River. Similar ideas are expressed by Anderson and others (1972) on the basis of geologic and geochronologic studies at the southeastern end of the Las Vegas shear zone in the vicinity of Lake Mead. Finally, Hamilton (1971) has proposed a transform model for the right-lateral strike-slip Agua Blanca fault of northern Baja California, that is strikingly similar to the Garlock model proposed here. Hamilton believes that recent displacements along the fault of 5 to 10 km can be accounted for by extensional (normal) faulting south of the Agua Blanca fault and near and at its eastern end.

Thompson (1971) and Scholz and others (1971) have recently proposed that development of the Basin and Range province is due to intracontinental spreading east of an east-dipping subduction zone along the Cenozoic margin of western North America. Both have suggested that more active east-west distension within the Basin and Range province occurred with the changeover documented by Atwater (1970) from a compressive stress regimen of subduction type to the present strike-slip tectonic regimen exemplified by the San Andreas fault system. Scholz and others (1971) and Garfunkel (1966) have also concluded that the Garlock fault may delineate the southern boundary of a more rapidly spreading portion of the Basin and Range province than is found to the south, an idea that agrees well with the geometry and genesis of the Garlock fault as developed in this paper.

## ACKNOWLEDGMENTS

Discussions with Donald F. Palmer have contributed to the development of ideas expressed in this paper. Earlier versions of this manuscript were critically reviewed and significantly improved by Allen M. Bassett, John C. Crowell, Warren Hamilton, Thomas L. Henyey, and Bennie W. Troxel, to whom our sincere appreciation is extended. This study is the outgrowth of regional tectonic studies in the southern Cordillera supported by grants from the National Science Foundation—GA-21401 and GA-1562 (Davis), and GA-21375 and GA-1079 (Burchfiel).

## REFERENCES CITED

Allen, C. R., St. Amand, P., Richter, C. F., and Nordquist, J. M., 1965, Relationship between seismicity and geologic structure in the southern California region: Seismol. Soc. America Bull., v. 55, p. 753–797.
Anderson, D. L., 1971, The San Andreas fault: Sci. American, November, p. 52–67.
Anderson, R. E., 1971, Thin-skin distension in Tertiary rocks of southeastern Nevada: Geol. Soc. America Bull., v. 82, p. 43–58.
Anderson, R. E., Longwell, C. R., Armstrong, R. L., and Marvin, R. F., 1972, Significance of K-Ar ages of Tertiary rocks from the Lake Mead region, Nevada-Arizona: Geol. Soc. America Bull., v. 83, p. 273–288.
Armstrong, R. L., 1970, Geochronology of Tertiary igneous rocks, eastern Basin and Range province, western Utah, eastern Nevada, and vicinity, U.S.A.: Geochim. et Cosmochim. Acta, v. 34, p. 203–232.
Atwater, Tanya, 1970, Implications of plate tectonics for Cenozoic tectonic evolution of western North America: Geol. Soc. America Bull., v. 81, p. 3513–3536.
Bucher, W. H., 1955, Deformation in orogenic belts: Geol. Soc. America Spec. Paper 62, p. 343–368.
Burchfiel, B. C., and Davis, G. A., 1971, Clark Mountain thrust complex in the Cordillera of southeastern California: Geologic summary and field trip guide: Riverside, Univ. California, Campus Museum Contr., no. 1, p. 1–28.
Burchfiel, B. C., and Stewart, J. H., 1966, "Pull-apart" origin of the central segment of Death Valley, California: Geol. Soc. America Bull., v. 77, p. 439–442.
Clements, T., and Emery, K. O., 1947, Seismic activity and topography of the sea floor off southern California: Seismol. Soc. America Bull., v. 37, p. 307–313.
Dibblee, T. W., Jr., 1967a, Areal geology of the western Mojave Desert, California: U.S. Geol. Survey Prof. Paper 522, 153 p.
—— 1967b, Evidence of major lateral displacement on the Pinto Mountain fault, southeastern California: Geol. Soc. America, Abs. for 1967, Spec. Paper 115, p. 188.
—— 1968, Displacements on the San Andreas fault system in the San Gabriel, San Bernardino, and San Jacinto Mountains, southern California, in Dickinson, W. R., and Grantz, A., eds., Proceedings of the conference on geologic problems of San Andreas fault system: Stanford Univ. Pubs. Geol. Sci., v. 11, p. 260–278.
Eaton, J. E., 1932, Decline of Great Basin, southwestern United States: Am. Assoc. Petroleum Geologists Bull., v. 16, p. 1–49.
Fleck, R. J., 1970a, Age and tectonic significance of volcanic rocks, Death Valley area, California: Geol. Soc. America Bull., v. 81, p. 2807–2816.
—— 1970b, Age and possible origin of the Las Vegas Valley shear zone, Clark and Nye Counties, Nevada: Geol. Soc. America, Abs. with Programs (Rocky Mtn. Sec.), v. 2, no. 5, p. 333.
Garfunkel, Zvi, 1966, Problems of wrench faults: Tectonophysics, v. 3, p. 457–473.
Gastil, R. G., DeLisle, Mark, and Morgan, JR, 1967, Some effects of progressive metamorphism on zircons: Geol. Soc. America Bull., v. 78, p. 879–906.
Grose, L. T., 1959, Structure and petrology of the northeast part of the Soda Mountains, San Bernardino County, California: Geol. Soc. America Bull., v. 70, p. 1509–1548.
Hamilton, Warren, 1969, Mesozoic California and the underflow of Pacific mantle: Geol. Soc. America Bull., v. 81, p. 2409–2430.
—— 1971, Recognition on space photographs of structural elements of Baja California: U.S. Geol. Survey Prof. Paper 718, p. 26.
Hamilton, Warren, and Myers, W. B., 1966, Cenozoic tectonics of the western United States: Rev. Geophysics, v. 4, p. 509–549.
Hazzard, J. C., 1937, Paleozoic section in the Nopah and Resting Springs Mountains, Inyo County, California: California Div. Mines and Geology, Rept. of State Mineralogist, XXXIII, p. 273–339.
Hess, F. L., 1910, Gold mining in the Randsburg quadrangle, California: U.S. Geol. Survey Bull. 430, p. 23–47.
Hewett, D. F., 1955, Structural features of the Mojave Desert region: Geol. Soc. America Spec. Paper 62, p. 377–390.
—— 1956, Geology and mineral resources of the Ivanpah quadrangle, California and Nevada: U.S. Geol. Survey Prof. Paper 275, 172 p.
Hill, M. L., and Dibblee, T. W., Jr., 1953, San Andreas, Garlock, and Big Pine faults, California—a study of the character, history, and tectonic significance of their displacements: Geol. Soc. America Bull., v. 64, p. 443–458.
Hill, M. L., and Troxel, B. W., 1966, Tectonics of Death Valley region, California: Geol. Soc. America Bull., v. 77, p. 435–438.
Hope, R. A., 1970, The Blue Cut fault, southeastern California: U.S. Geol. Survey Prof. Paper 650-D, p. D116–D121.
Hulin, C. D., 1925, Geology and ore deposits of the Randsburg quadrangle of California: California Div. Mines and Geology Bull. 95, 152 p.
Jahns, R. H., and Wright, L. A., 1960, Garlock and Death Valley fault zones in the Avawatz Mountains, California [abs.]: Geol. Soc. America Bull., v. 71, p. 2063.
Jahns, R. H., Troxel, B. W., and Wright, L. A., 1971, Some structural implications of a late Precambrian-Cambrian section in the Avawatz Mountains, California: Geol. Soc. America, Abs. with Programs (Cordilleran Sec.), v. 3, no. 2, p. 140.
Jennings, C. W., compiler, 1958, Death Valley sheet: California Div. Mines and Geology Map Sheet, scale, 1:250,000.
—— 1961, Kingman sheet: California Div. Mines and Geology Map Sheet, scale, 1:250,000.
Jennings, C. W., and Strand, R. G., compilers, 1969, Los Angeles sheet: California Div. Mines and Geology Map Sheet, scale, 1:250,000.
Jennings, C. W., Burnett, J. L., and Troxel, B. W., compilers, 1962, Trona sheet: California Div. Mines and Geology Map Sheet, scale,

Krause, D. C., 1965, Tectonics, bathymetry, and geomagnetism of the southern Continental Borderland west of Baja California, Mexico: Geol. Soc. America Bull., v. 76, p. 617–650.

Kupfer, D. H., 1968, A proposed deformation diagram for the analysis of fractures and folds in orogenic belts: Internat. Geol. Cong., XXIII, Prague, Sess. Rept., Proc. of Sec. 13, p. 219–232.

McKee, E. H., 1971, Tertiary igneous chronology of the Great Basin of western United States—implications for tectonic models: Geol. Soc. America Bull., v. 82, p. 3497–3502.

Moore, D. G., 1969, Reflection profiling studies of the California Continental Borderland: Structure and Quaternary turbidite basins: Geol. Soc. America Spec. Paper 107, 142 p.

Moore, J. G., 1960, Curvature of normal faults in the Basin and Range province of the western United States: U.S. Geol. Survey Prof. Paper 400-B, p. B409-B411.

Noble, L. F., 1926, The San Andreas rift and some other active faults in the desert region of southeastern California: Carnegie Inst. Washington Year Book no. 25, p. 415–428.

Noble, L. F., and Wright, L. A., 1954, Geology of the central and southern Death Valley region, California: California Div. Mines and Geology Bull. 170, p. 143–160.

Prodehl, Claus, 1970, Seismic refraction study of crustal structure in the western United States: Geol. Soc. America Bull., v. 81, p. 2629–2646.

Proffett, J. M., Jr., 1971, Late Cenozoic structure in the Yerington district, Nevada, and the origin of the Great Basin: Geol. Soc. America, Abs. with Programs (Cordilleran Sec.), v. 3, no. 2, p. 181.

Ross, D. C., 1970, Quartz gabbro and anorthositic gabbro: Markers of offset along the San Andreas fault in the California Coast Ranges: Geol. Soc. America Bull., v. 81, p. 3647–3662.

Scholz, C. H., Barazangi, Muawia, and Sbar, M. L., 1971, Late Cenozoic evolution of the Great Basin, western United States, as an ensialic interarc basin: Geol. Soc. America Bull., v. 82, p. 2979–2990.

Smith, A. R., compiler, 1964, Bakersfield sheet: California Div. Mines and Geology Map Sheet, scale, 1:250,000.

Smith, G. I., 1962, Large lateral displacement on Garlock fault, California, as measured from offset dike swarm: Am. Assoc. Petroleum Geologists Bull., v. 46, p. 85–104.

Smith, G. I., and Ketner, K. B., 1970, Lateral displacement on the Garlock fault, southeastern California, suggested by offset sections of similar metasedimentary rocks: U.S. Geol. Survey Prof. Paper 700-D, p. D1-D9.

Smith, G. I., Troxel, B. W., Gray, C. H., Jr, and von Huene, R. E., 1968, Geologic reconnaissance of the Slate Range, San Bernardino and Inyo Counties, California: California Div. Mines and Geology Spec. Rept. 96, 33 p.

Stewart, J. H., 1967, Possible large right-lateral displacement along fault and shear zones in the Death Valley-Las Vegas area, California and Nevada: Geol. Soc. America Bull., v. 78, p. 131–142.

—— 1971, Basin and Range structure: A system of horsts and grabens produced by deep-seated extension: Geol. Soc. America Bull., v. 82, p. 1019–1044.

Thompson, G. A., 1959, Gravity measurements between Hazen and Austin, Nevada: A study of basin-range structure: Jour. Geophys. Research, v. 64, p. 217–229.

—— 1971, Cenozoic basin range tectonism in relation to deep structure: Geol. Soc. America, Abs. with Programs (Cordilleran Sec.), v. 3, no. 2, p. 209.

Troxel, B. W., 1967, Sedimentary rocks of late Precambrian and Cambrian age in the southern Salt Springs Hills, southeastern Death Valley, California: California Div. Mines and Geology Spec. Rept. 92, p. 33–41.

Wilson, J. T., 1965, A new class of faults and their bearing on continental drift: Nature, v. 207, July 24, p. 343–347.

Wright, L. A., and Troxel, B. W., 1966, Strata of late Precambrian-Cambrian age, Death Valley region, California-Nevada: Am. Assoc. Petroleum Geologists Bull., v. 50, p. 846–857.

—— 1967, Limitations on right-lateral strike-slip displacement, Death Valley and Furnace Creek fault zones, California: Geol. Soc. America Bull., v. 78, p. 933–950.

—— 1969, Chaos structure and basin and range normal faults: Evidence for a genetic relationship: Geol. Soc. America, Abs. with Programs (Ann. Mtg.), v. 1, no. 7, p. 242.

Yeats, R. S., 1968, Rifting and rafting in the southern California Borderland, *in* Dickinson, W. R., and Grantz, A., eds., Proceedings of the conference on geologic problems of San Andreas fault system: Stanford Univ. Pubs. Geol. Sci., v. 11, p. 307–322.

Yerkes, R. F., and Campbell, R. H., 1971, Cenozoic evolution of the Santa Monica Mountains-Los Angeles Basin area: I. Constraints on tectonic models: Geol. Soc. America, Abs. with Programs (Cordilleran Sec.), v. 3, no. 2, p. 222–223.

MANUSCRIPT RECEIVED BY THE SOCIETY MARCH 27, 1972
REVISED MANUSCRIPT RECEIVED JULY 31, 1972

# Tectonic development of the southern California deserts

**J. Douglas Walker**
*Department of Geology, University of Kansas, Lawrence, Kansas 66045; e-mail: jdwalker@kuhub.cc.ukans.edu*

**Allen F. Glazner**
*Department of Geology, University of North Carolina, Chapel Hill, North Carolina 27599; e-mail: afg@unc.edu*

## INTRODUCTION

The southern California deserts constitute one of the most geologically and tectonically diverse parts of California. The area developed in part upon Precambrian crust of North America and has seen significant modifications over its geologic history, which encompasses much of the Phanerozoic. The California deserts are bounded by several major geologic structures (e.g., the San Andreas fault system and the eastern Sierran frontal escarpment) and are divided into several subprovinces by geologic and physiographic boundaries (Fig. 1).

One of the major geologic boundaries within the desert province is the Garlock fault (Fig. 1). This fault was named by Hess (1910). Hulin (1925) suggested left-lateral displacement along it of as much as 10 km. Recent left-lateral displacement was confirmed by Hill and Dibblee (1953) in an aerial survey of offset drainages, and Smith (1962) established that there had been ~65 km of left-lateral displacement since intrusion of the Mesozoic Independence dike swarm.

In spite of this evidence for large and recent displacement, the geologic significance of the Garlock fault, which forms the principal physiographic and geologic boundary between the Basin and Range province and the Mojave Desert, was poorly understood. One of the first papers to address this complexity and present a new interpretation for the deserts as a whole was "Garlock fault: An intracontinental transform structure, southern California" by Gregory A. Davis and B. Clark Burchfiel, published in 1973.

The main focus of this work was the nature of the Garlock fault as a significant tectonic feature (and not one of just local geologic importance) in the California deserts. Davis and Burchfiel presented new interpretations of the Garlock fault as a major strike-slip fault that separates the actively extending Basin and Range province from the unextending Mojave Desert region.

This paper is significant in many ways. Probably of greatest interest is that the paper applied newly developing concepts of plate tectonics, building on the stimulating work of Atwater (1970), to the landlocked California deserts. Although earlier papers (e.g., Hamilton, 1969) had used tectonic concepts to interpret the geologic development of California and the Cordillera as a whole, Davis and Burchfiel applied ideas of plate tectonics gained from the oceans to explain a specific problem of intracontinental tectonics.

The work was based on a growing knowledge of the geology of the California deserts. The authors used seven major tie points across the Garlock fault to demonstrate that there is little variation in slip across its exposed length. They then made the important interpretation, based on work in areas east of the exposed Garlock fault, that the fault must terminate before reaching the Nevada border. This led them to the interpretation that the Garlock fault was accommodating the strain gradient between extensional deformation in the Basin and Range and nonextensional strike-slip faulting in the Mojave Desert; thus the analogy between the Garlock fault and an oceanic transform fault (Fig. 4 of Davis and Burchfiel).

This paper stimulated much new research in the California deserts—in fact, much of our own research has been aimed at resolving problems and processes of continental deformation raised in the Davis and Burchfiel paper. We discuss some of the new findings here. We focus mostly on developments in strike-slip faulting and extensional deformation in the deserts, as these points bear most closely on the problems developed in the paper. In addition, we discuss briefly new findings related to other issues raised in the paper.

## EXTENSIONAL DEFORMATION IN THE BASIN AND RANGE

One of the bold predictions made in the Davis and Burchfiel paper is that there is significant extension north of the Garlock fault in the area between Death Valley and the Clark Mountains

---

Walker, J. D., and Glazner, A. F., 1999, Tectonic development of the southern California deserts, *in* Moores, E. M., Sloan, D., and Stout, D. L., eds., Classic Cordilleran Concepts: A View from California: Boulder, Colorado, Geological Society of America Special Paper 338.

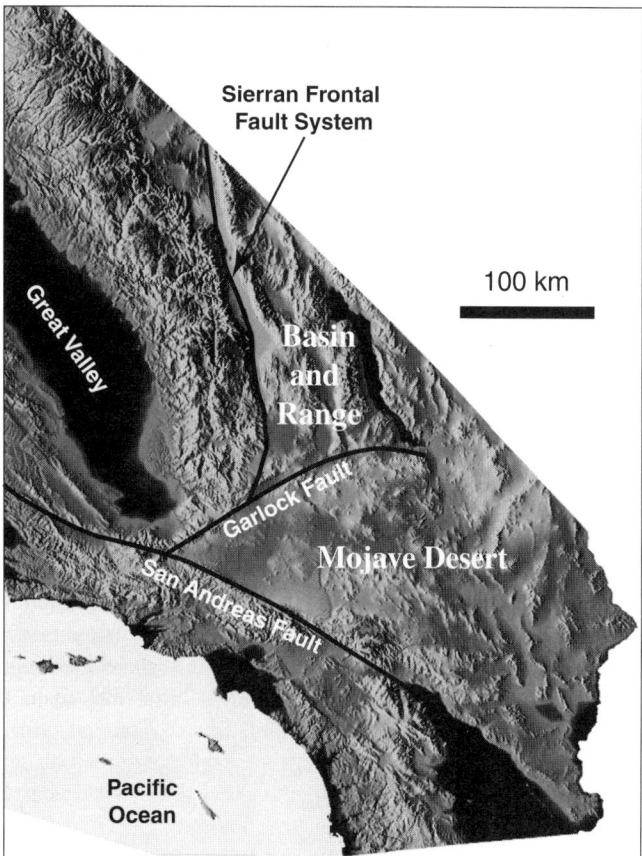

Figure 1. Shaded relief map of southern California showing major physiographic and structural boundaries. The deserts are bounded on the east by the state border and on the west by the Sierra Nevada frontal fault system and the San Andreas fault.

of Nevada (Fig. 3 of Davis and Burchfiel, 1973). They estimated extension in this area to be about 64 km, a stunningly large prediction for the time. Since the publication of Davis and Burchfiel (1973), studies of extension in the California deserts have exploded, both for the areas covered in their paper and throughout the deserts as a whole.

Extensional deformation in the Death Valley area was known for some time (e.g., Burchfiel and Stewart, 1966; Hunt and Mabey, 1966; Wright and Troxel, 1973). Wernicke et al. (1988) have since quantified the extension vector for the Basin and Range at the latitude of Las Vegas, Nevada, and showed that there is large-magnitude extension within the Death Valley area, substantiating the interpretation of Davis and Burchfiel (Fig. 2). (For more extensive discussions on California extension and Death Valley geology, see Wernicke and Spencer, and Troxel and Wright, this volume.) Although Wernicke et al. (1988) interpreted extensional magnitudes and vectors for the Basin and Range as a whole, their study shows amounts of extension in individual segments. For the area around and north of the Garlock fault, these authors show 125 km of extension directed toward N65°W. Because the trend of the Garlock fault is S75°W, this resolves into 90 km of extension parallel to the Garlock fault (e.g., the component of slip on the Garlock fault that would be apparent from extension occurring in a west-northwest orientation). If extension east of the Nopah Range is ignored (as it was by Davis and Burchfiel), then the estimate of extension reduces to 95 km with 70 km of Garlock-parallel extension. This figure is very close to that interpreted by Davis and Burchfiel.

## EXTENSIONAL DEFORMATION IN THE MOJAVE DESERT

For the California deserts as a whole, study of extensional deformation has probably been the most active area of recent research. Much work had been done on the nature of Basin and Range extension (e.g., Burchfiel and Stewart, 1966; Wright and Troxel, 1973; Wernicke et al., 1988). Indeed, the view at the time of Davis and Burchfiel (1973) was that the Basin and Range was an extensional province, whereas the other parts of the deserts, specifically the Mojave Desert, were largely unaffected. Starting with the seminal work of Davis et al. (1980), it became clear that the Mojave Desert had seen similar magnitudes of extension, but at a somewhat earlier time.

Davis et al. (1980) reported on the style and significance of low-angle normal faulting in the Whipple Mountains of southeastern California (Fig. 2). Although these authors initially interpreted the deformation in this area to be the result of gravity sliding, their descriptions of structures and style of deformation are classic examples of extensional deformation. It is clear that the Whipple Mountains are part of a larger area of metamorphic core complexes, which are common in the deserts of California and Arizona (e.g., Coney, 1980). Metamorphic core complexes are areas where ductilely deformed footwall rocks are juxtaposed against brittlely deformed hanging-wall rocks along a low-angle normal fault that is commonly referred to as a detachment fault. Detachment faults are interpreted to be rooted into the crust, so that large amounts of extension are accommodated along shallowly dipping faults. Extensional deformation in the Whipple Mountains and Colorado River area generally occurred between 25 and 15 Ma, with most of the extension occurring from 25 to 20 Ma (Spencer et al., 1995). The detachment faults rooted to the northeast, and extension is interpreted to be roughly northeast-southwest in orientation. The magnitude of extension is considered to be of the order of 70 km or more (Spencer and Reynolds, 1991).

It is clear that extension of metamorphic core complex style is present across the Mojave Desert (Fig. 2). Dokka (1986, 1989), Glazner et al. (1989), and Walker et al. (1990a, 1995) have interpreted various areas of the Mojave Desert to have undergone this style of deformation. The main area of well-exposed deformation is in the Barstow area, and is referred to as the central Mojave metamorphic core complex (Bartley et al., 1990a). Ductilely deformed footwall rocks of Mesozoic and Cenozoic age are separated from hanging-wall rocks by the Waterman Hills detachment fault. The hanging wall consists of lower Miocene synextensional strata, as well as undeformed Mesozoic plutonic

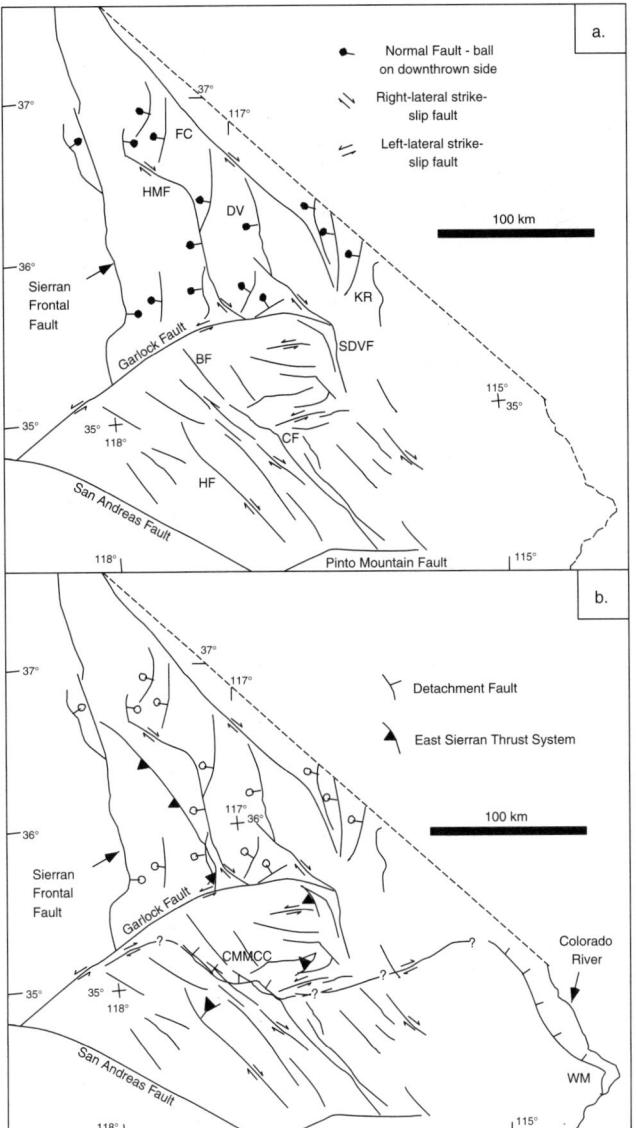

Figure 2. Main structural features of the southern California deserts. A: Major normal faults in the Basin and Range province and strike-slip faults in the Mojave Desert are shown. BF—Blackwater fault; CF—Calico fault; DV—Death Valley; FC—Furnace Creek fault zone; HF—Helendale fault; HMF—Hunter Mountain fault; KR—Kingston Range detachment; SDVF—southern Death Valley fault system. B: Miocene extensional system of the Mojave Desert and the east Sierran thrust system. Gray lines are normal faults and strike-slip faults of Miocene to recent age. CMMCC—central Mojave metamorphic core complex; WM—Whipple Mountains.

rocks (Bartley et al., 1990a; Glazner and others, 1989; Dokka, 1989; Walker et al., 1995). Extension roots toward the northeast. The magnitude of extension across the central Mojave is considered to be 70 km. Martin et al. (1993) and Bartley and Glazner (1991) considered the extension in the central Mojave Desert to be linked to that in the Whipple Mountains and environs (Fig. 2), on the basis of its magnitude, timing, and overall kinematics.

Many studies have shown that the extension in the Mojave Desert is early to middle Miocene in age (Dokka, 1986, 1989; Walker et al., 1995). The age of this extension is clearly older than that in the Basin and Range, and structures of this age in the Basin and Range province north of the Garlock fault are not significant. This led Walker et al. (1995) to propose that extension was transferred westward along a transform fault out of the deserts and into the southern Sierra Nevada Mountains and/or into the southern Great Valley.

## STRIKE-SLIP ACTIVITY IN THE MOJAVE DESERT

The feature that led Davis and Burchfiel to focus on the Garlock fault was the strain incompatibility between the extensional deformation in the Basin and Range and the lack of it in the Mojave Desert. It is clear that the Mojave Desert currently is undergoing overall transcurrent (Garfunkel, 1974) or transpressional (Bartley et al., 1990b) motions. One need only remember the 1992 Landers earthquake to appreciate strike-slip activity in the Mojave Desert (Hauksson et al., 1993; Nur et al., 1993).

Hewett (1954) first considered this set of faults as a regionally extensive set of dip-slip faults. Dibblee (1961) interpreted northwest-striking faults in the Mojave as right-lateral strike-slip faults. The presence of the northwest-trending strike-slip faults was put into a plate tectonic context by Atwater (1970). Finally, the overall role of these faults was assessed by Garfunkel (1974), who interpreted them as linked sets of strike-slip faults that produce area-conservative deformation and vertical-axis rotation. This deformation was interpreted as a link between extension in the Salton Trough–California borderlands and the Basin and Range. This model has formed the basis for much of the research on these faults in the past 25 years.

Garfunkel (1974) recognized that the model should be testable using paleomagnetic techniques on the widespread Miocene to recent volcanic rocks of the Mojave Desert. This prompted a large number of paleomagnetic and other studies aimed at looking for vertical-axis rotation in the Mojave Desert to test and refine the Garfunkel model (e.g., Luyendyk et al., 1980; Carter et al., 1987; Ross et al., 1989; Luyendyk, 1991; Valentine et al., 1993; Schermer et al., 1996). These studies showed that there is some systematic rotation in paleomagnetic declinations, but that local geologic complications are also important.

Results of these studies suggest that although the overall geometry of the Garfunkel model is accurate, the tectonic development of faulting is somewhat different. Garfunkel (1974) interpreted strike-slip faults to have formed mostly as a response to extension in the Basin and Range. Garfunkel's favored interpretation was that the Mojave Desert undergoes east-west–directed sinistral shear. However, it is clear that these faults have developed more in response to transcurrent motions along the Pacific–North American plate boundary. In other words, the Mojave Desert is undergoing northwest-southeast directed dextral shear.

One of the main factors leading to Garfunkel's interpretation (and Davis and Burchfiel's as well), was the view that extension immediately north of the Garlock fault is east-west directed. It is clear from other studies (e.g., Burchfiel and Stewart, 1966; Wernicke et al., 1988) that extension there is currently directed to the northwest. Given this extension vector, the Garlock fault, as well as the Mojave Desert, must rotate and shear in response to Basin and Range extension. Right-lateral shear of ~60 km is required through the Mojave Desert in order to resolve an average trend of S75°W for the Garlock and N65°W for Basin and Range extension, and a magnitude of extension of 95 km (see discussion above for orientation and magnitude of extension). This is consistent with some estimates for actual fault slip across the Mojave Desert (e.g., Dokka, 1983; Dokka and Travis, 1990; Schermer et al., 1996). Because the length of the Garlock fault at the boundary between the Basin and Range and the Mojave Desert is about 100 km, this means that the Garlock fault has rotated ~30° clockwise.

## GARLOCK FAULT

The Garlock fault is at the heart of the work by Davis and Burchfiel, and in the overall connection of the Basin and Range with the Mojave Desert. If the Davis and Burchfiel interpretation is correct, then initiation of motion on the Garlock must coincide with the beginning of Basin and Range extension. As with many strike-slip faults, understanding the timing of motion along this fault has proven elusive. Previous workers have considered Garlock fault motion possibly to have been initiated in the early Tertiary (Nilsen and Clarke, 1975; Cox, 1982), although evidence for significant motion of this age is unclear.

Recent work suggests that the Garlock fault initiated during the late Miocene. Monastero et al. (1997) provided evidence that rocks as young as 17 Ma are offset the entire 64 km of Garlock slip, implying that the fault initiated after that time. Burbank and Whistler (1987) placed initiation of the Garlock fault at 10 Ma on the basis of paleomagnetic results from Miocene rocks in the El Paso Mountains. Hence, it appears that the Garlock began movement at 10 to 11 Ma.

When did extension commence in the Basin and Range north of the Garlock fault? Detachment-fault–related extension in the eastern Kingston Range area began shortly after 13.4 Ma and continued for some time afterward (Fowler et al., 1995). This early phase of extension continues southward past any possible eastward trace of the Garlock fault (Davis et al., 1993) and does not, therefore, require the Garlock to be active as a transform structure. However, extension across the 12.4 Ma Kingston Peak pluton and in the Nopah and Resting Spring Ranges to the north and northwest, respectively, appears to have been most active between ca. 9 and 12 Ma (Wright et al., 1983; Davis et al., 1993; Topping, 1993). All of these areas are clearly within the Death Valley and southern Basin and Range extensional system and require some sort of Garlock fault to the south of them. Davis et al. (1993, their Fig. 4, event 5) and Davis and Burchfiel (1993) reported evidence for sinistral strike-slip faulting beneath the alluviated Kingston Wash of an appropriate age to represent the eastern, geometrically complex terminus of the Garlock fault.

It is clear, therefore, that the new discoveries on the Garlock fault and extensional deformation in the Basin and Range area of Figure 1 are compatible. The Garlock fault started moving at about the time extensional deformation became active in the Basin and Range of the Death Valley area (10 to 11 Ma). Hence, the supposition by Davis and Burchfiel that the Garlock could serve as a transform between the Basin and Range and the Mojave Desert is consistent with new data on the areas.

## OTHER DEVELOPMENTS

Davis and Burchfiel (1973) relied on several markers to examine the offset along the Garlock fault. These included some features of Tertiary age, but also markers of Mesozoic and Paleozoic age. Here we describe briefly progress made on understanding these other markers and their significance for tectonic development of the southern California deserts.

### Mesozoic thrusts

One of the tie points used by Davis and Burchfiel was a thrust fault in the southern Slate Range north of the Garlock fault and a similar structure in the Granite Mountains to the south (tie B-B' in Fig. 1 of Davis and Burchfiel, 1973). It is now recognized that these thrusts are part of a regionally extensive belt of shortening referred to as the east Sierran thrust system (Dunne et al., 1983). This system extends from the Sierra Nevada southward into the Mojave Desert (Fig. 2) (Dunne, 1986; Stevens et al., 1997; Walker et al., 1990b, 1995). Thrusts of this system place tectonized Mesozoic plutonic rocks over coeval volcanic rocks and older strata along moderately west-dipping faults. Activity was mostly during Middle to Late Jurassic time (Dunne, 1986; Dunne and Walker, 1993).

### Paleozoic strata

Davis and Burchfiel also used a distinctive group of Paleozoic rocks called the Garlock Formation as a tie across the Garlock fault (tie C-C'). It is now known that these rocks constitute outcrops of eugeoclinal strata similar to rocks of the Roberts Mountain allochthon (Burchfiel and Davis, 1975). These originally tectonically outboard rocks are juxtaposed against miogeoclinal to cratonal strata and are thus out of place in their current position. Their position is interpreted to have resulted from late Paleozoic truncation of the continental margin by strike-slip faulting (Davis et al., 1978; Walker, 1988; Stone and Stevens, 1988).

Miogeoclinal rocks were also used to determine Garlock fault displacement (marker E-E' of Davis and Burchfiel). It is clear that the California deserts are in a critical position for understanding the extent of the Precambrian and Paleozoic margin of the western part of the Cordillera. Facies trends in southern Nevada and eastern California continue westward across the Mojave Desert. Cratonal and miogeoclinal strata, along with Precambrian basement rocks of North America, extend essentially to the San Andreas fault (Stewart and Poole, 1975; Martin and Walker, 1992). This adds further to the out-of-place interpretation for the eugeoclinal rocks discussed here.

## CONCLUSIONS

Davis and Burchfiel (1973) provided new insights into the Garlock fault and the overall geologic configuration of the California deserts. Their work led to an appreciation of the significance of out-of-place sedimentary rocks and pointed to the important geologic relations exposed in relatively inaccessible military bases. Work since that time has tended to support and extend the interpretations made by those authors. In addition, some of the bold predictions made by Davis and Burchfiel, such as large-magnitude extension in the Death Valley area, have been confirmed.

## ACKNOWLEDGMENTS

Our understanding of the geology of the California deserts has grown from interactions with many geologists. We particularly wish to thank John Bartley, Stefan Boettcher, Robert Fillmore, John Fletcher, Jonathan Linn, Mark Martin, David Miller, Jonathan Miller, Francis Monastero, Elizabeth Schermer, and, of course, Gregory Davis and Clark Burchfiel for sharing ideas, data, and interpretations with us. Davis and Burchfiel also provided helpful reviews of this paper.

## REFERENCES CITED

Atwater, T., 1970, Implications of plate tectonics for the Cenozoic tectonic evolution of western North America: Geological Society of America Bulletin, v. 81, p. 3513–3536.

Bartley, J. M., and Glazner, A. F., 1991, En echelon Miocene rifting in the southwestern United States and model for vertical-axis rotation in continental extension: Geology, v. 19, p. 1165–1168.

Bartley, J. M., Fletcher, J. M., and Glazner, A. F., 1990a, Tertiary extension and contraction of lower-plate rocks in the central Mojave metamorphic core complex, southern California: Tectonics, v. 9, p. 521–534.

Bartley, J. M., Glazner, A. F., and Schermer, E. R., 1990b, North-south contraction of the Mojave block and strike-slip tectonics in southern California: Science, v. 248, p. 1398–1401.

Burbank, D. W., and Whistler, D. P., 1987, Temporally constrained tectonic rotations derived from magnetostratigraphic data: Implications for initiation of the Garlock fault, California: Geology, v. 15, p. 1172–1175.

Burchfiel, B. C., and Davis, G. A., 1975, Nature and controls of Cordilleran orogenesis, western United States: Extensions of an earlier synthesis: American Journal of Science, v. 275-A, p. 363–396.

Burchfiel, B. C., and Stewart, J. H., 1966, "Pull-apart" origin of the central segment of Death Valley, California: Geological Society of America Bulletin, v. 77, p. 439–442.

Carter, J. N., Luyendyk, B. P., and Terres, R. R., 1987, Neogene clockwise tectonic rotation of the eastern Transverse Ranges, California, suggested by paleomagnetic vectors: Geological Society of America Bulletin, v. 98, p. 199–206.

Coney, P. J., 1980, Cordilleran metamorphic core complexes: An overview, in Crittenden, M. D., Jr., Coney, P. J., and Davis, G. H., eds., Cordilleran metamorphic core complexes: Geological Society of America Memoir 153, p. 7–31.

Cox, B. F., 1982, Stratigraphy, sedimentology, and structure of the Goler Formation (Paleocene), El Paso Mountains, California: Implications for Paleogene tectonism on the Garlock fault [Ph. D. thesis]: Riverside, University of California, 248 p.

Davis, G. A., and Burchfiel, B. C., 1973, Garlock fault: An intracontinental transform structure, southern California: Geological Society of America Bulletin, v. 84, p. 1407–1422.

Davis, G. A., and Burchfiel, B. C., 1993, Tectonic problems revisited: The eastern terminus of the Miocene Garlock fault and the amount of slip on the southern Death Valley fault zone: Geological Society of America Abstracts with Programs, v. 25, no. 5, p. 28.

Davis, G. A., Monger, J. W. H., and Burchfiel, B. C., 1978, Mesozoic construction of the Cordilleran "collage," central British Columbia to central California, in Howell, D. G., and McDougall, K. A., eds., Mesozoic paleogeography of the western United States: Pacific Section, Society of Economic Paleontologists and Mineralogist Book 8, p. 1–32.

Davis, G. A., Anderson, J. L., Frost, E. G., and Shackelford, T. J., 1980, Mylonitization and detachment faulting in the Whipple-Buckskin-Rawhide Mountains terrane, southeastern California and western Arizona in Crittenden, M. D., Jr., Coney, P. J., and Davis, G. H., eds., Cordilleran metamorphic core complexes: Geological Society of America Memoir 153, p. 79–129.

Davis, G. A., Fowler, T. K., Bishop, K. M., Brudos, T. C., Friedmann, S. J., Burbank, D. W., Parks, M. A., and Burchfiel, B. C., 1993, Pluton pinning of an active Miocene detachment fault system, eastern Mojave Desert, California: Geology, v. 21, p. 627–630.

Dibblee, T. W., Jr., 1961, Evidence of strike-slip movement on northwest-trending faults in Mojave Desert, California: U.S. Geological Survey Professional Paper 424-B, p. 197–198.

Dokka, R. K., 1983, Displacements on late Cenozoic strike-slip faults of the central Mojave Desert, California: Geology, v. 11, p. 305–308.

Dokka, R. K., 1986, Patterns and modes of early Miocene crustal extension, central Mojave Desert, California, in Mayer, L., ed., Extensional tectonics of the southwestern United States: A perspective on processes and kinematics: Geological Society of America Special Paper 208, p. 75–95.

Dokka, R. K., 1989, The Mojave extensional belt of southern California: Tectonics, v. 8, p. 363–390.

Dokka, R. K., and Travis, C. J., 1990, Late Cenozoic strike-slip faulting in the Mojave Desert, California: Tectonics, v. 9, p. 311–340.

Dunne, G. C., 1986, Geologic evolution of the southern Inyo Range, Darwin Plateau, and Argus and Slate ranges, east-central California—An overview, in Dunne, G. C., compiler, Mesozoic and Cenozoic structural evolution of selected areas, east-central California: Los Angeles, Geological Society of America Cordilleran Section, p. 3–21.

Dunne, G. C., and Walker, J. D., 1993, Age of Jurassic volcanism and tectonism, southern Owens Valley region, east-central California: Geological Society of America Bulletin, v. 105, p. 1223–1230.

Dunne, G. C., Moore, S. C., Gulliver, R. M., and Fowler, J., 1983, East Sierran thrust system, eastern California: Geological Society of America Abstracts with Programs, v. 26, no. 7, p. 386.

Fowler, T. K., Friedmann, S. J., Davis, G. A., and Bishop, K. M., 1995, Two-phase evolution of the Shadow Valley Basin, southeastern California: A possible record of footwall uplift during extensional detachment faulting: Basin Research, v. 7, p. 165–169.

Garfunkel, Z., 1974, Model for the late Cenozoic tectonic history of the Mojave Desert, California, and its relation to adjacent regions: Geological Society of America Bulletin, v. 85, p. 141–188.

Glazner, A. F., Bartley, J. M., and Walker, J. D., 1989, Magnitude and significance of Miocene crustal extension in the central Mojave Desert, California: Geology, v. 17, p. 50–53.

Hamilton, W., 1969, Mesozoic California and the underflow of the Pacific mantle: Geological Society of America Bulletin, v. 80, p. 2409–2430.

Hauksson, E., Jones, L. M., Hutton, K., and Eberhart-Phillips, D., 1993, The 1992 Landers earthquake sequence—Seismological observations: Journal of Geophysical Research, v. 98, p. 19,835–19,858.

Hess, F. L., 1910, Gold mining in the Randsburg quadrangle, California: U.S. Geological Survey Bulletin 430, p. 23–47.

Hewett, D. F., 1954, General geology of the Mojave Desert region, California, in Jahns, R. H., ed., Geology of southern California: California Division of Mines and Geology Bulletin 170, p. 5–20.

Hill, M. L., and Dibblee, T. W., Jr., 1953, San Andreas, Garlock and Big Pine faults, California: Geological Society of America Bulletin, v. 64, p. 443–458.

Hulin, C. D., 1925, Geology and ore deposits of the Randsburg quadrangle, California: California Mining Bureau Bulletin, v. 95, 152 p.

Hunt, C. B., and Mabey, D. R., 1966, Stratigraphy and structure, Death Valley, California: U.S. Geological Survey Professional Paper 494A, 162 p.

Luyendyk, B. P., 1991, A model for Neogene crustal rotations, transtension, and transpression in southern California: Geological Society of America Bulletin, v. 103, p. 1528–1536.

Luyendyk, B. P., Kamerling, M. J., and Terres, R., 1980, Geometric model for Neogene crustal rotations in southern California: Geological Society of America Bulletin, v. 91, p. 211–217.

Martin, M. W., and Walker, J. D., 1992, Extending the western North American Proterozoic and Paleozoic craton through the Mojave Desert: Geology, v. 20, p. 753–756.

Martin, M. W., Glazner, A. F., Walker, J. D., and Schermer, E. R., 1993, Evidence for right-lateral transfer faulting accommodating en echelon Miocene extension, Mojave Desert, California: Geology, v. 21, p. 355–358.

Monastero, F. C., Sabin, A. E., and Walker, J. D., 1997, Evidence for post-early Miocene initiation of movement on the Garlock fault from offset of the Cudahy Camp Formation, east-central California: Geology, v. 25, p. 247–250.

Nilsen, T. H., and Clarke, S. H. Jr., 1975, Sedimentation and tectonics in the early Tertiary continental borderland of central California: U.S. Geological Survey Professional Paper 925, 64 p.

Nur, A., Ron, H., and Beroza, G. C., 1993, The nature of the Landers-Mojave earthquake line: Science, v. 261, p. 201–203.

Ross, T. M., Luyendyk, B. P., and Haston, R. B., 1989, Paleomagnetic evidence for Neogene clockwise tectonic rotations in the central Mojave Desert, California: Geology, v. 17, p. 470–473.

Schermer, E. R., Luyendyk, B. P., and Cisowski, S., 1996, Late Cenozoic structure and tectonics of the northern Mojave Desert: Tectonics, v. 15, p. 905.

Smith, G. I., 1962, Large lateral displacement on the Garlock fault, California, as measured from an offset dike swarm: American Association of Petroleum Geologists Bulletin, v. 46, p. 85–104.

Spencer, J. E., and Reynolds, S. J., 1991, Tectonics of mid-Tertiary extension along a transect though west-central Arizona: Tectonics, v. 10, p. 1204–1221.

Spencer, J. E., Richard, S. M., Reynolds, S. J., Miller, R. J., Shafiqullah, M., Gilbert, W. G., and Grubensky, M. J., 1995, Spatial and temporal relationships between mid-Tertiary magmatism and extension in southwestern Arizona: Journal of Geophysical Research, v. 100, p. 10,321–10,351.

Stevens, C. H., Stone, P., Dunne, G. C., Greene, D. C., Walker, J. D., and Swanson, B. J., 1997, Paleozoic and Mesozoic evolution of east-central California: International Geology Review, v. 39, p. 788–829.

Stewart, J. H., and Poole, F. G., 1975, Extension of the Cordilleran miogeocline belt to the San Andreas fault, southern California: Geological Society of America Bulletin, v. 86, p. 205–212.

Stone, P., and Stevens, C. H., 1988, Pennsylvanian and Early Permian paleogeography of east-central California: Implications for the shape of the continental margin and timing of continental truncation: Geology, v. 16, p. 330–333.

Topping, D. A., 1993, Paleogeographic reconstruction of the Death Valley extended region: Evidence for Miocene large rock-avalanche deposits in the Amargosa Chaos basin, California: Geological Society of America Bulletin, v. 105, p. 1190–1213.

Valentine, M. J., Brown, L. L., and Golombek, M. P., 1993, Cenozoic crustal rotations in the Mojave Desert from paleomagnetic studies around Barstow, California: Tectonics, v. 12, p. 666–677.

Walker, J. D., 1988, Permian and Triassic rocks of the Mojave Desert and their implications for timing and mechanisms of continental truncation: Tectonics, v. 7, p. 685–709.

Walker, J. D., Bartley, J. M., and Glazner, A. F., 1990a, Large-magnitude Miocene extension in the central Mojave Desert: Implications for Paleozoic to Tertiary paleogeography and tectonics: Journal of Geophysical Research, v. 95, p. 557–569.

Walker, J. D., Martin, M. W., Bartley, J. M., and Coleman, D. S., 1990b, Timing and kinematics of deformation in the Cronese Hills, California, and implications for Mesozoic structure of the southwestern Cordillera: Geology, v. 18, p. 554–557.

Walker, J. D., Fletcher, J. M., Fillmore, R. P., Martin, M. W., Taylor, W. J., Glazner, A. F., and Bartley, J. M., 1995, Connection between igneous activity and extension in the central Mojave metamorphic core complex: Journal of Geophysical Research, v. 100, p. 10,477–10,494.

Wernicke, B., Axen, G. J., and Snow, J. K., 1988, Basin and Range extensional tectonics at the latitude of Las Vegas, Nevada: Geological Society of America Bulletin, v. 100, p. 1738–1757.

Wright, L. A., and Troxel, B. W., 1973, Shallow-fault interpretation of Basin and Range structure, southwestern Great Basin, in deJong, K. A., and Scholten, R., eds., Gravity and tectonics: New York, John Wiley and Sons, p. 397–407.

Wright, L. A., Troxel, B. W., and Drake, R. E., 1983, Contrasting space-time patterns of extension-related, late Cenozoic faulting, southwestern Great Basin, California: Geological Society of America Abstracts with Programs, v. 15, p. 287.

MANUSCRIPT ACCEPTED BY THE SOCIETY NOVEMBER 23, 1998.

Geological Society of America
Special Paper 338
1999

# Chapter 19
# DEATH VALLEY: THE ULTIMATE DESERT VALLEY

**Lauren A. Wright**
*500 E. Marylyn Avenue, E 69*
*State College, Pennsylvania 16801*

**Bennie W. Troxel**
*2961 Redwood Road*
*Napa, California 94558*

(Center) A painting by Russell Chaney showing Noble as he appeared in 1930; (1) at age 12 months already with shovel and bucket in hand; (2) age 27, upon joining the U.S. Geological Survey; (3) in midlife; (4) in the later years with his contented cat; (5) his famous Model A field car; (6) at rest at the Noble's railroad tie cabin in Shoshone; (7) one of the two Jaguars owned concurrently by the Nobles in the 1950s and early 1960s; (8) Levi and Dorothy Noble with close friend and colleague Henry Ferguson; (9) at work at his private retreat in Valyermo; (10) in deep field conference with friends, Henry "Fergy" Ferguson and James "Jim" Gilluly.

Collage by L. A. Wright

## LEVI F. NOBLE

Levi Noble was born in Auburn, New York, in 1882 and spent his childhood there in his family's Victorian mansion. He attended Yale University. For his doctorate, awarded in 1909, he mapped the Shinumo quadrangle of the Grand Canyon and defined the Precambrian and Cambrian formations of that classic area. In 1909 he also began a lifetime association with the U.S. Geological Survey (USGS). A year later he married the former Dorothy Evans. They received, as a wedding gift from her parents, a fruit ranch at Valyermo in the northern foothills of the San Gabriel Mountains, and, not by coincidence, athwart the San Andreas fault. An independent income permitted Noble to negotiate a freedom to choose his own geological activities in exchange for a token salary from the USGS.

During much of the rest of their lives, the Nobles claimed three concurrent addresses. While at the Valyermo ranch, Noble engaged in his well-known investigation of the 30 mile San Gabriel segment of the San Andreas fault. His pioneer investigations of the geology of the Death Valley region were pursued from their voting address in Shoshone. Their summer months were ordinarily spent at the Auburn mansion. Noble died in 1965, a Department of Interior distinguished service medalist and an acknowledged major contributor to the geology of California. Dorothy followed him in 1969. Their graves in the Auburn cemetery are only a few yards from Noble's boyhood home.

Lauren A. Wright

## DEATH VALLEY IN 1929

Death Valley is a trench 165 to 200 miles long flanked on both sides by mountains of complex structure and varied stratigraphy. It has been visited casually by many geologists but has never been comprehensively studied. Indeed no single part of the district has been mapped and elaborated in detail. The origin and history of the valley are at best but imperfectly known. Nevertheless the region is one of great geologic interest and as it is now fast becoming accessible, it is being visited frequently by geologists and others interested in its origins. For this reason it seems worth while to offer a summary of the observations made by the writer in the course of four trips to the valley and to present a tentative outline of the geologic history.

Eliot Blackwelder

# STRUCTURAL FEATURES OF THE VIRGIN SPRING AREA, DEATH VALLEY, CALIFORNIA

BY LEVI F. NOBLE

## ABSTRACT

A flat thrust fault of middle or later Tertiary age, believed to have followed roughly the contact of later pre-Cambrian sediments with earlier pre-Cambrian metamorphic rocks is well exposed throughout the Virgin Spring area, about 10 miles square, in the Black Mountains east of Death Valley. On this thrust later pre-Cambrian, Cambrian, and Tertiary rocks have moved relatively westward for an unknown distance. The rocks of the overthrust plate are broken into innumerable blocks and slices, which are thrust over one another to form an extremely complex mosaic. This assemblage of blocks is named the Amargosa chaos, and the flat fault upon which the chaos lies is named the Amargosa thrust. The chaos is divided into the Virgin Spring, Calico, and Jubilee facies, each characterized by certain kinds of rock. The Amargosa thrust and chaos are folded into several plunging anticlines, of northwesterly trend, along whose crests the earlier pre-Cambrian rocks below the thrust are exposed.

Lying unconformably upon the folded thrust and chaos is the Funeral fanglomerate, probably of late Pliocene age, which consists of fanglomerates and basaltic lava flows. These rocks are deformed by folds and faults so recent that they are still reflected in the topography. Death Valley in the Virgin Spring area is primarily a syncline, modified by faulting, in the Funeral fanglomerate.

Features similar to the Amargosa thrust and chaos occur throughout an area of at least 8000 square miles that borders and includes the Death Valley trough.

The present paper deals mainly with the structure of the Virgin Spring area but it also notes related structures elsewhere in the Death Valley region, outlines the stratigraphy of the region, and describes a geologic section across it.

## INTRODUCTION AND ACKNOWLEDGMENTS

The fantastic disorder of the rock masses that form the precipitous mountain ranges bordering Death Valley imparts a quality of strangeness to the scenery that is felt even by the casual visitor. It is a strangeness perhaps equal in degree to that which the visitor feels at the Grand Canyon but it is of a wholly opposite kind. The scenery of the Grand Canyon has an orderly strangeness that is the expression of a rock structure whose order and simplicity are unequaled elsewhere. The scenery of the Death Valley region is the expression of a rock structure so chaotically disordered that even its general nature was but recently recognized.

The rocks throughout the Death Valley region are deformed by folding and faulting on a grand scale, but the areas of greatest disorder are associated with large thrust faults, many of which can be dated as middle or later Tertiary. Some of the features accompanying these faults appear to be without counterpart in North America; and they are so amazingly complex that the writer, in his endeavors to interpret them, has proceeded only slowly from bewilderment to partial understanding. The purpose of this paper is to describe some of these features as exemplified in a typical area of 100 square miles, designated the Virgin Spring area, which lies near the center of what will be called the southern Death Valley region....

C. Lewis Gazin rendered valuable assistance in the field for 2 months in the spring of 1933. During the years 1935, 1936, and 1937 H. Donald Curry co-operated in a reconnaissance of the region. This association was of invaluable aid; all the important discoveries of fossils that throw light on the age of the Tertiary rocks in the region were made by him, and the interpretation of many structural features was worked out with his help. In 1937 John Hazzard measured a section of the Paleozoic rocks in the Nopah Range which has become indispensable in the study of the region. During the years 1938 and 1939, Thomas P. Thayer of the Geological Survey made a study of the Tertiary rocks in the northern Black Mountains, where he mapped an area of 100 square miles gordering Furnace Creek. The results of Thayer's work are of fundamental importance in the interpretation of the Tertiary stratigraphy and structure of the region.

The existence of structural features like the most remarkable of those to be described in the present paper was probably first suspected at least 32 years ago by Ransome (1910) as a result of his work, during 1905 and 1908, in the Bullfrog district, which is 60 miles north of the Virgin Spring area. Ransome believed that the Tertiary rocks had there been thrust over the older rocks on a flat fault but he was unable to find proof of this hypothesis, and no such fault is shown in his structure sections....

Hewitt (1928), on the basis of observations begun in 1924, demonstrated the existence in the area east of the Death Valley region of flat thrust faults of middle or late Tertiary age upon which, in general, younger rocks have been pushed over older rocks. The faults described by Hewitt are in the Shadow Mountains, 35 miles southeast of the Virgin Spring area. Similar faults in the Johnnie district, 35 miles northeast of this area, were described by Nolan (1929), who correlated them with those studied by Hewitt in the Shadow Mountains.

Notwithstanding these pioneer studies, a great deal remained to be learned regarding both the stratigraphy and the structure of the region when, in 1931, the writer began his work. Six seasons of reconnaissance were therefore given to determining the age, character, and distribution of the rocks as a foundation for the study of the structure. This reconnaissance included not only the entire extent of the Furnace Creek and Avawatz Mountains quadrangles but adjacent parts of the Ballarat and Searles Lake quadrangles, the total area covered being more than 8000 square miles. The results of the first 2 years of the reconnaissance have been recorded in a brief progress report (Noble, 1934).

Areal mapping of the region and systematic study of the structure were begun in 1937. Since then about 500 square miles have been mapped by the writer and about 200 square miles each by Curry and Thayer. The ground covered by the writer includes the Virgin Spring area, which is the chief subject of the present paper, and a strip 50 miles long and 3 miles wide extending from Death Valley to the Kingston Range. A geologic section along this strip is described in this paper, which also embodies observations made in the Black, Owlshead, and Avawatz Mountains and in the Ibex Hills, where other small areas were mapped.

During the early years of reconnaissance, Professors William Morris Davis, H. E. Gregory, and John E. Wolff each made several visits to the region with the writer, who derived much help and stimulation from his field contacts with these eminent men.

During the later stages of the work, H. G. Ferguson, James Gilluly, D. F. Hewett, G. F. Loughlin, W. C. Mendenhall, G. W. Stose, and Anna Jonas Stose each spent a week or more in the Virgin Spring and adjacent areas. The constructive field criticism offered by these geologists and their personal interest in the problem have been of immeasurable value. The writer is indebted to H. G. Ferguson for reading the text of this paper and for giving helpful advice regarding its arrangement and to G. W. Stose for preparing the maps and illustrations, devising methods of representing the structure on the maps, and for helpful suggestions concerning the text. F. C. Calkins also has criticized the text with special regard to composition.

It is now an easy matter to visit the Death Valley region; the Death Valley National Monument has been systematically developed for travel, and its improved highways are annually traversed by thousands of tourists. The more remote parts of the region, on the other hand, are difficult of access except to the experienced desert traveler, though much of it can be reached by automobile over unimproved desert roads.

The region is a paradise for the geologist. Owing to the great irregularity of the topographic relief, the extreme aridity of the climate, and the almost complete absence of vegetation, the rocky structure of the mountain ranges is laid bare to an extent that is unusual even in the desert. Geologic work is possible almost everywhere throughout the year, except that the crest of the Panamint Range is snow-covered for part of the winter and the deeper parts of the Death Valley trough are intolerably hot in summer.

\* \* \* \* \* \*

## SOUTHERN DEATH VALLEY REGION

### STRATIGRAPHY

\* \* \* \* \* \*

The southern Death Valley region contains rocks of at least eight geologic systems—pre-Cambrian, Cambrian, Ordovician, Silurian, Devonian, Carboniferous, Tertiary, and Quaternary—whose aggregate thickness certainly exceeds 45,000 feet for the stratified rocks alone. The earlier pre-Cambrian rocks are intensely metamorphosed; the later preCambrian, Paleozoic, and Tertiary rocks are only locally altered, and their characteristic features are recognizable almost everywhere. Widespread angular unconformities separate the earlier pre-Cambrian from the later pre-Cambrian rocks, the later pre-Cambrian from the Cambrian rocks, and the Tertiary rocks from all older rocks. The greatest stratigraphic break of the region is the unconformity between the intensely metamorphosed earlier pre-Cambrian rocks and the relatively unmetamorphosed later pre-Cambrian rocks, which, though separated from the Cambrian rocks by an unconformity, are hardly more altered than even the youngest of the Paleozoic rocks. The unconformity at the base of the Tertiary rocks, however, is a major break, second only to that between the earlier and later pre-Cambrian.

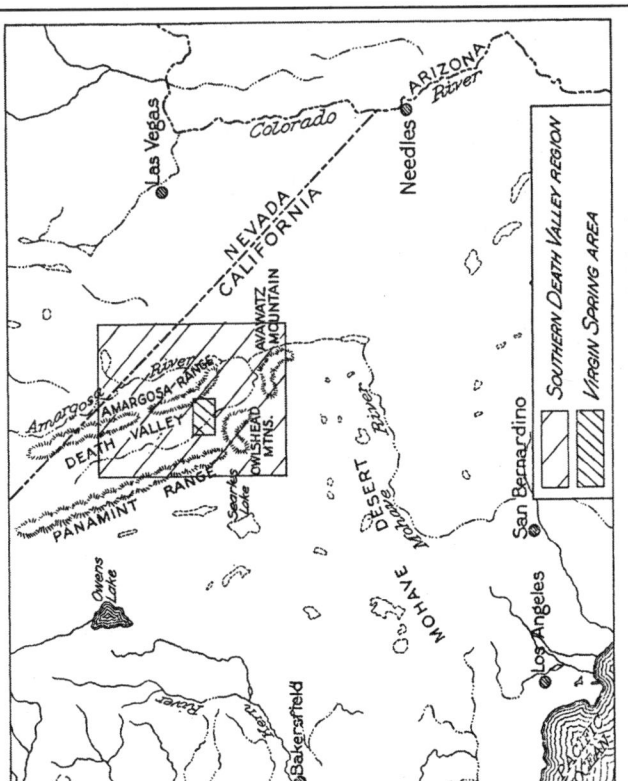

FIGURE 1.—*Index map of eastern California*
Showing location of southern Death Valley region and Virgin Spring area.

*Earlier pre-Cambrian metamorphic rocks.*—The earlier pre-Cambrian metamorphic rocks are widespread. They form the cores of the Panamint Range, the Funeral, Black, and Avawatz mountains, and the Ibex and Silurian hills and are exposed in the Greenwater and Kingston ranges and in the Owlshead Mountains.

In the Panamint Range the pre-Cambrian metamorphic rocks, as described by Murphy (1932, p. 337), comprise an older series, largely schists and gneisses, which he has named the Panamint metamorphic complex, and an upper series of metamorphosed sediments—largely schists, slates, quartzites, and dolomitic limestones—which he has subdivided under the names Marvel limestone, Surprise formation, and Telescope group. In the Funeral Mountains the pre-Cambrian metamorphic rocks include phyllites, schists, dolomites, quartzites, and graywackes that in part resemble Murphy's Surprise formation. Schists that resemble those of the Panamint Range are exposed in the Quail Mountains and the western part of the Avawatz Mountains.

Elsewhere in the southern Death Valley region the pre-Cambrian metamorphic rocks are chiefly granitic and dioritic gneisses. No rocks comparable to the Surprise formation and Telescope group described by Murphy are known east of the Death Valley trough, south of the Funeral Mountains. Only the gneisses are present in the Virgin Spring area.

*Later pre-Cambrian sedimentary rocks.*—The later pre-Cambrian sedimentary rocks in the southern Death Valley region were first described by Gilbert (1875, p. 170), who measured a section at Saratoga Springs (Pl. 1) 15 miles southeast of the Virgin Spring area. Later Campbell (1902, p. 14) described the same section. Both Gilbert and Campbell believed the rocks in this section to be pre-Cambrian. The writer of the present paper has described them in a general way and has referred to them as Algonkian strata (Noble, 1934). Hewett (1940) studied these rocks in the Kingston Range, just east of the southern Death Valley region, and grouped them into three formations—the Crystal Spring formation, Beck Spring dolomite, and the Kingston Peak formation—which together constitute the Pahrump series . . .

\*   \*   \*   \*

Hewitt's section of the Pahrump series in the Kingston Range (1940) is summarized as follows:

*Paleozoic rocks.*—Paleozoic rocks in the southern Death Valley region were first described by Spurr (1903, p. 194-205) in a brief reconnaissance report. The writer of the present report has also described them in a very general way (Noble, 1934).

The Paleozoic has been studied in detail in only two parts of the region. Nolan (1929, p. 461-472) described three Lower Cambrian formations, which occur in the Spring Mountain Range, near Johnnie, in the extreme northeast corner of the mapped area (Pl. 1). These he named the Johnnie formation, Stirling quartzite, and Wood Canyon formation. In the Resting Springs and Nopah Ranges, near Shoshone, in the eastern part of the area, Hazzard (1937, p. 273-339) has measured a detailed section, from Lower Cambrian to Pennsylvanian inclusive, which has become the standard for the region….

Hazzard's section of the Paleozoic in the Resting Springs and Nopah Ranges showing the formation names and thicknesses (1937, p. 273-339) is summarized as follows:

```
                                                Feet
Paleozoic
  Pennsylvanian
    Bird Spring (?) formation ...............   780+
  Mississippian ..............................  2167
    Monte Cristo (?) limestone ...............   987+
           UNCONFORMITY
    Stewart Valley limestone ................. 1180+
           UNCONFORMITY
  Devonian ...................................   890
    Middle (?) Devonian
      Sultan dolomite ........................   890
           UNCONFORMITY
  Silurian (?) ...............................   335
    Ordovician ...............................  2105
  Upper Ordovician
    Ely Springs (?) dolomite .................   800
  Middle Ordovician
    Eureka quartzite .........................   265
  Lower Ordovician
    Pogonip (?) dolomite ..................... 1040
  Cambrian ................................... 16,598+
    Upper Cambrian ........................... 1740
      Nopah formation ........................ 1740
    Middle Cambrian .......................... 5182
      Cornfield Springs formation ............ 2975
      Bonanza King formation .................  515
      Cadiz formation ........................  692
    Lower Cambrian ........................... 9676+
      Wood Canyon formation .................. 3033
      Stirling quartzite ..................... 2593
           DISCONFORMITY (?)
      Johnnie (?) formation .................. 2550
      Noonday dolomite ....................... 1500

  Total thickness of Paleozoic ............... 22,875+
           UNCONFORMITY
Pre-Cambrian rocks
          *       *       *       *       *

                                                Feet
Lower Cambrian
  Dolomite (Recently named Noonday dolomite)
           UNCONFORMITY
Later pre-Cambrian
  Pahrump series
    Kingston Peak formation .................. 1900
    Beck Spring dolomite ..................... 1000
    Crystal Spring formation ................. 2000
           UNCONFORMITY ....................... 4900
Earlier pre-Cambrian
  Granite gneiss
          *       *       *       *       *
```

*Granitic intrusive rocks.*—Granite is exposed at many places in the southern Death Valley region, where it covers altogether at least 600 square miles but it does not outcrop within the large area of Paleozoic rocks that occupies all the northeastern part of the region, or in the Funeral Mountains.

In the Owlshead Mountains outcrops of granite cover at least 150 square miles. Other large masses lie in the Black, Greenwater, Avawatz, Granite, and southern Panamint ranges (Pl. 1). The small part of the Kingston Range that lies within the area is composed of granite which extends far southward in the Ivanpah quadrangle to the Cima Dome.

Much of the granite is near quartz monzonite in composition. In some areas it is all coarsely crystalline; elsewhere it is porphyritic or porphyritic exhibits much variation in texture. None of it is gneissoid.

The age and relations of most of the granite are unknown. That in the Owlshead Mountains cuts the later pre-Cambrian Pahrump series, and the granite in the Panamint Range cuts rocks as young as Carboniferous. According to Thayer (personal communication), porphyritic granite in the Greenwater Range and Black Mountains cuts volcanic rocks that are believed to be a part of the Artist Drive formation of the earlier Tertiary. According to Hewett (in manuscript) the granite in the Kingston Range is early Tertiary. Although the granite in all the areas may be roughly contemporaneous it more probably represents two or more widely separated periods of intrusion.

A body of granite that will be described later is present in the Virgin Spring area and extends north of the area to Gold Valley.

*Tertiary rocks.*— ...The Tertiary rocks are widely and irregularly distributed. In the 3800 square miles shown on the map (Pl. 1) it would be difficult to select any area 10 miles square that does not contain them. They occur irregularly on the mountain ranges, at some places covering the slopes and at others forming the highest summits. Death Valley and all other large topographic depressions are wholly or in large part underlain by Tertiary rocks.

Thayer's section of the Tertiary rocks in the northern Black Mountains and Greenwater Range is summarized as follows:

Quaternary deposits
UNCONFORMITY
Tertiary rocks
    Funeral fanglomerate (Pliocene?) .................... 3000
    Greenwater volcanics (Pliocene?) .................... 2500
    UNCONFORMITY
    Furnace Creek formation (Upper Miocene or Pliocene) ... 2500
    UNCONFORMITY
    Artist Drive formation (Oligocene or Miocene) ........ 5000
    Total ............................................... 13,000
UNCONFORMITY
Paleozoic rocks

...It is not yet possible to correlate most of these Tertiary formations throughout the southern Death Valley region. The difficulty is not due so much to incomplete knowledge of the distribution and character of the rocks as to their great horizontal and vertical variability, which makes lithologic character almost valueless for correlation. The Funeral fanglomerate with its included basalt, however, is readily identifiable wherever it occurs in the region...

### STRUCTURE

The outstanding structural features of the general region are large low-angle thrust faults, which appear to be middle or late Tertiary in age. The thrust planes have been folded, and the folding has been accompanied or followed by large-scale movement on steep faults, which has continued to the present. For the most part the thrust faults follow unconformities between the earlier and the later pre-Cambrian rocks, between the later pre-Cambrian and the Cambrian rocks, and between the Tertiary and all older rocks....

Throughout the region the later pre-Cambrian and Paleozoic rocks are disordered by faulting. In some areas they are so intricately broken that the disorder is chaotic; structural blocks are measurable in hundreds or mere tens of feet and obviously not mappable on a scale of 1: 62,500. In other areas they are more coarsely broken, and individual blocks are mappable. In still other areas structural blocks in which the stratigraphic sequence is unbroken are measurable in thousands of feet or even miles. The degree of disorder varies from place to place in the region but in general it increases progressively westward and is greatest near the Death Valley trough. Imbricate, or "schuppen," structure is conspicuous at many places. Competent rocks such as massive dolomites and quartzites tend to override weaker rocks such as shales and to form a monoclinally dipping series in which the present succession of the rocks differs widely from their original stratigraphic sequence. Throughout the region, even in areas of chaotic disorder, the prevailing dip of the strata is easterly-ranging from southeast to northeast-and is commonly as steep as 45 degrees, so that in crossing the region from east to west one descends, broadly speaking, in the stratigraphic column, but parts of the section are repeated again and again in separate blocks and in separate areas, and this repetition gives the larger blocks the appearance of monoclinal fault blocks...

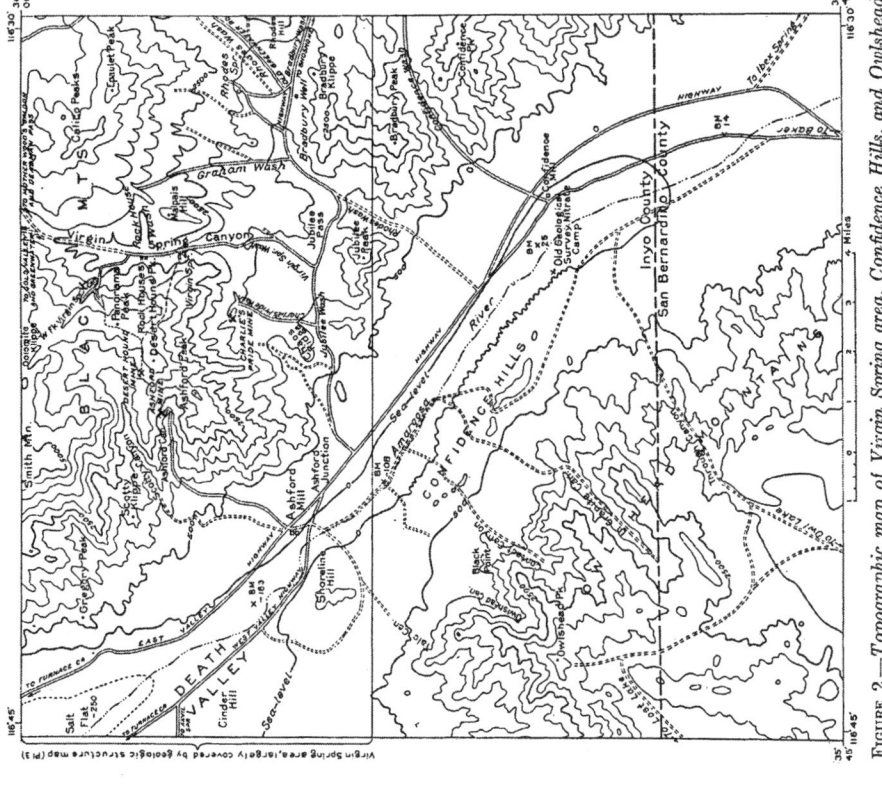

FIGURE 2.—*Topographic map of Virgin Spring area, Confidence Hills, and Owlshead Mountains*
Contour interval 500 feet.

## VIRGIN SPRING AREA

### GENERAL STATEMENT

...A flat thrust fault, whose plane of rupture is believed to have followed roughly the contact of the later pre-Cambrian sediments with the earlier pre-Cambrian metamorphic rocks, is well exposed throughout the area. On this thrust the later pre-Cambrian and Tertiary rocks that make up the overthrust plate have been carried relatively westward for an unknown distance.

The rocks of the overthrust plate are broken into innumerable blocks and slices, which are thrust over one another to form a mosaic so complex that it could hardly be mapped satisfactorily on any scale....

The name Amargosa chaos is applied to this broken up assemblage of rocks of various stratigraphic horizons, and the name Amargosa thrust is applied to the flat fault upon which the chaos lies (Pl. 5).

Although individual geologic formations in the Amargosa chaos are not mappable, the chaos as a whole is divisible into three mappable units, each characterized by certain kinds of rock. The three units are termed the Virgin Spring, Calico, and Jubilee phases of the Amargosa chaos. (See Plate 3.) The Virgin Spring phase consists almost wholly of blocks of later pre-Cambrian and Cambrian sedimentary rocks representing a stratigraphic range of many thousands of feet; the Calico phase consists chiefly of blocks of Tertiary volcanic rocks; and the Jubilee phase largely of blocks of Tertiary red conglomerate and tectonic breccias of granite. The stratigraphic range represented by the rocks in the Calico and Jubilee phases is unknown, but that of the Calico phase is certainly much less than that represented by the Virgin Spring. The Virgin Spring phase lies everywhere directly on the Amargosa thrust. The Calico and Jubilee phases lie on the thrust in some places and on the Virgin Spring phase in others.

The date of the thrusting is believed to be middle or later Tertiary. After the thrusting the Amargosa thrust was folded so that the older pre-Cambrian metamorphic rocks of the autochthonous block form the central cores of several plunging anticlines, which trend northwestward. The trace of the Amargosa thrust outlines these anticlines conspicuously on the geologic structure map (Pl. 3), and they are no less conspicuous in the field.

Lying unconformably upon the folded Amargosa chaos of the overthrust plate and, where the chaos was eroded from the anticlines, upon the pre-Cambrian gneiss of the autochthonous block, is the Funeral fanglomerate with its included lava flows, which are largely basalt. These Pliocene (?) rocks are themselves folded and are broken by large normal faults, the movement on some of which has continued into Recent time.

sharply truncated at the boundary of the block. Commonly the bedding, even of incompetent beds, is not greatly distorted....

The Amargosa chaos or features resembling it are widespread in the southern Death Valley region, and if they occur in other regions the term chaos, as a common noun, may prove to be a useful geological term....

The basal unit, the Virgin Spring phase, consists essentially of later pre-Cambrian and Cambrian formations but contains a few blocks of Tertiary volcanic rocks; the second unit, the Calico phase, consists almost wholly of Tertiary volcanic rocks; and the third unit, the Jubilee phase, consists of Tertiary volcanic and sedimentary rocks, granite, and pre-Cambrian and Cambrian rocks.

The three units are superposed. The Virgin Spring phase lies everywhere directly on the Amargosa thrust and nowhere overlies either of the other phases; the Calico phase lies at most places on the Virgin Spring phase but at two places forms the sole of the thrust; and the Jubilee phase lies at one place on the Virgin Spring phase and at another place forms the sole of the thrust. The Calico and Jubilee phases are not in contact within the area, and their relative position is not known.

*Virgin Spring phase*—The Virgin Spring phase, named from Virgin Spring, in Virgin Spring Canyon, covers about 20 square miles in the area mapped....

As seen in cross section (Pls. 7, 8, 9), the shapes of the blocks in the chaos are widely varied but are for the most part elongate. Lozenge shapes are particularly characteristic; many blocks have blunt or rounded ends, whereas others taper to a point; the forms of some are so irregular and odd as to defy description. Fishlike forms that suggest whales or huge tadpoles or lizards are common (Pl. 8, fig. 2).

The sizes of the blocks range widely, but one dimension commonly exceeds 200 feet, and many blocks are not far from 100 by 200 by 600 feet. A few blocks are as much as half a mile long.

The rocks in the Virgin Spring phase represent a stratigraphic range of at least 18,000 feet. This range extends upward from the base of the later pre-Cambrian Pahrump series at least as high as the Cornfield Springs formation of the Middle Cambrian, although no blocks of the basal Middle Cambrian Cadiz formation of the Resting Springs and Nopah section have been identified. The later pre-Cambrian rocks,

## AUTOCHTHONOUS BLOCK

Essentially all the rock below the thrust in the Virgin Spring area belongs to the earlier pre-Cambrian group. The principal rock is a dark-weathering coarse-grained augen gneiss, which consists essentially of feldspar, quartz, and brown biotite; the augen are of potash feldspar. Pegmatite dikes are abundant in some parts of the gneiss; they lie parallel with its foliation and are sheared and partly altered. Small masses of quartzite, mica schist, and hornfels are associated with the gneiss at a few places, but their relation to it is unknown. All these rocks are mashed, contorted, and thoroughly recrystallized, and all are regarded as earlier pre-Cambrian....

Dikes of quartz latite porphyry cut the gneiss in all directions and in places form complex networks. A remarkable feature of these dike rocks is that the feldspar phenocrysts have cores of sanidine and rims of albite. The albite is thought to be secondary, for the dikes contain much chlorite and other clearly secondary minerals. The cause of the albitization and other alteration phenomena is unknown. The fact that the orthoclase of the granite also is albitized suggests that the dike rocks and the granite may be genetically related. Some of these dikes intrude the Virgin Spring phase of the chaos and are believed to be Tertiary.

Dikes of unaltered fine-grained or glassy rhyolitic rock cut the gneiss at several places. These dikes are obviously related to the Tertiary extrusive rocks of the region.

## AMARGOSA CHAOS

...The characteristic features of the chaos shown in Plate 6 are:

(1) The arrangement of the blocks is confused and disordered—chaotic.

(2) The blocks, though mostly too small to map, are vastly larger than those in anything that could be called a breccia; most of them are more than 200 feet in length, some are as much as a quarter of a mile, and a few are more than half a mile in length.

(3) They are tightly packed together, not separated by much finer-grained material.

(4) Each block is bounded by surfaces of movement—in other words, each is a fault block.

(5) Each block is minutely fractured throughout, yet the original bedding in each block of sedimentary rock is clearly discernible and is

including the diabase sills associated with the sedimentary beds, constitute at least 6500 feet of the sequence, and the Cambrian rocks at least 11,500 feet....

Notwithstanding the appearance of great disorder, there is a rude system in the arrangement of the larger blocks in the chaos. In general, they are elongate slabs whose long axes lie parallel to the Amargosa thrust, and the chaos is essentially a mass of these slabs, lying one upon the other (Pl. 7). Viewed from a distance, their attitude suggests the scales of a fish or shingles on a roof. In a general way, these slabs are arranged in the normal stratigraphic sequence. Blocks of younger rocks rest, as a rule, on blocks of older rocks, though here and there the reverse is true....

The dikes of altered quartz latite porphyry with zoned feldspars and of rhyolite that cut the gneiss of the autochthonous block also cut the Virgin Spring phase of the chaos at many places. West of the Calico Peaks several dikes of both types of rock may be seen cutting the chaos, the Amargosa thrust, and the pre-Cambrian gneiss.

*Calico phase.*—Resting at most places upon the Virgin Spring phase of the Amargosa chaos but at two places upon the older pre-Cambrian gneiss of the autochthonous block is a mosaic of fault blocks that is here named the Calico phase....

Two outstanding features distinguish the Calico phase from the Virgin Spring phase—the fact that it is composed almost wholly of Tertiary volcanic rocks, and its color. The color differences enable the observer readily to distinguish the two phases even from a distance. The colors of the Virgin Spring phase are prevailingly dark shades of brown and gray, whereas those of the Calico phase are a kaleidoscopic mixture of light and dark—the light shades are yellowish-white to buff, and the dark shades dull pinkish to red. Such assemblages of parti-colored Tertiary volcanic rocks are called calico by the prospectors throughout the desert region, and the hills formed from them are called calico hills, the Calico Peaks being themselves an example of this usage. The name "Calico phase" is thus descriptive as well as otherwise appropriate.

Most of the blocks in the Calico phase of the chaos are glassy lavas and tuffs, which occur in nearly equal amounts. Although the flows and tuffs are now greatly disordered, it is evident that they once constituted a volcanic sequence which was piled up about one or more centers of eruption. The lavas are rhyolitic in appearance, and the tuffs are composed of rhyolitic material similar to that in the flows. The lavas are commonly dull pinkish or red, and the tuffs are yellowish white or buff. Blocks of black obsidian are present but not abundant....

The chaotic arrangement of the blocks in the Calico phase is shown in Plates 5, 6, and 13. It does not appear to differ much, particularly in the lower part of the phase, from that in the Virgin Spring phase, although the two phases are composed of very different types of rock, which may have behaved differently under deformation; few blocks in the Calico phase consist of bedded rocks. As in the Virgin Spring phase, all contacts between blocks are surfaces of movement (Pls. 11-13).

In most of the Calico phase west of the Calico Peaks and Epaulet Peak (Pl. 3) the blocks are relatively small and the pattern complex, but at places under and east of these peaks the rhyolitic flows and tuffs form sheets as much as half a mile long and several hundred feet thick in which an orderly sequence is preserved. These parts of the Calico phase suggest a stratified volcanic series that is intricately broken up by faulting but not entirely chaotic. This relative order becomes more pronounced eastward away from the Virgin Spring area until at Sheephead Peak (Pl. 1), 6 miles east of the area, the blocks in the chaos would be mappable on a map of the same scale as the geologic structure map of the Virgin Spring area (Pl. 3), although the disorder is still great....

In a distant view (Pl. 5), the contact of the Virgin Spring and Calico phases is conspicuous, because the two phases differ strikingly in color and lithologic character. On closer view (Pl. 13), the contact is seen to be at some places irregular and at some places obscure. But, wherever the contact is traced laterally for any considerable distance, different rocks are found to form successively the bottom of the Calico phase, a relationship which indicates that the contact is either an unconformity or, more probably, a fault. This fault is not nearly so well defined, however, as the Amargosa thrust. At some places just above the contact, Virgin Spring blocks are mixed with Calico blocks, and a few blocks contain rhyolite or tuffs in apparent depositional contact with Cambrian rock. But it is evident almost everywhere that the Calico and Virgin Spring phases are separated by a surface on which there has been movement....

In the Bullfrog district, 60 miles north of the Virgin Spring area, a body of intricately broken Tertiary volcanic rocks overlies early Paleozoic and pre-Cambrian rocks comprising limestone, quartzite,

Most of the blocks of granite and pre-Tertiary rocks are so thoroughly shattered that they are properly termed breccia.

These breccias are interpreted as slices of rock formations that have been thrust bodily over other rock formations and have been shattered in the process. They have not the internal structure characteristic of talus deposits or landslides. They will be called accordingly tectonic breccias.

The tectonic breccia of granite is composed of irregular fragments, some of them several hundred feet long, so tightly compacted and fitted together that at first sight the breccia resembles solid granite; elsewhere it is composed of innumerable smaller fragments in a matrix of pulverized granite (Pl. 15, fig. 2). All the material in the breccia is granite. Outcrops have a characteristic cavernous appearance (Pl. 14, fig. 1; Pl. 15, fig. 1) that renders the granite breccia easily distinguishable from other parts of the Jubilee phase....

At many places, masses of tectonic breccia, some of them as much as 1000 feet long and 300 feet thick, override conglomerate, andesite, tuff, and other rocks on thrust planes that commonly dip 35 degrees or less to the east (Pl. 14). At some places, material resembling gravel or talus appears to be overridden by advancing thrust sheets. Elsewhere, blocks of the igneous and sedimentary rocks are thrust over the breccias. In some blocks of sedimentary rock the bedding is folded. The prevailing dip of the bedding in the blocks in which it is not folded is easterly, as is also the prevailing dip of the thrust slices. Neither the folds in the blocks nor the imbricate faults that impart the shingled structure to the Jubilee phase penetrate the surface of movement upon which the Jubilee phase lies....

The Jubilee and Calico phases of the chaos are not in contact in the Virgin Spring area, so that the relation of one to the other is doubtful. Both phases overlie the Amargosa thrust and the Virgin Spring phase; both are overlain unconformably by the Pliocene (?) Funeral fanglomerate. In general the two phases differ in lithologic composition, but the Jubilee phase contains volcanic rocks resembling some of those in the Calico phase. On the other hand, the Calico contains no blocks of the tectonic breccias of granite and other rocks that are abundant in the Jubilee. The deformation that caused the original formation of these breccias possibly antedated the thrusting that produced the Jubilee. The Jubilee and Calico phases may conceivably be more or less equivalent to each other, but the writer

shale, and schist and is separated from them by a low-angle fault (Ransome, Emmons, and Garrey, 1910, p. 101-103). Ransome believed that this fault is an overthrust upon which the volcanic rocks have been pushed over the older rocks for an unknown distance and he suggests that the overthrusting of brittle volcanic rocks, under light load, over an irregular prevolcanic topography is the proximate cause of their complex dislocation, though he was unable to find proof of this hypothesis in the Bullfrog district....

*Jubilee phase*—Resting at one place upon the older pre-Cambrian metamorphic rocks of the autochthonous block and at another place upon the Virgin Spring phase of the chaos is another mosaic of fault blocks which is termed the Jubilee phase. The name is derived from Jubilee Wash, where this mosaic is exposed on both sides of the Shoshone-Death Valley highway at Chaos Ridge (Pl. 3; fig. 2).

Broadly speaking, the Jubilee phase is a complex mosaic of irregular blocks and thrust slices, which differ widely in size and composition, resting discordantly on a surface or surfaces of movement. It contains a much larger proportion of poorly consolidated and broken-up material and presents a more confused picture than the other two phases.

Scattered outcrops of the Jubilee phase occur in two areas-each about 3 square miles in extent-one in Jubilee Wash and the other at the mouths of Ashford and Scotty canyons. This phase, in contrast to the Virgin Spring and Calico phases, is in general poorly exposed, for it lies mostly in topographic depressions veneered with Quaternary alluvium, above which low hills of the chaos project like islands; but it is well exposed near Scotty Klippe, where the Jubilee phase forms hills rising above the pre-Cambrian gneiss.

The original thickness of the Jubilee phase is unknown, as it is overlain unconformably by the Funeral fanglomerate. Its maximum observed thickness is not over 1000 feet.

The blocks that constitute the Jubilee phase are much more varied lithologically than those in the Virgin Spring and Calico phases. Perhaps half of them consist of granite and red Tertiary conglomerate; others consist of rhyolite, rhyolitic tuff, porphyritic andesite, quartz latite porphyry, gypsiferous shale, fresh-water limestone and fanglomerate of Tertiary age, older pre-Cambrian gneiss and greenstone, and later pre-Cambrian and Cambrian sedimentary rocks.

believes that the Jubilee is the younger. It seems to have been the product of a movement that occurred later than or in the late stages of the Amargosa overthrusting and in general at a higher level but to have used, here and there, the plane of the older and lower thrust. Where this occurred, the Virgin Spring and Calico phases overlying the thrust were pushed away.

### ORIGIN OF AMARGOSA CHAOS

The evidence that the Amargosa chaos as a whole has moved over the underlying floor of earlier pre-Cambrian gneiss is everywhere clear and convincing. The blocks in the chaos that lie directly on the gneiss range from later pre-Cambrian to Tertiary; the broad sides of the blocks and the bedding within the blocks that show bedding lie at all possible angles to this floor. The contact between the chaos and the gneiss is obviously an enormous thrust fault, which is here called the Amargosa thrust.

Several features of this great fault, though each of them has already been more or less frequently mentioned, must be recapitulated here with special emphasis. One of these is that the upper plate of the overthrust—in other words the Amargosa chaos—consists of younger rocks than the mass below the thrust. It is more usual, of course, for the upper plate to consist of older rocks. Another significant feature is that the faults between blocks in the chaos, even between blocks that have come from strata originally hundreds of feet apart, do not dislocate the surface of the underlying gneiss; many have been observed to meet this surface at an acute angle and to steepen upward as they diverge from it. Both these facts, together with the fundamental fact of extreme disorder throughout the chaos, indicate that the thrust faulting took place at shallower depth and under lighter load than the movements that have produced thrust faults of the more usual type. . . .

### AMARGOSA THRUST FAULT

*Description.*—The Amargosa thrust fault (Pls. 5, 7, 11) is easily traced because of the great contrast in lithologic character and structure between the rocks above and below it and the excellent exposures of the fault contact afforded by the steep slopes of the area. Because of later folding of the fault plane and the ruggedness of the topography the outcrop of the fault is sinuous in the extreme. . . .

At many places where the chaos has moved over massive unaltered pre-Cambrian gneiss the fault surface is grooved and striated and the contact is sharp. At other places, where the gneiss is disintegrated at the fault surface and less resistant, the contact is ragged, or serrate, on a small scale. The serrate contacts appear to be due to the plowing of the overthrust rock into soft material.

Clay gouge is present at many places along the thrust but is commonly only a few inches thick (Fig. 4). Some of the gouge contains fragments of rock different from that in the adjacent overridden and underridden block; these fragments may have been dragged from some distance. Most of the gouge is white, owing to leaching of its iron content by circulating water.

The fault surface is uneven. The major irregularities are the anticlines and synclines into which the thrust surface is folded (Pl. 3), but the surface is also irregular in detail and is wavy (Pl. 10, fig. 1). Commonly this minor relief does not exceed 10 feet in a distance of a hundred but at some places it is much greater. It is not clear whether these minor irregularities are due to folding or to the original unevenness of the surface over which the chaos rode. . . .

*Direction and magnitude of movement.*—No direct measure of the magnitude and direction of the thrust movement is possible, because the roots of the thrust fault have not been found; but several lines of evidence indicate that the amount of movement was large and that the relative movement of the overthrust plate was toward the west.

The distance between the known outcrops of the thrust, in the Virgin Spring area, is 11 miles from east to west and 6 miles from north to south (Pl. 3), and, from outcrops of what is believed to be the same thrust, its apparent minimum extent is 75 miles from northwest to southeast.

Further evidence is afforded by the regional distribution of the Cambrian and pre-Cambrian rocks. The Virgin Spring phase of the chaos consists of later pre-Cambrian and Cambrian sedimentary rocks. Most Cambrian rocks in the southern Death Valley region lie east and northeast of the Virgin Spring area, and most later pre-Cambrian sedimentary rocks lie southeast of the area. The only exposures of Cambrian and later pre-Cambrian rocks lying westward from the area are in narrow strips along the bases of the Panamint and Owlshead ranges, where they are extremely disordered. The core and western slope of the Panamint Range consist of older pre-Cambrian metamorphic rocks. The known distribution of the rocks therefore suggests an eastern rather than a western source for the material in the Virgin Spring phase of the chaos.

Some facts, indeed, appear to preclude a western source. The blocks of later pre-Cambrian sedimentary rocks in the chaos include Beck Spring dolomite and the Kingston Peak formation. The writer has found no outcrops of either of these formations anywhere in a wide region lying westward from the Virgin Spring area. Conversely, he has found in the chaos no blocks of Murphy's Telescope group of pre-Cambrian rocks, which occurs in the Panamint Range, west and northwest of the Virgin Spring area. Furthermore, the granite in the Jubilee phase of the chaos resembles granite that is exposed in the Black Mountains and Kingston Range but is not found in the Panamint Range and Owlshead Mountains, westward from the Virgin Spring area.

The arrangement of the blocks in the Amargosa chaos is also believed to give evidence as to the direction of movement. The rude shingling, or imbricate structure, which is, very broadly, a piling up of tabular blocks that for the most part dip eastward, indicates that the relative njovement of the upper plate was from east to west (Fig. 5).

*Date of thrusting.*—The date of the Amargosa thrust is probably post-Miocene, for the Amargosa chaos contains some Tertiary rocks that are believed to be not older than Miocene and are probably lower Pliocene. If the three phases of the chaos are the result of independent movements, it is not possible to assign definite dates in the Tertiary to these movements until more is learned about the Tertiary formations from which the blocks in the phases are derived.

ROCKS LATER THAN THRUSTING

*Funeral fanglomerate.*—Lying unconformably upon the older pre-Cambrian gneiss and the three phases of the Amargosa chaos is a series of poorly consolidated sediments, largely fanglomerate in the upper part, but partly fanglomerate and partly talus or landslide breccia in the lower part. Basalt flows are interbedded with the fanglomerate. The beds are somewhat folded, but their prevailing dips are eastward at angles between 25 and 35 degrees (Pl. 3). The series is also cut by faults, most of them normal.

The basalt flows and associated fanglomerate are continuously traceable northward along the Black Mountains and Greenwater Range from Epaulet Peak, in the Virgin Spring area, to the type locality of the Funeral fanglomerate north of Ryan (Ph 1) and they are therefore correlated with this formation, which is believed to be mainly later Pliocene but possibly in part early Pleistocene.

The surface of erosion upon which the Funeral fanglomerate rests is very uneven and is overlain in some places by the breccias in the lower part of the formation, in other places by the basalt and fanglomerate of the upper part. The thickness of the formation thus varies widely from place to place. Where greatest, it is as much as 2500 feet. . . .

The breccias form lenticular beds which range in thickness from a few feet to more than 200 feet and commonly extend as much as half a mile along the strike. Each bed consists of angular fragments from a few inches to many feet in diameter. Some larger blocks are as much as 100 feet long. The material is wholly unstratified and jumbled. Each bed consists almost entirely of one type of rock, commonly a rock that underlies the Funeral formation somewhere near-by. At Jubilee Pass, for example, the underlying rock is a coarse augen gneiss, and the breccia is composed wholly of fragments of this gneiss; whereas, east of upper Virgin Spring Canyon, where the formation overlies the Calico phase of the chaos, the breccia beds are composed of fragments of rhyolite and tuff derived from that phase (Pl. 18, fig. 2) . . . .

It is clear that the breccias were deposited while topographic relief was great and a vast amount of broken-up rock material remained in the region from the faulting that produced the Amargosa chaos and, apparently, while the region was still tectonically active. But, in contrast with the breccias of the Jubilee phase of the chaos, which are regarded for the most part as broken-up thrust slices of rock formations, the breccias of the Funeral fanglomerate are interpreted for the most part as deposits of talus or of coarse fanglomerate. Some of them resemble landslide breccias described by Longwell (1936 p. 147) and by Woodford and Harris (1928, p. 279-283, 287-290) and may represent landslides.

There is no sharp boundary between the part of the Funeral formation that consists chiefly of breccia and the part that consists chiefly of fanglomerate. The breccia beds are most numerous and thickest near the base, gradually decreasing in number upward until all the rock is fanglomerate. The maximum thickness of the part that contains the breccia beds is about 1500 feet.

* * * * *

## FOLDING LATER THAN AMARGOSA THRUST FAULT

*General statement.*—There were two periods of folding in the Virgin Spring area later than the Amargosa thrust. The earlier folding took place before the deposition of the Funeral fanglomerate, which appears to have been deposited in tectonic depressions largely determined by the folding. This earlier folding is along axes that trend northwest and is believed to have been accompanied by faulting of unknown extent. The later folding involves the Funeral fanglomerate, and the axes of these later folds in part follow the axes of the earlier folds and in part are oblique to them. The folding was accompanied or followed by largescale normal faulting.

*Pre-Funeral folding.*—The Amargosa thrust, together with the overlying chaos, has been folded into three main anticlines, which trend between N. 30° W. and N. 50° W. and plunge to the southeast; they have been named the Graham, Rhodes, and Desert Hound anticlines. Erosion has removed the chaos from the crests of these anticlines and has exposed the pre-Cambrian gneiss below the thrust plane, so that the anticlines appear on the map as areas of gneiss surrounded by the Amargosa chaos (Pl. 3). The exhumed smooth, broadly convex surfaces of gneiss are the "turtleback" surfaces that are the outstanding topographic features of the slopes of the Black Mountains bordering Death Valley (Curry, 1938) . . .

*Post-Funeral folding.*—The folds that involve the Funeral fanglomerate are all gentle. Some of them follow the older folds and trend northwest; others cut across the older axes of folding.

The Death Valley syncline, the Confidence anticline, the shallow syncline in Rhodes and Bradbury washes (Pl. 12, fig. 2), and a part of the Malpais Hill syncline (Pls. 2, 3, 4), are all the result of the later folding and follow the axes of the older folding. The Death Valley syncline coincides with the present Death Valley trough. The Confidence Hills anticline is a local fold in the Death Valley syncline and is parallel with its axis (Pl. 7, fig. 1; Pl. 16, fig. 2) . . .

*Post-Funeral faulting.*—The entire Virgin Spring area is cut by a network of faults, which are so numerous that only the larger ones are shown on the map. Many faults were observed whose throw is only a few tens of feet, but some of the large faults are measurable in thousands of feet. Most of the faults are normal, but at least two are reverse faults, and some recent faults in the Death Valley trough appear to be strike-slip faults. Most of the larger faults follow the limbs of the later folds. This association suggests that the faults are surficial tensional features resulting from the folding. Some of the larger normal faults follow the plane of the Amargosa thrust where the plane dips steeply but diverge from it or cut it where it dips gently.

* * * * *

## ORIGIN OF DEATH VALLEY

The structure of the southern part of Death Valley, in the Virgin Spring area, is thought to be essentially a syncline modified by normal faulting rather than a graben bounded by normal faults. The western slope of Smith Mountain, which dips steeply toward Death Valley, consists mainly of the pre-Cambrian gneiss, overlain near Scotty Canyon by several klippen of the Amargosa chaos (Pl. 16, fig. 1). This general scope of the mountain is essentially the somewhat dissected floor of the Amargosa thrust fault, which dips under the Quaternary deposits of Death Valley and apparently rises to the surface in the Panamint Mountains west of the valley. West of Ashford mill the Funeral fanglomerate including the basalt member, which is exposed on both sides and in the bottom of Death Valley, is bent into a shallow syncline which coincides with the Death Valley trough (Pl. 3; Pl. 4, C-C'). This part of Death Valley is therefore interpreted as primarily a syncline in the Funeral fanglomerate, which follows the trend of the folding of the Amargosa thrust fault and chaos. Normal faults cut the Funeral fanglomerate on the east limb of the syncline and drop the fanglomerate stepwise toward Death Valley (Pl. 3; Pl. 4, B-B'). These faults are therefore later than the Pliocene (?) Funeral fanglomerate.

Very recent scarps in Quaternary alluvium show moderate displacements on these faults in the same direction.

It may be that faulting on a larger scale than is now evident occurred after the folding of the Amargosa thrust and chaos and prior to the deposition of the Funeral fanglomerate, and that these faults are concealed by the fanglomerate and by Quaternary deposits. The talus and landslide breccias at the base of the Funeral fanglomerate east of Jubilee Wash may have been deposited along such pre-Funeral fault scarps.

# Levi Noble's Death Valley, a 58-year perspective

**Lauren A. Wright**
*500 E. Marilyn Avenue, E69, State College, Pennsylvania 16801*
**Bennie W. Troxel**
*2961 Redwood Road, Napa, California 94558*

## INTRODUCTION

Levi Noble, in his 1941 classic paper entitled "Structural features of the Virgin Spring area, Death Valley, California," included an overview of what was then known of the geology of the Death Valley region. The overview served as background for his descriptions and interpretation of the complex and puzzling structural features that he termed the "Amargosa chaos." As Noble's paper is 59 pages long and includes six figures and 20 plates, the version reprinted here is necessarily much abridged. The interested reader is thus referred to the original paper and to a later paper and accompanying map (Noble and Wright, 1954). We frequently refer to each.

Since 1930 Noble had been engaged in a reconnaissance mapping of the Death Valley region for the U.S. Geological Survey. Before then the major faults that have shaped the present landscape remained either unplotted or only broadly delineated. Radiometrically determined ages of the igneous and metamorphic rocks were not yet available, and only the 1°-1:250,000 quadrangles were available as base maps. Also at that time, thrust faults were in vogue; extensive metamorphism was commonly cited as evidence for a Precambrian age; and the role of monolithologic sedimentary breccias in sedimentary successions was only beginning to be appreciated. The existence of low-angle normal faults in the Basin and Range province had been noted by W. M. Davis (1922), but he attributed them to "underdrag" in the upper plate of an advancing thrust fault, rather than to an extending crust.

By 1941, however, the general features of the Proterozoic and Paleozoic formations of the Death Valley and bordering regions had been described by Nolan (1929), Noble (1934), Hazzard (1937), and Hewett (1940), and the central part of the Panamint Range had been mapped in reconnaissance by Murphy (1932). Also by 1941, Noble and his younger coworkers, Don Curry and Tom Thayer, had plotted on their field maps most of the major geologic features of the region. Indeed, Noble's 1941 paper includes a basically accurate cross section, 60 km long, from the Black Mountains eastward to the Kingston Range. In our brief presentation we emphasize the features in the Death Valley terrane that were especially obvious to Noble and his immediate contemporaries and discuss how their early perceptions of these features have fared.

## THE 1954 MAP

In the early 1950s, Richard Jahns, editor of California Division of Mines Bulletin 170, *Geology of Southern California*, invited Noble to contribute his reconnaissance mapping in Death Valley. In this endeavor he also sought the assistance of Lauren Wright, who had visited with Noble frequently in connection with Wright's work on the talc deposits of the region. Noble complied, at first hesitantly and then enthusiastically. Aided by Wright, he produced, on a 1:250,000 scale, the first geologic map of the Death Valley region. It was accompanied by a text and a somewhat revised version of the earlier cross section (Noble and Wright, 1954). The map included generalized versions of the mapping of the Quail Mountains by Muehlberger (1954), of the Silurian Hills by Kupfer (1954), and of various talc-bearing areas by Wright. Wright accompanied Noble on a two-week reconnaissance in which blank areas were filled in. Otherwise the mapping and most of the interpretations are Noble's. We refer to the product as "the 1954 map."

The 1954 map faithfully shows the distribution of the unmetamorphosed occurrences of the Proterozoic Pahrump Group and of the overlying succession of later Proterozoic and Paleozoic formations. Noble divided the Cenozoic sedimentary rocks into a unit that later proved to predate the Basin and Range extensional event and another that accompanied it. The map also

---

Wright, L. A., and Troxel, B. W., 1999, Levi Noble's Death Valley, a 58-year perspective, *in* Moores, E. M., Sloan, D., and Stout, D. L., eds., Classic Cordilleran Concepts: A View from California: Boulder, Colorado, Geological Society of America Special Paper 338.

showed the components of the igneous terrane now known as the central Death Valley plutonic-volcanic field. In addition, it was the first map to show the major faults of the Death Valley region in a geologic setting, although they were shown on earlier fault maps (Lawson et al., 1908; Willis and Wood, 1922).

Without radiometrically determined ages Noble and Wright (1954) were unable to distinguish between plutons since assigned to either Mesozoic or Tertiary time; nor could they date the metamorphic events recorded in various parts of the Death Valley terrane. They thus combined under "undifferentiated pre-Cambrian rocks" units as diverse as: (1) the pre-Pahrump quartzofeldspathic crystalline complex later dated as about 1.7 Ga (Wasserburg et al., 1959; Silver et al., 1961); (2) the strongly metamorphosed succession composed of the Pahrump Group and the overlying Johnnie Formation and Stirling Quartzite in the northern Funeral Mountains (Troxel and Wright, 1968; Labotka et al., 1980; Wright and Troxel, 1993); (3) the Mesozoic plutons and associated metamorphosed pendants of Proterozoic and Paleozoic formations of the Avawatz Mountains (Spencer, 1990); and (4) the Willow Spring diorite-gabbro that underlies most of the central Black Mountains and is now dated as 11 to 12 Ma (Asmerom et al., 1990).

Noble and Wright did not view the Death Valley terrane as significantly extended. The Amargosa "thrust" and other elements of the Death Valley terrane are shown on the 1954 map as displaced by normal faults, but these are few and tend to be moderately to steeply dipping. Although Chester Longwell (1945) had identified downward-flattening faults in the Las Vegas quadrangle of southern Nevada, he stopped short of suggesting that they join along single low-angle regional faults.

## AREAL GEOLOGY

Since 1954 most of the Death Valley region has been covered by numerous maps at scales of 1:96,000 and larger (Table 1B; see especially Hunt and Mabey, 1966). These maps and numerous others add a wealth of detail to Noble's original observations, but also generally confirm them. The interpretations, however, have continued to evolve. Since then the late Cenozoic features, in particular, have been attributed to an intensely extending crust, a possibility that escaped Noble's original speculations but toward which he seemed to be moving in our later conversations with him. Subsequent estimates of the degree of crustal extension within the central Death Valley area, however, have been strikingly disparate, ranging from several tens (Serpa and Pavlis, 1996) to several hundreds (Wernicke, 1992; Topping, 1993) percent.

## PRE-CENOZOIC SEDIMENTARY GEOLOGY

In dealing primarily with the Cenozoic features of the Death Valley region, Levi Noble necessarily viewed the Proterozoic and Paleozoic formations mainly as building blocks in his Cenozoic terrane. Beginning in the 1970s, however, following an investigation by John Stewart (1970) that included the description of the late Proterozoic and Cambrian formations at various localities of the Death Valley region, most of the Proterozoic and Paleozoic formations have been studied individually, in various degrees of detail. The emphasis has been on sedimentology and depositional environments, but also includes paleontology and, in recent years, sequence stratigraphy (Table 1C).

As summarized by Wright and Prave (1993), the Middle Proterozoic through Lower Cambrian formations of the Death Valley region portray a transition from an early extensional, aulacogen-like terrane of basins and highlands to the formation of the northwestward-thickening clastic-dominated prism of the Cordilleran miogeocline. Perspectives on the timing and stratigraphic expression of rift to drift transition of the Cordilleran miogeocline are also provided in Bahde et al. (1997) and Cooper and Fedo (1998). The dominantly carbonate younger (Middle Cambrian–Carboniferous) formations of the Death Valley region have proven to be part of the Las Vegas branch of the Great American Bank of the Cordilleran miogeocline. Coverage of many of the carbonate and other units was provided by Cooper (1989, 1995), Cooper and Stevens (1991), Cooper et al. (1982), Osleger and Montanez (1996), Palmer and Halley (1979), and Levy and Christie-Blick (1991).

## INTERPRETING THE CENOZOIC TECTONIC FRAMEWORK

A large and growing literature on the Cenozoic tectonics of the Death Valley region was spawned with the representation on the 1954 map of the major faults in a geologic setting (Table 1B). Most of these investigations treat one or more aspects of the history of the triangular crustal block that is between the northwest-trending Furnace Creek fault zone and the mostly west-trending Garlock fault zone (Fig. 1). Although the evolution of this block has been interpreted differently by different workers, all now agree that its crust has been much more extended than the bordering crust to the northeast and south. The following are brief treatments of three features that have attracted special attention.

### Furnace Creek fault zone

We quote Noble and Wright (1954, p. 153) from the text that accompanied the 1954 map.

> The Furnace Creek fault zone is poorly exposed and the details of structure are unknown, but it is believed to be a major line of dislocation that has been recurrently active since late Mesozoic or early Tertiary time, with reversals of apparent vertical movement. Horizontal striae on vertical fault surfaces west of Echo Canyon suggest that it may have had a large horizontal as well as vertical component.

Curry (1938) was, in fact, the first to report this evidence of strike-slip displacement. Subsequent investigators have confirmed the earlier suspicion of large offset on the Furnace Creek

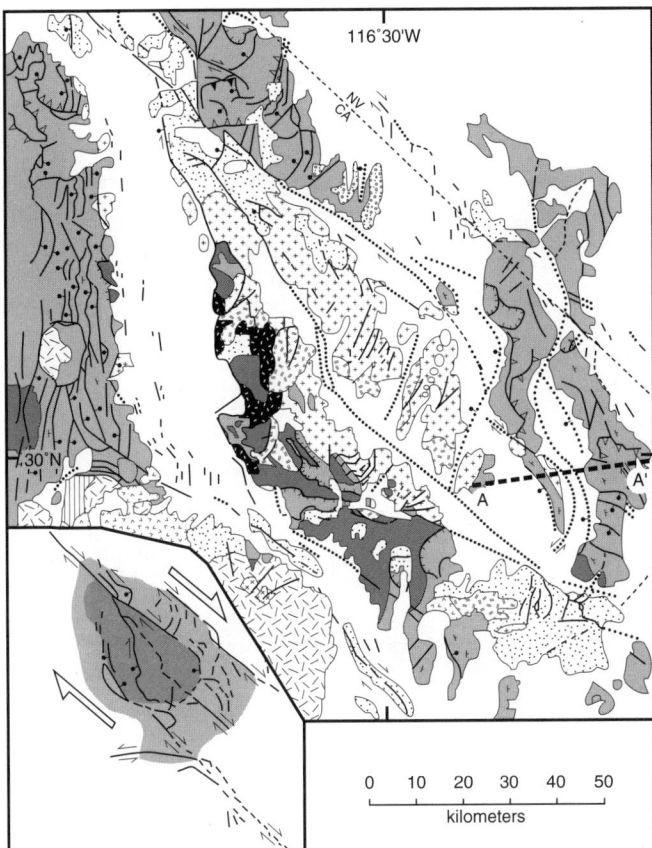
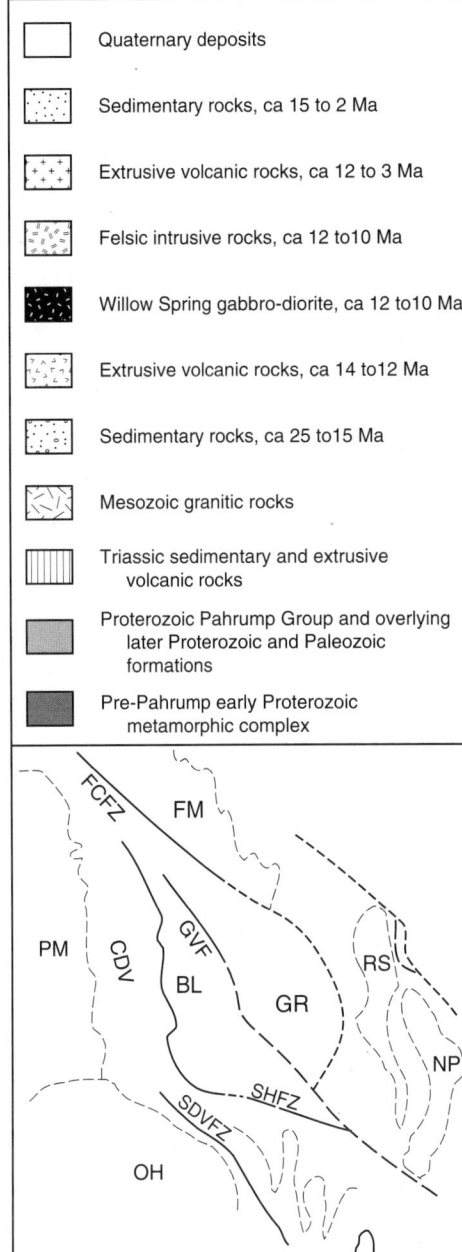

Figure 1 (above and to the right). Tectonic map of the central and southern parts of the Death Valley region. Extensional detachment faults are identified by hachures, Mesozoic thrust faults by open barbs, and Cenozoic thrust faults by solid barbs. Inset shows the principal faults of the Death Valley region interpreted as defining a large-scale dextral pull-apart between the Furnace Creek and southern Death Valley fault zones. Heavy stippling shows area of major extension between the Furnace Creek and Sheephead faults; lighter stippling shows a bordering area of strong, but less-severe extension. Index map beneath the explanation shows the principal topographic features and the major strike-slip faults of the region of the tectonic map: BL = Black Mountains; CDV = central Death Valley; FCFZ = Furnace Creek fault zone; FM = Funeral Mountains; GR = Greenwater Range; GVF = Grand View fault; NP = Nopah Range; OH = Owlshead Mountains; PM = Panamint Range; RS = Resting Spring Range; SHFZ = Sheephead fault zone; SDVFZ = southern Death Valley fault zone.

fault zone. The earliest of their reports, reviewed by Stewart et al. (1968), cite right-lateral offsets of tens of kilometers in isopachs and facies trend lines in Paleozoic formations exposed on opposite sides of the fault. Also in 1968 McKee described a 20 km offset of a Mesozoic pluton. Later workers (e.g., Snow and Wernicke, 1989) have stressed offsets on Mesozoic thrust faults. The kinematics of displacement along the 200 km length of the fault, however, have remained in dispute and are briefly referred to in the following.

Again we quote (italics are ours) Noble and Wright (1954, p. 153).

The earliest movements [*on the Furnace Creek fault*] raised the Black Mountain wedge, from which most of the Paleozoic rocks were removed during the Mesozoic-early Tertiary interval. Later in the Tertiary, a reversal of movement elevated the terrane of Paleozoic rocks northeast of the wedge and determined a northwest-trending synclinal trough that extended from North Death Valley through the site of the present Furnace Creek Valley to Amargosa Valley. The Furnace Creek formation was deposited in this trough.

The notion that the missing Paleozoic rocks in the Black Mountains and Greenwater Range had been eroded during uplift on the southwest side of the Furnace Creek fault has proven only

## TABLE 1. SELECTED REFERENCES BY CATEGORY (cf. References Cited)

**A. Levi F. Noble**
- Bradley, 1966
- Wright and Troxel, 1992

**B. Areal and structural geology and regional tectonics**
- Albee et al., 1981
- Applegate and Hodges, 1995
- Asmerom et al., 1990
- Brady, 1986
- Bucher, 1956
- Burchfiel and Stewart, 1966
- Burchfiel et al., 1983
- Butler et al., 1988
- Curry, 1938
- Curry, 1954
- Davis, 1922
- Davis and Burchfiel, 1973
- Denny and Drewes, 1965
- Dooley and McClay, 1996
- Drewes, 1963
- Fowler and Calzia, 1999
- Greene, 1997
- Haefner, 1972
- Hamilton, 1988
- Hill and Troxel, 1966
- Hodges et al., 1985
- Holm, 1992
- Holm et al., 1992
- Holm et al., 1993
- Holm et al., 1994
- Hopper, 1947
- Hunt and Mabey, 1966
- Jennings, 1985
- Johnson, 1957
- Kupfer, 1954
- Kupfer, 1960
- Labotka et al., 1980
- Lawson et al., 1908
- Longwell, 1945
- Mancktelow and Pavlis, 1994
- Mason, 1948
- McAllister, 1970
- McAllister, 1971
- McAllister, 1973
- McKee, 1968
- McKenna and Hodges, 1990
- Miller, 1992
- Miller, in press
- Muehlberger, 1954
- Murphy, 1932
- Noble, 1941
- Noble and Wright, 1954
- Otton, 1976
- Otton, 1977
- Pavlis, 1994
- Prave and Wright, 1986
- Ransome et al., 1910
- Reynolds, 1976
- Sears, 1953
- Serpa, 1990
- Serpa and Pavlis, 1996
- Serpa et al., 1988
- Smith, 1962
- Snow, 1990
- Snow and Wernicke, 1989
- Spencer, 1990
- Stewart, 1983
- Stewart et al., 1968
- Topping, 1993
- Troxel and Butler, 1998
- Wagner and Hsü, 1987
- Wernicke, 1981
- Wernicke, 1992
- Wernicke et al., 1988
- Wernicke et al., 1989
- Wernicke et al., 1993
- Wilhelms, 1963
- Willis and Wood, 1922
- Wright, 1954
- Wright, 1968
- Wright, 1987
- Wright and Troxel, 1969
- Wright and Troxel, 1971
- Wright and Troxel, 1973
- Wright and Troxel, 1984
- Wright and Troxel, 1993
- Wright et al., 1981
- Wright et al., 1984
- Wright et al., 1991

**C. Pre-Cenozoic sedimentary geology**
- Adams and Grotzinger, 1996
- Albright, 1989
- Albright, 1991
- Bahde et al., 1997
- Cooper and Edwards, 1991
- Cooper and Fedo, 1998
- Cooper and Keller, 1995
- Cooper et al., 1982
- Diehl, 1976
- Gutstadt, 1968
- Hazzard, 1937
- Hewett, 1940
- Levy and Christie-Blick, 1991
- Maud, 1979
- McCutcheon and Cooper, 1989
- Miller, 1985
- Nolan, 1929
- Osleger and Montanez, 1996
- Osleger et al., 1976
- Palmer and Halley, 1979
- Prave, 1991
- Prave, 1996
- Roberts, 1976
- Roberts, 1982
- Silver et al., 1961
- Stewart, 1970
- Stewart, 1972
- Stewart, 1982
- Summa, 1993
- Troxel, 1982a
- Troxel, 1982b
- Troxel and Wright, 1968
- Wasserburg et al., 1959
- Wertz, 1982
- Williams et al., 1976
- Wright and Prave, 1993
- Wright and Troxel, 1966
- Wright et al., 1976
- Wright et al., 1978

**D. Cenozoic sedimentary geology**
- Beratan et al., 1999
- Blair and Raynolds, 1999
- Brady and Troxel, 1999
- Cemen et al., 1985
- Cemen et al., 1999
- Fridrich, 1999
- Friedmann, 1999
- Knott et al., 1999
- Morrison, 1999
- Prave and McMackin, 1999
- Snow and Lux, 1999
- Stock and Bode, 1935
- Wright et al., 1999

partly supportable by the geologic record. Direct evidence of such erosion has been found only in the conglomerate member of the Bat Mountain Formation (ca. 16 Ma) at the southern end of the Funeral Mountains (Cemen et al., 1985, 1999). We now attribute the scarcity of exposures of late Proterozoic and Paleozoic formations in the Greenwater Range and Black Mountains to a combination of: (1) erosion both before and after the inception of the Furnace Creek fault; (2) engulfment and covering by the plutons and extrusive bodies of the central Death Valley plutonic-volcanic field; and (3) tectonic denudation.

### The type pull-apart basin

The 1954 map clearly showed the subdivision of Death Valley into (1) a northwest-trending northern part, controlled by the Furnace Creek fault zone and related faults; (2) a north-trending central part bordered on the east by the frontal fault of the Black Mountains; and (3) a northwest-trending southern part with several northwest-trending faults (Fig. 1). It also showed the faults of southern Death Valley as collectively extending arcuately northward into the central part of the valley, the entire feature being designated the "Death Valley fault zone." The trace of the Furnace Creek fault zone is shown by solid lines that extend southwestward into the lower part of Furnace Creek Wash and as a dotted line between there and the Amargosa Valley.

Burchfiel and Stewart (1966) observed (1) the indicators of strong right-lateral movement on the Furnace Creek fault zone; (2) the paucity of ground ruptures in the southeasternmost part of the Furnace Creek fault zone; (3) the dominantly dip-slip nature of the frontal fault of the Black Mountains; and (4) the parallelism between the Furnace Creek fault zone and the faults that occupy southern Death Valley. They then reasoned that central Death Valley is essentially a half-graben pulled apart between the en echelon terminations of the active part of the Furnace Creek fault zone and the fault zone that occupies southern Death Valley. Because theirs was the initial description of such a basin, central Death Valley qualifies as the type "pull-apart basin" and the starting point for later field and theoretical studies of transtensional and transpressional features along zones of strike-slip displacement. The pull-apart concept was foreshadowed in a succession of block diagrams by Drewes (1963), but he fell short of identifying central Death Valley as an extensional basin.

### The central Death Valley plutonic-volcanic field: An igneous rhombochasm?

The 1954 map clearly showed the terrane of extrusive and intrusive igneous bodies that underlies most of the Black Mountains and Greenwater Range. Unaware that these bodies share a late Cenozoic age, Noble assigned the Willow Creek gabbro-diorite (Fig. 1) to Precambrian time. He also designated all of the granitic bodies on the map as "mostly Mesozoic" but added that they "may be in part Tertiary." Even in the 1941 paper (p. 954), he observed that although the emplacement of granitic rock in the Death Valley region "may be roughly contemporaneous, it more probably represents two or more widely separated periods of intrusion." Noble also recognized an older-younger subdivision of the extrusive units. Subsequent detailed mapping, outcrop observations, radiometric dating, and related laboratory investigations have documented the late Cenozoic age for both the intrusive and extrusive bodies (e.g., Drewes, 1963; McAllister, 1971, 1973; Haefner, 1974; Otton, 1976, 1977; Wright and Troxel, 1984; Holm, 1992; Holm et al., 1992; Miller, 1992, 1999). The Black Mountains, in particular, have been shown to contain an exceedingly complex array of extension-related plutons and dikes (see Holm et al., 1994, for a review of these investigations).

The post-1954 investigations also have yielded a coherent, multiunit, volcanic stratigraphy consistent with Noble's younger-older subdivision and dominated by mafic to felsic magmatic flows and air-fall tuffs (Wright et al., 1991). The magmatic flows have been fed through closely spaced and widely distributed dikes and stocks driven by local extensional events within the field. Only one extensive ash-flow tuff—the Rhodes Tuff—has been detected.

As early as the 1930s Levi Noble must have wondered why the dominantly igneous terrane of the Black Mountains and Greenwater Range is surrounded by ranges composed mostly of Proterozoic and Paleozoic sedimentary formations (Fig. 1). Much later, Wright and Troxel (1971) suggested that the volcanic bodies were emplaced in preferentially extended crust which, like that of the central Death Valley pull-apart basin, is between the en echelon terminations of two strike-slip faults, specifically, the southeasternmost part of the Furnace Creek fault and the northwesternmost part of the Sheephead fault (Fig. 1). This tectonic setting later led to the designation of the entire feature as an igneous rhombochasm (Wright et al., 1991).

Wright and Troxel's (1971) initial assumption that the Sheephead fault has moved in a right-lateral sense was based, in part, upon its parallelism with the Furnace Creek fault and upon the direction of extension theoretically required for a true rhombochasm. However, the field evidence for right-lateral movement has proven inconclusive. Indeed, on the 1954 map, as well as in a recent three-dimensional tectonic model by Serpa and Pavlis (1996), the Sheephead fault has been represented as left lateral. A further investigation of this fault is obviously needed.

## CENOZOIC SEDIMENTARY GEOLOGY

On the 1954 map Noble designated the Tertiary sedimentary and volcanic rocks as "older" or "younger" mainly on the basis of stratigraphic position, shared lithologies, and scanty paleontologic evidence. Since then, the sedimentary deposits have yielded numerous radiometrically determined ages, and their depositional environments have been addressed in numerous papers (Table 1D). The later studies have corrected Noble's correlation of the Oligocene Titus Canyon Formation of the Funeral and Grapevine Mountains with the Artist Drive Formation (ca. 14 to 6 Ma) of the Black Mountains–Furnace Creek Wash area, and the succession (ca. 25 to 16 Ma) at the southern end of the Funeral Mountains (Cemen et al., 1985). Also, at various localities, deposits that Noble correlated with the Pliocene Funeral Formation are now known to be Miocene in age.

The later studies, however, have supported Noble's premonition that there is also a tectonic basis for the subdivision of the Tertiary sedimentary record into older and younger deposits. Specifically, they reveal stratigraphic evidence of the change, about 16 m.y. ago, from the broad terrain that preceded the initiation of major crustal extension to the formation of the present basins and ranges (Cemen et al., 1985, 1999; Snow and Lux, 1999).

## ORIGIN OF THE AMARGOSA CHAOS

At the end of his 1941 paper, Noble cited numerous localities, well beyond the Black Mountains, where he had observed features like the Amargosa chaos. He concluded, "It is thus evident that the Amargosa thrust and chaos, or structural features similar to them, are present within a northwest-trending belt, at least 150 miles long and in places at least 75 miles wide, which borders and includes the Death Valley trough" (Noble, 1941, p. 997).

For the next 32 years, Noble and later workers attributed the chaos to mechanisms other than crustal extension. However, he also was gradually moving away from the "regional thrust" interpretation. The 1954 map, cross section, and text represented most or all of the Jubilee phase of the chaos as being composed of sedimentary breccias. Although the Virgin Spring and Calico phases continued to be viewed as the upper plate of a thrust fault, the thrusting was attributed to compressional arching in the Black Mountains and was apparently related to major movement on bounding strike-slip faults. Noble's growing uneasiness with the "thrust fault" interpretation was recorded in an unpublished field trip guide, written in the late 1950s after he had frequently accompanied us to the Virgin Spring area. He wrote "In the Virgin Spring paper I assumed that the Virgin Spring chaos, Jubilee rocks, and Calico rocks formed a composite thrust plate of Pliocene age. This hypothesis doesn't look so good right now, but it still remains to be completely knocked out."

Others also were questioning the thrust-fault interpretation. Landsliding off a rising structural block was favored by Sears (1953) and Bucher (1956). Kupfer (1960) suggested near-surface bifurcation beneath a rising thrust plate. Sears attributed the adjacent uplift to injection of felsic magma; Bucher suggested that the landsliding was favored by the injection of hot fluids beneath the chaos; Kupfer extrapolated his interpretation from observations in the Silurian Hills southeast of the Black Mountains.

However, the thrust fault interpretation lived on. Drewes (1963) reviewed the three interpretations and favored the one by Noble and Wright (1954). Two publications in 1966, Hunt and Mabey (1966) and Hill and Troxel (1966), referred to the "Amargosa thrust." Hunt and Mabey accepted Noble's initial suggestion that the fault extended well beyond the limits of the Black Mountains.

We interpreted the Virgin Spring and Calico phases as extensional features (Wright and Troxel, 1969, 1973) after Noble's death in 1965. While completing our mapping of the Virgin Spring area, we observed that the Calico phase, although much less attenuated than the Virgin Spring phase, had acquired its chaotic appearance by movement along innumerable normal faults, most of which terminate in the underlying Virgin Spring phase. We also observed that other ranges of the region are bounded and cut by normal faults, and we questioned the association of extreme attenuation with crustal compression. In addition, we were aware that Anderson (1971), while mapping in the Lake Mead area of southern Nevada, had observed the downward termination of numerous normal faults against a single low-angle fault. It is now obvious that Noble provided an exceptionally detailed description of one or more extensional detachment surfaces long before this term entered the geological vocabulary.

## EXTENSIONAL TECTONIC MODELS FOR THE DEATH VALLEY REGION

A detailed reconstruction of the extensional tectonic evolution of the Death Valley region obviously requires the evaluation of a very large array of geological, geophysical, and geochemical data. This formidable task remains to be undertaken and complete objectivity is virtually impossible. Predictably, the existing interpretations have proven controversial, and the subject is too complex to be reviewed adequately here. In brief, the proposed tectonic models have been broadly grouped into the "shallow fault" and the "large-scale dextral shear" models. The proponents of each assign major roles to both low-angle extensional faults and strike-slip faults, the differences being mainly in the geometry and kinematics of the faulting, especially in the lateral extent of individual detachment faults, the degree of implied crustal thinning, and the depths to which the range-bounding faults penetrate the crust.

The "rolling hinge" model (Fig. 2), the most frequently mentioned of the shallow fault models (Wernicke, 1992; Holm et al., 1993), apparently requires the presence of one or more detachment surfaces at depths of 5 km or less beneath the entire Death Valley region. The application of the model is constrained, however, if range-bounding normal faults, such as the frontal faults of the Resting Spring and Nopah Ranges, can be demonstrated to penetrate the crust deeply.

Again we detect Levi Noble's influence. We paraphrase a proponent of the shallow fault model: "I am dealing with Noble's regional Amargosa thrust, but it is an extensional rather than a compressional feature." Those who favor the large-scale dextral shear model, however, recall that, in his later years, Noble assigned progressively greater importance to faults now recognized as right lateral.

## Large-scale dextral shear model

In the earliest version of this model Hill and Troxel (1966) proposed that the Cenozoic tectonic features of the Death Valley region are compatible with a strain system involving right-lateral movement on northwest-striking faults, northeast-southwest shortening, and moderate northwest-southeast extension. Although they observed the presence of normal faults, by referring to the Amargosa "thrust," they implicitly attributed the chaos to shortening. That the extensional component is nonuniformly distributed was implied in the proposal that the Central Death Valley plutonic-volcanic field formed in a pull-apart setting within the larger extensional framework (Wright and Troxel, 1971). Nonuniformity is also implied in the association of chaos with extreme extension (Wright and Troxel, 1973) and in the interpretation of the Amargosa fault as also limited to a part of the larger framework.

Paradoxically, the role of range-bounding normal faults in the large-scale dextral shear model was addressed in a simplistic cross section (Wright and Troxel, 1973) drawn from the Pahrump Valley to the Panamint Range, illustrating what we called a "shallow fault" interpretation. In the eastern part of this cross section (Fig. 2A), the frontal faults of the east-tilted Nopah and Resting Spring Ranges and Dublin Hills are shown as terminating downward at about 7 km, a depth viewed as compatible with the characteristic 45° to 55° dips of the constituent formations. By theoretically restoring these fault-bounded blocks to their original positions, we estimated an extension of 30% to 50%. Not included in Figure 2A is the central part of the original cross section in which the Amargosa Chaos and detachment fault, as well as Cenozoic intrusive bodies, are shown. There the percentage of extension must be much greater.

Evidence that the frontal faults of the Resting Spring and Nopah Ranges extend to a regional detachment zone at a depth of about 15 km was later observed in the deep seismic reflection profiles of the Death Valley COCORP project (Fig. 2b; Serpa et al., 1988). The implied geologic cross section (Fig. 2C) was viewed by Serpa et al. (1988) as a modification of our earlier one. We continue to agree with this interpretation.

We observe additional evidence for the deep rooting of the range-bounding faults in the presence of basaltic to intermediate volcanic flows and stocks (ca. 10 to 11.5 Ma) that flank the east side of the southern part of the Resting Spring Range (Fig. 2E). The flows onlap east-tilted Cambrian strata and are, in turn, east tilted, indicating the emplacement of mantle-derived material contemporaneously with the tilting of the principal fault block. Because the flows are south of a right-lateral strike-slip fault that cuts diagonally through the range (Fig. 1), we view the volcanic event as related to preferential extension of the southern part of the range.

In the latest version of the dextral shear model as applied to the Death Valley fault system, Serpa and Pavlis (1996) invoked measured and hypothetical vertical axis rotations and suggested a two-stage, three-dimensional tectonic history. The early stage is dominated by conjugate displacements on the Garlock and Furnace Creek fault zones; the later stage by large-scale dextral shear along major northwest-trending fault zones. Serpa and Pavlis also underscored the role of northeast-southwest shortening in the dextral shear model, especially in the formation of the well-known turtleback folds (Curry, 1954), and in partly compensating for the extension-related crustal thinning.

## Shallow fault model

Stewart (1983), in proposing that the Panamint Range block has moved 80 km westward from an original position immediately west of the site of the present Resting Spring Range, was the first to suggest that a single, low-angle normal fault once extended through a large part of the Death Valley region. The proposal also implied a right-lateral offset of comparable magnitude along the Furnace Creek fault zone and a several hundred percent extension in the area between the two

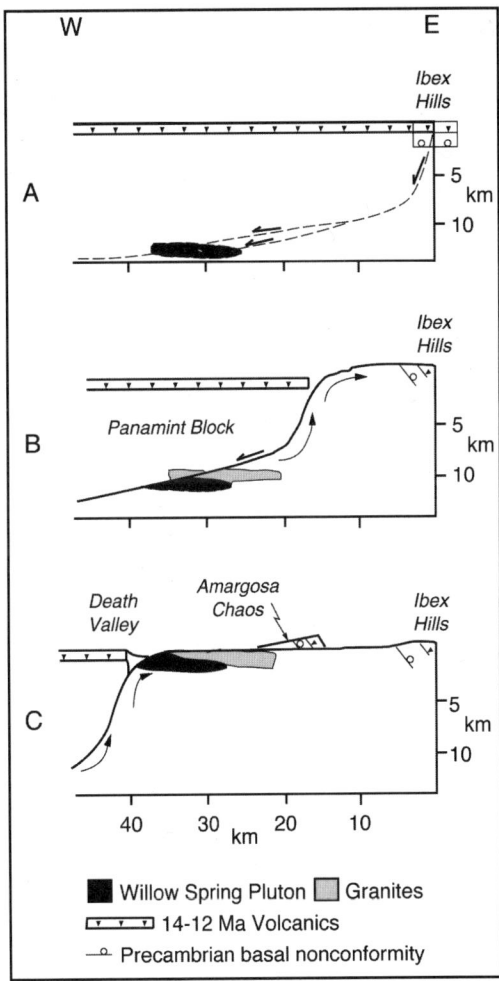

Figure 2. Sequential, idealized cross sections showing the Cenozoic tectonic development of the Death Valley region according to the rolling hinge model. Modified from Holm et al. (1993).

blocks. Stewart based this proposal mainly on (1) the nature and distribution of the Amargosa fault, (2) the presence of a west-dipping detachment fault low along the east slope of the Panamint Range, and (3) the observation that the thick succession of Proterozoic and Paleozoic sedimentary rocks that underlies most of the Panamint Range is "largely absent in the Black Mountains." He speculated that "the fault may be the upper part of a crustal scale fault, similar to the type described by Wernicke (1981) that cuts downward to the northwest rooted deep in the crust" (Stewart, 1983, p. 154).

Subsequently, Wright et al. (1984) cited evidence that, beginning about 13 Ma, the locale of the cessation of intensive extension has migrated westward from the site of the Kingston Range to at least as far as Death Valley. Still later and consistent with Stewart's suggested dislocation of the Panamint block, Wernicke et al. (1988) proposed a correlation of thrust faults that included the matching of the Panamint thrust at the northern end of the Panamint Range with the Chicago Pass thrust at the northern end of the Nopah Range, thereby theoretically placing the Resting Spring and Nopah Ranges within an area of extreme extension. These three proposals are central to the rolling hinge model of Wernicke (1992) and his coworkers. In the application of the rolling hinge to the Death Valley region, the Nopah and Resting Spring Ranges and the Black Mountains are viewed as thin slivers of crust detached sequentially, from east to west, above a migrating flexure in highly thinned crust. Wernicke's interpretation of the structural setting of the Nopah and Resting Spring Ranges is shown in Figure 2D. The basic kinematics of the rolling hinge model, as applied by Holm et al. (1994) to the area of the Black Mountains, are shown in Figure 3.

The matching of the two thrust faults, however, continues to be debated. An alternative correlation of thrusts was incorporated in the dextral shear model of Serpa and Pavlis (1996). If real, the latter correlation implies a much smaller percentage of extension in the central Death Valley region than that favored by Wernicke and his coworkers.

## CONCLUSIONS

The history of geologic investigations in the Death Valley region has been one of constantly changing hypotheses, but the region has remained and will continue to be a classic natural laboratory for the study of transtensional-extensional systems.

In reflecting upon the organization and content of the Virgin Spring paper and the 1954 map and also upon recalling our numerous conversations with Levi Noble, we remember especially the clarity of his prose, his precise separation of observation from interpretation, his prescience in sensing the findings of later workers, and his desire to be as objective as possible. We quote the final sentences in Bradley's (1966) memorial: "In the hurrying, somewhat blatant world, the value of a quiet, unassuming man is apt to be minimalized. Let that error never be made where Levi Noble is concerned, either as a geologist or as a human being."

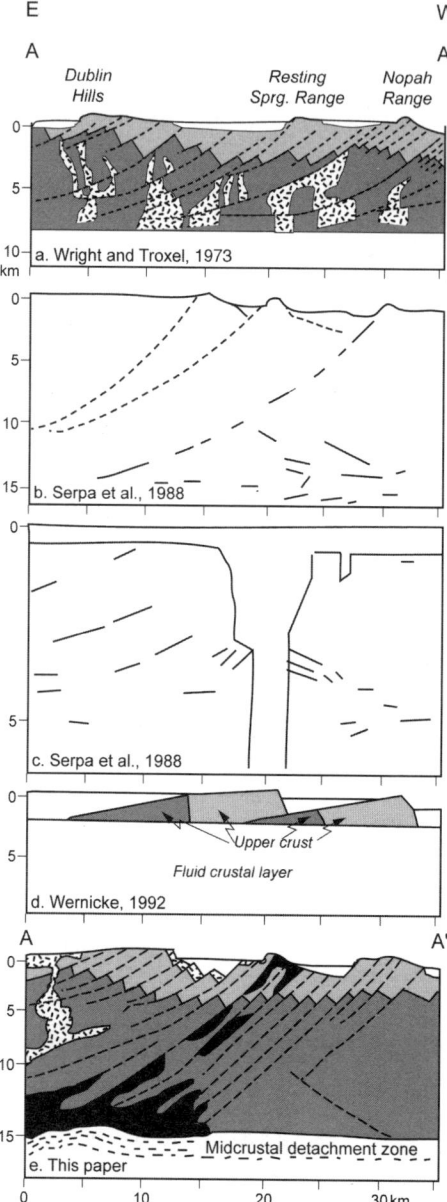

Figure 3. Successive interpretive cross sections along and near the dashed line A-A′ of the tectonic map of Figure 1. In A and C the dark shading is the Precambrian basement complex, light shading is the Proterozoic and Paleozoic cover, scattered dashes show Miocene felsic plutons and dikes, and black indicates Miocene mafic to intermediate igneous bodies exposed on the east side of the southern part of the Resting Spring Range. See text for further explanation. A: Initial cross section modified from Wright and Troxel (1973). B: Reinterpretation by Serpa et al. (1988) of the faulting shown in cross section A and based on data from the Consortium for Continental Reflection Profiling (COCORP). C: Lines drawn by Serpa et al. (1988) on COCORP profiles that served as the basis for the construction of cross section B. D: Part of a larger, simplistic hypothetical crustal section by Wernicke (1992) showing an application of the rolling hinge model to the area of the preceding cross sections. E: Our currently preferred cross section embracing the data from the COCORP profiling, the field observations of section A and others made subsequently.

## ACKNOWLEDGMENTS

We are indebted to John Cooper, John Crowell, Terry Pavlis, and Steve Rowland for their thorough and helpful reviews of the initial version of this paper. Jeff Knott helpfully and patiently supplied the computer graphics.

## REFERENCES CITED

Adams, R. D., and Grotzinger, J. P., 1996, Lateral continuity and parasequences in Middle Cambrian platform carbonates, Carrara Formation, southeastern California, U. S. A.: Journal of Sedimentary Research, v. 66, p. 1079–1091.

Albee, A. L., Labotka, T. C., Lanphere, M. A., and McDowell, S. D., 1981, Geologic map of the Telescope Peak quadrangle, California: U.S. Geological Survey Quadrangle Map GQ-1532, scale 1:62,500.

Albright, G., 1989, Stratigraphy and sedimentology of the Quartz Spring Sandstone and Tin Mountain Limestone, Funeral Mountains, California, *in* Cooper, J. D., ed., Cavalcade of carbonates: Los Angeles, Pacific Section, Society of Economic Paleontologists and Mineralogists, p. 37–44.

Albright, G. R., 1991, Late Devonian and Early Mississippian paleogeography of the Death Valley region, California, *in* Cooper, J. D., and Stevens, C. H., eds., Paleozoic paleogeography of the western United States: Los Angeles, Pacific Section, Society of Economic Paleontologists and Mineralogists, p. 253–270.

Anderson, R. E., 1971, Thin skin distension in Tertiary rocks of southeastern Nevada: Geological Society of America Bulletin, v. 82, p. 43–58.

Applegate, J. D. R., and Hodges, K. V., 1995, Mesozoic and Cenozoic extension recorded by metamorphic rocks in the Funeral Mountains, California: Geological Society of America Bulletin, v. 107, p. 1063–1076.

Asmerom, Y., Snow, J. K., Holm, D. K., Jacobson, S. B., Wernicke, B. P., and Lux, D., 1990, Rapid uplift and crustal growth in extensional environments: An isotopic study from the Death Valley region, California: Geology, v. 8, p. 223–226.

Bahde, J., Barretta, C., Cederstrand, L., Flaugher, M., Heller, R., Irwin, M., Swartz, C., Traub, S., Cooper, J., and Fedo, C., 1997, Neoproterozoic–Lower Cambrian sequence stratigraphy, eastern Mojave Desert, California: Implications for the base of the Sauk Sequence, craton margin hinge zone, and the evolution of the Cordilleran continental margin, *in* Girty, G. H., Hanson, R. E., and Cooper, J. D., eds., Geology of the western Cordillera: Perspectives from undergraduate research: Pacific Section, SEPM (Society for Sedimentary Geology) Special Publication 82, p. 1–20.

Beratan, K. K., Hsieh, J., and Murray, B., 1999, Pliocene-Pleistocene stratigraphy and depositional environments, southern Confidence Hills, Death Valley, California, *in* Wright, L. A., and Troxel, B. W., eds., Cenozoic basins of the Death Valley region: Geological Society of America Special Paper 333 (in press).

Blair, T. C., and Raynolds, R. J., 1999, Sedimentology, stratigraphy and paleotectonic implications of lower Pliocene fan-delta and lacustrine deposits, Hole in the Wall and Wall Front members, Furnace Creek basin, Death Valley, California, *in* Wright, L. A., and Troxel, B. W., eds., Cenozoic basins of the Death Valley region: Geological Society of America Special Paper 333 (in press).

Bradley, W. H., 1966, Memorial to Levi Fatzinger Noble (1882–1965): Geological Society of America Bulletin, v. 77, p. 49–52.

Brady, R. H., III, 1986, Cenozoic geology of the northern Avawatz Mountains in relation to the intersection of the Garlock and Death Valley fault zones, San Bernardino County, California [Ph.D. thesis]: Davis, University of California, 292 p., scale 1:24,000.

Brady, R. H., III, and Troxel, B. W., 1999, The Military Canyon Formation: Depositional evolution and constraints on lateral faulting, southern Death Valley, California, *in* Wright, L. A., and Troxel, B. W., eds., Cenozoic basins of the Death Valley region: Geological Society of America Special Paper 333 (in press).

Bucher, W. H., 1956, Role of gravity in orogenesis: Geological Society of America Bulletin, v. 67, p. 1295–1318.

Burchfiel, B. C., and Stewart, J. H., 1966, "Pull-apart" origin of the central segment of Death Valley, California: Geological Society of America Bulletin, v. 94, p. 439–441.

Burchfiel, B. C., Hamill, G. S., and Wilhelms, D. E., 1983, Structural geology of the Montgomery Mountains and northern half of the Nopah and Resting Spring Ranges, Nevada and California: Geological Society of America Bulletin, v. 94, p. 1359–1376, scale 1:62,500.

Butler, P. L., Troxel, B. W., and Verosub, K. B., 1988, Late Cenozoic history and styles of deformation along the southern Death Valley fault zone: Geological Society of America Bulletin, v. 100, p. 402–410.

Cemen, I., Wright, L. A., Drake, R. E., and Johnson, F. C., 1985, Cenozoic sedimentation and sequence of deformational events at the southeastern end of the Furnace Creek fault zone, *in* Biddle, K. T., and Christie-Blick, N., eds., Strike-slip deformation, basin formation, and sedimentation: Society of Economic Paleontologists and Mineralogists Special Publication 37, p. 127–141.

Cemen, I., Wright, L. A., and Prave, A. R., 1999, Stratigraphy and tectonic implications of the latest Oligocene and early Miocene sedimentary succession southernmost Funeral Mountains, Death Valley region, California, *in* Wright, L. A., and Troxel, B. W., eds., Cenozoic basins of the Death Valley region: Geological Society of America Special Paper 333 (in press).

Cooper, J. D., ed., 1989, Cavalcade of carbonates: Los Angeles, Pacific Section, Society of Economic Paleontologists and Mineralogists, p. 37–44.

Cooper, J. D., ed., 1995, Ordovician of the Great Basin: Field trip guidebook and volume for the Seventh International Symposium on the Ordovician System: Pacific Section, SEPM (Society for Sedimentary Geology), 151 p.

Cooper, J. D., and Edwards, J. C., 1991, Cambro-Ordovician craton-margin carbonate section, southern Great Basin: A sequence-stratigraphic perspective, *in* Cooper, J. D., and Stevens, C. H., eds., Paleozoic paleogeography of the western United States: Los Angeles, Pacific Section, Society of Economic Paleontologists and Mineralogists, p. 237–252.

Cooper, J. D., and Fedo, C. M., 1998, Anatomy of a craton margin hinge zone: Sequence stratigraphy of upper Neoproterozoic–basal Cambrian succession, eastern Mojave Desert, California, USA, *in* Guidebook to field trip 1, Geological Society of America, Cordilleran Section, p. 1.1–1.26.

Cooper, J. D., and Keller, M., 1995, Ordovician craton margin-miogeoclinal transition, southern Great Basin, *in* Cooper, J. D., ed., Ordovician of the Great Basin: Field trip guidebook and volume for the Seventh International Symposium on the Ordovician System: Pacific Section, SEPM (Society for Sedimentary Geology), p. 107–132.

Cooper, J. D., and Stevens, C. H., eds., 1991, Paleozoic paleogeography of the western United States: Los Angeles, Pacific Section, Society of Economic Paleontologists and Mineralogists, p. 253–270.

Cooper, J. D., Miller, R. H., and Sundberg, F. A., 1982, Environmental stratigraphy of the upper part of the Nopah Formation (Upper Cambrian), *in* Cooper, J. D., Troxel, B. W., and Wright, L. A., eds., Geology of selected areas in the San Bernardino Mountains, western Mojave Desert and southern Great Basin (Geological Society of America Cordilleran Section guidebook): Shoshone, California, Death Valley Publishing, p. 97–116.

Curry, H. D., 1938, Strike-slip faulting in 1938 [abs.]: Geological Society of America Bulletin, v. 49, p. 1874–1875.

Curry, H. D., 1954, "Turtlebacks" in the central Black Mountains, Death Valley, California, *in* Jahns, R. H., ed., Geology of southern California: California Division of Mines Bulletin 170, p. 53–59.

Davis, G. A., and Burchfiel, B. C., 1973, Garlock fault: An intracontinental transform structure, southern California: Geological Society of America Bulletin, v. 84, p. 1407–1422.

Davis, W. M., 1922, Faults, underdrag, and landslides of the Great Basin ranges: Geological Society of America Bulletin, v. 33, p. 92–96.

Denny, C. S., and Drewes, H., 1965, Geology of the Ash Meadows quadrangle: U.S. Geological Survey Bulletin 1181L, 56 p., scale 1:62,500.

Diehl, P., 1976, Stratigraphy and sedimentology of the Wood Canyon Formation, *in* Troxel, B. W., and Wright, L. A., eds., Geologic features of Death Valley: California Division of Mines and Geology Special Report 106, p. 51–62.

Dooley, T. P., and McClay, K. R., 1996, Strike-slip deformation in the Confidence Hills, southern Death Valley fault zone, eastern California, USA: Geological Society of London Journal, v. 153, p. 375–387.

Drewes, H., 1963, Geology of the Funeral Peak quadrangle on the east flank of Death Valley: U.S. Geological Survey Professional Paper 413, 78 p., scale 1:62,500.

Fowler, K. T., and Calzia, J. P., 1999, The Kingston Range detachment fault, southwestern Death Valley region: Relation to Tertiary deposits and reconstruction of initial dip, *in* Wright, L. A., and Troxel, B. W., eds., Cenozoic basins of the Death Valley region: Geological Society of America Special Paper 333 (in press).

Fridrich, C. J., 1999, Tectonic evolution of the Crater Flat basin, Yucca Mountain region, Nevada, *in* Wright, L. A., and Troxel, B. W., eds., Cenozoic basins of the Death Valley region: Geological Society of America Special Paper 333 (in press).

Friedmann, S. J., 1999, Sedimentology and stratigraphy of the Shadow Valley basin, eastern Mojave Desert, California, *in* Wright, L. A., and Troxel, B. W., eds., Cenozoic basins of the Death Valley region: Geological Society of America Special Paper 333 (in press).

Greene, R. C., 1997, Geology of the northern Black Mountains, Death Valley, California: U. S. Geological Survey Open-File Report OF 97-79, 110 p., scale 1:24,000.

Gutstadt, A. M., 1968, Petrology and depositional environments of the Beck Spring Dolomite (Precambrian), Kingston Range, California: Journal of Sedimentary Petrology, v. 38, p. 1280–1288.

Haefner, R., 1972, Igneous history of a rhyolite lava flow series near Death Valley, California [Ph.D. thesis]: University Park, Pennsylvania State University, 281 p.

Haefner, R., 1974, Geology of the Shoshone Volcanics, Death Valley region eastern California, *in* Troxel, B. W., and Wright, L. A., eds., Geologic features of Death Valley, California: California Division of Mines and Geology Special Report 106, p. 59–64.

Hamilton, W. B., 1988, Detachment faulting in the Death Valley region, California and Nevada, *in* Carr, M. D., and Yount, J. C., eds., Geologic and hydrologic investigations of a potential nuclear waste disposal site at Yucca Mountain, southern Nevada: U.S. Geological Survey Bulletin 1790, p. 51–85.

Hazzard, J. C., 1937, Paleozoic section in the Nopah and Resting Springs Mountains, Inyo County, California: California Journal of Mines and Geology Report 33, p. 273–339.

Hewett, D. F., 1940, New formation names to be used in the Kingston Range of the Ivanpah quadrangle, California: Washington Academy of Science Journal, v. 30, p. 239–240.

Hill, M. L., and Troxel, B. W., 1966, Tectonics of the Death Valley region, California: Geological Society of America Bulletin, v. 77, p. 435–438.

Hodges, K. V., McKenna, L. W., and Harding, M. B., 1985, Structural unroofing of the central Panamint Range, southeastern California, *in* Wernicke, B. P., ed., Basin and Range extensional tectonics near the latitude of Las Vegas, Nevada: Geological Society of America Memoir 176, p. 377–390.

Holm, D. K., 1992, Structural, thermal and paleomagnetic constraints on the tectonic evolution of the Black Mountains crystalline terrane, Death Valley, California, and implications for extensional tectonism [Ph.D. thesis]: Cambridge, Massachusetts, Harvard University, 237 p.

Holm, D. K., Snow, J. K., and Lux, D. R., 1992, Thermal and barometric constraints on the intrusive and unroofing history of the Black Mountains: Implications for timing, initial dip, and kinematics in the Death Valley region, California: Tectonics, v. 11, p. 507–522.

Holm, D. K., Geissman, J. W., and Wernicke, B., 1993, Tilt and rotation of the footwall of a major normal fault system: Paleomagnetism of the Black Mountains, Death Valley extended terrane, California: Geological Society of America Bulletin, v. 105, p. 1373–1387.

Holm, D. K., Pavlis, T. L., and Topping, D. J., 1994, Black Mountains crustal section, Death Valley region, California, *in* McGill, S. G., and Ross, T. M., eds., Geological investigations of an active margin: Tulsa, Oklahoma, Geological Society of America Cordilleran Section Guidebook, p. 31–54.

Hopper, R. H., 1947, Geologic section from the Sierra Nevada to Death Valley, California: Geological Society of America Bulletin, v. 58, p. 393–432.

Hunt, C. B., and Mabey, D. R., 1966, General geology of Death Valley, California: U. S. Geological Survey Professional Paper 494-A, 162 p., scale 1:96,000.

Jennings, C. W., 1985, An explanatory text to accompany the 1:750,000 scale fault and geologic maps of California: California Division of Mines and Geology Bulletin 201, 197 p.

Johnson, B. K., 1957, Geology of a part of the Manly Peak quadrangle, southern Panamint Range: University of California Publications in the Geological Sciences, v. 30, p. 353–424, scale 1:50,000.

Knott, J. R., Sarna-Wojcicki, A. M., Meyer, C. E., Tinsley, J. C., Wells, S. G., and Wan, E., 1999, Late Neogene stratigraphy and tephrochronology of the western Black Mountains piedmont, Death Valley, California: Implications for the tectonic development of Death Valley, *in* Wright, L. A., and Troxel, B. W., eds., Cenozoic basins of the Death Valley region: Geological Society of America Special Paper 333 (in press).

Kupfer, D. H., 1954, Geology of the Silurian Hills, San Bernardino County, California, *in* Jahns, R.H., ed., Geology of southern California: California Division of Mines Bulletin 170, map sheet 19.

Kupfer, D. H., 1960, Thrust faulting and chaos structure, Silurian Hills, San Bernardino County, California: Geological Society of America Bulletin, v. 71, p. 181–214, scale 1:9600.

Labotka, T. C., Albee, A. L., Lanphere, M. A., and McDowell, S. D., 1980, Stratigraphy, structure, and metamorphism in the central Panamint Range (Telescope Peak quadrangle), Death Valley area, California, part II: Geological Society of America Bulletin, v. 91, p. 843–933.

Lawson, A. C., ed., 1908, The California earthquake of April 18, 1906, Report of the State Earthquake investigation Commission: Carnegie Institution of Washington Publication 87, 451 p.

Levy, M., and Christie-Blick, N., 1991, Tectonic subsidence of the early Paleozoic passive continental margin in eastern California and western Nevada: Geological Society of America Bulletin, v. 103, p. 1590–1606.

Longwell, C. R., 1945, Low-angle normal faults in the Basin and Range province: Eos (Transactions, American Geophysical Union) v. 26, p. 107–118.

Mancktelow, L. W., and Pavlis, T. L., 1994, Fold-fault relationships in low-angle detachment systems: Tectonics, v. 13, p. 668–685.

Mason, J. F., 1948, Geology of the Tecopa area, southeastern California: Geological Society of America Bulletin, v. 59, p. 333–352.

Maud, R. L., 1979, Stratigraphy and depositional environments of the carbonate-terrigenous member of the Crystal Spring Formation, Death Valley, California [Ph.D. thesis]: University Park, Pennsylvania State University, 221 p.

McAllister, J. F., 1970, Geology of the Furnace Creek borate area, Death Valley, Inyo County, California: California Division of Mines and Geology Map Sheet 14, scale 1:24,000.

McAllister, J. F., 1971, Preliminary geologic map of the Funeral Mountains in the Ryan quadrangle, Death Valley region, Inyo County, California: U.S. Geological Survey Open-File Map, scale 1:62,500.

McAllister, J. F., 1973, Geologic map and sections of the Amargosa borate area—southeast continuation of the Furnace Creek area—Inyo County, California: U.S. Geological Survey Miscellaneous Investigations Map I-782, scale 1:24,000.

McCutcheon, K. P., and Cooper, J. D., 1989, Environmental carbonate stratigraphy and cyclic deposition of the Smoky Member of the Nopah Formation (Upper Cambrian), Nopah Range, southern Great Basin, *in* Cooper, J. D., ed., Cavalcade of carbonates: Los Angeles, Pacific Section, Society of Economic Paleontologists and Mineralogists, p. 87–100.

McKee, E. H., 1968, Age and rate of movement on the Death Valley–Furnace Creek fault zones: Geological Society of America Bulletin, v. 79, p. 509–512.

McKenna, L. W., and Hodges, K. V., 1990, Constraints on the kinematics and timing of late Miocene–recent extension between the Panamint and Black Mountains, southeastern California, *in* Wernicke, B. P., ed., Basin and Range extensional tectonics near the latitude of Las Vegas, Nevada: Geological Society of America Memoir 176, p. 363–376.

Miller, J. M. G., 1985, Glacial and syntectonic sedimentation: The Upper Proterozoic Kingston Peak Formation, southern Panamint Range, eastern California: Geological Society of America Bulletin, v. 96, p. 1537–1553.

Miller, M. G., 1992, Brittle faulting induced by ductile deformation of a rheologically stratified rock sequence, Badwater turtleback, Death Valley, California: Geological Society of America Bulletin, v. 104, p. 1376–1385.

Miller, M. G., 1999, Implications of ductile strain on the Badwater turtleback for pre-14 Ma extension, *in* Wright, L. A., and Troxel, B. W., eds., Cenozoic basins of the Death Valley region: Geological Society of America Special Paper 333 (in press).

Morrison, R. B., 1999, Quaternary geology of Tecopa Valley, California: A multi-million-year record and its relevance to the proposed nuclear waste repository at Yucca Mountain, Nevada, *in* Wright, L. A., and Troxel, B. W., eds., Cenozoic basins of the Death Valley region: Geological Society of America Special Paper 333 (in press).

Muehlberger, W. R., 1954, Geology of the Quail Mountains, San Bernardino County, California, *in* Jahns, R. H., ed., Geology of southern California: California Division of Mines Bulletin 170, map sheet 16, scale: 1 inch = 4000 ft.

Murphy, F. M., 1932, Geology of a part of the Panamint Range, California: Miners in California: California Division of Mines Report 28, p. 329–356.

Noble, L. F., 1934, Rock formations of Death Valley, California: Science, p. 173–178.

Noble, L. F., 1941, Structural features of the Virgin Spring area, Death Valley, California: Geological Society of America Bulletin, v. 52, p. 941–999.

Noble, L. F., and Wright, L. A., 1954, Geology of the central and southern Death Valley region, California, *in* Jahns, R. H., ed., Geology of southern California: California Division of Mines Bulletin 170, p. 143–160.

Nolan, T. B., 1929, Notes on the stratigraphy and structure of the northeast portion of Spring Mountain, Nevada: American Journal of Science, v. 17, p. 461–472.

Osleger, D. A., and Montenez, I. P., 1996, Cross-platform architecture of a sequence boundary in mixed siliciclastic-carbonate lithofacies, Middle Cambrian, southern Great Basin, USA: Sedimentology, v. 43, p. 197–218.

Osleger, D. A., Montanez, I. P., Martin-Chivelet, J., and Lehman, C., 1976, Cycle and sequence stratigraphy of Middle to Upper Cambrian carbonates, *in* Abbot, P. L., and Cooper, J. D., eds., American Association of Petroleum Geologists field conference guide, San Diego, CA: Society of Economic Paleontologists and Mineralogists Book 80, p. 55–85.

Otton, J. K., 1976, Geologic features of the central Black Mountains, California, *in* Troxel, B. W., and Wright, L. A., eds., Geologic features Death Valley, California: California Division of Mines and Geology Special Report 106, p. 27–34.

Otton, J. K., 1977, Geology of the central Black Mountains, Death Valley, California: The turtleback terrane [Ph.D. thesis]: University Park, Pennsylvania State University, 155 p.

Palmer, A. R., and Halley, R. B., 1979, Physical stratigraphy and trilobite biostratigraphy of the Carrara Formation (Lower and Middle Cambrian) of the southern Great Basin: U.S. Geological Survey Professional Paper 1047, 131 p.

Pavlis, T. L., 1994, Kinematics of extensional, ductile to brittle structures in the Death Valley extended terrane, *in* McGill, S. F., and Ross, T. M., eds., Geological investigations of an active margin: Tulsa, Oklahoma, Geological Society of America Cordilleran Section Guidebook, p. 32–42.

Prave, A. R., 1991, Depositional and sequence stratigraphic framework of the Lower Cambrian Zabriskie Quartzite: Implications for regional correlations and the Early Cambrian paleogeography of the Death Valley region of California and Nevada: Geological Society of America Bulletin, v. 104, p. 505–515.

Prave, A. R., 1996, Neoproterozoic carbonate successions, Death Valley, California: Comparison and contrast with Phanerozoic examples, *in* Abbot, P. L., and Cooper, J. D., eds., American Association of Petroleum Geologists field conference guide, San Diego, CA: Society of Economic Paleontologists and Mineralogists Book 80, p. 35–55.

Prave, A. R., and McMackin, M. R., 1999, Depositional framework of mid to late Miocene strata in the Dumont Hills and along the margin of the Kingston Range: Tentative implications for the tectonostratigraphic evolution of the southern Death Valley region, *in* Wright, L. A., and Troxel, B. W., eds., Cenozoic basins of the Death Valley region: Geological Society of America Special Paper 333 (in press).

Prave, A. R., and Wright, L. A., 1986, Isopach pattern of the Lower Cambrian Zabriskie Quartzite, Death Valley region, California-Nevada: How useful in tectonic reconstructions?: Geology, v. 14, p. 251–254.

Ransome, F. L., Emmons, W. H., and Garry, G. H., 1910, Geology and ore deposits of the Bullfrog district, Nevada: U.S. Geological Survey Bulletin 407, 130 p.

Reynolds, M. W., 1976, Geology of the Grapevine Mountains, Death Valley, California: A summary, *in* Troxel, B. W., and Wright, L. A., eds., Geologic features of Death Valley, California: California Division of Mines and Geology Special Report 106, p. 19–25.

Roberts, M. T., 1976, Stratigraphy and depositional environments of the Crystal Spring Formation, southern Death Valley region, California, *in* Troxel, B. W., and Wright, L. A., eds., Geologic features of Death Valley, California: California Division of Mines and Geology Special Report 106, p. 35–43.

Roberts, M. T., 1982, Depositional environments and tectonic setting of the Crystal Spring Formation, southern Death Valley region, California, *in* Cooper, J. D., Troxel, B. W., and Wright, L. A., eds., Geology of selected areas in the San Bernardino Mountains, western Mojave Desert, and southern Great Basin, California: Shoshone, California, Death Valley Publishing Company, p. 143–154.

Sears, D. H., 1953, Origin of the Amargosa chaos, Virgin Spring area, Death Valley, California: Journal of Geology, v. 61, p. 182–187.

Serpa, L., 1990, Structural styles across an extending orogen: Results from the COCORP Mojave and Death Valley seismic transects, *in* Wernicke, B. P., ed., Basin and Range extensional tectonics near the latitude of Las Vegas, Nevada: Geological Society of America Memoir 176, p. 335–344.

Serpa, L., and Pavlis, T., 1996, Three-dimensional model of the late Cenozoic history of the Death Valley region, southeastern California: Tectonics, v. 15, p. 1113–1128.

Serpa, L., deVoogd, B., Wright, L., Willemin, J., Oliver, J., Hauser, E., and Troxel, B., 1988, Structure of the central Death Valley pull-apart basin and vicinity from COCORP profiles in the southern Great Basin: Geological Society of America Bulletin, v. 100, p. 1437–1450.

Silver, L. T., McKinney, C. R., and Wright, L. A., 1961, Some Precambrian ages in the Panamint Range, Death Valley, California: Geological Society of America Special Paper 68, 55 p.

Smith, G. I., 1962, Large lateral displacement on the Garlock fault, California, as measured from offset dike swarm: American Association of Petroleum Geologists Bulletin, v. 46, p. 85–104.

Snow, J. K., 1990, Cordilleran orogenesis, extensional tectonics, and geology of the Cottonwood Mountains area, Death Valley region, California and Nevada: Cambridge, Massachusetts, Harvard University 463 p.

Snow, J. K., and Lux, D. R., 1999, Tectono-sequence stratigraphy of Tertiary rocks in the Cottonwood Mountains and northern Death Valley area, California and Nevada, *in* Wright, L. A., and Troxel, B. W., eds., Cenozoic basins of the Death Valley region: Geological Society of America Special Paper 333 (in press).

Snow, J. K., and Wernicke, B., 1989, Uniqueness of geological correlations: An example from the Death Valley extended terrain: Geological Society of America Bulletin, v. 101, p. 1351–1362.

Spencer, J. E., 1990, Late Cenozoic extensional and compressional tectonism in the southern and western Avawatz Mountains, southeastern California, *in* Wernicke, B. P., ed., Basin and Range extensional tectonics near the latitude of Las Vegas, Nevada: Geological Society of America Memoir 176, p. 317–333.

Stewart, J. H., 1970, Upper Precambrian and Lower Cambrian strata, southern Great Basin: U.S. Geological Survey Professional Paper 620, 206 p.

Stewart, J. H., 1972, Initial deposits in the Cordilleran geosyncline: Evidence of a Late Precambrian (<850 m.y.) continental separation: Geological Society of America Bulletin, v. 83, p. 1345–1360.

Stewart, J. H., 1982, Regional relationships of Proterozoic Z and Lower Cambrian rocks in the western United States and northern Mexico, in Cooper, J. D., Troxel, B. W., and Wright, L. A., eds., Geology of selected areas in the San Bernardino Mountains, western Mojave Desert, and southern Great Basin, California: Geological Society of America, Cordilleran Section guidebook, p. 171–181.

Stewart, J. H., 1983, Extension tectonics in the Death Valley area, California: Transport of the Panamint Range structural block 80 km northwestward: Geology, v. 11, p. 153–157.

Stewart, J. H., Albers, J. P., and Poole, F. G., 1968, Summary of regional evidence for right-lateral displacement in the western Great Basin: Geological Society of America Bulletin, v. 79, p. 1407–1414.

Stock, C., and Bode, F. D., 1935, Occurrence of lower Oligocene mammal-bearing beds near Death Valley, California: California Academy of Natural Sciences Proceedings, v. 21, p. 571–579.

Summa, C. L., 1993, Sedimentologic, stratigraphic, and tectonic controls of a mixed carbonate-siliciclastic succession: Neoproterozoic Johnnie Formation, southeastern California [Ph.D. thesis]: Cambridge, Massachusetts Institute of Technology, 615 p.

Topping, D. J., 1993, Paleogeographic reconstruction of the Death Valley extended region: Evidence from large rock-avalanche deposits in the Amargosa Chaos basin, California: Geological Society of America Bulletin, v. 105, p. 1190–1213.

Troxel, B. W., 1982a, Basin facies (Ibex Formation) of the Noonday Dolomite, southern Saddle Peak Hills, southern Death Valley, California, in Cooper, J. D., Troxel, B. W., and Wright, L. A., eds., Geology of selected areas in the San Bernardino Mountains, western Mojave Desert, and southern Great Basin, California: Geological Society of America, Cordilleran Section guidebook, p. 43–46.

Troxel, B. W., 1982b, Description of the uppermost part of the Kingston Peak Formation, Amargosa River Canyon, Death Valley region, California, in Cooper, J. D., Troxel, B. W., and Wright, L. A., eds., Geology of selected areas in the San Bernardino Mountains, western Mojave Desert, and southern Great Basin, California: Geological Society of America, Cordilleran Section guidebook, p. 61–70.

Troxel, B. W., and Butler, P. R., 1998, Tertiary and Quaternary fault history of the intersection of the Garlock and Death Valley fault zones, southern Death Valley, California, in Calzia, J. P., and Reynolds, R. E., eds., Finding faults in the Mojave: San Bernardino County Museum Association Quarterly, v. 45, p. 91–98.

Troxel, B. W., and Wright, L. A., 1968, Precambrian stratigraphy of the Funeral Mountains, Death Valley, California [abs.]: Geological Society of America Special Paper 121, p. 574–575.

Wagner, D. L., and Hsu, E. Y., 1987, Reconnaissance geologic map of parts of the Wingate Wash, and Manly Peak quadrangles, San Bernardino and Inyo Counties, California: California Division of Mines and Geology Open-File Report 87-10, scale 1:62,500.

Wasserburg, G. J,. Wetherill, G. W., and Wright, L. A., 1959, Ages in the pre-Cambrian terrane of Death Valley, California: Journal of Geology, v. 67, p. 702–708.

Wernicke, B. P., 1981, Low-angle normal faults in the Basin and Range Province: Nappe tectonics in an extending orogen: Nature, v. 291, p. 645–648.

Wernicke, B., 1992, Cenozoic, extensional tectonics of the U. S. Cordillera, in Burchfiel, B. C., Lipman, P. W., and Zoback, M. L., eds., The Cordilleran orogen; conterminous United States: Boulder, Colorado, Geological Society of America, Geology of North America, v. G-3, p. 553–581.

Wernicke, B., Axen, G. J., and Snow, J. K., 1988, Basin and Range extensional tectonics, at the latitude of Las Vegas, Nevada: Geological Society of America Bulletin, v. 100, p. 1738–1757.

Wernicke, B. P., Snow, G. J., Axen, G. J., Burchfiel, B. C., Hodges, K. V., Walker, J. D., and Guth, P. L., 1989, Extensional tectonics in the Basin and Range province between the Sierra Nevada and the Colorado Plateau: 28th International Geological Congress field trip guidebook T138: Wahington, D.C., American Geophysical Union, 80 p.

Wernicke, B., Snow, J. K., Hodges, K. V., and Walker, J. D., 1993, in Lahren, M. M., Trexler, J. H., Jr., and Spinoza, C., eds., Crustal evolution of the Great Basin and the Sierra Nevada: (Geological Society of America field trip guidebook, Cordilleran/Rocky Mountain Sections): Reno, Nevada, University of Nevada, p. 453–479.

Wertz, W. E., 1982, Stratigraphy and sedimentology of the Stirling Quartzite, Death Valley area, California and Nevada, in Cooper, J. D., Troxel, B. W., and Wright, L. A., eds., Geology of selected areas in the San Bernardino Mountains, western Mojave Desert, and southern Great Basin, California: Geological Society of America, Cordilleran Section, guidebook, p. 165–170.

Wilhelms, D. E., 1963, Geology of part of the Nopah and Resting Spring Ranges, Inyo County, California [Ph.D. thesis]: Los Angeles, University of California, 224 p., scale 1: 48,000.

Williams, E. G., Wright, L. A., and Troxel, B. W., 1976, The Noonday Dolomite and equivalent units, southern Death Valley region, California, in Troxel, B. W., and Wright, L. A., eds., Geologic features, Death Valley, California: California Division of Mines and Geology Special Report 106, p. 45–50.

Willis, B., and Wood, H. O., 1922, Fault map of the State of California: Seismological Society of America, scale 1:506,800.

Wright, L. A., 1954, Geology of the Alexander Hills area, Inyo and San Bernardino Counties, California, in Jahns, R. H., ed., Geology of southern California: California Division of Mines Bulletin 170, map sheet 17, scale 1:32,000.

Wright, L. A., 1968, Talc deposits of the southern Death Valley–Kingston Range region, California: California Division of Mines and Geology Special Report 95, 79 p.

Wright, L. A., 1987, Overview of the role of strike-slip and normal faulting in the Neogene history of the region northeast of Death Valley, California-Nevada, in Ellis, M. A., ed., Late Cenozoic evolution of the southern Great Basin: Nevada Bureau of Mines and Geology Open File 89-1, p. 1–11.

Wright, L. A., and Prave, A. R., 1993, Proterozoic–Early Cambrian tectonostratigraphic record in the Death Valley region, California-Nevada, in Reed, J., and others, eds., Precambrian rocks of the conterminous United States: Boulder, Colorado, Geological Society of America, Geology of North America, v. C-2, p. 529–533.

Wright, L. A., and Troxel, B. W., 1966, Strata of late Precambrian–Cambrian age, Death Valley region, California-Nevada: American Association of Petroleum Geologists Bulletin, v. 50, p. 846–857.

Wright, L. A., and Troxel, B. W., 1969, Chaos structure and basin and range normal faulting: Evidence for a genetic relationship: Geological Society of America Abstracts with Programs:, v. 7, p. 242.

Wright, L. A., and Troxel, B. W., 1971, Evidence for the tectonic control of volcanism, Death Valley region, California: Geological Society of America Abstracts with Programs, v. 3, p. 221.

Wright, L. A., and Troxel, B. W., 1973, Shallow-fault interpretation of Basin and Range structure, southwest Great Basin, in DeJong, K. A., and Scholten, R., eds., Gravity and tectonics: New York, John Wiley and Sons, p. 933–950.

Wright, L. A., and Troxel, B. W., 1984, Geology of the northern half of the Confidence Hills 15-minute quadrangle, Death Valley region, eastern California; the area of the Amargosa chaos: California Division of Mines and Geology Map Sheet 34, scale 1:24,000.

Wright, L. A., and Troxel, B. W., 1992, Levi Noble: Pioneer geologist of the Death Valley area, in Pisarowicz, J., ed., Proceedings, third Death Valley conference on history and prehistory: Death Valley, California, Death Valley Natural History Association, p. 143–157.

Wright, L. A., and Troxel, B. W., 1993, Geologic map of the central and northern Funeral Mountains, Death Valley region, southern California: U.S. Geological Survey Miscellaneous Investigations Series Map I-2305, scale 1:48,000.

Wright, L. A., Troxel, B. W., Williams, E. G., Roberts, M. T., and Diehl P. E., 1976, Precambrian sedimentary environments of the Death Valley region, eastern California, *in* Troxel, B. W., and Wright, L. A., eds., Geologic features Death Valley, California: California Division of Mines and Geology Special Report 106, p. 7–14.

Wright, L. A., Williams, E. G., and Cloud, P., 1978, Algal and cryptalgal structures and platform environments of the late pre-Phanerozoic Noonday Dolomite, eastern California: Geological Society of America Bulletin, v. 89, p. 321–333.

Wright, L. A., Troxel, B. W., Burchfiel, B. C., Chapman, R. H., and Labotka, T. C., 1981, Geologic cross section from the Sierra Nevada to the Las Vegas Valley, eastern California to southern Nevada: Geological Society of America Map and Chart Series MC 28-M, scale 1:250,000.

Wright, L. A., Drake, R. E., and Troxel, B. W., 1984, Evidence for the western migration of severe Cenozoic extension, southwestern Great Basin, California: Geological Society of America Abstracts with Programs, v. 16, p. 701.

Wright, L. A., Thompson, R. A., Troxel, B. W., Pavlis, T. L., DeWitt, E. H., Otton, J. K., Ellis, M. A., Miller, M. G., and Serpa, L. F., 1991, Cenozoic magmatic and tectonic evolution of the east-central Death Valley region, *in* Walawender, M. J., and Hannan, B. B., eds., Geological excursions in southern California and Mexico: Geological Society of America Cordilleran Section Guidebook, San Diego, California, p. 93–127.

Wright, L. A., Greene, R. C., Cemen, I., Johnson, F. C., and Prave, A. R., 1999, Tectonostratigraphic development of the Miocene-Pliocene Furnace Creek basin, Death Valley region, California, *in* Wright, L. A., and Troxel, B. W., eds., Cenozoic basins of the Death Valley region: Geological Society of America Special Paper 333 (in press).

MANUSCRIPT ACCEPTED BY THE SOCIETY NOVEMBER 23, 1998.

Section VI
# THE MODERN LANDSCAPE—MOUNTAINS TO COAST

# The modern landscape: Introduction

> Go where you may within the bounds of California, mountains are ever in sight, charming and glorifying every landscape.
>
> John Muir, 1894 [1991, p. 1]

## INTRODUCTION

This section contains three diverse chapters, each of which addresses in some aspect the origin of the California landscape. This landscape dominated the early settlement, and early geologic description, and is itself a reflection of its geology, including its location along an active plate boundary. In a real sense, all chapters of this volume address some aspect of the California landscape. California's inspiring beauty goes hand in hand with its inherent geologic hazards. Periodic events, such as earthquakes, volcanic eruptions, landslides, mudflows, and floods, are all agents that shape the landscape of California and that affect the safety and well-being of its inhabitants.

We begin this chapter with the quote from Muir, because for him, the "Sierra Nevada" included not only the Sierra Nevada as currently designated, but also peaks of the Klamath Mountains and the volcanoes of the Cascade Range. The aspects of the landscape touched upon in this section include the modern and ancient volcanic deposits exposed throughout California, the glacial features present in the Sierra Nevada and ranges to the east, and the landslide features of California, particularly in its coastal ranges and valleys.

In the first chapter, the paper on Mount Lassen by H. Williams provides the entree for the essay "Volcanism in California" by R. L. Christiansen and M. A. Clynne. Active volcanism is related to active subduction along the northern California coast, to processes along the San Andreas fault system, and along the eastern edge of the Sierra Nevada and western margin of the Basin and Range province in east-central California. This active volcanism is a direct result of the complex plate tectonic interaction between the Pacific and North American plates. Older volcanic activity in the Sierra Nevada and southeastern California deserts was a product of an earlier plate tectonic interaction between the Farallon and North American plates. Thus features discussed in this chapter overlap with previous ones, particularly those by Prentice, Wakabayashi, and Saleeby (see also Atwater, this volume). Christiansen's and Clynne's essay should be read with the previous essays in mind.

In the second chapter of this section, the essay by Gillespie, Clark, and Burke, entitled "Eliot Blackwelder and the alpine glaciations of the Sierra Nevada" focuses on Blackwelder's 1931 documentation of glacial features in the Sierra Nevada and ranges to the east. This glacial activity is superimposed upon the older rocks, including the Paleozoic-Cenozoic deposits and the modern volcanoes. Glacial features form much of the striking scenery of the high mountains of California and the Basin and Range mountains. Gillespie and his coauthors summarize current ideas about this glacial history. These features are of interest not only for their intrinsic value, but also because of the relationship between the development of humans and ice ages, and the uncertainty about anthropogenic versus natural influences on future climatic fluctuations. As the historian Will Durant said, "civilization is an interlude between ice ages" (quoted *in* Clarke, 1998). Understanding the causes and results of past glacial epochs enriches our discussion of this important issue.

The final chapter of this section and of this volume is entitled "Commentary on Karl Terzaghi's 'Mechanism of Landslides, 1950'," by Roy J. Shlemon, based on the reprinted "classic" paper by Terzaghi. Shlemon summarizes the pioneering contribution of Terzaghi and the present state of knowledge of landslides, particularly in urban regions of California. Rapid uplift of poorly consolidated rocks in California's mountains and ranges in response to present and recent plate interactions makes landslides an important landscape-forming force, and one that is of prime engineering importance in California's rapidly expanding mountain and coastal urban regions.

Finally, it is worth considering the origin of that most unique feature of California's landscape "...the Coast Range on the west side [and] the Sierra Nevada on the east...coming together in curves on the north and south inclose [sic] a magnificent basin, with a level floor more than 400 miles long and from 35 to 60

---

### A 1978 Cordilleran Section Symposium: Debating Ideas Relevant to the West

The following memories are from a Cordilleran section meeting early in my professional career. The memories involve a symposium that, like many other Cordilleran symposia, provided an opportunity to explore a problem of specific interest to the West.

In 1975, I was hired by the U.S. Geological Survey (USGS) to investigate a geologic hazard that had arisen during the siting of some nuclear power reactors. Several proposed power reactor sites were located in unconsolidated sedimentary basins from which ground water was either being pumped or might be pumped. Because a few ground-water basins had developed large ground cracks that appeared related to land subsidence and ground-water withdrawal, concern for the safety of reactors on these "soft sites" had been expressed by regulatory officials. If such cracks occurred beneath a power reactor, they could threaten the safety of the reactor. The USGS hired me to develop a methodology to predict the occurrence of these cracks. Ironically, selection of "soft sites" as locations for Western power reactors was motivated by a desire to avoid rock sites where issues of active faulting seemed commonly to arise and delay and jeopardize construction.

Because not much literature on these ground cracks was published, and they were an unfamiliar phenomenon to me, I spent much of 1975 and 1976 examining ground cracks in the field with scientists who had reported them. These scientists included Malcolm Clark, Martin Mifflin, Douglas Morton, Ralph Patt, Herbert Schumann, and Carl Winikka. I quickly realized that none of the scientists whom I was visiting were in contact with each other. In fact, within a couple of years, I was the only one who appreciated how pervasive these features were in the western United States. Because most of the basins with these features were in the Cordilleran section, I decided in 1977 to organize a full-day symposium at the 1978 Cordilleran section meeting in Tempe, Arizona. The purposes of the symposium, "Ground Failure Caused by Ground-water Withdrawal from Tectonic Basins," were both to demonstrate how widespread these features were and to provide a forum on their mechanism of formation. There were 16 papers presented — 6 described specific geographic occurrences of ground failure and 10 discussed aspects of potential causes of the failure.

For many of the presenters, it was the first time that they had met and discovered a common interest. Debate at the symposium was lively about the causes of ground failure. Four mechanisms were proposed. While all agreed that the cracks were caused by tension, the source of the tension was unclear. Proposed sources of the tension included localized differential compaction, regional horizontal movements associated with regional subsidence, horizontal seepage stresses, and soil desiccation. While some investigators entertained multiple hypotheses, a few insisted that theirs alone was the correct explanation.

For me, it was the first symposium that I had ever organized, and it started me down a path that would eventually lead to a volume in the GSA Reviews in Engineering Geology (Holzer, 1984) and a National Academy of Sciences report (NRC, 1991). Some participants would become lifelong friends or professional acquaintances and one, Ken Lee, would die tragically shortly before the symposium was held.

I would also learn that some scientists never abandon hypotheses regardless of field evidence to the contrary. By 1984, field monitoring and studies done by others and me would show that these cracks were caused by localized differential compaction. One study, published in the GSA *Bulletin* (Jachens and Holzer, 1982) was so detailed that the differential subsidence caused by the topographic relief of fields that had originally been graded flat for farming. The study also revealed the time history of the process of failure and the strains at failure. A second study, by Holzer and Pampeyan (1981), reported measurements with first-order leveling techniques of the differential subsidence at ground cracks. Ironically, the theoretical underpinnings of the localized differential compaction mechanism had been developed by Ken Lee.

Thomas L. Holzer
U.S. Geological Survey

**REFERENCES CITED**

Jachens, R. C., and Holzer, T. L., 1982, Differential compaction mechanism for earth fissures near Casa Grande, Arizona: Geological Society of America Bulletin, v. 93, p. 998–1012

Holzer, T. L., 1984, Man-induced land subsidence: Geological Society of America Reviews in Engineering Geology, v. VI, 221p.

Holzer, T. L., and Pampeyan, E. H., 1981, Earth fissures and localized differential subsidence: Water Resources Research, v. 17, p. 223–227.

NRC, 1991, Mitigating losses from land subsidence in the United States: National Academy of Sciences, National Research Council, 58 p.

miles wide...the grand Central Valley of California" (Muir, 1894, p. 1). In other words, why have the Sierra Nevada and Coast Ranges been uplifted rapidly and recently, leaving the valley near sea level? We may finally have an answer in the documentation by Godfrey et al. (1997) and Godfrey and Klemperer (1998), by geophysical means, of a large allochthonous slab of oceanic crust and mantle, named by them the Great Valley ophiolite, underlying the entire Great Valley. Godfrey and others suggest that this ophiolite was emplaced on the western edge of the North American continent in Late Jurassic–Early Cretaceous time. The Great Valley ophiolite is one of the largest detached ocean crust-mantle slabs known in the world. Its presence prevents the valley from rising along with the surrounding Coast Ranges and Sierra Nevada (see also Moores and others, this volume).

## REFERENCES CITED

Clarke, A. C., 1998, Presidents, experts, and asteroids: Science, v. 280, p. 1531–1532.

Godfrey, N., and Klemperer, S., 1998, Ophiolitic basement to a forearc basin and implications for continental growth: The Coast Range/Great Valley ophiolite, California: Tectonics, v. 17, p. 558–570.

Godfrey, N. J., Beaudoin, B. C., Klemperer, S. L., and the Mendocino Working Group, 1997, Ophiolitic basement to the Great Valley forearc basin, northern California, from seismic and gravity data: Implications for crustal growth at the North American continental margin: Geological Society of America Bulletin, v. 109, p. 1536–1562.

Muir, J., 1991, The mountains of California (new and enlarged edition): Berkeley, California, Ten Speed Press, 389 p.

MANUSCRIPT ACCEPTED BY THE SOCIETY NOVEMBER 23, 1998.

Geological Society of America
Special Paper 338
1999

# Chapter 20
# MODERN AND ANCIENT VOLCANOES

**Robert L. Christiansen and Michael A. Clynne**
*U.S. Geological Survey*
*345 Middlefield Road, MS 910*
*Menlo Park, California 94025*
*E-mail: rchris@mojave.wr.usgs.gov;*
*mclynne@mojave.wr.usgs.gov*

# HOWEL WILLIAMS

Howel Williams introduced a truly modern geologic approach to problems of volcanism in California and the Cordilleran region. Williams, born and reared in Liverpool, came to geology through earlier interests in geography and archaeology. Work in the Ordovician strata of Wales had brought him into contact with the abundant volcanic rocks there and had led him to visit classic volcanic areas in Germany and France. That experience, in turn, set his lifelong interest in problems of volcanism and brought him to California, eager to work on younger volcanic fields. In 1926 Williams arrived in Berkeley to study under A. C. Lawson at the University of California and immediately began his first investigation of California's volcanic geology at the Sutter (Marysville) Buttes in the Sacramento Valley.

The summer of 1927 found Williams' attention drawn to the area of the 1914–1917 eruptions of Lassen Peak, and he became deeply interested in geologic problems of the surrounding young volcanic field. Over the next several years he produced several important papers on that region, including a monograph on the geology of Lassen Volcanic National Park.

It was difficult to select just one of these papers for the purpose of this volume, but it is possible to reproduce here a single complete paper that represents fully the characteristically insightful, beautifully illustrated, and broadly applicable aspects of his work. Thus, we have chosen to reproduce his 1929 paper on the volcanic domes of the Lassen Peak area, in which he demonstrated the existence and origins of a large cluster of dacitic domes on the northern flank of an eroded older andesitic stratovolcano.

Robert L. Christiansen
Michael A. Clynne

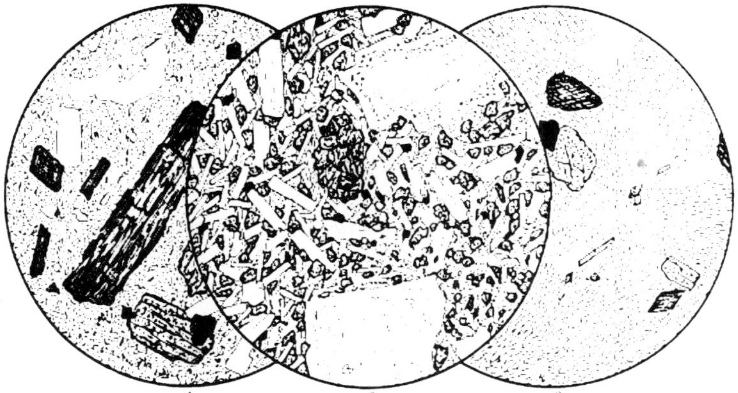

# THE VOLCANIC DOMES OF LASSEN PEAK AND VICINITY, CALIFORNIA.

HOWEL WILLIAMS.

## INTRODUCTION

Since the spectacular rise of the spine of Mont Pelée in 1902, the attention of geologists has been drawn repeatedly to the fact that volcanoes are able to force more or less solid lava upwards within their conduits. In 1904, Russell[1] discussed the diagnostic characters of such "massive-solid eruptions"; in 1907, under the caption 'Staukuppen' (plug-domes), Bergeat[2] also described protrusions of Peléan type; and in 1916, Powers[3] briefly reviewed the literature relating to Pacific domes. More recently, the rise of the dacite plug in the crater of Lassen Peak has been ably considered by Day and Allen;[4] the phenomenal growth of the Santa Maria dome in Guatemala during 1922-25 has been recorded by Sapper[5] and the upheaval of the Fouqué Kaméni dome at Santorini in 1925 has been dealt with by Washington[6] and others.

The purpose of this paper is to attract notice to the remarkable development of similar domes in the Lassen Volcanic National Park, California. Here, within the space of 50 square miles, there are no less than 13 domes, chiefly of glassy dacite. Most of them are clustered on the flanks of the old Brokeoff volcano; the largest is Lassen Peak itself and the most recent is only 200 years old. For the present purpose, it will be best to describe the recent domes first.

## THE DOMES OF THE CHAOS CRAGS.

The northern slope of Lassen Peak sweeps down 2,500 feet to a confused cluster of rocky peaks, expressively termed the Chaos Crags. These peaks attain an elevation of 8,545 feet

---

[1] Russell, I. C., Criteria relating to massive-solid volcanic eruptions, this Journal, **17**, 253-268, 1904.
[2] Bergeat, A., Staukuppen, Neues Jahrb., Festband, 310-329, 1907.
[3] Powers, S., Volcanic Domes in the Pacific, this Journal, **42**, 261-274, 1916.
[4] Day, A. L., and Allen, E. T., The Volcanic Activity and Hot Springs of Lassen Peak, Carnegie Instit., Washington No. 360, 1925.
[5] Sapper, K., Der vulkanische Tätigkeit in Mittelamerika im 20. Jahrhundert, Zeitschr. für Vulkanologie, Band IX, Heft 4, 158-200, 1926.
[6] Washington, H. S., Santorini Eruption of 1925, Bull. Geol. Soc. America, **37**, 349-384, 1926.

Fig. 17. Andesites of the Brokeoff cone. A. South slope of cone, on fire trail at 8050 feet (298). Phenocrysts of medium andesine, basaltic hornblende and augite, in a base composed of glass and feldspathic residuum. B. Typical augite andesite of the Huckleberry series, near Huckleberry Lake (304). Phenocrysts and microliths of acid labradorite with intersertal granules of augite and magnetite in a feldspathic matrix. C. Glassy andesite at base of Brokeoff cliff, elevation 8500 feet (310). Phenocrysts of andesine, brownish green hornblende, augite and hypersthene in a pale yellowish brown glass crowded with belonites, chiefly of plagioclase.

Example of Howel Williams' thin section sketches. (From Williams, H., 1932b)

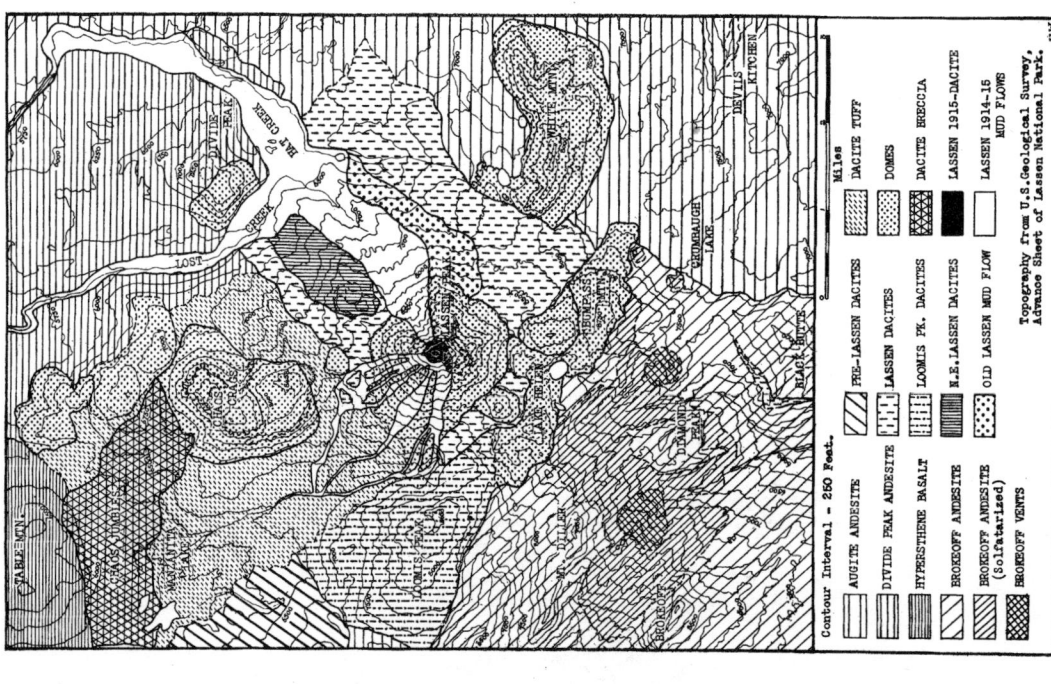

Fig. 1. Geologic sketch-map of Lassen Peak and vicinity.

and stretch northwards for a mile and a half. In an east-west direction, they extend for about a mile (Fig. 1). Their upper surface is hardly more than a wilderness of dacite blocks, and their sides are almost wholly encompassed by great aprons of talus.

It has long been known that these crags represent one of the latest volcanic eruptions of the Lassen area. Brewer and King,[7] who visited them in 1863, cite the testimony of several persons, that, from 1854 to 1857, these crags "were constantly emitting large quantities of steam and gases." Diller[8] mentions "Chaos," as having been formed about two centuries ago, and the writer[9] has given independent evidence to the same effect. From the northern base of these crags, a gentle slope descends toward Manzanita Lake and Lake Reflection. Upon this slope rests a hummocky pile of angular dacite boulders, about two and half square miles in extent. The surface of this deposit, like that of the Chaos Crags, is almost entirely devoid of vegetation, and resembles the bare moraine of an Alpine glacier. This remarkable accumulation of rocks is known locally as the Chaos Jumbles. It was formed, as we shall see, by the partial collapse of the Chaos Crags.

It is not only the recency and hence the almost perfect preservation of the original features of these domes that invites attention; the structures here displayed by the dacites are much more apparent than elsewhere and permit us to draw conclusions regarding the mode of upheaval. Only by analogy with the Chaos Crags is it possible to understand many of the obscure features of Lassen Peak and adjacent domes.

(a) *Pyroclastic eruptions*: It has often been remarked that the upheaval of volcanic domes and spines is initiated and may be followed by explosive eruptions. Ash-explosions had occurred at Lassen Peak for a period of one year before the lava-plug rose to the surface of the crater in May, 1915. Similarly, pumice eruptions preceded the rise of the obsidian spines in the Mono Craters, California; the formation of the domes of Novarupta (Alaska), of Santa Maria (Guatemala), and of Tarumai (Japan), were also heralded by pyroclastic

---

[7] In Whitney, J. D., Geological Survey of California, **1**, 315, 1865.
[8] Diller, J. S., Lassen Peak—our most active volcano, Bull. Seismol. Soc. America, **6**, 2, 1916.
[9] Williams, H., A Recent Volcanic Eruption near Lassen Peak, Univ. Calif. Publ. Bull. Dept. Geol. Sci., **17**, 262, 1928.

eruptions. The protrusion of the dacite domes of the Chaos Crags likewise succeeded a series of tuff-explosions. The line of vents formerly existed on the present site of the Chaos Crags. At the southern end of the crags, a well-preserved remnant of one of these lapilli-cones may still be seen (Fig. 2a). The freshness of the biotite in the bread-crust bombs suggests that the temperature of the eruptions was comparatively low.[10]

At the close of this initial explosive phase, there must have been a vast, hilly field of lapilli passing outward into a wide plain of sand-like tuff. Over much of this area the vegetation was destroyed, so that the barren, white tuff-sheet must have resembled the desolate Pumice-Flats about Mono Lake.

(b) *Dacite protrusions*: Two dacite domes, each of them originally a square mile in extent, were then protruded through the tuff- and lapilli-cones, and one of them was elevated at least 1,800 feet. The earlier, southern mass has an approximately arcuate form and partly encloses the later dome, from which it is separated by a conspicuous, moat-like depression (Figs. 2a and 2b). It is possible that the earlier dome was once circular in plan and that its crescentic form results from explosions that destroyed its northern side, just as they are known to have destroyed the northern face of the later dome. A somewhat similar sequence of events occurred during the rise of McCulloch and Metcalf domes in the Bogoslof Islands, Alaska.[11]

Except along the periphery, the banding of the dacite in the earlier protrusion is invariably steep and is usually vertical. Marginally, the banding has a low inward dip, as if the lava had just begun to spread laterally before it was chilled. Analogous structures were observed in the dome of Novarupta, Alaska,[12] and may be studied in the Panum spine near Mono Lake. That the Chaos dacite was largely solid when it reached its present position is apparent not only from the attitude of the banding, but from the fact that the surface of the dome is chiefly made up of a tumultuous pile of loose, angular blocks. These rock piles may result from the fracture of a solid crust owing to differential upward movement in the

[10] Cf. Day, A. L., and Allen, E. T., op. cit. 49, 1925.
[11] Jaggar, T. A., The Evolution of Bogoslof Volcano, Bull. Amer. Geog. Soc., **40**, 385-400, 1908.
[12] Sapper, K., Vulkankunde, Stuttgart, 1927, footnote 199-200. "Die Neubildung eines Doms am Novarupta-Vulkan (nahe Katmai) 1912 zeigte vermöge der Bandstruktur des Gesteins deutlich, dass die Bänderung den Dom umgibt, nahe der Basis flach einwärts fällt, nahe dem Gipfel aber steil, fast senkrecht ist." Dr. C. N. Fenner has kindly sent me photographs illustrating this phenomenon.

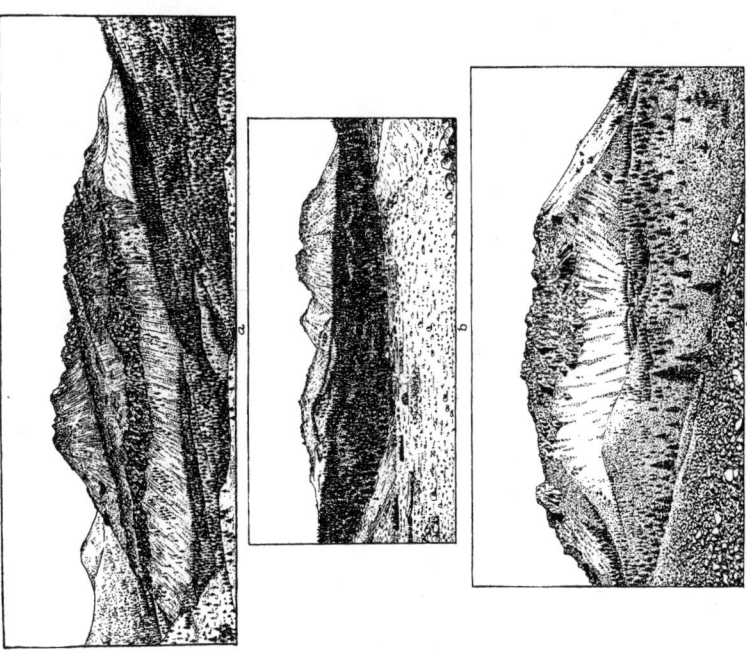

FIG. 2. *The Chaos Crags. a*: looking east. Shows ruined lapilli-cone and tuff-mantled hills in the foreground; also, the two dacite domes, the earlier well-banded and the later structureless. *b*: looking west. Shows the moat between the domes and the long talus banks. In the foreground is part of the 1915 mud-flow from Lassen Peak. *c*: looking south. Shows the craters at the northern base of the later dome and part of the avalanche-deposit, the Chaos Jumbles.

products of these explosions are widespread (Fig. 1). From a study of the distribution of lapilli and bread-crust bombs in the dacite-tuff, it is manifest that a more or less north-south

still viscous lava beneath, and in part also may be due to quick changes of temperature, especially sudden chilling, and to steam explosions. The dacite is abnormally rich in glass, which must have been subject to much strain both during and after consolidation.

The superficial rock piles are not devoid of regular arrangement; on the contrary, they are disposed in great, arcuate ridges concentric with the outer margin of the dome, as if they express some internal structure. Viewed from the west (Fig. 2a), they appear as a series of terraces, inclined northwards at a low angle. If the dome were originally circular and the ridges were continuous, they must then have resembled the rims of a group of hollow cylinders one within another, the highest at the center. If such were the case, it is reasonable to suppose that the outermost cylinder solidified first and that the others rose successively, the innermost last as a narrow pencil of lava. Fouqué[13] noticed circular cracks on the surface of the Georgios dome, Santorini, in 1866, and attributed them to violent explosions. Washington[14] saw similar fissures among the rock piles at the base of the Fouqué Kaméni dome, Santorini, in 1925, his explanation being that they were "caused by a differential sinking of the dome itself relatively to the lava plateau on which it stands." Such a sinking might be facilitated by the outpouring of lava from below and by the loose, blocky character and weight of the dome itself. It is worthy of note that the loose blocks comprising much of the surface of the domes at the Chaos Crags are exceptional. Over large areas, the blocks average a yard in length; elsewhere it is uncommon to find blocks less than 10 feet in dimensions. Many blocks measure as much as 20 feet long and a few reach a length of 30 feet. They are generally smooth-faced and angular, and frequently their sides are slightly concave, suggesting conchoidal fracture on a large scale.

The high talus banks that partly surround the domes are, in large measure, an original character. To some extent the domes rose, as it were, through their own talus. In this respect, as in many others, they resemble the twin domes of Santa Maria, Guatemala, so beautifully illustrated in Sapper's work.

[13] Fouqué, F., Santorin et ses Éruptions, 49, and 75, Paris, 1879.
[14] Washington, H. S., op. cit., 370, 1926.

The later, northern protrusion at the Chaos Crags, like the earlier, consists of a glassy hornblende-mica dacite but the lava here is almost wholly free from banding and is more abundantly charged with purplish, basic, hornblendic secretions. Rarely can any structure be detected within the mass, and it is then restricted to a faint, vertical banding near the core and to a low, inward-dipping banding at the edges. This marginal arrangement of the flow-planes is thus similar to that observable in the earlier dome; it is due in part to the weight of superincumbent lava and suggests an abortive attempt at lateral flow. There appears to be no regular distribution of the basic secretions nor do they exhibit any common direction of elongation. Here, as in the earlier dome, the surface is largely made up of loose blocks in irregular heaps. Most of the blocks have suffered superficial oxidation by steam, for the hornblende and mica on their outer faces are of a deep red color, their normal color being green.

(c) *Explosive eruptions.* The later dome had perhaps attained its present elevation when the northern face of the mass collapsed, producing a gigantic avalanche or rock stream, the Chaos Jumbles. The immediate cause of the collapse is uncertain, but two possibilities suggest themselves: either the dome had become so high and unstable that its collapse was inevitable, or its breakdown was precipitated by the blasting of craters at or near its base. At the northern base of the dome there are two deep, crateriform depressions (Fig. 2c), around which there are signs of solfataric activity. These depressions must have been partly formed after the collapse of the dome, for they have been blasted through the loose debris. It is possible, however, that they originated immediately before the collapse, and that their formation hastened that inevitable event. About 150 million cubic yards of dacite blocks were thus hurled onto the tuff-sheet that had been formed by the initial pyroclastic eruptions. The avalanche rushed northward and rose 400 feet up the opposite slopes of Table Mountain, almost two miles distant from the craters. Elsewhere, the writer[15] has offered evidence to show that the old tuff-sheet over which the avalanche passed had been largely converted into mud by rains that attended the steam explosions. The mobility of the avalanche must have been increased considerably by this muddy surface. The size and character of these explosions are almost precisely similar to

[15] Williams, H., op. cit., 257-258, 1928.

those of the eruption that occurred at Bandai-san, Japan, in 1888.[16] In the case of the Chaos dome, we may suppose that a major cause of its upheaval was the vapor tension in a subjacent crystallizing magma, and we may postulate, as Day and Allen have done in the case of the plug that rose in the crater of Lassen Peak in 1915, that the lava continued to rise until the propelling force was violently released by the blasting of vents at or near its base.

## WHITE MOUNTAIN DOME.

We are now in a position to discuss the older domes of the Lassen region. Of particular interest is the dome of White Mountain, situated about two miles southeast of Lassen Peak (map, Fig. 1). At first sight, the mountain appears to be little else than an enormous pile of blocks. In a few places, crags and pinnacles of 'solid' rock do project through the talus banks, but they are of insignificant proportions by comparison with the volume of superficial debris (Fig. 3c). It is this extraordinary development of talus that first impresses the onlooker. It is not a simple product of weathering, and by analogy with the talus banks of the Chaos Crags, it seems to be due to the fracturing of the lava during its actual emplacement. In a word, it appears to be an original feature of the dome. Probably the loose jumble of blocks passes downward into brecciated andesite, in which the fragments are only semi-detached, and this in turn grades into massive, unbroken lava.

Griggs,[17] writing of the dome of Novarupta, states that in cooling, the upper portion became "a confusion of broken blocks, heaped topsy turvy over the surface, concealing for the most part the bedrock and giving the lava plug the appearance of a huge stone pile." Washington[18] vividly describes the rapid changes in the form of the Santorini dome of 1925. He notes:

These changes were obviously caused by the continuous redistribution of the constituent loose blocks by the violent eruptions. .... The dome was not covered by a more or less continuous "carapace," but .. was (and probably now is) simply a mass of

[16] Bonney, T. G., Volcanoes, 35-36. London, 1912.
[17] Griggs, R. F., The Valley of Ten Thousand Smokes, Nat. Geog. Mag., 249, 1922.
[18] Washington, H. S., op. cit, 361-365, 1926.

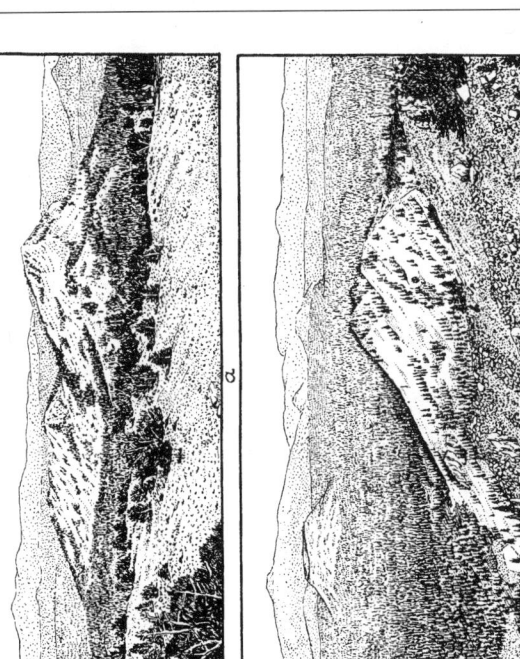
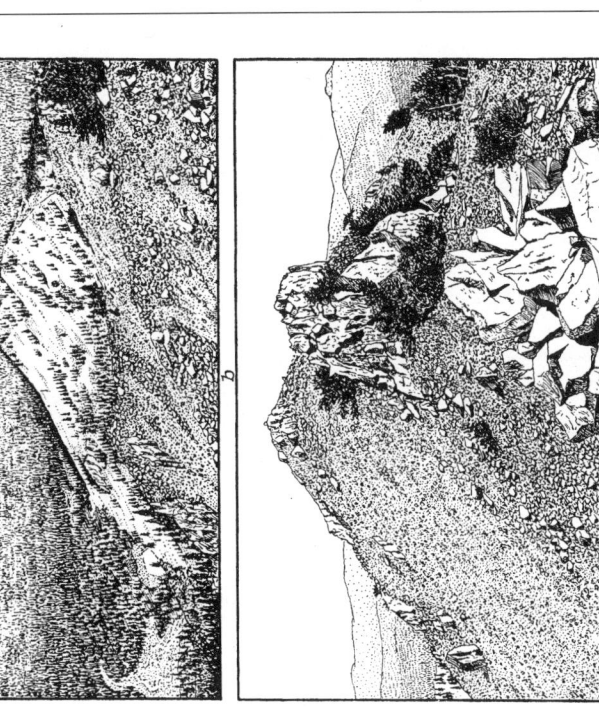

Fig. 3. *The White Mountain Dome.* a: looking east from the slopes of Lassen Peak, showing the great talus banks. b: east end of the ridge. The Central Plateau of augite andesite and Hat Mountain cone in the middle distance. In the background, from left to right, the dome and cone of Prospect Peak, the quartz basalt Cinder Cone and Fairfield Cone. c: part of the summit ridge, showing vertically jointed rock pinnacles protruding through coarse, angular talus blocks.

loose blocks. The dome showed no definite crater. . . . The great vulcanian explosions took place from different places on the upper part of the dome. . . . Occasionally the dome, so to speak "blew its whole head off."

Fouqué[19] refers to the rise of the Georgios dome in 1866 thus:

L'accroissement de l'îlot se faisait sans secousse, sans projection, silencieusement, mais avec une telle rapidité, que M. de Cigalla le compare au développement d'une bulle de savon. Il s'opérait du dedans au dehors, comme par un mouvement d'expansion, les blocs semblaient partir du centre de la surface et progresser de là vers la périphérie; on avait peine à suivre du regard la marche de tous ces blocs pierreux et leurs déplacements incessants. . . Tout autour de nous, nous entendons un craquement incessant dû à la rupture des blocs pendant la contraction causée par leur refroidissement; ce bruit est accompagné d'un cliquetis semblable à celui que donne le choc de débris de porcelaine; il est produit par la chute des fragments détachés des cassures. récentes.

Again, Jaggar[20] describes McCulloch Peak in the Bogoslof Islands in these words:

(It was) of rounded cone shape . . ; and appeared at first sight like a steaming heap of bowlders. The top in profile was flattish, but lumpy, made of hard rock, and without any crater. There were rocky crags at the core and summit of the mass, with slide-rock about its foot-slopes fed by the tumble from the rising cliffs above. Not that the tumbling or rising could be seen—it was too slow for that. All was deathly quiet—but dangerous looking.

It is needless to cite other instances of similar growth among volcanic domes; the fracturing of their crusts and the contemporaneous development of talus banks appear to be almost diagnostic features of massive protrusions.

At White Mountain, one is also impressed by the scarcity of any regular structures within the 'solid' portions. In this respect, it resembles the northern dome at the Chaos Crags, and, as we shall see, the dome of Lassen Peak. Only at the western end of White Mountain do the andesites exhibit flow-banding. Here, close to the margin of the dome, the lava assumes a slabby or even platy structure and the banding lies flat or has a low inward dip. Further within the mass the flow-planes become increasingly steep and at 400 yards from the margin they are generally vertical. It would seem, therefore, that here, as at the Chaos Crags, Novarupta and other domes, the flattening of the flow-planes toward the base and periphery suggests an abortive attempt at lateral spread. Throughout most of White Mountain, however, banding is entirely wanting.

Along the northern edge of the dome there are curved 'lager' joints, convex outwards. Elsewhere, the joints are usually vertical (Fig. 3c), but the joint-planes are oriented without apparent regularity. Over large areas, particularly near the western end of the dome, the andesite blocks are deeply oxidized and have been converted locally into a mixture of opal and kaolin by solfataric action. Hereabouts, the joint faces are commonly covered with brilliant white, yellow and red incrustations of opal, sulphur and iron oxide.

Petrographically, the lava is a pyroxene andesite, poor in glass, rarely vesicular and carries abundant hornblendic secretions. Save for this last character and for the scarcity of flow-banding, the lava does not differ materially from the pale andesites that form the whole region about the head of the Warner Valley to the east. The dips of these andesite flows point to a vent on or near the site of White Mountain. It is therefore suggested that the White Mountain dome rose in the old crater after the cessation of the quiet, liquid effusions.

### THE DOME OF LASSEN PEAK.

The edifice of Lassen Peak rises conspicuously above the surrounding country. Like White Mountain, its sides are almost entirely covered with banks of talus, some of which have a vertical sweep of over 3,000 feet. The exact limits of the peak are poorly defined, but its volume cannot be less than three-fifths of a cubic mile and it must have been elevated about 2,500 feet above the old Lassen crater. Except for the cliffs illustrated in Fig. 4, 'solid' exposures are few and small, and are characterized by an almost complete lack of any regular structure. To many, this must seem the most remarkable feature which the mountain presents, though heretofore no particular record seems to have been made concerning the fact. Excluding the products of the 1914-15 eruptions, the peak is composed of a grey or pink, glassy hornblende-mica dacite carrying an abnormal abundance of dark,

---

[19] Fouqué, F., op. cit, 42 and 54, 1879.
[20] Jaggar, T. A., op. cit, 392, 1908.

hornblendic secretions. Viewed from a few yards' distance, the rocks might well be mistaken for unbedded volcanic breccias. A crude banding may be seen among the crags on the southeast side of the peak; it rolls irregularly at low angles with a general pitch toward the interior of the dome.

If the flow-structures be considered vague, the joint-system is no less obscure. Vertical joints predominate far over all others (Fig. 4c), but the orientation of the planes is so haphazard as to defy any attempt at mapping. On the south side of the peak, the joint-planes may be curved. The high and beautifully smooth cliff overlooking Lake Helen affords an excellent instance of this (Fig. 4b). The smoothness of this cliff-face is not due to glacial polishing. Proof is impossible, but the impression is strong that this is part of the original face of the dacite dome, and that its broad vertical fluting and narrow striae and polishing are due to friction of the solid or almost solid crust of the mass as it rose to the surface through a circular orifice. The curved joints that lie parallel to the face are thus attributed to contraction on cooling. Such a polishing of the sides of massive protrusions has been described elsewhere; for example, one face of the Pelée spine was smooth, and Jaggar[21] says of the Metcalf dome at Bogoslof that its

"resemblance to a beak was increased by regular markings on the rounded surface as though the horn had been shoved up at intervals. Apparently, it had been squeezed out, like paste from a tube, in a semi-plastic condition, and had gradually shaped a roundish cavity for itself in the lava about it."

Possibly much of the outer face of the Lassen dome now concealed by talus may be smooth like the cliff above Lake Helen, and it is likely that if the talus were stripped away the solid core of the Lassen protrusion, like those of the Chaos Crags, would be found to be more or less cylindrical.

Reference to the map (Fig. 1) shows that Lassen Peak is partly surrounded by flows of dacite. Most of these dacites are black, glassy lavas, mineralogically similar to that of Lassen itself, and likewise rich in hornblendic secretions. One of the flows, that of Loomis Peak, consists of a lava that can only be distinguished from the dacite of Lassen Peak by the prominence of its banding. The attitude of the flows indi-

[21] Jaggar, T. A., op. cit., 394, 1908.

AM. JOUR. SC.—FIFTH SERIES, VOL. XVIII, No. 106, OCTOBER, 1929.
22

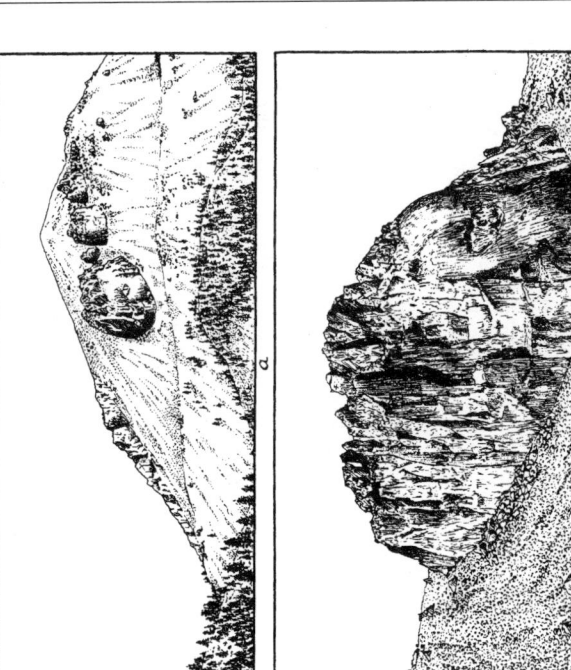
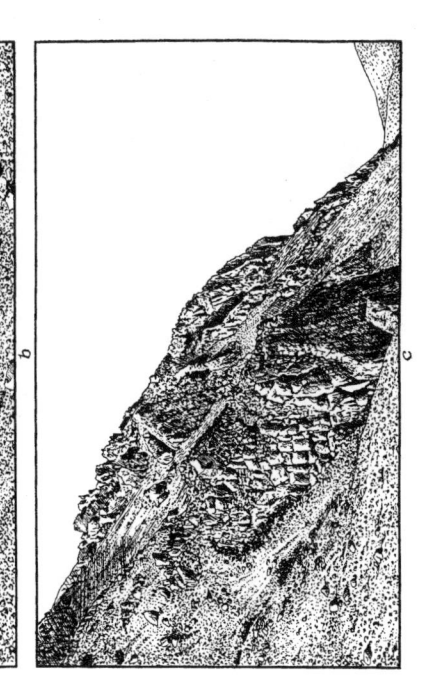

FIG. 4. *The Dome of Lassen Peak.* *a*: south slope from above Lake Helen, showing smooth sides of the plug-dome and concentric banks of talus. *b*: nearer view of the smooth face of the dome, showing the rounded form and fine vertical striation; height of cliff, about 500 feet. *c*: crags near the summit, southeast side, showing vertical joints and long talus slides.

cates that they issued from a crater on or near the site of Lassen. It seems reasonable to infer, therefore, that the almost structureless mass of Lassen Peak represents the upheaved plug of this crater.

Plug-domes of similar type are far from uncommon: Santa Maria (Guatemala), Galunggung (Java), Ko-usu, O-usu and Tarumai (Japan) may be mentioned as examples. In a measure, Lassen Peak may have risen as an endogenous dome, as did the trachyte *puys* of the Auvergne and Ascension Island, but chiefly, it seems, the mass was propelled from below as a homogeneous body or 'stopper' by the continued accession of fresh magma and by the pressure of aqueous vapor. Its rise appears to have been attended by steam explosions and by mud flows comparable with those that accompanied the rise of the plug in the summit crater of Lassen in 1915. The signs of some of these earlier mud flows may be seen along the eastern margin of the recently "Devastated Area," and along Lost Creek. Indeed, the upheaval of the 1915 lava into the summit crater of Lassen was a miniature copy of the rise of Lassen itself through the pre-Lassen crater.

The inner structure of Lassen Peak is, of course, unknown. It is probable, however, that an ill-defined central conduit exists and that the lava becomes increasingly vesicular toward it. Should fresh magma rise beneath the dome, this central portion would suffer fusion first and tend to be uplifted. This is what may have happened during 1914-15. Central spines rose from the summits of the Santa Maria, Pelée and Metcalf domes, and probably the mechanism of emplacement of the 1915 lava of Lassen Peak was the same. There, however, the dacite did not have the properties requisite to the formation of a narrow spine.

### DOMES SOUTH OF LASSEN PEAK.

A cluster of small domes lies at the southern base of Lassen Peak, separating it from the exposed parts of the Brokeoff volcano. These domes were protruded through the black, glassy dacite flows that issued from the old Lassen crater. For instance, Eagle Peak is made up of a pink, massive and poorly banded dacite like that of Lassen Peak, whereas its base is almost encircled by a nearly horizontal flow of black, columnar dacite. Five hundred yards north of Lake Helen, the long talus slope of Lassen abuts on a small dacite pro-

trusion, almost circular in plan, which shows a marginal, concentric banding that stands vertically or dips inwards at steep angles and encloses an unbanded core. Bumpass Mountain, again, is for the major part a structureless mass of gray dacite the surface of which is littered with great piles of angular blocks. Closely-set, steep or vertical joints and low, inward-dipping flow-planes may be distinguished at the margins of the dome. Along its south side, adjacent to the hot springs of Bumpass Hell, the dacite has suffered hydrothermal alteration. This protrusion and that lying immediately to the north closely resemble the talus-mantled dome of White Mountain. The fantastic crags of Vulcan's Castle, one mile northwest of Lake Helen, appear to represent part of another dacite dome, but there is reason to suppose that hereabouts the domal eruptions were accompanied by short, thick flows. It must also be stated that the forms of these domes have been seriously modified by glaciation.

Finally, about 8 miles south of Lassen Peak, close to the Park Highway, there are two other domes, composed of acid andesite rich in hornblendic secretions, namely Morgan Mountain and the hill west of the 'Boy Scout Camp.' Each measures approximately two miles long and a mile wide. They exhibit marginal banding and massive cores like the domes already mentioned and are associated with thick, stumpy flows. They lie near the southern base of the old Brokeoff volcano.

### DOMES NORTH OF THE CHAOS CRAGS.

The irregular hills that lie between the Chaos Crags and Table Mountain are largely covered by glacial drift, but their forms suggest them to be steep lava-domes. Here, as on other domes, the surfaces are chiefly made up of angular dacite blocks lying about in great disorder. The dacite is not of the usual hornblende-mica variety, but is a pyroxene-bearing type, rich in hypersthene phenocrysts set in a glassy base, and is almost devoid of hornblendic secretions. Banding is seldom developed and seems to be oriented haphazard.

### SUMMARY CONCERNING THE LASSEN DOMES.

The domes referred to in the preceding pages consist chiefly of glassy, acid andesites and dacites, rich in hornblendic secretions. Most of them are singularly devoid of well-defined structures, except marginally where they reveal a low, inward-dipping banding. In the cores of the domes banding is either absent or is vertical; the dominant joint-planes are also vertical but of irregular orientation. All the domes are characterized by loose, blocky surfaces, due to the break-up of their solid crusts during protrusion. Two domes—those of the Chaos Crags—were protruded through a line of tuff-cones; two others—Lassen Peak and White Mountain—rose through old craters that had previously erupted fluid dacite and andesite, respectively. These four domes may thus be termed plug-domes (Staukuppen) and appear to represent the upheaved infillings of volcanic pipes.

It is worthy of note that although so many of these domes are clustered on the flanks of the old, hypersthene andesite, Brokeoff volcano, no domes rose into the Brokeoff caldera itself. It must be remembered, however, that the Brokeoff caldera is now the center of an extensive area of fumaroles and hot springs, and it is possible that this solfatarism is the surficial expression of a concealed intrusion, perhaps of dacitic character. That the Brokeoff caldera was faulted and collapsed after the emission of some of the glassy dacites from the Lassen crater is clear from the fact that a flow of Lassen dacite occurs on the opposite side of the Brokeoff caldera, on Black Butte (map, Fig. 1). Perhaps the suggestion may be allowed that the collapse of the Brokeoff caldera was occasioned by the protrusion of the adjacent domes of dacite. Such a wholesale migration of magma as that involved during the dacite protrusions may well have withdrawn the support from beneath the Brokeoff cone.

By analogy with the observed rate of protrusion of such domes as those of Santa Maria, Tarumai and Santorini, it can hardly be doubted that the rise of the Lassen domes was a rapid process. Two years after its first appearance, the Santa Maria dome was 500 meters high and 1,200 meters wide at the base. Considered solely on this basis, it seems that Lassen Peak may have reached its present elevation during a period of about five years, and that some of the smaller domes may have grown in a year or so.

Excepting the domes of the Chaos Crags, which are about 200 years old, all the Lassen protrusions are preglacial. It should be borne in mind, however, that glaciation persisted among these mountains until a very late date.

## THE DIVIDE PEAK DOME.

The form of the augite andesite cone of Divide Peak is surprisingly little modified by glaciation. The same is true of other large cones in the Lassen National Park, especially of Red Mountain, Prospect Peak and Mount Harkness. On the summit of Divide Peak, there still remains a well-preserved cinder cone. On the southwest side of Divide Peak, the andesite is intruded by a large body of dacite, finely revealed in the long line of cliffs overlooking the head of Lost Creek (Figs. 1 and 5). Toward the base, the dacite has a strongly developed columnar structure, the columns standing vertically. Flow-banding is conspicuous throughout, especially at higher levels; throughout most of the dome it lies flat, but near the top it is vertical or dips to the east and southeast at angles of from 40° to 70°. The dacite is a glassy, hornblende-mica variety, not unlike that of the Chaos Crags and Lassen Peak, and is charged with abundant hornblendic secretions, particularly toward the base.

Although the actual contacts between the dacite and andesite cannot be observed, there is reason to suppose that the margin of the dacite has a steep outward dip and may be vertical in places. Thus, toward the southern end of the protrusion the horizontal banding of the dacite bends over to the vertical, and at several localities the dips of the andesites are seen to have been reversed and tilted away from the dacite at high angles.

Since the form of the Divide Peak cone has been so little affected by glaciation, it is probable that the present form of the dacite dome is also not very different from its original shape. If such be the case, then the dacite cannot have risen far above the surface of the andesite cone. The area of reversed dips among the andesites near the margin of the dome is not extensive; again, there cannot have been more than a thin cover of andesite over the dacite, and it is more likely that no such cover ever existed; and, further, there is no marginal assimilation of andesite. It follows that the dacite did not displace much andesite and is probably of no great thickness. Perhaps it is of mushroom form and has an internal structure such as that depicted in Fig. 5. A viscous lava rising through a narrow orifice and spreading in this manner would develop a more or less concentric structure (Zwiebelstruktur), as Reyer[22] has shown experimentally. The dome of Divide Peak thus presents a close similarity to many of the trachyte domes described by Daly[23] from Ascension Island; that is, it is a true endogenous dome that grew

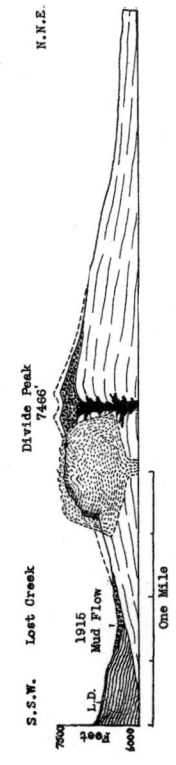

Fig. 5. Section through Divide Peak andesite cone and dacite dome. The summit of the former consists chiefly of bedded pyroclastic rocks. Flows of glassy dacite from the crater on the N. E. flank of Lassen Peak (L.D.) overlie the andesites on the south.

by expansion from within. It was not upheaved as a virtually solid plug, as were most of the other Lassen domes. Its location on the slope of the Divide Peak cone is comparable, however, with that of the other domes on the flanks of the older Brokeoff volcano.

UNIVERSITY OF CALIFORNIA,
BERKELEY, CALIFORNIA.

[22] Reyer, E., in Sapper, K., Vulkankunde, 199, 1927.
[23] Daly, R. A., The Geology of Ascension Island, Proc. Amer. Acad. Arts and Sci., 60, 23-38, 1925.

# Volcanism in California

**Robert L. Christiansen and Michael A. Clynne**
*U.S. Geological Survey, 345 Middlefield Road, MS 910, Menlo Park, California 94025;*
*e-mail: rchris@mojave.wr.usgs.gov; mclynne@mojave.wr.usgs.gov*

## INTRODUCTION

A rich literature exists on volcanism and volcanic geology in California, although only a single eruptive episode—that of Lassen Peak in the second decade of the twentieth century—has been witnessed closely during the period of written history. Without doubt the most influential early investigator to bring a modern geologic approach to bear on problems of California and west coast volcanism was Howel Williams, and among his earliest studies was his work in the Lassen region. This paper briefly reviews some of Williams' relevant work and its influences, then summarizes later studies of volcanism in California.

In 1926 Williams came to study under A. C. Lawson at the University of California, and the rest of his career remained centered at Berkeley. Following studies at the University of Liverpool and Imperial College on Ordovician strata in Wales, with their abundant volcanic rocks, and visits to classic volcanic areas of Germany and France (McBirney, 1985), he arrived in California eager to work on volcanic problems and soon began a study at the Sutter (Marysville) Buttes in the Sacramento Valley (Williams, 1929a).

Interest in the 1914–1917 eruptions of Lassen Peak, then still in recent memory, attracted Williams there in the summer of 1927, where he found his attention drawn to a number of interesting issues. In particular, he recognized the Chaos Crags, aptly named piles of broken lava blocks that rise as much as 430 m above the north base of Lassen Peak, to be a cluster of volcanic domes—protrusions of viscous lava that had mounded above their vents with only limited lateral flowage. He also recognized the Chaos Jumbles, a hummocky field of lava blocks spreading northwestward from the Crags, as a catastrophic avalanche deposit rather than morainal debris, as thought by earlier workers. Williams recognized that Lassen Peak itself is a much larger volcanic dome, one of the largest known, and made detailed petrographic observations of the clustered dacitic domes of the Lassen area and their abundant mafic inclusions. These investigations, together with systematic geologic mapping, constituted one of the first detailed modern studies of a volcanic area in California; they were published in a series of topical papers (Williams, 1928, 1929b, 1931) and were summarized, together with a colored geologic map, in a volume of the important publication series produced over many decades by the Department of Geological Sciences of the University of California (Williams, 1932b).

## LASSEN PEAK ERUPTIONS OF 1914–1917

An extended series of eruptions began at Lassen Peak in May 1914, climaxed in May 1915, and waned through June 1917, occasioning great scientific as well as public interest. Although these volcanic events were not scientifically well studied during their activity, they were among the first eruptive episodes to be documented through an extensive photographic record.

The first scientist to visit Lassen in order to observe its activity was J. S. Diller, a pioneering geologist who had mapped much of southern Oregon and northernmost California in geologic reconnaissance for the U.S. Geological Survey during the late nineteenth and early twentieth centuries. Diller, by then no longer an active field geologist, accompanied the local businessman and enthusiastic photographer B. F. Loomis to Lassen Peak and the surrounding area in June 1914. The climactic explosions of May 19–22, 1915, were not witnessed directly by scientists, but following those eruptions Diller visited again in 1915 and 1921. Although he published some significant observations (e.g., Diller, 1914a, 1914b, 1916, 1917), Diller did not carry out a systematic study. During the summer of 1915 and for the next several seasons, A. L. Day and E. T. Allen of the Carnegie Institution of Washington undertook a detailed study of the ongoing volcanic activity, as well as the hot springs of the Lassen Peak area. Their

---

Christiansen, R. L., and Clynne, M. A., 1999, Volcanism in California, *in* Moores, E. M., Sloan, D., and Stout, D. L., eds., Classic Cordilleran Concepts: A View from California: Boulder, Colorado, Geological Society of America Special Paper 338.

ensuing paper (Day and Allen, 1925) summarized the 1914–1915 eruptions and offered the first integrated interpretation of the devastation produced on the northeast flank of Lassen Peak in May 1915 and the resulting far-traveled debris flows. B. F. Loomis published his own ideas of the explosive eruptions and their interpretation in a small book featuring some of his photographs (Loomis, 1926). These events and studies formed the background against which Williams set his more comprehensive study of the Lassen region.

## "THE VOLCANIC DOMES OF LASSEN PEAK AND VICINITY, CALIFORNIA"

It is somewhat arbitrary to select just one of the several papers Howel Williams produced from his work in the Lassen Peak region for the purpose of this volume. In order to reproduce a single complete paper that represents some of the characteristically insightful and generally applicable aspects of his work, however, we have chosen to highlight his 1929 paper on the volcanic domes of the Lassen area. In that paper Williams demonstrated the existence of at least 13 dacitic domes clustered on the northern flank of an eroded older andesitic stratovolcano, the Brokeoff cone.

His observations of Chaos Crags and Chaos Jumbles had led Williams to new conclusions about their origins and had focused his attention on the origin and emplacement of volcanic domes. The importance of domes had begun to become more widely appreciated through a number of early twentieth-century studies. In particular, the catastrophic 1902 eruption of Mont Pelée on Martinique and subsequent activity there (Lacroix, 1904), and the 1922–1925 growth of the Santiaguito dome at Santa María volcano in Guatemala (Sapper, 1926) had demonstrated important aspects of the growth mechanisms of volcanic domes and their relation to violently explosive eruptions. In the paper reproduced here, Williams (1929b) brought these lessons to the domes of the Lassen region and extended them with his careful descriptions and interpretations of the variously exposed domes and related flows and pyroclastic deposits there.

The Chaos Crags are the youngest domes of the Lassen region. Williams described pyroclastic ejecta that form a partly buried crater at the south end of the Crags and a blanket over much of the surrounding terrain. The chaotically blocky surfaces of the domes were demonstrated to have formed by the fracturing of a brittle crust as the domes grew by swelling during emplacement while their interiors retained structures related to viscous flowage. Interpretation of these features in terms of dome-emplacement mechanisms allowed Williams to extend his interpretations to the emplacement of the older domes, Reading Peak (known at the time as White Mountain), Lassen Peak, and others nearby. The domes, at least initially glassy in texture, are dacites. The initial enveloping of the domes by talus-like breccia was shown to be characteristic, although glacial erosion had removed such a carapace from many of the older bodies. The degrees of preservation of carapace breccia provided estimates of the relative ages of various domes. Spines or other smooth-surfaced protrusions characteristically rise above the brecciated surfaces to expose the interior lithologies and shear surfaces related to emplacement. Individual domes represent monogenetic episodes of growth over periods of months to a few years. The dome on the west side of Raker Peak (then known as Divide Peak) was interpreted to have forcibly displaced the lavas of the older andesitic cone that it intruded and partly overgrew.

The detailed descriptions and interpretations of these bodies in the Lassen Peak region led logically to a comprehensive review of volcanic domes and their origins (Williams, 1932a). The Lassen work and its outgrowths became highly influential publications, setting a high standard for geologic work in areas of young volcanism, beginning a course of study that carried Williams to several other significant volcanoes in the Cascades, and helping to establish his worldwide reputation in volcanology and volcanic geology.

## LATER STUDIES AND THE EXTENDED INFLUENCE OF HOWEL WILLIAMS

Williams followed his Lassen work with important investigations at other Cascades volcanoes. A brief study of Mount Shasta (Williams, 1932c, 1934) led to a reconnaissance of the northernmost California Cascades (Williams, 1949). What undoubtedly became his most influential study, that at Crater Lake, Oregon (Williams, 1942), demonstrated the collapse origin of the Crater Lake caldera and resulted in another highly influential comprehensive review, analyzing calderas at volcanoes around the world (Williams, 1941). Meanwhile, Williams visited and studied volcanoes elsewhere, including Tahiti (Williams, 1933) and the Navajo-Hopi country of Arizona (Williams, 1936). Furthermore, he began to attract other young members of the Berkeley faculty to volcanic studies, notably C. A. Anderson, who made significant contributions to understanding volcanic mudflow deposits (Anderson, 1933b) and carried out geologic studies of youthful volcanism in the Medicine Lake Highland (Anderson, 1933a, 1941) and the Clear Lake volcanic field (Anderson, 1936) in northern California.

Williams also gained a prominent academic reputation. Among those who came under his influence and went on to advance volcanological knowledge were Gordon Macdonald, who later made major contributions to the geology of Hawaii (e.g., Macdonald, 1968) and the Lassen region (e.g., Macdonald, 1963). Williams' student Garniss Curtis, after a study of Tertiary volcanism in the central Sierra Nevada (Curtis, 1954), began an important investigation of the 1912 eruption in the Valley of Ten Thousand Smokes, Alaska (Curtis, 1968), although his efforts were soon diverted from a volcanological emphasis to development and application of the newly emerging technology of K-Ar dating (e.g., Curtis et al., 1958). Curtis did, however, join in revising the earlier results of Williams' first California study, at Sutter Buttes (Williams and Curtis, 1977). Following the 1943–1952 eruptions of Parícutin (Williams, 1950), Williams' major volcanological interests turned to Central America. It was there that he encountered the person who was to become probably his most notable volcanological student and long-time colleague, A. R. McBirney. Separately, together, and with others, Williams and McBirney produced a series of important studies of Guatemalan, Nicaraguan, and Salvadoran volcanoes (e.g., McBirney and Williams, 1965) as well as the Galápagos Islands (McBirney and Williams, 1969). Williams collaborated with fellow Berkeley professors F. J. Turner and C. M. Gilbert to produce a classic textbook on petrography, illustrated with his thin-section drawings (Williams et al., 1954), and later with McBirney on a volcanology textbook (Williams and McBirney, 1979).

## OTHER STUDIES OF CALIFORNIA VOLCANISM

Volcanic rocks in California from Precambrian to Holocene have provided numerous important studies, but a brief review of advances made in the path blazed by Howel Williams and his contemporaries at Berkeley inevitably focuses on Cenozoic volcanism. Figure 1 shows the locations of important localities mentioned in the following discussion.

### The Cascades–Modoc Plateau–Warner Mountains region

When Howel Williams undertook studies at Lassen, R. J. Russell (1928) was establishing a Tertiary volcanic stratigraphy in the Warner Mountains, at the edge of the Great Basin. The andesitic "lower Cedarville series" of Russell, now known to be Eocene, is part of an early continental-margin arc that ended southward in northwestern Nevada (Christiansen and Yeats, 1992); the "upper Cedarville series" of Russell, 32–17 Ma (Duffield and McKee, 1986), is a southeastward continuation of the mid-Tertiary arc designated in Oregon as the Western Cascades.

Basalts, mainly late Miocene but as young as Quaternary, mantle the Modoc Plateau and much of the region between the Warner Mountains and the Cascade Range (McKee et al., 1983; Duffield and McKee, 1986). The Quaternary Medicine Lake volcanic shield and caldera adjacent to the Cascades is predominantly mafic with lesser amounts of silicic lavas (Anderson, 1941; Donnelly-Nolan, 1988); Holocene rhyolite erupted high on the shield, and Holocene basalt is extensive on its flanks (Anderson, 1933a; Heiken 1978; Donnelly-Nolan et al., 1990).

The southern High Cascades arc, younger than about 7 Ma, extends from Oregon to south of Lassen Peak. Patterns of volcanic vents and faults are strongly influenced by basement structures and crustal properties (Blakely et al., 1997). Mount Shasta, the largest stratocone of the Cascades (it has nearly double the volume of Mount Rainier although it has a summit elevation about 250 feet lower), represents 600 k.y. of episodic cone building (Williams, 1934; Miller, 1980; Christiansen, 1982); an ancestral cone was largely destroyed by sector collapse about 300 ka (Crandell, 1989). A nearly continuous chain of mafic shields is broken by a gap of about 50 km between the Shasta and Lassen areas (Guffanti and Weaver, 1988). In the south, the oldest High Cascades eruptions produced volcanic debris-flow deposits of the Tuscan Formation, mainly about 2.5–3.5 Ma (Anderson, 1933b; Harwood et al., 1981), that interfinger with several volcanic centers (Lydon, 1968) and include rhyolitic ash-flow tuff that once covered most of the Sacramento Valley (Anderson and Russell, 1939; Sarna-Wojcicki, 1976). The southernmost active center of the High Cascades is the cluster of late Quaternary dacitic domes, including Lassen Peak, on the eroded flank of the 600–400 ka Brokeoff stratovolcano (Williams, 1929b, 1932b; Clynne, 1990).

Volcanic rocks of the southern Cascades, Modoc Plateau, and adjacent parts of the Great Basin were reviewed by Macdonald (1966).

### Sierra Nevada

Rhyolitic sediments and lavas, mainly of 23–21 Ma, occupy former west-flowing drainages in the northern and central Sierra Nevada (Piper et al., 1939; Dalrymple, 1964; Durrell, 1987) that headed farther east during a 28–22 Ma period of voluminous rhyolitic ash-flow eruptions in western Nevada (Slemmons, 1966). High in the Sierra, the rhyolites include several 28–21 Ma welded ash-flow tuffs that originated from sources farther east (Dalrymple, 1963, 1964; Deino, 1994).

During the middle to late Miocene, intermediate calc-alkalic magmas erupted in eastern California and western Nevada, many of them as eruptive and debris-flow breccias (Piper et al., 1939; Curtis, 1954; Chesterman, 1968; Gilbert et al., 1968; Durrell, 1987) that repeatedly filled the Sierran drainages before the range was uplifted and separated from the Great Basin (Bateman and Wahrhaftig, 1966). The source volcanoes, mainly older than 9 Ma, erupted near the present range crest and farther east (Slemmons, 1966), but a few dacitic domes occur in the western foothills (Rose, 1959). These predominantly andesitic volcanic rocks are parts of a mid-Miocene arc that extended southward through eastern California and western Nevada to about the latitude of Las Vegas (Christiansen and Yeats, 1992).

Latitic to basaltic 11–9 Ma lavas and ash flows, including a caldera center, occur high in the mid-Miocene arc sequence of the central Sierra Nevada (Dalrymple, 1963, 1964; Slemmons, 1966; Gilbert et al., 1968; Noble et al., 1974). Young volcanic rocks from the northern Sierra Nevada southward to about Sonora Pass, Bodie Hills, and Mono Lake include small areas of 8–3 Ma basalt, trachyandesite, and rhyolite (Dalrymple, 1964; Thompson and White, 1964; Chesterman, 1968; Gilbert et al., 1968; Silberman and Chesterman, 1972).

A notable contribution from the Berkeley group was the recognition by C. M. Gilbert (1938) of voluminous, mainly welded rhyolitic ash flows, the Bishop Tuff, east of the Sierran front. The source of the tuff is the Long Valley caldera, 30 km across (Bailey et al., 1976; Bailey, 1989). By 1.1 Ma rhyolitic magma had filled a large chamber (Metz and Mahood, 1985) that was disrupted at 760 ka by eruption of the 700 km$^3$ Bishop Tuff (Dalrymple et al., 1965; Bailey et al., 1976; Hildreth, 1979; Wilson and Hildreth, 1997), depositing fallout ash over much of the western United States (Izett et al., 1970). The rhyolitic Inyo and Mono domes and flows, younger than 30 ka, extend 50 km north-

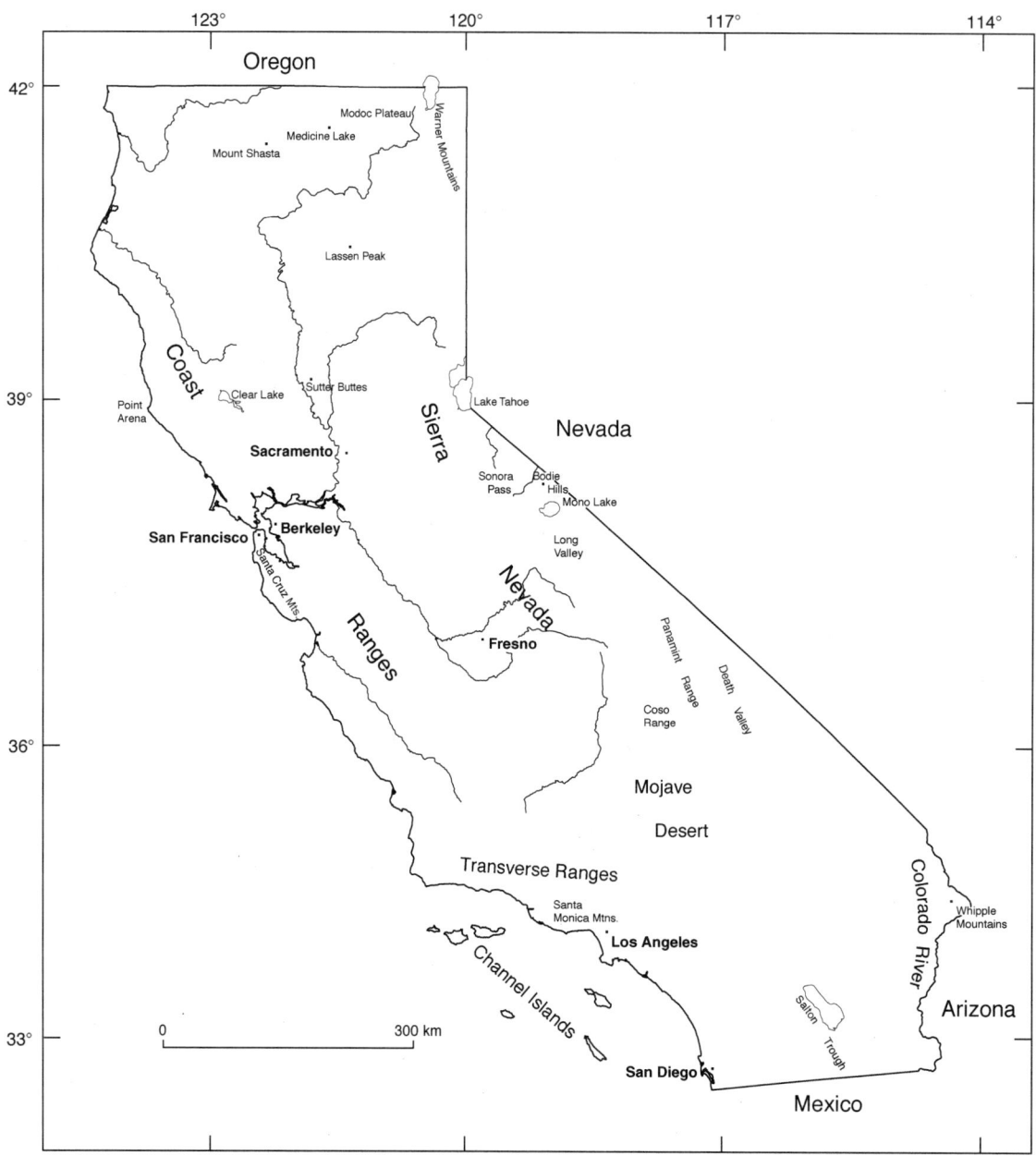

Figure 1. Map of California showing major localities mentioned in the text.

ward from the caldera to Mono Lake (Dalrymple, 1967; Bailey et al., 1976; Wood, 1977); their most recent eruptions were about 600 years ago (Miller, 1985; Sieh and Bursik, 1986). Since 1980, intense seismicity, uplift, and $CO_2$ emissions have signaled continuing activity of the Long Valley magmatic system (Hill, 1996).

*California deserts*

Miocene and younger volcanism between the southern Sierra Nevada and central Death Valley produced bimodal assemblages of basalt-trachyandesite and rhyolite. An area near the north end of Death Valley is pocked by basaltic Holocene maar volcanoes (Crowe and Fisher, 1973). In the Coso Range, bimodal volcanism has been especially intense since about 6 Ma (Duffield et al., 1980; Bacon et al., 1980), and the youngest eruptions are roughly time predictable from the volume of the preceding eruption (Bacon, 1982).

In the central Mojave Desert region, ca. 25–22 Ma volcanic and sedimentary rocks are generally dominated by andesite and are overlain by 20–17 Ma basalts and silicic tuffs from local centers. The Peach Springs Tuff, a distinctive and voluminous 18 Ma rhyolitic ash-flow sheet, extends from the western Colorado Plateau to the central Mojave Desert (Young and Brennan, 1974; Nielson et al., 1990). Middle and upper Miocene andesites and dacites from Death Valley to the northern Mojave Desert (Dibblee, 1967; Smith, 1964) may be parts of the mid-Miocene arc, but a gap in the arc extends southward from the central Mojave and southern Nevada to beyond the Mexican border, marking initiation of a transform boundary between the Pacific and North America plates. Local 10 Ma and younger basalts to basanites, commonly with conspicuous ultramafic mantle inclusions, dot the Mojave region (Parker, 1963; Wise, 1969; Dohrenwend et al., 1984; Farmer et al., 1995).

In a corridor of highly extended upper crustal rocks along the lower Colorado River, magmatism generally followed the 25–26 Ma earliest dated local extension, although in the Whipple Mountains (Davis et al., 1980), andesite and basalt of 32–26 Ma that predate detachment faulting are overlain by 19–17 Ma synextensional andesite. Early to middle Miocene calc-alkalic volcanism in this corridor was succeeded by later Miocene basalts and bimodal rhyolite-basalt fields associated with regionally distributed basin-range faulting (Crowe, 1978).

The Gulf of California oceanic spreading center penetrates the continent at the Salton Trough. Small-volume volcanism (Robinson et al., 1976) associated with opening of the Gulf of California since about 4–5 Ma produced rhyolite domes as well as tholeiitic basalts that occur as inclusions in the rhyolite and as subsurface sills, dikes, and flows. The magmatism sustains several high-temperature geothermal systems that are actively metamorphosing the sedimentary fill of the Salton trough (Muffler and White, 1969).

## Coastal California

Two lower Miocene, dominantly silicic or bimodal volcanic fields at the western margin of the California desert area are offset more than 300 km northwest along the San Andreas fault system (Dibblee, 1967; Turner, 1970; Crowell, 1973; Matthews, 1976; Frizzell and Weigand, 1993; Sims, 1993). Late Oligocene to early Miocene volcanism also produced dacitic volcanic and shallow intrusive rocks in the southern Coast Ranges (Ernst and Hall, 1974) and submarine lavas in the Santa Cruz Mountains and near Point Arena (Turner, 1970; Fox et al., 1985). Restoring mid-Miocene and younger right slip places these volcanic rocks between the paleolatitudes of Las Vegas and Tucson, suggesting that they were initially parts of a continuous largely rhyolitic mid-Tertiary magmatic belt across the southern basin-range region (Christiansen and Yeats, 1992). Dickinson (1997), however, related them to mantle upwelling that occurred as the Pacific plate initially encountered North America to end subduction of part of the Vancouver-Farallon plate.

Predominantly intermediate ca. 20–12 Ma calc-alkalic rocks are widespread in coastal southern California, some of them interlayered in marine sedimentary rocks, including in the Los Angeles basin (Woodring et al., 1946; Shelton, 1955; Yerkes, 1957; Schleicher, 1974; Weigand, 1982), the western Transverse Ranges (Yerkes and Campbell, 1979; Weigand, 1982; Weigand and Savage, 1993), the northern Channel Islands (Crowe et al., 1976; Vedder et al., 1979), and near the Mexican border (Hawkins, 1970; Gastil et al., 1979). Volcanic-arc rocks of 24-8 Ma on both sides of the Gulf of California (Gastil et al., 1979; Sawlan and Smith, 1984), separated as much as 480 km by oceanic spreading since 5 Ma, are relatively little rotated. By contrast, the northern Channel Islands and Santa Monica Mountains are laterally disrupted and may have been rotated 70°–130° clockwise since the mid-Miocene (Luyendyk et al., 1985). Restoration would make mid-Miocene andesitic rocks of southern California contiguous with the continental-margin arc in northwestern Mexico, south of the transform-related gap (Christiansen and Yeats, 1992), although Dickinson (1997) related them to a second pulse of mantle upwelling following a rise-trench encounter that ended subduction of the Monterey-Arguello plate.

With widening of the subduction gap and northward retreat of the southern termination of the arc since about 15 Ma, large magmatic systems have propagated northward through the central Coast Ranges, lagging migration of the Mendocino triple junction by about 80–100 km (Dickinson and Snyder, 1979; Weigand, 1982; Fox et al., 1985; Dickinson, 1997). The youngest of these is the basaltic to rhyolitic Clear Lake volcanic field of 2 Ma to 10 ka (Anderson, 1936; Hearn et al., 1981, 1995). The important geothermal field at the Geysers is sustained by heat from this magmatic system. To the east, rhyolitic domes of the Pleistocene Sutter Buttes and contemporaneous buried intrusives (Williams and Curtis, 1977) were emplaced during a wave of tectonic deformation in the Sacramento Valley as the Mendocino triple junction passed offshore (Harwood and Helley, 1987).

Small-volume basaltic or bimodal magmatism has occurred intermittently since the early Miocene along the lengthening coastal transform fault system, including the California Continental Borderland (Hawkins et al., 1971), the southern Coast Ranges and westernmost Transverse ranges (Vedder, 1975; Hall, 1981), and the San Francisco Bay area (Fox et al., 1985; McLaughlin et al., 1996).

## PERSPECTIVE

Studies of volcanism and volcanic rocks in California have advanced considerably in the past few decades. It is important to note, however, that the conceptual advances are built upon a foundation of detailed and systematic geologic mapping, stratigraphic study, geochronology, petrologic investigation, and volcanological interpretation of individual volcanic areas. Such intensive studies were pioneered in California by Howel Williams and a notable group that he influenced during his long tenure at the University of California at Berkeley. Great strides taken since then follow paths laid out in those first modern approaches to problems of California volcanism.

## ACKNOWLEDGMENTS

This brief review benefited considerably from reviews by Bill Dickinson, Julie Donnelly-Nolan, Wes Hildreth, and Pete Weigand; to all of whom we are indebted for their careful readings and thoughtful comments.

# REFERENCES CITED

Anderson, C. A., 1933a, Volcanic history of Glass Mountain, northern California: American Journal of Science, v. 226, p. 485–506.

Anderson, C. A., 1933b, The Tuscan Formation in northern California, with a discussion concerning the origin of volcanic breccias: California University Publications, Department of Geological Sciences Bulletin, v. 23, p. 215–276.

Anderson, C. A., 1936, Volcanic history of the Clear Lake area, California: Geological Society of America Bulletin, v. 47, p. 629–664.

Anderson, C. A., 1941, Volcanoes of the Medicine Lake Highland: California University Publications, Department of Geological Sciences Bulletin, v. 25, p. 347–422.

Anderson, C. A., and Russell, R. D., 1939, Tertiary formations of northern Sacramento Valley, California: California Journal of Mines and Geology, v. 35, p. 219–253.

Bacon, C. R., 1982, Time-predictable bimodal volcanism in the Coso Range, California: Geology, v. 10, p. 65–69.

Bacon, C. R., Duffield, W. A., and Nakamura, K., 1980, Distribution of Quaternary rhyolite domes of the Coso Range, California: Implications for the extent of the geothermal anomaly: Journal of Geophysical Research, v. 85, p. 2425–2433.

Bailey, R. A., 1989, Geologic map of the Long Valley caldera, Mono-Inyo craters volcanic chain, and vicinity, eastern California: U.S. Geological Survey Miscellaneous Investigations Map I-1933, scale 1:62,500.

Bailey, R. A., Dalrymple, G. B., and Lanphere, M. A., 1976, Volcanism, structure, and geochronology of Long Valley caldera, Mono County, California: Journal of Geophysical Research, v. 81, p. 725–744.

Bateman, P. C., and Wahrhaftig, C., 1966, Geology of the Sierra Nevada, *in* Bailey, E. H., ed., Geology of northern California: California Division of Mines and Geology Bulletin 190, p. 107–172.

Blakely, R. J., Christiansen, R. L., Guffanti, M., Wells, R. E., Donnelly-Nolan, J. M., Muffler, L. J. P., Clynne, M. A., and Smith, J. G., 1997, Gravity anomalies, Quaternary vents, and Quaternary faults in the southern Cascade Range, Oregon and California: Implications for arc and backarc evolution: Journal of Geophysical Research, v. 102, p. 22,513–22,527.

Chesterman, C. W., 1968, Volcanic geology of the Bodie Hills, Mono County, California, *in* Coats, R. R., Hay, R. L., and Anderson, C. A., eds., Studies in volcanology: A memoir in honor of Howel Williams: Geological Society of America Memoir 116, p. 45–68.

Christiansen, R. L., 1982, Volcanic hazard potential in the California Cascades, *in* Martin, R. C., and Davis, J. F., eds., Status of volcanic prediction and emergency response capabilities in volcanic hazard zones of California: California Division of Mines and Geology Special Publication 63, p. 41–59.

Christiansen, R. L., and Yeats, R. S., 1992, Post-Laramide geology of the U.S. Cordilleran region, *in* Burchfiel, B. C., Lipman, P. W., and Zoback, M. L., eds., The Cordilleran orogen: Conterminous U.S.: Boulder, Colorado, Geological Society of America, Geology of North America, v. G-3, p. 261–406.

Clynne, M. A., 1990, Stratigraphic, lithologic, and major element geochemical constraints on magmatic evolution of the Lassen volcanic center, California: Journal of Geophysical Research, v. 95, p. 19,651–19,669.

Crandell, D. R., 1989, Gigantic debris avalanche of Pleistocene age from Mount Shasta Volcano, California, and debris-avalanche hazard zonation: U.S. Geological Survey Bulletin 1861, 32 p.

Crowe, B. M., 1978, Cenozoic volcanic geology and probable age of inception of basin-range faulting in the southeasternmost Chocolate Mountains, California: Geological Society of America Bulletin, v. 89, p. 251–264.

Crowe, B. M., and Fisher, R. V., 1973, Sedimentary structures in base-surge deposits with special reference to cross-bedding, Ubehebe Craters, Death Valley, California: Geological Society of America Bulletin, v. 84, p. 663–682.

Crowe, B. M., McLean, H., Howell, D. G., and Higgins, R. E., 1976, Petrography and major element chemistry of the Santa Cruz Island volcanics, *in* Howell, D. G., ed., Aspects of the geologic history of the California Continental Borderland: American Association of Petroleum Geologists Miscellaneous Publication 24, p. 196–215.

Crowell, J. C., 1973, Problems concerning the San Andreas fault system in southern California, *in* Kovach, R. L., and Nur, A., eds., Proceedings, Conference on Tectonic Problems of the San Andreas Fault System: Stanford University Publications in the Geological Sciences, v. 13, p. 125–135.

Curtis, G. H., 1954, Mode of origin of pyroclastic debris in the Mehrten Formation of the Sierra Nevada: University of California Publications in Geological Sciences, v. 29, p. 453–502.

Curtis, G. H., 1968, The stratigraphy of the ejecta from the 1912 eruption of Mount Katmai and Novarupta, Alaska, *in* Coats, R. R., Hay, R. L., and Anderson, C. A., eds., Studies in volcanology: A memoir in honor of Howel Williams: Geological Society of America Memoir 116, p. 153–210.

Curtis, G. H., Evernden, J. F., and Lipson, J., 1958, Age determination of some granitic rocks in California by the potassium-argon method: California Division of Mines Special Report 54, 16 p.

Dalrymple, G. B., 1963, Potassium-argon dates of some Cenozoic volcanic rocks of the Sierra Nevada, California: Geological Society of America Bulletin, v. 74, p. 379–390.

Dalrymple, G. B., 1964, Cenozoic chronology of the Sierra Nevada, California: University of California Publications in the Geological Sciences, v. 47, 41 p.

Dalrymple, G. B., 1967, Potassium-argon ages of recent rhyolites of the Mono and Inyo Craters, California: Earth and Planetary Science Letters, v. 3, p. 289–298.

Dalrymple, G. B., Cox, A. V., and Doell, R. R., 1965, Potassium-argon age and paleomagnetism of the Bishop Tuff, California: Geological Society of America Bulletin, v. 76, p. 665–673.

Davis, G. A., Anderson, J. L., Frost, E. G., and Shackelford, T. J., 1980, Mylonitization and detachment faulting in the Whipple-Buckskin-Rawhide Mountains terrane, southeastern California and western Arizona, *in* Crittenden, M. D., Coney, P. J., and Davis, G. H., eds., Cordilleran metamorphic core complexes: Geological Society of America Memoir 153, p. 79–183.

Day, A. L., and Allen, E. T., 1925, The volcanic activity and hot springs of Lassen Peak: Carnegie Institution of Washington Publication 360, 190 p.

Deino, A. L., 1994, Nine Hill Tuff near Soda Springs, California, *in* John, D. A., ed., Field guide to Oligocene-Miocene ash-flows and source calderas in the Great Basin of Nevada: U.S. Geological Survey Open-File Report OF 94-0193, p. 40–41.

Dibblee, T. W., Jr., 1967, Areal geology of the western Mojave Desert, California: U.S. Geological Survey Professional Paper 522, 153 p.

Dickinson, W. R., 1997, Tectonic implications of Cenozoic volcanism in coastal California: Geological Society of America Bulletin, v. 109, p. 936-954.

Dickinson, W. R., and Snyder, W. S., 1979, Geometry of triple junctions related to San Andreas transform: Journal of Geophysical Research, v. 84, p. 561–572.

Diller, J. S., 1914a, The eruptions of Lassen Peak, California: Seismological Society of America Bulletin, v. 4, p. 103–107.

Diller, J. S., 1914b, The Lassen eruption: Science, n.s., v. 40, p. 49–51.

Diller, J. S., 1916, Lassen Peak—Our most active volcano: Seismological Society of America Bulletin, v. 6, p. 1–7.

Diller, J. S., 1917, Was the new lava from Lassen Peak viscous at the time of its eruption?: Washington Academy of Sciences Journal, v. 7, p. 82.

Dohrenwend, J. C., McFadden, L. D., Turrin, B. D., and Wells, S. G., 1984, K-Ar dating of the Cima volcanic field, eastern Mojave Desert, California: Late Cenozoic volcanic history and landscape evolution: Geology, v. 12, p. 163–167.

Donnelly-Nolan, J. M., 1988, A magmatic model of Medicine Lake Volcano, California: Journal of Geophysical Research, v. 93, p. 4412–4420.

Donnelly-Nolan, J. M., Champion, D. E., Miller, C. D., Grove, T. L., and Trimble, D. A., 1990, Post-11,000 year volcanism of Medicine Lake Volcano, Cascade Range, northern California: Journal of Geophysical Research, v. 95, p. 19693–19704.

Duffield, W. A., and McKee, E. H., 1986, Geochronology, structure, and basin-range tectonism of the Warner Range, northeastern California: Geological Society of America Bulletin, v. 97, p. 142–146.

Duffield, W. A., Bacon, C. R., and Dalrymple, G. B., 1980, Cenozoic volcanism, geochronology, and structure of the Coso Range, Inyo County, California: Journal of Geophysical Research, v. 85, p. 2381–2404.

Durrell, C., 1987, Geologic history of the Feather River country, California: Berkeley, University of California Press, 337 p.

Ernst, W. G., and Hall, C. A., Jr., 1974, Geology and petrology of the Cambria Felsite, a new Oligocene formation, west-central California Coast Ranges: Geological Society of America Bulletin, v. 85, p. 523–532.

Farmer, G. L., Glazner, A. F., Wilshire, H. G., Wooden, J. L., Pickthorn, W. J., and Katz, M., 1995, Origin of late Cenozoic basalts at the Cima volcanic field, Mojave Desert, California: Journal of Geophysical Research, v. 100, p. 8399–8415.

Fox, K. F., Jr., Fleck, R. J., Curtis, G. H., and Meyer, C. E., 1985, Implications of the northwestwardly younger age of the volcanic rocks of west-central California: Geological Society of America Bulletin, v. 96, p. 647–654.

Frizzell, V. A., and Weigand, P. W., 1993, Whole-rock K-Ar ages and geochemical data from middle Cenozoic volcanic rocks, southern California: A test of correlations across the San Andreas fault, in Powell, R. E., Weldon, R. J., II, and Matti, J. C., eds., The San Andreas fault system: Displacement, palinspastic reconstruction, and geologic evolution: Geological Society of America Memoir 178, p. 273–287.

Gastil, G., Krummenacher, D., and Minch, J., 1979, The record of Cenozoic volcanism around the Gulf of California: Geological Society of America Bulletin, v. 90, part I, p. 839–857.

Gilbert, C. M., 1938, Welded tuff in eastern California: Geological Society of America Bulletin, v. 49, p. 1829–1862.

Gilbert, C. M., Christensen, M. N., Al-Rawi, Y., and Lajoie, K. R., 1968, Structural and volcanic history of Mono basin, California-Nevada, in Coats, R. R., Hay, R. L., and Anderson, C. A., eds., Studies in volcanology: A memoir in honor of Howel Williams: Geological Society of America Memoir 116, p. 275–329.

Guffanti, M., and Weaver, C. S., 1988, Distribution of late Cenozoic volcanic rocks in the Cascade Range: Volcanic arc segmentation and regional tectonic considerations: Journal of Geophysical Research, v. 93, p. 6513–6529.

Hall, C. A., Jr., 1981, San Luis Obispo transform fault and middle Miocene rotation of the western Transverse Ranges, California: Journal of Geophysical Research, v. 86, p. 1015–1031.

Harwood, D. S., and Helley, E. J., 1987, Late Cenozoic tectonism of the Sacramento Valley, California: U.S. Geological Survey Professional Paper 1359, 46 p.

Harwood, D. S., Helley, E. J., and Doukas, M. P., 1981, Geologic map of the Chico monocline and northeastern part of the Sacramento Valley, California: U.S. Geological Survey Miscellaneous Investigations Map I-1238, scale 1:62,500.

Hawkins, J. W., 1970, Petrology and possible tectonic significance of late Cenozoic volcanic rocks, southern California and Baja California: Geological Society of America Bulletin, v. 81, p. 3323–3338.

Hawkins, J. W., Allison, E. C., and Macdougall, D., 1971, Volcanic petrology and geologic history of Northeast Bank, Southern California Borderland: Geological Society of America Bulletin, v. 82, p. 219–228.

Hearn, B. C., Jr., Donnelly-Nolan, J. M., and Goff, F. E., 1981, The Clear Lake Volcanics: Tectonic setting and magma sources, in McLaughlin, R. J., and Donnelly-Nolan, J. M., eds., Research in the Geysers–Clear Lake geothermal area, northern California: U.S. Geological Survey Professional Paper 1141, p. 25–46.

Hearn, B. C., Jr., Donnelly-Nolan, J. M., and Goff, F. E., 1995, Geologic map and structure sections of the Clear Lake Volcanics, Northern California: U.S. Geological Survey Miscellaneous Investigations Map I-2362, scale 1:24,000.

Heiken, G., 1978, Plinian-type eruptions in the Medicine Lake Highland, California, and the nature of the underlying magma: Journal of Volcanology and Geothermal Research, v. 4, p. 375–402.

Hildreth, W., 1979, The Bishop Tuff: Evidence for the origin of compositional zonation in silicic magma chambers, in Chapin, C. E., and Elston, W. E., eds., Ash-flow tuffs: Geological Society of America Special Paper 180, p. 43–75.

Hill, D. P., 1996, Earthquakes and carbon dioxide beneath Mammoth Mountain, California: Seismological Research Letters, v. 67, p. 8–15.

Izett, G. A., Wilcox, R. E., Powers, H. A., and Desborough, G. A., 1970, The Bishop ash bed, a Pleistocene marker bed in the western United States: Quaternary Research, v. 1, p. 121-132.

Lacroix, A., 1904, La montagne Pelée et ses éruptions: Paris, Masson, 662 p.

Loomis, B. F., 1926, Pictorial history of the Lassen volcano: Anderson, California, Valley News Press, 142 p.

Luyendyk, B. P., Kamerling, M. J., Terres, R. R., and Hornafius, J. S., 1985, Simple shear of southern California during Neogene time suggested by paleomagnetic declinations: Journal of Geophysical Research, v. 90, p. 12,454–12,466.

Lydon, P. A., 1968, Geology and lahars of the Tuscan Formation, California, in Coats, R. R., Hay, R. L., and Anderson, C. A., eds., Studies in volcanology: A memoir in honor of Howel Williams: Geological Society of America Memoir 116, p. 441–475.

Macdonald, G. A., 1963, Geology of the Manzanita Lake quadrangle, California: U.S. Geological Survey Geologic Quadrangle Map GQ-248, scale 1:62,500.

Macdonald, G. A., 1966, Geology of the Cascade Range and Modoc Plateau, in Bailey, E. H., ed., Geology of northern California: California Division of Mines and Geology Bulletin 190, p. 65–96.

Macdonald, G. A., 1968, Composition and origin of Hawaiian lavas, in Coats, R. R., Hay, R. L., and Anderson, C. A., eds., Studies in volcanology: A memoir in honor of Howel Williams: Geological Society of America Memoir 116, p. 477–522.

Matthews, V., III, 1976, Correlation of the Pinnacles and Neenach Volcanic Formations and their bearing on the San Andreas fault problem: American Association of Petroleum Geologists Bulletin, v. 60, p. 2128–2141.

McBirney, A. R., 1985, Memorial to Howell Williams, 1898–1979: Geological Society of America Memorials, v. 15, 3 p.

McBirney, A. R., and Williams, H., 1965, Volcanic history of Nicaragua: University of California Publications in Geological Sciences, v. 55, 73 p.

McBirney, A. R., and Williams, H., 1969, Geology and petrology of the Galápagos Islands: Geological Society of America Memoir 118, 197 p.

McKee, E. H., Duffield, W. A., and Stern, R. J., 1983, Late Miocene and early Pliocene basaltic rocks and their implications for crustal structure, northeastern California and south-central Oregon: Geological Society of America Bulletin, v. 94, p. 292–304.

McLaughlin, R. J., Sliter, W. V., Sorg, D. H., Russell, P. C., and Sarna-Wojcicki, A. M., 1996, Large-scale right-slip displacement on the East San Francisco Bay region fault system: Implications for location of the late Miocene to Pliocene Pacific plate boundary: Tectonics, v. 15, p. 1–18.

Metz, J. M., and Mahood, G. A., 1985, Precursors to the Bishop Tuff eruption: Glass Mountain, Long Valley, California: Journal of Geophysical Research, v. 90, p. 11,121–11,126.

Miller, C. D., 1980, Potential hazards from future eruptions in the vicinity of Mount Shasta volcano, northern California: U.S. Geological Survey Bulletin 1503, 43 p.

Miller, C. D., 1985, Holocene eruptions at the Inyo volcanic chain, California: Implications for possible eruptions in Long Valley caldera: Geology, v. 13, p. 14–17.

Muffler, L. J. P., and White, D. E., 1969, Active metamorphism of upper Cenozoic sediments in the Salton Sea geothermal field and the Salton Trough, southeastern California: Geological Society of America Bulletin, v. 80, p. 157–182.

Nielson, J. E., Lux, D. R., Dalrymple, G. B., and Glazner, A. F., 1990, Age of the Peach Springs Tuff, southeastern California and western Arizona: Journal of Geophysical Research, v. 95, p. 571–580.

Noble, D. C., Slemmons, D. B., Korringa, M. K., Dickinson, W. R., Al-Rawi, Y., and McKee, E. H., 1974, Eureka Valley Tuff, east-central California and adjacent Nevada: Geology, v. 2, p. 139–142.

Parker, R. B., 1963, Recent volcanism at Amboy Crater, California: California Division of Mines and Geology Special Report 76, 23 p.

Piper, A. M., Gale, H. S., Thomas, H. E., and Robinson, T. W., 1939, Geology and ground-water hydrology of the Mokelumne area, California: U.S. Geological Survey Water-Supply Paper 780, 230 p.

Robinson, P. T., Elders, W. A., and Muffler, L. J. P., 1976, Quaternary volcanism in the Salton Sea geothermal field, Imperial Valley, California: Geological Society of America Bulletin, v. 87, p. 347–360.

Rose, R. L., 1959, Tertiary volcanic domes near Jackson, California: California Division of Mines Special Report 60, 21 p.

Russell, R. J., 1928, Basin and Range structure and stratigraphy of the Warner Range, northeastern California: California University Publications, Department of Geological Sciences Bulletin, v. 17, p. 387–496.

Sapper, K., 1926, Der vulkanische Tätigkeit in Mittelamerika im 10. Jahrhundert: Zeitschrift für Vulkanologie, v. 9, p. 158–200.

Sarna-Wojcicki, A. M., 1976, Correlation of late Cenozoic tuffs in the central Coast Ranges of California by means of trace- and minor-element chemistry: U.S. Geological Survey Professional Paper 972, 30 p.

Sawlan, M. G., and Smith, J. G., 1984, Petrologic characteristics, age and tectonic setting of Neogene volcanic rocks in northern Baja California Sur, Mexico, in Frizzell, V. A., Jr., ed., Geology of the Baja California peninsula: Los Angeles, Pacific Section, Society of Economic Paleontologists and Mineralogists, p. 237–251.

Schleicher, D., 1974, Emplacement mechanism of the Miraleste Tuff Bed, Palos Verdes Hills, California: Geological Society of America Bulletin, v. 85, p. 505–512.

Shelton, J. S., 1955, Glendora volcanic rocks, Los Angeles basin, California: Geological Society of America Bulletin, v. 66, p. 45–89.

Sieh, K., and Bursik, M., 1986, Most recent eruption of the Mono Craters, eastern central California: Journal of Geophysical Research, v. 91, p. 12,539–12,571.

Silberman, M. L., and Chesterman, C. W., 1972, K-Ar age of volcanism and mineralization, Bodie mining district and Bodie Hills volcanic field, Mono County, California: Isochron/West, no. 3, p. 13–22.

Sims, J. D., 1993, Chronology of displacement on the San Andreas fault in central California: Evidence from reversed positions of exotic rock bodies near Parkfield, California, in Powell, R. E., Weldon, R. J., II, and Matti, J. C., eds. The San Andreas fault system: Displacement, palinspastic reconstruction, and geologic evolution: Geological Society of America Memoir 178, p. 231–256.

Slemmons, D. B., 1966, Cenozoic volcanism of the central Sierra Nevada, California, in Bailey, E. H., ed., Geology of northern California: California Division of Mines and Geology Bulletin 190, p. 199–208.

Smith, G. I., 1964, Geology and volcanic petrology of the Lava Mountains, San Bernardino County, California: U.S. Geological Survey Professional Paper 457, 97 p.

Thompson, G. A., and White, D. E., 1964, Regional geology of the Steamboat Springs area, Washoe County, Nevada: U.S. Geological Survey Professional Paper 458-A, 52 p.

Turner, D. L., 1970, Potassium-argon dating of Pacific Coast Miocene foraminiferal stages, in Bandy, O. L., ed., Radiometric dating and paleontologic zonation: Geological Society of America Special Paper 124, p. 91–129.

Vedder, J. G., 1975, Juxtaposed Tertiary strata along the San Andreas fault in the Temblor and Caliente Ranges, California, in Crowell, J. C., ed., San Andreas fault in southern California: California Division of Mines and Geology Special Report 118, p. 234–240.

Vedder, J. G., Howell, D. G., and Forman, J. A., 1979, Miocene strata and their relation to other rocks, Santa Catalina Island, California, in Armentrout, J. M., Cole, M. R., and TerBest, H., Jr., eds., Cenozoic paleogeography of the western United States: Pacific Section, Society of Economic Paleontologists and Mineralogists, Pacific Coast Paleogeography Symposium 3, p. 239–256.

Weigand, P. W., 1982, Middle Cenozoic volcanism of the western Transverse Ranges, in Fife, D. L., and Minch, J. A., eds., Geology and mineral wealth of the California Transverse Ranges (Mason Hill volume): Santa Ana, California, South Coast Geological Society, p. 170–188.

Weigand, P. W., and Savage, K. L., 1993, Review of the petrology and geochemistry of the Miocene Conejo Volcanics of the Santa Monica Mountains, California, in Weigand, P. W., Fritsche, A. E., and Davis, G. E., eds., Depositional and volcanic environments of middle Tertiary rocks of the Santa Monica Mountains, southern California: Pacific Section, SEPM (Society for Sedimentary Geology) Book 72, p. 93–112.

Williams, H., 1928, A recent volcanic eruption near Lassen Peak: California University Publications, Department of Geological Sciences Bulletin, v. 17, p. 241–263.

Williams, H., 1929a, Geology of the Marysville Buttes, California: California University Publications, Department of Geological Sciences Bulletin, v. 18, p. 103–220.

Williams, H., 1929b, The volcanic domes of Lassen Peak and vicinity, California: American Journal of Science, v. 218, p. 313–330.

Williams, H., 1931, The dacites of Lassen Peak and vicinity, California, and their basic inclusions: American Journal of Science, v. 222, p. 385–403.

Williams, H., 1932a, The history and character of volcanic domes: California University Publications, Department of Geological Sciences Bulletin, v. 21, p. 51–146

Williams, H., 1932b, Geology of the Lassen Volcanic National Park, California: California University Publications, Department of Geological Sciences Bulletin, v. 21, p. 195–383.

Williams, H., 1932c, Mount Shasta, a Cascade volcano: Journal of Geology, v. 45, p. 417–429.

Williams, H., 1933, Geology of Tahiti, Moorea, and Maiao: Bernice P. Bishop Museum Bulletin 105, 89 p.

Williams, H., 1934, Mount Shasta, California: Zeitschrift für Vulkanologie, v. 15, p. 225–253.

Williams, H., 1936, Pliocene volcanoes of the Navajo-Hopi country: Geological Society of America Bulletin, v. 47, p. 111–172.

Williams, H., 1941, Calderas and their origin: University of California Publications, Department of Geological Sciences Bulletin, v. 25, p. 239–346.

Williams, H., 1942, The geology of Crater Lake National Park, Oregon, with reconnaissance of the Cascade Range southward to Mount Shasta: Carnegie Institution of Washington Publication 540, 162 p.

Williams, H., 1949, Geology of the Macdoel quadrangle: California Division of Mines Bulletin 151, p. 7–60.

Williams, H., 1950, Volcanoes of the Parícutin region, Mexico: U.S. Geological Survey Bulletin 965-B, p. 165–279.

Williams, H., and Curtis, G. H., 1977, The Sutter Buttes of California: University of California Publications in Geological Sciences, v. 116, 56 p.

Williams, H., and McBirney, A. R., 1979, Volcanology: San Francisco, Freeman, Cooper, 397 p.

Williams, H., Turner, F. J., and Gilbert, C. M., 1954, Petrography: San Francisco, Freeman, 406 p.

Wilson, C. J. N., and Hildreth, W., 1997, The Bishop Tuff: New insights from eruptive stratigraphy: Journal of Geology, v. 105, p. 407–439.

Wise, W. S., 1969, Origin of basaltic magmas in the Mojave Desert area, California: Contributions to Mineralogy and Petrology, v. 23, p. 53–64.

Wood, S. H., 1977, Distribution, correlation, and radiocarbon dating of late Holocene tephra, Mono and Inyo craters, eastern California: Geological Society of America Bulletin, v. 88, p. 89–95.

Woodring, W. P., Bramlette, M. N., and Kew, W. S. W., 1946, Geology and paleontology of the Palos Verdes Hills, California: U.S. Geological Survey Professional Paper 207, 145 p.

Yerkes, R. F., 1957, Volcanic rocks of the El Modeno area, Orange County, California: U.S. Geological Survey Professional Paper 274-L, p. 313–334.

Yerkes, R. F., and Campbell, R. H., 1979, Stratigraphic nomenclature of the central Santa Monica Mountains, Los Angeles County, California: U.S. Geological Survey Bulletin 1457-E, 31 p.

Young, R. A., and Brennan, W. J., 1974, The Peach Springs Tuff: Its bearing on structural evolution of the Colorado Plateau and development of Cenozoic drainage in Mohave County, Arizona: Geological Society of America Bulletin, v. 85, p. 83–90.

MANUSCRIPT ACCEPTED BY THE SOCIETY NOVEMBER 23, 1998.

Printed in U.S.A.

Geological Society of America
Special Paper 338
1999

# Chapter 21
# MOUNTAINS SHAPED BY ICE

**Alan R. Gillespie**
*Department of Geological Sciences*
*University of Washington*
*Seattle, Washington 98195*
*E-mail: alan@oz.geology.washington.edu*

**Malcolm M. Clark**
*26135 Altadena Drive*
*Los Altos Hills, California 94022*
*E-mail: mclark@isdmnl.wr.usgs.gov*

**Raymond M. Burke**
*Department of Geology*
*Humbolt State University*
*Arcata, California 95221*
*E-mail: rmb2@axe.humboldt.edu*

## ELIOT BLACKWELDER

Eliot Blackwelder (1880–1969), president of the Geological Society of America in 1940, was a remarkably productive and influential geologist. Although he specialized in alpine glaciation and desert processes in the western United States, he also wrote "easily and well about fossils, climate, structural geology, Precambrian rocks, and phosphate deposits" (Krauskopf, 1976). After training in geology at the University of Chicago and holding successive faculty positions there and at Wisconsin, Illinois, Stanford, and Harvard Universities, Blackwelder accepted appointments as professor and head of the Geology Department at Stanford in 1922. He retired from these positions in 1945.

## HIGHLIGHTED ARTICLE

In an era when the "shelf life" of most research articles is a few years, a classic paper is one whose methodology and results are viewed as largely correct decades after its writing, and which has had a profound influence on subsequent generations of researchers. By this definition, "Pleistocene glaciation in the Sierra Nevada and Basin Ranges" is a classic paper on mountain glaciation in the Cordillera of North America. In it Blackwelder set the tone and content for modern studies of alpine glaciation in North America. He focused discussion on the number of glacial stages and their relative durations, the relative weathering and geomorphic criteria by which glacial deposits and landforms could be assigned ages and correlated, and the actual distribution and correlation of glacial features from the Sierra Nevada east to the Rocky Mountains and beyond. Blackwelder anticipated modern concerns by trying to correlate not just deposits from one valley to another, but glacial stages from one part of the world to another. His stated goal was to reconstruct Pleistocene history, which he viewed as dominated by global climatic fluctuations, driven perhaps by astronomical causes.

Alan R. Gillespie

Autographed reprint cover page from Blackwelder (1931).

# Eliot Blackwelder and the alpine glaciations of the Sierra Nevada

**Alan R. Gillespie**
*Department of Geological Sciences, University of Washington, Seattle, Washington 98195-1310*
**Malcolm M. Clark**
*26135 Altadena Drive, Los Altos Hills, California 94022*
**Raymond M. Burke**
*Department of Geology, Humboldt State University, Arcata, California 95521*

## INTRODUCTION

Eliot Blackwelder read "Pleistocene glaciation in the Sierra Nevada and Basin Ranges" before the Geological Society of America on December 27, 1929. The article was published two years later (Blackwelder, 1931), 68 years after the Whitney Survey in 1863 discovered evidence of Pleistocene alpine glaciation in the Sierra Nevada of California (Whitney, 1865; Brewer, 1966). J. D. Whitney and his colleagues were inspired by Louis Agassiz, who had earlier inferred the existence of vast, long-vanished glaciers in Europe and beyond. The naturalist John Muir (1872) soon thereafter found active cirque glaciers in the Sierra Nevada, for example on Merced Peak. Muir argued eloquently that vanished glaciers had played a dominant role in sculpting the stunning topography of broad valleys and steep cliffs of the Sierra Nevada, as exemplified by the canyons of the Merced and Tuolumne Rivers. Later, François Matthes (1930) argued that most of the erosion was actually fluvial, and the glaciers only modified the landscape.

By the turn of the century, the existence of large Pleistocene glaciers was widely accepted, and attention was directed toward filling in the details of their fluctuations. How many times did they advance? How long were the glacial periods, and what was the interval between them? How did the glacial history of the Sierra Nevada accord with the presumably better-understood history of the Laurentide ice sheets? These were the questions that motivated Eliot Blackwelder. The summary of his experience and observations in attempting to answer them formed the core of his 1929 address to the society. He began his research into alpine glaciations immediately after receiving his undergraduate degree, working with R. D. Salisbury in the mountains of Wyoming (Blackwelder and Salisbury, 1903), and by 1929 he had nearly three decades of experience.

"Pleistocene glaciation in the Sierra Nevada and Basin Ranges" was a remarkable paper, one that set the tone and topics for most subsequent investigations of alpine glaciations in the American West. It occupies the present midpoint of the history of Sierran glacial studies and is a true "classic," if a classic paper is one that has survived, largely intact, the test of time and has been the nucleus for a de facto school of geological inquiry.

Although Blackwelder (1931) galvanized much subsequent research, and even though his research has a modern feel to it, Blackwelder was very much a product of his times. Most obviously, he worked when access was more difficult than it is today. His geographic coverage is remarkable when viewed from that perspective, but some observations that have proven important were missing. As Blackwelder (1931, p. 866) predicted, even in the eastern Sierra Nevada, which was his geographic focus, he had "mapped only a few of [the glaciated drainages] in the detail their interest merits. It will take many years to complete the mapping, and that work will necessarily be completed by others." Blackwelder studied the effects of the great Pleistocene climatic oscillations, before the marine and ice-core records showing sea-level and climatic fluctuations were measured and analyzed. Blackwelder wrote before the great discoveries in geochronology —radiocarbon, potassium-argon, and cosmogenic-nuclide analysis—that allowed numerical dating of some of the deposits he studied. Working as he did against these handicaps, Blackwelder's insights are all the more impressive.

Blackwelder's writing provides an interesting footnote. At its best it was evocative, almost lyrical; Blackwelder's concern was evidently to convey a powerful impression of the generalized observations he had made. He wrote well. Modern scientific writers might view his style as needlessly wordy; today's science is presented in a more terse, matter-of-fact style. We might also criticize Blackwelder as being occasionally imprecise, or as overgeneralizing. Viewed against his predecessors and contemporaries, however, Blackwelder was notable for his "clear, direct thinking and for expressing complex ideas in simple words" (Krauskopf, 1976).

Gillespie, A. R., Clark, M. M., and Burke, R. M., 1999, Eliot Blackwelder and the alpine glaciations of the Sierra Nevada, *in* Moores, E. M., Sloan, D., and Stout, D. L., eds., Classic Cordilleran Concepts: A View from California: Boulder, Colorado, Geological Society of America Special Paper 338.

In this article we first briefly set the stage for Blackwelder's 1929 speech by reviewing the work of his colleagues in the Sierra Nevada, and recapitulate the main points of his talk and article. We then examine the influence this has had on his intellectual heirs and comment on the changes that have occurred in his field since his talk.

## SCIENTIFIC SETTING

Blackwelder focused his studies on the eastern escarpment of the Sierra Nevada and the ranges to the east, sharing the scientific stage with Israel Russell (1885, 1889, 1898), G. K. Gilbert (1890), and Adolf Knopf (1918). Much of the pioneering research on Sierran glaciation had been conducted on the less-arid west side of the Sierra Nevada, and there the work of Clarence King (1878), Henry Turner (1900), and François Matthes (1930) was paramount. In the ranges of the Great Basin to the east, King (1878) and Russell (1885) had reported evidence of former glaciation.

Blackwelder (1931, p. 867) noted that Willard D. Johnson, working with Israel Russell near Mono Lake, was the first to recognize evidence of multiple glacial stages in the Sierra Nevada. Johnson, however, left his findings unpublished, and credit for first publishing on distinct glacial stages in the Sierra Nevada went to Turner and Ransome (1898), although Gilbert (1890) had earlier inferred at least two distinct glaciations throughout the Cordillera. Russell (1889) had recognized multiple advances, but had not claimed that these necessarily represented different stages. To Blackwelder, this was an important distinction: deposits from different stages differed enough to be mapped and correlated regionally, and stages were first-order climatic events, possibly global in scope. Curiously, as we shall see, modern dating techniques may now have blurred some of the boundaries between what had been, to Blackwelder, obviously different glaciations. Nevertheless, in 1931 Blackwelder probably felt he was on solid scientific ground.

Blackwelder (1931, p. 866) focused on the "dry eastern slope of the Sierra Nevada, where the glacial features are best displayed free from obscuring forests" (Fig. 1), but his interest in regional correlation of glacial deposits led him also to consider evidence from the Central Plains, the Rocky Mountains, and the Great Basin, and especially from the western slopes of the Sierra Nevada. Blackwelder carefully examined evidence of past glaciations in the west-slope valleys of the Merced, Stanislaus, American, and Yuba Rivers. Here he crossed paths with Matthes, who was aggressively pursuing similar lines of research, especially in Yosemite Valley (Matthes, 1929, 1930). There Blackwelder and Matthes met to sort out their differences in the field, in September, 1930, while Blackwelder was preparing his manuscript. As Blackwelder (1931, p. 907) wrote: "At that time certain correlations and interpretations were agreed upon. These were later supplemented by a re-examination by the writer in March, 1931." Blackwelder's article was submitted to the Geological Society of America *Bulletin* six weeks later; it must have benefited from Matthes' input. However, Blackwelder's (1931, p. 907) correlations and interpretations were "accepted only in part by Mr. Matthes."

Blackwelder regarded the Central Plains as providing the "standard section" against which the less-developed Cordilleran evidence would be compared and interpreted. Antevs (1925) had suggested that five glacial stages, corresponding to the then-recognized five stages of the Laurentide ice sheet, would ultimately be recognized in the Cordillera, and Blackwelder felt he was well on the way to realizing Antevs' prediction. Blackwelder (1931, p. 869) assumed that the advance and retreat of the Laurentide ice sheet was driven by "general changes of climate of worldwide influence," and he never questioned that fluctuations in different localities were probably synchronous. Thus, he displayed a predilection for incorporating field observations and interpretations into a neat, theoretical package, in this case supplied by his presumptions on how the global climate system actually behaved. This explains his concern with correlation: the existence of global climatic fluctuations gave correlation exercises a strong rationale, because glaciations really were synchronous. As Blackwelder (1931, p. 919) put it:

### PURPOSE OF THE CORRELATIONS

The writer is interested in the glacial stages of the West, not so much for their own sakes as for their utility in reconstructing the Pleistocene history of the Cordilleran region. Many individual events in that history have been determined, but serious difficulties are encountered when the attempt is made to arrange these in consecutive historical order and still more when one tries to fit such a sequence into the general scheme of the period. Fossils are rare, and even when found they give but little comfort, because the evolutionary changes among organisms were too slow to bring about very distinct faunas within the Pleistocene period–especially in the later part.

The best and most convenient criteria at present available for working out the sequence of Pleistocene events are physiographic. For general correlations there seems to be no basis as good as that afforded by the climatic pulsations for which the Pleistocene period is noted. Such climatic variations, if due to general atmospheric or perhaps astronomic causes, must have affected all parts of the region and impressed their record upon its topographic forms and deposits. Basins held fresh lakes during the cooler epochs, but only playas or salinas in the intervening dry times. Alluvial fans grew larger in the more arid ages and were intrenched under the influence of the next cool moister régime. Such illustrations serve to indicate how the establishment of a series of climatic ages may facilitate the integration of a continuous history of the Pleistocene period in western United States.

Already tentative correlations have been extended out from the glacial area of the Sierra Nevada to the desert basins eastward as far as Death Valley and thence to the Colorado River. The great pediments and alluvial fans as well as the lake terraces and salt beds can thus be assigned to approximate geologic dates and still older features referred to them in turn.

Figure 1. "Moraines along the east base of the Sierra Nevada southwest of Mono Lake. In the center of the view a large smooth lateral moraine of the Tahoe stage extends out from Bloody Canyon. The scarcity of boulders and the absence of a terminal moraine are typical. To the right, in front of Mount Dana, is a maturely eroded mass of till of the Sherwin stage resting upon granite; the white lines mark the contact. The Tioga stage moraines of Bloody Canyon are scarcely visible" (Blackwelder, 1931, Plate 23). Bloody Canyon has been studied repeatedly since Blackwelder's visit (Sharp and Birman, 1963; Burke and Birkeland, 1979; Gillespie, 1982; Birkeland and Burke, 1988; Phillips et al., 1990).

It should not be supposed, just because of his seeming preoccupation with distinguishing stages from advances (which might occur several times within a stage), that Blackwelder was naive or entranced by geologic nomenclature. Instead, his was a more pragmatic, modern concern. Commenting that "the difference of opinion is more apparent than real" Blackwelder (1931, p. 869), seemed to regard the issue of the number of glacial stages as of interest in that their recognition enabled correlation across vast regions. Blackwelder employed the terms *age*, *stage*, and *epoch* interchangeably, all to indicate a time in which alpine glaciers advanced to occupy valleys far from their cirques, although he specifically claimed to "use at present the terms *age* and *stage* for the times and deposits of the major individual glacial and interglacial advances and retreats" (Blackwelder, 1931, p. 869).

## "PLEISTOCENE GLACIATION IN THE SIERRA NEVADA AND BASIN RANGES"

Blackwelder (1931) divided his article into two main parts: the glaciations of the eastern slopes of the Sierra Nevada, and correlation with the ranges to the east and the Central Plains. In the first section, he focused on three topics: the number of glacial stages, criteria by which deposits and landforms could be assigned relative ages, and the duration of the glacial stages. In addition, Blackwelder discussed the characteristics associated with each of the four glacial stages that he recognized.

In the second section, Blackwelder correlated the Sierran glacial stages with other alpine glacial stages, and with the "standard section" offered by the tills of the Central Plains. Essentially, Blackwelder used two approaches: landforms and till were correlated on the basis of their degree of weathering, and sequences of stages were correlated by their number and relative duration or spacing in the different regions.

### "Eastern slope of the Sierra Nevada"

One of Blackwelder's contributions (Blackwelder, 1930) was to add a fourth glacial stage to the three advances noted by Russell (1889). Blackwelder regarded all four as stages and gave them local geographic names, to preclude unnecessary literature revision should anticipated correlations with the "standard section" not hold up. He commented that Matthes (1930) had adopted the name "Wisconsin" for the youngest stage in Yosemite, because the correlation seemed unimpeachable. Yet, Blackwelder conservatively resisted what must have been, for him also, a reasonable thing to do. His four stages were, in order of increasing age, the Tioga, Tahoe, Sherwin, and McGee stages. Blackwelder suspected that, to match the "standard section," a fifth stage would ultimately be detected as well, falling between the Tahoe and Sherwin stages. He correctly predicted that the evidence would be found near Mono Lake, as it was 32 years later (Sharp and Birman, 1963).

Blackwelder expressed dissatisfaction with the application to Cordilleran deposits of relative dating criteria commonly used in the Central Plains. Soil development in the semiarid West was more difficult to decipher, for Blackwelder, than soil development in the more humid East and Midwest. Relevant fossils and loess were notably absent in the Sierra Nevada.

To fill this void, Blackwelder contrived a long list of relative weathering criteria for use in the West. Chief among these were degree of landform erosion (Fig. 2), boulder weathering (Fig. 3), and degree of stream dissection. Stratigraphic relations between till and tephra were also employed, as were geographic sequences

Figure 2. "A granite Roche Moutonnée of the Tahoe Stage. The view, taken in Leavitt Meadow, shows how weathering and erosion have changed its original smooth, rounded contour to one that is more angular and ragged" (Blackwelder, 1931, Fig. 13). Leavitt Meadow is east of Sonora Pass, in California. Blackwelder was evidently misled by the "ragged" appearance of the outcrop, which was buried deep under a Tioga glacier (Clark, 1967) and either never possessed a "smooth, rounded contour," or lost it to postglacial erosion.

Figure 3. "Typical isolated granite boulder on the Sherwin Till, Rock Creek Plateau. The rock is cavernous and crumbly (rule 6 inches long indicates size of boulder)" (Blackwelder, 1931, Fig. 23). Blackwelder pioneered the systematic estimation of boulder weathering for the relative dating and correlation of moraines.

within valleys. Possibly the most widely quoted and distinctive measure was Blackwelder's GWR (granite weathering ratio). This involved counting granodiorite boulders exposed on moraine crests, for example, and classifying them into three categories according to their degree of weathering: "(a) almost unweathered; (b) notably decayed on the surface but still solid; and (c) greatly weathered, cavernous, or rotten" (Blackwelder, 1931, p. 877). Characteristic ratios, which admittedly varied with climate, were associated with deposits of the different stages. Blackwelder recognized the effects of forest fires on the GWR, but somehow thought that the effects of fire-spalling could be differentiated from those of chemical weathering, all over a time he recognized as spanning tens of thousands of years!

Blackwelder discussed more than 20 relative-weathering criteria. Perhaps because of problematic situations posed by spatially variable fire effects, wind exposure, and rock jointing, he cautioned against reliance on any single one. Realistically, not all criteria could be used in each locality, but even a small set of criteria could strengthen the conclusions greatly. Blackwelder seems to have trusted boulder counts and even systematized field observations over subjective judgment of age.

Blackwelder and his contemporaries lacked the necessary tools to come to grips with questions about the duration of glaciations and the intervals between advances of the glaciers. The more quantitative relative-weathering criteria afforded a scheme for assessing duration, but the scheme was necessarily speculative and unreliable. The interval between the Tioga and Tahoe stages, for example, was estimated by Knopf (1918) as five units and by Russell (1925) as two and a half units, where a single unit represented the time since retreat of the Tioga glaciers. To Blackwelder, this would have meant that the Tahoe stage was as old as 125 ka, if the Tioga deglaciation was taken to have coincided with retreat of the Wisconsin glaciers in the Central Plains, estimated by Antevs (1925) to have begun 25 ka and ended 10 ka years ago. Argued similarly, the Sherwin stage would have been ~50 units (<1.25 Ma) and the McGee stage perhaps 150–250 units (<6 Ma). Blackwelder did not *believe* these age estimates; he thought that they helped clarify the "historical view."

*"The Tioga stage."* Blackwelder completed his discussion of the eastern Sierra Nevada by giving specifics on locations and characteristics of deposits and glaciated terrain of his four stages. He named the youngest after Tioga Pass, where he regarded Tioga-age landforms and deposits as well displayed. To Blackwelder (1931, p. 882):

> The glacial features that were made by the ice tongues of the Tioga epoch are even now almost as fresh and unaltered as at the time of their formation. The cirques are still as bare and ragged as if recently abandoned by the ice. Talus cones are few and small where the rocks are not closely jointed, and have not grown to large size even where the closer spacing of joints is favorable to frost action. The original polished and striated surface is still rather generally intact, even on such easily weathered rocks as coarse granite. Acres of polished and grooved rock are a familiar sight near Tenaya Lake (Yosemite National Park) and many other places in the high Sierra. If it were not for the destructive effects of forest fires, such surfaces would be even more completely preserved.
>
> The lateral moraines generally stand out as bold embankments, marred only by few landslides and by sharp ravines where tributary brooks descend the canyon sides. The terminal moraines are still complete, except for V-shaped notches through which the main streams tumble down to the plains beyond. As the distinctive glacial topography of the moraines is almost entirely preserved, it is generally an easy task to map them continuously.

Tioga moraines were bouldery, and the boulders were largely fresh. Lakes were commonly found behind end moraines.

*"The Tahoe stage."* Tahoe moraines were characterized by Blackwelder as more eroded than Tioga moraines; axial streams were deeply incised; lakes were filled or marshy; and boulders were less abundant and more heavily weathered. Some boulders were "decidedly rotten" In Mono basin, tephra blanketed Tahoe moraines, but not adjacent Tioga moraines. Although he concluded that the Tioga and Tahoe glaciers belonged to different stages, Blackwelder clearly agonized over this judgment. In the end, he may have been swayed by the appearance of some cirques that he supposed had been occupied by Tahoe, but not Tioga, glaciers; here the contrast between the subdued forms and deep talus of the "Tahoe" cirques contrasted dramatically with the "ragged" Tioga cirques.

*"The Sherwin stage."* The Sherwin stage was named for the area north of Sherwin Hill, near the town of Bishop. The original topographic expression of Sherwin deposits has long been lost, and main streams have cut deeply into underlying bedrock. Exposed granitic boulders are thoroughly weathered, and even subsurface boulders are grusy. In some valleys, Sherwin-age deposits are in close proximity to younger Tahoe and Tioga moraines from the same drainages, although considerably more extensive. Yet, sufficient time had elapsed since the Sherwin stage that, elsewhere, erosion or "diastrophism" had eliminated the lower-elevation Sherwin deposits, leaving the Tahoe and Tioga moraines unaffected.

*"The McGee stage."* Evidence of the McGee stage is preserved only in unusual localities, most notably the top of McGee Mountain, 800 m above the modern McGee Creek and its flanking moraines. Original topographic expression was entirely destroyed. Blackwelder expressed concern that McGee "tills" were glacial, and not mudflow deposits.

**Extent of glaciers.** Blackwelder regarded Tahoe moraines as longer than Tioga moraines of the same drainage, terminating in some cases as much as 500 ft (152 m) lower in elevation. They were also more voluminous, by an estimated factor of 50 in the Bridgeport Basin (Fig. 4). This conclusion was significant, because it drove Blackwelder to regard the Tioga glaciers, even in their cirques, as thinner and less extensive than their Tahoe ancestors, an inference no doubt supported by the subdued nature of some of his "Tahoe" cirques. Blackwelder viewed Tahoe glaciers west of the Sierra crest as confluent, in contrast to the presumably isolated glaciers of Tioga time. Large tracts of bare, heavily weathered granitic bedrock in the Sierran uplands were attributed to Tahoe-stage abrasion, consistent with the supposed small size of the Tioga glaciers.

*"Correlations"*

*"Western slope of the Sierra Nevada."* Blackwelder's immediate concern, and most of his emphasis in the article, was correlating his glacial sequence from the eastern Sierra with that of Matthes to the west. The challenge here was that, because the climatic conditions and vegetation differed across the crest of the Sierra, the relative weathering criteria were difficult to apply. Blackwelder considered two drainages, Yosemite and Yuba Valleys. In Yosemite, he concluded that Matthes' Wisconsin stage included both the Tioga and the Tahoe stages; Matthes' El Portal and the Sherwin stages were the same; and evidence for the McGee stage was lacking (Table 1). Blackwelder considered the moraines on the floor of Yosemite Valley near El Capitan to mark the maximum extent of the Tahoe glaciers and relegated the Tioga glaciers to positions far upvalley, for example, above Nevada Falls. Blackwelder's interpretation of Yosemite accorded well with his preconception from the eastern slope that Tioga glaciers were much smaller than Tahoe glaciers, but both were probably in error. Today, the Tioga end moraines, not the Tahoe, are thought to be the ones near El Capitan, and it is possible that remnants of Tahoe moraines were mistakenly identified as El Portal till by Matthes.

For all of his emphasis on the GWR, Blackwelder seems to have relied exclusively on subjective assessments of his relative weathering criteria in drawing his conclusions in Yosemite. However, he did use GWR values and boulder frequencies in correlations with the Yuba Valley glaciations, coming to similar

Figure 4. "Robinson Canyon near Bridgeport [California]. The view shows a frontal moraine of the Tioga stage (XX) only 30 ft high, with a 700-ft lateral moraine of the earlier (Tahoe) stage on the left" (Blackwelder, 1931, Fig. 11). Evidence such as this persuaded Blackwelder that Tioga moraines were commonly much smaller than Tahoe moraines; yet, at this locality the Tioga end moraines could be traced to high lateral moraines deposited against the Tahoe moraines that caught Blackwelder's eye.

*Tentative Correlation Table of Glacial Stages in Western United States*

| Iowa and Illinois | Montana | Wyoming (Wind River) | Colorado (San Juan) | Utah (Wasatch) | Nevada (Ruby Mts.) | Sierra Nevada East | Yosemite Matthes |
|---|---|---|---|---|---|---|---|
| Wisconsin | Wisconsin | Pinedale | Wisconsin | "Younger" | Angel Lake | Tioga | Wisconsin |
| Iowan | Iowan (?) | Bull Lake | Durango | "Older" | Lamoille | Tahoe | |
| Illinoisan | ......... | ......... | ......... | ......... | ......... | —?— | |
| Kansan | Kennedy | Buffalo | Cerro | ......... | —?— | Sherwin | El Portal / Glacier Pt. |
| Nebraskan | ......... | ......... | ......... | ......... | ......... | McGee | |

Table 1. "Tentative Correlation Table of Glacial Stages in Western United States"(Blackwelder, 1931, p. 918).

substantive conclusions that Tioga ice descended only to 6500 ft (1980 m), compared to 4000 ft (1220 m) for larger and more vigorous Tahoe glaciers.

**Ranges of the Great Basin.** Blackwelder's discussion of glaciation in the Basin Ranges was more cursory than for the Sierra Nevada, probably because his reconnaissance there was briefer. Yet the same was true of his correlation with the Rocky Mountains, where he had worked extensively on the Big Horn Range and later in the Tetons and Wind River Mountains (Blackwelder, 1915).

In the Ruby Mountains, Nevada, Blackwelder recognized two glacial stages, which he correlated to his Tioga and Tahoe stages. He claims he did so "by applying there the methods developed on the dry eastern slope of the Sierra Nevada" (Blackwelder, 1931, p. 911), but if this is the case, there is little sign of it in his writing; instead, he resorted to generalizations and evocative prose ("wild crags of the freshly torn cirques") to make his case. Perhaps he collected data but was frustrated in their interpretation, for he reported concerning the Tahoe moraines "the boulders are somewhat more decayed than is usual on moraines of this age, but the value of the comparison is diminished because the rocks... were gneisses of sedimentary origin" (Blackwelder, 1931, p. 911). Regardless, he came to the conclusion that here, as in the Sierra, Tioga glaciers were diminutive, barely reaching down to 8000 ft (2440 m) elevation, in contrast to the massive Tahoe glaciers that crossed the range front. Throughout the remainder of the Great Basin, the story was much the same: Tioga and Tahoe evidence abounded, with hints here and there of older (and therefore Sherwin) tills.

In the Wasatch Range, Utah, Blackwelder was back on more solid ground, partly because Atwood (1909) had independently made the claim for two distinct stages there, and partly because the Sierran relative weathering data—GWRs and boulder frequency counts—seemed to work once more, on the granitic rocks. The story was again the same: an early, bulky Tahoe equivalent, and smaller Tioga equivalents. Blackwelder connected Lake Bonneville shorelines with Tioga-equivalent moraines at Little Cottonwood Creek, demonstrating that the latest glaciation and lake high stand were coeval.

**"Rocky Mountains."** In the Rocky Mountains of Wyoming, Blackwelder (1931, p. 916–917) was content to observe "an obvious parallelism between the three stages of Wyoming and the Sherwin, Tahoe, and Tioga stages in California," citing in a general way the decay of boulders, the extent and depth of post-glacial stream erosion, and the "relations of moraines to valleys" to support this opinion.

The story in Colorado and Montana was much the same. In the San Juan Mountains, Colorado, Blackwelder (1931, p. 916) found that "the Durango stage is represented by large moraines inside the canyons and by gravel valley-train terraces 150–250 ft above the creeks. This indicates the Tahoe stage. The Wisconsin moraines are lower, lie within the Durango moraines, and are connected with gravel terraces only 20–30 ft high, thus conforming closely to the Tioga stage of the Far West." For Montana, Blackwelder relied on Alden, who recognized three stages (e.g., Alden, 1932), which he related to the three Blackwelder recognized in the Tetons.

**Laurentide ice sheet.** Blackwelder (1931, p. 917–918) did not feel competent to extrapolate his results to the plains, but relied on W. C. Alden, who had "traced [the glacial formations of the prairie states] more or less continuously from Iowa across Nebraska, South Dakota, and Montana to the Rocky Mountains," for correlation to the "standard section." Evidently, Blackwelder at one point also had a field conference with G. F. Kay, who had studied the Laurentide sequence in Iowa (Kay, 1931). According to Blackwelder, Alden concluded that the Pinedale and Bull Lake stages of Wyoming corresponded to the continental Wisconsin and Iowan stages (Table 1). "Although realizing the liability of error," Blackwelder (1931, p. 918) forwarded his correlations as a "tentative scheme for continuing studies."

In particular, correlations of older tills were uncertain. Nevertheless, "The absence of stages [in the West] equivalent to the Illinoisian stands out... as a challenge" (Blackwelder, 1931, p. 918). Blackwelder claimed some "rather definite evidence" of such a stage in the eastern Sierra Nevada, presumably in Mono basin. He felt stymied by the difficulties in applying relative weathering criteria to the older tills, concluding that drift of the "missing" stage was difficult to distinguish from the older Sherwin till, and that the evident absence of the moraines was because the glaciers were smaller than the later Tahoe glaciers, and thus exposed only locally.

## CHANGES IN THE INTERPRETATION OF SIERRAN GLACIAL GEOLOGY SINCE 1931

### Number of stages

Blackwelder predicted the discovery of a fifth, pre-Tahoe glacial stage in the Sierra Nevada, which was realized with Sharp and Birman's (1963) analysis of moraines at Bloody Canyon (Fig. 1). Their "Mono Basin" moraines were preserved accidentally, because the path of the longer Tahoe glacier shifted to the north (the Mono Basin moraines are out of view, behind the moraine complex in Fig. 1). The Mono Basin stage was the first of many new, named Pleistocene glaciations. Joseph Birman's dissertation, inspired by Blackwelder and later published in 1964, correlated glaciations across the Sierra crest and also recognized a pre-Tioga, post-Tahoe stage, later named "Tenaya" at Bloody Canyon by Sharp and Birman (1963). Birman (1964) also turned attention to cirques with two Holocene "Neoglaciations" in addition to Matthes' Little Ice Age glacial events. Sharp (1968) reported evidence of a pre-Sherwin till in Rock Creek, and Clark (1967) found evidence for two other tills between Tahoe and Sherwin(?) deposits above the West Walker River. Gillespie (1982) noticed that the "Tahoe" moraine at Bloody Canyon was composite, consisting of a younger crest and inner flank partly burying a distinctly older moraine that was best exposed near the glacier terminus. In addition, two or three pre-Mono Basin moraine remnants were found up-canyon. In the face of this newfound complexity, some glacial advances were notably missing, in particular, the latest-Pleistocene Younger Dryas tills found in Europe (Clark and Gillespie, 1997).

### Geochronology

Numerical dating, unknown in 1931, became feasible three decades later. Radiocarbon dating and dendrochronologic techniques have now made possible quantitative age estimates throughout the Holocene and back into the late Pleistocene. Lichenometry has allowed ages to be estimated for otherwise undatable deposits, at least for geographic areas and age ranges for which independent data have made calibration feasible (e.g., Curry, 1968; Scuderi, 1984). K/Ar analysis has permitted the dating of lavas intercalated with glacial deposits (Dalrymple, 1964; Gillespie et al., 1984). Recently, analysis of exotic nuclides produced by bombardment of rocks by cosmic rays has yielded exposure ages under a wide range of circumstances (Bierman, 1994).

The Tioga advances are now firmly correlated to the late Wisconsin glaciation. In Mono basin, Tioga glaciers began advancing before ~30,000 yr B.P., with the maximum extent of the glaciers achieved between 25,000 and 20,000 yr B.P. and retreat to the cirques completed before 13,200 yr B.P. (Clark, 1997, and written comm., 1998; Benson et al., 1998; Bursik and Gillespie, 1993). The Tenaya glaciers may have retreated after ~31,000 yr B.P. (Bursik and Gillespie, 1993).

The Tahoe stage is more difficult to date, largely because of the wide range of deposits assigned to it (for a recent review, see Gillespie and Molnar, 1995). Phillips et al. (1990) used cosmogenic $^{36}Cl$ to arrive at a minimum age estimate of ~49,000 yr B.P. (recalculated with production rates of Swanson et al., 1993) for the younger "Tahoe" moraine crest at Bloody Canyon, establishing it as mid-Wisconsin in age. The $^{36}Cl$ ages of boulders from the older "Tahoe" part of that compound moraine were <185,000 yr B.P. In a different drainage, another "Tahoe" or pre-Tahoe moraine has been bracketed at 130,000–460,000 yr B.P. by $^{39}Ar/^{40}Ar$ dates on interbedded lava (Gillespie et al., 1984). Thus, Blackwelder's Tahoe stage appears to include both Wisconsin and pre-Wisconsin advances. His Tioga and the younger of the Tahoe stages (as well as the intermediate Tenaya stage) now appear to be substages of the Wisconsin glaciation, as he had thought all along (Table 1). Perhaps the older till from Blackwelder's Tahoe moraine at Bloody Canyon, or the Mono Basin moraines it must have buried, are evidence of Blackwelder's sought-for "Illinoisian" stage.

The Sherwin glaciation where it is best exposed, in a roadcut on U.S. Highway 395 north of Bishop, California, underlies Bishop Tuff now dated as ~760,000 yr B.P. (Sarna-Wojcicki and Pringle, 1992) and is probably 40–100 k.y. older (Sharp, 1968; Birkeland et al., 1980). The McGee stage has not been dated.

### Paleoclimatology and stable isotope analysis

Blackwelder could not have foreseen the estimation of ice-volume and sea-level fluctuations from the $^{18}O/^{16}O$ analysis of foraminifers from sea-floor cores, yet the marine isotope stages, corresponding to periods of high-latitude glaciation and interglaciation, provide the complete global record he sought to use as a reference for his alpine glaciations. The picture is probably far more complex than he or his contemporaries imagined, but his suspicions about astronomical causes now appear to have been well founded. Dated sediment cores from Mono Lake (Benson et al., 1998) and lakes of the Owens River system (e.g., Smith, 1984) provide sedimentological and isotopic evidence that pluvial and glacial events both were temporally correlated in a general way with the marine record. Perhaps most significant, climate change has supplanted Pleistocene history as a focus of study.

### Relative weathering criteria

Recent decades have seen significant advances in relative dating. Sharp (1969) gave it a more numerical, "semiquantitative" foundation; Birkeland (1984) and Burke and Birkeland (1979) added soil development, which had frustrated Blackwelder, to the list of successful measures of relative age. Colman and Pierce

(1981) measured distributions of clast weathering-rind thicknesses; Crook (1986) measured acoustic wave speeds through clasts.

### Glacier physics

Work on ice flow rules out the possibility of thin Tioga glaciers flowing down the modest slopes where Blackwelder had them, for example, along the Yuba River (60 m thick; Blackwelder, 1931, p. 909). Tioga glaciers, which Blackwelder showed in the East Walker drainage to be 80%–90% of the length of their Tahoe predecessors, were probably proportionately large everywhere else along the eastern slope of the Sierra (e.g., Clark, 1967, Table 3), and to the west. Such large Tioga glaciers would necessarily have nearly the same thicknesses, widths, accumulation areas, and cirques as the Tahoe glaciers at the same places, contrary to Blackwelder's conclusions.

### Geomorphology

The equilibrium-line altitude (ELA), not the terminus elevation, is now used as the best measure of climatic conditions (e.g., Porter, 1964). The ELA, which can be reconstructed from the outline and elevations of long-vanished glaciers, is the contact between the zones of ice accumulation and ablation. It is controlled by a gamut of climatic parameters, including temperature and precipitation (e.g., Leonard, 1989). Gillespie (1991) demonstrated that the ELA depression of Tioga and Tahoe glaciers differed by only 10%, and therefore that their surface elevations and widths must have been about the same except near their termini.

In addition, the composite nature of many moraines was recognized at Convict Creek and Bridgeport basin (Sharp, 1969, 1972), Bloody Canyon (Gillespie, 1982), and many other localities. Many Tahoe moraines, for example, were deposited on top of older moraines, whereas the slightly shorter Tioga glaciers did not overtop the Tahoe moraines near their ends, but plastered till on their inside slopes. In the Bridgeport basin, Tioga moraines did locally overtop Tahoe moraines a few kilometers upstream from their ends (e.g., Crook and Gillespie, 1986). This finding cast further doubt on Blackwelder's opinion that the Tahoe moraines were so bulky, and therefore also on some of his correlations.

### Major corrections

Blackwelder's conclusion that Tioga glaciers were much smaller than Tahoe glaciers may have been reinforced by the compound nature of Tahoe moraines, with a volume of till deposited over two or more glaciations, but it probably originated from comparing the mass and height of Tioga end moraines to those of nearby Tahoe lateral moraines, and his failure to recognize continuity of Tioga end moraines with many of those big lateral moraines (see Blackwelder, 1931, p. 884). Blackwelder compared small Tioga end moraines to large Tahoe lateral moraines at both Robinson Creek (Fig. 4) and Fallen Leaf Lake (Blackwelder, 1931, p. 889), and at both sites those small Tioga end moraines lead to big Tioga lateral moraines (admittedly easier to see with aerial photographs unavailable to Blackwelder). As a result, Blackwelder (1931, p. 887) mislabeled many Tioga moraines as Tahoe, which in turn led to his incorrect statements that most of the bare granite in the Sierra is Tahoe (rather than Tioga), and that "many [Tahoe] cirques... were not occupied by more recent glaciers."

Blackwelder failed to appreciate the possibility of significant subglacial (or interglacial) fluvial incision of channels into bedrock, and survival of some of those channels under sliding glaciers. This led him to describe the deep bedrock channels on either side of Sonora Pass as remarkable examples of great post-Tahoe fluvial incision (Blackwelder, 1931, p. 893), which erroneously seemed to require the absence of Tioga glaciers at those places. In fact, the channel in granite west of Sonora Pass (near Dardanelles Station) was buried by a Tioga glacier, but might predate it; and channels of Leavitt Creek and the West Walker River, east of Sonora Pass, were both deep under the Tioga glacier (Clark, 1967). Blackwelder's identification of these channels as post-Tahoe required his Tioga glaciers on either side of Sonora Pass to be unrealistically small.

## LEGACY

Blackwelder (1931) strongly influenced succeeding Quaternary geologists working in the American West by his emphasis on standardized criteria of relative age, by his determined quest to determine the number of glacial stages and their relative durations, and by his focus on the regional distribution and correlation of glacial features across vast regions. His ultimate goal to describe Pleistocene history may have been supplanted by the effort to understand the processes underlying climatic change, and the ways in which climate can change, but the necessity to know the history of climatic fluctuations is unchanged. For all his relative-weathering criteria, Blackwelder did occasionally fail to distinguish Tioga from Tahoe moraines, and this has led to persistent misconceptions about the relative sizes of the moraines, the extent of the glaciers, and possibly the durations of the glaciations. Regardless of his relatively few serious errors, Blackwelder's work and integration of the research of his colleagues and predecessors inspired the very work that permitted his flaws to be recognized. All of his successors owe him a debt of gratitude for that.

## ACKNOWLEDGMENTS

We thank Judy Terry Smith, School of Earth Sciences, Stanford University, for supplying Eliot Blackwelder's portrait and information about his career from the school's Geohistory Archives. Dorothy Stout helped us bring this manuscript to completion with her tolerant view of our lapses as successive editorial deadlines receded into history. Critical readings by Don Easterbrook, Eldridge Moores, and Doris Sloan improved our writing. Conversations over many years with dozens of colleagues, most of whom were influenced and inspired by Eliot Blackwelder's work, have added to our appreciation and understanding of Cordilleran geology.

## REFERENCES CITED

Alden, W. C., 1932, Physiology and glacial geology of eastern Montana and adjacent areas: U.S. Geological Survey Professional Paper 174, 133 p.

Antevs, E., 1925, On the Pleistocene history of the Great Basin: Carnegie Institution of Washington Publication no. 352, p. 51–114.

Atwood, W. W., 1909, Glaciation of the Uinta and Wasatch Mountains: U.S. Geological Survey Professional Paper 61, 96 p.

Benson, L. V., Lund, S. P., Burdett, J. W., Kashgarian, M., Rose, T. P., Smoot, J. P., and Schwartz, M., 1998, Correlation of Late-Pleistocene lake-level oscillations in Mono Lake, California, with North Atlantic climate events: Quaternary Research v. 49, p. 1–10.

Bierman, P., 1994, Using in situ cosmogenic isotopes to estimate rates of landscape evolution: A review from the geomorphic perspective: Journal of Geophysical Research, v. 99, no. B-7, p. 13,885–13,896.

Birkeland, P. W., 1984, Soils and geomorphology: New York, Oxford University Press, 372 p.

Birkeland, P. W., and Burke, R. M., 1988, Soil catena chronosequences on eastern Sierra Nevada moraines, California, U.S.A: Arctic and Alpine Research, v. 20, p. 473–484.

Birkeland, P. W., Burke, R. M., and Walker, A. L., 1980, Soils and subsurface rock-weathering features of Sherwin and pre-Sherwin glacial deposits, eastern Sierra Nevada, California: Geological Society of America Bulletin, v. 91, p. 238–244.

Birman, J. H., 1964, Glacial geology across the crest of the Sierra Nevada: Geological Society of America Special Paper 75, 80 p.

Blackwelder, E., 1915, Post-Cretaceous history of the mountains of central Wyoming, Part III, Quaternary cycles of stream erosion: Journal of Geology, v. 23, p. 307–340.

Blackwelder, E. 1929, Glacial history of the east side of the Sierra Nevada (abst): Geological Society of America Bulletin, v. 40, p. 127.

Blackwelder, E., 1930, Correlation of glacial epochs in Western United States [abs.]: Geological Society of America Bulletin, v. 41, p. 91.

Blackwelder, E., 1931, Pleistocene Glaciation in the Sierra Nevada and Basin Ranges: Geological Society of America Bulletin, v. 42, p. 865–922.

Blackwelder, E., and Salisbury, R. D., 1903, Glaciation in the Big Horn Mountains, Wyoming: Journal of Geology v. ii, p. 216–233.

Brewer, W. H., 1966, Up and down California in 1860–1864 (third edition): Los Angeles, University of California Press, 583 p.

Burke, R. M., and Birkeland, P. W., 1979, Reevaluation of multiparameter relative dating techniques and their application to the glacial sequence along the eastern escarpment of the Sierra Nevada, California: Quaternary Research, v. 11, p. 21–51.

Bursik, M. I., and Gillespie, A. R., 1993, Late Pleistocene glaciation of Mono Basin, California: Quaternary Research, v. 39, p. 24–35.

Clark, D. H., 1997, A new alpine lacustrine sedimentary record from the Sierra Nevada: Implications for late Pleistocene paleoclimate reconstructions and cosmogenic isotope production rates: Eos (Transactions, American Geophysical Union) v. 78, p. F249.

Clark, D. H., and Gillespie, A. R., 1997, Timing and significance of late-glacial and Holocene glaciation in the Sierra Nevada, California: Quaternary International, v. 38/39, p. 1–38.

Clark, M. M., 1967, Pleistocene glaciation of the drainage of the West Walker River, Sierra Nevada, California [Ph.D. thesis]: Stanford, California, Stanford University, 170 p.

Colman, S. M., and Pierce, K. L., 1981, Weathering rinds on andesitic and basaltic stones as a Quaternary age indicator, western United States: U.S. Geological Survey Professional Paper 1210, 56 p.

Crook, R. Jr.,1986, Relative dating of Quaternary deposits based on P-wave velocities in weathered granitic clasts: Quaternary Research, v. 25, p. 281-292.

Crook, R., Jr., and Gillespie, A. R., 1986, Weathering rates in granitic boulders measured by P-wave speeds, in Colman, S. M., and Dethier, D., eds., Rates of chemical weathering of rocks and minerals: Orlando, Florida, Academic Press, p. 395–417.

Curry, R. R., 1968, Quaternary climatic and glacial history of the Sierra Nevada, California [Ph.D. thesis]: Berkeley, University of California, 204 p.

Dalrymple, G. B., 1964, Potassium-argon dates of three Pleistocene interglacial basalt flows from the Sierra Nevada, California: Geological Society of America Bulletin, v. 75, p. 753–758.

Gilbert, G. K., 1890, Lake Bonneville: U.S. Geological Survey Monograph 1, 438 p.

Gillespie, A. R., 1982, Quaternary glaciation and tectonism in the southeastern Sierra Nevada, Inyo County, California [Ph.D. thesis]: Pasadena, California Institute of Technology, 705 p.

Gillespie, A. R., 1991, Testing a new climatic interpretation for the Tahoe glaciation, in Hall, C. A., Jr., Doyle-Jones, V., and Widawski, B., eds., Natural history of eastern California and high-altitude research: White Mountain Research Station Symposium 3, p. 383–398.

Gillespie, A. R., and Molnar, P., 1995, Asynchronous maximum advances of mountain and continental glaciers: Reviews of Geophysics, v. 33, p. 311–364.

Gillespie, A. R., Huneke, J. C., and Wasserburg, G. J., 1984, Eruption age of a ~100,000-year-old basalt from $^{40}Ar$-$^{39}Ar$ analysis of partially degassed xenoliths: Journal of Geophysical Research, v. 89, p. 1033–1048.

Kay, G. F., 1931, The relative ages of the Iowan and Wisconsin drift sheets: Iowa Geological Survey, v. 21, p. 158–172.

King, C., 1878, Report of the geological exploration of the Fortieth Parallel: Systematic Geology, v. 1, p. 459–529.

Knopf, A., 1918, A geologic reconnaissance of the Inyo Range and the eastern slope of the southern Sierra Nevada, California; with a section on the stratigraphy of the Inyo Range by Edwin Kirk: U.S. Geological Survey Professional Paper 110, 130 p.

Krauskopf, K. B., 1976, Eliot Blackwelder, June 4, 1880–January 14, 1969: Biographical Memoirs XLVIII, National Academy of Sciences: New York, Columbia University Press, p. 83–103.

Leonard, E. M., 1989, Climatic change in the Colorado Rocky Mountains: Estimates based on modern climate at late Pleistocene equilibrium lines: Arctic and Alpine Research, v. 21, p. 245–255.

Matthes, F., 1929, Multiple glaciation in the Sierra Nevada: Science, v. 70, p. 75–76.

Matthes, F., 1930, Geologic history of the Yosemite Valley: U.S. Geological Survey Professional Paper 160, 137 p.

Muir, J., 1872, Living glaciers of California: Overland, v. 9, p. 547–549.

Phillips, F. M., Zreda, M. G., Smith, S. S., Elmore, D., Kubik, P. W., and Sharma, P., 1990, Cosmogenic Chlorine-36 chronology for glacial deposits at Bloody Canyon, eastern Sierra Nevada: Science, v. 248, p. 1529–1532.

Porter, S. C., 1964, Composite Pleistocene snowline of Olympic Mountains and Cascade Range, Washington: Geological Society of America Bulletin, v. 75, p. 477–482.

Russell, I. C., 1885, Geological history of Lake Lahontan, a Quaternary lake of northwestern Nevada: U.S. Geological Survey Monograph 11, 288 p.

Russell, I. C., 1889, Geological history of Mono Valley, California: U.S. Geological Survey Eighth Annual Report, Part I, p. 261–394.

Russell, I. C., 1898, The glaciers of North America: Journal of Geology. v. 12, p. 553–564.

Russell, J., 1925, Glaciation in the King's River Canyon of the Sierra Nevada Mountains [Ph.D. thesis]: Stanford, California, Stanford University.

Sarna-Wojcicki, A. M., and Pringle, M. S., Jr., 1992, Laser fusion $^{40}Ar$-$^{39}Ar$ ages of the tuff of Taylor Canyon and Bishop Tuff, E. California–W. Nevada [abs.]: Eos (Transactions, American Geophysical Union), v. 73, p. 633.

Scuderi, L. A., 1984, A dendroclimatic and geomorphic investigation of late-Holocene glaciation, southern Sierra Nevada, California [Ph.D. thesis]: Los Angeles, University of California, 247 p.

Sharp, R. P., 1968, Sherwin Till–Bishop Tuff relationship, Sierra Nevada, California: Geological Society of America Bulletin, v. 79, p. 351–364.

Sharp, R. P., 1969, Semiquantitative differentiation of glacial moraines near Convict Lake, Sierra Nevada, California: Journal of Geology, v. 77, p. 68–91.

Sharp, R. P., 1972, Pleistocene glaciation, Bridgeport Basin, California: Geological Society of America Bulletin, v. 83, p. 2233–2260.

Sharp, R. P., and Birman, J. H., 1963, Additions to the classical sequence of Pleistocene glaciations, Sierra Nevada, California: Geological Society of America Bulletin, v. 74, p. 1079–1086.

Smith, G. I., 1984, Paleohydrologic regimes in the southwestern Great Basin 0–3.2 m.y. ago compared with other long records of "global" climate: Quaternary Research, v. 22, p. 1–17.

Swanson, T. W., Caffee, M., Finkel, R., Harris, L., Southon, J., Zreda, M. G., and Phillips, F. M., 1993, Establishment of new production parameters for Chlorine-36 dating based on the deglaciation history of Whidbey Island, Washington: Geological Society of America Abstracts with Programs, v. 25, no. 6, p. 461.

Turner, H. W., 1900, The Pleistocene geology of the south-central Sierra Nevada with especial reference to the origin of Yosemite Valley: California Academy of Sciences Proceedings, Geology, v. 3, p. 261–321.

Turner, H. W., and Ransome, F. L., 1898, Description of the Big Trees quadrangle, California: U.S. Geological Survey Geological Atlas, Folio 51, 8 p.

Whitney, J. D., 1865, Geology: Geological Survey of California, v. 1, p. 450–455.

MANUSCRIPT ACCEPTED BY THE SOCIETY NOVEMBER 23, 1998.

Geological Society of America
Special Paper 338
1999

# Chapter 22
# CALIFORNIA'S SHIFTING SLOPES

**Roy J. Shlemon**
*Roy J. Shlemon and Associates, Inc.*
*P.O. Box 3066*
*Newport Beach, California 92659-0620*
*E-mail: rshlemon@ix.netcom.com*

Karl Terzaghi, an outstanding contributor to the field of engineering geology (photograph taken in 1957; courtesy of George A. Kiersch, 1991a, p. 33).

# KARL TERZAGHI

Karl Terzaghi, the author of the 1950 "Mechanism of Landslides," a classic paper in engineering geology, was a distinguished professor at Harvard University and a world-renowned consultant. He is regarded as the "Father of Soil Mechanics," and he received several honorary doctorates and many awards from various universities and professional societies.

Among his many contributions, he brought the concepts of soil mechanics into engineering geologic practice; he emphasized that best results can only be obtained by integration of theory and professional judgment; and he identified the technical background needed both by the engineer and geologist to work successfully on engineering-geologic projects.

The 1950 Terzaghi paper "Mechanism of Landslides" is a classic because of its synthesis of landslide-analysis principles, its presentation of pertinent case studies, its integration of soil mechanics and geologic principles, and its philosophical expression about the cooperative roles of the geologist and engineer. Some 50 years later, it is still required reading, the hallmark of a truly classic paper.

Roy J. Shlemon

*The Geological Society of America*

# APPLICATION OF GEOLOGY TO ENGINEERING PRACTICE

## BERKEY VOLUME

SIDNEY PAIGE, *Chairman*

*Papers By*

JOHN L. SAVAGE, ROGER RHOADES, EDWARD B. BURWELL, JR., GEORGE D. ROBERTS, BERLEN C. MONEYMAKER, JAMES F. SANBORN, KARL TERZAGHI, GEORGE D. LOUDERBACK, O. E. MEINZER, E. F. BEAN, W. C. KRUMBEIN, DUNCAN MCCONNELL, RICHARD C. MIELENZ, WILLIAM Y. HOLLAND, KENNETH T. GREENE, K. C. HEALD, MURL H. GIDEL, AND CHAS. B. HUNT

PUBLISHED BY THE SOCIETY
1950

THE GEOLOGICAL SOCIETY OF AMERICA
ENGINEERING GEOLOGY (BERKEY) VOLUME               NOVEMBER 1950

# MECHANISM OF LANDSLIDES

By KARL TERZAGHI

*Harvard University, Cambridge, Mass.*

## CONTENTS

| | Page |
|---|---|
| Varieties of slope movements | 84 |
| Definitions | 84 |
| Differences between landslides and creep | 84 |
| Varieties of landslides | 87 |
| Processes leading to landslides | 88 |
| Causes of landslides | 88 |
| External changes of stability conditions | 88 |
| Earthquake shocks | 89 |
| Lubricating effect of water | 91 |
| Rise of piezometric surface | 92 |
| Progressive structural changes in slope-forming material | 94 |
| Rapid drawdown | 97 |
| Spontaneous liquefaction | 100 |
| Seepage from artificial sources of water | 100 |
| Periodicity of landslides | 102 |
| Review of slide-producing processes | 105 |
| Dynamics of landslides | 110 |
| Surface movements preceding the slide | 110 |
| Movements during the slide | 111 |
| Movements after the slide | 112 |
| Landslide problems | 115 |
| Preventive measures | 115 |
| Landslide correction | 119 |
| Co-operation between geologist and engineer on landslide problems | 121 |
| References | 121 |

### ILLUSTRATIONS

| Figure | | Page |
|---|---|---|
| 1. | Hallowell valley near Northampton, England | 87 |
| 2. | Diagram illustrating conventional method for computing effect of earthquake on stability of a slope | 89 |
| 3. | Section through a slide which was caused by an excess hydrostatic pressure in the silt layers of a stratum of varved clay | 92 |
| 4. | Diagrammatic section through site of rock slide of Goldau (1806) prior to slide | 93 |
| 5. | Turtle Mountain after the Great Frank Alberta Slide in 1903 | 95 |
| 6. | Diagram showing gradual decrease of shearing resistance of stiff, fissured London clay | 98 |
| 7. | Diagrams illustrating effect of speed of drawdown on stability of temporarily submerged slope | 98 |
| 8. | Folkestone Warren, Channel Coast, England | 103 |
| 9. | Section through the Hudson slide of 1915 | 104 |
| 10. | Diagram illustrating variations of the factor of safety of different slopes prior to a landslide | 109 |
| 11. | Diagram illustrating the ground movements which precede a landslide | 111 |
| 12. | Sketch map of the site of the Kenogami slide south of the Saguenay River, province of Quebec, Canada | 113 |
| 13. | Artesian conditions | 117 |
| 14. | Ore loading | 118 |
| 15. | Diagram showing relation between position of the water table with reference to a slope after failure and the horizontal component of the corresponding downhill movement of the surface of the slope | 120 |

| Plate | | Following page |
|---|---|---|
| 1. | Deformation of strata | 96 |
| 2. | Slides | 96 |
| 3. | Slides | 96 |

## VARIETIES OF SLOPE MOVEMENTS

### DEFINITIONS

The term *landslide* refers to a rapid displacement of a mass of rock, residual soil, or sediments adjoining a slope, in which the center of gravity of the moving mass advances in a downward and outward direction. A similar movement proceeding at an imperceptible rate is called *creep*. The velocity of the masses involved in a typical landslide increases more or less rapidly from almost zero to at least 1 foot per hour. Then it again decreases to a small value. By contrast, typical creep is a continuous movement which proceeds at an average rate of less than 1 foot per decade. Higher rates of creep movements are rather uncommon.

Slides on the slopes of man-made cuts are sometimes referred to as *slope failures*. In this paper, for the sake of convenience, the term landslide will be retained for failure on both natural slopes and slopes of cuts.

### DIFFERENCES BETWEEN LANDSLIDES AND CREEP

A landslide is an event which takes place within a short period of time as soon as the stress conditions for the failure of the ground located beneath the slope are satisfied. By contrast, creep is a more or less continuous process. A landslide represents the movement of a relatively small body of material with well-defined boundaries, whereas creep may involve the ground located beneath all the slopes in a whole region and no sharp boundary exists between stationary and moving material. Most landslides are produced only by the force of gravity, whereas creep movements can also be due to the combined action of the force of gravity and various other agents.

Within the zone of seasonal changes of moisture and temperature, at least part of the horizontal component of the ground movement is produced by thermal expansion and contraction, by swelling and shrinking, freezing and thawing, and other seasonal processes (Sharpe and Dosch, 1942, p. 46–48). These processes result in a downhill movement of a sheet of earth, with depth equal to or smaller than the depth of seasonal variations in the condition of the ground. Below this depth, creep can be produced only by the force of gravity, unaided by other agents. Since the force of gravity does not change with the seasons, the rate of the resulting gravity creep is fairly constant. This type of creep will be referred to as *continuous creep* in contrast to the *seasonal creep* which can occur in the top layer of the ground only. In the following comparison between landslides and continuous creep the seasonal creep is not considered.

If the difference between landslides and continuous creep resided only in the velocity of the movement, it would hardly be justifiable to consider landslides and creep as different types. However, experience has disclosed another much more significant distinguishing feature. It consists in the difference between the pattern of the deformations produced by these two processes. Patterns of equally different character can be obtained in the laboratory by loading tests on blocks of a nonhomogeneous mass composed of materials such as asphalt which have the properties of a very viscous liquid. A heavy load on such a block causes an almost sudden failure by separation along one or more surfaces of rupture, which cut across the boundaries between the strong and weak portions of the block. By contrast, the instantaneous effect of a very small load on a similar block is imperceptible; but if the load acts on the block for many years the block undergoes very important and very intricate deformations which reflect all the details of the internal structure of the block.

The striking difference between the resulting deformation patterns is due to the fact that the laws which determine the deformations are as different as those of hydraulics and of the mechanics of elastic solids. If a system composed of strata with very different elastic properties is acted upon for a long time by shearing stresses which are smaller than the average shearing strength of the system, the most rigid members only will behave like solids, whereas the balance will be deformed like a very viscous liquid. The deformation of the system will be like that of a sheet of asphalt containing layers of a brittle material. As the shearing stresses increase, a higher and higher percentage of the members of the system will perform like solids, and, if the stresses are rapidly increased to the point of failure, the entire system will behave like a solid.

The load per unit of area under which a block fails by shear or splitting is commonly known as the *compressive strength* of the material of which the block is composed. The load at which creep begins is very much smaller. It is called the *fundamental strength* (Griggs, 1936, p. 364). As long as the shearing stresses in the material beneath a slope are smaller than the "fundamental" shearing resistance of the material, the slope is at rest. If they exceed this value the slope creeps, and if they become equal to the stress required to produce a shear failure a landslide occurs.

\* \* \* \* \*

The known manifestations of creep suggest that creep is nothing but a small-scale and superficial replica of what takes place at depth under the influence of tectonic forces. Both processes go on continuously over vast areas, and the mechanism of both is essentially the same although the driving forces are different.

\* \* \* \* \*

From time to time, at geographically widely separated points, the intensity of the shearing stresses, within a zone of creep, becomes equal to the shearing resistance of the material, or the shearing resistance of the material decreases until it becomes equal to the shearing stresses. Under such circumstances, a landslide takes place.

\* \* \* \* \*

Creep, like tectonic movements, may lead to intricate deformations, revealing and accentuating the resistance pattern of the masses subject to deformation. In contrast to creep deformations, landslides are characterized by sliding movements along well-defined surfaces, which cut across the boundaries between competent and incompetent strata like the shear planes produced by a shear test of short duration. In spite of these radical differences, no sharp boundaries between these two groups of slope movements can be drawn. On any one slope located above unconsolidated material such as residual soil or sedimentary clay, creep may develop into a slide, and the slide may be followed by creep in the material which has moved out of the slope (Sharpe and Dosch, 1942).

Both creep and landslides require the attention of the engineer. Creep deformation of strata, located beneath the bottom of erosion valleys like those shown in Figure 1, may have a profound effect on the foundation conditions at the site of proposed storage dams (Lapworth, 1911). Lugeon (1922) has published a brief account of repeated failures of a water main near Lausanne due to the rapid creep of weathered Flysch shales underlying the slope which supported the piers of the pipe line. Haefeli (1944) described serious damage to a railroad viaduct caused by creep. It required expensive underpinning operations. The foundations for the piers of the bridge rest on talus. Similar creep phenomena, leading to bridge defects in Switzerland were thoroughly investigated and described by Mohr et al. (1947). However, the implications of creep are beyond the scope of this paper, which deals exclusively with typical landslides.

## VARIETIES OF LANDSLIDES

Landslides may involve materials of any kind, ranging between hard rock and soft clay, or any combination of materials. A similar variety prevails among the processes which may lead to landslides. They include undercutting by river erosion,

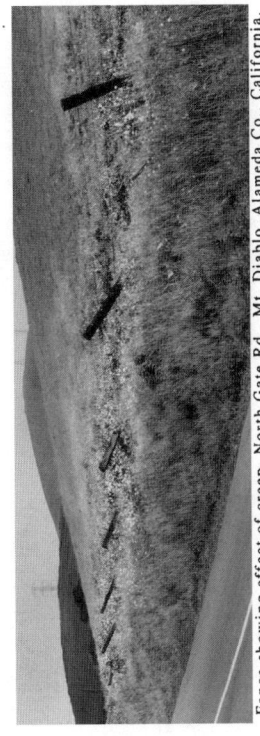

Fence showing effect of creep. North Gate Rd., Mt. Diablo, Alameda Co., California. Photo by Doris Sloan

man-made excavation, change in the ground water regime, and progressive structural changes in the material adjoining the slopes.

A phenomenon involving such a multitude of combinations between materials and disturbing agents opens unlimited vistas for the classification enthusiast. The result of the classification depends quite obviously on the classifier's opinion regarding the relative importance of the many different aspects of the classified phenomenon.

In this paper no new classification will be added to the numerous existing ones. In exchange, an attempt will be made to discriminate between the processes which may conceivably lead to landslides, and to analyze each one of them.

## PROCESSES RESPONSIBLE FOR LANDSLIDES

### CAUSES OF LANDSLIDES

The causes of landslides can be divided into external and internal ones. External causes are those which produce an increase of the shearing stresses at unaltered shearing resistance of the material adjoining the slope. They include a steepening or heightening of the slope by river erosion or man-made excavation. They also include the deposition of material along the upper edge of slopes and earthquake shocks. If an external cause leads to a landslide, we can conclude that it increased the shearing stresses along the potential surface of sliding to the point of failure.

Internal causes are those which lead to a slide without any change in surface conditions and without the assistance of an earthquake shock. Unaltered surface conditions involve unaltered shearing stresses in the slope material. If a slope fails in spite of the absence of an external cause, we must assume that the shearing resistance of the material has decreased. The most common causes of such a decrease are an increase of the pore-water pressure, and progressive decrease of the cohesion[1] of the material adjoining the slope. Intermediate between the landslides due to external and internal causes are those due to rapid drawdown, to subsurface erosion, and to spontaneous liquefaction.

### EXTERNAL CHANGE OF STABILITY CONDITIONS

One of the most common and most obvious causes of landslides consists in the undercutting of the foot of a slope or the deposition of earth or other materials along the upper edge of the slope. Both operations produced an increase of the shearing stresses in the ground beneath the slope. If and as soon as the average shearing stress on the potential surface of sliding[2] becomes equal to the average shearing resistance, a landslide occurs.

A slope failure on a man-made slope may occur during or at any time after construction. If the slope fails several weeks after construction or later, the slide can be ascribed only to an internal cause which reduced the shearing resistance of the slope material after the completion of the construction operations. Delayed slides occur most commonly during heavy rainstorms.

---

[1] The term cohesion indicates the resistance of a material, rock, or sediment against shear along a surface which is under no pressure.

[2] The term *potential surface of sliding* indicates that surface located beneath a slope for which the ratio between average shearing stress and average shearing resistance is a maximum. If the material adjoining the slope is fairly homogeneous, the cross section of the surface of sliding resembles a cycloid. Stability computations are commonly based on the simplifying assumption that the profile has the shape of an arc of a circle; the error due to this assumption is unimportant.

## EARTHQUAKE SHOCKS

Earthquake shocks are considered external causes of landslides because they increase the shearing stresses along the potential surface of sliding, whereas the shearing resistance remains unchanged. The conventional method for evaluating the effect of an earthquake shock on the stability of a slope is illustrated by Figure 2, represent-

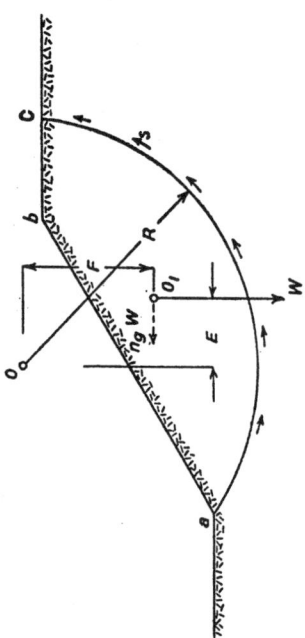

FIGURE 2.—*Diagram illustrating conventional method for computing effect of earthquake on stability of a slope*

The earthquake produces a horizontal acceleration $n_g$-times the acceleration of gravity, which increases the moment tending to rotate the wedge *abc* clockwise by $n_g FW$.

ing a cross section of a slope. The section through the potential surface of sliding is assumed to be an arc of a circle, $ac$, with the center $O$. Let

$W$ = weight of the earth (water and solid combined) located above the surface of sliding, per unit of length of the slope measured at a right angle to the plane of the section shown in Figure 2,
$l$ = length of the arc $aC$,
$s$ = average shearing resistance per unit of area of surface of sliding,
$g$ = acceleration due to gravity,
$n_g$ = ratio between the greatest horizontal acceleration produced by the earthquake and the acceleration $g$ due to gravity, and
$O_1$ = center of gravity of the slice $abC$.

The weight $W$, acting at the lever arm $E$, tends to produce a rotation of the slice $abc$ about the axis $O$ and the rotation is resisted by the shearing resistance, $sl$, acting at a lever arm $R$. Hence, prior to the earthquake the safety factor $G_s$ of the slope with respect to sliding is

$$G_{s_c} = \frac{\text{resisting moment}}{\text{driving moment}} = \frac{slR}{EW} \qquad (1)$$

An earthquake with an acceleration equivalent $n_g$ produces a mass force acting in a horizontal direction of intensity $n_g$ per unit of weight of the earth. (*See*, for instance, Terzaghi, 1943b, p. 473–479.) The resultant of this mass force, $n_g W$, passes, like the weight $W$, through the center of gravity $O_1$ of the slice $abc$. It acts at a lever arm with length $F$ and increases the moment which tends to produce a rotation of the slice $abc$ about the axis $O$ by $n_g FW$. Hence the earthquake reduces the factor of safety of the slope with respect to sliding from $G_a$, equation (1), to

$$G'_s g'_s = \frac{sR}{EW + n_a FW} \quad (2)$$

The numerical value of $n_a$ depends on the intensity of the earthquake. Independent estimates (Freeman, 1932) have led to the following approximate values:

Severe earthquakes, Rossi-Forel scale IX, $n_a = 0.1$
Violent, destructive, Rossi-Forel scale X, $n_a = 0.25$
Catastrophic $n_a = 0.5$

The earthquake of San Francisco in 1906 was violent and destructive (Rossi-Forel scale X), corresponding to $n_a = 0.25$.

Equation (2) is based on the simplifying assumptions that the horizontal acceleration $n_a g$ acts permanently on the slope material and in one direction only. Therefore the conception it conveys of earthquake effects on slopes is very inaccurate, to say the least. Theoretically a value of $G'_s = 1$ would mean a slide, but in reality a slope may remain stable in spite of $G'_s$ being smaller than unity and it may fail at a value of $G'_s > 1$, depending on the character of the slope-forming material.

The most stable materials are clays with a low degree of sensitivity, in a plastic state (Terzaghi and Peck, 1948, p. 31), dense sand either above or below the water table, and loose sand above the water table. The most sensitive materials are slightly cemented grain aggregates such as loess and submerged or partly submerged loose sand.

If a violent earthquake shock strikes a slope on plastic clay with low sensitivity, it will hardly have any effect beyond the formation of tension cracks along the upper edge associated with a slight bulging of the slope, because the viscosity of the clay interferes with more extensive displacements under impact. The slopes of earth dams or dikes, consisting of sand, may bulge slightly, and the crest of the fills may settle, but the slopes will not fail, provided the fills rest on a rough and stable base. After the earthquake, the fills will be more stable than before, because the earthquake vibrations tend to compact the material. Toward the end of the last century two earth dams have been built across the San Andreas fault in California—the San Andreas Dam, 95 feet high, and the Upper Crystal Springs Dam, 85 feet high. During the earthquake of San Francisco in 1906 the horizontal displacement along the fault at the site of the dams amounted to more than 10 feet. Yet the slopes of the dams remained intact except at the point where they were warped by shear (Eng. News-Rec., 1932).

If a mass of stable material, such as dense sand, rests on a slippery base, like the surface of a layer of soft clay, a slope failure may occur by sliding of the stable material on its base.

The destructive effect of earthquakes on slightly cemented grain aggregates, such as loess, and on submerged loose sand, seems to be chiefly due to the rapid vibratory movement of the particles with reference to each other and not to the quasi-static effects described by equation (2). In loess these movements are likely to break the connection between the grains, whereupon the material assumes the character of cohesionless sand. In December 1920 a catastrophic earthquake occurred in the heart of the loess district of the province of Kansu in China. "In each case the earth which came down bore the appearance of having shaken loose clod from clod and grain from grain, and then cascaded like water, forming vortices, swirls, and all the convolutions into which a torrent might shape itself" (Close and McCormick, 1922, p. 463). During the earthquake of New Madrid in 1811, many loess slopes failed (Fuller, 1912). Submerged masses of loose sand may, under the impact of an earthquake shock, temporarily assume the character of a suspension which flows like a viscous liquid. (See the subheading "Spontaneous Liquefaction.")

### LUBRICATING EFFECT OF WATER

If a slide takes place during a rainstorm at unaltered external stability conditions, most geologists and many engineers are inclined to ascribe it to a decrease of the shearing resistance of the ground due to the "lubricating action" of the water which seeped into the ground. This explanation is unacceptable for two reasons.

First of all, water in contact with many common minerals, such as quartz, acts as an anti-lubricant and not as a lubricant. Thus, for instance, the coefficient of static friction between smooth, dry quartz surfaces is 0.17 to 0.20 against 0.36 to 0.41 for wet ones (Terzaghi, 1925, p. 42–64).

Second, only an extremely thin film of any lubricant is required to produce the full static lubricating effect characteristic of the lubricant. Any further amount of lubricant has no additional effect on the coefficient of static friction between them (Hardy, 1919). In humid regions such as the eastern United States and within less than 1–2 feet from the sloping surface every sediment—sand included—permanently contains far more than the quantity of water needed for "lubricating" the surfaces of the grains (Terzaghi, 1942, Fig. 15, p. 356). Yet in humid regions rainstorms start landslides as often as they do in arid ones. In other words, since practically all the sediments located beneath slopes are permanently "lubricated" with water, a rainstorm cannot possibly start a slide by lubricating the soil or boundaries between soil strata.

However, the rain water which seeps into a slope affects the stability of the slope in various other ways. If the voids of the ground are partly filled with air, the water eliminates the surface tension which imparts to fine-grained, cohesionless soils a considerable amount of apparent cohesion (Terzaghi and Peck, 1948, p. 114–128). The water which enters the voids also increases the unit weight of the soil though, as a rule, this increase is commonly unimportant.

Some soils, such as typical loess, owe their cohesion to a soluble binder. If a slope on such a soil is submerged for the first time, or if the soil becomes saturated by seepage from a newly created, artificial source of water, the binder is removed by solution and the soil loses its cohesion.

Last—but not least—water which enters the ground beneath a slope always causes a rise of the piezometric surface[3], which, in turn, involves an increase of the pore-water pressure and a decrease of the shearing resistance of the soil. Since water can affect the stability of slopes in several radically different ways, its actions will be discussed under different subheadings.

---

[3] The piezometric surface is the locus of the points to which the water would rise in piezometric tubes. If the permeability of a soil, such as a soft clay, is too low to permit locating the position of the piezometric surface by means of observation wells, pressure gages must be used.

## RISE OF PIEZOMETRIC SURFACE

Throughout a saturated mass of jointed rock, soil, or sediment, the water which occupies the voids is under pressure. Let

$p$ = pressure per unit of area at a given point $P$ of a potential surface of sliding, due to the weight of the solids and the water located above the surface,
$h$ = the piezometric head at that point,
$w$ = the unit weight of the water, and
$\phi$ = the angle of sliding friction for the surface of sliding.

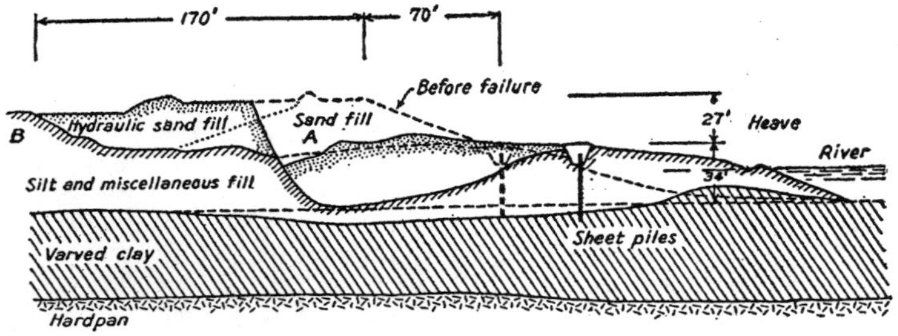

Regarding the relation between these four quantities, soil mechanics has led to the following conclusions (Terzaghi and Peck, 1948, p. 51–55). If the potential surface of sliding is located in a layer of sand or silt, the shearing resistance $s$ per unit of area at the observation point is equal to

$$s = (p - hw) \tan \phi \qquad (3)$$

Hence, if the piezometric surface rises, $h$ increases, and the shearing resistance $s$ decreases. It can even become equal to zero. The action of the water pressure $hw$ can be compared to that of a hydraulic jack. The greater $hw$, the greater is the part of the total weight of the overburden which is carried by the water, and as soon as $hw$ becomes equal to $p$ the overburden "floats." If a material has cohesion, $c$ per unit of area, its shearing resistance is equal to the sum of $s$, equation (3), and the cohesion value $c$, whence

$$s = c + (p - hw) \tan \phi \qquad (4)$$

The effect of a decrease of the shearing resistance $s$ on the stability of slopes on stratified sediments is illustrated by Figure 3. The figure shows a vertical section through sand dikes which were constructed along a river for flood-protection purposes. The dikes rest on a layer of soft silt and miscellaneous fill which covers the surface of a horizontal stratum of varved clay with a thickness of about 50 feet. The dash line shows the position of the dike and of the boundaries between the underlying strata prior to the slide.

Some time after the dike $A$ was completed by depositing and compacting moist

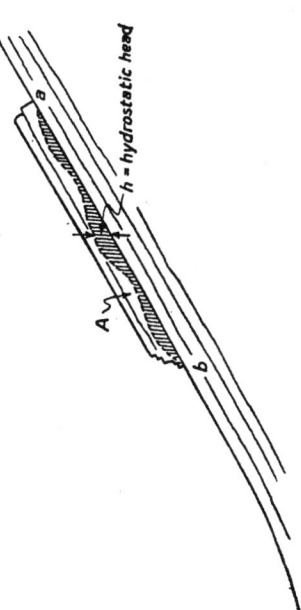

FIGURE 3.—*Section through a slide which was caused by an excess hydrostatic pressure in the silt layers of a stratum of varved clay*

The row of sheetpiles advanced in a few minutes over a maximum distance of up to 60 feet toward the river.

sand in layers, the space between the landward slope of this dike and the outer slope of an older dike, $B$, was filled with sand. The sand was excavated by means of a hydraulic dredge and deposited in a semiliquid condition. The sluicing operations raised the piezometric level in the pore water of the silt layers in the varved clay to a considerable height $h$ above the original water table. As a consequence the resistance $s$, equation (3), against sliding decreased.

During the construction of the hydraulic fill the dike $A$ suddenly subsided, and the row of sheet piles, together with the foreland, moved over a distance up to 60 feet and over a length of about 1200 feet toward the river. The row of sheet piles remained perfectly intact. This fact showed that the failure had occurred by sliding along one or more horizontal surfaces of sliding located in the varved clay. If the shearing resistance along these surfaces had not been extremely low, the wedge-shaped body of silt, located on the river side of the sheet piles, could not possibly have advanced over a distance up to 60 feet without undergoing intense compression and shortening in the direction of the movement.

After a slide has occurred the excess pore-water pressure in the zone of shear always decreases, on account of progressive consolidation, and approaches a value zero. In order to get information on the rate of decrease of the pressure which prevailed in the varved clay, a great number of pressure gages were installed. The first readings were made more than 3 months after the slide occurred. Yet, even beneath the banks of the river, the piezometric elevation $h$, equation (3), still amounted to more than 10 feet, with reference to the river level.

The relation expressed by equation (3) also applies to stratified or jointed rocks. To illustrate its bearings on rockslides, the classical slide of Goldau in Switzerland will be discussed. This slide has always been ascribed to the "lubricating action" of the rain- and meltwater. Figure 4 is a diagrammatic section through the slide area. It shows a slope oriented parallel to the bedding planes of a stratified mass of Tertiary Nagelflue (conglomerate with calcareous binder) which rises at an angle of 30° to the

FIGURE 4.—*Diagrammatic section through site of rock slide of Goldau (1806) prior to slide*

Slab $A$ was separated from its base by a thin layer of weathered rock. The dashed line represents the piezometric surface in this layer during a heavy rainstorm.

horizontal. On this slope rested a slab of Nagelfiue 5000 feet long, 1000 feet wide, and about 100 feet thick. It was separated from its base by a porous layer of weathered rock.

The fact that the slab had occupied its position since prehistoric times indicates that the shearing force, which tended to displace the slab, never exceeded the shearing strength, in spite of the effects of whatever hydrostatic pressures, $hw$, in equation (3), may have temporarily acted on the base of the slab in the course of its existence.

On September 2, 1806, during heavy rainstorms, the slab moved down the slope, wiped out a village located in its path, and killed 457 people (Heim, 1882). This catastrophe can be explained in at least three ways. One explanation is that the angle of inclination of the slope had gradually increased on account of tectonic movements, until the driving force which acted on the slab became equal to the resistance against sliding. A second explanation is based on the assumption that the resistance of the slab against sliding was due not only to friction, but also to a cohesive bond between the mineral constituents of the contact layer. The total shearing resistance due to the bond was gradually reduced by progressive weathering, or by the gradual removal of cementing material, either in solution or by the erosive action of water veins. The third explanation is that $h$ in equation (3) or (4) assumed an unprecedented value during the rainstorm, whereas the cohesion $c$, in equation (4), remained unchanged, provided cohesion existed. In Figure 4 the value $h$ is equal to the vertical distance between the potential surface of sliding, $ab$, and the dash line interconnecting $ab$ which represents the piezometric line. During dry spells $h$ is equal to zero. In other words, the piezometric surface is located at the slope. During rainstorms the rain water enters the porous layer located between slab and slope at $a$ and leaves it at $b$. Since the permeability of this layer is variable, the piezometric line descends from $a$ to $b$ in steps, and the value $h$ in equations (3) and (4) is equal to the average vertical distance between the piezometric line and the slope.

The maximum value of $h$ changes from year to year, and if the exits of the water veins at $b$ are temporarily closed by ice formation while rain- or melt-water enters at $a$, $h$ assumes exceptionally high values. However, the seasonal variations of $h$, the corresponding variations of $s$, equations (3) and (4), and the occasional obstruction of the exits at $b$ have occurred in rhythmic sequence for thousands of years, without catastrophic effects. It is very unlikely that $h$ assumed a record value in 1806, in spite of unaltered external conditions. Therefore it is more plausible to assume that the slide was caused by a process which worked only in one direction, such as a gradual increase of the slope angle or the gradual decrease of the strength of the bond between slab and base. In no event can the slide be explained by the "lubricating effect" of the rain water. One might as well ascribe a theft to some mysterious effects of the presence of the thief in the house instead of inquiring about his physical actions.

## PROGRESSIVE STRUCTURAL CHANGES IN THE SLOPE-FORMING MATERIALS

Every rainstorm causes an increase of the value $h$ in equations (3) and (4) and, as a consequence, a decrease of the shearing resistance along potential surfaces of sliding. Therefore the factor of safety $G_s$, equation (1), of every slope with respect to sliding is subject to cyclic changes. The minor variations have a period of a few weeks or

months, and the major ones of many years. (See section on "Periodicity of Landslides"). These variations are part of the routine of the slopes. Hence the probability that an old slope should be exposed in our lifetime to unfavorable conditions without

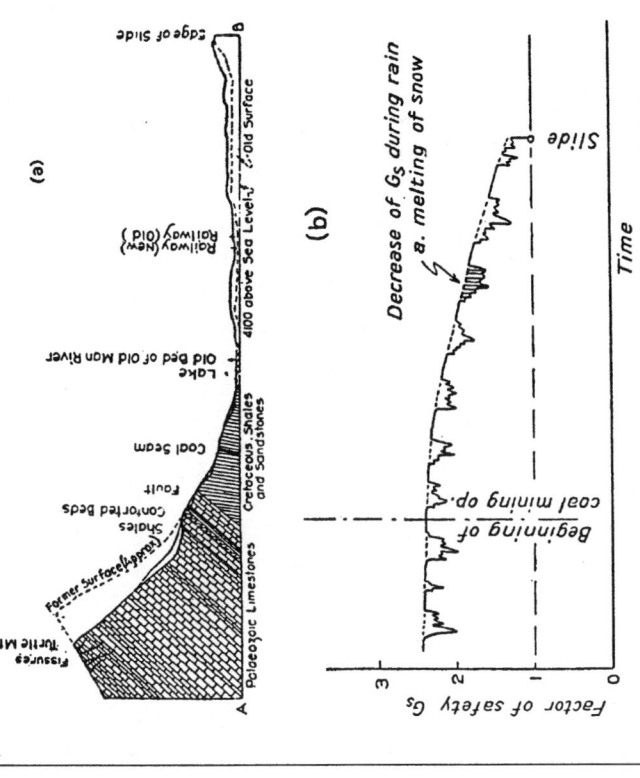

FIGURE 5.—*Turtle Mountain after the Great Frank Alberta Slide in 1903*
After McConnell and Brock (1904). (a) Cross section (b) Diagram illustrating the writer's concept of the changes of the safety factor of the slope prior to the slide.

any precedent is almost nil. If such a slope fails without external provocation it is much more probable that it failed on account of a gradual decrease of the cohesion of the slope-forming materials.

In hard, jointed rocks, resting on softer rocks, a decrease of the cohesion of the rock adjoining a slab may occur on account of creep of the softer rocks forming their base. The great Turtle Mountain slide of 1903 near Frank, Alberta (Fig. 5a), seems to belong to this category. Percolating waters and frost action have contributed to the breakdown (Sharpe, 1938, p. 79). They always do, but they have done it for many thousands of years. Percolating waters cannot move blocks located between joints at great depth, and the frost action is only skin deep. Hence neither water nor frost could have altered the stability conditions in the rock adjoining the slope beyond a distance of a few feet from the slope. However, the limestones, forming the bulk of the peak, rested on weaker strata which certainly "crept" under the influence of the unbalanced pressure produced by the weight of the limestone, and the rate of creep was accelerated by coal-mining operations in the weaker strata.

The total cohesion along the potential surface of sliding in a jointed rock is equal to the combined shearing strength of all those blocks of rock which interfere, like dowels, with the sliding movement. The yield of the base of the limestone caused an increase of the shearing stresses; the increases of the stresses caused one dowel after another to "snap," and the slope failed when it was ripe for failure, at a time when the factor of safety assumed one of its periodic minimum values. Figure 5 b is a graph illustrating the process which led to the slide.

Another incident of a similar kind was the collapse of the Pulverhörndl, an isolated limestone tower in the northern Alps, which rested on a bed of shale. The tower was a favorite training ground for mountaineers. It had a volume of about 260,000 cu. yds. In the fall of 1920 the tower suddenly collapsed, without any provocation. The fragments struck the shale, whereupon the shale assumed the character of a mudstream, and about ten million cubic yards of shale advanced on a gentle slope toward the mouth of a valley (Lehmann, 1926). This catastrophe, too, was probably caused by the gradual yield of the base of the tower, whereas all the other circumstances attending the slide were only a repetition of what had happened many times before.

The writer had an opportunity to investigate a flow slide which was indirectly caused by the swelling of a homogeneous clay stratum. The slide occurred on the side of an open excavation for one of the locks adjoining the powerhouse of the hydroelectric power development, Swir III east of Leningrad.

The slope rose at an angle of about 35° and intersected three horizontal strata. The uppermost stratum consisted of a well-compacted and slightly cohesive glacial till, the middle one of a stiff, greenish Devonian clay, which could be cut with a knife, and the lower one of a very stiff, reddish, sandy clay. The slope made a perfectly sound impression, and nobody, the writer included, felt the slightest concern about its stability.

In the summer of 1930, during a heavy rainstorm, the till turned into a mudflow and descended into the lock excavation. The only explanation which appeared to be acceptable was that the breakdown of the till was due to a horizontal expansion of its base. Since the till is too rigid to participate in a lateral expansion of its base it is torn into fragments by tension. The rain water accumulates in the open spaces between the fragments. The fragments disintegrate, and the supersaturated mixture of water and till flows into the cut.

In order to find out whether this explanation is correct, a niche was carved out of the green clay, a few hours after the slide had started. Figure 1 of Plate 2 is a photographic view of the niche 1 hour after the excavation was completed. One could already see that the distance between the sides of the niche, above the base of the green clay, had decreased and that the rear face had advanced with reference to the underlying red clay. A few hours later the relative displacements were very conspicuous, and the overhanging parts of the expanding green clay started to crumble (Pl. 2, fig. 2).

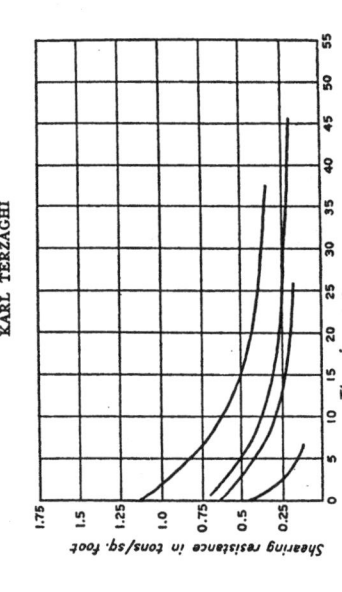

FIGURE 6.—*Diagram showing gradual decrease of shearing resistance of stiff, fissured London clay*. The curves are based on the results of a statistical study of slope failures in the London area. Each curve represents a different locality. (After A. W. Skempton 1948.)

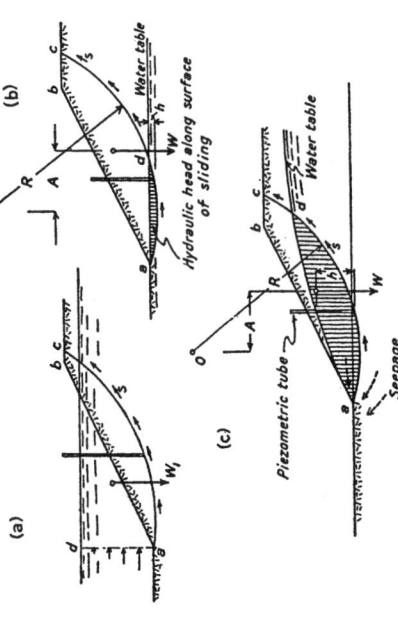

FIGURE 7.—*Diagrams illustrating effect of speed of drawdown on stability of temporarily submerged slope*

(a) Section through the slope prior to drawdown. (b) Forces acting on ground above potential surface of sliding after very slow drawdown, and (c) after rapid drawdown. Dashed arrows in (c) indicate directions of seepage toward foot of slope.

\* \* \* \* \*

### RAPID DRAWDOWN

The term *rapid drawdown* refers to the lowering of the water level in a storage reservoir or to the descent of the water level in a river after a flood at a rate of at least several feet per day. The effect of this process on the stability of the slopes forming the sides of the reservoir or the river is illustrated by Figure 7. Figure 7a shows a vertical section through a partly submerged slope $ab$. The potential surface of sliding is indicated by the arc $ac$ of a circle. Let

$W$ = Weight of slice $abc$, solid and water combined, per unit of length of the slope,
$l$ = length of the arc $ac$,
$c$ = cohesion of the slope-forming material,
$\phi$ = angle of internal friction of this material,
$p$ = average unit pressure on the surface of sliding $ac$ due to the weight $W$ of the slice $abc$,
$h$ = piezometric head for any point of the potential surface of sliding, at any time,
$h_1$ = average of the piezometric heads $h$ for the surface of sliding after a very slow drawdown, and
$h_2$ = as before, after a rapid drawdown.

The following analysis is based on the assumption that the voids of the soil are completely filled with water both below and above the piezometric surface. The error due to this assumption is very small unless the slope material consists of very coursegrained sediments such as coarse sand or gravel without an admixture of finer fractions. The effect of the capillary forces on the stability of the slope will be disregarded. These forces increase the stability of the slope under any circumstances.

If the level of the body of water adjoining the slope goes down very slowly, the water table remains horizontal and descends at the same rate as the water level of the reservoir. After the drawdown is complete, the piezometric surface is a horizontal surface passing through the foot of the slope (Fig. 7b). The average shearing resistance $s$ of the material adjoining the surface of sliding is determined by equation (4), and the factor of safety of the slope with respect to sliding is

$$G_s = \frac{Rlc + (p - h_1w)\tan\phi}{AW} \quad (5)$$

On the other hand, if a drawdown takes place very rapidly the descent of the piezometric surface lags behind the descent of the free water level, and at the end of the drawdown the piezometric surface rises from the foot of the slope as indicated in Figure 7c and intersects the potential surface of sliding at a point $d$ which is located high above $d$ in Figure 7b. The corresponding factor of safety with respect to sliding is

$$G'_s = \frac{Rlc + (p - h_2w)\tan\phi}{AW} \quad (6)$$

In Figures 7b and 7c the total water pressure on the surface of sliding $ac$ is indicated by shaded areas. Since the total water pressure on $ac$ in Figure 7b (slow drawdown) is very much smaller than that on $ac$ in Figure 7c (rapid drawdown), $h_1$ is very much smaller than $h_2$, and, as a consequence, $G'_s$, equation (6), is smaller than $G_s$, equation (5). Hence, even if a slope has survived a great number of slow drawdowns it may fail after a rapid drawdown, because $G'_s$ is smaller than $G_s$.

Landslides caused by rapid drawdown are very common, and many records of such slides have been published. (See, for instance, Pollack, 1912). The sediments which are most seriously affected by a rapid drawdown are those intermediate between sand and clay.

As long as the piezometric surface in the ground beneath the slope has a gradient, the water percolates through the ground toward the surfaces adjoining the foot $a$ of the slope, as indicated in Figure 7c by dashed arrows. On account of its viscosity the percolating water exerts on the soil particles a pressure known as *seepage pressure*.

This pressure acts in the direction of the flow, and its intensity increases in simple proportion to the seepage velocity (Terzaghi and Peck, 1948, p. 54). At the foot of the slope the seepage velocity and the corresponding seepage pressure are much greater than higher up, and the seepage pressure tends to move the soil particles along the flow lines which are directed toward the foot of the slope. As a consequence, at the foot of the slope, the point of failure is reached much earlier than at higher elevations, and once the lower part of the slope has failed, the upper part follows because it has lost its support. The mechanics of this process and the means of preventing its detrimental effects have been investigated by Reinius (1948).

## SPONTANEOUS LIQUEFACTION

The arrangement of the grains of fine sand or coarse silt can be so unstable that a slight disturbance of the equilibrium of the grains may cause a rearrangement of the grains, whereby the grains settle into more stable positions, and the porosity of the sediment decreases.

If this process takes place above the water table it has no noticeable effect other than a settlement of the ground surface. By contrast, if it occurs below the water table its consequences can be catastrophic, because the viscosity of the water, which occupies the voids of the sand, prevents a rapid decrease of the porosity. During the time between the collapse of the structure and the reconsolidation under the new conditions of equilibrium, the sediment has the properties of a thick viscous liquid which spreads laterally, until its surface becomes almost horizontal. The transformation into the liquid state is known as *spontaneous liquefaction*. (See, for instance, Terzaghi and Peck, 1948, p. 100–105.)

The liquefaction of an unstable sediment can be caused by vibrations such as those produced by pile driving or quarry blasts. It can also be produced by the rapid rise or fall of the water table. The best-known slides, due to spontaneous liquefaction, are those which occur from time to time at the coast of the province of Zeeland (Holland), on sand slopes whose rise may be as gentle as 1:4. Between 1881 and 1946 no less than 229 slides have been reported. The quantity of sand which moves out during a slide ranges between about 80 cubic yards and 3 million cubic yards.

The slides are probably preceded by erosion caused by shore currents, associated with an increase of the average slope angle. The slides commonly occur after exceptionally high tides, particularly if the tide coincides with a heavy gale. The moving sand spreads out like a fan on the bottom of the sea, and after the slide the slope angle of the surface of the slide material may be as small as 3° to 4° (Koppejan et al., 1948).

## SEEPAGE FROM ARTIFICIAL SOURCES OF WATER

Seepage from artificial sources of water, such as storage reservoirs or unlined canals, may compromise the stability of existing slopes in at least four ways, depending on the character of the slope-forming material and on the conditions of stratification. It may reduce the shearing resistance of the ground by increasing the item $hw$ in equations (3) and (4); it may eliminate apparent cohesion produced by the surface tension in drained soils; it may eliminate real cohesion by removing cementing materials in solution; it may also cause a slope failure by retrogressive underground erosion by water veins emerging at the foot of the slope. The term "artificial source of water" implies that the source is of recent origin. Otherwise the slope failure caused by the action of the seepage derived from the source would have occurred long ago.

Figure 3 illustrates a slope failure caused by an increase of the pore-water pressure, bw. The mechanics of this slide are explained in the figure caption. The seepage water came from a freshly deposited hydraulic fill located between the dikes $A$ and $B$.

Moist, fine, silty sand can form permanent vertical slopes with a height of several tens of feet. This can be seen in any sand pit. The cohesion required for maintaining the equilibrium of such slopes is due to the friction produced by the surface tension of the contact moisture (water particles, surrounding the points of contact of the grains; see Terzaghi and Peck, 1948, p. 127). Hence the stability of such slopes requires the existence, within the slope-forming material, of a large area of contact between air and soil moisture.

Experience shows that the water, which seeps toward steep slopes during rainstorms, does not displace enough air to destroy the apparent cohesion of sand or silt. However, if water percolates through the ground toward the slope in large quantities and without any intermissions, the air is almost completely expelled, the apparent cohesion is eliminated, and the slope fails. A similar failure would occur if a steep slope on fine sand or silty sand is submerged for the first time in its history, for instance by the creation of a storage reservoir.

Slope failures due to the removal of a binder by solution are a common phenomenon in loess regions. Loess owes its cohesion to a soluble binder which consists chiefly of calcium carbonate. Since typical loess contains numerous vertical root holes, it commonly forms vertical cliffs, and the cliffs remain stable for years or decades, provided the water table is permanently located below the level of the base of the cliffs. Furthermore many artificial caves in loess are known to have existed for centuries without their unsupported roofs showing any signs of deterioration. These facts lead to the conclusion that the water, which percolates through the voids of loess during rainstorms, does not perceptibly weaken the bond between the loess particles. However, if loess is submerged or if a permanent flow of seepage through loess is established, the bond between the loess particles perishes within a few weeks or months and the loess assumes the character of supersaturated rock flour which flows like molasses. After a loess flow has come to rest, the excess water gradually drains out, and the final product of the process of drainage has the density and the properties of a very fine, loose sand.

The intensity of the effects of saturation on the physical properties of loess increases with increasing initial porosity, which ranges between about 40 and 60 per cent. The results of a comprehensive survey of the physical properties of loess and their engineering implications were presented by Scheidig (1934).

The effects of saturation on loess were impressively demonstrated by a large-scale test which was performed some 15 years ago in southeastern Turkestan in connection with an irrigation project. In order to investigate the reaction of loess to seepage through the bottom of a proposed irrigation canal, a bowl-shaped excavation 13 feet deep was made. The bottom of the excavation was rectangular, and it covered an area of 70 by 35 feet. The rise of the slopes was 1:1.5. The base of the loess stratum was located at a depth of about 80 feet below the bottom of the excavation. After the excavation was finished it was filled with water, and the water level was maintained at a height of about 10 feet above the bottom, by pumping.

A short time after the bottom was flooded it started to subside, and the slopes began to slough. The subsidence continued, first at an accelerated and later at a decreasing rate. At the end of 6 weeks the bottom of the excavation was located at a depth of about 2½ feet below its original position. The sloughing had spread about 20 feet beyond the original upper edge of the slope, and within the zone of sloughing the loess was so soft that it was not possible to walk on its surface. In its original state the loess, in the vicinity of the site of the test, is so coherent that vertical cliffs, 60 feet high, remain permanently stable.

Seepage coming from a storage reservoir or an unlined canal may also decrease the stability of a slope by subsurface erosion, proceeding from the exit of water veins toward the source of water supply. As the length of the underground conduit increases, the quantity of water seeping into the conduit and the cross section of the conduit also increase. Finally the width of the conduit becomes so great that the roof collapses, whereupon the mass of sediment above the roof breaks up and a slide ensues.

Under natural conditions this process may produce, in the course of time, deep gullies with a considerable length. The *Bossorocas* in the State of Sao Paulo (Brazil) are an example. A spring undermines the foot of the slope forming the rear wall of the niche which surrounds the spring. The slope fails, the slide mass is removed by erosion, and the process starts again. Some of the resulting gullies have a depth of about 160 feet. The upper part of the formation exposed on the walls of the gullies, with a thickness of about 50 feet, consists of a loesslike sediment and the lower part of a fairly coarse sand with an argillaceous binder. The mechanics of the formation of the Bossorocas has been investigated by E. Pichler (results not yet published). Figure 1 of Plate 3 shows the rear wall of one of the gullies. The water which performs the subsurface erosion leaves the ground through a cave at the foot of the rearwall (not visible in the photograph).

Subsurface erosion is also the most common cause of the dreaded failure of storage dams by piping. The erosion works back from springs emerging close to the foot of the downstream slope of the dam (Terzaghi and Peck, 1948, p. 507–510). If part of the enclosure of a storage reservoir consists of a natural ridge composed of unconsolidated sediments or decomposed rock, subsurface erosion may cause a failure of the downstream slope of the ridge, followed by a discharge of the contents of the reservoir through the gap. The failure of the Cedar Reservoir, Washington, in December 1918, is a typical example. A quantity of morainal material, estimated between 800,000 and 2,000,000 cubic yards, moved out of the slope. The slide initiated a flood which wrecked the tracks of the Milwaukee Railroad and destroyed the town of Edgewick and several industrial plants (Mackin, 1941).

* * * *

TABLE 1.—*Processes leading to landslides*

| A | B | C | D | E | F |
|---|---|---|---|---|---|
| Name of agent | Event or process which brings agent into action | Mode of action of agent | Slope materials most sensitive to action | Physical nature of significant actions of agent | Effects on equilibrium conditions of slope |
| Transporting agent | Construction operations or erosion | 1. Increase of height or rise of slope | Every material | Changes state of stress in slope-forming material | Increases shearing stresses |
| | | | Stiff, fissured clay, shale | Changes state of stress and causes opening of joints | Increases shearing stresses and initiates process 8 |
| Tectonic stresses | Tectonic movements | 2. Large-scale deformations of earth crust | Every material | Increases slope angle | Increase of shearing stresses |
| Tectonic stresses or explosives | Earthquakes or blasting | 3. High-frequency vibrations | Every material | Produces transitory change of stress | |
| | | | Loess, slightly cemented sand, and gravel | Damages intergranular bonds | Decrease of cohesion and increase of shearing stresses |
| | | | Medium or fine loose sand in saturated state | Initiates rearrangement of grains | Spontaneous liquefaction |
| Weight of slope-forming material | Process which created the slope | 4. Creep on slope | Stiff, fissured clay, shale, remnants of old slides | Opens up closed joints, produces new ones | Reduces cohesion, accelerates process 8 |
| | | 5. Creep in weak stratum below foot of slope | Rigid materials resting on plastic ones | | |

## REVIEW OF LANDSLIDE-PRODUCING PROCESSES

Table 1 contains a review of the processes which may cause landslides. It is intended as an aid to memory and as a guide in landslide investigation.

In connection with landslide investigations the following fact should be remembered. Most slope failures take place during periods of exceptionally heavy rainfall or in spring, when the snow melts. However, exposure to rain or melting snow belongs to the normal existence of a slope. Hence, if a slope is old, heavy rainstorms or rapidly melting snow can hardly be the sole cause of a slope failure, because it is most unlikely that they are without any precedent in the history of the slope. They can only be considered contributing factors.

The circumstances attending the Hudson slide of 1915 (Fig. 9) are an example. There can be no doubt that the slope failed on account of exceptionally high pore-water pressures in the proximity of the boundary between clay and gravel. Yet the slope was very old. Therefore the slide must have been preceded by an unprecedented change in the conditions of the existence of the slope. One of them consisted in the accumulation of stockpiles of crushed rock, with a total weight of about 25,000 tons, along the upper edge of the slope (Newland, 1916, p. 104). Furthermore, it is conceivable that the deforestation of the outcrops of the gravel or of an adjacent aquifer has produced an unprecedented increase of the highest elevation of the water table, associated with unprecedented pore-water pressures at the base of the varved clay. Either one of the two changes may account for the catastrophe. Therefore the slide can be ascribed to action 1 (column D, Table 1), action 14, or to a combination of both. In connection with the rock slide of Goldau (Fig. 4), it was also impossible to decide whether the slide was caused by tectonic movements (action 2), by progressive weathering (action 8), or by a combination of both.

In order to facilitate discrimination between causes and contributing factors, Figure 10 has been prepared. It shows the factor of safety $G_s$ with respect to sliding as a function of time. Each curve represents an individual slope. Curve A, (Fig. 10a) refers to the failure of an old slope due to progressive weathering of the slope-forming material or to tectonic movements. Curve B illustrates the failure of an old slope

| Water | Rains or melting snow | 6. Displacement of air in voids | Moist sand | Increases pore-water pressure | Decrease of frictional resistance |
|---|---|---|---|---|---|
| | | 7. Displacement of air in open joints | Jointed rock, shale | | |
| | | 8. Reduction of capillary pressure associated with swelling | Stiff, fissured clay and some shales | Causes swelling | Decrease of cohesion |
| | | 9. Chemical weathering | Rock of any kind | Weakens intergranular bonds (chemical weathering) | |
| | Frost | 10. Expansion of water due to freezing | Jointed rock | Widens existing joints, produces new ones | |
| | | 11. Formation and subsequent melting of ice layers | Silt and silty sand | Increases water content of soil in frozen top-layer | Decrease of frictional resistance |
| | Dry spell | 12. Shrinkage | Clay | Produces shrinkage cracks | Decrease of cohesion |
| | Rapid drawdown | 13. Produces seepage toward foot of slope | Fine sand, silt, previously drained | Produces excess pore-water pressure | Decrease of frictional resistance |
| | Rapid change of elevation of water table | 14. Initiates rearrangement of grains | Medium or fine loose sand in saturated state | Spontaneous increase of pore-water pressure | Spontaneous liquefaction |
| | Rise of water table in distant aquifer | 15. Causes a rise of piezometric surface in slope-forming material | Silt or sand layers between or below clay layers | Increases pore-water pressure | Decrease of frictional resistance |

TABLE 1.—*Continued*

| A | B | C | D | E | F |
|---|---|---|---|---|---|
| Name of agent | Event or process which brings agent into action | Mode of action of agent | Slope materials most sensitive to action | Physical nature of significant actions of agent | Effects on equilibrium conditions of slope |
| Water—(*Cont.*) | Seepage from artificial source of water (reservoir or canal) | 16. Seepage toward slope | Saturated silt | Increases pore-water pressure | Decrease of frictional resistance |
| | | 17. Displaces air in the voids | Moist, fine sand | Eliminates surface tension | Decrease of cohesion |
| | | 18. Removes soluble binder | Loess | Destroys intergranular bond | |
| | | 19. Subsurface erosion | Fine sand or silt | Undermines the slope | Increase of shearing stress |

which occurred during a heavy rainstorm several years after the foot of the slope was undercut. The curves C to G in Figure 10b refer to slopes of recent origin, such as the sides of open cuts, and to old slopes which were exposed to unprecedented conditions (curves D and G). Slope C failed on account of spontaneous liquefaction

(a) Old slopes

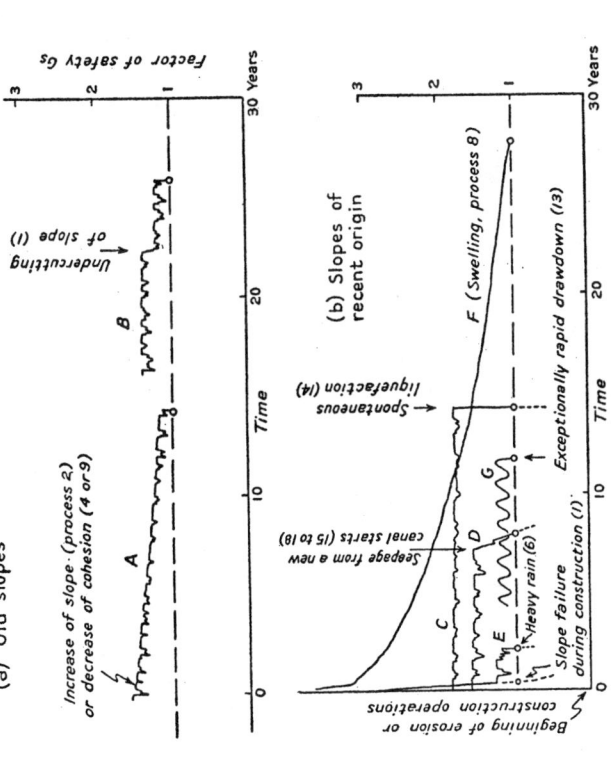

FIGURE 10.—*Diagram illustrating the variations of the factor of safety of different slopes prior to a landslide*

The ups and downs of the lines A to E are due to the changes in the pore-water pressure in the slope-forming material, associated with the alternation of dry and wet spells. Those of line G represent the effects of filling and emptying of the reservoirs adjoining the slope. The numerals in parentheses indicate the slide-producing processes listed in column C of Table 1.

caused by a near-by blast; D on account of seepage through the bottom of a recently constructed unlined canal located beyond the upper edge of the slope; E as a result of the heaviest rainfall since the time when the slope was formed; F on account of the gradual softening of stiff, fissured clay on which the slope was located, and G on account of an exceptionally rapid drawdown produced by a partial failure of the storage dam or the failure of an outlet valve.

The ups and downs of the curves A to E reflect the sequence of dry and wet spells, and those of curve G the normal variations of the free water level adjoining the slope of a storage reservoir. Special attention is called to curve B in Figure 10a, which represents a slope failure caused by a slight undercutting of the foot of the slope. The effects of this operation were not important enough to cause immediate failure. Yet, without it, the heavy rains that fell on the slope several years later would not have produced a slide. Hence the undercutting was the real cause of the failure, but the rainstorm was equally essential. This is one of many instances in which a slope failure can be accounted for only by a combination of two of the processes listed in column D of Table 1.

The degree of stability of an existing or a proposed slope cannot be reliably estimated, unless the process or processes, which may conceivably lead to a failure, are clearly understood and quantitative information regarding the controlling factors is available. The means for securing this information will be discussed in the last part of this paper.

## LANDSLIDE PROBLEMS
### PREVENTIVE MEASURES

The practical importance of a thorough investigation of the degree of stability of existing and of proposed slopes is illustrated by the following statistics published by Ladd (1934):

"Within the last three years landslides have resulted in more than 3,000 deaths and very great material losses. Since the spring of 1931 landslides have led to at least thirteen railroad disasters, four of which were in foreign countries and nine in the United States. By these, 227 people were killed and 31 were injured."

The foremost requirements for slide prevention are reliable information on the geologic structure of the ground adjoining the slope under consideration, to be obtained by test borings combined with a geologic survey and a clear conception of the processes which may conceivably lead to a failure of the slope. The processes that produce slides are listed in Table 1 (Nos. 1 to 19).

The first step toward slide prevention is to take all the measures required to make the slide-producing processes as ineffective as conditions permit. The rise of the piezometric surface behind the slope, associated with a displacement of air during heavy rainstorms (process 7), can be reduced by covering the slope and a broad strip of the area beyond the crest of the slope with a layer or lining having a low permeability. The formation of ice layers (process 11) and subsequent sloughing, known as solifluction, can be counteracted by drainage and various other means. (See, for instance, Terzaghi and Peck, 1948, p. 131–134.) The formation of deep shrinkage cracks (process 12) can be avoided by covering the slope with sod or a thick layer of sand. The danger of spontaneous liquefaction (processes 3 and 14) can be eliminated by compaction. The available technical means for accomplishing it was described by Terzaghi and Peck (1948, p. 379–381). The risk of slope failures, due to the concentration of flow lines at the foot of a slope after rapid drawdown (process 13), can be avoided by covering the lower part of the slope with an inverted filter of a weight sufficient to counteract the seepage pressure exerted by the percolating water (Reinius, 1948). Seepage from an artificial source of water, such as a canal or a storage reservoir (process 16–18), can be prevented by a water-tight lining of the bottom of the body of open water, a cut off, deep drains, or a combination of these. The danger of subsurface erosion (process 19) can be removed by covering the exit area of the water veins with an inverted filter or by adequate drainage (Terzaghi and Peck, 1948, p. 502–514). If the soil located beneath the slope contains layers or pockets of relatively permeable material such as sand or silt which may conceivably communicate with a distant aquifer (process 15), drainage is imperative. If the process which led to the Hudson slide of 1915 (Fig. 14) had been recognized, the catastrophe could have been avoided at a moderate expense, by "bleeding" the gravel beneath the varved clay.

By means of these and similar provisions it is commonly possible to prevent a decrease of the degree of stability of the slope. There are, as a matter of course, exceptions to this rule. Slides on slopes on stiff, fissured clay, due to the gradual softening of the slope-forming material (process 8), are an example. However, the effect of time on the stability of such slopes may be evaluated on the basis of experience. Figure 6 is a graphic representation of what experience can teach a competent observer.

The next step is to estimate the degree of stability of the existing or proposed slope under the conditions prevailing at the time of the investigation. If a slope is located on clay, which does not contain layers or pockets of relatively permeable materials, a fairly accurate computation of the factor of safety of the slope, with respect to sliding, can be made on the basis of the results of laboratory tests on undisturbed samples. This investigation belongs in the realm of soil mechanics.

The stability of slopes on clay containing layers of soil or sand and that of slopes on relatively permeable materials, such as silt or coarser sediments, depends not only on the physical properties of the slope-forming material but also to a large extent on the pore-water pressure in the most permeable members of the geologic system. In such instances it is necessary to secure reliable information concerning the pore-water pressure.

Standard observation wells can be relied upon only if their lower ends are located in relatively permeable material, not finer than fine sand. Otherwise it is necessary to use pore-pressure gages (Terzaghi, 1943a). The relative merits of the existing types of gages are being investigated by the Subcommittee on the Measurement of Pore Water Pressures in Earth Dams and their Foundations (chairman Prof. R. B. Peck, Univ. of Illinois), Committee on Earth Dams, Soil Mechanics & Foundations Division, A.S.C.E.

\* \* \* \* \*

## LANDSLIDE CORRECTION

If a slope has started to move, the means for stopping the movement must be adapted to the processes which started the slide (Table 1, column C). It is hardly an exaggeration to say that most slides are due to an abnormal increase of the pore-water pressure in the slope-forming material or in a part of its base (processes 6, 7, 13, 15, 16). In such instances radical drainage is indicated.

The extraordinary efficacy of drainage has recently been demonstrated by the following observation. During a tropical cloudburst, involving a precipitation of 9 inches in 24 hours, a slide occurred on a slope rising at an average angle of 30°.

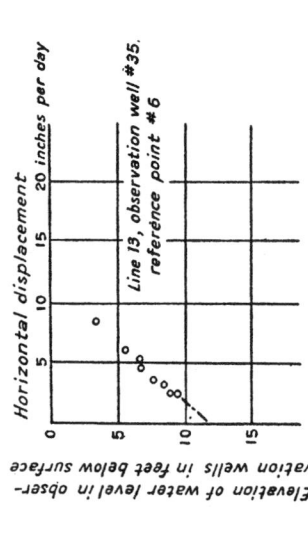

FIGURE 15.—*Diagram showing the relation between the position of the water table with reference to a slope after failure (ordinates) and the horizontal component of the corresponding downhill movement of the surface of the slope*

The slope is located on deeply weathered metamorphic rocks, and the deepest part of the surface of sliding was about 130 feet below the surface. The slide area was about 500 feet wide, 1000 feet long, and the quantity of material involved in the slide exceeded half a million cubic yards.

Since the slide occurred in the close proximity of a hydro-electric power station, immediate action was indicated. In order to get quantitative information concerning the ground movements and the factors which determine the rate of movement, reference points were established on several horizontal lines across the slide area, and observation wells were drilled in the proximity of the reference points. By plotting the vertical distance between slope and the water level in the wells as ordinates, and the corresponding rate of movement of the adjoining reference points as abscissas, diagrams like Figure 15 were obtained. Although the moving mass had a depth up to 130 feet, the diagrams showed that the lowering of the water table by not more than about 15 feet would suffice to stop the movement.

Drainage was accomplished by means of toe trenches, drainage galleries, and horizontal drill holes extending from the headings into water-bearing zones of the jointed rock. The movements ceased while the drainage was still in an initial state. The following rainy season brought record rainfalls; yet the ground movements in the slide area remained imperceptible.

Drainage can also be used to advantage if the water seeps through open joints between chunks of relatively impermeable material such as shale (Forbes, 1947).

If the permeability of the slide material is too low to permit drainage by pumping from wells or bleeding through galleries, the resistance against sliding can be increased and the ground movements can be stopped by means of the vacuum method or the electro-osmotic method (Terzaghi and Peck, 1948, p. 337–340). Both procedures create a reduction of the pore-water pressure associated with a permanent decrease of the water content of the drained strata which, in turn, increases the cohesion and shearing resistance of the ground. Similar effects can be obtained by circulating hot, dry air through galleries located within the unstable material (Hill, 1934).

If drainage is difficult or if its success is doubtful, the ground movements can be stopped either by reducing the slope angle or by constructing artificial barriers, such as heavy retaining walls or rows of piles across the path of the moving material. A list of the current procedures and comments on their efficacy have been published by Ladd (1935). (*See also* Seaton, 1938).

Every landslide or slope failure is the large-scale experiment which enables competent investigators to draw reliable conclusions regarding the shearing resistance of the materials involved in the slide. Once a slide has occurred on a construction job, the data derived from the failure may permit reliable computation of the factor of safety of proposed slopes on the same job and modification of the design in accordance with the findings. This procedure was successfully used during the construction of several of the German superhighways on treacherous ground (Gottstein, 1936).

### CO-OPERATION BETWEEN GEOLOGIST AND ENGINEER ON LANDSLIDE PROBLEMS

If a geologist is called upon to report on the degree of stability of an existing or a proposed slope, he is likely to furnish an adequate account of the geology of the site and of the hydrologic conditions. However, his understanding of the physical processes, which may impair the stability of the slope, is commonly deficient because he has not been trained to think in terms of exact physical concepts. This is demonstrated by the indiscriminate use of the term "lubrication" and other misnomers. Very few geologists have a clear conception of the difference between total and effective pressure, of the effect of the pore-water pressure and of surface tension on the shearing resistance of sediments, and of the relation between stress, strain, and time for cohesive soils. Yet, an opinion concerning the means for increasing the stability of a slope is merely guesswork unless it is based on a knowledge of fundamental physical relationships, and the guess may be wrong.

A civil engineer, trained in soil mechanics, may have a better grasp of the physical processes leading to slides. However, he may have a very inadequate conception of the geologic structure of the ground beneath the slopes, and he may not even suspect that the stability of the slope may depend on the hydrologic conditions in a region at a distance of more than a mile from the slope.

On account of the wide range of specialized knowledge and experience required for judging the stability of slopes, important landslide problems call for co-operation between geologist and engineer. To get satisfactory results the geologist should be familiar with the fundamental principles of soil mechanics, and the engineer should know at least the elements of physical geology.

### REFERENCES

Close, U., and McCormick, E. (1922) *Where the mountains walked*, Nat. Geog. Mag., vol. 41, p. 445–464.

Eng. News-Rec. (1932) *Three dams on San Andreas fault have resisted earthquakes*, vol. 109, p. 218–219.

Eng. Rec. (1909), *Landslides*, vol. 59, p. 737–740.

Forbes, H. (1947) *Landslide investigations and correction*, Am. Soc. Civil Eng., Tr., vol. 112, Paper 2303, p. 377–442.

Freeman, John R. (1932), *Earthquake damage and earthquake insurance*, New York, McGraw-Hill Book Co., Inc.

Fuller, M. L. (1912) *The New Madrid earthquake*, U. S. Geol. Survey, Bull. 494, 119 pages.

Gottstein, E. v. (1936) *Two examples concerning underground sliding caused by construction of embankments*, 1st Inter. Conf. Soil Mech., Pr. Cambridge, Mass., vol. III, p. 122–128.

Griggs, David T. (1936) *Deformation of rocks under high confining pressures*, Jour. Geol., vol. 44, p. 541–577.

Haefeli, R. (1944) *Zur Erd- und Kriechdruck Theorie*, Schweiz. Bauztg., vol. 124.

Hardy, W. B. and T. V. (1919) *Note on static friction and on the lubricating properties of certain chemical substances*, Philos. Mag. London, vol. 39, no. 223, p. 32–35.

Heim, A. (1882) *Über Bergstürze*, Naturf. Gesell. Zürich, Neujahrsblatt 84.

Hill, R. A. (1934) *Clay stratum dried out to prevent landslips*, Civil Eng., vol. 4, p. 403–407.

Hollingworth, S. E., Taylor, J. H., and Kellaway, G. A. (1944) *Large-scale superficial structures in the Northampton Ironstone Field*, Geol. Soc. London, Quart. Jour., vol. C, p. 1–44.

Jarnvagrastyrelsen (1922), *Statens Jarnvagars Geotekniska Commission, 1914–1922*, Final Rep. to Royal Bd. State Railroads, May 31.

Koppejan, A. W., van Wamelen, B. M., and Weinberg, L. J. H. (1948) *Coastal flow slides in the Dutch Province of Zeeland*, 2d Inter. Conf. Soil Mech. Found. Eng., Pr., Rotterdam, Holland, vol. V, p. 89–96.

Ladd, G. E. (1934), *Bank slide in deep cut caused by draught*, Eng. News-Record, vol. 112, p. 324–326.

——— (1935), *Landslides, subsidences and rockfalls*, Am. RR. Eng. Assoc., Pr., vol. 36, p. 1091–1162.

Lapworth, H. (1911) *The geology of dam trenches*, Inst. Water Eng., Tr., vol. 16, p. 25.

Lehmann, O. (1926) *Die Verheerungen in der Sandlingsgruppe*, Denkschrift Ak. Wiss. Wien, Math. natw. Klasse, vol. 100, p. 263–299.

Lugeon, M., and Oulianoff, N. (1922) *Sur le balancement des couches*, Univ. de Lausanne, Lab. Géol. Géog. Bull., no. 32.

Mackin, J. H. (1941) *A geologic interpretation of the failure of the Cedar Reservoir, Washington*, Eng. Exper. Sta., Bull. 107. Univ. Washington.

Marmer, H. A. (1930) *Chart datums*, U. S. Coast Geod. Survey, Spec. Pub. 170.

McConnell, R. G., and Brock, R. W. (1904) *Report on the great landslide at Frank, Alberta, Canada* Dept. Interior, Ann. Rept. 1902–1908, pt. 8, 17 pages.

Mohr, C., Haefeli, R., Meisser, L., Waltz, F. and Schaad, W. (1947) *Umbau der Landquaribrücke der Rhätischen Bahn in Klosters*, Schweiz. Bauztg., vol. 65, p. 5–8, 20–24, 32–37.

Newland, D. H. (1916) *Landslides in unconsolidated sediments*, N. Y. State Mus., Bull. 187, p. 79–105.

Pollack, V. (1913) *Über Seeuferbewegungen*, Österr. Wochenschr. für öffentl. Baudienst, vol. 19, Heft 35, p. 595–603.

Raschka, H. (1912) *Die Rutschungen im Abschnitt Ziersdorf-Eggenburg der Kaiser-Franz-Josefs-Bahn*, Zitschr. Österr. Ing. Arch. Ver., p. 561.

Reinius, E. (1948) *On the stability of the upstream slope of earth dams*, Doctor's Thesis, Kungl. Tekniska Högskolan, Stockholm, Sweden.

Scheidig, A. (1934) *Der Löss und seine geotechnischen Eigenschaften*, Theodor Steinkopf, Dresden and Leipzig, 233 pages.

Seaton, T. H. (1938) *Engineering problems associated with clay, with special reference to clay slips*, Inst. Civil Eng. Jour. (London), Paper 5170, p. 457–498.

Sharpe, C. F. S. (1938) *Landslides and related phenomena*, Columbia Univ. Press, New York, 136

Sharpe, C. F. S., and Dosch, E. F. (1942) *Relation of soil-creep to earth-flow in the Appalachian Plateaus*, Jour. Geomorph., vol. 5, p. 312-324.

Skempton, A. W. (1948) *The rate of softening of stiff, fissured clays*, 2d Inter. Conf. Soil Mech. Found. Eng. Pr., Rotterdam, vol. II, p. 50-53.

Terzaghi, K. (1925) *Erdbaumechanik*, Franz Deuticke, Wien, 399 pages.

——— (1936) *Stability of slopes on natural clay*, 1st Inter. Conf. Soil Mech. Found. Eng. Pr., Cambridge, Mass., vol. I, p. 161-165.

——— (1942) *Soil moisture and capillary phenomena in soils*, Physics of the Earth, vol. IX (Hydrology), McGraw-Hill Co., New York, p. 331-363.

——— (1943a) *Measurements of pore-water pressure in silt and clay*, Civil Eng., vol. 13, p. 33-36.

——— (1943b) *Theoretical soil mechanics*, John Wiley and Sons, New York, 510 pages.

——— and Peck, R. B. (1948) *Soil mechanics in engineering practice*, John Wiley and Sons, New York, 566 pages.

Toms, A. H. (1946) *Folkestone Warren landslips: Research carried out in 1939 by the Southern Railway Company*, Inst. Civ. Eng., RR. Eng. Div., Railway Paper no. 19.

### Commentary on Karl Terzaghi's 1950 "Mechanism of Landslides"

**Roy J. Shlemon**
*Roy J. Shlemon & Associates, Inc., P.O. Box 3066, Newport Beach, California 92659-0620*

## INTRODUCTION

Engineering geology in North America has come a long way since the trail-blazing work of William Otis Crosby 100 years ago (Kiersch, 1991a), about the same time as the founding of the Cordilleran Section of the Geological Society of America (GSA). Now regarded as a distinct scientific discipline, engineering geology has been variously defined, but is generally regarded as the application of geosciences to the solution of engineering problems (Kiersch, 1991b; Kiersch and Hatheway, 1991).

Modern engineering geology encompasses all manner of applied scientific research, ranging from geological characterization for siting large dams, bridges, canals, and waste repositories to hazard assessments of subsidence, faults, and mass movements. The Cordilleran Section of the GSA extends across a wide variety of terrain that provides abundant slope stability problems and related opportunities for analysis and recommendations for mitigation. Every winter brings new bedrock and surficial failures; highways are disrupted, homes are destroyed, and increasingly litigation abounds. In fact, the direct and indirect cost of landslide damage in the United States is about $1.5 to $2.0 billion annually (Schuster and Fleming, 1986; Slosson and Cronin, 1999). Indeed, it is the now all-too-common urban landslide that provides the engineering geologist with much work, as well as with much potential liability (Fig. 1).

Engineers and geologists have long recognized and, in most cases, successfully analyzed and mitigated landslide movement (see summaries in Kiersch, 1991b; Fleming and Varnes, 1991; Turner and Schuster, 1996; Slosson et al., 1992). However, it was the GSA "Berkey Volume" of 1950 that brought the attention of the geological community to the evolving "practice of soil mechanics" as set forth by Karl Terzaghi in his now-classic "Mechanism of Landslides," a paper short in title, and long in technical concepts and analytical philosophy.

The 1950 GSA "Berkey Volume" (Paige, 1950), named in honor of the distinguished American engineering geologist, Charles Peter Berkey, president of the GSA in 1941 and first chairman of the Engineering Geology Division (Kiersch and Hatheway, 1991), contains 12 papers written by a "who's who" of American geologists. These topics are subsumed within the field of engineering geology, and many are specifically pertinent to the Cordilleran Section. In addition to the Terzaghi paper, the Berkey volume contains such classics as "Geology of Dam Construction" by E. B. Burwell, Jr., and B. C. Moneymaker; "Engineering Geology and Dam Construction" by J. F. Sanborn; "Faults and Engineering Geology" by G. D. Louderback; "Geology and Engineering in the Production and Control of Ground Water" by O. E. Meinzer; "Geological Aspects of Beach Engineering" by W. C. Krumbein; and "Military Geology" by C. B. Hunt. A classic book in its own right, the Berkey volume reached its sixth printing by 1967.

## KARL TERZAGHI

Karl Terzaghi was born in Prague on October 2, 1883, and he died 80 years later in Winchester, Massachusetts, on October 25, 1963. Engineers regard him as "The Father of Soil Mechanics." More familiarly, among engineering geologists, he is known as "The Great Man." In fact, he has attained almost cult status in both communities. He received the Norman Medal of the American Society of Civil Engineers in 1930, 1943, 1946, and 1955. Terzaghi was a distinguished professor at the Technical University in Vienna and later at Harvard University. In addition, he was a much-honored adjunct professor at the University of Illinois and a world-acclaimed consultant. Terzaghi published 7 books, almost 300 papers, and received 9 honorary doctorates from United States and international universities (Lamb and Whitman, 1969; Richard E. Goodman, 1998, personal commun.). Aspects of his life are now routinely posted on the Internet, where it is reported that museums and institutes housing collections of his works can be found in Canada, Nor-

Shlemon, R. J., 1999, Commentary on Karl Terzaghi's 1950 "Mechanism of Landslides," *in* Moores, E. M., Sloan, D., and Stout, D. L., eds., Classic Cordilleran Concepts: A View from California: Boulder, Colorado, Geological Society of America Special Paper 338.

Figure 1. Typical urban landslide in southern California: Via Estoril failure in Laguna Niguel took place 19 March 1998; photograph taken on 20 March 1998. The two bordering houses collapsed a few days later, and litigation abounds. (Photograph courtesy of Geo-Tech Imagery International, Oceanside, CA).

way, and Turkey. Tributes to Karl Terzaghi are many; they range from accolades for his method of teaching and scientific problem solving (Bjerrum, 1960) to plaudits from Ralph Peck (1997), the esteemed coauthor of the first and second editions of Terzaghi's impressive textbook *Soil Mechanics in Engineering Practice* (Terzaghi and Peck, 1948, 1967). Well-deserved homage to Terzaghi continues, and a complete biography of his life and work has recently been written by Richard Goodman (1999).

Trained both as an engineer and geologist with an astute appreciation of field observations, Terzaghi's passion for geology is well exemplified by his personal characterization as "an engineering geologist." In fact, he has been described as being not only the founder of geotechnical engineering, but also one of the early teachers of engineering geology (Redlich et al., 1929; Einstein, 1991; Kiersch, 1991b). Terzaghi always emphasized that geological thinking was an important part of any engineering investigation, particularly with respect to understanding landslides. Alas, his calling to engineering colleagues to acquire basic training in geological principles and techniques, and thereby strengthen their engineering capability, has never been fully accepted, much to his consternation in later life (Terzaghi, 1963).

## THE "MECHANISM OF LANDSLIDES" PAPER

The classic 1950 "Mechanism of Landslides," portions of which are excerpted here, is elegant in its simplicity, in its synthesis of landslide-analysis principles, in its presentation of pertinent case studies, and in its philosophical expression about the respective roles that the engineer and geologist play as they ostensibly work toward a common goal. Moreover, owing to its publication by the GSA, the paper was made available to a great number of geologists, many of whom were not familiar with the soil mechanics concepts then being developed by Terzaghi, nor with the relatively small body of related engineering literature. The Terzaghi paper is 40 pages in length, the longest in the Berkey volume. Even so, a few years later, Terzaghi still found it necessary to acquaint the geologist with "engineering properties of sediments," which he defined as ". . . those properties of sediments that have a significant influence on the performance of sediments during and after the completion of construction operations" (Terzaghi, 1955, p. 558). However, these concepts and terminology were published in "Economic Geology," then a broad-based journal that served the four branches of applied geology from the 1930s to the mid-1950s. Unfortunately, this journal was not readily appreciated by most engineering geologists at the time (Kiersch, 1991b).

A particular "gem" in Terzaghi's 1950 paper is the subtle way in which he chided some geologists for attributing rainfall-induced landslides to the "lubricating effect of water" (p. 91). Pointing out the fallacy of such thinking, Terzaghi, presented examples from his earlier work (1925, 1943; Terzaghi and Peck, 1948) to show that water in contact with minerals such as quartz may well act as an anti-lubricant; that water displaces air in voids to reduce soil cohesion and hence slope stability; and, importantly, that water increases pore pressure and thus decreases soil shearing resistance. Fundamental and almost intuitive as they now seem to be, these impacts of soil water were often not appreciated by many geologists prior to Terzaghi's 1950 paper.

Terzaghi's 1950 paper also called attention to the impact of seismic shaking on slope stability (p. 89), on liquefaction (p. 100), and on subsidence (p. 101), subjects extremely relevant to contemporary engineering-geologic problems. He also emphasized the relationship of landslide periodicity to fluctuations in pore-water pressure and to related changes in slope factor-of-safety (p. 102), the basis for installing dewatering devices and hence mitigating many landslide movements.

The 1950 paper similarly brought attention to the detrimental impacts of rapid drawdown of reservoir water levels, of changes in piezometric surfaces, and of seepage pressure on slope stability (p. 97), and to the importance of effective stress, concepts previously elucidated in his soil mechanics textbooks (Terzaghi, 1925; Terzaghi and Peck, 1948; Skempton, 1960). But the "meat" of Terzaghi's 1950 paper is Table 1: "Processes Leading to Landslides" (p. 106-108). Here he provided an all-encompassing synthesis of landslide description, of variability of natural materials, and of events and agents that produce slope movement. The Table incorporates traditional geological description and perspective, as well as soil mechanics methods, as a framework to analyze equilibrium of slopes. But in his unassuming way, Terzaghi reminded us that Table 1". . . is intended [merely] as an aid to memory and as a guide in landslide investigation" (p. 105).

The shortest section, and perhaps most thought-provoking of Terzaghi's 1950 paper, is the last few paragraphs. Entitled "Co-operation Between Geologist and Engineer on Landslide Problems" (p. 121), Terzaghi set forth the philosophical basis for differences and commonality of engineering and geology, the backbone of the modern discipline of engineering geology. Here, on the one hand, Terzaghi admonished many geologists for not having a clear concept of the difference between total and effective stress, of the effect of pore-water pressure and of surface tension on sediment shear strength, and of the relationship of stress, strain and time on cohesive soils. On the other hand, Terzaghi was not loathe similarly to chasten the soils engineer, as they were then known, who often did (and still do) site-specific landslide analysis without the benefit of adequate regional geologic and hydrologic information. Indeed, in his many writings, Terzaghi emphasized that soils, as natural sediments, are highly variable, and engineering models, based on extrapolation from too-few sampling points, have the potential for disastrous results. He further pointed out that theory alone, without the application of mature judgment, may similarly lead to problems. Terzaghi succinctly summed up his belief about the respective role of the engineer and geologist on a project (1950, p. 121): "To get satisfactory results, the geologist should be familiar with the fundamental principles of soil mechanics, and the engineer should at least know the elements of physical geology."

## ENGINEERING GEOLOGY SINCE THE 1950 TERZAGHI PAPER

As in every scientific field of inquiry, engineering geology has changed in the 50 years since publication of the "Mechanism of Landslides" paper (Kiersch, 1991b). To be sure, analytical tools are more abundant: piezometers and slope indicators are now routinely emplaced in landslides; downhole logging to document the location and geometry of slip surfaces is standard procedure; and abundant samples are normally collected for routine geotechnical laboratory analysis. Further, there is increasing recognition about the complexity — if not multiplicity — of landslide slip surfaces, and the potential rejuvenation of ancient ruptures, geological factors that heretofore have not always been accounted for in traditional engineering modeling. Also more appreciated is the influence of time and of Pleistocene climatic change on slope stability. And, too, the impact of direct and indirect anthropic activities is brought home by the apparent increasing frequency of urban slope failures. Nevertheless, the basic principles of landslide analysis, the importance of earth material properties and their stratigraphic variability, and the various cause-and-effect relationships that give rise to landslides, still remain the same as in 1950 when Terzaghi's concepts were brought to the attention of many in the geological community. Accordingly, though now almost 50 years later, the 1950 "Mechanism of Landslides" paper is still required reading for engineering geologists, the hallmark of a truly classic technical and philosophical work-of-art.

## ACKNOWLEDGEMENTS

I thank Richard E. Goodman, Allen W. Hatheway, George A. Kiersch, Murray Levish, J. David Rogers, and James E. Slosson for providing their much-appreciated time, literature, photographs and insights about the impact of the Terzaghi paper on the engineering-geological profession. I am likewise grateful for their review of a draft manuscript and for their valued suggestions for improvement.

## REFERENCES CITED

Bjerrum, L., 1960, Some notes on Terzaghi's method of working, in Bjerrum, L., Casagrande, A., Peck, R. B., and Skempton, W. W., From theory to practice in soil mechanics, New York, John Wiley and Sons, p. 22-25.

Einstein, H. H., 1991, Observation, quantification, and judgment: Terzaghi and engineering geology: Journal of Geotechnical Engineering, American Society Civil Engineers, v. 117, no. 11, paper no. 26379, p. 1172-1178.

Fleming, R. W., and Varnes, J. D., 1991, Slope movements: in Kiersch, G. A., ed., The heritage of engineering geology; the first hundred years: Boulder, Colorado, Geological Society of America, Centennial Special Volume 3, p. 201-218.

Goodman, R. E., 1999, The engineer as artist — the life and work of Karl Terzaghi, 1883 to 1963: Reston Virginia, American Society of Civil Engineers Press (in press).

Kiersch, G. A., 1991a, The heritage of engineering geology; changes through time, in Kiersch, G. A., ed., The heritage of engineering geology; the first hundred years: Boulder, Colorado, Geological Society of America, Centennial Special Volume 3, p. 51-85.

Kiersch, G. A., 1991b, Modern practice, training and academic endeavors, 1940s to 1980s, in Kiersch, G. A., ed., The heritage of engineering geology; the first hundred years: Boulder, Colorado, Geological Society of America, Centennial Special Volume 3, p. 51-85.

Kiersch, G. A, and Hatheway, A. W., 1991, History and heritage of Engineering Geology Division, Geological Society of America, 1940s to 1990, in

Kiersch, G. A., ed., The heritage of engineering geology; the first hundred years: Boulder, Colorado, Geological Society of America, Centennial Special Volume 3, p. 109-147.

Lamb, T. W., and Whitman, R. V., 1969, Soil mechanics: New York, John Wiley & Sons, 553 p.

Paige, S., ed., 1950, Application of geology to engineering practice, Berkey Volume: Engineering Geology, Geological Society of America, 327 p.

Peck, R. B., 1997, Gaining ground: Civil Engineering, v. 67, no. 12, p. 54-56.

Redlich, K., Terzaghi, K., and Kampe, R., 1929, Ingenieurgeologie: Vienna, Julius Springer, 708 p.

Schuster, R. L., and Fleming, R. W., 1986, Economic losses and fatalities due to landslides: Association of Engineering Geologists Bulletin, v. 23, no. 1, p. 11-28.

Skempton, A. W., 1960, Significance of Terzaghi's concept of effective stress, in Terzaghi, E., From theory to practice in soil mechanics: New York, John Wiley and Sons, p. 42-53.

Slosson, J. E., and Cronin, V. S., in press, Lessons relearned: Geological Society of America, Abstracts with Program, Annual Meeting, Toronto, Canada.

Slosson, J. E., Keene, A. G., and Johnson, J. A., 1992, Landslides/landslide mitigation: Boulder, Colorado, Geological Society of America, Reviews in Engineering Geology, v. IX, 120 p.

Terzaghi, K., 1925, Erdbaumechanik auf bodenphysikalischer grundlage: Vienna, Franz Deuticke, 399 p.

Terzaghi, K., 1943, Theoretical soil mechanics: New York, John Wiley and Sons, 510 p.

Terzaghi, K., 1950, Mechanism of landslides: in Paige, S. (Chairman), Application of geology to engineering practice, Berkey Volume: Geological Society of America, p. 83-123.

Terzaghi, K., 1955, Influence of geological factors on the engineering properties of sediments: in Bateman, A. M., ed., Economic Geology 50th Anniversary Volume: Lancaster, Pennsylvania, Economic Geology Publishing Company, Part 2, p. 557-618.

Terzaghi, K. 1963, Karl Terzaghi's last writing on soils: Engineering News Record, v. 171, no. 21, p. 1-2.

Terzaghi, K., and Peck, R. B., 1948, Soil mechanics in engineering practice: New York, John Wiley and Sons, 566 p.

Terzaghi, K., and Peck, R. B., 1967, Soil mechanics in engineering practice, 2nd edition, New York, John Wiley and Sons, 729 p.

Turner, A. K., and Schuster, R. L., eds., 1996, Landslides, investigations and mitigation: Natural Research Council, Transportation Research Board Special Report 247, Washington, D.C., National Academy Press, 673 p.

MANUSCRIPT ACCEPTED BY THE SOCIETY NOVEMBER 23, 1998

> # EPILOGUE

# *Epilogue*

# *Past, present, and future context of Cordilleran geology*

**Eldridge M. Moores, Doris Sloan, and Dorothy L. Stout**

*Those who can't remember the past are condemned to repeat it.*

G. Santayana

## HISTORIC CONTEXT

Throughout this volume we have emphasized the excitement of geologic discovery in the Cordillera and the conceptual revolutions that have swept geology over the past century. The romance and fascination of the Gold Rush were part of a tremendous outpouring of enthusiasm for exploration and discovery of the wide open and "unpeopled" great American West. This enthusiasm comes through in the writings of early geologists, explorers, and pioneers.

We tend to forget, however, that these heady developments in geology did not occur in a sociopolitical vacuum. During the late nineteenth century, the early development of the state of California occurred at a time when California society was marked by nearly complete discrimination against people of color. The discovery of gold coincided with the 1848 Treaty of Guadalupe Hidalgo, in which Mexico ceded claims to approximately half its territory, now the states of Arizona, California, Colorado, New Mexico, Nevada, and Texas, to the United States. The treaty's stipulation that prior property rights would be respected was never enforced, and Mexican landholders such as M. Vallejo and J. A. Sutter lost their holdings in the overwhelming inrush of people. In response to pressure from California lawmakers, the United States reneged on treaties signed in 1851-1852 that promised California Native Americans substantial land rights in return for peace (Magagnini, 1997). Until 1870, no person of color could give evidence in a court, which meant that they effectively had no rights (Chen, 1980). A tax on foreign miners was in 1853 the single largest source of state income (25-50%; Chen, 1980, p. 48). For most of the late nineteenth century, it was legal to kill Native Americans, and the California Legislature reimbursed volunteer militias for ammunition expended. Women had few rights in the late nineteenth century in California, despite having gained the vote in Wyoming in 1869 when it became a territory.

Although the outcome of the Civil War and the Fourteenth and Fifteenth Amendments to the Constitution granted broad rights to all citizens, the United States largely remained a massively segregated society until the civil rights movement in the 1960s (a fact reflected in personnel patterns in geology, as discussed below). Lindgren's (1895; this volume) paper and the founding of the Cordilleran Section correspond in time with the 1896 *Plessy vs. Ferguson* decision, enshrining segregation in the "separate but equal" doctrine, and with the Spanish American War, the foray of United States into nineteenth century imperialism. Lawson's (1908; this volume) paper appeared one year after the "Gentleman's Agreement" between the Theodore Roosevelt administration and the Japanese government effectively ended for years Japanese emigration to the United States; one year before the discovery of reversals in the Earth's magnetic field by the French geophysicist B. Bruhnes, and of the Moho discontinuity by the Croatian geophysicist A. Mohorovičič; and two years after publication of the German geologist G. Steinmann's recognition of what became known as "Steinmann's trinity,"–the ubiquitous association of serpentines, pillow lava, and radiolarian chert. Although his article principally concerned the Alpine orogen, he made some observations about the San Francisco Bay area. Steinmann stated in his article (1906, p. 30):

"Auch die in der Küstenkordillere *Californiens* in Gesellschaft ophiolithischer Eruptive weit verbreiteten 'Phthanite' waren von BECKER für Umwandlungsprodukte erklärt, ebenso wie die Ophiolithe selbst. Als ich im Jahre 1891 nach dem Kongress in *Washington* teils allein, teils unter Führung der Herrn LAWSON einige Exkursionen in der Umgebung von *San Francisco* machte, fiel es mir nicht schwer, die Radiolaritnature der "Phthanite" und ihre gesetzmassige Verknüpfung mit den Ophiolithen zu erkennen. Die späteren Untersuchungen von LAWSON, PALACHE, FAIRBANKS, RANSOME, und HINDE haben beides vollauf bestätigt."

Also the widely distributed 'phthanites' in the California Coastal Cordillera in association with ophiolitic eruptives were interpreted by BECKER as metamorphic products, as well as the ophiolites themselves. When in 1891 after the Congress in Washington I made some excursions in the vicinity (environment) of San Francisco, partly alone and partly under the guidance of Mr. LAWSON, I found it not difficult to recognize the radiolarite nature of the "phthanite" and its regular linkage with the ophiolites. The more recent investigations of LAWSON, PALACHE, FAIRBANKS, RANSOME, and HINDE have fully confirmed both (observations).

I suspect that no California geologist took note of Steinmann's comments. Some 60 years would pass before anyone else again recognized ophiolites in the California Coast Ranges (e.g., Bezore, 1969, Moiseyev, 1970).

During the first two decades of the twentieth century, western Americans were isolated from most of the rest of the United States, let alone Europe. The isolation of western geologists continued until the Second World War (Stout, this volume). Gilbert's (1914; this volume) paper appeared as World War I began, three years after the Austrian geologists Ampferer and Hammer (1911) invoked the process of "verschluchung" or subduction to explain the origin of the Alps, the same year as B. Gutenberg's discovery of the Earth's core, and one year before publication (in German) of the first edition of Wegener's book on continental drift (Wegener, 1915). The period between Gilbert's paper and the next one (Williams, 1929; this volume) saw the end of World War I, the Nineteenth Amendment giving women the right to vote (1920), and the Great Kanto Earthquake (1923), which ruined the Japanese economy and brought about military rule in that country. Wegener first published an account in English of continental drift in *Discovery* in 1922, and the English translation of the third edition of his book, *The Origin of Continents and Oceans* appeared in 1924 (cf. Vine, 1977; Schwarzbach, 1986).

The publication of Wegener's book in English gave western American geologists the opportunity to incorporate European ideas into their own work, but they paid little attention, except to reject them. Wegener's ideas, generally rejected at the 1926 meeting of the American Association of Petroleum Geologists, the proceedings of which were published in 1928 (van Waterschoot van der Gracht, 1928), one year before Williams' (1929; this volume) paper and the same year as Wadati's recognition of a dipping seismic zone beneath Japan.

Three "classic" articles appeared in the 1930s in the years leading up to World War II. This time also saw the first stirrings of organized research in marine geology and geophysics in the United States. California and neighboring states saw the pursuit of several regional studies, e.g. the first wave of major geologic explorations of the California and neighboring deserts (Wright and Troxel; this volume), progress on a major early synthesis of the geology of the Coast Ranges (Taliaferro, 1943) and of California in general (Reed, 1933). Blackwelder's (1931; this volume) paper appeared as the Dutch geophysicist F. A. Vening Meinesz began his gravity measurements at sea using a borrowed American submarine of World War I vintage, and coincided with the depths of the Great Depression, with major bank failures in Germany (leading in 1933 to Adolph Hitler's accession to power), and with the Japanese invasion of China. Kleinpell's (1938; this volume) paper appeared one year after the South African geologist, A. DuToit, kept continental drift alive with his book *Our Wandering Continents*, the same year as Hitler's invasion of Czechoslovakia and the Munich agreement.

Only one "classic" article appeared during the 1940s. Publication of Noble's (1940; this volume) work coincided with the Battle of Britain, one year after publication of Hess' (1939) first attempt at a global tectonic synthesis (given at the International Geological Congress in Moscow in 1937) entitled "Island arcs, gravity anomalies, and serpentine intrusions: A contribution to the ophiolite problem."

During 1941–45, western geologists were at war along with the rest of their fellow citizens. No classic papers appeared during the years of World War II. Many American geologists in the armed forces found themselves in parts of the world they otherwise would never have visited, which enriched their appreciation of global geologic features. Technological advances enabled the surge in exploration and knowledge of the oceans that came after the War, as scientists strived to apply all the new technology to scientific questions, and as government support for basic research was galvanized by the start of the Cold War.

Several papers appeared in the decade or so after World War II. Bramlette's (1946; this volume) paper coincided with publication of Hess' paper on guyots, the latter arguably the first to suggest a role for surficial processes on terrains preserved beneath the oceans. Churchill delivered his "Iron Curtain" speech the same year. Larsen's (1948; this volume) paper appeared as Truman was re-elected President, and as R. Raitt collected the first seismic data for a uniformly thin (5 km) oceanic crust (Menard, 1986). Terzaghi's (1950; this volume) paper appeared as the Korean War began and the National Science Foundation was established, and one year after publication of H. Benioff's (1949) global summary of dipping seismic zones (see also Benioff, 1955). Hill and Dibblee (1953; this volume) appeared as the Korean War ended, and as the U.S. Supreme Court handed down the *Brown vs. Board of Education* decision outlawing segregation in the schools, as M. Ewing and B. Heezen published evidence of the mid-Atlantic rift valley. The last classic paper of this volume to appear in the 1950's, the California Water Plan (1957; this volume), appeared the same year as the "Space Race" began with the successful launch by the U.S.S.R. of Sputnik.

The plate tectonics revolution resulted largely from the marine geological, geophysical and seismic research of the 1950s and early 1960s made possible by technological developments fueled largely by the Cold War. Young Cordilleran geoscientists practiced their science in an atmosphere of political upheaval and the challenging of authority. This carried over into science and gave the young scientists the courage to challenge their fixist elders. In the 1960s the gathering momentum of the sea floor spreading–plate tectonics revolution occurred at the height of the Cold War and the climax of the Civil Rights movement. Bateman and Wahrhaftig's (1966; this volume) paper appeared the same year as the GSA Annual Meeting in San Francisco, at which talks by F. J. Vine and A. Cox, R. Doell, and G. B. Dalrymple led to widespread acceptance of sea floor spreading by the geologic community. Publication of Ojakangas's (1968; this volume) paper coincided with civil unrest over the Vietnam war, the assassinations of Martin Luther King and Robert F. Kennedy, and the establishing papers of the plate tectonics revolution.

Finally, in the 1970s, the papers by Ernst, Moores, and Evernden and Kistler (all 1970; this volume) appeared the same year as the Kent State incident. Armstrong's (1972; this volume) paper appeared as the plate tectonics revolution extended to paleontology, about the same time as the Watergate burglary; and publication of Burchfiel and Davis' (1973; this volume) paper coincided with the U.S.–North Vietnam agreement ending direct U.S. participation in Vietnam, and with Watergate hearings in Congress. The oil crisis of 1973 quadrupled the price of oil, but also resulted in a boom in jobs for geologists and for enrollments in geology departments. In the 1970s, geological research began to be hampered in a major way by the shift in funding practices from rather informal and short grant applications, quickly reviewed and funded, to the stultifyingly time-consuming process subsequently in effect. One might doubt whether the founding research of the plate tectonics revolution could ever have been possible two or three decades subsequently.

## THE SITUATION TODAY

Since the 1970s, plate tectonics has been in a "mopping-up" phase, in the words of Kuhn (1970). We have experienced, however, revolutions in planetary geology, imaging of planetary surfaces and the Earth's interior, and in our understanding of Earth's place in space, epitomized by the Cretaceous-Tertiary boundary impact hypothesis.

We are still faced with a pronounced lack of diversity that has characterized the field of geology through the past century. In 1899, all GSA fellows but one were men. As recently as 1970, women were discouraged from taking geology; some faculty members readily stated that they "didn't want any females in my class," and even in the 1980s, there were practically no women on any geological faculty, and women graduate students faced discouraging discrimination in the field and the mines. For ethnic minorities, the situation has been even more pronounced, and there are few non-Euro-American geoscientists. The legacy of centuries of discrimination lingers. A noted California geologist, who is African-American, was told by one of his professors in the 1950s that he "didn't belong in the class." This same person faced such overt bias in the California oil industry that employment there proved impossible, despite tens of applications.

Participation in the geosciences among underrepresented minorities has been very slight, and there is little change. Geoscience educators have traditionally found it difficult to attract underrepresented minority students into courses or majors, despite their growing numbers in the student population. A recent survey found that in geosciences, fewer than 3% of recent Ph.D.'s at any institution were granted to underrepresented minorities (Mervis, 1998). One problem may be that in some cases students of color who reside in inner cities have had little or no contact with wilderness, mountains, the coast, or other aspects of field work or the natural landscape that many geologists have found so attractive (S. E. Hirschfield, oral communication, 1998). A second problem is certainly the general lack of geoscience in K-12 education. Implementation of the new NRC K-12 standards (National Research Council, 1996), which include geoscience, would go a long way towards eventually solving this latter problem, but any such action is necessarily a state-by-state and district-by-district matter and will require long-term efforts and involvement by a broad representation of the earth science community.

California and the entire Cordilleran section region is changing rapidly with regard to ethnic composition. Some school districts in California have as many as 80-100 native languages. The entire Cordilleran region, therefore, is becoming a diverse, multi-ethnic international society, much as San Francisco became an international city during the nineteenth century.

The situation is changing with regard to women. According to data from the American Geological Institute, in 1995-1996 approximately one-third of students enrolled in geoscience classes were women, and in 1993, women employed as geologists numbered 9550, or 15% of a total of 62,150 geologists. At an August, 1998, committee meeting at GSA headquarters, fully 50% of the participants were women. Clearly there is progress, but much more needs to be done.

The patterns of funding for research in geology also have undergone a change, along with the rest of science. The "social contract" between science and society that grew out of World War II and the Manhattan project is ending (Byerly and Pilke, 1995). A new "contract," as yet unformulated, will probably require greater attention to societal needs, and closer attention to "societal good." Geologists, as well as the rest of science, face the challenge of adjusting to this new reality.

## THE FUTURE

*For now I see through a glass darkly*
—*I Corinthians, 13:12*

Any prediction of future events is tricky at best, and downright foolhardy at worst. Nevertheless, here are a few:
• Society along the Pacific margin of North America will become increasingly urban and multi-ethnic. Geologic hazard

> **REFLECTIONS OF A CALIFORNIA GEOLOGIST**
>
> My entry as a person of color into the geologic profession in the early 1950s was an interesting experience. As a young, recently discharged, Army Air Corps pilot trainee (Tuskegee Airmen), I found embarking on university life to be a new and exciting adventure. I was excited by the prospect of becoming a professional geologist in a time when job opportunities were bountiful, and the road to success seemed certain. With an undergraduate degree in hand and a few letters of recommendation from the geology faculty of a highly respected University of California campus, all looked rosy. However, after sending out 42 applications and having many interviews with petroleum and mining companies, my exuberance quickly waned. I received not one job offer, and only a few letters of rejection. After searching for many months, I landed a job with a small mining company with financial backing of an African-American businessman.
>
> This was not the most auspicious beginning for a career in geology, but it led to other positions in the private sector, the California Division of Mines and Geology, the USGS, academia, and as a private consultant. At present, although opportunities for women and people of color in geology have improved markedly, few people of color choose the field. There are many programs (including several with which I have been involved) that are actively attempting to encourage students of color to take geology, but these programs have had limited success. I hope that in the future we will find other avenues or approaches that will make the geological sciences become a profession of choice.
>
> Robert A. Matthews<br>Department of Geology<br>University of California, Davis

risk assessment, warning, and avoidance will become increasingly important. An increased awareness of geology by all citizens becomes ever more important.

- Geoscientists will learn that the most effective way to communicate with the public is in scenarios, rather than making specific predictions.
- Sometime in the next 50 years, the world-wide production of oil will start to decline. Predictions of that date range from by 2005 to 2040 (Kerr, 1998). Oil will become permanently expensive, and the need for alternative sources of energy will become acute.
- The increasingly multi-ethnic nature of California and society along the Pacific margin of North America will pose a special challenge to a field as lacking in diversity as geology, because any professional group ultimately depends for its acceptance and existence upon its credibility with the population as a whole.
- Professionals in all branches of geology will increasingly find themselves working outside the United States and Canada, in cooperation with local professionals. To be effective this work will require knowledge of the local language, culture, and a sensitivity to the professional and intellectual worth of the local co-workers.
- A great earthquake will strike a major population center somewhere along the Pacific Rim, with widespread economic consequences. Prime California candidates for such an event include the Hayward fault in the San Francisco Bay area, and the blind thrust fault system beneath downtown Los Angeles.
- A years-long drought will overtax California water supply systems to the breaking point.
- California will experience several so-called 100-year floods.
- A moderate earthquake in or near the Sacramento-San Joaquin Delta will lead to catastrophic levée failure and thereby to collapse of the California Water Project.
- The increasing use of satellite communications, imagery, geographic information systems (GIS), global positioning systems (GPS) will revolutionize field work, as money for field-based research will continue to dry up.
- The $CO_2$ content of the atmosphere will rise through the next century. Future geologists will be able to participate in a real-life experiment testing current conflicting climatic predictions and determine which, if any, were correct.
- Sometime in the next century, after the usual delay, NASA will achieve its goal for the next 25 years of imaging Earth-like planets out to 40 light-years (e.g. Dressler, 1996). The discovery of oceans, continents, and mountain ranges, on extra-terrestrial planets will have deeply profound geologic and philosophical implications.
- Increasingly detailed images of the Earth's interior coupled with surface historical geological information will lead to new insight into the history of global tectonic processes.
- There will be at least one impact of a large extraterrestrial body.

Finally, one can predict that the twenty-first century will be one of resource uncertainty and genuine scarcity. This, coupled with other geosocietal considerations should make the geosciences the premier science of the next century and an integral part of K-12 education.

What can we learn from the history of the past century? We should strive to look at events and our fellow citizens with new eyes. In American culture a century ago, it was acceptable to subjugate and ignore women, and to discriminate against immigrants and underrepresented minorities. There are afoot in the body politic today some disturbing tendencies today to revert to these attitudes. We should resist these regressive pressures! Instead, we should embrace our society's diversity and strive collectively to make society more inclusive. This will happen only with hard work and reaching out. It means that when one needs a student or an employee, there is a moral imperative to ask "Have I looked in all the places?" We geoscientists should also be less reticent than in the past to embrace and broadcast the central role of the geosciences in society. In other words, we collectively should become much more involved in the everyday policy debates at all levels from local to international. If we do these things, we will indeed leave our profession and the world in a better state than we have found it.

## REFERENCES CITED

Ampferer, O. and Hammer, W., 1911, Geologischer Querschnitt durch die Ostalpen vom Allgäu zum Gardasee: Jahrbuch Geologisches Reichanstalt, v. 61, no. 3-4, p. 531–710.

Benioff, H., 1949, Seismic evidence for the fault origin of oceanic deeps: Geological Society of America Bulletin, v. 60, p. 1837–1856.

Benioff, H., 1955, Seismic evidence for crustal structure and tectonic activity, in Poldervaart, A., ed., Crust of the Earth: Geological Society of America Special Paper 62, p. 61-74.

Bezore, S. P., 1969, The Mount Saint Helena ultramafic-mafic complex of the northern California Coast Ranges: Geological Society of America Abstracts with Programs, 1969, part 3, Cordilleran Section, p. 5–6.

Byerly, R., and Pilke, R. A., 1995, The changing ecology of United States science: Science, v. 269, p. 1531–1532.

Chen, Jack, 1980, The Chinese of America, San Francisco: Harper, 274 p.

Dressler, A., 1996, Exploration and the search for origins: a vision for ultraviolet-optical-infrared space astronomy: Washington, D.C., Association of Universities for Research in Astronomy, Report of the HST and Beyond Committee, 89 p.

Hess, H. H., 1939, Island arcs, gravity anomalies, and serpentine intrusions: A contribution to the ophiolite problem: International Geological Congress, Moscow, 1937, v. 2, p. 262–283.

Kerr, R. A., 1998, The next oil crisis looms large–and perhaps close: Science, v. 281, p. 1128–1130.

Kuhn, T. S., 1970, The structure of scientific revolutions: Cambridge, MA, Harvard University Press, 210 p.

Maganini, S., 1997, California's lost tribes, Sacramento Bee, June 29-July 2, 1997.

Menard, H. W., 1986, The Ocean of Truth; a Personal History of Global Tectonics: Princeton, N.J., Princeton University Press, 353 p.

Mervis, J., 1998, Wanted: a better way to boost numbers of minority Ph.D.s: Science, v. 281, p. 1268–1270.

Moiseyev, A. N., 1970, Late Serpentinite Movements in the California Coast Ranges: New evidence and its implications, Geological Society of America Bulletin, v. 81, p. 1721–1732.

National Research Council, 1996, National Science Education Standards, Washington, D. C., National Academy Press, 262 p.

Reed, R. D., 1933, Geology of California: Tulsa, Oklahoma, The American Association of Petroleum Geologists, 355 p.

Schwarzbach, M., 1986, Alfred Wegener, the father of continental drift: Madison, Wisconsin, Science Tech, Inc., 241 p.

Steinmann, G., 1906, Geologische Beobachtungen in den Alpen. Die Schartsche Ueberfaltungstheorie und die geologische Bedeutung der Tiefseeabsätze und der ophiolithischen Massengesteine, Berichte der Naturforschenden Gesellschaft zu Freiburg I. BR., Freiburg: C. A. Wagners Universitäts-Buchdruckerei, v. 16, p. 44–65.

Taliaferro, N. L., 1943, Geologic history and structure of the central Coast Ranges of California: California Division of Mines and Geology Bulletin 118, p. 119–163.

Van Waterschoot Van der Gracht, W. A. J. M., ed., 1928, Theory of continental drift: A symposium: Tulsa, Oklahoma, American Association of Petroleum Geologists, 240 p.

Wegener, A., 1915, Die Entstehung der Kontinente und Ozeane: Braunschweig, F. Vieweg and Sohn Aktien-Gesellschaft, 94 p.

MANUSCRIPT ACCEPTED BY THE SOCIETY ON NOVEMBER 23, 1998

# Index

## A

Adams, Leason, photograph of, 40
adjudicated basin management, 335
adjudicated ground-water basins, 335
Agassiz, Louis, 18, 443
Alamitos gap, fresh-water injection barrier at, 329
Alaska-Juneau gold belt, gold mineralization in, 59
Alden, W. C., 448
Alleghany district, California, 55, 56, 57, 60
    fluid inclusions from, 64
Allen, Clarence, 30
Allen, E. T., 431
alpine glaciations, of Sierra Nevada, 443–452
Amargoa Valley, 403
Amargosa chaos, 399
    origin of, 404
American River Valley, 444
Ampferer, O., 478
Anderson, Charles A., 25, 29, 432
Anderson, John, photograph of, 40
arc-continent collision, 229
Armstrong, Richard L., 339, 479
    biography, 342
    classic paper, 343–356
    commentary on work by, 357–362
    photograph of, 342
Arnold, Ralph, 25, 27
    photograph of, 40
Arnold, Z. M., 270
artificial recharge, 323
Artist Drive Formation, 403
Auriferous Gravels, 39, 107, 237
Avawatz Mountains, 400

## B

Balk, Robert, 35
bank vegetation, importance in channel form, 255
Bartleson-Bidwell expedition (1841), 339
basement rocks, western Klamath Mountains, 231
Basin and Range, extensional deformation, 375–376, 377, 378
Basin Ranges, glaciation in, 445–449
Basin-and-Range province, large-scale extension in, 182
Bat Mountain Formation, 403
Bateman, Paul C., 107, 479
    biography, 162
    classic paper, 165–172
    commentary on work by, 173–184
    photograph of, 162
batholithic complexes, chronology and emplacement of, 189–191, 193–200

batholiths, 203–215, 217–218
    age of, 195–197
    depth patterns in, 195–197
    granitoid emplacement in, 194, 197
    origin of, 197
    relation between development and sedimentation, 197
Becker, H., 35, 477–478
bed-load transport, 253–255
Benioff, Hugo, 25, 478
    photograph of, 40
Berkeley Hills, 302
Berkey, Charles Peter, 471
Berry, William L., photograph of, 316
Bezore, Stephen, 29
Big Pine fault, 101
biostratigraphic principles, 270
Birman, Joseph, 449
Bishop Tuff, 433
Black Mountains, 399, 400, 401, 403
Blackerby, Bruce, 29
Blackwelder, Eliot, 25, 29, 30, 415, 478
    biography, 440
    commentary on work of, 443–452
    photograph of, 440
    quoted, 383
Blake, William P., 24
Bloody Canyon, 445, 449, 450
Bodie Hills, 433
Bonsall tonalite, 217
Bramlette, Milton Nunn, 238, 270, 478
    biography, 274
    classic paper, 275–299
    commentary on work by, 301–313
    photograph of, 274
Branner, John C., 19, 22, 24, 26, 28, 30, 80
    photograph of, 19, 23
Branner Club, 26
Bridgeport basin, 447, 450
Brokeoff cone, 432
Brokeoff stratovolcano, 433
Brooks, Howard, 16
Brown, Harrison, 193
Bruhnes, B., 477
Burchfiel, B. Clark, 35, 340, 479
    biography, 364
    classic paper, 364, 365–373
    commentary on work by, 375–380
    photograph of, 364
    quoted, 108
Buwalda, John P., 25, 26, 28, 29, 80
    photograph of, 40
Byerly, Perry, photograph of, 40

## C

Calaveras Complex, 175
CALFED Bay-Delta Program, 331–332, 334, 336

Calico phase, Amargosa chaos, 404
California Aqueduct, 324, 330, 331, 334
California Cascades, 432
California Water Code, 335
California Water Plan, 238, 317–322, 323–332, 333–334, 478
California Water Plan team, 316
Calkins, F. C., 26
Campbell, Ian, 25, 30
Carrizo Plain, offset on, 80
Carson, Kit, 339
Cascades, 433
Cathedral Range epoch, 195–197
Central basin, South Coast region, 328
Central Coast region, water supplies in, 329
Central Metamorphic Belt, 229
central Mojave metamorphic core complex, 376
Central Valley Improvement Act, 334
Central Velley Project, 324
Chaffee, George, 27
Chainman Formation, 229
Chaney, R. W., 29
Channel Islands, 435
Chaos Crags, 431, 432
Chaos Jumbles, 431, 432
Chetco ophiolite, 231
Chicago Pass thrust, Nopah Range, 406
Chuckwalla Valley basin, 331
Clark, Malcolm, 416
Clark Mountains, extension, 375
Claypole, E. W., 24
Clear Lake volcanic field, 432, 435
climate, 237, 333
Cloos, Ernst, 30
Cloos, Hans, 35
Coachella Valley basin, 331
Coachella Valley segment, San Andreas fault, 101–102
Coast Range ophiolite, 123–124, 126, 231
    emplacement of, 231
Coast Ranges, California, 80, 101–102, 126, 194, 197
    mercury deposits, 63–64
    Salinian block of, 180, 193
    uplift of, 157
Coast Ranges segment, San Andreas fault, 101
coastal California, volcanism, 435
Coats, Robert, 194
Colorado Desert region, water supplies in, 331
Colorado River, water imported from, 237, 328–329
Colorado River region, water supplies in, 331
Colorado River trough, 360

Comstock Lode, 238
conjunctive use, defined, 334
continental extension, 357
Convict Creek, 450
Coombs, D. S., quoted, 227
Cordilleran Geological Club, formation of, 22
Cordilleran Section
 first meeting, 22–23
 locations of meetings, 18, 20–21 (table)
 officers of, 18, 20–21(table)
Cordilleran tectonic studies, European contributions to, 33–36
Coso Range, volcanism, 434
Cotter, photograph of, 18
Cox, A., 479
Cragin, F. W., 23
Crater Lake, Oregon, 432
Crater Lake caldera, collapse origin of, 432
Crosby, William Otis, 471
Crowell, John, 101
 John, quoted, 17
crustal evolution, of Sierra Nevada, 180–182
crustal thickening, 360
crustal thinning, 360
Curry, Don, 399
Curtis, Garniss, 432
Cushman, Joseph, 269
Cuyama basin, 302

## D

Dalrymple, G. B., 479
Dana, James Dwight, 35
Davidson, G., 80
Davis, Gregory A., 29, 35, 340, 479
 biography, 364
 classic paper, 364, 365–373
 commentary on work by, 375–380
 photograph of, 364
 quoted, 29
Davis, Kingsley, 270
Davis, William Morris, 25, 26
Day, Arthur L., 431
 photograph of, 40
Death Valley fault zone, 403
Death Valley area, 340, 383, 385–397, 399–411
 extension in, 375, 379
 extensional tectonic models for, 404–406
 major faults of, 400
 tectonic map of, 401
debris, transported by water, 238, 244–252, 253–256
Deep Creek Mountains, 359
deep well injection, 323
denudation faults, 339, 343–356, 357–362
deserts, volcanism, 434–435
detachment faults, 360–361

Dewey, John F., quoted, 220
Diablo Range, 126
Diamond Peak Formation, 229
Dibblee, Thomas W., Jr., 9, 80, 478
 biography, 88
 classic paper, 89–99
 commentary on work by, 101–103
 photograph of, 88, 103
Dibblee, Thomas, Sr., 88
Dickinson, William R., 30, 155, 227
 biography, 6
 contributions to plate tectonic revolution, 6
 photograph of, 6
Diller, J. S., 27, 39, 431
direct recharge, defined, 335
Divide Peak, 432
Doell, R., 479
Dominguez gap, fresh-water injection barrier at, 329
Donner Pass region, 179
d'Orbigny, Alcide, 268
Drewes, Harold, 359
drought, 237
drought of 1987–1992, 334
Dublin Hills, faults, 405
DuToit, A., 478

## E

Eakle, Arthur S., 24
Egan Range, 358, 359
El Capitan, 447
El Paso Mountains, 378
elastic rebound theory, development of, 80
Ellis, Gertrude, 269
engineering geology, 29, 455–470, 471, 473
environmental consequences, of gold mines, 55
epochs, intrusive, 195–196
Ernst, W. Gary, 30, 107, 479
 biography, 112
 classic paper, 113–121
 commentary on work by, 123–133
 contribution to plate tectonics revolution, 112, 123
 photograph of, 112
Evernden, Jack Foord, 479
 biography, 186
 classic paper, 189–191
 commentary on work by, 193–200
 photograph of, 186
Ewing, M., 478
extension, 357, 360
 Death Valley area, 379
 in Salton Trough–California borderlands, 377
extensional deformation
 Basin and Range, 375–376, 377, 378
 Mojave Desert, 376, 377
extensional tectonic models, for Death Valley region, 404–406

## F

Fairbanks, H. W., 22, 23, 488
Fairchild, Herman LeRoy, 22, 23
Fallen Leaf Lake, 450
Fandango Pass, 339
Farallon plate
 interactions with Pacific plate, 81
 relative motion to North American plate, 59, 81, 229
 subduction of, 57, 59
faults, 57, 357. See also specific faults
Feather River, 324
Feather River–Devil's Gate ophiolite complex, formation of, 230
Feather River Project, 334
Feather River watershed, 323
Ferguson, Henry, photograph of, 382
field trips, at GSA meetings, 26
flume experiments, 253–255
fluvial geomorphology, 255
Folsom dam, 324
Foothills belt, 175, 179, 180
foothills fault system, 56
fossil faunas, use in oil exploration, 268
fossil foraminifers, 258, 267, 269
Fox, J. P., 28
Franciscan assemblage, 197. See also Franciscan Complex
Franciscan Complex, 107, 112, 123–133, 197
 development of, 231
 evolution of, 229
 geochronologic data, 126–128
 inception of subduction of, 231
 melanges, 125, 126
 metamorphism, 124–125
 related to subduction processes, 123
 strike-slip faults through, 129
Frémont's expedition (1843–1844), 39, 339
fresh-water injection barriers, 329
Friant dam, 324
Funeral Mountains, 400, 403
Furnace Creek fault zone, 400–403, 405
Furnace Creek Wash, 403

## G

Gardner, photograph of, 18
Garfunkel, Z., 377
Garlock fault, 101, 180, 339, 340, 364, 365–373, 375–378
 rotation of, 378
Garlock fault zone, 400, 405
Garlock Formation, 378
geochronological data, worldwide production of, 193
geochronology
 radioisotopic, 193
 Sierra Nevada, 449
geomagnetic polarity reversals, 193

geomorphology, Sierra Nevada, 450
geoscience, departments of, 11
Gilbert, C. M., 433
Gilbert, Grove Karl, 17, 26, 28, 34, 79, 80, 81, 82, 238, 478
    biography, 242
    classic paper, 244–252
    commentary on work by, 253–256
    photograph of, 242
Gilluly, James, 9, 30, 35, 36
    photograph of, 382
glacial epochs, 415
glacial stages, 445–447
glaciation
    relation to pluvial lake levels, 29
    of Sierra Nevada, 415, 443–452
glide-decollement models, 359
Golconda allochthon, 175–176
    emplacement of, 231
gold
    amount extracted in California, 55
    Auriferous Gravels as source of, 39
    discovery of at Sutter's Mill (1848), 39
    initial documentation of geology of, 39
    Mother Lode, 55–67
    ore deposition, 62–63
    origin of, 59–62
    in western Sierra Nevada foothills, 55
*Gold Belt Folios*, 27
Gold Hill, 359
gold mineralization
    age of, 56–59
    tectonic environment of, 56–59
gold-quartz veins, in Sierra Nevada foothills, 55–67
Gold Rush (1849–1852), 55, 238, 339
Granite Mountains, thrust fault in, 378
granitoids, emplacement in batholiths, 194, 197
Grant Range, 359
Grapevine Mountains, 403
Grass Valley district, California, 55, 57, 60, 136
Great Valley, 238
    crust-mantle slab beneath, 232
    flood (1861–1862), 237
    settlement of, 237
Great Valley basement, 180
Great Valley forearc, 123
Great Valley Group, 123–124, 155–160
    basin analysis, 156–157
    petrology, 155
    provenance, 155
    sedimentology, 155
    stratigraphy, 156
Great Valley ophiolite, 231, 417
    origin of, 156
Great Valley sequence, 107, 197
Green Valley tonalite, 217
Greenwater Range, 401, 403
ground water, contamination of, 329
ground-water development, 326–327(table)

ground-water extraction, 239
ground-water management, 335–336
ground-water models, 328
ground-water overdraft, defined, 335
ground-water recharge, 238
Grove, Karen, quoted, 99
Gulf of California, oceanic spreading center, 435
Gutenberg, Beno, 25, 478
    photograph of, 40

## H

Haller, C. R., 270
Hamilton, W., 176, 194, 227
Hammer, W., 478
Haner, Barbara E., quoted, 4
Hanna, G. Dallas, 270
Harcuvar Complex, lower Colorado River trough, 360
Hart, S. R., 195
Hayward fault, 80
Hedberg, Hollis, 270
Heezen, B., 478
Henry Mountains, 253
Hess, F. L., 375
Hess, H. H., 478
Hetch-Hetchy project, 329
Hilgard, Ernest W., 19, 22, 24
    photograph of, 23
Hill, H. Stanton, quoted, 25
Hill, Mason L., 9, 25, 28, 29, 80, 478
    biography, 88
    classic paper, 89–99
    commentary on work by, 101–103
    photograph of, 88
Hinds, Norman E.A., 25
Hollister, J. J., photograph of, 19
Holzer, Thomas L., quoted, 416
Honey Lake, 331
Hoover, Herbert C., 19, 27
    photograph of, 19
Hose, Richard, 358
Howell, David G., quoted, 8
Humboldt complex, 231
    emplacement in western Nevada, 231
Humboldt River, 339
Humboldt route, 339
Humboldt Sink, 339
Huntington Lake epoch, 195–196
hydraulic mining, ban on, 237–238
hydrothermal fluids responsible for gold deposits, origin of, 59–62
hypothesis of velocity reversal, 255

## I

Imperial Valley, settlement of, 237
in-lieu recharge, defined, 335
incidental conjunctive use, defined, 334–335
individual water management, 335
Ingersoll, Raymond V., 155
intrusive epochs, 195
Inyo Mountains epoch, 195–196

irrigation, 237
Irwin, W. P., quoted, 227

## J

Jahns, Richard, 30, 399
    quoted, 28
Jeffreys, Harold, photograph of, 40
Johnnie Formation, 400
Johnson, Willard, D., 28, 444
Jordan, David Starr, 18, 19
    photograph of, 19
Josephine ophiolite, 231
Jubilee phase, Amargosa chaos, 404

## K

Kay, G. F., 448
Kay, Marshall, 34
Kern River, 330
Kettleman Hills oil field, 274
Kew, W. S. W., 29
King, Clarence, 18
    photograph of, 18
King, Philip B., 34
Kingston Peak pluton, 378
Kingston Range area, 378
Kistler, Ronald Wayne, 479
    biography, 187
    classic paper, 189–191
    commentary on work by, 193–200
    photograph of, 187
Klamath Mountains, 56, 60, 63, 155, 194, 197, 229
    decoupling from Nevada, 232
    radioisotopic dating, 193
    vein mineralization in, 59
    volcanic arcs in, 231
    westward thrusting in, 231
Kleinpell, Robert Minssen, 238, 478
    biography, 258
    classic paper, 259–266
    commentary on work by, 267–271
    photograph of, 258
Knight, Wilbur C., 24
Knopf, Adolf, 30
Kober, Leopold, 33
Krauskopf, Konrad, 30
Kula plate, relative motion to North American plate, 59

## L

land subsidence, 238, 324, 328
landslides, 415–416, 454, 455–470, 471–474
Lapworth, Charles, 269
large-scale dextral shear model, Death Valley region, 405
Larsen, Esper Signius, Jr., 108, 478
    biography, 202
    classic paper, 203–215
    commentary on work by, 217–218
    photograph of, 202
Lassen "cut-off," 339

Lassen Peak, 420, 421–430
　eruptions (1914–1917), 27, 431–432
Laurentide ice sheet, advance and retreat of, 444
Lawson, Andrew Cowper, 17, 18, 19, 22, 24, 25, 28, 29, 30, 101, 420, 431, 477–478
　biography, 70
　classic paper, 71–76
　commentary on work by, 79–85
　mapping by, 70
　photograph of, 22, 23, 70
　postearthquake observations, 70
　quoted, 18
Le Conte Club, 22
Leavitt Creek, 450
Leavitt Meadow, 446
LeConte, Joseph L., 17, 18, 22, 23, 30
　*Elements of Geology*, 22
　photograph of, 23
Lee, Ken, 416
Lee Vining epoch, 195–196
Leopold, Luna, 30
Lindgren, Waldemar, 19, 27, 30, 39, 477
　biography, 42
　classic paper, 43–53
　commentary on work by, 55–67
　photograph of, 42
Livermore Valley basin, 329
local agency water management, 335
local water management ordinances, 335–336
Loma Prieta earthquake (1989), 83
Lone Pine, earthquake rupture at, 80
Long Valley caldera, 433
Longwell, Chester, 7
Loomis, B. F., 431, 432
Los Angeles basin, 435
Louderback, George D., 19, 22, 25, 28, 29
　photograph of, 23
lower Colorado River, volcanism, 435
Lyell, Charles, 268

## M

Macdonald, Gordon, 432
Macelwane, Father J. B., photograph of, 40
Mackin, J. Hoover, quoted, 11
Mallory, V. Standish, 270
Manley-Brier party (December 1849–January 1850), 340
mapping
　first north California geologic, 70
　of Sierra Nevada, 39
Matthes, François E., 26, 80, 82, 443, 444
Matthews, Robert A., quoted, 480
Maxson, John, 25
McBirney, A. R., 433
McGee stage, 447
McKee, Bates, 29

McMath, 254
McPhee, John, quoted, 53, 143
Medicine Lake Highland, volcanism in, 432
Medicine Lake volcanic shield, 433
Melones fault zone, 55, 56, 57, 175
Merced Peak, 443
Merced River, 443
Merced River Valley, 444
Merriam, John C., 19, 22, 24, 30
　photograph of, 23
Mifflin, Martin, 416
Miller, W. J., 25
mineral wealth, of California, 27
Misch, Peter, 358
Modoc Plateau, 433
Mohorovicic, A., 477
Mojave Desert area, 180, 197
　extensional deformation, 376, 377
　strike-slip activity in, 377–378
　volcanism, 434–435
Mojave River basin, 331
Mojave segment, San Andreas fault, 101–102
Mojave-Snow Lake fault, displacement on, 231
Mojave-Sonora megashear, 231
Mokelumne project, 329
Mono Basin
　export of water from, 331
　moraines, 449
Mono Lake, 331, 433, 434, 444, 449
Mont Pelée, Martinique, eruption (1902), 432
Montebello forebay, Central basin, Coastal Plain, 328
Monterey–Arguello plate, subduction of, 435
Monterey Formation, 238, 275–299, 301–313
　age of, 302–303
　biostratigraphy, 302–303
　carbonate diagenesis, 301, 306
　composition, 303–304
　depositional environments, 304
　diagenesis, 274, 301, 304–306
　lithology, 303–304
　microfossils, 302
　mineralogic alteration, 304, 305
　oceanographic context of deposition of, 301
　as oil reservoir, 307
　as oil source, 307
　organic matter diagenesis, 301
　paleoceanographic significance, 308–309
　paleoenvironmental significance, 308–309
　petroleum generation, 301, 306–307
　petroleum potential, 301
　petroleum reservoirs, 301, 307–308
　as petroleum source, 306–307
　phosphate diagenesis, 306
　silica diagensis, 274, 301, 305–306

Moores, Eldridge M., 29, 30, 108, 479
　biography, 220
　classic paper, 221–226
　commentary on work by, 227–234
　contributions to plate tectonics revolution, 228
　photograph of, 220
Morton, Douglas, 416
Mother Lode gold, 55–67, 195
Mount Lassen, 415. *See also* Lassen Peak
Mount Morrison, 175
Mount Shasta, 432, 433
Mount Whitney region, 179
Muir, John, 18, 107, 443
　photograph of, 18
　quoted, 107, 415
Murphy, E. C., 253
Murphy Wash fault, 359

## N

Napa-Sonoma Valley basin, 329
Naples Beach, 302
Neinesz, F. A. Vening, 478
Noble, Dorothy, photograph of, 382
Noble, Levi F., 80, 101, 340, 478
　biography, 382
　classic paper, 385–397
　commentary on work of, 399–411
　photograph of, 382
Nolan, Thomas B., 30
Nopah Range, 378, 404, 405, 406
North American plate
　relative motion to Farallon plate, 59, 81, 229
　relative motion to Kula plate, 59
　tectonic interaction with Pacific plate, 81, 227, 415
North Coast region, water supplies in, 330
North Lahontan region, water supplies in, 330–331
Northern Sierra terrane, 230

## O

Oakeshott, Gordon, 29
oil, Monterey Formatin, 307
oil deposits, development of, 238
oil exploration, 27
oil recovery, 267
Ojakangas, Richard W., 107, 479
　biography, 136
　classic paper, 137–153
　commentary on work by, 155–160
　photograph of, 136
Olema, San Andreas surface rupture near, 79
Omori, F., 80
ophiolites
　Coast Range, 123–124, 126
　formation of, 229
　Great Valley, 156

ophiolites (continued)
  Sierra Nevada, 231
  significance in tectonics, 228, 232
  used as a tool to interpret orogenic belts, 220
Oppel, Albert, 268, 269
Orange County, fresh-water injection barriers in, 329
Orange County Water District, 328
Oriental mine, Alleghany district, 62, 63
orogenic belts, relation to plate tectonics, 228
Oroville Dam, 324, 334
Owens River system, 449
Owens Valley, 182
  export of water from, 331
  importation of water from, 237

## P

Pacific plate
  interactions with Farallon plate, 81
  tectonic interaction with North American plate, 81, 227, 415
Page, Benjamin M., photograph of, 109
Pahrump Group, 399, 400
Palache, Charles, 30, 478
Palmdale, California, 80
Palos Verdes Hills, 274, 302
Panamint Range, 406
  mapping of, 399
Panamint thrust, Panamint Range, 406
Parícutin, Central America, eruptions (1943–1952), 432
Paso Robles Formation, 268
Patt, Ralph, 416
Peach Springs Tuff, volcanism, 434
Peninsular Ranges, 194
Petaluma Valley basin, 329
petroleum, Monterey Formation, 301–309
petroleum exploration, 268–269
petroleum production, 27
Philippines, oil production in, 269
Pilger, Andreas, 34
plate tectonics, 107–109
  relation to orogenic belts, 228
  role in Cordilleran evolution, 357
  and San Andreas fault, 102
plate tectonics paradigm, applied to Sierra Nevada batholith, 176
plate tectonics revolution, 11, 107, 156, 478–479
  contributions of Eldridge M. Moores, 228
  contributions of W. Gary Ernst, 112, 123
  contributions of William R. Dickinson, 6
  and San Andreas fault, 81
Point Arena, submarine lavas in, 435
Point Delgada
  earthquake rupture at, 80

Point Delgada (continued)
  San Andreas fault near, 83, 101
Point Sal Formation, Santa Maria basin, 303
Poncho Rico, 267
postearthquake investigations, 70, 79
Powell, J. W., 26, 237
precipitation, 237–238, 323, 333
  North Coast region, 330
  North Lahontan region, 330
  South Lahontan region, 331
Preston Peak ophiolite, 231

## Q

Quail Mountains, mapping of, 399
Quartz Hill gold deposits, Klamath Mountains, 63
Quenstedt, August, 268

## R

railroads, construction of, 238
rain-shadow effect, 237, 339
Raitt, R., 478
Raker Peak, 432
Ransome, F. L., 25, 27, 29, 488
Raymond basin, South Coast region, 328
Reading Peak, 432
Reed, Ralph D., 29, 33, 268, 269
Reid, Harry Fielding, 28
  classic paper, 76–77
  commentary on work by, 79–85
  photograph of, 40
Resting Spring Range, 378, 404, 405, 406
Rhodes Tuff, 403
Richter, Charles F., 25
  photograph of, 40
rifting, of western North American margin, 229
Ritter Range, 175
Ritter Range caldera, 197
Ritter Range pendant, 179
Roberts, Ralph J., 227
Roberts Mountain allochthon, 175, 229–230, 378
Robinson Canyon, 447
Robinson Creek, 450
Rocky Mountains, Wyoming, 448
rolling hinge model, 404, 406
Rubey, William, 30
Ruby Mountains, Nevada, 448
Russell, Israel C., 27, 444
Russell, R. J., 433

## S

Sacramento River system, impacts of hydraulic mining for gold on, 253
Sacramento River watershed, 323
Sacramento–San Joaquin Delta, 238, 239, 331

Sacramento Valley area, 137–153, 197
  water imports to, 334
  water supplies in, 330
Salinas Valley, seawater-intrusion problem at, 329
Salinian block, Coast Ranges, 180, 193
Salisbury, R. D., 443
Salton Trough, 435
Salton Trough–California borderlands, extension in, 377
Saltos Shale, 302
San Andreas fault, 28, 39, 79–85, 101–103, 129, 379
  Coachella Valley segment, 101–102
  Coast Ranges segment, 101
  evolution of concepts about, 80–83
  first publications on, 19, 70
  horizontal displacement along, 29, 40, 80–81
  Mojave segment, 101–102
  and plate tectonics revolution, 81
  and plate tectonics theory, 102
  strike-slip displacement across, 80
San Andreas fault system, 415, 435
  development of, 231
San Diego County, water import to, 329
San Fernando Reference, 323
San Fernando Valley, South Coast region, 328, 329
San Francisco Bay area, water imports to, 334
  water supplies in, 329–330
San Francisco earthquake (1906), 28, 39, 70, 79–85
San Gabriel basin, South Coast region, 328
San Gabriel Mountains, Transverse Ranges, 197
San Gabriel Valley, 329
San Jacinto fault, 80
San Jacinto Mountains, 218
San Joaquin basin, 303
  oil production in, 307
San Joaquin Valley, 197
  artificial recharge in, 323
  water imports to, 334
  water supplies in, 330
San Juan Bautista, earthquake rupture near, 80
San Juan Mountains, 448
San Marcos gabbro, 217
San Pedro Formation, 329
Santa Ana gap, fresh-water injection barrier at, 329
Santa Ana River basins, South Coast region, 328
Santa Clara Valley basin, 329
Santa Cruz Mountains, submarine lavas in, 435
Santa Lucia region, central Coast Ranges, 197
Santa Maria basin, 303
  reservoirs in, 308
Santa Maria Valley, 274
Santa María volcano, Guatemala, 432

Santa Monica Mountains, 435
Schell Creek Range, 359
Schenck, Hubert, 268
Schumann, Herbert, 416
sea-floor spreading, 11
  in Troodos complex, 227
sea-floor spreading revolution, 479
seawater conversion, 331
sediment transport, in rivers, 254
Seeber, L., quoted, 227
seismic hazards analysis, 81–83
Seismological Society of America, establishment of, 28
Sengör, A. M. Celâl
  biography, 32
  photograph of, 32
shallow fault model, Death Valley region, 405–406
Shasta dam, 324
Sheephead fault, 403
Sherwin stage, 447
Sherwin Till, Rock Creek Plateau, 446
Shoo Fly Complex, 175, 230
Sieh, K. E., 83
Sierra Club, founding of (1892), 18
Sierra Nevada, 163, 165–172, 173–184, 415
  active cirque glaciers in, 443
  alpine glaciations of, 443–452
  decoupling from Nevada, 232
  deformation of, 231
  geochronology, 449
  geomorphology, 450
  glaciation in, 415, 445–449
  radioisotopic dating, 193
  roof pendants, 231
  volcanic arcs in, 231
  volcanism, 433–434
Sierra Nevada batholith, 107, 162, 176
  age of, 194–195
  basement as source component for, 178
  framework rocks of, 173–176
  origin of, 176–179, 194–195
  petrogenesis, 176–179
  unroofing of, 179
Sierra Nevada Eastern Belt, development of, 230
Sierra Nevada foothills, gold-quartz veins in, 55, 56, 58
Sierra Nevada metamorphic belt, 55
Sierran thrust system, 378
Silurian Hills, mapping of, 399
Silver, Leon, 30
Silverado aquifer, 329
Sinclair, J. W., 22
  photograph of, 23
Sixteen-to-One Mine, Alleghany district, 62
Skibitzke, 328
Slate Creek ophiolite, 231
Slate Range, thrust fault in, 378
Smartville ophiolite, 231
Smith, James P., 19, 22
  photograph of, 19

Smith, W. S. Tangier, 23
Smith, Ward, photograph of, 103
Smith, William, 267
Sonora Pass, 433, 450
South Coast region, 323
  artificial recharge facilities, 328
  basin management, 328
  ground-water basins, 328
  spreading grounds, 328
  water supplies in, 328–329
South Lahonton region, water supplies in, 331
Stanford University, 19, 22, 20, 30
Stanislaus River Valley, 444
State Water Project, 324
  construction of, 323
Steinmann, G., 477
Stevens Sands, San Joaquin basin, 303
Stille, Hans Wilhelm, 33, 34, 35, 36
  photograph of, 34
Stirling Quartzite, 400
Stock, Chester, 25, 30
Stout, Dorothy L.
  biography, 16
  photograph of, 16
Stout, Martin, 29
stratigraphic correlation, 301
stratigraphic paleontology, 258, 259–266, 267–271
stratigraphy, 238, 259–266, 267–271
stream capacity, 254
stream power, 254–255
strike-slip activity, in Mojave Desert, 377–378
strike-slip faults, 80, 102
  through Franciscan Complex, 129
subduction, 227
Subjacent series, 107
Suess, Eduard, 33
Suisun-Fairfield Valley basin, 329
Sullivan, Frank, 270
Superfund sites, 329
Superjacent series, 107
Sutter Buttes, 435
Sutter's Mill, discovery of gold at (1848), 39
SWEAT (southwest U.S.–east Antarctic) connection, 220
synclinorium concept, 175, 178

## T

Tahoe stage, 447
Taliaferro, Nicholas, 25
Talmage, J. E., 22, 23
  photograph of, 23
tectonic development, southern California deserts, 375–380
tectonic studies, European contributions to, 33–36
tectonics, 227–228
  and crustal evolution, 180–182
  in California Desert, 339–340
  See also plate tectonics

Teichert, Curt, 35
Tenaya stage, 449
terran analyses, 229
terrane concept, 228
terrane maps, 228
terranes, California, 221–226, 227–234
  and tectonic development, 229–231
Terzaghi, Karl, 415, 478
  biography, 454, 471–472
  classic paper, 455–470
  commentary on works by, 471–474
  photograph of, 454
Thayer, Tom, 10, 399
Thomas, R. G., 328
Thompson, George, 30
Tioga stage, 446–447
Tipton, Ann, 270
Titus Canyon Formation, 403
Tolman, Cyrus F., 29
  photograph of, 260
total stream power, 254
Transverse Ranges, 102, 194, 197, 435
treated waste water, 323
Troodos complex, Cyprus, 227
Troxel, Bennie W., 358, 478
Truckee River, 339
Trümpy, Rudolf, 36
Tuolumne Intrusive Suite, Cathedral Range epoch, 197
Tuolumne River, 443
Turekian, Karl, 342
Turner, F. J., 433
Turner, H. W., 24, 27, 29, 39
Tuscan Formation, volcanic debris-flow deposits of, 433

## U

University of California, Berkeley, 19, 22, 29, 30
Urey, Harold, 193

## V

Valley of Ten Thousand Smokes, Alaska, eruption (1912), 432
VanderHoof, V. L., 29
Vaughan, T. Wayland, 30
Ventura County, ground-water basin in, 329
Vine, F. J., 479
Virgin Spring phase, Amargosa chaos, 404
volcanic arcs
  eastern Klamath Mountains, 231
  Sierra Nevada, 231
volcanic deposits, alteration of, 274
volcanism, 420, 421–430, 431–438
  California deserts, 434–435
  coastal California, 435
  elated to subduction, 415
  Sierra Nevada, 433–434

## W

Wadati, 478
Wagner, Norm, 16
Wahrhaftig, Clyde, 107, 479
   biography, 163
   classic paper, 165–172
   commentary on work by, 173–184
   photograph of, 163
   sketches by, 10, 162, 166, 170, 186
Warner Mountains, 433
Warner Range, 339
Warren, A. D., 270
Wasatch fault, Utah, 81
Wasatch Range, Utah, 448
waste water treatment, 323, 328
water
   distribution of, 323
   importation of, 237
   possibility of shortage, 331
   scarcity of, 237
   sources for California, 27
   storage, 239
water banking, defined, 335
water development, history of, 333
water management agencies, 335
water marketing, defined, 335
water rights, 323
water transfer, defined, 335
Waterman Hills detachment fault, 376
weathering criteria, 445–447, 449–450
Weaver, Donald, 270
Weber, E. M., 328
Wegener, A., 478
West Coast basin, South Coast region, 328
West Walker River, 450
Wheeler, Greg, quoted, 16
Whipple Complex, lower Colorado River trough, 360
Whipple Mountains, 376
White-Inyo Mountains, 231
White Mountain, 432
White Pine Range, 359
Whiterock Bluff Shale, 302
Whitewater Canyon, San Andreas fault at, 80
Whitney, Josiah D., 39, 443
   quoted, 42
Williams, Howel, 25, 415, 478
   biography, 420
   classic paper, 421–430
Williams, Howel (continued)
   commentary on work by, 431–438
   photograph of, 420
Willis, Bailey, 24, 25, 30
   photograph of, 28
Willow Spring diorite-gabbro, 400
Winikka, Carl, 416
Wood, Ethel M. R., 269
Wood, Harry O., 25, 28, 29, 80
   photograph by, 84
   photograph of, 40
Woodford, A. O., 25
   photograph of, 25
Woodring, W. P., 274
Woodson Mountain granodiorite, 217
Wornardt, Walter W., 270
Wright, Lauren, 399, 478
Wrights tunnel, offset of, 83

## Y

Yosemite epoch, 195–196
Yosemite Valley, 444, 447
Yuba River, impacts of hydraulic mining for gold on, 253
Yuba River Valley, 444, 447